Stereochemical Applications of Gas-Phase Electron Diffraction

Edited by
**István Hargittai and
Magdolna Hargittai**

Part B
**Structural Information for
Selected Classes of Compounds**

István Hargittai
Magdolna Hargittai
Structural Chemistry Research Group
of the Hungarian Academy of Sciences
Eötvös University
Budapest, VIII., Puskin utca 11 – 13
Pf. 117, H-1431, Hungary

Library of Congress Cataloging-in-Publication Data

Stereochemical applications of gas-phase electron diffraction/edited by István Hargittai and Magdolna Hargittai.
p. 000 cm.—(Methods in stereochemical analysis: v. 10)
Bibliography: p.
Includes indexes.
Contents: pt. A. The electron diffraction technique—pt. B. Structural information for selected classes of compounds.
ISBN 0-89573-337-4 (pt. A).   ISBN 0-89573-292-0 (pt. B)
1. Electrons—Diffraction.   2. Chemistry, Analytic—Qualitative.   3. Molecular structure.
I. Hargittai, István. H. II. Hargittai, Magdolna. III. Title: Gas-phase electron diffraction.
IV. Series.
QD79.E4S74   1988   87-37169
541.2'23—dc19

ISBN 0-89573-292-0 VCH Publishers
ISBN 3-527-26790-5 VCH Verlagsgesellschaft
ISBN 0-89573-719-1 (set) VCH Publishers
ISBN 3-527-27820-6 (set) VCH Verlagsgesellschaft

Distributed in North America by:

VCH Publishers, Inc.
220 East 23rd Street
Suite 909
New York, New York 10010

Distributed Worldwide by:

VCH Verlagsgesellschaft mbH
P.O. Box 1260/1280
D-6940 Weinheim
Federal Republic of Germany

Contents

Part B

STRUCTURAL INFORMATION FOR SELECTED CLASSES OF COMPOUNDS

1. **Boron and Silicon Compounds** 1
   Vladimir S. Mastryukov

2. **Nitrogen and Phosphorus Compounds** 35
   Lev V. Vilkov and Nina I. Sadova

3. **Oxygen and Sulfur Compounds** 93
   Victor A. Naumov

4. **Fluorine Derivatives** 147
   Heinz Oberhammer

5. **Saturated Organic Molecules** 209
   Lawrence K. Montgomery

6. **Unsaturated Organic Molecules** 239
   Marit Traetteberg

7. **Substituted Benzene Derivatives** 281
   Aldo Domenicano

8. **Organometallic Compounds of Main Group Elements** 325
   Arne Haaland

9. **Metal Halides** 383
   Magdolna Hargittai

10. **Interaction of Theoretical Chemistry with Gas-Phase Electron Diffraction** 455
James E. Boggs

Author Index 477

Formula Index 497

Part A
THE ELECTRON DIFFRACTION TECHNIQUE

1. **A Survey: The Gas-Phase Electron Diffraction Technique of Molecular Structure Determination** 1
István Hargittai

2. **Status of Electron Scattering Theory with Respect to Accuracy in Structure Analyses** 55
Lawrence S. Bartell

3. **Small-Angle Electron Scattering by Gas Molecules: Experimental and Theoretical Aspects** 85
Shigehiro Konaka

4. **Information on Electron Density Distribution from High-Energy Electron Scattering** 107
Shuzo Shibata and Fumihiko Hirota

5. **Temperature Dependence of Electron Diffraction Structural Parameters: Theory and Experiment** 139
Manfred Fink and Denis A. Kohl

6. **Gas Electron Diffraction Experiment** 191
János Tremmel and István Hargittai

7. **Joint Use of Electron Diffraction and High-Resolution Spectroscopic Data for Accurate Determination of Molecular Structure** 227
Kozo Kuchitsu, Munetaka Nakata, and Satoshi Yamamoto

8. **Spectroscopic Information from Electron Diffraction** 265
Victor P. Spiridonov

9. **Molecular Orbital Constrained Electron Diffraction (MOCED) Studies: The Concerted Use of Electron Diffraction and Quantum Chemical Calculations** 301
Lothar Schäfer, John D. Ewbank, Khamis Siam, Ning-Shih Chiu, and Harrell L. Sellers

10. **Self-Consistent Molecular Models from a Combination of Electron Diffraction, Microwave, and Infrared Data Together with High-Quality Theoretical Calculations**. 321
Herman J. Geise and W. Pyckhout

11. **Conformational and Thermodynamic Properties from Electron Diffraction** 347
Kenneth Hedberg

12. **Investigation of Large-Amplitude Motion** 367
Alfred H. Lowrey

13. **Common Ground between EXAFS and GED?** 413
Brian Beagley

14. **Combined Application of Electron Diffraction and Liquid Crystal NMR Spectroscopy** 451
David W. H. Rankin

15. **Low-Resolution Microwave Spectroscopy and Its Combination with Electron Diffraction** 483
Robert K. Bohn

16. **Gas-Phase X-ray Diffraction** 511
Takao Iijima, Keiko Nishikawa, and Toshiyuki Mitsuhashi

Author Index 539

Formula Index 553

Subject Index 559

Foreword

The field of fast electron diffraction of gaseous molecules was initiated by the work of Herman Mark and Raimund Wierl in 1930. It had its origins in the revolutionary concepts of the early 1920s which contained the idea of wave/particle duality for material particles proposed by Louis de Broglie and the new quantum mechanics introduced by Erwin Schrödinger, Werner Heisenberg, Max Born, and Pascual Jordan for describing the characteristics and behavior of elementary substances such as electrons. The possibility of performing practical experiments was enhanced by technological developments in high-voltage and high-vacuum techniques. Successful demonstrations of electron diffraction from metal foils, thin slices of mica, and powder samples suggested that electron diffraction of gaseous molecules could be feasible, particularly in view of the relatively high scattering power of atoms for electrons compared to X-rays, which had already been shown by Peter J. W. Debye to form a diffraction pattern from a gaseous sample of carbon tetrachloride. The long period of time required to obtain a diffraction pattern from X-rays made the technique undesirable. By comparison, the first experiments of Mark and Wierl produced a photograph of an electron diffraction pattern of carbon tetrachloride in a fraction of a second.

A strong motivation for pursuing the gas electron diffraction technique was the opportunity to gain deep insights into the nature of the chemical bond by investigating the detailed geometric arrangements of atoms in gaseous molecules. Interest in the technique proliferated very quickly in several laboratories throughout the world, even though the apparatus required was fairly elaborate and usually was constructed within the investigator's institution.

A remarkable aspect of gas electron diffraction is the accuracy and detail to which the geometric structure and internal motion can be determined in view of the fact that the diffraction data of interest are generated in the form of a rather modest interference signal superimposed on a much stronger and steeply falling background intensity. The eye perceives such radially symmetric patterns as diffuse rings, and this response greatly facilitated the measurements of the diameters of the maxima and minima of the rings and the estimation of the intensity pattern. Such evaluations of the interference patterns characterized the data that were used to determine molecular structure by the so-called visual method. Progress was so considerable that by 1936 Lawrence Brockway was able to write an extensive review article in *Review of Modern Physics* which included structural information determined for 146 different molecules. Lawrence Brockway, as a graduate student of Linus Pauling, initiated gas electron diffraction in the United States, a project that Pauling brought to the United States, on his return from study in Germany, with the help and cooperation of Herman Mark.

Continuous progress followed, in particular the development of technical means for extracting interference data quantitatively, thus obviating the use of visual estimates. The method became known as the sector–microphotometer method, evidently because such apparatus played a key role. A sector can level considerably the steeply falling background intensity, whereupon the microphotometer can be readily applied to the evaluation of the diffracted intensities. Sector devices for gas electron diffraction were developed independently and described by Christian Finbak in 1938 and P. P. Debye (P.J.W. Debye's son) in 1939. Initial impetus to the use of the sector and microphotometer was given by the work of Finbak and Odd Hassel and the school that developed around them in Norway, starting in the late 1930s.

The term "sector–microphotometer procedure," although initially based on the special apparatus employed, came to imply, in addition, the many improvements incorporated into molecular structure determination which have been developed from deeper theoretical insights, considerably improved analytical procedures, and the enhanced accuracy of the experimental data. An insightful theory that sparked many of the improvements was published by P.J.W. Debye in 1941. It concerned the interpretation of the Fourier integral transform of interference intensities and suggested that if the measured experimental data were properly prepared, there was the possibility of obtaining not only enhanced accuracy in the structural parameters, but also much information about the internal motion of the molecules.

Isabella Karle and I were considerably influenced by the 1941 paper of P.J.W. Debye and, on arriving at the Naval Research Laboratory in Washington, DC, in 1946, we set about further development of the theory, treatment of the data, and establishment of analytical procedures to fulfill the implications of this paper. We learned many things from working in this field. Among them was the great value of introducing mathematical and physical constraints into the theoretical aspects of the analyses. One constraint was a very useful nonnegativity criterion applied to the Fourier transform of the intensity data from the interference signal. The good results obtained from this criterion stimulated the development of the implications of the nonnegativity of electron density distributions in crystals. This nonnegativity property has played a valuable role in the development of theoretical formulas and practical procedures for crystal structure analysis.

After the sector–microphotometer method was firmly established, a number of additional valuable developments afforded deeper insights and greater analytical facility and accuracy. Many workers have contributed to these developments, and their numerous contributions can be represented by those of particular individuals. There was the development of least-squares analyses to refine structural parameters by Lise and Kenneth Hedberg, the enhanced insight into scattering theory that led to the introduction of complex atomic scattering coefficients by Verner Schomaker and Roy Glauber, and the study of molecular motion and its role in electron diffraction and spectroscopy by Yonezo Morino, Kozo Kuchitsu, and Takehiko Shimanouchi. Special features such as interatomic multiple scattering, isotope effects and anharmonic vibrations have been investigated by Lawrence Bartell and Kozo Kuchitsu; molecular force constants, shrinkage effects, and connections with spectroscopy have been studied by Sven Cyvin; and the thermodynamic implications of conformational changes have been calculated by Otto Bastiansen and Kenneth Hedberg. Structural studies of inorganic materials have been pursued with special high-temperature devices by Victor Spiridonov and Nicholas Rambidi. New possibilities in the high-temperature area as well as in the study of unstable species have been opened by the combined mass spectrometry–electron diffraction approach of István Hargittai and János Tremmel. Small-angle electron diffraction combined with appropriate theoretical considerations have produced information about electronic wave functions as described by Russel Bonham and Manfred Fink. Investigations in theoretical chemistry facilitated by structural information have been carried out by James Boggs, Richard Hilderbrandt, and Lothar Schäfer. Large-amplitude motions have been studied by Victor Spiridonov, and atomic clusters have been investigated by Marcel Rouault, Philippe Audit, Gerard Torchet, Jean Farges, Bernard Raoult, M. F. de Feraudy, and Lawrence Bartell. Finally structural studies of large numbers of related compounds have been pursued by many workers, eg, Magdolna and István Hargittai, Kozo Kuchitsu, Vladimir Mastryukov, Nina Sadova, Heinz Oberhammer, Marit Traetteberg, and Lev Vilkov, revealing many subtle structural interrelationships.

As a result of the dedicated efforts in a relatively small number of laboratories, gas electron diffraction has served as a valuable tool in the investigation of molecular structure. Much information has been obtained concerning molecular configuration, bond distances and angles, internal motion (including hindered internal rotation and barrier heights), preferred orientation in conformers, and conjugation and aromaticity. Investigations have also concerned mixtures in equilibrium, including evaluations of thermodynamic quantities, free radicals, a wealth of high-temperature studies, clusters, isotope effects, and the joint use of other techniques such as laser excitation, microwave spectroscopy, and mass spectrometry.

We thus have the view of gas electron diffraction as a technique of wide application to many aspects of molecular structure, and when it is combined with various spectroscopic techniques, the value of each may be considerably enhanced. The broad range of topics that have been under recent investigation is readily seen by noting the chapter titles of this book. They concern in Part A special studies such as small-angle electron scattering, electron density distributions, combination techniques including spectroscopy, and calculations in theoretical chemistry. Part B gives a broad summary of investigations concerning various classes of compounds.

Structural research not only has inherent scientific interest, it also stimulates and provides the conceptual basis on which many related fields of science that depend upon structural information can progress. It is therefore especially fitting to have the up-to-date and extensive reviews of forefront topics in the electron diffraction of gases that are presented in this volume.

Jerome Karle

Preface

Gas electron diffraction is a "direct" physical method for stereochemical analysis. Its main application is the determination of molecular geometry. The technique has been around for decades and has had such spectacular successes as, eg, Odd Hassel's conformational studies. However, only recently has it become a rigorous tool rather than a type of scientific art. Up-to-date computational methods, the understanding of the physical meaning of the determined parameters, and the combined applications of this technique with other techniques have together contributed to this development.

Gas electron diffraction is one of the two principal experimental tools for the determination of vapor-phase molecular structures (the other being high-resolution spectroscopy). With several controversies behind it, today GED is a reliable and, for many purposes, a unique tool. It complements, rather than competes with, microwave spectroscopy, X-ray and neutron crystallography, and quantum chemical calculations. During the past several years the capabilities of gas electron diffraction have expanded in several directions. One is the increased accuracy and reliability of the structural parameters determined. Another is the increasing amount of vibrational information extracted from the electron scattering data. Yet another is the investigation of unstable molecules, excited states, and other "exotic" species.

There is hesitation sometimes in accepting electron diffraction results, especially on the part of those who have witnessed certain earlier fiascos. Reports of structure studies still vary in depth and in the reliability of the analysis and error estimations. However, there is a wide audience, which is interested in a source of accessible information on molecular structure. Most readers of electron diffraction papers, while not specialists in the technique, are consumers of structural information and should be assisted in judging critically the electron diffraction papers. This wide readership should also be informed of the present potentialities of the electron diffraction technique. The main function of this book is to present modern gas electron diffraction to the nonspecialist, to convey a real sense of the importance of this field of research to the nonexpert. It may be considered an offering by the electron diffractionists to their fellow chemists.

The potential users of structural information often ask such seemingly simple questions as, Which compounds are suitable for electron diffraction stereochemical analysis and which are not? What are the limitations of the size of a molecule for such an analysis? How reliable are the parameters determined, and how reliable are the error limits given in the original works? In which directions is the technique developing? What are the advantages in using GED compared with other techniques? These are obviously not the questions of electron diffraction specialists, but the questions of spectroscopists, crystallographers, physical organic chemists, and other chemists who are trying to relate their findings to other physicochemical and structural information. It appears that a large body of experimental data has been accumulated over the years, and some guidelines will be offered as to the meaning and reliability of these data with intention of facilitating the utilization of this body of information. Furthermore, it is hoped that the book will help to direct many chemists toward the application of electron diffraction stereochemical analysis to problems for which it is uniquely qualified.

The book is organized into two major parts. Part A discusses the development and the present capabilities of gas electron diffraction. Several contributions deal with the combined application of gas electron diffraction with other techniques. The interrelationship between geometry and motion is discussed with emphasis on large-amplitude motion. The possibilities of electron diffraction for determining vibrational and thermodynamic information are also described.

Part B contains chapters on structural results for selected classes of compounds. These critical chapters serve as reliable sources of structural information. Beyond merely surveying the respective areas, they present empirical trends and interpretation of structural variations as well.

As editors of these volumes, we feel privileged to have been given the enthusiastic cooperation of our electron diffractionist colleagues. The project owes much to the trust and help provided by the series editor, Professor Alan P. Marchand. Most of the editorial work was carried out in the summer and fall of 1986 during our visit to the Department of Chemistry of the University of Texas at Austin. Professor James E. Boggs provided the necessary conditions for this work, and facilitated its successful completion. We appreciate the pleasant and efficient cooperation with the Staff of VCH Publishers, and the work of Director of Production/Manufacturing Alan Winick in bringing out these volumes.

The appearance of this book coincides with the 70th birthday of Professor Otto Bastiansen. As a token of our appreciation and affection we dedicate our work in this project to him.

István Hargittai
Magdolna Hargittai

Introduction

Sparked and sustained by significant progress in vacuum techniques and by improvements of calorimetry, a new branch of physics developed around the turn of the century, namely *cryogenics*, and with it physics and chemistry at very low temperatures. Two laboratories in Europe were leading in this discipline: that of Kammerling-Onnes in Leiden and that of Walter Nernst in Berlin; both contributed enormously important experimental observations to the start of modern physics in its infancy. Kammerling-Onnes discovered super-conductivity, a strange and incredible phenomenon which resisted a rational explanation for almost 50 years.[1] Nernst,[2] on the other hand, established through careful and systematic experiments the anomalous behavior of ideal gases at very low temperatures, which also posed a startling enigma, because it could not be explained by the application of Boltzmann's statistic which treats each molecule as a particle maintaining its individuality during its distribution over the cells in the phase space. A correct interpretation was only possible through a different statistical treatment which was first introduced by Bose[3] and later expanded by Einstein.[4] In his article Einstein comments that in the degenerated state the gas molecules do not behave like permanently distinguishable particles, but more like waves which interfere with each other and do not simply collide like solid particles. This was the first published inkling concerning the wave character of moving particles; it occurred in 1924, one year before de Broglie's[5] fundamental paper of 1925 which described in full detail the particle-wave correlation by the equation:

$$\lambda = \frac{h}{mv} \tag{1}$$

where m and v = mass and velocity of the moving particle, h = Planck's constant, and λ = wavelength coordinated to the particle.

The first experimental verification of this relation came for slow electrons from Davisson and Germer[6] and for fast electrons from Thompson[7] in 1927. Immediately thereafter electron diffraction phenomena were used for two different research purposes: first to obtain information on the basic properties of the electron, particularly its spin, and second to use the electron waves to study the structure of matter in the same manner which had been in use with X-rays for 15 years.[8] Electron diffraction, however, offered new opportunities for structural studies in comparison with X-ray diffraction. X-rays are scattered by the electrons which surround the atoms of the scattering materials, whereas electrons are deviated— scattered—by the electric field of the positively charged nuclei. The main consequence is a much stronger interaction of the electron beam with the irradiated material which clearly manifests itself in the much larger absorption of electrons—beta radiation—compared with X-rays— gamma radiation. Electron diffraction, therefore, provides excellent means to study the *surface* structure of crystalline materials; it also makes it possible to localize hydrogen atoms—protons—in a lattice which cannot be "seen" by X-rays because their only electron is used for bonding them to the structure.[9] If one proceeds from the scattering by a crystalline body to that by isolated molecules, the strong interaction between the electron beam and the irradiated system gives much higher scattered intensities and permits one to shorten the exposure times by a factor between 1,000 and 10,000 in comparison with X-rays. Debye[10] had already established that X-ray scattering from gases produces certain characteristic patterns but that the exposure times of several days rendered this approach difficult and impractical. It occurred to me that the scattering of fast electrons from gases should lead to much shorter exposure times, to a much better control of the experimental conditions, and to the possibility of longer systematic studies of the interatomic distances. When I asked Dr. Raimund Wierl, one of the three high-level physicists in our laboratory, of his opinion he agreed that such experiments would be very interesting but certainly not easy. Wierl had received his Ph.D. summa cum laude with Professor Willy Wien in Munich and had an excellent training in the physics of high vacuum and high voltage. The difficulties of the new experiment were rather formidable. The electron beam had to be *narrow* and well *collimated,* which is difficult to achieve because the negatively charged electrons repel each other as they travel close to each other over a distance of a few centimeters. The beam has to be *monochromatic*; according to eq. (1) that means that all electrons should have the same velocity. In a strict sense this is impossible and there is always a certain velocity distribution which, for the purpose of this test, would have to be narrowed as much as possible. This electron beam would have to impinge perpendicularly on a jet stream of the gas which would have to be as narrow and as dense as possible in order to give a maximum of interaction between the electron and the scattering molecules. All this had to happen in a vacuum camera in order to avoid any scattering of a gas which did not belong to the jet stream. Evidently this experiment required inventiveness in instrument construction and extreme care and skill in its execution. Fortunately Wierl had both to an admirable degree and only a few weeks after our first conversation he showed me a beautiful photograph of carbon tetrachloride produced with 45-kV electrons in one tenth of a second. What a tremendous difference between this test and the daylong exposure with X-rays with much more diffuse patterns of lower contrast. We had selected CCl_4 for these first experiments because the heavy highly charged chlorine nuclei would be responsible for most of the deviation of the electron beam and the pattern would simply reflect the Cl—Cl distance in the tetrahedral molecule; knowing this distance it is easy to calculate the C—Cl distance, which was found to be 1.82 Å. This method, therefore, permits the direct experimental determination of the distance between atoms which are joined together by a covalent bond. In the late 1920s such "interatomic distances" were of considerable interest to arrive at *quantitative* models of organic molecules; some of them such as the C—C and C=O distance had been determined with some difficulty from X-ray crystal analysis and were known as the classic "Bragg atomic radii," but they had to be somewhat laboriously separated from the much more pronounced effect of the lattice scattering and the results of different independent sources did not at all agree in a satisfactory manner. Now electron diffraction of molecules in the gas phase offered the chance to *directly* measure these interesting and important quantities by the study of simple molecules in their natural state, namely in the gas phase. Thus the C—H distance was determined from CH_4, the C—F from CF_4, the C—C from ethane C_2H_6 and propane C_3H_8, and more complicated bonds like C=C from ethylene, C≡C from acetylene, and so on. Wierl[11] demonstrated the utility and reliability in a few typical cases. For instance, he found for the aromatic C—C distance in benzene 1.41 Å, but for the aliphatic C—C distances in cyclohexane 1.58; the latter agrees very well with the Bragg value for aliphatic chain molecules obtained from X-ray studies of aliphatic fatty acids. Meyer and I[12] had proposed in 1928 quantitative chain models for cellulose, silk, rubber, chitin, and starch with the use of the then available approximate values for C—C, C=C, C—O, and C—N; it was evident that better data for these fundamental bond distances would permit the construction of much more reliable models for these and other, even more complicated systems such as myosin, globulin, and keratin. Obviously a large field for fundamental research had been opened by the new method which was inviting elaborate systematic work. However, our laboratory was part of an industrial organization and not part of a university. The members of our top management, C. Bosch, K. H. Meyer, and A. Mittasch, progressive and enlightened as they were, had viewed with satisfaction the development of the new method, but certainly would disagree with extended studies of interatomic distances. In fact, Wierl already was aiming his electron beams on the surfaces of catalysts and magnetic tapes.

Fortunately at that time Linus Pauling, who had spent several months with Professor Arnold Sommerfeld in Munich, visited our laboratory and was told and shown all that we had established on the structure of polymers and on X-ray and electron diffraction. His own interest was focused on quantum chemistry with all its ramifications and he was, therefore, looking for all available background of quantitative character. Electron diffraction was a very promising method to provide some of this background. Pauling immediately liked the new method and we gladly gave him everything we had: construction of the camera, operating conditions, precautions, possible sources of errors, etc.

It is well known from the literature how much Pauling improved our original procedure in his own laboratory in Pasadena, how he collected essential new data, and how much all this improved and extended information ultimately contributed to the elucidation of protein and nucleic acid structure.

Since then enormous progress was made in the experimental technique of electron diffraction[13]: producing and maintaining very high vacuums in relatively large containers, exact regulation and control, velocity of electron beams, monochromatization, and collimation of the beam. As for the collection and indexing of the individual scattered beams, all refinements developed for X-rays were also used for electrons and later for protons and neutrons. All important novel developments of this character are presented by a group of distinguished experts in Part A.

Part B presents the results obtained up to date, including systems which contain B, Si, N, P, O, S, F, and various metals.

Thus, this book is both a textbook and an encyclopedic presentation of the existing information.

Herman F. Mark

References

1. An excellent survey on superconductivity can be found in Tinkham, M. "Modern Physics", Gordon & Breach: New York, 1965.
2. Nernst, W. *Z. Electrochem.* **1914**, *20*, 357.
3. Bose, S. N. *Z. Phys.* **1924**, *26*, 178.
4. Einstein, A. *Berl. Ber.* **1924**, p. 261; *Berl. Ber.* **1925**, p. 3.
5. de Broglie, L. *Ann. Phys.* **1925**, *3*, 22.
6. Davisson, C. J.; Germer L. H. *Phys. Rev.* **1927**, *30*, 705.
7. Thompson, G. P.; Reid, A. *Nature* **1927**, *119*, 890.
8. Cf. Mark, H. "Use of X-rays in Chemistry and Technology", J. A. Barth: Leipzig, 1926.
9. Cf. Mark, H.; Wierl, R. *Naturw.* **1930**, *18*, 778.
10. Debye, P. Lecture at "The Bunsen Society", 1930.
11. Cf. Wierl, R. *Phys. Z.* **1930**, *31*, 366; 1028; *Ann. Phys.* **1931**, *8*, 521; Mark, H.; Wierl R. *Fort. Phys. Chem.* **1931**, *21*, 1.
12. Meyer, K. H.; Mark, H. "The Structure of High Molecular Weight Compounds", Leipzig, 1930.
13. For a recent comprehensive treatment see Cowley, J. M. "Diffraction Physics", North-Holland: New York, 1985.

1

BORON AND SILICON COMPOUNDS

Vladimir S. Mastryukov

DEPARTMENT OF CHEMISTRY
MOSCOW STATE UNIVERSITY
MOSCOW, USSR

CONTENTS

INTRODUCTION   1
STEREOCHEMISTRY OF BORON COMPOUNDS   2
   Historical Notes   2
   Molecules with Two-Coordinated Boron   3
   Molecules with Three-Coordinated Boron   4
   Molecules with Four-Coordinated Boron   8
   Molecules with Four- to Six-Coordinated Boron   9
STEREOCHEMISTRY OF SILICON COMPOUNDS   16
   Historical Notes   16
   Molecules with Two- and Three-Coordinated Silicon   17
   Molecules with Four-Coordinated Silicon   18
ACKNOWLEDGMENTS   29
REFERENCES   29

Introduction

In a brief review chapter like this it is nearly impossible to discuss or even mention all the compounds containing boron and silicon that have been studied so far by gas-phase electron diffraction (GED). Therefore, it is not our intention to cover the literature thoroughly; instead, we prefer to select about 70 % of the compounds studied and to use them for discussion of some problems and trends of stereochemical interest. We apologize for having to leave out some interesting research papers.

All the discussion hinges on the three types of geometrical parameter, ie, internuclear distances, bond angles, and dihedral (torsion) angles. The classification and comparison of these parameters implies the existence of a certain reference point and such standard values usually do exist, but they can be valid only within a special class of compounds.

Stereochemically, boron and silicon have little in common, but there exists some resemblance in their bonding properties that distinguish them from carbon. In fact, B and Si usually have no tendency to form double bonds, but at the same time they have *enhanced capability for double-bond character*. This fact frequently is interpreted by invoking the *vacant p* and *d* atomic orbitals (AO) on boron and silicon atoms, respectively. Although there are still some uncertainties as to their proper description, the arguments mentioned form the main theoretical basis that has stimulated many studies and is used in discussion of the results obtained.

All the compounds treated in this chapter are classified according to "coordination number" (c.n.) of the silicon or boron atom. This characteristic, in spite of its approximate nature, is very useful and is still widely used in modern stereochemical literature (see, eg, a recent book by Hargittai[1]).

The sections devoted to boron and silicon each begin with a brief historical introduction that enumerates the most important milestones. This is greatly facilitated by the publication of Bibliography of Electron Diffraction, 1930–1979.[2] Nearly all these pioneer investigations were repeated subsequently. Therefore, the actual discussion is based on more recent data.

Stereochemistry of Boron Compounds

Historical Notes

Stock and Wierl were the first to investigate the molecular structure of a boron compound, borazine,[3] as early as 1932. In 1937 Bauer and Brockway reported data on a number of boron compounds. This work concerned topics that later attracted much interest, ie, *boranes* (diborane),[4] *boron halides* BX_3 (X = F, Cl, Br),[5] and *donor–acceptor complexes* (H_3BNMe_3).[6] Among other substances, boranes, due to their complexities, clearly demonstrated the weak points of GED, thereby providing some skepticism with respect to the method. In fact, for diborane, B_2H_6, the ethanelike structure first reported[4] later proved to be incorrect. The difficulties encountered in the course of the study of borane structures at the California Institute of Technology were very vividly described by Schomaker and Hedberg.[7] Research on boron halides has initiated numerous investigations on interaction of boron *p* AOs with lone pairs of electrons. From the 1970s, this field has been studied systematically by the Norwegian school by dealing with compounds that contain B—N, B—O, B—S, and B—Se bonds.

In early 1960s a new class of compounds, the *carboranes*, was discovered. Polyhedral frames of these molecules are formed by boron and carbon atoms linked together. The first representative of the class of icosahedral carboranes, $1,2\text{-}C_2B_{10}H_{12}$, was studied in 1965 by Vilkov and co-workers.[8] This field developed further in the 1970s.

Molecules with Two-Coordinated Boron

Two-coordinated boron exists in a handful of molecules of purely inorganic origin. We begin our discussion with the diboron trioxide molecule, B_2O_3, which plays a particularly important part in the stereochemistry of boron. Studies of molecular properties of B_2O_3 have attracted great interest during the past two decades; several contradictory reports appeared. In the early 1980s, a revival of interest in this system occurred due to improvements in both experimental and computational techniques.

Recently, new GED data were obtained[9] for B_2O_3 by using the modernized high-temperature equipment at Moscow State University. Initially, the data were treated in terms of the vibrationally averaged internuclear distances r_g, which is commonly used today. In agreement with earlier electron diffraction (ED) studies, the new data were found to be consistent with a V-shaped molecule of C_{2v} symmetry. Then, the data were reinterpreted in terms of a "harmonic equilibrium structure" r_e^h, by using a recently developed scheme of analysis. This new structure was found to deviate slightly from the simple V-shaped geometry by the nonlinearity of the BO_2 fragments (**1**).

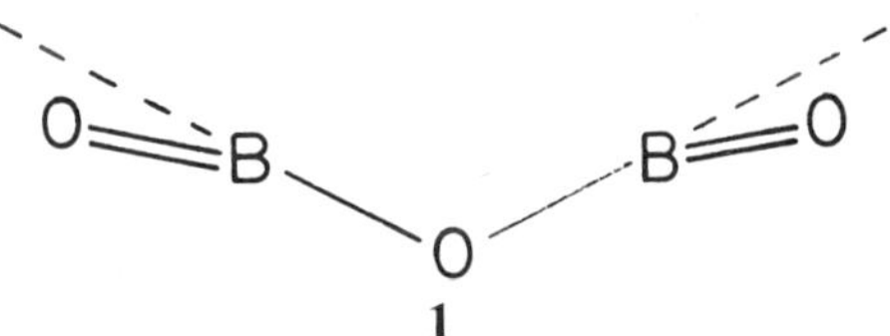

1

The main features of this structure were confirmed by *ab initio* calculations[10] as shown in Table 1-1, which summarizes the r_g, r_e^h, and r_e parameters. In conclusion, it may be stated that at the present time the principal structural aspects of the B_2O_3 molecule seem to be reliably established. On the other hand, its sulfur analog, B_2S_3, has not been studied in so much detail.[11]

The second group of compounds with two-coordinated boron consists of the metal metaborates MBO_2 [M = Li, Na, K, Rb, Cs, Tl(I); see reference 12 and references cited therein]. In general, the metaborates contain linear symmetric OBO groups, although differences between the two types of B—O bond lengths (up to 0.05 Å) could not be ruled out. The average distance, approximately

Table 1-1. EXPERIMENTAL AND CALCULATED
STRUCTURAL PARAMETERS OF DIBORON TRIOXIDE[a,b]

Parameter	r_g	r_e^h	r_e
B=O, Å	1.219(7)	1.195(6)	1.191
B—O, Å	1.323(8)	1.329(10)	1.330
∠BOB, deg	137.5(63)	134.2(50)	137.1
∠OBO, deg	180	173.4(44)	178.8

[a] In tables and throughout this chapter: parenthesized numbers
are uncertainties referring to the last digit; data without uncertainties represent assumed values.

[b] Data for r_g and r_e^h from reference 9, data for r_e from reference 10.

1.251(1) Å, is close to the arithmetic mean of r_g(B=O) and r_g(B—O) of B_2O_3 (ie, 1.27 Å).

The following M—O distances and the effective MOB bond angles for a series of molecules MOBO were found to be: 3.26(5) Å and 112(5)° (M = Rb); 3.55(5) Å and 130(6)°(M = Cs); 3.39(5) Å and 132(5)° (M = Tl).

Molecules with Three-Coordinated Boron

Compounds of the Type BX_3, BX_2Y, and B_2X_4. The molecules BX_3 have three electron pairs in the valence shell of the central atom, and according to the valence shell electron-pair repulsion (VSEPR) concept,[13] they are planar, with ∠XBX = 120°. Table 1-2 presents the internuclear distances B—X in the molecules studied; some additional examples are also given in Table 1-3.

The GED study of BF_3 by Kuchitsu and Konaka[14] is remarkable because it helped to reveal an interpretational error in the earlier spectroscopic work, a fact that subsequently was confirmed experimentally.[15] Quite recently, the original GED data were used by Spiridonov and Gershikov[16] in a combined analysis together with detailed spectroscopic information (vibrational frequencies, Coriolis constants, rotational constants, and centrifugal constants). The authors reported r_e^h (B—F) = 1.31074(8) Å, measured with a high accuracy.

The bonding and structure of organoboranes are of particular interest when an electron-rich organic moiety is bonded to an electron-deficient boron atom (**2**). In this case one can expect specific substituent effects and conformations. The trivinylborane molecule $B(CH=CH_2)_3$,[17] serves as a good example in this regard. The C=C bond length of 1.370(6) Å in trivinylborane was found to be considerably longer than the corresponding C=C bond in ethylene (where

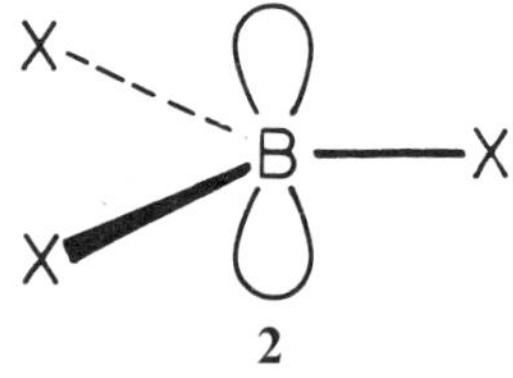

2

Table 1-2. BOND LENGTHS FOR BX_3 MOLECULES[a]

Molecule	B—X (Å)
BF_3	1.313(1)
BCl_3	1.742(4)
BBr_3	1.893(5)
BI_3	2.118(5)
$B(CH_3)_3$	1.578(1)
$B(CH{=}CH_2)_3$	1.558(3)

[a] Individual references may be found in Landolt-Börnstein.[11]

$r_g = 1.337(2)$ Å). The best agreement with experiment was obtained for a planar dynamic model. Both these facts are interpreted as providing evidence for π-electron delocalizations from the $C{=}C$ double bond to the vacant boron p AO in trivinylborane.

Similar problems have been treated by Norwegian scientists[11,18-20] in a series of investigations of compounds $Me_{3-n}B(XMe)_n$ where X = NH, O, S. The B—N, B—O, and B—S distances (see Table 1-3) are considered in terms of the possible transfer of electrons from a π donor (NHMe, OMe, SMe) to the boron $2p$ orbital. The double-bond character of the link between B and X depends on the π-electron donor capacity of the ligand and on the number of strong π donors attached to each boron atom. Since the CH_3 group is a weak π donor, one can expect the B—X bonds to become shorter as the number of methyl groups increases. In fact, such an expectation is verified for the B—N and B—S bonds but not for the B—O bonds. In the last case, due to the high electronegativity of oxygen, an opposing effect is possible, since bonds to a central atom are known to become shorter when the number of electronegative ligands is increased. In all molecules studied, the planar or nearly planar

Table 1-3. BOND DISTANCES FOR SOME BX_3, BX_2Y, AND BXY_2 MOLECULES

Molecule	Bond distances (Å)		Ref.
	B—X	B—C	
Me_2BNHMe	1.397(2)	1.586(2)	18
$MeB(NHMe)_2$	1.418(2)	1.586(3)	18
$B(NHMe)_3$	1.432(2)		18
Me_2BOMe	1.361(2)	1.575(2)	19
$MeB(OMe)_2$	1.375(4)	1.571(6)	19
$B(OMe)_3$	1.368(2)		19
Me_2BSMe	1.779(5)	1.570(4)	11
$MeB(SMe)_2$	1.796(7)	1.567(10)	20
$B(SMe)_3$	1.805(2)		11

conformation of heavy-atom skeletons is attributed to substantial amounts of double-bond characters. *Ab initio* calculations[21,22] performed for similar model compounds reproduced fairly well the trends observed in the experimental studies.

The diboron tetrahalides B_2X_4 consist of two BX_2 groups joined by a B—B single bond. These molecules are the simplest representatives of conformational systems, since rotation around the B—B bond gives rise to different conformers. The equilibrium conformer for B_2F_4 is planar (D_{2h} symmetry:3) as opposed to B_2Cl_4 and B_2Br_4, for which the equilibrium conformer is staggered (D_{2d}:4). This result has been rationalized in terms of different balancing of the effects of conjugation (favoring the planar form) and steric repulsion. In the crystal phase all of these molecules are planar.

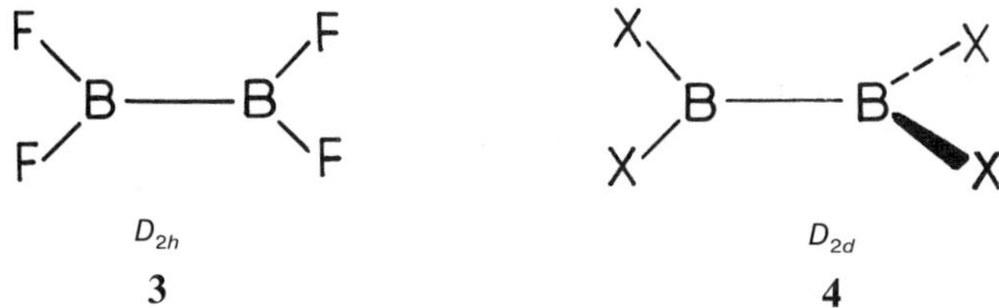

The B_2X_4 molecules were studied by Hedberg and co-workers[11,23,24] at three to five different temperatures, to measure the potential barrier (treated parametrically). The potential of the form $V = \frac{1}{2}V_0(1 - \cos 2\varphi)$ was assumed, and the V_0 values thereby obtained are shown in Table 1-4.

In the last paper[24] of this series, Danielson and Hedberg conclude: "The structural work on the three lower diboron tetrahalides provides a clear picture of trends in the bond distances and bond angles and allows one to predict the properties of the very unstable iodine compound with considerable confidence." These predicted characteristics are also found in Table 1-4.

Cyclic Compounds. There are only a few examples of cyclic compounds in which one or more boron atoms are contained within a five- or six-membered ring. Their main geometrical parameters are displayed in Table 1-5. In spite of the diversity of their composition, these molecules are planar or nearly planar, except for 1,3-dimethyl-2-chloro-diazaboracyclohexane.[25]

Table 1-4. STRUCTURAL PARAMETERS FOR B_2X_4 MOLECULES

Molecule	Bond distances (Å)		∠XBX (deg)	V_0 (kJ mol^{-1})	Ref.
	B—B	B—X			
B_2F_4	1.720(4)	1.317(2)	117.2(2)	1.76(67)	23
B_2Cl_4	1.702(69)	1.750(11)	118.7(7)	7.74(21)	11
B_2Br_4	1.689(16)	1.902(4)	120.7(3)	12.84(1.38)	24
B_2I_4	1.69	2.10	123	18.4	24

Table 1-5. SELECTED GEOMETRICAL PARAMETERS OF CYCLIC BORON-CONTAINING COMPOUNDS[a]

Molecule	Distance (Å)		Bond angle (deg)	
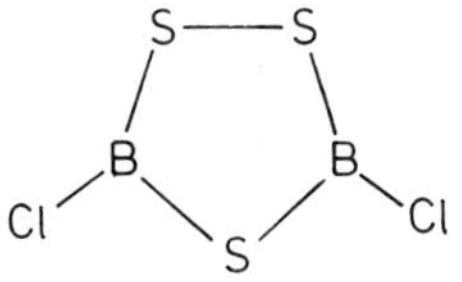	B—S$_{av}$	1.794(5)	SBS	121.7(5)
	B—Cl	1.756(9)	ClBS(B)	120.8(5)
	S—S	2.069(3)	BSB	96.9(6)
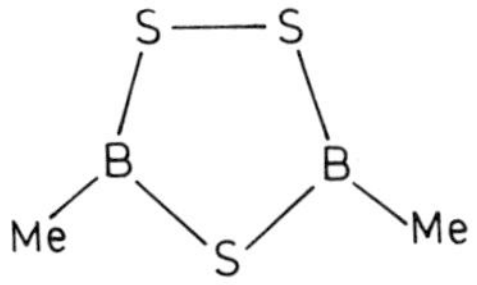	B—S$_{av}$	1.803(3)	SBS	117.7(2)
	B—C	1.569(5)	CBS(B)	122.8(16)
			BSB	101.6(4)
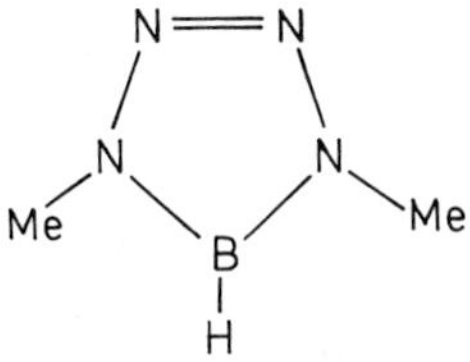	B—N	1.413(10)	NBN	101.8(6)
	N—N	1.375(5)	CNN	115.8(3)
	N=N	1.291(6)		
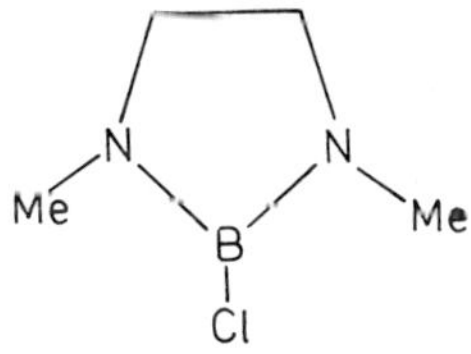	B—N	1.413(3)	NBN	110.8(3)
	B—Cl	1.770(4)	BNC(C)	108.6(3)
	C—N$_{av}$	1.455(2)	NCC	105.7(3)
(HBO)$_3$	B—O	1.376(2)	OBO = BOB	120.00(64)
(HBNH)$_3$	B—N	1.455(21)	NBN	117.7(12)
	B—H	1.258(14)	BNB	121.1(12)
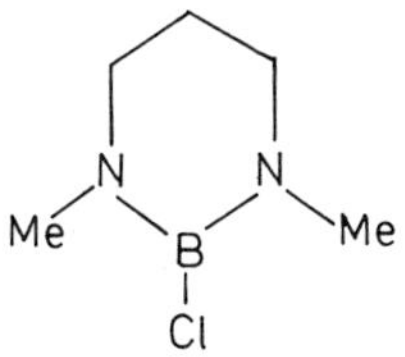	B—N	1.417(5)	NBN	120.8(5)
	B—Cl	1.782(5)	BNC	121.6(5)
			BNC(C)	124.1(5)

[a] References to individual molecules may be found in Landolt-Börnstein,[11] except for the last molecule listed (see Seip and Seip[25]).

In borazine,[26] $B_3N_3H_6$, the boron-nitrogen analog of benzene, the B—N distance is close to that found in $B(NHMe)_3$ (see Table 1-3) and is about 0.04 Å larger than the aromatic carbon–carbon bond in benzene.

Molecules with Four-Coordinated Boron

The ability of the boron atom to use the formally empty p orbital is demonstrated especially clearly by its ability to form complex compounds of the type X_3BDY_n. The elements of Groups V and VI act as electron donors (D) in such compounds.

Table 1-6 lists the geometrical parameters of the donor–acceptor complexes studied. Traditionally,[38,39] these parameters are compared with those of the free donor and acceptor molecules to reveal the structural consequences of complex formation. With this in mind, at the bottom of Table 1-6 are collected the data for the donors, while the internuclear distances of acceptor halides are found in Table 1-2.

The changes in the bond angle that are observed upon complexation can be at least partly rationalized in terms of VSEPR theory, in which the molecule X_3BDY_3 is formulated in the following manner.[38,39] When the lone electron pair in the free donor is replaced by a bonding pair in the complex, a smaller repulsion for the ligands Y is obtained and, therefore, the $\angle$YDY bond angle is expected to *increase*. The appearance of a fourth electron pair on an acceptor atom leads to *a decrease* in the $\angle$XBX bond angle. In addition to these two consequences of the VSEPR concept, there appear three new nonbonded

Table 1-6. Main Geometrical Parameters of Donor–Acceptor Complexes and Free Donor Molecules

Molecule	Bond distances (Å)			Bond angles (deg)		Ref.
	B—D	B—X	D—Y	X—B—X	Y—D—Y	
F_3BOMe_2	1.73(5)	1.361(8)	1.440(13)	117(2)	110	27
H_3BNMe_3	1.656(2)	1.261(6)	1.485(1)	112.8(4)	109.2(2)	28
F_3BNMe_3	1.664(11)	1.354(6)	1.468(10)	113.1(9)	108.5(7)	29
	1.674(4)	1.374(2)	1.485(2)	112.6(3)	109.2(4)	30
Cl_3BNMe_3	1.659(6)	1.839(4)	1.495(4)	110.8(3)	108.7(5)	29
	1.652(9)	1.836(2)	1.497(3)	110.9(2)	108.1(3)	31
Cl_3BPMe_3	1.941(16)	1.851(7)	1.800(4)	109.4(4)	109.3(3)	32
Br_3BNMe_3	1.663(13)	2.001(3)	1.500(5)	110.3(3)	107.8(5)	31
Br_3BPMe_3	1.946(29)	2.010(9)	1.804(4)	111.7(7)	108.0(7)	33
I_3BNMe_3	1.663(13)	2.245(4)	1.497(5)	108.6(4)	106.0(8)	34
I_3BPMe_3	1.947(11)	2.233(3)	1.809(3)	111.6(3)	106.0(5)	35
OMe_2			1.415(1)		111.8(2)	36
NMe_3			1.458		110.9	37
PMe_3			1.844		98.8	37

interactions D...X, B...Y, and X...Y, which tend to *decrease* all the aforementioned bond angles. Thus, this new factor acts in the same direction as the VSEPR arguments applied to the acceptor part of a complex, whereas in the donor part there arises a competition between two opposing effects. In practice, the fact that trimethylphosphine and trimethylamine acting as donor molecules behave differently illustrates a predominance of one of the two factors mentioned.

In all the complexes studied, the properties of the donor–aceptor bond are of special concern. In some cases the heights of the potential barrier around this bond were measured. Additionally, several complexes were studied both in the solid phase and in the gas phase. It is concluded that the B—N bonding is enhanced in the solid phase while for the B—P bond there is no such trend: while the B—P bond is *shorter* in crystalline Br_3BPMe_3, it is *longer* in crystalline Cl_3BPMe_3 vis-à-vis the corresponding gas-phase data.[32,33]

Molecules with Four- to Six-Coordinated Boron

Boranes. The boron hydrides (ie, boranes) form a specific group of compounds; several decades of research were required to understand their structural properties.[40] W. N. Lipscomb was especially active both in the experimental structural studies and in the development of the theory, and in 1976 he was awarded the Nobel Prize in chemistry for his contributions. The major part of the structural work has been accomplished by using the X-ray technique, while GED helped to refine the geometries of some lower boranes. We shall start our discussion with the simplest borane (B_2H_6), the structure of which is common to several derivatives of the Group III elements.

There have been many experimental and theoretical studies concerning the structure and chemical bonding of diborane. The bridge model of B_2H_6 possesses two distinct structural types of hydrogen atom, the bridge (H_b) and terminal (H_t) hydrogens, as shown in Figure 1-1. It is seen from the data presented in Table 1-7 that the bridge B—H_b distance is about 0.14 Å longer than the terminal distance. As shown in Figure 1-1, there is no direct B—B bond in diborane, but this distance is usually reported (as B—B) and discussed. According to Bartell and Carrol[41] (1.775 Å), this B...B distance is very similar to the normal B—B bond lengths in B_2X_4 molecules, which vary from 1.69 to 1.72 Å (see Table 1-4) or from the sum of the covalent radii reported by Pauling[46] (1.62 Å) or by Slater[47] (1.70 Å).

Bartell and his co-workers systematically studied the influence of isotopic substitution on molecular structure.[48] As a part of this program, deuterodiborane and diborane were investigated. The *primary* isotope effect (the difference between B—H_b and B—D_b) is 0.005 Å, while the *secondary* isotope effect on the B—B bond length is about 0.01 ± 0.005 Å. The geometrical parameters of diborane were subsequently revised by Kuchitsu[49] when the rotational constants of this molecule became available.

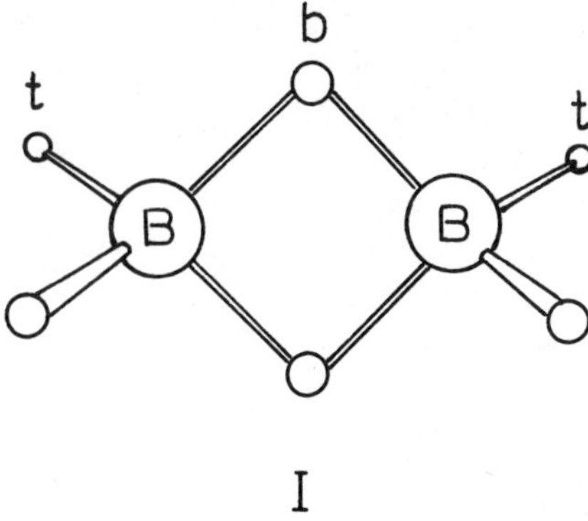

Figure 1-1. Structure of B_2H_6 (I) and diborane–like molecules (II–V).

Figure 1-1 also shows different types of the diborane-like structure produced by substitution. The structures of type II and III are obtained by substitution at terminal and bridge positions, respectively. In the terminal-substituted compounds like 1,2-dimethyldiborane two stereoisomers, cis and trans, are possible. Finally, the substitution of boron (and hydrogen) atoms leads to models IV and V in Figure 1-1. The geometrical parameters for some of these systems are given in Table 1-7. The structures of dimeric metal trihalides A_2X_6 are described in Chapter 9.

Table 1-7. MOLECULAR STRUCTURE OF DIBORANE AND RELATED COMPOUNDS

Molecule	Bond distances (Å)			Bond angle (deg)		Ref.
	B—B or B—M	B—H$_t$	B—H$_b$			
B_2H_6	1.775(3)	1.196(8)	1.339(6)	BBH$_t$	120.5(4)	41
				H$_b$BH$_b$	97.0(3)	
B_2D_6	1.771(3)	1.198(6)	1.333(4)	BBH$_t$	119.3(9)	41
				H$_b$BH$_b$	96.8(3)	
H_2BH_2BHCl	1.775(15)	1.205(13)	1.331(15)	BBCl	120.9(3)	42
	1.775(5)(BCl)			H$_t$BH$_t$	125.0(60)	
trans-MeHBH$_2$BMeH	1.799(8)	1.241(10)	1.365(8)	BBC	121.8(6)	43
cis-MeHBH$_2$BMeH	1.798(7)	1.239(8)	1.358(6)		122.6(5)	43
Me$_2$BH$_2$BMe$_2$	1.840(13)	^{a}	1.364(45)		120.0(13)	11
$H_2BH(NH_2)BH_2$	1.93(9)	1.15(9)	^{b}	BNB	76.2(2.8)	44
Me$_2$AlH$_2$BH$_2$	2.128(8)	1.218(16)av		CMC	118.4(7)	45
Me$_2$GaH$_2$BH$_2$	2.163(8)	1.205(19)av			118.8(12)	45

^{a} B—C = 1.590(3) Å.
^{b} B—N = 1.504(26) Å.

The diborane type of bonding sometimes occurs in metal borohydrides, ie, compounds that contain BH_4 groups linked to the metal center. The molecular structures of Be(BH$_4$)$_2$,[11] Al(BH$_4$)$_3$,[11] Zr(BH$_4$)$_4$,[11] Ti(BH$_4$)$_3$,[50] Cp$_2$Ti(BH$_4$),[51] and Ga(BH$_4$)$_2$H (see reference 52 and other references cited therein) have been reported. Of these compounds, beryllium borohydride is of particular interest. In spite of numerous experimental[53] and theoretical[54] investigations, there is no clear-cut decision regarding its structure. The possibility of Be(BH$_4$)$_2$ comprising different species in the gas phase (depending on the state of the sample) has been discussed.

The BH_4 groups in metal (M) borohydrides act as bidentate (**5**) or tridentate (**6**) ligands. Structure **5** and model IV (Figure 1-1) represent essentially the same type of bonding, which is exemplified by Al(BH$_4$)$_3$.[11] The unusual bidentate BH_4 groups with nonequivalent hydrogen bridges occur in HGa[($\mu - H)_2BH_2]_2$.[52] Structure **6** (with trihydrogen-bridged ligands) was found in Zr(BH$_4$)$_4$[11] and in Ti(BH$_4$)$_3$.[50]

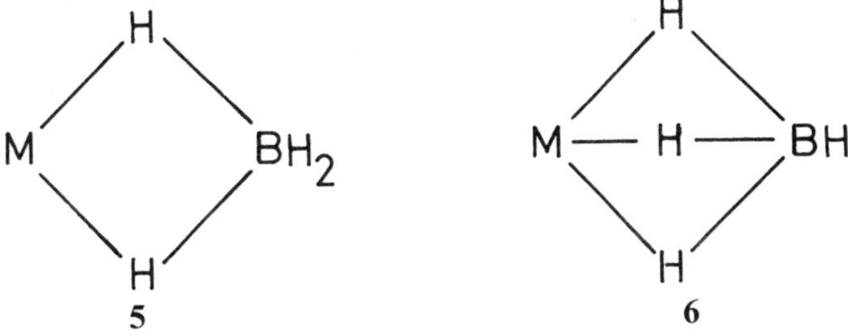

Few higher boranes have been studied by GED, and we mention only B_4H_{10},[55] B_5H_9,[56] and $B_{10}H_{14}$.[57] The most up-to-date structural results are

available for tetraborane(10).[55] The reliability of the visual measurements by Hedberg and co-workers for pentaborane(9)[56] was justified later by the investigation of methyl- and silylpentaboranes[58] and finally by direct microwave study of this molecule.[59]

From the studies carried out on boranes, it is known that there are three main structural types: *closo*, *nido*, and *arachno*. *closo*-Boranes $B_n H_n^{2-}$ have the structure of an *n*-vertex polyhedron, which also serves as the basis for *nido*-boranes ($B_n H_{n+4}$) and *arachno*-boranes ($B_n H_{n+6}$). The *nido*- and *arachno*-boranes are produced via removal of one and two vertices from the parent polyhedron, respectively. According to these considerations,[60–64] $B_4 H_{10}$ and $B_5 H_9$ are the fragments of *closo*-borane $B_6 H_6^{2-}$, which has octahedral geometry as determined by the X-ray technique.[65] This is illustrated in Figure 1-2, which also shows the B—B bond lengths. The structural similarity of these molecules is revealed by a remarkable constancy of the B—B bond length, which belongs to the type common to the whole triad (1.705, 1.700 and 1.69 Å). The second type of B—B bond (ie, that which contains the hydrogen bridge) disappears at the final stage of polyhedron construction.

There is a strong correlation between the type of skeletal structure (complete or incomplete polyhedra) and the number of the skeletal-bonding electrons in the system. Such a conclusion given in terms of molecular orbital can be applied to other systems, including carboranes, metallocarboranes, metal carbonyl clusters, and metal–hydrocarbon π complexes. Wade[62] has stated: "Although boranes may not have launched any satellites to brighten our night skies, they have nevertheless shed much light on several important and developing areas of chemistry."

In concluding this section, we would like to mention several important theoretical papers,[66–71] which discuss the structures of well-known boranes together with structural predictions for molecules that have not yet been prepared.

Carboranes. Carboranes may be regarded as being derivatives of boranes in which carbon atoms are an integral part of the skeletal framework. This is exemplified by *closo*-carboranes, $C_2 B_{n-2} H_n$, which retain the same type of polyhedron that occurs in the isoelectronic *closo*-boranes $B_n H_n^{2-}$. For general structural data on carboranes studied in the gas phase as well as in crystal, the reader is referred to the review by Mastryukov, Dorofeeva, and Vilkov.[72] Since *closo*-carboranes have been studied more thoroughly than other families, we have confined these considerations to the most symmetrical members of the series. These compounds are shown in Figure 1-3, and their geometrical parameters are collected in Table 1-8.

Triangles of boron atoms in polyhedral boranes are also retained as the most favorable building blocks for carboranes. The bond angles in these triangular faces are usually close to 60°, being, however, *larger* (up to 73.1°) when an apical carbon atom appears and correspondingly *smaller* (up to 53.4°) for the angles CBB.[75] Carboranes are particularly interesting in that they frequently contain carbon atoms with coordination number greater than four.[77] Analysis of the

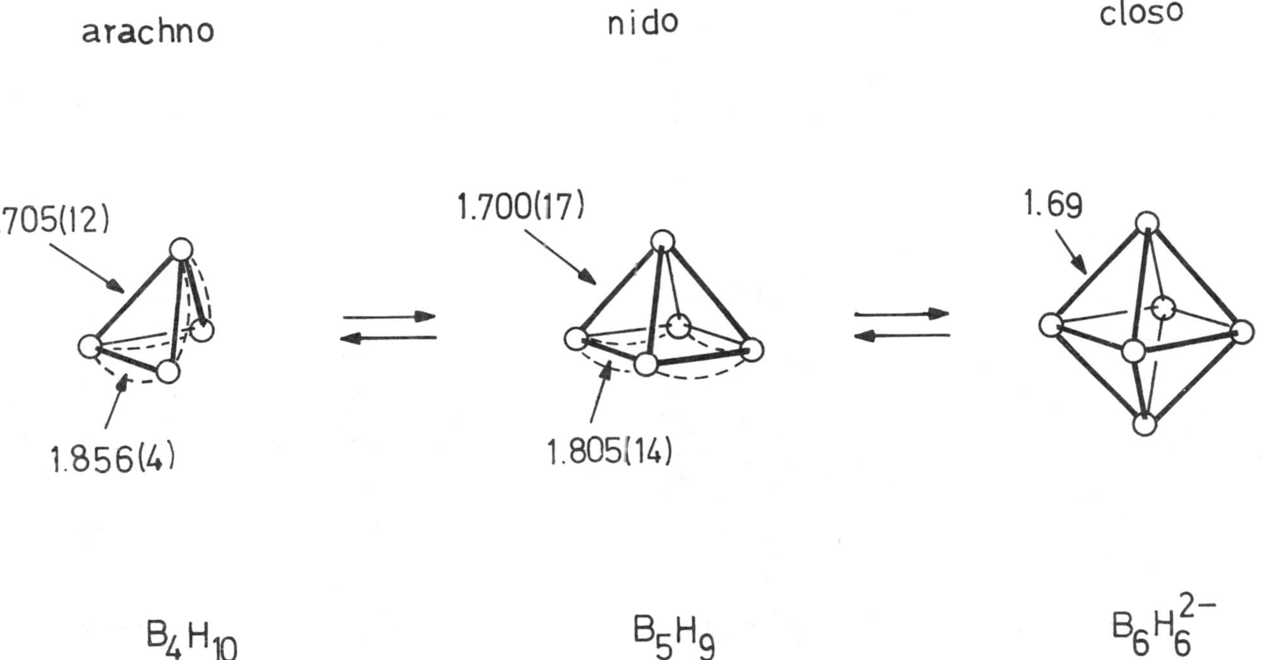

Figure 1-2. The "arachno-nido-closo" relationship for boranes. Terminal hydrogens are omitted for clarity, while the B—H—B links are shown by dotted lines. (Adapted from reference 64.)

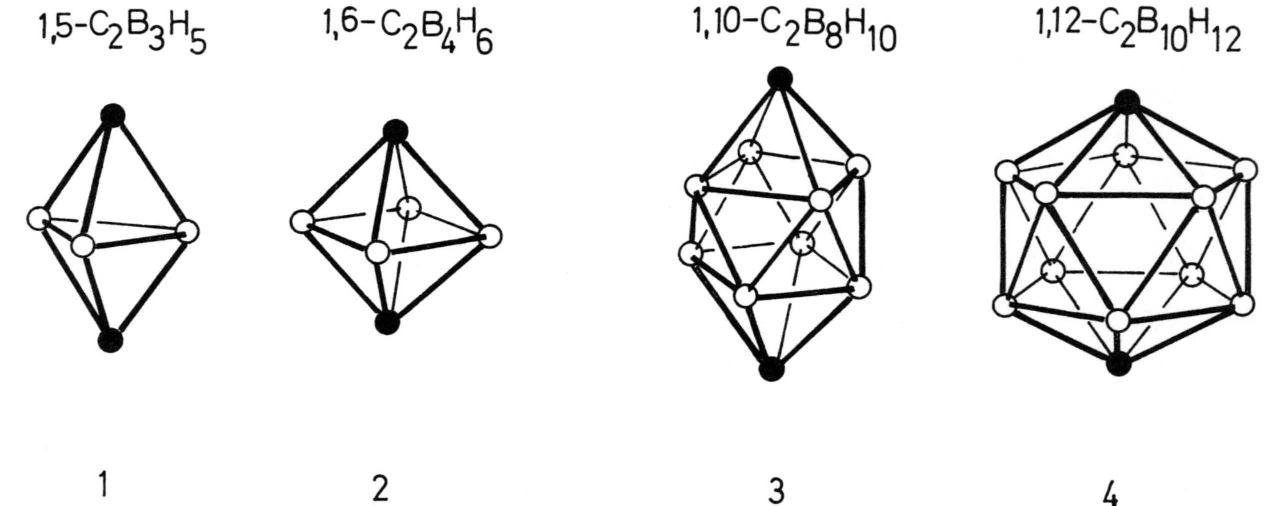

Figure 1-3. Carborane geometries: (1) trigonal bipyramid, (2) octahedron, (3) bicapped square antiprism, (4) icosahedron. Carbon atoms are indicated by solid circles; hydrogen atoms are omitted for clarity.

Table 1-8. COMPARISON OF GEOMETRICAL PARAMETERS
OF CARBORANE MOLECULES

Molecule	B—C (Å)	B—B (Å)		Ref.
		Basal	Equatorial	
$1,5\text{-}C_2B_3H_5$	1.556(2)	1,853(2)		73
$1,6\text{-}C_2B_4H_6$	1.633(4)	1.720(4)		73
	1.635(4)	1.725(12)		74
$1,10\text{-}C_2B_8H_{10}$	1.602(2)	1.850(5)	1.829(4)	75
$1,12\text{-}C_2B_{10}H_{12}$	1.710(11)	1.792(7)	1.772(13)	76

B—C distances in carboranes $1,5\text{-}C_2B_3H_5$, $1,6\text{-}C_2B_4H_6$, and $1,12\text{-}C_2B_{10}H_{12}$ suggests that the distance varies linearly with the coordination number of the carbon atom (n_C)[72,75]:

$$r(B\text{—}C) = 1.248 + 0.077 n_C \ (\text{Å}) \qquad n_C = 4, 5, 6$$

A similar empirical relationship was established for the C—C distances that occur in other carboranes:

$$r(C\text{—}C) = 1.074 + 0.058n \ (\text{Å}). \qquad n = 6, 8, 9, 10$$

Here, n is the sum of carbon coordination numbers minus 2. This pattern resembles Stoicheff's rule, well known in the stereochemistry of organic compounds.[79] On the other hand, the third type of polyhedral distance, B—B, does not reveal any distinct dependence on the coordination number of boron.[72,75] The B—B bond length is likely to be controlled by electronic factors in the polyhedral system. This point may be illustrated at least in part by the results of an *ab initio* calculation for $1,5\text{-}C_2B_3H_5$.[80] The long B—B bond in this molecule (see Table 1-8) can be rationalized in terms of a very small overlap population between the boron atoms contrasted to a very strong bonding between B and C. Therefore, as an extreme case, one can imagine this molecule as having a bicyclo [1.1.1] pentane-like structure, ie, one that lacks B—B bonds.

From the material presented above, one can draw a general conclusion: the larger the coordination numbers of the boron and carbon atoms, the longer are the polyhedral B—C and C—C bonds in carboranes. By contrast, exopoly-hedral bonds tend to be *shorter* than the strandard distances. Initially, this was illustrated by the C—I distance of 2.095(15) Å in $1,12\text{-}C_2I_2B_{10}H_{10}$.[81] This bond is shorter than the "aliphatic" C—I bond, being indistinguishable from the "olefinic" one, since $r(C\text{—}I) = 2.106(5)$ Å in $CI_2\text{=}CI_2$.[11] Furthermore, it was found[72] that the exopolyhedral C—C bond lengths depended little on the environment, which is described by the empirical equation:

$$r(C\text{—}C)_e = 1.510 + 0.002n \ (\text{Å}) \qquad n = 6, 7, 8, 10$$

These distances vary on average from 1.52 to 1.53 Å while n increases from 6 to 10; ie, they are nearly invariant. Our recent results on $1,12\text{-}(CCH_3)_2B_{10}H_{10}$[82]

and $(1,12\text{-}CB_{10}H_{10}CH)_2$[83] confirm this finding. It is of particular interest to compare two formally equivalent C—C distances formed by six-coordinated carbons:

$$r(\text{C}—\text{C}) = 1.653(49) \text{ Å in } 1,2\text{-}C_2B_{10}H_{12} \qquad \text{(reference 76)}$$

$$r(\text{C}—\text{C})_e = 1.553(26) \text{ Å in HCB}_{10}H_{10}\text{C}—\text{CB}_{10}H_{10}\text{CH} \qquad \text{(reference 83)}$$

Their difference is 0.1 Å.

The earlier electron diffraction studies of carboranes usually ignored the theoretical account of vibrational effects because of the lack of force fields for these polyhedral molecules. In 1980 the first attempt to calculate an r_α structure was undertaken for $1,10\text{-}C_2B_8H_{10}$,[75] based on the unpublished force field worked out by Urevig for a similar molecule, $1,6\text{-}C_2B_4H_6$ (the corresponding paper appeared later[84]). The result of the calculation was particularly rewarding because it revealed an interesting feature: the amplitudes of the bonded and nonbonded cage distances (which differ by a factor of 1.5) are close to one another. This finding is at variance with the empirically established relations of Mastryukov and Cyvin,[85] who showed that frequently the amplitudes of vibration are roughly proportional to the corresponding internuclear distances.

Electron diffraction studies of p-phospha- and p-arsacarboranes, $1,12\text{-}XCHB_{10}H_{10}$ (X = P, As),[86] provided the first structural determination of heteroatomic carboranes in the gas phase. These molecules, together with the parent carborane $1,12\text{-}C_2B_{10}H_{12}$, are regarded as being three-dimensional aromatic systems. The authors[86] discuss the geometrical parameters in comparison with those for their two-dimensional counterparts: phosphabenzene, arsabenzene, and benzene.

Carboranes like boranes represent a field of chemistry of unusual theoretical interest in view of the widespread occurrence of multicenter bonds. There are a number of excellent theoretical papers (see, eg, references 80, 87, and 88) and previous publications), which contain valuable structural information that pertains to these systems. Many of the structural features of carboranes have been properly described, while others still await clarification.

Stereochemistry of Silicon Compounds

Historical Notes

The bases of stereochemistry of silicon compounds can be found in papers by Brockway and associates that appeared in 1934 and 1938 and referred to *silicon halides* (SiCl_4)[89] and *silanes* (Si_2H_6).[90]

In 1955 Hedberg published his classical paper[91] on $\text{N(SiH}_3)_3$ followed by a study[92] of $\text{O(SiH}_3)_2$ in 1963. The impact of these articles was felt in the late 1960s, when numerous investigations of the participation of silicon and germanium *d* orbitals in chemical bonding with nitrogen, phosphorus, arsenic, antimony, oxygen, sulfur, and selenium were reported by British investigators.

In the period 1955–1958 American[93] and Japanese[94] research groups used Si_2Cl_6 as a model compound for studying hindered internal rotation. The influence of these contributions on the subsequent development of the field is illustrated in an excellent review by Clark.[95]

In 1962 a paper appeared that presented the structure of silacyclopentane,[96] the first representative of the cyclic organosilicon compounds. The structure of the first cyclic polysilane (cyclopentasilane, Si_5H_{10})[97] was not reported until 1976.

Wells,[98] in his authoritive treatise "Structural Inorganic Chemistry," writes that "In its simple molecules and ions Si does not exhibit a covalency of less than four except (possibly) in the silyl ion." However, in 1980 the molecular structure of $Me_2Si{=}CH_2$ with *three-coordinated Si* was reported.[99] Slightly later, in 1983, *two-coordinated Si* was reported for the carbene analogs $SiCl_2$ and $SiBr_2$.[100] On the other hand, an attempt to reach the elevated coordination number of silicon was not so successful. In fact, silicon in methyl silatrane exhibits coordination number 5 according to single-crystal X-ray data, whereas in the gas phase silicon appears to retain an approximately tetrahedral arrangement.[101] Selected aspects of stereochemistry of silicon compounds are described in a number of review articles.[102–105]

Molecules with Two- and Three-Coordinated Silicon

Compounds of the type SiX_2 that contain two-coordinated silicon (ie, silicon analogs of carbenes) are highly reactive; therefore, they polymerize readily at ordinary temperatures. The difficulties associated with their structural investigation lie in procedures for generating these reactive intermediates and in the control of vapor composition. Both problems were solved by Hargittai and his co-workers[100] in a combined electron diffraction/mass spectrometric study using a high-temperature reactor nozzle system. Dihalides $SiCl_2$ and $SiBr_2$ were produced in the molybdenum reactor nozzle at about 1200°C by the reaction of $Si(solid)$ with Si_2Cl_6 and $SiBr_4$, respectively. The following molecular parameters were obtained from the structural analysis:

$$r_a(Si{-}Cl) = 2.083(4)\ \text{Å} \quad \text{and} \quad \angle\,ClSiCl = 102.8(6)° \quad \text{for } SiCl_2$$

$$r_a(Si{-}Br) = 2.243(5)\ \text{Å} \quad \text{and} \quad \angle\,BrSiBr = 102.7(3)° \quad \text{for } SiBr_2$$

Furthermore, from the amplitude of vibration $l(Br...Br)$, the bending frequency of $SiBr_2$ could be estimated ($v_2 = 122.5\ \text{cm}^{-1}$). It may be noted that the resulting bond angles in SiX_2 are only 1.5–2.5° larger than those in the germanium analogs: 100.3(4)° in $GeCl_2$[106] and 101.2(9)° in $GeBr_2$[107] (for further details, see Chapter 9).

Much experimental work was also involved in the electron diffraction study of 1,1-dimethylsilaethylene, $CH_2=SiMe_2$, the unique representative of molecules containing a carbon–silicon double bond.[99] This short-lived intermediate was generated via pyrolysis of 1,1-dimethylsilacyclobutane, and its percentage in the gas phase varied from 23(8)% to 49(4)%. The average $r_g(Si=C) = 1.83(4)$ Å seems to be too long according to the previous theoretical estimates (1.63–1.75 Å) and to the subsequent response[108,109] to this work. This conclusion is partly supported by the X-ray measurement leading, eg, to the value $r(Si=C) = 1.764(3)$ Å.[110]

Molecules with Four- Coordinated Silicon

Compounds of Type SiX_4, SiX_3Y, and SiX_2Y_2. The structural data on SiX_4 molecules are summarized in Table 1-9. These molecules have an exactly tetrahedral environment, and owing to the high symmetry, they are characterized by one Si—X distance which has been measured accurately. Structural studies of various SiX_3Y species have been made (see Table 1-10). However, only a few examples of SiX_2Y_2 molecules have been studied by GED (Table 1-11).

The Si—C bond length will be considered first, since it is the most important parameter for many organosilicon compounds. Unlike various types of $C(sp^n)$—$C(sp^n)$ bonds,[78] one can select only three for the Si—C bond, ie, $Si—C(sp^3)$. $Si—C(sp^2)$, and $Si—C(sp)$, since the formation of multiple bonds is not typical for silicon. These Si—C bonds occur in the following triads of molecules:

$$H_3Si—CH_3 \qquad H_3Si—CH=CH_2 \qquad H_3Si—C\equiv CH$$

or

$$Si(CH_3)_4 \qquad Si(CH=CH_2)_4 \qquad Si(C\equiv CH)_4$$

Table 1-9. Bond Lengths for SiX_4
Molecules

Molecule	Si—X (Å)
SiF_4	1.552(2)
$SiCl_4$	2.019(3)
$SiMe_4$	1.875(2)
$Si(CH=CH_2)_4$	1.855(2)[a]
$Si(SiMe_3)_4$	2.361(3)
	1.889(3) (Si—C)
$Si(NCO)_4$	1.688(3)
$Si(OMe)_4$	1.614(1)[b]

[a] Reference 111.
[b] Reference 112.

Table 1-10. COMPARISON OF GEOMETRICAL PARAMETERS
IN SOME SiX_3Y MOLECULES

Molecule	Bond distances (Å)		Bond angles XSiX, [XSiY] (deg)	Ref.
	Si—X	Si—Y		
H_3SiNCO	1.470(9)	1.703(4)	110	11
H_3SiNCS	1.486(22)	1.704(6)	110	11
H_3SiCH_2—CH=CH_2	1.479	1.875(4)	[107]	113
H_3SiN=S=O	1.486	1.762(6)	109.5	114
H_3SiN=N=N	1.485	1.719(8)	109.5	11
H_3SiNMe_2	1.485	1.713(5)	[109.5]	115
H_3SiC≡C—Cl	1.488(12)	1.812(5)	[109.4(20)]	116
H_3SiC≡C—Me	1.506(10)	1.802(4)	[110.7(10)]	117
$H_3SiC_4H_9$	1.474(9)	1.873(3)	110.9(20)	118
$H_3SiCo(CO)_4$	1.480	2.381(7)	109.5	11
$H_3SiMn(CO)_5$	1.490	2.407(5)	110	119
$H_3SiRe(CO)_5$	1.514(37)	2.562(13)	108	120
F_3SiNCO	1.553(4)	1.648(10)	107.9(2)	11
F_3SiCH_2—CH=CH_2	1.582(2)	1.837(10)	105.9(7)	121
$F_3SiC_6H_5$	1.572(6)	1.822(31)	105.0(21)	122
$F_3SiRe(CO)_5$	1.583(4)	2.360(7)	112.5(4)	123
Cl_3SiNCO	2.014(7)	1.646(8)	109.5(4)	11
Cl_3SiCH_2Cl	2.019(1)	1.905(10)	[109.8(3)]	124
	2.024(4)	1.910(9)	[110.6(3)]	125
Me_3SiN=N=N	1.854(3)	1.734(7)	111.4(15)	11
Me_3Si—C≡C—Cl	1.855(6)	1.825(8)	110.1	11
$Me_3SiC_3H_5$	1.872(4)$_{av}$		108.7(13)	126
Me_3SiCN	1.871(8)	1.844(22)	[107.0(15)]	11
Me_3SiNCO	1.864(2)	1.740(4)	108.8(25)	127
Me_3SiOMe	1.864(4)	1.639(4)	[108.6(2)]	128
Me_3SiOCH=CH_2	1.864(3)	1.663(5)	[107.0(7)]	129
Et_3SiH	1.886(4)	1.48	110.5(40)	130
$(Me_3C)_3SiH$	1.934(6)	1.49	[105.3(13)]	131
$(MeO)_3SiMe$	1.632(4)	1.842(13)	[109.6(5)]	132
$(NCO)_3SiCl$	1.684(5)	2.020(9)	[109.6(6)]	11

Table 1-11. STRUCTURAL PARAMETERS FOR SiX_2Y_2 MOLECULES

Molecule	Bond distances (Å)		Bond angles (deg)		Ref.
$(NCO)_2SiCl_2$	Si—Cl	2.024(8)	ClSiCl	107.7(8)	11
	Si—N	1.687(8)	NSiN	113.0(2)	
Me_2SiCl_2	Si—Cl	2.055(2)	CSiC	114.2(2)	133
	Si—C	1.849(4)	ClSiCl	107.5(1)	
$Me_2Si(CH_2Cl)_2$	Si—C_{av}	1.875(9)	CSiC	109.5	134
$Me_2Si(OMe)_2$	Si—C	1.859(4)	CSiC	114.5(16)	135
	Si—O	1.641(3)	CSiO	108.1(13)	

The first series was studied by microwave spectroscopy,[11] and the Si—C distances were reported, in angstrom units, as follows:

$$1.867(3) \quad (r_0) \qquad 1.853(3) \quad (r_s) \qquad 1.826(3) \quad (r_s)$$

Two molecules from the second triad can be found in Table 1-9, while the Si—C(sp) bond length is borrowed from H_3Si—C≡C—CH_3 (Table 1-10). The corresponding Si—C distances are as follows:

$$1.875(2) \qquad 1.855(2) \qquad 1.802(4) \text{ Å}$$

In general, a good parallel exists between the spectroscopic and electron diffraction measurements; in the case of Si—C(sp^2) the results are even indistinguishable within the error limits despite differences in physical meaning among the parameters.

Steric strain noticeably influences the Si—C bond length. Overcrowding in a molecule arises from the introduction of bulky trimethylsilyl groups $SiMe_3$ and manifests itself in the lengthening of the bond, eg, up to 1.889(4) Å in $CH_2(SiMe_3)_2$.[136] For additional examples, see Tables 1-9 and 1-10. An interesting lack of symmetry from the point of view of the Si—C distance may be observed in two similar molecules $HSi(CMe_3)_3$[131] and $HC(SiMe_3)_3$.[137] The Si—C bond is more strained in the first molecule [1.934(6) Å] than in the second [1.886(6) Å], and this feature is well reproduced by molecular mechanics calculations.[137]

Earlier we touched upon $B(CH=CH_2)_3$,[17] and now we mention its analog, $Si(CH=CH_2)_4$.[111] As in the previous case, the C=C bond length is increased [up to 1.355(2) Å, which exceeds by 0.019(3) Å the C=C distance in ethylene[11]]. This fact is interpreted in terms of $(p - d)\pi$ interaction between the vinyl group and the silicon atom. A smaller degree of lengthening of the C=C bond [to 1.347(5) Å] is also found in $H_2C=CHSiClMe_2$.[138]

The concept of $(p - d)\pi$ back-bonding is also used for qualitative explanation of the shortening of the Si—Cl bond. This distance is 2.02 Å in $SiCl_4$, while application of the Schomaker–Stevenson rule[139] (the sum of the covalent radii of silicon and chlorine corrected for the differences in their electronegativities) predicts a Si—Cl bond length of 2.08 A; according to a recent modification,[140] the estimate has been decreased to 2.06 Å.

The Si—Cl distance depends on the number of these bonds in a molecule, being largest when the number is *small* and vice versa [eg, Si—Cl = 2.065(4) Å in $CH_2BrSi(CH_3)_2Cl$[141]; for additional examples see reference 105]. Such regularity was previously observed in investigations of carbon–halogen bond lengths, and this result was interpreted in terms of electronegativity.[142] The development of these ideas can be found in the study of the series of $CH_nCl_{3-n}SiCl_3$ molecules. There is a small decrease in the Si—Cl distance on going from the methyl to the trichloromethyl derivative. This variation parallels an increase in substituent electronegativity ($CH_3 \rightarrow CCl_3$), which shortens the Si—Cl bonds. On the other hand, the neighboring unsaturated groups have a much more pronounced effect on the Si—Cl distances. The longest bonds were found in the following molecules: $Cl_3SiCH=CH_2$ [2.060(4) Å],[143] $ClMe_2SiCH=CH_2$ [2.078(2) Å],[138] and $ClH_2SiC_6H_5$ [2.076(10) Å].[144]

The pseudohalogens, NCO and NCS, are of particular interest among the ligands in the molecules presented in Table 1-11 [additional information on these compounds can be found in Table 2-1 of Chapter 2]. The molecular structures of pseudohalide derivatives of carbon, silicon, and germanium have been extensively investigated by a variety of techniques. We only outline the problem, since this subject has been reviewed.[105,145] The main question in these studies is whether the fragment $Si—N{=}C{=}O$ is linear or bent at the N atom. The answer to this question is: "It depends on the physical method selected and the aggregation state." Much relevant information can be extracted via spectroscopy.[146] In an electron diffraction study, a linear molecule with a low-frequency bending vibration may appear to have a nonlinear skeleton. The pseudohalide group exhibits large-amplitude motion (see Chapter 12 of Part A), and care should be taken in interpreting the results because of the problems caused by the shrinkage effect. The introduction of accurate vibrational corrections, however, is not always possible even in a modern electron diffraction study.[127]

In the past decade there was considerable interest in compounds that contain highly branched trimethylsilyl groups, $SiMe_3$, and many of them were characterized structurally. When two or more such groups are bonded to the same atom, considerable steric strain will result, which is reflected in all the geometrical parameters. The bond lengths become longer as mentioned above, and the bond angles open to values that are significantly larger than the tetrahedral angle of 109.5° [for comparison, $\angle SiCSi = 123.2(9)°$ in $H_2C(SiMe_3)_2$].[136] Additionally, dihedral angles differ by 10–15° from those that correspond to the staggered conformation. In a molecule with several $SiMe_3$ groups, silicon loses its role as the central atom, and classification of such compounds becomes more difficult. Accordingly, we present only a list of the molecules studied $(R = SiMe_3)$: CH_2R_2,[136] CR_4,[147] R_2CCO,[148] $C_5H_4R_2$,[11] $C_6H_4R_2$,[149] R_3CPH_2,[150] $LiHCR_2$,[151] $(LiNR_2)_2$[152]; $A(NR_2)_2$, where $A = Be$,[11] Mg,[153] Zn,[154] Hg,[155] Cd,[156] Ge, Sn, Pb[157]; $A(NR_2)_3$, where $A = Ce$, Pr,[158] Sc[159]; $A(CHR_2)_2$, where $A = Ge$, Sn.[160] Most of these complicated species were studied by Fjeldberg during the period 1981–1985 [Thesis, Trondheim, 1985; For more details, see Chapter 8].

Cyclic Organosilicon Compounds. The simplest cyclic organosilicon molecules studied by GED are shown below, (7):

7

The structural data for some of these compounds are included in Table 1-12. The molecular parameters of Cl_2SiMe_2 are presented in Table 1-11, while those of H_2SiMe_2 are as follows: Si—C 1.867(2) Å and $\angle$ CSiC 110°59′ ± 10′.[11] Comparison of the Si—C bond distances reveals no significant differences in four-, five-, and six-membered ring systems [1.895(2), 1.892(2), and 1.885(3) Å, respectively]. This behavior contrasts with the patterns that are characteristic of cycloalkanes[170] and other heterocyclic compounds,[171] where bond distances are largest in four-membered rings and become gradually shorter as the ring size increases.

Aside from this feature, the C—Si—C and Cl—Si—Cl bond angles appear to be unusual. Both these angles are less than the tetrahedral value, 109.5°, and this is especially conspicuous for the four-membered ring system.[162] This finding does not conform to the structural pattern that is typical for other organic compounds. Here, the variations among the CCC and HCH bond angles in the fragments C—CH_2—C are such that one of the angles is *smaller* than 109.5° while the other is *larger*.[172] Furthermore, there exists a similar relationship for other types of fragment, ie, C—CX_2—X (X = Hal, Me) and C—AX_2—C (A = Ge, Sn).

Conformational characteristics of the cyclic silicon compounds in question are also noteworthy. The four-membered ring in silacyclobutane[161] is more puckered (dihedral angle $\varphi = 34 \pm 2°$) than in the parent cyclobutane[173] ($\varphi = 26 \pm 3°$). Silacyclopentane[163] differs from cyclopentane in that pseudorotation is destroyed in the silacyclopentane and the molecule exists in a half-chair conformer of C_2 symmetry. The chair conformer, which is common to many saturated six-membered rings, is also preserved in silacyclohexane.[163] However, it is flattened at the silicon and puckered at the opposite carbon atom relative to cyclohexane. The conformational features of silacyclopentane and silacyclohexane are well reproduced by molecular mechanics calculations.[163]

Table 1-12. COMPARISON OF GEOMETRICAL PARAMETERS OF
CYCLIC ORGANOSILICON COMPOUNDS

| | Bond distances (Å) | | Bond angles (deg) | | |
Molecule	Si—C	Si—X	CSiC	XSiX	Ref.
$(CH_2)_3SiH_2$	1.895(2)	1.496(18)	80.8(5)	116(9)	161
$(CH_2)_3SiCl_2$	1.886(4)	2.032(2)	84.0(6)	105.1(3)	162
$(CH_2)_4SiH_2$	1.892(2)	1.497(8)	96.3(3)	112.3(29)	163, 164
$(CH_2)_4SiCl_2$	1.852(3)	2.046(2)	99.2(3)	108.7(2)	165
$(CH_2)_4SiF_2$	1.853(3)	1.582(6)	a		166
$C_4H_6SiH_2$	1.894(3)	1.491	95.7(12)	105	167
$C_4H_6SiCl_2$	1.872(11)	2.047(3)	96.9(11)	104.8(33)	168
$(CH_2)_5SiH_2$	1.885(3)	1.465(22)	104.2(14)	105(14)	163
$(CH_2)_5SiCl_2$	1.861(5)	2.055(2)	106.5(12)	106.0(15)	169

^{a} CSiF = 113,4(3)°.

Hilderbrandt and co-workers[174–178] investigated a number of strained poly-cyclic compounds (**8–13**) to study, in particular, the geometry at the bridgehead silicon (which has a marked effect on the relative reactivities of these species).

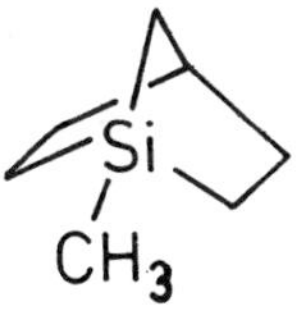

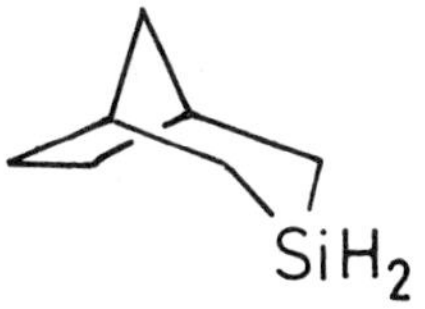

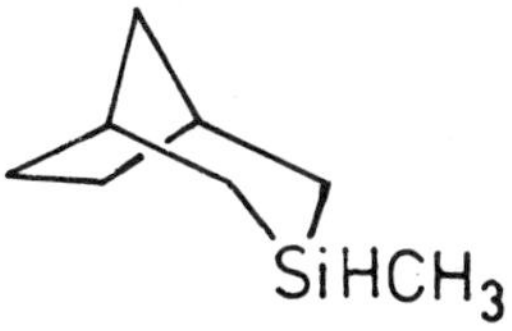

Reference: 174   175   175

8   **9**   **10**

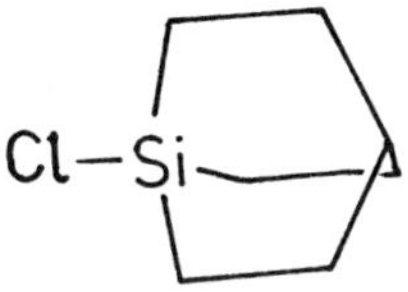

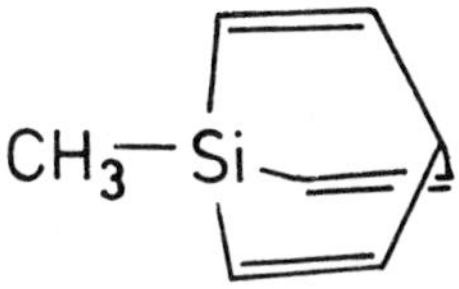

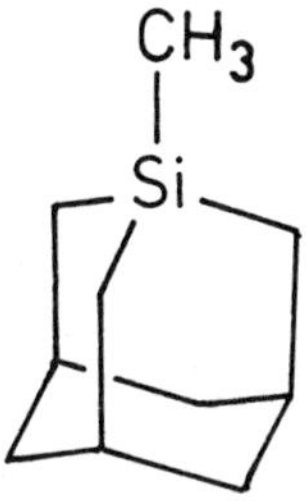

Reference: 176   177   178

11   **12**   **13**

Of these interesting compounds we focus only on 3-silabicyclo[3.2.1]octane,[175] (**9**) which contains a six-membered fragment. The conformation of this structural unit may be viewed as being a consequence of substitution of a carbon by a silicon atom in the cyclohexane[179] ring and bridging the same ring to produce bicyclo[3.2.1]octane (**14**).[180] All the changes in the conformational parameters

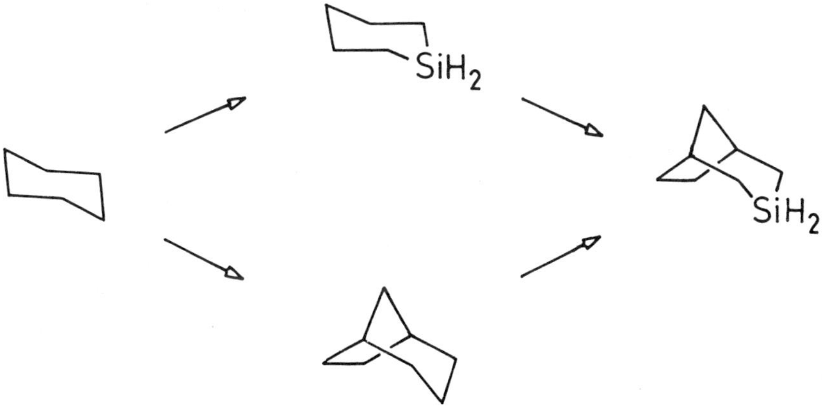

14

of the six-membered ring turned out to be predictable within experimental error as being the sum of the individual effects of silicon substitution and bridging.

Silanes and Their Derivatives. Recent developments in the stereochemistry of silanes were initiated by Bartell and associates in 1970 by their investigation of $Si(SiMe_3)_4$.[181] Later, open-chain silanes and their derivatives containing Si—Si and Si—Si—Si linkages were studied. The geometrical parameters are listed in Table 1-13A. The largest cyclosilane that has been studied to date is Si_6H_{12} (see Table 1-13B). On the other hand, the lower members of this family, Si_3H_6 and Si_4H_8, are unstable; therefore, structural data are available only for their derivatives. It should be stressed that the prime reason for studying related silanes and hydrocarbons is to determine their structural similarities and instructive differences and thereby increase our understanding of bonding and structure in general.

Again, as in the preceding section, we start with a discussion of internuclear distances. The relevant basic types of bonds are summarized below.

C—C	Si—C	Si—Si
1.534(1) Å	1.875(2) Å	2.331(4) Å

C=C	Si=C	Si=Si
1.337(2) Å	1.834(4) Å	2.151 Å

The list includes the Si=Si bond distances so far measured only in crystal; here we present the average value found for two molecules.[189]

Actually, it is possible to compare in more detail the Si—Si and C—C distances using cyclic systems as an example. Figure 1-4 illustrates how these distances change in three- to six-membered rings. The average Si—Si distance for the Si_3 ring is calculated from the X-ray results of Masamune and colleagues: 2.407 Å,[190] 2.428 Å,[191] and 2.416 Å.[191] The C—C distances have been discussed earlier elsewhere.[170] From Figure 1-4 it is clear that the principal difference occurs with the three-membered system: the corresponding C—C bond distance is *shortest* in cycloalkanes, while the Si—Si distance is *longest* in cyclosilanes. Masamune[190] ascribes this elongation to the effect of steric interactions between the ligands. On the other hand, one cannot expect a complete geometric similarity for silanes and hydrocarbons because of differing electronic factors, which are operative in these systems. The structural features of the silicon-silicon double bond may serve as a good illustration in this regard. The silicon atoms have a pyramidal configuration, and the Si=Si bond is slightly twisted.[189] Therefore, we may conclude that the geometrical parameters of three-membered silanes provide challenges for both experiment and theory.

As we have pointed out,[79] the CCC bond angle 112.7° is one of the important geometrical characteristics of *n*-alkanes (together with the C—C distance of 1.533 Å) when used as a reference in discussing the stereochemistry of organic substances. Note the similarity between this value and that found in cyclohexane[179] (111.4 ± 0.2°), which is regarded as being nearly strainless.

Table 1-13. COMPARISON OF GEOMETRICAL PARAMETERS OF SILANES AND
THEIR DERIVATIVES

A. ACYCLIC SILANES

Molecule	Bond distances (Å)		Bond angles (deg)		Ref.
	Si—Si	Si—H, Si—C, or Si—Hal			
Si_2H_6	2.331(3)	1.492(3)	SiSiH	110.3(4)	11
Si_2Cl_6	2.324(30)	2.009(4)	ClSiCl	109.7(6)	11
Si_2F_6	2.324(6)	1.569(2)	FSiF	110.6(3)	182
	2.317(6)	1.564(2)		108.6(3)	183
Si_2Me_6	2.340(9)	1.887(3)	CSiC	110.5(4)	11
$Si_2(Me_2Cl)_2$	2.338(13)	1.860(3)	SiSiC	109.8(7)	184
		2.077(2)	SiSiCl	107.7(6)	
Si_3Me_8	2.325(12)	1.887(3)	SiSiSi	118.0(25)	185
Si_3Cl_8	2.329(7)	2.026(7)($SiCl_3$)	SiSiSi	118.7(16)	186
		2.034(22)($SiCl_2$)			
$Si(SiMe_3)_4$	2.361(3)	1.889(3)	CSiC	107.9(5)	181

B. CYCLIC SILANES

Molecule	Bond distances (Å)		Bond angles (deg)		Ref.
	Si—Si	Si—H, Si—C	SiSiSi	HSiH, CSiC	
Si_4Me_8	2.362(6)	1.894(6)	89.3(9)	109(3)	187
Si_5H_{10}	2.342(2)	1.496(6)	104.3(7)	105.3(29)	97
Si_6H_{12}	2.342(5)	1.484(8)	110.4(14)	103.0	188

Turning now to silanes, we find no standard value for the SiSiSi bond angle, since Si_3H_8 and higher homologs remain unstudied. Arguments derived from hydrocarbon stereochemistry enable us to assess the range (rather wide, however) for this parameter: 110.3–118.0° (the first figure is taken from cyclohexasilane[188] while the second is from Si_3Me_8[185]).

The main features worth noting in the results on cyclic silanes concern their respective conformations. The four-membered ring in Si_4Me_8* is less puckered than in cyclobutane,[173] as illustrated by values for their dihedral angles φ (**15**).

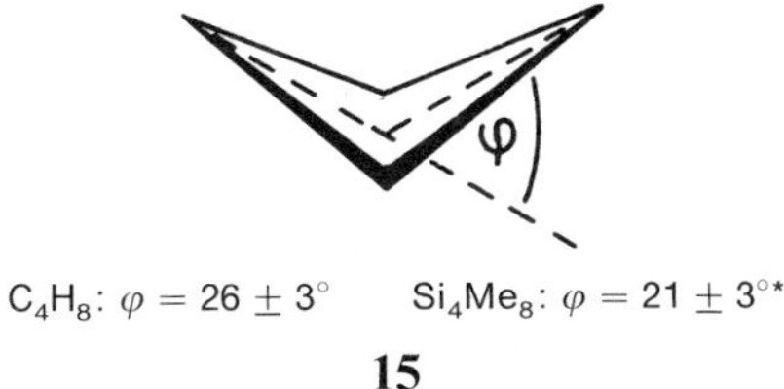

15

* In our preliminary data on this compound,[187] an error was found that affected mainly only the dihedral angle. The value reported here was obtained after the correction.

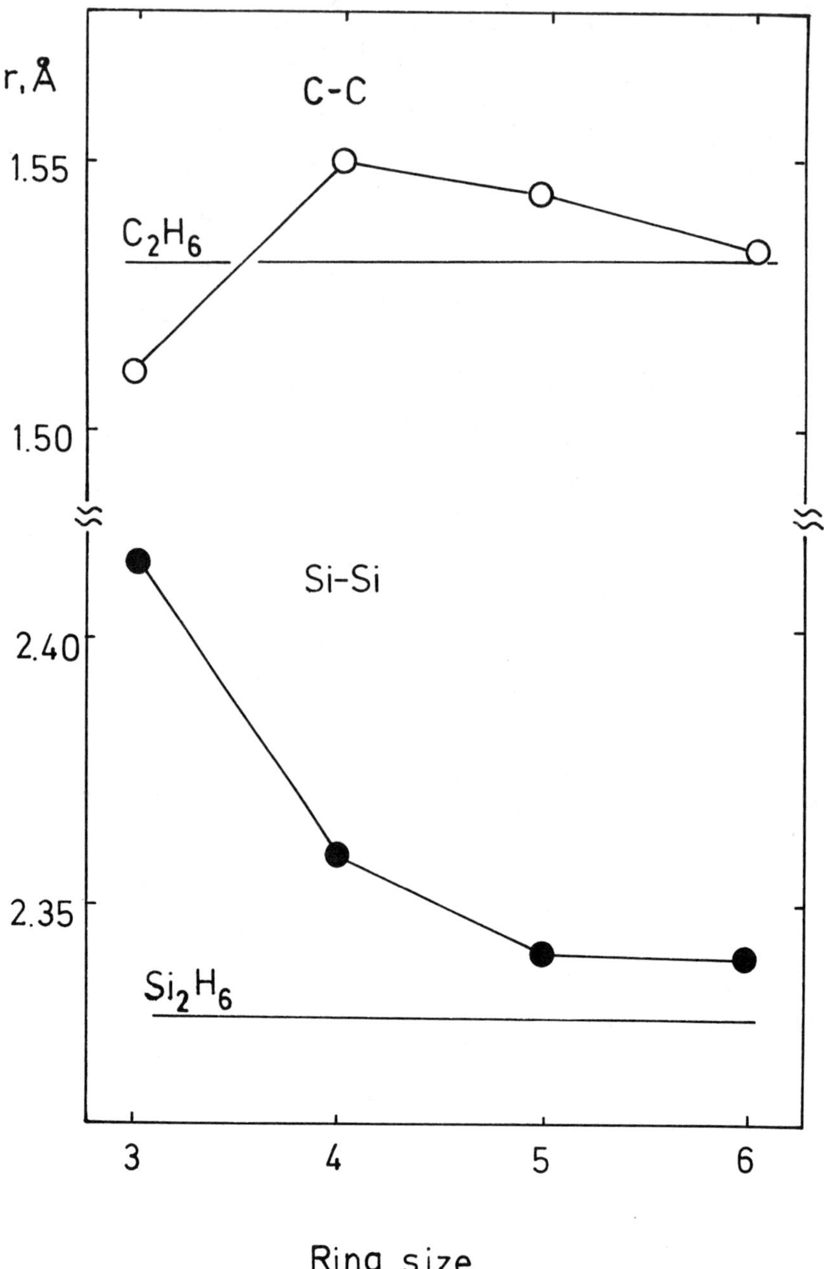

Figure 1-4. Comparison of the C—C and Si—Si bond lengths in cyclic systems as a function of ring size. Horizontal lines correspond to the "unstrained" distances taken from C_2H_6 and Si_2H_6.

Since the nonplanarity of cyclobutane is caused by torsion strain,[79] one may conclude that the flattening of the Si_4 ring is in accord with a decrease in the torsional potential on proceeding from the C—C to the Si—Si bond.

Reduction of torsion (and angular) strain produces no effect on the five-membered ring system and, accordingly, cyclopentasilane is remarkably similar in structure to cyclopentane.[97,192] On the other hand, an essential difference between C_6H_{12} and Si_6H_{12} consists in the appearance of the twist conformer in the latter case. It is generally agreed that for cyclohexane the twist conformer (17) lies about 21–25 kJ mol^{-1} above the chair (16) and it is, therefore, accessible only at high temperatures.[193] The chair–twist energy gap ΔE in cyclohexasilane is dramatically decreased due to the reduction of torsion strain. Thus, up to $37 \pm 8\%$ of the twist form has been detected by GED at 130°C in vapor of Si_6H_{12}. This value corresponds to the energy difference $\Delta E = 5.4$ kJ mol^{-1}. The chair itself is less flattened than in cyclohexane, a fact that may be attributed to weaker gauche repulsions in Si_6H_{12}.

Chair   Twist

16   **17**

Siloxanes and Silazanes. In the compounds considered above, silicon was mainly a central atom. The structures of silyl compounds (with Si being treated as a "ligand" atom) are not strictly relevant to this chapter. However, since these compounds are of particular interest, it was decided to discuss briefly the bond angle data summarized in Table 1-14. (For additional information see Chapters 2 and 3.

The silyl compounds bear little resemblance to their methyl analogs when the central atom A contains lone electron pairs. Indeed, the calculated differences in the bond angles for A = N and O are 7–13 times greater than that for A = C. This result was ascribed traditionally to π bonding involving the donation of nitrogen and oxygen lone electron pairs to empty silicon d orbitals. (However,

Table 1-14. COMPARISON OF THE $\angle$ B—A—B BOND ANGLES (A = C, N, O; B = CH_3, SiH_3) FOR SOME RELATED COMPOUNDS

	A = C	A = N	A = O
$\angle$C—A—C	$H_2C(CH_3)_2$: 112(1)°	$HN(CH_3)_2$: 111.8(8)°	$O(CH_3)_2$: 111.8(2)°^{a}
$\angle$Si—A—Si	$H_2C(SiH_3)_2$: 114.4(2)°^{b}	$HN(SiH_3)_2$: 127.7(1)°^{c}	$O(SiH_3)_2$: 144.1(9)°
Difference	2.4°	15.9°	32.3°

^{a} Reference 36.
^{b} Larger bound angles found in $H_2C(SiF_3)_2$ [117.7(4)°][194] and in $H_2C(SiMe_3)_2$ [123.2(9)°][136] can be accounted for by steric interactions.
^{c} The largest $\angle$Si—N—Si bond angle of 131.3(15)° is found in $HN(SiMe_3)_2$.[195]

ab initio calculations by Oberhammer and Boggs[196] do not support this widely held view.) The increase in the bond angles that occurs upon CH_3/SiH_3 substitution was accounted for alternatively in terms of nonbonded Si...Si interactions, which force the silyl groups apart, since the "one-angle nonbonded radius" of silicon is 1.55 Å.[196] Note that the Si...Si distances are remarkably constant despite the large difference in bond angles:

$$3.149 \text{ Å} \qquad 3.097 \text{ Å} \qquad 3.107 \text{ Å}$$
$$H_2C(SiH_3)_2 \qquad HN(SiH_3)_2 \qquad O(SiH_3)_2$$

However, the "one-angle nonbonded radii" concept is also open to criticism.[198]

Strain in the four-membered rings of hexamethylcyclodisilazane[199] and of tetramethylcyclodisiloxane[200] (X-ray study) can definitely be seen from inspection of the following values:

	$(MeNSiMe_2)_2$	$(OSiMeS_2)_2$
∠ Si—A—Si	93.5(3)°	86°
Si...Si	2.527(2) Å	2.31 Å

A striking feature of the cyclic siloxane structure[200] is the Si...Si distance of 2.31 Å, which is shorter than the normal Si—Si distance in disilane (2.331 Å)!

As expected, the six-membered rings in hexamethylcyclotrisilazane[201] and hexamethylcyclotrisiloxane[202] are strained to a much lesser extent:

	$(HNSiMe_2)_3$	$(OSiMe_2)_3$
∠ Si—A—Si	126.8(8)°	131.6(4)°
Si...Si	3.090 Å	2.983 Å

The molecular structure of $(SiH_3)_2NBF_2$ is of particular interest, since the molecule contains both boron and silicon, and their vacant orbitals can compete for the nitrogen lone pair.[203] The Si_2NB and NBF_2 units are found to be planar, and there is some evidence that these planes are slightly twisted. The bond angles are: ∠ SiNSi = 123.9(3)° and ∠ FBF = 123.2(8)°.

The concept of $(p - d)\pi$ back-bonding, irrespective of its true value, significantly impacted many areas of structural chemistry and stimulated numerous interesting studies. In particular, it was found that the presence of electronegative substituents on silicon enhances the effect of widening the angle at oxygen [up to 156(2)° in $(SiF_3)_2O$]. Furthermore, the manner in which SiH_3 groups influence the stereochemistry of other Group V and Group VI elements and the difference that occur upon SiH_3/GeH_3 substitution have been studied (cf. Table 1-15).

In conclusion, it may be noted that the main object of this survey was to demonstrate the wide range of boron- and silicon-containing substances studied

Table 1-15. COMPARISON OF BOND ANGLES FOR SOME RELATED
COMPOUNDS WITH THE CENTRAL ATOM FROM GROUP V AND
GROUP VI

Bond angle	Group V	Group VI
$\angle$ Si—A—Si	$P(SiH_3)_3$: 96.4(5)°[a]	$S(SiH_3)_2$: 97.4(5)°[b]
$\angle$ Si—A—Si	$As(SiH_3)_3$: 93.8(2)°[c]	$Se(SiH_3)_2$: 96.6(5)°[d]
$\angle$ Ge—A—Ge	$N(GeH_3)_3$: 120°	$O(GeH_3)_2$: 126.5(4)°
$\angle$ Ge—A—Ge	$P(GeH_3)_3$: 95.4(5)°[a]	$S(GeH_3)_2$: 98.9(3)°[b]

[a] Compare with $\angle$ C—P—C = 98.8° in PMe_3.[37]
[b] Compare with $\angle$ C—S—C = 99.05(4)° in SMe_2.[204]
[c] Compare with $\angle$ C—As—C = 96.2° in $AsMe_3$.[37]
[d] Compare with $\angle$ C—Se—C = 98.9(2)° in $SeMe_2$.[205]

by gas-phase electron diffraction. This is illustrated by using a partial sampling, which includes about 70 and 120 molecules that contain B and Si, respectively. Our understanding of the stereochemical patterns of these compounds continues to progress, and we foresee further development of this important area of chemistry.

Acknowledgments

I thank Professors O. Bastiansen and L. V. Vilkov for stimulating discussions, and L. M. Shkolnikova for typing the manuscript.

References

1. Hargittai, I. "The Structure of Volatile Sulphur Compounds". Akadémiai Kiadó: Budapest; Reidel: Dordrecht, 1985.
2. Buck, I.; Maier, E.; Mutter, R.; Seiter, U.; Spreter, C.; Starck, B.; Hargittai, I.; Kennard, O.; Watson, D. G.; Lohr, A.; Pirzadeh, T.; Schirdewahn, H. G.; Majer, Z. "Bibliography of Gas Phase Electron Diffraction 1930–1979". Fachinformationszentrum Energie-Physik-Mathematik GmbH: Karlsruhe, 1981.
3. Stock, A.; Wierl, R. *Z. Anorg. Allg. Chem.* **1932**, *203*, 228.
4. Bauer, S. H. *J. Am. Chem. Soc.* **1937**, *59*, 1096.
5. Lévy, H. A.; V. Brockway, L. O. *J. Am. Chem. Soc.* **1937**, *59*, 2085.
6. Bauer, S. H. *J. Am. Chem. Soc.* **1937**, *59*, 1804.
7. Schomaker, V.; Hedberg, K. In "Fifty Years of Electron Diffraction", Goodman, P. Ed.; Reidel: Dordrecht, 1981, pp. 208–221.
8. Vilkov, L. V.; Mastryukov, V. S.; Akishin, P. A.; Zhigach, A. F. *Zh. Strukt. Khim.* **1965**, *6*, 447.
9. Spiridonov, V. P.; Gershikov, A. G.; Zasorin, E. Z.; Ivanov, A. A.; Ermolayeva, L. I. *High Temp. Sci.* **1983**, *16*, 325.
10. Sellers, H.; Boggs, J. E.; Nemukhin, A. V.; Almlöf, J. *J. Mol. Struct.* **1981**, *85*, 195.
11. Landolt-Börnstein: Numerical Data and Functional Relationships in Science and Technology, New Series, Hellwege, K.-H., Editor-in-Chief. Vol. 7: "Structure Data of Free Polyatomic Molecules", Collomon, J. H.; Hirota, E.; Kuchitsu, K., et al; Springer-Verlag: Berlin, 1976.

12. Ezhov, Y. S.; Komarov, S. A. *J. Mol. Struct.* **1978**, *50*, 305.
13. Gillespie, R. J. "Molecular Geometry". Van Nostrand Reinhold: London, 1972.
14. Kuchitsu, K.; Konaka, S. *J. Chem. Phys.* **1966**, *45*, 4342.
15. Ginn, S. G.; Kenney, J. K.; Overend, J. *J. Chem. Phys.* **1968**, *48*, 1571.
16. Gershikov, A. G.; Spiridonov, V. P. *Vestn. Mosk. Univers., Ser. Khim.*, **1985**, *26*, 15.
17. Foord, A.; Beagley, B.; Reader, W.; Steer, I. A. *J. Mol. Struct.* **1975**, *24*, 131.
18. Almenningen, A.; Gundersen, G.; Mangerud, M.; Seip, R. *Acta Chem. Scand.* **1981**, *A35*, 341.
19. Gundersen, G.; Jonvik, T.; Seip, R. *Acta Chem. Scand.* **1981**, *A35*, 325.
20. Lindøy, S.; Seip, H. M.; Seip, R. *Acta Chem. Scand.* **1976**, *A30*, 54.
21. Fjeldberg, T.; Gundersen, G.; Jonvik, T.; Seip, H. M.; Saebø, S. *Acta Chem. Scand.* **1980**, *A34*, 547.
22. Gundersen, G. *Acta Chem. Scand.* **1981**, *A35*, 729.
23. Danielson, D. D.; Patton, J. V.; Hedberg, K. *J. Am. Chem. Soc.* **1977**, *99*, 6484.
24. Danielson, D. D.; Hedberg, K. *J. Am. Chem. Soc.* **1979**, *101*, 3199.
25. Seip, R.; Seip, H. M. *J. Mol. Struct.* **1975**, *28*, 441.
26. Harshbarger, W.; Lee, G.; Porter, R. F.; Bauer, S. H. *Inorg. Chem.* **1969**, *8*, 1683.
27. Iijima, K.; Yamada, Y.; Shibata, S. *J. Mol. Struct.* **1981**, *77*, 271.
28. Iijima, K.; Adachi, N.; Shibata, S. *Bull. Chem. Soc. Japan* **1985**, *57*, 3269.
29. Hargittai, M.; Hargittai, I. *J. Mol. Struct.* **1977**, *39*, 79.
30. Iijima, K.; Shibata, S. *Bull. Chem. Soc. Japan* **1979**, *52*, 711.
31. Iijima, K.; Shibata, S. *Bull. Chem. Soc. Japan* **1980**, *53*, 1908.
32. Iijima, K.; Shibata, S. *Bull. Chem. Soc. Japan* **1979**, *52*, 3204.
33. Iijima, K.; Koshimizu, E.; Shibata, S. *Bull. Chem. Soc. Japan* **1981**, *54*, 2255.
34. Iijima, K.; Shibata, S. *Bull. Chem. Soc. Japan* **1983**, *56*, 1891.
35. Iijima, K.; Koshimizu, E.; Shibata, S. *Bull. Chem. Soc. Japan* **1984**, *55*, 2551.
36. Tamagawa, K.; Takemura, M.; Konaka, S.; Kimura, M. *J. Mol. Struct.* **1984**, *125*, 131.
37. Beagley, B.; Medwid, A. R. *J. Mol. Struct.* **1977**, *38*, 229.
38. Hargittai, M.; Hargittai, I. "The Molecular Geometries of Coordination Compounds in the Vapour Phase". Akadémiai Kiadó: Budapest; Elsevier: Amsterdam, 1977.
39. Haaland, A. *Top. Curr. Chem.* **1975**, *53*, 1.
40. Lipscomb, W. N. "Boron Hydrides". Benjamin: New York, 1963.
41. Bartell, L. S.; Carroll, B. L. *J. Chem. Phys.* **1965**, *42*, 1135.
42. Iijima, T.; Hedberg, L.; Hedberg, K. *Inorg. Chem.* **1977**, *16*, 3230.
43. Hedberg, L.; Hedberg, K.; Kohler, D. A.; Ritter, D. M.; Schomaker, V. *J. Am. Chem. Soc.* **1980**, *102*, 3430.
44. Hedberg, K.; Stosick, A. J. *J. Am. Chem. Soc.* **1952**, *74*, 954.
45. Barlow, M. T.; Downs, A. J.; Thomas, P.D.P.; Rankin, D.W.H. *J. Chem. Soc. Dalton Trans.* **1979**, 1793.
46. Pauling, L. "The Nature of the Chemical Bond", 3rd ed. Cornell University Press: Ithaca, NY, 1963.
47. Slater, J. C. *J. Chem. Phys.* **1964**, *41*, 3199.
48. Bartell, L. S. In "Physical Methods of Chemistry", Weissenberger, A.; Rossiter, B. W., Eds.; Wiley-Interscience: New York, 1971, pp. 125-158.
49. Kuchitsu, K. *J. Chem. Phys.* **1968**, *49*, 4456.
50. Dain, C. J.; Downs, A. J.; Rankin, D.W.H. *Angew. Chem.* **1982**, *94*, 557.
51. Mamaeva, G. I.; Hargittai, I.; Spiridonov, V. P. *Inorg. Chim. Acta* **1977**, *25*, L123.
52. Barlow, M. T.; Dain, C. J.; Downs, A. J.; Laurenson, G. S.; Rankin, D.W.H. *J. Chem. Soc. Dalton Trans.* **1982**, 597.
53. Brendhaugen, K.; Haaland, A.; Novak, D. P. *Acta Chem. Scand.* **1975**, *A29*, 801.
54. Marynick, D. S. *J. Chem. Phys.* **1976**, *64*, 3080.
55. Dain, C. J.; Downs, A. J.; Laurenson, G. S.; Rankin, D.W.H. *J. Chem. Soc. Dalton Trans.* **1981**, 472.
56. Hedberg, K.; Jones, M. E.; Schomaker, V. *Proc. Natl. Acad. Sci. US* **1952**, *38*, 679.
57. Mastryukov, V. S.; Dorofeeva, O. V.; Vilkov, L. V. *Zh. Strukt. Khim.* **1975**, *16*, 128.
58. Wiser, J. D.; Moody, D. C.; Huffman, J. C.; Hilderbrandt, R. L.; Schaeffer, R. *J. Am. Chem. Soc.* **1975**, *97*, 1074.
59. Schwoch, D.; Burg, A. B.; Beaudet, R. A. *Inorg. Chem.* **1977**, *16*, 3219.
60. Williams, R. E. *Inorg. Chem.* **1971**, *10*, 210.
61. Rudolph, R. W. *Acc. Chem. Res.* **1976**, *9*, 446.
62. Wade, K. *New Sci.* **1974**, *62*, 615.

63. Wade, K. *Chem. Br.* **1975**, *11*, 177.
64. Wade, K. *Adv. Inorg. Radiochem.* **1976**, *18*, 1.
65. Schaeffer, R.; Johnson, Q.; Smith, G. S. *Inorg. Chem.* **1965**, *4*, 917.
66. Lipscomb, W. N. *Science* **1977**, *196*, 1047.
67. Dixon, D. A.; Kleier, D. A.; Halgren, T. A.; Hall, J. H.; Lipscomb, W. N. *J. Am. Chem. Soc.* **1977**, *99*, 6226.
68. Dewar, M.J.S.; McKee, M. L. *J. Am. Chem. Soc.* **1977**, *99*, 5231.
69. Aihara, J. *J. Am. Chem. Soc.* **1978**, *100*, 3339.
70. Brown, L. D.; Lipscomb, W. N. *Inorg. Chem.* **1977**, *16*, 2989.
71. Bicerano, J.; Marynick, D. S.; Lipscomb, W. N. *Inorg. Chem.* **1978**, *17*, 3443.
72. Mastryukov, V. S.; Dorofeeva, O. V.; Vilkov, L. V.; *Russ. Chem. Rev.* **1980**, *49*, 1181 (*Usp. Khim.* **1980**, *49*, 4377).
73. McNeill, E. A.; Gallaher, K. L.; Scholer, F. R.; Bauer, S. H. *Inorg. Chem.* **1973**, 2108.
74. Mastryukov, V. S.; Dorofeeva, O. V.; Vilkov, L. V.; Golubinskii, A. V.; Zhigach, A. F.; Laptev, V. T.; Petrunin, A. B. *Zh. Strukt. Khim.* **1975**, *16*, 171.
75. Atavin, E. G.; Mastryukov, V. S.; Golubinskii, A. V.; Vilkov, L. V. *J. Mol. Struct.* **1980**, *65*, 259.
76. Bohn, R. K.; Bohn, M. D. *Inorg. Chem.* **1971**, *10*, 350.
77. Mastryukov, V. S.; Vilkov, L. V.; Dorofeeva, O. V. *J. Mol. Struct.* **1975**, *24*, 217.
78. Stoicheff, B. P. *Tetrahedron* **1962**, *17*, 135.
79. Vilkov, L. V.; Mastryukov, V. S.; Sadova, N. I. "Determination of the Geometrical Structure of Free Molecules". Mir: Moscow, 1983.
80. Fitzpatrick, N. J.; Fanning, M. O. *J. Mol. Struct.* **1977**, *40*, 271.
81. Dorofeeva, O. V.; Mastryukov, V. S.; Golubinskii, A. V.; Vilkov, L. V.; Almenningen, A. *Zh. Strukt. Khim.* **1981**, *22*(5), 51.
82. Mastryukov, V. S.; Atavin, E. G.; Golubinskii, A. V.; Vilkov, L. V.; Stanko, V. I.; Goltyapin, Y. V. *Zh. Strukt. Khim.* **1982**, *23*, 51.
83. Dorofeeva, O. V.; Mastryukov, V. S.; Vilkov, L. V.; Karimov, A. S. Ninth Austin Symposium on Gas-Phase Molecular Structure, Austin, TX, 1982, p. 69.
84. Bragin, J.; Urevig, D. S.; Diem, M. *J. Raman Spectrosc.* **1982**, *12*, 86.
85. Mastryukov, V. S.; Cyvin, S. J. *J. Mol. Struct.* **1975**, *29*, 15.
86. Mastryukov, V. S.; Atavin, E. G.; Vilkov, L. V.; Golubinskii, A. V.; Kalinin, V. N.; Zhigareva, G. G.; Zakharkin, L. I. *J. Mol. Struct.* **1979**, *56*, 139.
87. Cheung, C.-C.S.; Beaudet, R. A.; Segal, G. A. *J. Am. Chem. Soc.* **1970**, *92*, 4158.
88. Dewar, M.J.S.; McKee, M. L. *Inorg. Chem.* **1980**, *19*, 2662.
89. Brockway, L. O.; Wall, F. T. *J. Am. Chem. Soc.* **1934**, *56*, 2373.
90. Brockway, L. O.; Beach, J. Y. *J. Am. Chem. Soc.* **1938**, *60*, 1836.
91. Hedberg, K. *J. Am. Chem. Soc.* **1955**, *77*, 6491.
92. Almenningen, A.; Bastiansen, O.; Ewing, V.; Hedberg, K.; Traetteberg, M. *Acta Chem. Scand.* **1963**, *17*, 2455.
93. Swick, D. A.; Karle, I. L. *J. Chem. Phys.* **1955**, *23*, 1499.
94. Morino, Y.; Hirota, E. *J. Chem. Phys.* **1958**, *28*, 185.
95. Clark, A. H. In "Internal Rotation in Molecules", Orville-Thomas W. J., Ed.; Wiley-Interscience: London, 1974, Chap. 10.
96. Dzhaparidze, K. G. *Soobshch. Akad. Nauk. Gruz. SSR* **1962**, *29*, 401.
97. Smith, Z.; Seip, H. M.; Hengge, E.; Bauer, G. *Acta Chem. Scand.* **1976**, *A30*, 697.
98. Wells, A. F. "Structural Inorganic Chemistry", 4th ed. Oxford University Press: London, 1975.
99. Mahaffy, P. G.; Gutoewski, R.; Montgomery, L. K. *J. Am. Chem. Soc.* **1980**, *102*, 2854.
100. Hargittai, I.; Schultz, G.; Tremmel, J.; Kagramanov, N. D.; Maltsev, A. K.; Nefedov, O. M. *J. Am. Chem. Soc.* **1983**, *105*, 2895.
101. Shen, Q.; Hilderbrandt, R. L. *J. Mol. Struct.* **1980**, *64*, 257.
102. Bokiy, N. G.; Struchkov, Y. T. *Zh. Strukt. Khim.* **1968**, *9*, 722.
103. Vilkov, L. V.; Mastryukov, V. S.; Oppenheim, V. D.; Tarasenko, N. A. In "Molecular Structures and Vibrations", Cyvin, S. J., Ed.; Elsevier: Amsterdam, 1972, pp. 310-329.
104. Vilkov, L. V.; Khaikin, L. S. *Top. Curr. Chem.* **1975**, *53*, 25.
105. Hargittai, I.; Rozsondai, B. *Kém. Közlem.* **1978**, *50*, 427.
106. Schultz, G.; Tremmel, J.; Hargittai, I.; Berecz, I.; Bohátka, S.; Kagramanov, N. D.; Maltsev, A. K.; Nefedov, O. M. *J. Mol. Struct.* **1979**, *55*, 207.
107. Schultz, G.; Tremmel, J.; Hargittai, I., Kagramanov, N. D.; Maltsev, A. K.; Nefedov, O. M. *J. Mol. Struct.* **1982**, *82*, 107.

108. Yoshioka, Y.; Goddard, J. D.; Schaefer, H. F. *J. Am. Chem. Soc.* **1981**, *103*, 2452.
109. Schaefer, H. F. *Acc. Chem. Res.* **1982**, *15*, 283.
110. Brook, A.; Nyburg, S. C.; Abdesaken, F.; Gutekunst, B.; Gutekunst, G.; Krishna, R.; Kallury, M. R.; Poon, Y. C.; Chang, Y.-M.; Wong-Ng, W. *J. Am. Chem. Soc.* **1982**, *104*, 5667.
111. Rustad, S.; Beagley, B. *J. Mol. Struct.* **1978**, *48*, 381.
112. Boonstra, L. H.; Mijlhoff, F. C.; Renes, G.; Spelbos, A.; Hargittai, I. *J. Mol. Struct.* **1975**, *28*, 129.
113. Beagley, B.; Foord, A.; Moutran, R.; Rozsondai, B. *J. Mol. Struct.* **1977**, *42*, 117.
114. Cradock, S.; Ebsworth, E. A. V.; Meikle, G. D.; Rankin, D.W.H. *J. Chem. Soc. Dalton Trans.* **1975**, 805.
115. Gundersen, G.; Mayo, R. A.; Rankin, D.W.H. *Acta Chem. Scand.* **1984**, *A38*, 579.
116. Cradock, S.; Fraser, A.; Rankin, D.W.H. *J. Mol. Struct.* **1981**, *71*, 209.
117. Cradock, S.; Koprowski, J.; Rankin, D.W.H. *J. Mol. Struct.* **1981**, *77*, 113.
118. Dakkouri, M.; Oberhammer, H. *J. Mol. Struct.* **1983**, *102*, 315.
119. Rankin, D.W.H.; Robertson, A. *J. Organomet. Chem.* **1975**, *85*, 225.
120. Rankin, D.W.H.; Robertson, A. *J. Organomet. Chem.* **1976**, *105*, 331.
121. Kuznetsova, T. M.; Alekseev, N. V.; Veniaminov, N. N. *Zh. Strukt. Khim.* **1979**, *20*, 336.
122. Il'enko, T. M.; Veniaminov, N. N.; Alekseev, N. V. *Zh. Strukt. Khim.* **1975**, *16*, 292.
123. Rankin, D.W.H.; Robertson, A.; Seip, R. *J. Organomet. Chem.* **1975**, *88*, 191.
124. Typke, V.; Dakkouri, M.; Schiele, M. *Z. Naturforsch.* **1980**, *A35*, 1402.
125. Vajda, E.; Székely, T.; Hargittai, M.; Maltsev, A. K.; Baskir, E. G.; Nefedov, O. M. *J. Mol. Struct.* **1981**, *73*, 243.
126. Kuznetsova, T. M.; Veniaminov, N. N.; Alekseev, N. V. *Zh. Strukt. Khim.* **1979**, *20*, 535.
127. Cradock, S.; Huntley, C. M.; Durig, J. R. *J. Mol. Struct.* **1985**, *127*, 319.
128. Csákvári, B.; Wagner, Zs.; Gömöry, P.; Hargittai, I.; Rozsondai, B.; Mijlhoff, F. C. *Acta Chim. (Budapest)* **1976**, *90* 149.
129. Shen, Q. *J. Mol. Struct.* **1979**, *51*, 61.
130. Csákvári, B.; Wagner, Z.; Hargittai, I. *Acta Chim. (Budapest)* **1976**, *90*, 141.
131. Doun, S. K.; Bartell, L. S. *J. Mol. Struct.* **1980**, *63*, 249.
132. Gergö, É.; Hargittai, I.; Schultz, G. *J. Organomet. Chem.* **1976**, *112*, 29.
133. Belyakov, A. V.; Zavgorodnii, V. S.; Mastryukov, V. S. Eleventh Austin Symposium on Gas-Phase Molecular Structure, Austin, TX, 1986, p. 85.
134. Shen, Q. *J. Mol. Struct.* **1978**, *49*, 337.
135. Gergö, É.; Hargittai, I. Sixth International Symposium on Organosilicon Chemistry, August 1981, Budapest, p. 139.
136. Fjeldberg, T.; Seip, R.; Lappert, M. F.; Thorne, A. J. *J. Mol. Struct.* **1983**, *99*, 295.
137. Beagley, B.; Pritchard, R. G. *J. Mol. Struct.* **1982**, *84*, 129.
138. Shen, Q. *J. Mol. Struct.* **1982**, *95*, 215.
139. Schomaker, V.; Stevenson, D. P. *J. Am. Chem. Soc.* **1941**, *63*, 37.
140. Blom, R.; Haaland, A. *J. Mol. Struct.* **1985**, *128*, 21.
141. Shen, Q. *J. Mol. Struct.* **1984**, *112*, 169.
142. Bent, H. A. *J. Chem. Phys.* **1960**, *33*, 1260.
143. Vilkov, L. V.; Mastryukov, V. S.; Akishin, P. A. *Zh. Strukt. Khim.* **1964**, *5*, 183.
144. Vilkov, L. V.; Mastryukov, V. S. *Dokl. Akad. Nauk SSSR* **1965**, *162*, 1306.
145. Hargittai, I.; Paul, I. C. In "Chemistry of Cyanates and Their Thio Derivatives", Patai, S. Ed.; Wiley: New York, 1977, pp. 69–129.
146. Halonen, L.; Mills, I. M. *J. Mol. Spectrosc.* **1983**, *98*, 484.
147. Beagley, B.; Pritchard, R. G., Tenth Austin Symposium on Gas-Phase Molecular Structure, Austin, TX, 1984, A3, p. 81.
148. Rozsondai, B.; Hargittai, I. *Acta Chim. (Budapest)* **1976**, *90*, 157.
149. Rozsondai, B.; Zelei, B.; Hargittai, I. *J. Mol. Struct.* **1982**, *95*, 187.
150. Cowley, A. H.; Kilduff, J. E.; Ebsworth, E.A.V.; Rankin, D.W.H.; Robertson, H. E.; Seip, R. *J. Chem. Soc. Dalton Trans.* **1984**, 689.
151. Fjeldberg, T.; Lappert, M. F.; Thorne, A. J. *J. Mol. Struct.* **1985**, *127*, 95.
152. Fjeldberg, T.; Lappert, M. F.; Thorne, A. J. *J. Mol. Struct.* **1984**, *125*, 265.
153. Fjeldberg, T.; Andersen, R. A. *J. Mol. Struct.* **1984**, *125*, 287.
154. Haaland, A.; Hedberg, K.; Power, P. P. *Inorg. Chem.* **1984**, *23*, 1972.
155. Alyea, E. C.; Fisher, K. J.; Fjeldberg, T. *J. Mol. Struct.* **1985**, *130*, 263.
156. Alyea, E. C.; Fisher, K. J.; Fjeldberg, T. *J. Mol. Struct.* **1985**, *127*, 325.

157. Fjeldberg, T.; Hope, H.; Lappert, M. F.; Power, P. P. Thorne, A. J. *J. Chem. Soc. Chem. Commun.* **1983**, 639.
158. Fjeldberg, T.; Andersen, R. A. *J. Mol. Struct.* **1985**, *129*, 93.
159. Fjeldberg, T.; Andersen, R. A. *J. Mol. Struct.* **1985**, *128*, 49.
160. Fjeldberg, T.; Haaland, A.; Lappert, M. F.; Seip, R.; Schilling, B.E.R.; Thorne, A. J.; Volden, H. V. *J. Chem. Soc. Dalton Trans.* **1986**, 1551.
161. Mastryukov, V. S.; Dorofeeva, O. V.; Vilkov, L. V.; Cyvin, B.; Cyvin, S. J. *Zh. Strukt. Khim.* **1975**, *16*, 473.
162. Cyvin, B. N.; Cyvin, S. J.; Strelkov, S. A.; Mastryukov, V. S.; Vilkov, L. V.; Golubinskii, A. V. *J. Mol. Struct.* **1986**, *144*, 385.
163. Shen, Q.; Hilderbrandt, R. L.; Mastryukov, V. S. *J. Mol. Struct.* **1979**, *54*, 121.
164. Mastryukov, V. S.; Atavin, E. G.; Golubinskii, A. V.; Vilkov, L. V.; Cyvin, B. N.; Cyvin, S. J. *Zh. Strukt. Khim.* **1979**, *20*, 726.
165. Dakkouri, M. Fourth Microwave Spectroscopy Conference in Tübingen, 1977.
166. Shen, Q.; Dakkouri, M. *J. Mol. Struct.* **1985**, *130*, 283.
167. Veniaminov, N. N.; Alekseev, N. V.; Bashkirova, S. A.; Komalenkova, N. G.; Chernyshev, E. A. *Zh. Strukt. Khim.* **1975**, *16*, 290.
168. Veniaminov, N. N.; Alekseev, N. V.; Bashkirova, S. A.; Komalenkova, N. G.; Chernyshev, E. A. *Zh. Strukt. Khim.* **1975**, *16*, 918.
169. Carleer, R.; Van den Enden, L.; Geise, H. J.; Mijlhoff, F. C. *J. Mol. Struct.* **1978**, *50*, 345.
170. Kozina, M. P.; Mastryukov, V. S.; Milvitskaya, E. M. *Usp. Khim.* **1982**, *51*, 1337 (*Russ. Chem. Rev.* **1982**, *51*, 765).
171. Mastryukov, V. S.; Osina, E. L.; Vilkov, L. V. *Zh. Strukt. Khim.* **1975**, *16*, 850.
172. Mastryukov, V. S.; Osina, E. L. *J. Mol. Struct.* **1977**, *36*, 127.
173. Takabayashi, F.; Kambara, H.; Kuchitsu, K. Seventh Austin Symposium on Gas-Phase Molecular Structure, Austin, TX, 1978, p. 63.
174. Hilderbrandt, R. L.; Homer, G. D.; Boudjouk, P. *J. Am. Chem. Soc.* **1976**, *98*, 7476.
175. Shen, Q.; Hilderbrandt, R. L.; Blankenship, C. S.; Cremer, S. E. *J. Organomet. Chem.* **1981**, *214*, 155.
176. Schei, H.; Shen, Q.; Cunico, R. F.; Hilderbrandt, R. L. *J. Mol. Struct.* **1980**, *63*, 59.
177. Shen, Q.; Hilderbrandt, R. L.; Burns, G. T.; Barton, T. J. *J. Organomet. Chem.* **1980**, *195*, 39.
178. Shen, Q.; Kapfer, C. A.; Boudjouk, P.; Hilderbrandt, R. L. *J. Organomet. Chem.* **1978**, *169*, 147.
179. Bastiansen, O.; Fernholt, L.; Seip, H. M.; Kambara, H.; Kuchitsu, K. *J. Mol. Struct.* **1973**, *18*, 163.
180. Mastryukov, V. S.; Osina, E. L.; Vilkov, L. V.; Hilderbrandt, R. L. *Zh. Strukt. Khim.* **1981**, *22* (N2), 57.
181. Bartell, L. S.; Clippard, F. B., Jr.; Boates, T. L. *Inorg. Chem.* **1970**, *9*, 2436.
182. Rankin, D.W.H.; Robertson, A. *J. Mol. Struct.* **1975**, *27*, 438.
183. Oberhammer, H. *J. Mol. Struct.* **1976**, *31*, 237.
184. Kveseth, K. *Acta Chem. Scand.* **1979**, *A33*, 453.
185. Almenningen, A.; Fjeldberg, T.; Hengge, E. *J. Mol. Struct.* **1984**, *112*, 239.
186. Almenningen, A.; Fjeldberg, T. *J. Mol. Struct.* **1981**, *77*, 315.
187. Strelkov, S. A.; Mastryukov, V. S.; Vilkov, L. V.; Rozsondai, B.; Schuster, H. G.; Hengge, E. Tenth Austin Symposium on Gas-Phase Molecular Structures, Austin, TX, 1984, p. 112.
188. Smith, Z.; Almenningen, A.; Hengge, E.; Kovar, D. *J. Am. Chem. Soc.* **1982**, *104*, 4362.
189. West, R. *Pure Appl. Chem.* **1984**, *56*, 163.
190. Masamune, S.; Hanzawa, Y.; Murakami, S.; Bally, T.; Blount, J. F. *J. Am. Chem. Soc.* **1982**, *104*, 1150.
191. Dewan, J. C.; Murakami, S.; Snow, J. T.; Collins, S.; Masamune, S. *J. Chem. Soc. Chem. Commun.* **1985**, 892.
192. Burkert, U.; Allinger, N. L. "Molecular Mechanics". American Chemical Society: Washington, DC, 1982.
193. Offenbach, J. L.; Fredin, L. S.; Strauss, H. L. *J. Am. Chem. Soc.* **1982**, *104*, 4362.
194. Vajda, E.; Kolonits, M.; Fritz, G.; Thomas, J.; Sattler, E. *J. Mol. Struct.* **1984**, *117*, 329.
195. Fjeldberg, T. *J. Mol. Struct.* **1984**, *112*, 159.
196. Oberhammer, H.; Boggs, J. E. *J. Am. Chem. Soc.* **1980**, *102*, 7241.
197. Glidewell, C. *Inorg. Chim. Acta* **1976**, *20*, 113.
198. Gundersen, G.; Rankin, D.W.H.; Robertson, H. E. *J. Chem. Soc. Dalton Trans.* **1985**, 191.
199. Gergö, E.; Schultz, G.; Hargittai, I. *J. Organomet. Chem.* **1985**, *292*, 343.

200. Fink, M. J.; Haller, K. J.; West, R.; Michl, J. *J. Am. Chem. Soc.* **1984**, *106*, 822.
201. Rozsondai, B.; Hargittai, I.; Golubinskii, A. V.; Vilkov, L. V.; Mastryukov, V. S. *J. Mol. Struct.*
     1975, *28*, 339.
202. Oberhammer, H.; Zeil, W.; Fogarasi, G. *J. Mol. Struct.* **1973**, *18*, 309.
203. Robiette, A. G.; Sheldrick, G. M.; Sheldrick, W. S. *J. Mol. Struct.* **1970**, *5*, 423.
204. Iijima, T.; Tsuchiya, S.; Kimura, M. *Bull. Chem. Soc. Japan* **1977**, *50*, 2564.
205. Pandey, G. K.; Dreizler, H. *Z. Naturforsch.* **1977**, *32a*, 482.

2

NITROGEN AND PHOSPHORUS COMPOUNDS

Lev V. Vilkov and Nina I. Sadova

DEPARTMENT OF CHEMISTRY
MOSCOW STATE UNIVERSITY
MOSCOW, USSR

CONTENTS

INTRODUCTION   36
NITROGEN DERIVATIVES   36
   One-Coordinated Nitrogen Atom   37
   Two-Coordinated Nitrogen Atom   38
   Three-Coordinated Nitrogen Atom   45
   Four-Coordinated Nitrogen Atom   59
PHOSPHORUS DERIVATIVES   60
   One-Coordinated Phosphorus Atom   60
   Two-Coordinated Phosphorus Atom   61
   Three-Coordinated Phosphorus Atom   62
   Four-Coordinated Phosphorus Atom   73
   Five-Coordinated Phosphorus Atom   79
DISCUSSION   80
   Bonds and Their Configurations   80
   Calculation of Internal Rotation Potential Function from Electron
     Diffraction Data   81
   Conformations of Five-Membered Rings Without Double Bonds   83
ACKNOWLEDGMENTS   86
REFERENCES   86

Introduction

The stereochemistry of nitrogen and phosphorus appears to occupy a rather special place in structural chemistry. These two elements occur in a variety of valence states; therefore, there are ample possibilities for comparing experimental data and for testing various structural theories. Elements of neighboring groups such as carbon and silicon or oxygen and sulfur offer less opportunity for this kind of treatment.

A distinguishing feature of nitrogen and phosphorus in the lower coordination states is the formation of complexes in which the elements occur in unusual valence states. In this chapter, these complexes will practically be omitted from consideration; in such instances, nitrogen and phosphorus derivatives act as ligands attached to a central atom as acceptor. Data on these substances are included elsewhere, eg, in Chapters 1 and 8 of this volume.

Coordination number five is rather characteristic of phosphorus. Practically all the five-coordinated phosphorus derivatives studied by gas-phase electron diffraction are fluorinated phosphoranes, which are discussed in Chapter 4.

This review discusses data on more than 350 nitrogen and phosphorus compounds and covers the literature through 1985.

To reduce the list of references, original works published before 1975 which are included in Landolt-Börnstein (LB)* are as a rule not cited, and the reader is referred to that handbook. As electron diffraction data frequently are compared with structural information obtained by using other techniques, the following abbreviations will be employed:

ED: gas-phase electron diffraction
LCNMR: liquid crystal nuclear magnetic resonance spectroscopy
MW: microwave spectroscopy
XR: X-ray diffraction

The parameter values and uncertainties were borrowed directly from original papers. They are mostly r_a structure parameters. In a number of instances, r_e, r_s, r_z, r_α and r_g parameters are given. Numbers in parentheses [eg, 1.0976(2) Å] are the limits of error of the last digits.

Nitrogen Derivatives

In chemical compounds nitrogen can be linked with one, two, three, or four other atoms.

* Landolt-Börnstein: Numerical Data and Functional Relationships in Science and Technology, New Series; Hellwege, K. H.; Hellwege, A. M., Eds.; Vol. 7: "Structure Data of Free Polyatomic Molecules". Springer-Verlag: Berlin, 1976.

One-Coordinated Nitrogen Atom

The simplest nitrogen compound is the N_2 molecule in which the equilibrium distance r_e ($N\equiv N$) = 1.0976(2) Å.

Cyano derivatives make up a large group of compounds containing one-coordinated nitrogen atoms. The $C\equiv N$ bond length in the cyano group is practically independent of the nature of the substituent. The $C\equiv N$ bond possesses the shortest carbon–nitrogen bond observed experimentally:

	HCN	CH₃CN	CH₂=CHCN	ClC≡CCN[1]	NCCN
r_g($C\equiv N$), Å:	1.158(3)	1.159(2)	1.167(4)	1.160(3)	1.162(2)

With nitriles, the linearity of the $C-C\equiv N$ group is subject to discussion. According to microwave data, the $C-C\equiv N$ group and the XCN molecules (X = H, F, Cl, Br, and I) are linear. Deviations from linearity caused by the shrinkage effect were observed in electron diffraction studies of $C(CN)_4$, $O=C(CN)_2$,[2] and $NCCH_2CH_2CN$.[3] Introducing shrinkage corrections for, eg, $NCCH_2CH_2CN$ (**1**) afforded a linear $C-C\equiv N$ fragment structure. Considerable departures from linearity were also reported for the $=N-C\equiv N$ group in $NC-N=N-CN$ and $NNN-CN$.[4] The effect can be explained by C_s symmetry of the $\overset{N}{\diagdown}N-C\equiv N$ fragment (see also Tables 2-6 and 2-2, below).

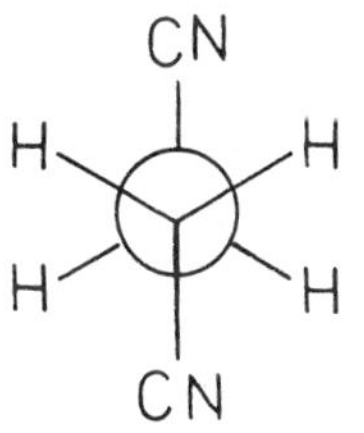

anti(74 ± 7 %)

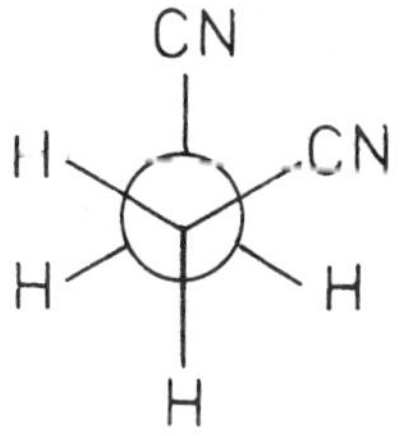

gauche

Electron diffraction data on conformations of 3-butenenitrile[5] (**2**) and benzyl cyanide[6] have been reported. Only the syn conformer occurs in the

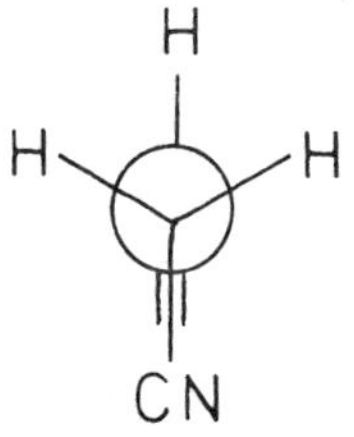

syn(94(18) %)

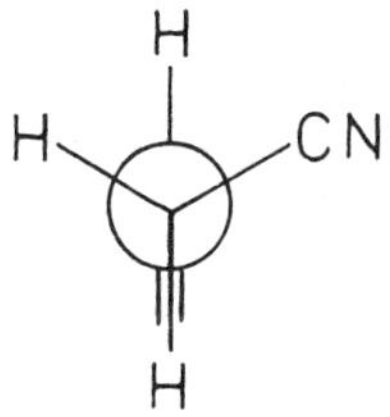

gauche

2

$CH_2=CHCH_2CN$ (**2**) vapor. The $C_6H_5CH_2CN$ molecule is nonplanar, with the ring and the CCN group making an angle of 42°.

The cyano group affects neighboring atomic groups as a strong σ-electron acceptor; in a number of molecules its structural effects are quite strong.

Examples are distortion of the benzene ring in NC—⟨○⟩—CN (see Chapter 7) and a substantial shortening of the central N—C bond [1.338(2) Å] in the nonplanar $(CH_3)_2N$—CN molecule[7] compared to amines.

Two-Coordinated Nitrogen Atom

Isocyanates, Isothiocyanates, Azides, and Diimides. The structural chemistry of these compounds is described in reviews,[8,9] where the problem of linearity versus nonlinearity of the $N=C=O$, $N=C=S$, and $N=N\equiv N$ chains has been analyzed in detail.

Combined electron diffraction and microwave data show that the $X—N=C=O(S)$* and $X—N=N\equiv N$ molecules and molecular frames are planar. For certain simple molecules, "planarity" refers linearity of the $N=C=O(S)$ and $N=N\equiv N$ groups.

In the X—NCO, X—NCS, and $X—N=N\equiv N$ (X = H, CH_3) molecules (Tables 2-1 and 2-2), the valence angles at nitrogen are smaller than 180°, and $\angle X—N=N < \angle X—N=C$. Substituting SiH_3 for H or CH_3 as a rule results in a substantial widening of the XNC and XNN angles.

In the X—NCO and X—NCS molecules (X = P, S, Se, Te, and Cl), the XNC angle is narrowed relative to the corresponding angle in species where X is a Group IV element (C, Si, or Ge). The angle SeNC is smallest in F_5SeNCO both in the F_5XNCO series (X = S, Se, Te) and among all the cyanates and thiocyanates studied. The longest $N=C$ bonds occur in isocyanates that contain P, S, Se, Te, and Cl atoms (Table 2-1).

The $N=N$ bond length in azides, $N—N=N\equiv N$ (1.25 Å on average), is comparable with that observed in azo compounds, while the $N\equiv N$ bond (1.13 Å on average) is elongated somewhat from N_2 (1.10 Å). The XNN valence angle values vary over a wide range, from about 109° in ClN_3 to 128° in $(CH_3)_3SiN_3$.

In addition to the compounds considered above, data on three diimides of the general formula $X—N=C=N—X$ are given in Table 2-3.

Imines. In imines $XN=C(X'X'')$ (Table 2-4), $—N=C\big\langle$ bond lengths are increased, and valence angles at nitrogen are narrowed compared to isocyanates

---

* One isoselenocyanate, F_2PNCSe, has also been studied (see Table 2-1).

Table 2-1. ISOCYANATES AND ISOTHIOCYANATES

Compound	$\angle X{-}N{=}C$ (deg)	$N{=}C$ (Å)	Method	Ref.
HNCO	128.1(0.5)	1.207(1)	MW	LB[a]
HNCS	135.0(0.2)	1.216(2)	MW	LB
CH_3NCO	140.3(0.4)	1.202(5)	ED	LB
CH_3NCS	141.6(0.4)	1.192(6)	ED	LB
SiH_3NCO	151.7(1.2)	1.216(9)⎫[b]	ED	LB
	152.7(1.2)	1.200(5)⎭		
	159	1.199(12)	MW	10
SiH_3NCS	163.8(2.6)	1.197(7)	ED	LB
$(CH_3)_3SiNCO$	156.9(3.0)	1.202(16)	ED	11
$(CH_3)_3SiNCS$	154(2)	1.18(1)	ED	LB
$Si(NCO)_4$	146.4(0.5)	1.209(2)	ED	LB
$ClSi(NCO)_3$	145(3)	1.213(8)	ED	LB
$Cl_2Si(NCO)_2$	136(2)	1.217(9)	ED	LB
Cl_3SiNCO	138.0(0.8)	1.219(9)	ED	LB
F_3SiNCO	160.7(1.2)	1.19[c]	ED	LB
GeH_3NCO	143.2(3.4)	1.168(27)	MW	12
GeH_3NCO	141.3(0.3)	1.190(7)	ED	13
F_2PNCO	130.6(0.8)	1.256(6)	ED	LB
F_2PNCS	138.7(0.6)	1.248(4)	ED + LCNMR	14
F_2PNCSe	149.0(1.5)	1.212(8)	ED	15
$Cl_2(O)PNCO$	120(1.5)	1.161(15)	ED	LB
$(SO_2Cl)NCO$	123.8(3.8)	1.159(11)⎫[b]	ED	16
		1.233(8) ⎭		
F_5SNCO	124.9(1.2)	1.234(8)	ED	17
F_5SeNCO	116.9(0.8)	1.260(11)	ED	17, 18
F_5TeNCO	126.5(2.4)	1.244(13)	ED	17
ClNCO	118.8(0.5)	1.228(10)	ED + MW	LB

[a] Landolt-Börnstein: See note, page 36.
[b] Two parameter sets were obtained.
[c] Assumed.

Table 2-2. AZIDES

Compound	$\angle X{-}N{=}N$ (deg)	Bond lengths (Å) $N{=}N$	$N{\equiv}N$	Method	Ref.
$HN{=}N{\equiv}N$	114.1(1.0)	1.237(3)	1.133(3)	MW	LB
$ClN{=}N{\equiv}N$	108.7(0.7)	1.252(10)	1.133(10)	MW	LB
$CH_3N{=}N{\equiv}N$	116.8(0.3)	1.216(4)	1.130(5)	ED	LB
$CF_3N{=}N{\equiv}N$	112.4(0.2)	1.252(5)	1.118(3)	MW + ED	19
$SiH_3N{=}N{\equiv}N$	123.8(1.0)	1.304(11)	1.125(7)	ED	LB
$(CH_3)_3SiN{=}N{\equiv}N$	128(3)	1.198(8)	1.150(11)	ED	LB
$GeH_3N{=}N{\equiv}N$	123.3(0.3)	1.217(20)	1.131(9)	MW	12
$NC{-}N{=}N{\equiv}N$	120.2(2.0)	1.252(20)	1.133(20)	MW	LB
$NC{-}N{=}N{\equiv}N$	114.5(2)	1.261(2)	1.121(2)	ED	4

Table 2-3. DIIMIDES

Compound	$N{=}C$ (Å)	Bond angles (deg)	
		$\angle XNC$	$\angle XNCNX$
$GeH_3N{=}C{=}NGeH_3{}^{a}$	1.184(9)	138.0(0.5)	75(cis)
$H_3SiN{=}C{=}NSiH_3$	1.206(5)	150.6(1.2)	180(trans)
$F_2PN{=}C{=}NPF_2$	1.240(5)	132.8(0.5)	55(cis)

[a] Reference 13.

and isothiocyanates. An exception is the cyclic molecule (cf. $ClN{=}C{=}O$, Table 2-1):

$$\text{(cyclic structure: } C_6F_4(CF Cl){=}NCl)$$

It is noteworthy that, in the series of methyleneimine derivatives $XN{=}CX_2(X = H, F, Cl)$, the $N{=}C$ bond length is constant (within experimental error) despite variations in substituent X electronegativities.

In compounds containing the $N{=}C{-}C{\equiv}C$ and $N{=}C{-}C{=}N$ conjugated systems, the $N{=}C$ bonds are longer by about 0.01 Å than in methyleneimines; the substituents are oriented trans with respect to the $N{=}C$ bond.

A coplanar conformation was only observed for $H_3CN{=}CHC_6H_5$ (**3**). Note that an isomer with eclipsed $N{=}C$ and $C{-}Cl$ bonds is characteristic for $CH_3N{=}CH(CCl_3)$. The $C_6H_5N{=}CHC_6H_5$ molecule **4** is nonplanar.

Table 2-4. IMINES

Compound	$N{=}C$ (Å)	$\angle X{-}N{=}C$ (deg)	Method	Ref.
$HN{=}CH_2$	1.273	110.4	MW[a]	20
$FN{=}CF_2$	1.274(6)	107.9(0.2)	MW + ED	21
$ClN{=}CCl_2$	1.266(5)	117.1(0.4)	ED	22
(cyclic structure: $C_6F_4(CFCl){=}NCl$)	1.224(10)	126.0	ED	23
$HN{=}C(CF_3)_2$	1.294(29)	110.0(0.9)	ED	LB
$H_3CN{=}CH(CCl_3)$	1.308(13)	115.2(2.0)	ED	24
$H_3CN{=}CHC_6H_5$	1.286(8)	113.6(1.6)	ED	25
$H_5C_6N{=}CHC_6H_5$	1.284(10)	115.0(2.0)	ED	26
$(CH_3)_3CN{=}CH{-}CH{=}NC(CH_3)_3$	1.283(6)	122.8(1.5)	ED	27
$HO{-}N{=}CHC(CH_3){=}O$	1.260(4)	112.3(0.4)	ED	28

[a] r_s/r_0 parameters.

The $(CH_3)_3CN{=}CH{-}CH{=}NC(CH_3)_3$ molecule has a gauche conformation (**5**) with respect to the $\mathrm{C{-}C}$ bond with a 65° rotation angle from the syn form; (the presence of a small amount of the trans conformer cannot be ruled out). A planar syn, s-trans conformation is characteristic of $HO{-}N{=}CH{-}C(CH_3){=}O$ (**6**).

Azines. Azines (Table 2-5) exemplify conjugated systems that contain a $={N}{-}N{=}$ central bond. The $N{=}C$ bond lengths in azines are the same as in imines, $XN{=}CX_2$. The $N{-}N$ bonds in $X_2C{=}N{-}N{=}CX_2$ (X = H, Br) are shortened substantially compared to hydrazine, N_2H_4 (Table 2-14). The $N{-}N{=}C$ valence angles are comparable with the $X{-}N{=}C$ angles listed in Table 2-4.

Interestingly, $H_2C{=}N{-}N{=}CH_2$ and $Br_2C{=}N{-}N{=}CBr_2$ possess different conformations. Whereas the trans conformer predominates in diazabutadiene (**7**) [its fraction is 0.91(7) at $-30°C$, 0.78(10) at $60°C$, and 0.75(11) at $225°C$], a single nonplanar conformer with a 72(5)° rotation angle of two planar $NCBr_2$ groups about the $N{-}N$ bond was observed in $Br_2C{=}N{-}N{=}CBr_2$ (**8**).

Table 2-5. AZINES

| Compound | Bond lengths(Å) | | $\angle$ NNC (deg) | Ref. |
	N=C	N—N		
CH_2=N—N=CH_2	1.277(2)	1.418(3)	111.4(0.2)	29
CBr_2=N—N=CBr_2	1.266(8)	1.381(18)	114.5(1.6)	30
CH_3CH=N—N=$CHCH_3$	1.277(4)	1.436(12)	110.5(0.7)	31

According to the authors of this study, the nonplanarity of the C=N—N=C frame can be explained by strong repulsive interactions between bromine atoms and the lone electron pair of nitrogen oriented cis to the bromines. A planar (or nearly planar) trans arrangement of bonds in the C=N—N=C group was observed in CH_3CH=N—N=$CHCH_3$, where the anti, trans, conformer **9** was found to predominate.

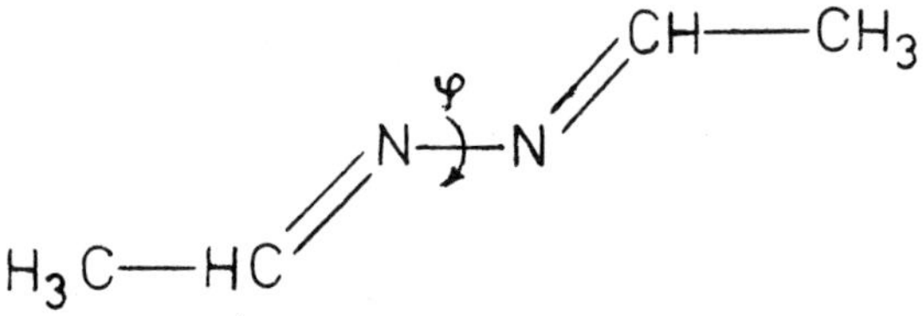

$$\varphi = 160\text{–}180°$$

9

Azo Compounds. The N=N bond length in azo compounds (Table 2-6) depends only slightly on the substituent (cf the N—N bond in hydrazines, Table 2-14). Its average value (1.24 Å) is close to that found in azides. The same N=N bond length is observed in cyclic molecules

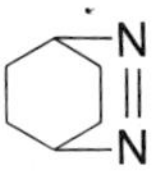

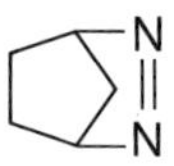

(see reference 36). The N=N bond is elongated in

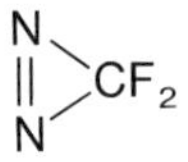

(see reference 37) to 1.293(9) Å.

The value of the NNC valence angle depends on the conformation: as a rule, the NNC valence angle is smaller in trans isomers than in cis isomers.

Table 2-6. AZO COMPOUNDS

Compound	N=N (Å)	∠N=NX (deg)	Conformation, symmetry	Ref.
$FN=NF$	1.230(10)	105.5(0.7)	s-trans, C_{2h}	32
$FN=NF$	1.214(12)	114.4(1.0)	s-cis, C_{2v}	32
$CH_3N=NCH_3$	1.247(3)	123.3(0.3)	trans	33
$CH_3N=NCH_3$	1.254(3)	119.9(0.5)	trans	34a
$CH_3N=NCF_3$	1.219(8)	110.5(4.0) [a] 126.2(2.0)	trans, ∠CNNC = 180°	34a
$CF_3N=NCF_3$ [b]	1.236(15)	133.0(08)	cis, CNNC = 0°	34a
$CF_3N=NCF_3$	1.235(10)	113.4(1.6)	trans, ∠CNNO = 180°	34b
$NC—N=N—CN$	1.261(2)	113.0(0.2)	s-trans	4
$H_5C_6N=NC_6H_5$	1.259(4) 1.268(4)	116.0(0.4) 114.5(0.4)	∠CNNC(C_2) = 31.0(1.6)° ∠CNNC(C_i) = 27.9(2.9)°	35

[a] The first value refers to the NCC_{CH_3} angle, and the second to NNC_{CF_3}.
[b] See updated data in reference 34b.

Trans conformers occur in $CH_3N=NCH_3$, $CH_3N=NCF_3$ **(10)**[34a] and $CF_3—N=N—CF_3$ **(11)**.[34b]

Two isomers were observed in N_2F_2, the trans isomer having a lower boiling point and the cis isomer a higher boiling point. The *trans*-azobenzene molecule is nonplanar, the rotation angle about the N—C bond being 30°.

Compounds Containing the N=P, N=V, and N=S Bonds. The Si—N—P angle in $(CH_3)_3SiN=P(CH_3)_3$[38] is 144.6(1.1)° [cf. $(CH_3)_3SiN=CO(S)$, Table 2-1]. The valence angle at N in $ClN=VCl_3$,[39] where the ClNV moiety is almost linear, is substantially larger: 169.7(4.2)°. This molecule contains a very short N—Cl distance of 1.597(8) Å.

The structure of molecules containing the N=S bond depends on the sulfur coordination number (Table 2-7). For compounds that contain sulfur in the same valence state, the valence angles at nitrogen vary under the influence of substituents: ∠ClN=S (or FN=S) < ∠CN=S, and ∠CN=S < ∠SiN=S.

Table 2-7. COMPOUNDS CONTAINING N=S
BONDS FOR $n = 2-5$[a]

Compound	$\angle XN{=}S$ (deg)	Ref.
	n = 2	
$ClN{=}SO$	116.3(0.8)	LB
$CH_3N{=}SO$	126(2)	40
$SiH_3N{=}SO$	129.2(0.2)	41
$CH_3N{=}S{=}NCH_3$	124.3(1.1)$_{cis}$	42
	116.5(1.2)$_{trans}$	
	n = 3	
$ClN{=}SF_2$	120.0(0.6)	LB
$ClN{=}S(CF_3)_2$	138.2(3.8)	43
$CF_3N{=}SF_2$	130.4(0.7)	44
	n = 4	
$(CH_3)_2S(O){=}NH$	114(7)	LB
$(CH_3)_2S\overset{/\!/ \text{NH}}{\underset{\backslash\!\backslash \text{NH}}{}}$	114(7)	LB
$F_2S(O) = NCl$	114.7(2.4)	LB
	n = 5	
$FN{=}SF_4$	117.6(1.2)	45
$CH_3N{=}SF_4$	127.2(1.1)	46

[a] n is the coordination number of sulfur.

Like the CNN angle in azo compounds, the CNS angle in $CH_3N{=}S{=}NCH_3$ depends upon the molecular conformation.

Nitroso Compounds. The structures of CF_3NO and $(CH_3)_2NNO$ were studied by using modern electron diffraction techniques (see Table 2-8).

The C—N=O and N—N=O groups are nonlinear in both molecules. In CF_3NO, the N=O bond is eclipsed with one of the C—F bonds, and the C—N bond is elongated compared to CH_3NO [1.49(3) Å, MW data]. All the nonhydrogen atoms in $(CH_3)_2NNO$ lie in the same plane.

Table 2-8. NITROSO COMPOUNDS

Compound	Bond lengths (Å)		$\angle XNO$[a] (deg)	Ref.
	$N{=}O$	$N{-}X$[a]		
CF_3NO	1.197(5)	1.546(8)	113.2(1.3)	LB
$(CH_3)_2NNO$	1.234(2)	1.344(2)	113.6(0.2)	LB

[a] X = C, N

Three-Coordinated Nitrogen Atom

The coordination number 3 is most characteristic of nitrogen compounds. The specific feature of the stereochemistry of trivalent nitrogen derivatives is the variation of the bond configuration about nitrogen in amines and amides depending on the substituents. A comparison of the experimental and calculated energies of formation of various molecules shows that the effects of mutual influence of atoms in molecules can exceed the barrier to inversion in nitrogen derivatives.[46] Consequently substantial influence of the groups attached to nitrogen can change the valence state and the bond configuration of nitrogen from pyramidal to a planar or almost planar configuration.

Amines and Their Derivatives

Ammonia and Alkylamines. According to the spectroscopic data, the equilibrium geometrical parameters of ammonia are $r_e(N—H) = 1.0116$ Å and $\angle HNH = 106.68°$. *Ab initio* calculations give $r_e(N—H) = 1.016$ Å and $\angle HNH = 106.2°$.[47]

In the alkylamine series (Table 2-9) a tendency toward C—N bond shortening can be observed in passing from the primary to the tertiary amines. The bond configuration at nitrogen is pyramidal. Substituting CH_3 for H in NH_3 results in an increase of about 3° in the nitrogen valence angles.

In $(CF_3)_3N$ the bond configuration approaches planarity ($\angle CNC = 118°$), and the C—N bond is shortened compared to methylamines.

Aromatic Amines. The valence angles at nitrogen in $C_6H_5N(CH_3)_2$ (**12**) and $(C_6H_5)_3N$ (**13**) are $116(2)°$; ie, the bond configuration at nitrogen is flattened significantly. The authors of the study on $(C_6H_5)_3N$ suggested that the equilibrium bond configuration is planar, and that a small departure from planarity may be caused by out-of-plane inversion-type vibrations.[52] In reference 53 the CNH_2 group in $\langle\bigcirc\rangle$—NH_2 is shown to be almost planar;

Table 2-9. ALKYLAMINES[a]

Compound	C—N (Å)	$\angle$CNC (deg)	Ref.
CH_3NH_2	1.471(2)	—	LB
$(CH_3)_2NH$	1.455(6)	111.8(0.8)	LB
$(CH_3)_3N$[b]	1.454(4)	110.6(0.6)	LB
$(CF_3)_3N$	1.426(6)	117.9(0.4)	49
$(CH_3)_2NC_2H_5$	1.452(6)	111.3(1.0)	50
$CH_2[N(CH_3)_2]_2$	1.449(4)	110.7(1.0)	51

[a] Methylamines were studied by both electron diffraction and microwave techniques; the results agree with each other. Listed here are electron diffraction data except for CH_3NH_2: for this compound only $r(C—N) = 1.465(2)$ Å was determined by electron diffraction.
[b] An additional analysis of data gave $r(C—N) = 1.461$ Å.[48]

gas-phase electron diffraction results, however, do not afford reliable information regarding planarity versus nonplanarity.

$$\varphi = 45°$$

12   **13**

When considering the structure of aromatic amines, the microwave data on aniline[54] should be cited: $r(\text{C—N}) = 1.402(2)$ Å, $\angle\text{HNH} = 113°$. The $C_6H_5NH_2$ molecule is nonplanar; the NH_2 group and the benzene ring make an angle of $37.5(0.2)°$. A comparison of aromatic and aliphatic amines shows that $r(\text{C—N})_{\text{aliphatic}}$ is greater than $r(\text{C—N})_{\text{aromatic}}$, and $\alpha_N(\text{aliphatic}) < \alpha_N(\text{aromatic})$ (α_N is an average valence angle at nitrogen).

Halogenated Amines. The XNX and XNY angles in molecules of the type $NF_2X(X = F, Cl, CF_3)$ are smaller than in NH_3 and in methylamines (Table 2-10). The C—N bond lengths decrease consistently along the series $CF_3NF_2 > (CF_3)_2NF > (CF_3)_3N$ (Tables 2-9 and 2-10). The N—F distance was reported[55] to be independent within experimental error of the number of CF_3 groups. The average valence angle at nitrogen increases in this series.

Analysis of the geometrical structure of chloromethylamines is hindered by the lack of sufficiently accurate data.

Organoelement Compounds. A substantial change in the bond configuration around the amine nitrogen atom is observed in silylamines compared to alkylamines (Table 2-11).

Amines containing one silyl and two methyl groups represent flattened pyramids with CNC and SiNC angles of about 112 and 120°, respectively; here, a silyl group may be SiH_3, SiH_2CH_3, or $SiH(CH_3)_2$.

Table 2-10. HALOGENATED AMINES: X = F, Cl

Compound	Bond lengths (Å)		Bond angles (deg)		Ref.
	N—X	C—N	$\angle$XNX	$\angle$XNY	
NF_3	1.365(2)	—	102.4(0.3)	—	LB
NF_2Cl	1.382(9)F	—	103(1)	105(1)	LB
	1.730(8)Cl				
NF_2CF_3	1.371(4)	1.476(5)	105.3(0.5)	104.1(0.7)	55
$NF(CF_3)_2{}^a$	1.378(12)	1.446(4)	—	104.8(0.7)	55
NCl_3	1.759(2)	—	107.1(0.5)	—	LB

^{a} $\angle$CNC $= 116.6(0.6)°$.

Table 2-11. SILYLAMINES

Compound	Bond lengths (Å)		Bond angles (deg)			Ref.
	Si—N	C—N	∠SiNSi	∠SiNC	∠CNC	
$(CH_3)_2N(SiH_3)$	1.713(5)	1.457(6)	—	120.9(0.3)	112.0(0.6)	56
$(CH_3)_2N(SiH_2CH_3)$	1.715(6)	1.455(3)	—	121.5(0.8)	112.7(0.8)	56
$(CH_3)_2N[SiH(CH_3)_2]$	1.719(5)	1.460(4)	—	119.3(0.8)	113.7(1.5)	56
$CH_3N(SiH_3)_2$	1.726(3)	1.465(6)	125.4(0.4)	117.3(0.2)	—	LB
$CH_3N(SiH_2CH_3)_2$	1.718(3)	1.492(12)	125.6(1.0)	117.2(0.6)	—	57
$CH_3N[SiH(CH_3)_2]_2$	1.727(4)	1.483(7)	126.1(0.5)	116.9(0.3)	—	57
$NH(SiH_3)_2$	1.725(3)	—	127.7(0.3)	—	—	LB
$NH[SiH(CH_3)_2]_2$	1.727(3)	—	130.4(1.5)	—	—	58
$NH[Si(CH_3)_3]_2$	1.738(5)	—	131.3(1.5)	—	—	59
$N(SiH_3)_3$	1.734(3)	—	119.7(0.3)	—	—	LB
$N(SiH_2CH_3)_3$	1.729(13)	—	120.0	—	—	60
$(CH_3)_2N(SiCl_3)$	1.657(12)	1.446(12)	—	123.1(0.8)	113.1(1.8)	LB
$(CH_3)_2N(SiF_3)$	1.654(15)	1.433(15)	—	119.7(1.6)	120.5(3.9)	LB

The $NCSi_2$ frame is planar in silylamines with two silyl groups and one methyl group. A planar bond configuration at nitrogen occurs in $N(SiH_3)_3$ and $N(SiH_2CH_3)_3$.

A change in the bond configuration at the amine nitrogen atom in silylamines takes place not only as the number of silyl groups increases, but also as H atoms in SiH_3 are replaced by fluorine or chlorine atoms, ie, in $SiF_3N(CH_3)_2$ and in $SiCl_3N(CH_3)_2$.

The Si—N bond length decreases only slightly along the trisilylamine-disilylamine-monosilylamine series, but it is reduced noticeably in $(CH_3)_2NSiX_3$ (X = F, Cl). In methyl(bis)silylamines, the C—N bonds are elongated somewhat compared to dimethyl(silyl) amines. The shortest C—N bonds were observed in $(CH_3)_2NSiX_3$ (X = F, Cl).

Amino derivatives of other Group IV elements also have planar bond configurations around nitrogen: $N(GeH_3)_3$ with $r(Ge-N) = 1.836(5)$ Å and $N[Sn(CH_3)_3]_3$[61] with $r(Sn-N) = 2.038(3)$ Å or nearly planar (eg, $Sn[N(CH_3)_2]_4$).

The structures of $M\{N[Si(CH_3)_3]_2\}_2$ (M = Be, Mg, Zn, Cd, Hg, Ge, Sn, Pb) and $M\{N[Si(CH_3)_3]_2\}_3$ (M = Sc, Ce, Pr) were all solved assuming a planar bond configuration at N. The silylamine groups were proved experimentally to be planar only for the Ge and Sn derivatives. Certain geometrical parameters of these compounds including M—N bond lengths are listed in Table 2-12.

The average valence angle at nitrogen (α_N) increases not only under the influence of Group IV elements but also under the influence of phosphorus. Thus, $\alpha_N = 117.5°$ in $P[N(CH_3)_2]_3$ and $120°$ in $ClP[N(CH_3)_2]_2$[63] and $(CH_3)_2NPCl_2$. The electron diffraction data on H_2NPF_2 and $(CH_3)_2NPF_2$ are at variance with the microwave results. According to the latter, $\alpha_N = 120°$ in both compounds, whereas electron diffraction experiments afford values of

Table 2-12. MOLECULES OF THE TYPE
$M\{N[Si(CH_3)_3]_2\}_n$ $(n = 2, 3)^a$

| | Bond lengths (Å) | | ∠SiNSi |
| | | | |
M	M—N	Si—N	(deg)
Be	1.562(24)	1.722(7)	129.3(0.5)
Mg	1.91(3)	1.703(6)	132(2)
Zn	1.824(14)	1.728(7)	130.4(2.0)
Cd	2.03(2)	1.727(7)	132.0(1.2)
Hg	2.01(2)	1.732(9)	128.0(2.0)
Ge	1.89(1)	1.743(6)	118(2)
Sn	2.09(1)	1.74	125.3
Pb	2.20(2)	1.75(1)	121.(2)
Sc	2.02(3)	1.715(8)	128.0(2.0)
Ce	2.33(4)	1.696(14)	129(4)
Pr	2.31(4)	1.705(15)	129(4)

^{a} The Be, Mg, Zn, Cd, and Hg atoms were assumed to
have linear bond configurations. Bond configurations around M in Ge, Sn, and Pb compounds are
characterized by the angles NGeN = 101.(1.5)°,
NSnN = 96°, and NPbN = 91(2)°. Sc has a
planar bond configuration (∠NScN = 119.5(1.5)°;
Ce and Pr form pyramids with the angles
∠NCeN = 112(3)° and ∠NPrN = 113(3)°.

115.3 and 116°. The bond configuration around the nitrogen atom was shown to
be planar in $N(PF_2)_3$,[64] $(H_3X)N(PF_2)_2$ (X = C,[65] Ge,[66] Sn[66]), $(SiH_3)_2NPF_2$[66]
and $(CH_3)_2NPF_4$.[67] The Ge—N bond in $(GeH_3)N(PF_2)_2$, 1.889(13) Å, is
longer by 0.05 Å than it is in $N(GeH_3)_3$. The Si—N bond lengths in
$(SiH_3)_2NPF_2$ [1.755(4) Å] and $(SiH_3)N(PF_2)_2$ [1.767(7) Å] are larger than
those found in most other silylamines (see Table 2-11). In general, the nature of
substituents at P and the valence state of phosphorus affect the structure of
phosphorus amino derivatives.

The geometry of the amino group in sulfur amino derivatives such as
$S[N(CH_3)_2]_2$, $OS[N(CH_3)_2]_2$, $O_2S[N(CH_3)_2]_2$, and $(CH_3)_2NSO_2X$ (X = Cl,
CH_3) is characterized by flattening of the nitrogen pyramid. The SNC and CNC
angles are in the range 115–118°. The SNS valence angle in $N(SCF_3)_3$ is
118.8(0.6)°.

A planar bond configuration at nitrogen is realized in boron amino derivatives $(CH_3)_2NBCl_2$ and $[(CH_3)_2N]_3B$. All the heavy atoms in the former
molecule are coplanar. The C—N bond length is the same as in dimethylamine.
The boron–nitrogen bond [1.379(6) Å] differs substantially from that observed
in $H_5B_2N(CH_3)_2$ and therefore can be regarded as being a double bond.

The B—N bond length in $[(CH_3)_2N]_3B$ is 1.431(12) Å. The parameters
$r(C—N) = 1.475(7)$ Å and ∠CNC = 118.1(1.0)° do not differ from the corresponding parameters in $(CH_3)_2NBCl_2$ within experimental error. Steric hin-

Table 2-13. HYDROXYLAMINES

Compound	Bond lengths (Å)			Bond angles (deg)		Ref.
	C—N	N—O	C—O	∠CNO	∠NOC	
NH_2OH	—	1.453(2)	—	—	—	70
NH_2OCH_3	—	1.463(2)	1.388(4)	—	108.7(0.3)	70
CH_3NHOH	1.420(4)	1.477(2)	—	107.7(0.2)	—	70
CH_3NHOCH_3	1.439(14)	1.496(9)	1.374(8)	104.5(0.2)	109[a]	70
$(CH_3)_2NOCH_3$	1.442(5)	1.513(9)	1.350(6)	102.6(0.7)	109.3(1.2)	70
$(CF_3)_2NOH$	1.435[a]	1.399(24)	—	111.3(0.8)	—	71
$HN(OCH_3)_2$	—	1.392(7)	1.431(2)	—	112.4(1.2)	72

[a] Assumed.

drance in $[(CH_3)_2N]_3B$ results in a 33(3)° rotation of the dimethylamino groups out of the BN_3 plane. In borazole, $H_6B_3N_3$, $r(B—N) = 1.435(2)$ Å.

Results of a study on compounds of the series $(CH_3)_{3-n}B(NHCH_3)_n$ ($n = 1, 2, 3$)[69] indicate that the B—N bond length depends substantially on n: 1.397(2) Å for $n = 1$, 1.418(2) Å for $n = 2$, and 1.439(2) Å for $n = 3$. The C—N bond length is constant over this series (1.456 Å).

Hydroxylamines. The most marked trend in hydroxylamines of the type $R_2R_3N—OR_1$ (R_1, R_2, R_3 = H, CH_3) is a tendency toward increasing N—O bond length with increasing number of methyl groups attached to nitrogen (see Table 2-13). The C—N bonds are shortened compared to methylamines and to dimethylhydrazine, and the C—O bonds are shorter than in dimethylether. The N—O bond is as a rule longer than the C—O bond. The N—O bond length in $(CF_3)_2NOH$ is shorter than in the methyl derivatives. The bond configuration at nitrogen is pyramidal in this molecule and also in $(CH_3)_2NOCH_3$. A reliable determination of the bond configuration at nitrogen in other hydroxylamines is hindered by the presence of hydrogen atoms.

Anti conformer **14** predominates in CH_3NHOCH_3 and $(CH_3)_2NOCH_3$ vapors; the percentage of syn conformers **15** is 30% and 20%, respectively.

The $HN(OCH_3)_2$ molecule **16** has a near gauche, gauche conformation, the CONO dihedral angles being 45 and 75°.

14 **15** **16**

Table 2-14. HYDRAZINES

| Compound | N—N (Å) | Bond angles (deg), X = F, C, Si | | | Ref. |
		∠XNX	∠XNN	∠XNNX	
NH_2NH_2	1.449(2)	—	—	91(2)	73
NF_2NF_2	1.489(7)	104.1(1.0)	101.4(1.0)	67.1(1.8)	74
NF_2NF_2	1.492(7)	103.1(0.6)	101.4(0.4)	64.2(3.7)	75
$CH_3NHNHCH_3$[a]	1.419(11)	—	112(1)	90(12)	76
$(CH_3)_2NN(CH_3)_2$[b]	1.410(4)	110.8(1.6)	113.5(0.6)	78.5	77
$(CF_3)_2NN(CF_3)_2$[c]	1.402(20)	121.2(1.5)	119.0(1.5)	88(4)	78
$(SiH_3)_2NN(SiH_3)_2$[d]	1.457(16)	129.5(0.7)	—	82.5(0.8)	79

[a] C—N 1.463(5) Å.
[b] C—N 1.463(1) Å.
[c] C—N 1.433(10) Å.
[d] Si—N 1.731(4) Å.

Hydrazines and Hydrazones. The study of the structure of hydrazine (N_2H_4) and its derivatives such as N_2F_4 and $CH_3NHNHCH_3$ has a long history. Sufficiently reliable data on hydrazine were obtained in a combined analysis of electron diffraction and microwave spectroscopic data (Table 2-14). It was shown experimentally that there is a difference between inner and outer NNH angles in N_2H_4 (**17**): $\angle NNH_i = 112(2)°$ and $\angle NNH_o = 106(2)°$.

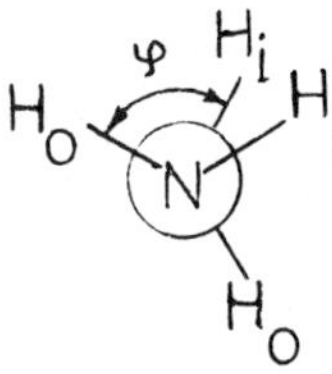

$\varphi = 90°$

17

In fully substituted hydrazines the N—N bond length varies within approximately 0.1 Å depending on the substituent (X):

X:	CF_3	CH_3	H	SiH_3	F
$r(N—N)$, Å:	1.40	1.41	1.45	1.46	1.49

Methylhydrazines show a tendency toward N—N bond shortening as the number of methyl substituents on nitrogen increases.

The bond configuration around nitrogen is nonplanar in N_2F_4, N_2H_4, and in the methyl derivatives of N_2H_4. It is, however, planar in $(CF_3)_2NN(CF_3)_2$ and $(SiH_3)_2NN(SiH_3)_2$.

Hydrazine derivatives (except for N_2F_4) mostly occur as gauche conformers with the XNNX dihedral angle equal or close to 90°. The trans conformer predominates in tetrafluorohydrazine.

Hydrazones (Table 2-15) possess nonplanar bond configurations around nitrogen with the average valence angle α_N of 114.9(0.7)° in

Table 2-15. Hydrazones

Parameter	$CH_3CH{=}N{-}N(CH_3)_2$	$(CH_3)_2C{=}N{-}N(CH_3)_2$
N—N, Å	1.402(12)	1.420(9)
C—N, Å	1.469(20)	1.476(19)
N=C, Å	1.286(6)	1.281(6)
C—C, Å	1.492(33)	1.492(11)
$\angle$ NNCH$_3$, deg	114.1(0.6)	109.2(2.6)
$\angle$ CNC, deg	117.5(1.0)	118.1(1.0)
$\angle$ C=NN, deg	122.2(2.5)	121.8(4.0)
$\angle$ N=C—CH$_3$, deg	114.0(1.8)	116.6(3.5)
$\angle \varphi$, deg[a]	19.2(7.1)	61.5(3.0)

[a] The angle $\varphi = 0°$ corresponds to the C_2 conformation in which both methyl groups at the amine nitrogen atom occur on the same side of the imine fragment.
Source: Reference 80.

$CH_3CH{=}N{-}N(CH_3)_2$ and 112.2(2.1) in $(CH_3)_2C{=}N{-}N(CH_3)_2$. The N—N bond lengths are the same as in methylhydrazines, and the N=C bonds have lengths typical for compounds containing two-coordinated nitrogen. The $N(CH_3)_2$ group is rotated from the imine fragment plane in both molecules; the rotation angles φ are, however, different. It is likely that in $(CH_3)_2C{=}N{-}N(CH_3)_2$ the equilibrium value is $\varphi = 90°$ and the deviation from 90° is caused by large-amplitude torsional motions.

Amides. The central fragment of organic acid amides contains the $\underset{O}{\overset{\diagdown}{C}}{-}N{\diagup}$ bond. A detailed analysis of the structural patterns for the simplest amides, $HC(O)NH_2$, $HC(O)NHCH_3$, $CH_3C(O)NH_2$, and $CH_3C(O)NHCH_3$ is given in the electron diffraction studies by Kuchitsu and co-workers.[81,82] These studies, in which microwave data were also employed, substantiated once more the notion that the bond configuration at the amide nitrogen atom is planar and that the central N—C(O) bond is shorter by about 0.1 Å than the N—C$_{Me}$ bond (Table 2-16). A planar bond configuration is also characteristic of

Table 2-16. Amides of Organic Acids and Their Derivatives

Compound	Bond lengths (Å)		Bond angles (deg)		Ref.
	C—N	C=O	$\angle$ NCC	$\angle$ NCO	
$HCONH_2$	1.368(3)	1.212(2)	—	125.0(0.4)	81a
$HCONHCH_3$	1.366(8)	1.219(5)	—	124.6(0.5)	81b
CH_3CONH_2	1.380(3)	1.220(3)	115.1(1.6)	122.0(0.6)	81c
$CH_3CONHCH_3$	1.386(4)	1.225(3)	114.1(1.5)	121.8(0.4)	82
CH_2FCONH_2	1.358(2)	1.226(2)	114.4(0.6)	126.7(0.7)	83a
$CH_2ClCONH_2$	1.362(2)	1.222(2)	118.0(0.2)	123.2(0.8)	83a
CH_2ICONH_2	1.370(3)	1.222(3)	116.9(0.4)	121.1(0.6)	83b
$(NC)CONH_2$	1.373(2)	1.219(2)	109.9(0.8)	126.6(1.0)	83c
$(CH_3CO)_2NH_2$	1.402(2)	1.210(2)	117.9(1.8)	123.7(1.5)	84

$CH_3C(S)NH_2$,[85] where the N—C bond is somewhat shorter [1.356(3) Å] than it is in acetamide.

The trans arrangement of the CH_3 groups with respect to the N—C bond was observed in $H_3CC(O)NHCH_3$. 2-Halosubstituted acetamides, $XCH_2C(O)NH_2$ (X = F, Cl), have various conformations. Steric hindrance in $CH_2ClCONH_2$ (**18**) and in CH_2ICONH_2 (**19**) causes a rotation of the $CXH_2(X = Cl, I)$ group from the syn configuration observed in CH_2FCONH_2 (**20**).

Although on the whole amides are characterized by a planar bond configuration at nitrogen, a flattened pyramid occurs in tetramethylurea and tetramethylthiourea,[86] where the $N(CH_3)_2$ groups are rotated somewhat about the N—C(X) bonds. The $(CH_3)_2N—C(O)Cl$[87] molecule is nonpolar; the rotation angle of the $N(CH_3)_2$ group about the N—C(O) bond is 26°.

Nitro Compounds. The presence of the nitro group NO_2 is the distinguishing feature of nitro compounds, whose geometrical structure is discussed in the review[88] that covers the literature to 1979.

Organic Nitro Compounds. Among aliphatic nitro compounds, differences in the geometry of the CNO_2 group between chloronitromethanes and methyl nitromethane derivatives should be mentioned first (Table 2-17). The ONO angle varies most: it increases by about 7° along the series CH_3NO_2—CH_2ClNO_2—CCl_3NO_2 and decreases by 3° on passing from $(CH_3)_2CHNO_2$ to $(CH_3)_3CNO_2$. Interestingly, $(CH_3)_2C(NO_2)_2$ (**21**) and $CCl_2(NO_2)_2$ (**22**) possess distinctly different conformations. In **21** the two nitro groups occupy nonequivalent positions relative to the methyl groups and are perpendicular to each other. In **22**, the nitro groups are equivalent relative to chlorines and are situated in a propellerlike arrangement.

Table 2-17. ALIPHATIC NITRO COMPOUNDS

Compound	Bond lengths (Å)		$\angle$ONO (deg)	Ref.
	N$=$O	C$-$N		
CH_3NO_2[a]	1.230(4)	1.488(5)	124.9(0.5)	89
CH_2ClNO_2	1.228(2)	1.506(9)	129.0(1.5)	LB
CF_3NO_2	1.21(1)	1.56(2)	132(2)	LB
CCl_3NO_2	1.190(6)	1.594(20)	131.7(2.6)	LB
CBr_3NO_2	1.22(1)	1.59(2)	134(3)	LB
$(CH_3)_2CHNO_2$	1.226(2)	1.518(10)[b]	125.4(0.3)	90
$(CH_3)_3CNO_2$	1.240(2)	1.533(15)[b]	122.2(0.6)	90
$(CH_3)_2C(NO_2)_2$	1.227(2)	1.517(15)[b]	124.5(0.6)	91
$CCl_2(NO_2)_2$	1.224(4)	1.506(12)	124.9(1.1)	LB
$CH(NO_2)_3$[c]	1.219(3)	1.505(5)	128.6(1.0)	LB
$CCl(NO_2)_3$[d]	1.213(3)	1.513(3)	128.3(0.6)	LB
$CBr(NO_2)_3$[e]	1.214(3)	1.514(6)	132.5(2.0)	LB
$C(NO_2)_4$[f]	1.218(3)	1.526(6)	129.3(1.0)	LB

[a] Microwave data.
[b] Average value of $r(C-N)$ and $r(C-C)$.
[c] Dihedral angle HCNO = 26.3(0.6)°.
[d] Dihedral angle ClCNO = 49.3(0.6)°.
[e] Dihedral angle BrCNO = 49.5(1.2)°.
[f] Dihedral angle NCNO = 47(1)°.

Nitro groups cause a shortening of adjacent C$-$Cl and C$-$Br bonds in halogenated nitroalkanes. Thus, the shortest C$-$X (X = Cl, Br) bonds are observed in $CCl(NO_2)_3$ [$r(C-Cl)$ = 1.712(4) Å] and in $CBr(NO_2)_3$ [$r(C-Br)$ = 1.885(9) Å].

Conformations of polynitroalkanes are characterized by rotation of the nitro groups from the N$-$C$-$X plane (X = H, Cl, Br, N).

Halopicrin molecules CX_3NO_2 (X = F, Cl, Br), have specific structures that cannot be considered to have been determined unambiguously because of the large discrepancies between microwave and electron diffraction data.

Passage from aliphatic to aromatic nitro compounds (Table 2-18) is accompanied by a change in the C$-$N bond length and in the ONO angle of the CNO_2 group. Both parameters are, as a rule, smaller in benzene derivatives than in saturated aliphatic compounds. The N$=$O bond length varies only slightly depending on the substituent.

The geometry of aromatic nitro compounds depends on the mutual influence of the NO_2 groups and the other substituents. Thus, in chloro- and bromonitrobenzenes, CNO_2 group parameters vary depending on the substituent position in the ring, eg, in m-$XC_6H_4NO_2$ and in p-$XC_6H_4NO_2$(X = Cl, Br), the N$=$O distance is increased somewhat compared to nitrobenzene. In p-nitroaniline and also in o- and p-chloronitrobenzenes, the C$-$NH$_2$ and C$-$Cl bonds are noticeably shorter (by up to ~0.03 Å) than the corresponding bonds in

Table 2-18. AROMATIC NITRO COMPOUNDS

Compound	Bond lengths (Å)			Bond angles (deg)				Ref.
	C—N	N=O	C—X^a	$\angle$ONO	$\angle CC_{NO_2}C$	$\angle CC_XC$	φ_{NO_2}	
$C_6H_5NO_2$	1.478(13)	1.218(4)	—	123.5(1.4)	125.1(1.4)	—	23(4)	92
o-$C_6H_4(NO_2)_2$	1.475(12)	1.224(2)	—	124.9(0.5)	120.4(1.2)	—	31(2)	93
m — $C_6H_4(NO_2)_2$	1.461(6)	1.225(2)	—	125.3(0.7)	121.8(1.0)	—	23(3)	94
p-$C_6H_4(NO_2)_2$	1.463(10)	1.221(2)	—	125.8(0.5)	122.9(0.9)	—	18(4)	93
sym-$C_6H_3(NO_2)_3$	1.475(5)	1.224(2)	—	125.7(0.5)	123.2(0.9)	—	21(2)	93
o-$ClC_6H_4NO_2$	1.462(12)	1.226(2)	1.721(3)	123.6(1.0)	121.3(1.2)	121.6(0.7)	34(3)	95
m-$ClC_6H_4NO_2$	1.442(10)	1.243(3)	1.746(6)	122.6(1.0)	123.0(1.5)	121.5(1.2)	13(6)	94
p-$ClC_6H_4NO_2$	1.469(3)	1.237(3)	1.707(4)	123.0(0.9)	122.6(1.1)	120.5(0.6)	19(5)	96
o-$BrC_6H_4NO_2$	1.494(14)	1.218(3)	1.894(6)	128.6(1.5)	120.4(2.4)	119.6(1.2)	43(3)	97
m-$BrC_6H_4NO_2$	1.448(14)	1.238(3)	1.865(8)	121.8(1.4)	121.3(1.0)	121.4(1.0)	25(5)	97
p-$BrC_6H_4NO_2$	1.465(12)	1.240(4)	1.896(7)	125.0(0.7)	121.6(0.6)	122.6(0.6)	19(3)	98
2,6-$(NO_2)_2C_6H_3Cl$	1.447(4)	1.229(2)	1.712(5)	123.0(0.3)	121.6(0.4)	118.9(0.6)	54(1)	99
2,6-$(NO_2)_2C_6H_3Br$	1.468(9)	1.229(3)	1.899(9)	125.6(0.9)	122.7(0.9)	117.1(1.2)	58(2)	97
p-$NH_2C_6H_4NO_2$	1.470(11)	1.225(2)	1.367(17)	124.7(0.7)	122.8(1.4)	119.8(1.7)	17(7)	100

^{a} X = Cl, Br, NH_2.

$C_6H_5NH_2$ and C_6H_5Cl,[101] respectively. No shortening of the C—Br bond is observed in o- and p-bromonitrobenzenes compared to C_6H_5Br[102]; the C—Br bond length is, however, decreased by 0.03 Å in m-BrC$_6$H$_4$NO$_2$.

The nitro group and also the Cl and Br atoms in m- and p-halonitrobenzenes cause benzene ring distortions (see Chapter 7). No noticeable ring angle deformations are observed in o-nitrobenzenes, probably because of steric factors. The $(C\text{---}C)_{benz}$ bond lengths are shorter than in benzene (1.397 Å) in a number of molecules.

Nonplanar conformations of m- and p-nitrobenzenes observed in electron diffraction experiments arise from torsional vibrations of the nitro groups. When these vibrations are taken into account, there result planar equilibrium models that contain a barrier to rotation V_2 of about 17–21 kJ mol^{-1}. The barrier in m- and o-XC$_6$H$_4$NO$_2$ (X = Cl, Br) was estimated by using a new technique that will be discussed later.

Orthoderivatives of nitrobenzene contain considerable nonbonded steric interactions. The angle of rotation of the NO$_2$ group is about 30–40° (see Table 2-18). A similar conformation is characteristic for o-nitroaromatic compounds in crystals except for certain classes of compounds. Thus o-nitroaromatic ethers and thioethers form planar structures with shortened O$_{NO_2}$...O and O$_{NO_2}$...S distances. In the gas phase, conformers of this type occur in o-NO$_2$C$_6$H$_4$SCl[103] and in o-NO$_2$C$_6$H$_4$SeBr.[104]

In nitroamines that contain the $\diagup\!\!\!\diagdown$N—NO$_2$ group, the bond configuration around the amine nitrogen atom and the N—N bond length depend substantially on the substituents on N$_{amine}$ (Table 2-19). A planar configuration was found for $(CH_3)_2NNO_2$, $CH_3(CH_2Cl)NNO_2$, and $CH_3(CH_2{=}CH)NNO_2$. Replacing one of the CH$_3$ groups in dimethylnitroamine by NO$_2$ or Cl results in a pyramidal amine nitrogen configuration and elongation of the N—N bond.

Table 2-19. NITROAMINES CH$_3$NXNO$_2$

			X			
Parameter	H	CH$_3$	CH$_2$Cl[a]	—CH=CH$_2$[b]	Cl[c]	NO$_2$
N=O, Å	1.228(3)	1.223(2)	1.212(3)	1.213(3)	1.209(2)	1.231(3)
N—N, Å	1.381(6)	1.382(3)	$\{$1.426(10)$_{av}$	1.385(8)	1.469(5)	1.480(5)
C—N, Å	1.452(5)	1.460(3)	$\}$	1.459(15)	1.478(5)	1.494(6)
∠ONO, deg	125.3(1.0)	130.4(1.5)	127.5(1.0)	128.0(2.1)	128.5(0.9)	132(1)
∠CNN, deg	109.0(1.3)	116.2(0.3)	116.6(0.6)	113.5(1.2)	112.9(1.5)	107.5(1.1)
∠CNX, deg	119(9)	127.6(0.6)	127.0(0.9)	124.1(1.0)	115.0(1.2)	107.5(1.1)
∠CNNO, deg	28	0	35	0	27	28

[a] Average value r(C—N, N—N); r(C—Cl) = 1.809(4) Å; ∠NCCl = 107.7(0.8)°.
[b] r(C=C) = 1.343(15) Å; r(C—N) = 1.432(5) Å; ∠NCC = 119.3(.1.5)°.
[c] r(N—Cl) = 1.720(4) Å; ∠ClNN = 108.4(1.3)°; ∠ClNNO = 24.4°.
Source: References 88 and 105.

Table 2-20. NITROGEN OXIDES

Compound	Bond lengths (Å)			Bond angles (deg)		Ref.
	N=O	N—N	N—O	$\angle$O=N=O	$\angle$N—O—N	
NO_2	1.202(3)	—	—	134.4(1.3)	—	106
N_2O_4	1.190(2)	1.782(8)	—	135.4(0.6)	—	106
N_2O_5	1.188(2)	—	1.498(4)	133.2(0.6)	111.8(1.6)	107

Additivity of fragments is observed in planar $CH_3(CH_2=CH)NNO_2$, ie, the geometrical parameters are not affected by conjugation.

Nitrogen Oxides, Nitrites, and Nitrates. The dimeric molecule N_2O_4 has a planar conformation of D_{2h} symmetry with a very long N—N bond whose length exceeds that of the corresponding N—N bonds in hydrazine and its derivatives by 0.3 Å. Comparison of the structural data on NO_2 and N_2O_4 (Table 2-20) does not reveal any substantial differences in the geometry of the NO_2 group between the dimeric and monomeric species.

Two NO_2 groups linked through the central oxygen atom in N_2O_5 (**23**) are involved in large-amplitude torsional motion, which has been analyzed in detail in reference 107. The potential energy minima have been shown to correspond to the equilibrium torsional angles $\tau_1 = \tau_2 = 30°$ ($\tau = 0$ for the planar conformation).

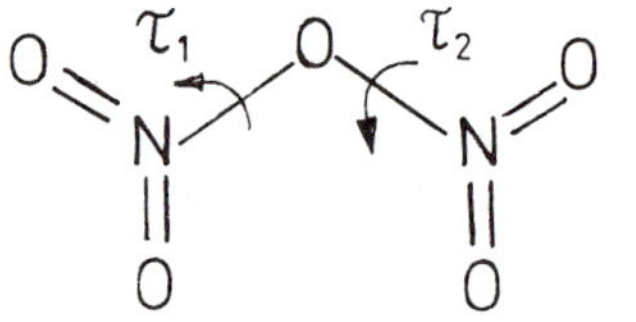

23

The torsional motion of each of the two groups is mutually correlated; thus, increase of one of the τ angles causes a decrease of the other. The nitro group parameters in N_2O_5 (Table 2-20) differ only slightly from those in NO_2 and in N_2O_4; the NON angle has a similar value to CON in CH_3ONO_2 [112.7(0.3)°, MW[108]] and COC in dimethyl ether.

The available electron diffraction data on alkali metal nitrites MNO_2 (M = Rb, Cs)[109] indicate that they possess cyclic structures (**24**) with equal M—O distances and NO_2 group parameters other than those in N_2O_4 (NO_2):

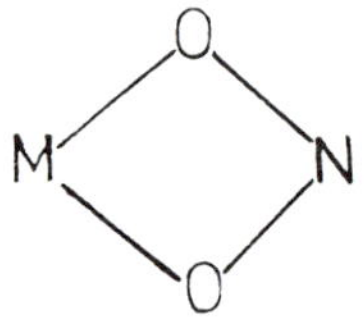

24

	N—O (Å)	$\angle$ONO (deg)
$RbNO_2$	1.252(5)	116(3)
$CsNO_2$	1.256(5)	118(3)

Table 2-21. NITRATES

Compound	Bond lengths (Å)		$\angle O_1NO_1'$ (deg)	Ref.
	$N—O_1$	$N{=}O_2$		
$RbNO_3$	$1.252(3)^a$		115(6)	110a
$CsNO_3$	$1.252(5)^a$		117(5)	110b
$TlNO_3$	$1.254(8)^a$		114(5)	110c
$Cu(NO_3)_2$	1.298(3)	1.205(5)	113.5(0.6)	111
$Zr(NO_3)_4$	1.284(7)	1.184(10)	—	112
$Sn(NO_3)_4$	1.281(5)	1.179(8)	—	112

^{a} Average value for $N—O_1$ and $N{=}O_2$.

Univalent element nitrates MNO_3 (M = Cs, Rb, Tl) (Table 2-21) form a bidentate structure (**25**) in the gas phase. The NO_3 group is planar, and the M atom lies in a plane bisecting the ONO angle. The planar structure of the $M(O_2NO)$ fragment was also observed in the following nitrates: $Cu(NO_3)_2$, $Zr(NO_3)_4$, and $Sn(NO_3)_4$ (Table 2-21). The copper atom in $Cu(NO_2)_2$, occupies the center of a planar square formed by oxygen atoms of two bidentate NO_3 groups. In the Zr and Sn compounds, the ligands are arranged tetrahedrally; the tetrahedron about Zr is distorted.

25

Studies of $Cu(NO_3)_2$, $Zr(NO_3)_4$, and $Sn(NO_3)_4$ showed the nonequivalence of the N—O bonds, the $N—O_2$ bond being shorter by about 0.1 Å than $N—O_1$. Average N—O distances were determined only for univalent metal nitrates. The ONO angle in the rings has the same value as in nitrites ($< 120°$).

Saturated Cyclic Nitrogen Compounds. Three-, four-, five-, and six-membered nitrogen-containing compounds have been studied (Table 2-22). Unlike the C—C distance, the C—N bond length is practically independent of ring size. A distinguishing feature of three-membered nitrogen rings is that the C—N bond is not shorter than in acyclic amines. The puckering angle in four-membered azetidine [29.7(1.4)°] is larger than that in cyclobutane (26°). Since the ring in trimethyleneoxide $(CH_2)_3O$ is nearly planar, one can see that, contrary to expectations, azetidine cannot be placed in between its two counterparts. A similar situation arises with pyrrolidine, which forms an envelope structure, while cyclopentane and tetrahydrofuran are characterized by pseudorotation. The axial N—H bond position agrees with the results of *ab initio* calculation,

Table 2-22. SATURATED CYCLIC NITROGEN COMPOUNDS

| Compound | Bond lengths (Å) | | ∠CNC (deg) | Cycle conformation | N—X bond configuration | Ref. |
	C—N	N—N C—C				
 ▷NX[a]　X = H	1.475(3)	1.481(3)	60.3(0.2)	—	—	LB
X = Cl	1.489(3)	1.484(3)	59.8(0.2)	—	—	LB
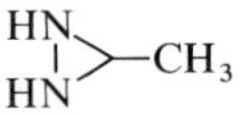 (HN)(HN)C—CH₃	1.489(9)	1.444(13)	—	—	—	113
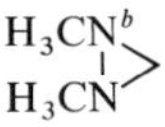 (H₃CN)[b](H₃CN)	1.489[c]	1.444[c]	112.0(0.5)	—	trans	113
 ◇NH	1.473(3)	1.563(3)	91.2(0.4)	Nonplanar	eq	114
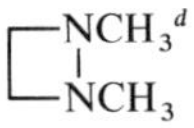 —NCH₃[d] / —NCH₃	1.481(8)	1.427(7) 1.537(11)	90.8(CNN)	Nonplanar	eq eq	115
 NX　X = H	1.469(10)	1.543(8)	105.2(3.5)	Envelope	ax	116
X = Cl[e]	1.476(5)	1.535(3)	108.0(0.8)	Envelope	eq	117
X = CH₃[f]	1.455(3)	1.542(4)	107.4(1.7)	Envelope	eq	117
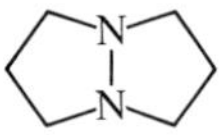 (N–N bicyclic)	1.486(8)	1.434(16) 1.531(10)	110.4(3.0)	Skewed Envelope	—	118
 NH (six-membered)	1.472(11)	1.531(6)	109.8(2.1)	Chair	eq	119
 HN NH	1.467(4)	1.540(8)	109.0(0.8)	Chair	—	LB
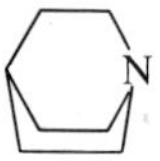 (bicyclic N)	1.469(3)	1.552(2)	107.3(0.5)	Boat	—	120

[a] MW data, r(N—Cl) = 1.738(3) Å.
[b] r(N—CH₃) = 1.445(3) Å.
[c] Assumed.
[d] r(N—CH₃) = 1.471(7) Å.
[e] r(N—Cl) = 1.736(3) Å.
[f] r(N—CH₃) = 1.446(3) Å.

while preferred equatorial orientations of the N—Cl and N—CH₃ bonds are at variance with theoretical predictions based on consideration of the corresponding chloro derivative.

The five-membered rings in 1,5-diazabicyclo[3.3.0]octane, possess skewed envelope conformation: the planes normal to the N—N bond and to the CCC plane make an angle of 18°, and the rotation angle about the N—N bond is about 38°.

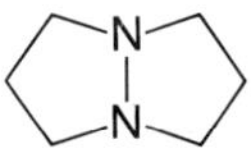

Table 2-23. DONOR-ACCEPTOR COMPLEXES OF
THE TYPE $BX_3 \cdot N(CH_3)_3$

| | Bond lengths (Å) | | $\angle CNC$ |
BX$_3$	N—B	N—C	(deg)
BH$_3$	1.656(2)	1.485(1)	109.2(0.2)
BF$_3$	1.674(2)	1.485(2)	109.2(0.4)
BCl$_3$	1.652(9)	1.497(3)	108.1(0.3)
BBr$_3$	1.663(13)	1.500(5)	107.8(0.5)
BI$_3$	1.663(13)	1.497(5)	106.0(0.8)

Six-membered monocyclic compounds usually have chair conformations.
The structure of 1-azabicyclo[2.2.2]octane is strained; the strain manifests itself in a 12° departure from coplanarity in the N—C—C—C bond system.

Four-Coordinated Nitrogen Atom

Four-coordinated nitrogen compounds include *inter alia*, donor–acceptor complexes of the type $AX_3 \cdot NY_3$, where A is a Group III element such as B, Al, or Ga (see Chapters 1, 8, and 9 and reference 121).

As shown in Table 2-23, the C—N bond length in $N(CH_3)_3$ increases upon the formation of the complexes, while the YNY angles are not much affected.

The molecular structure of the lithium complex **26** has also been studied. On the assumption that the bridge NSi_2 groups are perpendicular to the four-membered ring LiNLiN plane, the structure could be refined to the parameters $r(Li—N) = 1.99(3)$ Å, $r(Si—N) = 1.712(7)$ Å, $\angle LiNLi = 80(3)°$, and $\angle SiNSi = 129.8(1.7)°$. It follows that the bond lengths and the valence angles at nitrogen are essentially nonequivalent.

H$_3$Si   Li   SiH$_3$

N   N

H$_3$Si   Li   SiH$_3$

26

Trifluoroamine oxide F_3NO[122] is an interesting example of a four-coordinated nitrogen compound. The distances for N—F [1.432(3) Å] and N=O [1.158(4) Å] correspond to the formation of five covalent bonds around nitrogen. The molecular symmetry is C_{3v}. Four ligands occupy the vertices of an almost regular tetrahedron with nitrogen in the center. The O...F and F...F nonbonded distances are 2.21(2) Å.

Phosphorus Derivatives

Coordination of phosphorus may be more diverse than that of nitrogen. One- to five-coordinated phosphorus derivatives have been investigated by gas-phase electron diffraction. The most stable and therefore most thoroughly studied compounds are those containing three- and four-coordinated phosphorus. In the 1960s chemists succeeded in synthesizing one- and two-coordinated phosphorus derivatives, and this considerably extended the scope of phosphorus stereochemistry.

One-Coordinated Phosphorus Atom

Because of their low stability, one-coordinated phosphorus derivatives have been studied mostly by microwave spectroscopy[123–125] (Table 2-24). At this time only the structure of 2,2-dimethylpropylidynephosphine, $(CH_3)_3C-C{\equiv}P$, has been determined by electron diffraction[126] (Table 2-24). Since the $P{\equiv}C$ and $C-C$ bond lengths are very similar, a complete solution of the structural problem could be obtained only in a combined analysis of electron diffraction and microwave spectroscopic data.[126] Corrections for vibrational effects have been found from approximate force field calculations. The most important result was the determination of the $P{\equiv}C$ triple bond length. One can see that this parameter varies only insignificantly over the series of molecules included in Table 2-24. It can also be noticed that the $P{\equiv}C$ bond tends to shorten the $\equiv C-C\diagup_{\diagdown}$ bond compared to $HC{\equiv}C-C(CH_3)_3$ [1.495(5) Å] and $N{\equiv}C-C(CH_3)_3$ [1.495(15) Å]. However, for the latter molecule the large uncertainty of the parameter renders any comparison doubtful.

The $C-C{\equiv}P$ bond system has been demonstrated convincingly to be linear.

Stereochemically, the $P{\equiv}C$ distance does, in fact, correspond to a triple bond length, for it is equal to the sum of the carbon and phosphorus covalent radii

Table 2-24. METHYNEPHOSPHINES

	Bond lengths (Å)			
$P{\equiv}C-X$	$P{\equiv}C(1)$	$\equiv C(1)-X$	Method	Ref.
$P{\equiv}C-H^a$	1.540452(18)	1.06596(11)	MW	123
$P{\equiv}C-F^b$	1.541(5)	1.285(5)	MW	124
$P{\equiv}C-CH_3{}^c$	1.544(4)	1.465(3)	MW	125
$P{\equiv}C-C(CH_3)_3{}^d$	1.536(2)	1.473(4)	ED + MW	126

^{a} r_e structure.
^{b} r_0 structure.
^{c} r_0 value for $P{\equiv}C$ and r_s values for other parameters.
^{d} r_{av} structure; $C(2)-C(3) = 1.543(2)$ Å; $\angle C(3)$-$C(2)$-$C(3') = 109.0(2)°$.

calculated from $r(C\equiv C) = 1.20$ Å in $HC\equiv CH$ and $r(P\equiv P) = 1.890$ Å in P_2, ie, $r(P\equiv C) = \frac{1}{2}[r(C\equiv C) + r(P\equiv P)] = 0.60 + 0.95 = 1.545$ Å. The triple character of the $P\equiv C$ bond was also confirmed via the results of *ab initio* calculations generated with a double-zeta plus polarization basis.[127]

Two-Coordinated Phosphorus Atom

The distribution of bonds around trivalent two-coordinated phosphorus can be as follows:

1. A double bond and a single bond:

2. Two equivalent bonds of the order of 1.5:

3. Intermediate types

Until the 1960s, an opinion prevailed that a two-coordinated phosphorus compound should be unstable because of the instability of $p_\pi\!-\!p_\pi$ bonds formed by phosphorus, as in, eg, $P{=}C$, $P{=}N$, $P{=}P$. New two-coordinated phosphorus compounds were, however, prepared in the 1960s. Structural studies of these products enabled the determination of the stereochemical parameters of two-coordinated phosphorus including bond lengths and valence angles.

1,2,3-Phosphadiazoles. Electron diffraction studies of newly synthesized 2-phenyl-4-methyl phosphadiazole-1,2,3[128] and 2-acetyl-4-methyl-phosphadiazole-1,2,3,[129,130] and also an X-ray study of 2-cyanoethyl-4-phenylphosphadiazole-1,2,3[131] provided the earliest evidence that, together with other physicochemical investigations, proved the existence of a new class of heterocyclic compounds containing two-coordinated phosphorus. The principal geometrical parameters of these molecules are listed in Table 2-25, which also includes data generated via *ab initio* calculations.[133] The heterocycle has a planar structure.

Unfortunately the earliest electron diffraction results[128,129] cannot be considered to be very reliable, for the molecules include a large number of parameters and contain very similar distances such as $r(P{=}C)$ and $r(P{-}N)$. Nevertheless, the existence of a new class of compounds was established. Further studies including the results of quantum mechanical calculations confirmed this conclusion.

A reanalysis of the electron diffraction data of 2-acetyl-4-methylphosphodiazole-1,2,3 led to two models with two parameter sets giving equally good agreement with experiment.[130] Both parameter sets are quoted in Table 2-25. One of the sets is favored by comparison with quantum chemical calculations on the formyl derivative; however, the $P{-}N$ bond distance of this set seems to be too large. The other parameter set contains a more realistic $P{-}N$ bond distance. For comparison, the $P{-}N$ bond is about 1.70 Å long in phosphine amines.

Table 2-25. R—N ⟨N=C(R′), P=C(R″)⟩ 1,2,3-Phosphadiazoles[a]

| R | R′ | R″ | Bond lengths (Å) | | $\angle$C=P—N (deg) | Method | Ref. |
			P=C	P—N			
Ph	CH_3	H	1.759(5)	1.684(7)	87.5(1.0)	ED	128
$CH_3C(O)$	CH_3	H	1.75(1)	1.68(1)	89.0(1.5)	ED	129
			1.712(10)	1.744(8)	89.2(0.5)	ED	130
			1.751(7)	1.711(6)	89.2(0.4)	ED	130
$N\equiv CC_2H_4$	Ph	H	1.698(5)	1.775(5)	88.5(0.3)	XR	131
CH_3	CH_3	NHPh	1.746(3)	1.661(3)	88.2(0.2)	XR	132
H	H	H	1.683	1.683	89.2	Ab initio	133
HC(O)	H	H	1.673	1.695	88.9	Ab initio	133

[a] Single bond, P—C = 1.85 Å, P—N = 1.73 Å; double bond, P=C = 1.65 Å, P=N = 1.55 Å, calculated by the Schomaker–Stevenson equation.

It is of interest to compare the experimental data with phosphorus–carbon and phosphorus–nitrogen single- and double-bond lengths by using the Schomaker–Stevenson equation along with tables of covalent radii and electronegativities (Table 2-25). An elongation of the ring P=C bond from the calculated value of 1.65 Å can be explained by invoking electron delocalization. The ability of the three-coordinated nitrogen to participate in conjugation is substantiated by its planar bond configuration.

Phosphabenzene. In 1971 an analog of pyridine, ie, phosphorine, or phosphabenzene, was synthesized for the first time.[134] Its structure was determined via a combination of gas-phase electron diffraction and microwave spectroscopy[135] (Table 2-26). The same authors also determined the structure of arsabenzene (Table 2-26). A comparison of these structures—pyridine, phosphabenzene, and arsabenzene—shows that aromaticity probably remains intact along the series. The rings are planar, and the carbon–carbon bond lengths deviate insignificantly from that of benzene [r(C⋯C) = 1.397 Å]. Carbon–element separations are indicative of the multiple (intermediate between single and double) nature of the C⋯N, C⋯P, and C⋯As bonds. The C⋯P bond can be assigned a multiplicity of 1.5. The CXC valence angle decreases consistently from X=N to X=As.

Three-Coordinated Phosphorus Atom

Unlike amine derivatives, phosphines with single bonds to phosphorus adopt a rather peaked pyramidal configuration.

Table 2-26. PHOSPHABENZENE AND ITS ANALOGS

	N	P	As
X—C, Å	1.3376(4)	1.733(3)	1.850(2)
C(2)—C(3), Å	1.3938(5)	1.413(10)	1.391(9)
C(3)—C(4), Å	1.3916(4)	1.384(12)	1.400(10)
$(C \doteq C)_{av}$, Å	1.3927(4)	1.398(3)	1.396(2)
$\angle$C—X—C, deg	116.94(3)	101.1(3)	97.0(3))
$\angle$X—C(2)—C(3), deg	123.8	124.4(7)	125.3(7)
$\angle$C(2)—C(3)—C(4), deg	118.5	123.7(8)	123.9(13)
$\angle$C(3)—C(4)—C(5), deg	118.4	122.8(8)	124.5(11)
Single bond: X—C, Å[a]	1.48	1.85	1.97
Double bond: X=C, Å[a]	1.28	1.65	1.80
Method	MW	ED + MW	ED + MW
Ref.	LB	135	LB

[a] Calculated by the Schomaker-Stevenson equation.

Phosphine and Phosphorus Halides. The valence angle at phosphorus increases consistently from phosphine to halogenated phosphines (Table 2-27). In PH_3, this angle is minimal not only due to electronegativity considerations but, probably, also for steric reasons. In phosphorus halides, steric and electronegativity factors again act in the same direction.

In an electron diffraction study of PCl_3,[136] the temperature dependence of the geometrical parameters was determined. Thus, $r(P—Cl) = 2.039(1)$ Å and $\angle ClPCl = 100.3(0.1)°$ at room temperature $(T = 27°C)$, while $r(P—Cl) = 2.045(1)$ Å and $\angle ClPCl = 100.4(0.2)°$ at $T = 220°C$. It follows that temperature effects on the geometry of PCl_3 are relatively weak. A more recent microwave investigation of PCl_3 that included the study of a large number of isotopically substituted derivatives has afforded practically the same results.

Table 2-27. PHOSPHINES AND PHOSPHORUS HALIDES, PX_3

	X				
Parameters	H	F	Cl	Br	I[a]
P—X, Å	1.4115(6)	1.563(2)	2.043(1)	2.220(3)	2.43(4)
$\angle$XPX, deg	93.32(2)	96.9(0.7)	100.1(0.3)	101.0(0.4)	102(2)
Method	MW	ED + MW	MW	ED + MW	ED

[a] Obsolete data. A repeated investigation is required.

Source: LB.

Table 2-28. ALKYLPHOSPHINES CONTAINING BONDS OF THE

TYPE $\diagdown$P—C$\diagup$— (ED DATA)

Compound	P—C (Å)	∠CPC (deg)	Ref.
PH_2CH_3	1.858(3)	—	LB
$PH(CH_3)_2$	1.853(3)	99.2(0.6)	LB
$PH_2C[Si(CH_3)_3]_3$	1.808(9)	—	137
$P(CH_3)_3$	1.844(3)	98.8(0.3)	LB, 138
$P[C(CH_3)_3]_3$	1.919(5)	109.9(0.7)	139
$P(CF_3)_3$	1.904(7)	97.2(0.7)	140
$P(CH_3)_2CH_2P(CH_3)_2$	1.849(2)	99.9(1.0)	141

Alkylphosphines and Halogenated Alkylphosphines. Phosphorus valence angles vary within a rather narrow range in alkylphosphines (Table 2-28) and in halogenated alkylphosphines (Table 2-29). A substantial CPC angle broadening (to $\sim 110°$) is observed only in highly strained molecules such as $P[C(CH_3)_3]_3$. The P—C bond length then also increases by about 0.05 Å compared to H_2P—CH_3. Interestingly enough, the CPC angle in $P(CF_3)_3$ is not broadened, although the bond length in this compound is close to that in $P[C(CH_3)_3]_3$. The two effects, therefore, have different origins.

Substituting F or Cl for *t*-Bu groups in $P[C(CH_3)_3]_3$ or CH_3 groups in $P(CH_3)_3$ results in a P—C bond shortening, whereas no corresponding substantial change in halogen-phosphorus-carbon angles is observed. Table 2-29 shows

Table 2-29. HALOGENATED PHOSPHINES CONTAINING BONDS OF THE TYPE

$\diagdown$P—C$\diagup$—

Compound	Bond lengths (Å)		Bond angles (deg)[a]		Ref.
	P—C	P—X[a]	∠XPC	∠XPX, ∠CPC	
$FP[C(CH_3)_3]_2$	1.859(6)	1.619(7)	96.0(2.0)	113.8(1.9)[b]	139
$F_2PC(CH_3)_3$	1.822(12)	1.589(4)	99.0(0.6)	99.1(1.7)	139
$ClP(C_2H_5)_2$	1.818(6)	2.051(3)	98.9(0.6)	100.3(3.3)[b]	142
Cl_2PCH_3	1.831(10)	2.061(3)	98.8(0.6)	100.7(0.5)	143
$Cl_2PC(CH_3)_3$	1.873(17)	2.064(4)	100.9(0.8)	101.2(1.1)	144
$ClP[C(CH_3)_3]_2$	1.894(5)	2.079(4)	101.5(0.7)	109.7(1.9)[b]	145
$Cl_2PCH(CH_3)_2$	1.850(12)	2.058(2)	101.6(0.7)	100.6(0.4)	145
$Cl_2PCH_2PCl_2$	1.849(8)	2.056(2)	96.5(0.7)	101.1(0.6)	146

[a] X = F, Cl.

[b] ∠CPC

that the presence of methyl groups produces significant elongation of the P—F and P—Cl bonds compared to PF_3 and PCl_3, respectively.

A staggered conformation is characteristic of the bonds. In CH_2ClPCl_2,[147] $CH_3CH_2PCl_2$,[142] $(CH_3CH_2)_2PCl$,[142] $(CH_3)_2PCH_2P(CH_3)_2$,[141] and $Cl_2PCH_2PCl_2$,[146] conformational equilibria are observed. The last two compounds are the most interesting ones. Conformer III predominates in $Me_2PCH_2PMe_2$ (**27**). In $Cl_2PCH_2PCl_2$, however, conformers III and IV occur in a 56:44 ratio. It should be noted that in conformer IV, the Cl atoms are drawn toward each other. For similar derivatives of *n*-propane, conformers of type IV were not observed.[148].

I   II   III   IV   V

27

Variation in the carbon valence state in phosphines and halogenated phosphines containing bonds of the types $>P—C\langle$, $>P—C\langle$, and $/P—C\equiv$ (Table 2-30) result in a regular shortening of the phosphorus–carbon bond. The shortening in phenyl derivatives is, however, small, when compared with, eg, the

Table 2-30. PHOSPHINES AND HALOGENATED PHOSPHINES CONTAINING $>P—C\langle$, $>P—C\langle$, AND $>P—C\equiv$ BONDS

Compound	Bond lengths (Å)		Bond angles (deg)		
	P—C	P—X[a]	∠CPX[a]	∠XPX[a]	Ref.
$H_2PC_6H_5$	1.839(5)	—	—	—	143
$HP(C_6H_5)_2$	1.832(8)	—	100.8(1.7)	—	149
$(CH_3)_2PC_6H_5$	1.845(6)$_{av}$	—	103.4(1.0)	96.9(2.5)	150
$F_2PC_6H_5$	1.809(7)	1.580(3)	98.8(1.1)	102.3(1.2)	151
$Cl_2PC_6H_5$	1.83(3)	2.072(5)	100.6(2.0)	100.4(1.5)	152
$F_2PCH=CH_2$[b]	1.805	1.577	98.0	97.9	153
$Cl_2PCH=CH_2$	1.771(7)	2.056(2)	100.5(0.9)	100.2(0.7)	154
$Cl_2P—\langle S \rangle$	1.776(23)	2.059(3)	99.2(0.6)	97.6(0.6)	155
$(CH_3)_2P—C(O)CH_3$	1.863(2)$_{av}$	—	105.9(0.9)	99.3(2.0)	156
$(CH_3)_2P—C\equiv N$[b]	1.783	—	101	—	LB
$F_2P—C\equiv N$	1.792(9)	1.568(3)	98.3(0.3)	97.9(0.3)	157
$P(C\equiv N)_3$	1.806(4)	—	98.9(0.0)	—	158

[a] X = F, Cl.
[b] MW data.

P—C bond in $P(CH_3)_3$. Structures containing the $\diagdown$P—C$\diagup\diagup$ bond have not been determined with sufficient reliability. The available data are, nevertheless, indicative of a further decrease in the $\diagdown$P—C$\diagup\diagup$ bond length.

Dimethylacetylphosphine, $(CH_3)_2P$—$C(O)CH_3$,[156] seems to be an exception in that the $\diagdown$P—C$\diagup\diagup$ bond in this compound is longer than other $\diagdown$P—C$\diagup\diagup$ or even $\diagdown$P—C$\diagdown$ bonds. This is indicative of the absence of any significant interaction between the lone electron pair on phosphorus and the π bond of one carbonyl group (as occurs in amides of organic acids). In this respect the

$$\diagdown P-C\diagup^{O}_{\diagdown CH_3}$$

bond more closely resembles the C—Cl bond in

$$Cl-C\diagup^{O}_{\diagdown CH_3}$$

which is also elongated compared to

$$Cl-C\diagup^{CH_2}_{\diagdown H} .$$

The conformation of H_2PPh has not been determined. Two different angles of rotation of the phenyl groups have been observed in Ph_2PH (**28**): $\varphi_1 \cong 14(10)°$ and $\varphi_2 \cong 66(4)°$. Conformation I facilitates conjugation of the phenyl ring and the phosphorus lone electron pair. The overall molecular conformation, probably, also depends on interactions between the phenyl rings.

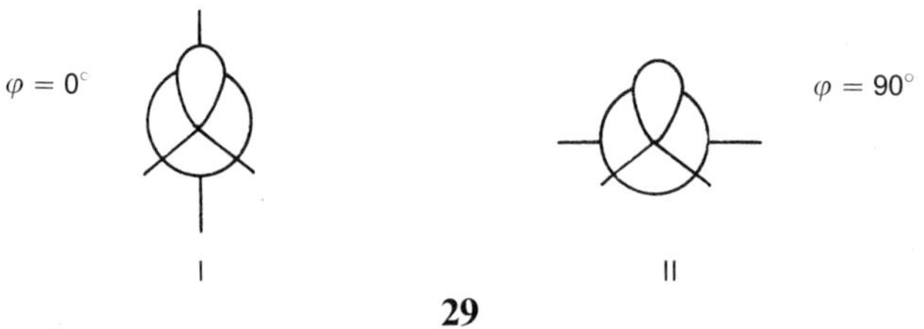

An electron diffraction study of $PhPMe_2$ (**29**)[150] has been carried out to obtain information about the conformation equilibrium. Vapors have been found to contain conformers I and II in a 70:30 ratio.

The $F_2PC_6H_5$ and $Cl_2PC_6H_5$ molecules occur in the bisector conformer I. The deviations from the symmetrical structure are about 31° for $F_2PC_6H_5$[151] and about 10° for $Cl_2PC_6H_5$[152].

A cis/trans conformational equilibrium occurs in $Cl_2PCH{=}CH_2$ (**30**).[154]

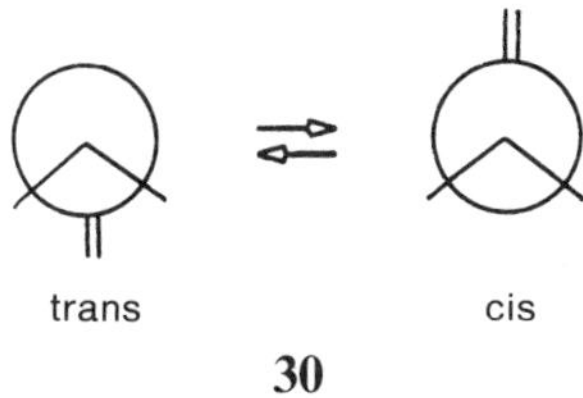

30

The microwave rotational spectra of $F_2PCH{=}CH_2$ are indicative of a trans conformation.[153] The Cl_2P— molecule has a somewhat unexpected configuration: the P—Cl and C—S bonds practically eclipse one another.[155] This result can be explained in terms of electrostatic attraction between Cl and S moieties whose effective charges have opposite signs.

Two models agree with the experimental data for $(CH_3)_2PC(O)CH_3$ (**31**). Model II provides somewhat better agreement. Large vibrational amplitude values for rotation-dependent C...O distances are indicative of only a small curvature of the potential surface that describes rotations about the P—C bond.

31

Ab initio calculations for $H_2PC(O)H$[159] as a model molecule and for $(CH_3)_2PC(O)CH_3$[159] reveal that the energy minimum occurs at 90°, thereby favoring interactions between the $C{=}O$ group and the lone electron pair on phosphorus.

In phosphines that contain the $\backslash P{-}C{\equiv}$ bond, this bond is shortened compared to the $\backslash P{-}C{-}$ bond in $P(CH_3)_3$. The CPC valence angles, however, are unaffected.

Phosphites. P—O bonds show a tendency to undergo shortening upon substitution of halogens for OR (Table 2-31). Bond lengths are similar for P—O and P—F. The OCH_3 group increases the P—Cl distance compared to PCl_3 to a certain extent. The valence angles at phosphorus vary only insignificantly. The phosphorus atom noticeably affects the bond angles at oxygen (see Chapter 3).

One of the most important points to consider when discussing the sterochemistry of phosphites is the molecular conformation about the P—O bond. The F_2POCH_3 and Cl_2POCH_3 molecules have structures in which the bisector of the XPX angle and the O—C bond are oriented mutually cis. Electron

Table 2-31. PHOSPHITES

	Bond lengths (Å)		Bond angles (deg)		
Compound	P—O	P—X[a]	$\angle$OPX[a]	$\angle$XPX[a]	Ref.
F_2POCH_3[b]	1.560(15)	1.561(6)	102.2(1.0)	94.8(0.6)	160
Cl_2POCH_3	1.585(12)	2.067(3)	101.4(1.1)	98.1(1.9)	161
F_2POPF_2	1.631(10)	1.568(4)	97.6(1.2)	99.2(2.4)	162
$P(OCH_3)_3$	1.620(3)	—	100.5(0.9)[c]	—	163
$P(OC_2H_5)_3$	1.600(6)	—	96.5(0.5)[c]	—	164
$P(OC_2H_3)_3$	1.600(6)	—	100.0(1.0)[c]	—	164

[a] X = F, Cl.
[b] MW data.
[c] $\angle$OPO

diffraction data on F_2POPF_2 can be fitted by a mixture of conformers of point groups $C_1(40\%)$, $C_2(20\%)$, $C_s(20\%)$, and $C_{2v}(20\%)$ (see type V in (27)).[162]

Unfortunately, because of complex molecular structures and the presence of several rotational forms with O—C as well as P—O axes, conformational analysis has not been performed for unsubstituted phosphites, $P(OR)_3$.

Thiophosphites. As with phosphites, the central P—S bond in the bisderivative $(F_2P)_2S$ is longer than in the mono derivatives. Substituting F and Cl for SCH_3 in $P(SCH_3)_3$ causes a decrease in the P—S bond length (Table 2-32). A certain tendency toward elongation of the P—F and P—Br bonds compared to PF_3 and PBr_3 can be observed. The P—Cl bond length practically coincides with that in PCl_3. The valence angles at phosphorus remain the same as in the simplest phosphines, except for the case of the $P(SCH_3)_3$ molecule.

The conformation of the F_2PSMe molecule has not been determined unambiguously; in F_2PSGeH_3, however the S—Ge bond and the bisector of the FPF angle are nearly mutually cis. For Cl_2PSMe, the authors of reference 167 rule out the eclipsed conformation. Other conformers (cis, gauche, and trans) fit the experimental data equally well. For Br_2PSMe,[167] the same authors eliminate the eclipsed and trans arrangements of the S—CH_3 bond relative to the BrPBr

Table 2-32. THIOPHOSPHITES AND SELENOPHOSPHITES

	Bond lengths (Å)		Bond angles (deg)		
Compound	P—S(Se)	P—X[a]	$\angle$SPX[a]	$\angle$XPX[a]	Ref.
F_2P—SCH_3	2.085(3)	1.589(3)	101.2(0.3)	95.6(0.6)	165
F_2P—$SGeH_3$	2.115(8)	1.590(9)	99.9(0.4)	97.0(1.0)	166
Cl_2P—SCH_3	2.082(12)	2.038(6)	100.7(0.6)	99.6(1.2)	167
Br_2P—SCH_3	2.116(16)	2.244(4)	101.3(0.6)	100.6(0.6)	167
$P(SCH_3)_3$	2.115(4)	—	94.0(0.6)	—	168
$(F_2P)_2S$	2.132(4)	1.572(2)	100.2(0.4)	97.4(0.5)	165
$(F_2P)_2Se$	2.273(5)	1.573(3)	98.7(0.4)	100.6(0.5)	165

[a] X = F, Cl, Br.

angle bisector. Two conformers predominate in $P(SMe)_3$[168] (**32**). The $(F_2P)_2X$ (X = S, Se) (**33**) molecules have near C_{2v} symmetry, with the lone electron pairs on P in an eclipsed (cis) orientation with respect to the P—S bond. The conformations of $(F_2P)_2X$ are thus different from $(F_2P)_2O$.

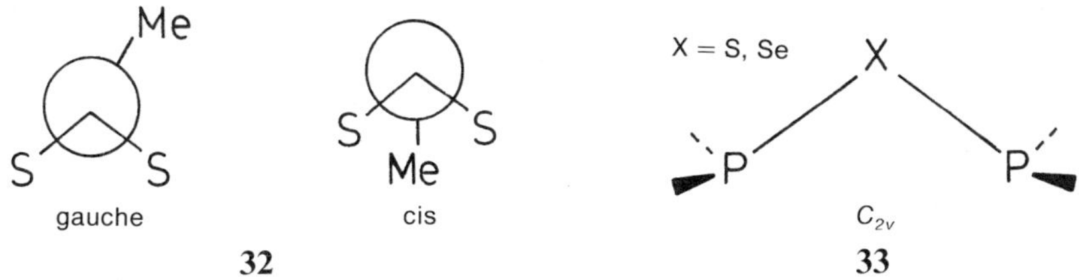

Problems pertaining to the sterochemistry of sulfur are discussed in Chapter 3.

Amino and Imidophosphines. With aminophosphines, the most important points to consider include: (a) patterns of P—N bond length variations, (b) the effects of P on the amine nitrogen atom configuration, and (c) rotational isomerism about the P—N bond.

Tables 2-33 and 2-34 show that P—N bond lengths vary only slightly over the series of compounds studied except for a noticeable increase in the case of $P[N(CH_2)_2]_3$ and a decrease in the case of $F_2PN(CH_3)_2$. A change in the valence state of N in isocyanates causes no significant shortening of the $\diagdown$P—N= bond compared to $\diagdown$P—N$\diagup$.

The bond configuration at N has been discussed in the first section. As mentioned previously, the nitrogen atom adopts a planar (or nearly planar) environment in aminophosphines. So far as conformations about the $\diagdown$P—N$\diagup$ bond are concerned, there is a certain ambiguity. Molecules are known with both bisector (**34**) and orthogonal (**35**) orientations of the plane containing

Table 2-33. AMINOPHOSPHINES AND CHLOROAMINOPHOSPHINES

| Compound | Bond lengths (Å) | | Bond angles (deg) | | |
	P—N	P—Cl	∠NPN	∠NPCl	Ref.
$P[N(CH_3)_2]_3$	1.700(5)	—	96.5(1.0)	—	169
$P[N(CH_2)_2]_3$	1.75(1)	—	97.5(1.5)	—	169
$ClP[N(CH_3)_2]_2$	1.730(5)	2.180(4)	96.2(2.6)	100.5(7)	170
$Cl_2PN(CH_3)_2$	1.69(3)	2.083(5)	—	100(1)	171

Table 2-34. DIFLUOROPHOSPHINES CONTAINING BONDS
$>$P—N$<$ AND $>$P—N= (ED DATA)

Compound	Bond lengths (Å)		Bond angles (deg)		
	P—N	P—F	$\angle$NPF	$\angle$FPF	Ref.
F_2P—NH_2	1.667(7)	1.581(3)	101.0(1.1)	93.5(1.1)	172
F_2P—$N(CH_3)_2$	1.648(8)	1.589(3)	97(4)	99(3)	172
F_2P—$N(SiH_3)_2$	1.680(4)	1.585(3)	99.4(0.7)	96.9(1.0)	173
$(F_2P)_2NH$	1.684(8)	1.584(3)	98.3(0.7)	95.6(1.0)	174
$(F_2P)_2NCH_3$	1.680(6)	1.583(2)	99.6(0.3)	95.1(0.3)	175
$(F_2P)_2NSiH_3$	1.691(4)	1.570(2)	99.3(0.3)	96.1(0.5)	173
$(F_2P)_2NGeH_3$	1.698(8)	1.592(5)	99.6(0.5)	96.5(1.1)	176
$(F_2P)_3N$	1.712(4)	1.574(2)	99.0(0.4)	97.1(0.6)	177
F_2P—N=C=O	1.683(6)	1.563(3)	99.5(0.7)	97.9(0.8)	178
F_2P—N=C=S	1.686(7)	1.566(3)	97.7(0.8)	99.4(0.9)	178
F_2P—N=C=Se	1.649(12)	1.530(4)	98.8(0.8)	97.9(1.4)	179
F_2P—N=C=N—PF_2	1.680(6)	1.562(2)	94.4(0.7)	103.4(0.8)	180

nitrogen bonds with respect to the PX_2 group (X = F, Cl). Most molecules occur as bisector conformers; the exceptions are $Cl_2PN(CH_3)_2$ and F_2PNHCH_3.[181]

Molecules of the type F_2PN=C=X (X = O, S, Se) are characterized by C_s symmetry and a trans orientation of the N=C=X group relative to the bisector of the FPF angle. It is of interest to note that in the F_2PN=C=NPF_2 cumulene system, the PNCNP dihedral angle is about 55° rather than 90°. The PF_2 groups are oriented trans with respect to the N=C bonds.

The presence of an amino group causes an increase in the P—F and P—Cl bond lengths compared to PX_3. The largest increase is observed for the P—Cl bond in $ClP(NMe_2)_2$: this bond length substantially exceeds the sum of the covalent radii (equal to 2.04 Å), which is indicative of a bond order smaller than unity.

Diphosphines. The P—P bond becomes 0.1 Å longer on passing from $F_2PP(GeH_3)_2$ to P_2F_4 (Table 2-35). A similar elongation is observed in hydrazine derivatives. Both phosphorus atoms have the usual pyramidal bond configuration.

Unlike hydrazine derivatives that possess gauche conformations, these compounds exist as trans or near trans conformers (**36**).

An increase in P—F bond lengths in P_2F_4 and $F_2PP(GeH_3)_2$ and P—C bond lengths in $P_2(CF_3)_4$ should be mentioned. The CPP and CPC angles are also increased substantially in the latter molecule. The relative angle values for

36

Table 2-35. DIPHOSPINES

Compound	Bond lengths (Å)		Angles (deg)			Ref.
	P—P	P—X	∠PPX	∠XPX	φ(P—P)	
P_2H_4	2.218(4)	1.451(5)	95.2(0.6)	91.3(1.4)	—	182
$P_2D_4{}^a$	2.219(4)	1.416(2)	94.3(0.2)	92.0(0.3)	74.0(2.2)	183
			99.1(0.1)			
P_2F_4	2.281(6)	1.587(3)	95.4(0.3)	99.1(0.4)	180	184
$F_2PP(GeH_3)_2$	2.177(10)	2.320(6)Ge	95.7(1.3)	98.6(1.6)	156.1(3.5)	185
		1.581(6)F		98.5(2.3)		
$P_2(CH_3)_4$	2.192(9)	1.853(3)	101.1(0.7)	99.6(1.0)	164(23)	186
$P_2(CF_3)_4$	2.182(16)	1.914(4)	106.7(0.7)	103.8(0.8)	180	184

^{a} MW data.

FPF versus PPF (in P_2F_4) and FPF versus PPGe (in $F_2PP(GeH_3)_2$) are at variance with those predicted by VSEPR theory.

Phosphines with P—Si and P—Ge Bonds. Unlike silylamines and germylamines, the corresponding phosphorus derivatives have usual pyramidal bond configurations around phosphorus (Table 2-36). The P—Si bond undergoes a noticeable shortening in H_2PSiF_3 under the influence of F atoms.

Phospholanes and Phospholenes Containing P—O, P—N, and P—S Bonds. Among phosphorus-containing ring compounds, those that contain the five-membered ring have been studied by electron diffraction most thoroughly. The results obtained show the ring to have an envelope conformation (37). Exo P—Cl or P—O bonds are in the axial position. The ring puckering angle τ is given in Table 2-37.

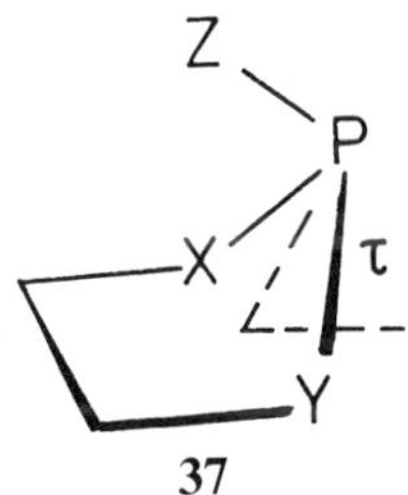

37

Table 2-36. PHOSPHINES CONTAINING P—Si AND P—Ge BONDS

Compound	P—El (Å)	∠El—P—El (deg)	Ref.
$H_2P—SiH_3$	2.249(3)	—	187
$H_2P—SiF_3$	2.207(3)	—	188
$(CH_3)_2P—SiH_3$	2.245(3)	99.0(0.5)SiPC	187
		100.8(1.2)CPC	
$P(SiH_3)_3$	2.248(3)	96.5(0.5)	189
$P(GeH_3)_3$	2.306(3)	95.7(1.0)	189

Table 2-37. PHOSPHOLANES AND PHOSPHOLENES CONTAINING P—O, P—N, AND P—S BONDS

Compound	Bond lengths (Å)		Angles (deg)[a,b]			Ref.
	P—X[a]	P—Y[b]	∠XPY	∠YPY	∠τ	
$(CH_2O)_2PCl$	2.105(7)	1.631(7)	99.6(0.5)	95.7	25.8	190
$[C(CH_3)_2O]_2PCl$	2.093(10)	1.630(10)	103.2(0.6)	95.9(2.0)	—	191
$[CH_2O]_2POCH_3$		1.624(12)	103.6(2.1)	93.6(1.2)	—	192
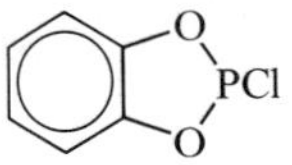 (benzodioxaphosphole)PCl	2.106			99.0	24.3	193
$X = CH_3$	2.174(9)	1.622(13)PO 1.692(13)PN	98.1 101.2(1.5)	93.9	13	194
$X = C_6H_5$	2.155(4)	1.620(5)PO 1.653(5)PN	100.4 103.5(0.6)	92.7	21.8	195
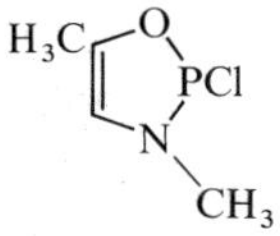 (methyl-oxazaphospholene)PCl	2.167(5)	1.630(6)PO 1.698(8)PN	101.0 101.0	91.2(0.5)	28	196
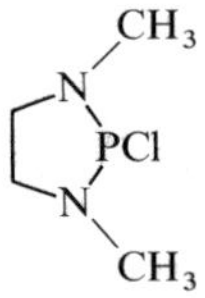 (dimethyl-diazaphospholane)PCl	2.19(2)	1.68	102.0(2.0)	89.8	28.8	197
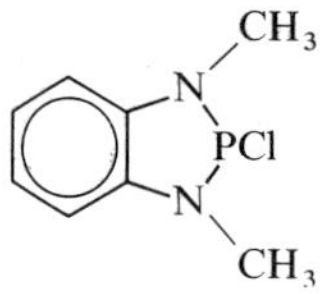 (dimethyl-benzodiazaphosphole)PCl	2.19(2)	1.701(3)	99.8(1.0)	87.4(0.8)	32(2)	198
$(CH_2S)_2PCl$	2.153(29)	2.077(15)	95.3(3.1)	83.3(2.1)	43.8(7.5)	199

[a] X = Cl, O.
[b] Y = O, N.

Table 2-38. METAPHOSPHATES

M⟨O⟩P=O	Distances (Å)			∠OPO	
	$(P—O)_{av}$	$(M—O)_{av}$	$(O\text{---}O)_{av}$	(deg)	Ref.
$NaPO_3$	1.480(3)	2.20(3)	2.58(1)	89(10)	200
$NaPO_3{}^a$	1.55	2.2	2.68	108	201
$CsPO_3$	1.474(4)	2.83(5)	2.57(2)	107(7)	202

[a] Quantum chemical calculation.

Considerable P—Cl bond elongation compared to PCl_3 is found in these compounds, especially in molecules containing endocyclic P—N bonds. Other parameters such as P—O, P—N, and P—S bond lengths have their usual values.

Alkali Metal Metaphosphates. Studies of sodium and cesium metaphosphates reveal that these compounds possess average P—O and M—O bond length values (Table 2-38). Their respective molecular configurations correspond to model **38**, with a planar arrangement of atoms. The P—O bond length was found to be far shorter than a P—O single bond (~ 1.60 Å).

$$O—P\langle{}^{O}_{O}\rangle M$$

M = Na, Cs

38

Four-Coordinated Phosphorus Atom

In terms of a classic theory of atomic valence states, there are two possible distributions of four-coordinated phosphorus valencies:

1. pentavalent: $\diagdown\!\!\diagup P{=}O$

2. Trivalent with a coordination bond $\diagdown\!\!\diagup P \rightarrow$

Acyclic Compounds

Phosphorus Oxo-, Sulfo-, and Selenium Halides. Four-coordinated pentavalent phosphorus halides are compounds that possess the simplest structure and composition (Table 2-39). Atoms attached to phosphorus comprise an irregular tetrahedron. The Hal—P—Hal valence angles are far below 109°28′ and increase along the series F to Br in symmetric $X{=}PY_3$ molecules for a given X atom (X = O or S). In compounds with X = O, the angle is larger than with X = S. Naturally, the X=P—Y valence angle is always larger than the YPY angle. These observations are in accord with expectations based on simple steric

Table 2-39. PHOSPHORUS OXOHALIDES

Compound	Bond lengths (Å)		$\angle$ Hal—P—Hal (deg)	Method	Ref.
	P=X	P—Hal			
P(O)F$_3$	1.436(6)	1.524(3)	101.3(0.2)	ED	LB
P(O)F$_3$	1.436	1.523	101.3	MW	203
P(O)HF$_2$	1.437(6)	1.539(3)	99.8(0.5)	MW	LB
P(S)F$_3$	1.866(5)	1.538(3)	99.6(0.3)	ED	204
	1.867	1.551	98.6	MW	205
P(Se)HF$_2$	2.026(4)	1.557(3)	98.1(0.7)	ED, NMR	206
P(O)Cl$_3$	1.449(5)	1.992(3)	103.3(0.2)	ED	LB
P(O)Cl$_3$	1.445(5)	1.989(2)	103.7(0.2)	MW	LB
P(S)Cl$_3$	1.884(5)	2.010(3)	101.8(0.2)	ED	LB
P(O)Br$_3$	1.455(7)	2.175(3)	104.1(0.2)	ED	207
P(S)Br$_3$	1.895(4)	2.193(3)	101.9(0.2)	ED	207
P(S)F$_2$Cl	1.864(8)	1.535(2)PF	100.5(0.8)FPF	ED	205
		1.895(8)PCl			
P(S)F$_2$Br	1.881(4)	1.543(3)PF	98.3(1.0)FPF	ED	205
		2.155(4)PBr			
P(S)F$_2$I	1.902(6)	1.546(5)PF	—	ED	205
		2.422(6)PI	—		

considerations. Interestingly, the Hal—P—Hal valence angles in PY$_3$ (Table 2-27) are smaller than in OPY$_3$.

Phosphorus-halogen bonds in X=PY$_3$ are somewhat shorter than in PY$_3$. This result probably reflects the operation of an increase in P—Y bond polarity. This explanation agrees with a lengthening of P—Y bonds on passing from X=O to X=S.

P=X (X = O, S, Se) bond lengths correspond to those expected for double bonds. A considerable shortening of the bonds from the sums of the covalent radii for double bonds (1.60 Å for P=O and 2.00 Å for P=S) is observed. The P=O and P=S bond lengths increase from F to Cl and to Br derivatives, ie, with decreasing P=O and P=S bond polarities. An alternative view holds that $\diagdown\!\!-\!\!\diagup$P=X bonds should be treated as being intermediate between $\diagdown\!\!-\!\!\diagup$P—X and $\diagdown\!\!-\!\!\diagup$P≡X bonds (X = O, S).[208]

Alkylphosphine Oxides, Sulfides, and Selenides. (Table 2-40). The bond configuration around phosphorus is close to that in phosphorus oxohalides (Table 2-39), although the CPC angles are larger than the Hal—P—Hal angles (Table 2-40). The P=O, P=S, and P=Se bonds are elongated noticeably compared to the halides. The P—C bonds are shorter, and the CPC valence angles larger than in PMe$_3$. Methyl group rotations are hindered. The most probable arrangement of the bonds about the P—C bond corresponds to a staggered conformation. Steric strain produced by *tert*-butyl groups increases the P=O and P—C bond lengths by about 0.1 Å and the CPC valence angle by about 10°.

Table 2-40. ALKYLPHOSPHINE OXIDES, SULFIDES, AND SELENIDES
$(CH_3)_3P(X)$; X = O, S, Se

| | Bond lengths (Å) | | Bond angles (deg) | | |
Compound	P=X	P—C	∠X—P—C	∠C—P—C	Ref.
$(CH_3)_3P(O)$	1.476(2)	1.809(2)	114.4(0.7)	104.1(0.8)	209
$[(CH_3)_3C]_3P(O)$	1.590(12)	1.888(6)	106.1(0.5)	112.9(0.5)	210
$(CH_3)_3P(S)$	1.940(2)	1.818(2)	114.1(0.2)	104.5(0.3)	209
$(CH_3)_3P(Se)$	2.091(3)	1.816(3)	113.8(0.4)	104.8(0.4)	211
$P(CH_3)_3$	—	1.846(3)	—	98.6(0.2)	LB

Also, the $—C(CH_3)_3$ group local symmetry is lowered from C_{3v}. The group axis and the P—C bond make an angle of 3.1(8)°, and the group is rotated from the staggered arrangement by 15.8(7)°.

Substituting a hydrocarbon radical for Cl in $Cl_3P=X$ (X = O, S) noticeably increases the P—Cl bond length (Table 2-41). Other bond lengths and valence angles are the same, within experimental error, as those listed in Tables 2-39 and 2-40. In the case of the compounds under consideration, the most important question is the conformation about the P—C bond. A trans–gauche equilibrium is observed in $CH_2ClP(X)Cl_2$ (X = O, S). The cis form predominates in $CH_2=CHP(O)Cl_2$, and in $C_6H_5P(O)Cl_2$, the P=O bond lies in the plane of the benzene ring. The predominant conformer of $(CH_2=CH)_2P(O)Cl$ has C_s symmetry and P—C=C dihedral angle of about 35°.

Phosphoranes Containing —P=C *Bonds.* Bonds of a new type, P=C, occur in phosphoranes (Table 2-42). Their length corresponds to that of a double bond, for they are shorter by about 0.2 Å than P—C bonds. In conformity to stereochemical patterns, the P=C bond is shortened in $Me_3P=C=PMe_3$ because of a change in the valence state of the central carbon atom. The valence angles at P are indistinguishable from those in the corresponding oxo derivatives.

The most interesting point is the conformation of molecule about the P=C bond. In $Me_3P=CHSiH_3$, the C—P=C—Si dihedral angle was found to be

Table 2-41. SUBSTITUTED PHOSPHORUS
OXOHALIDES

Compound	Y	P—Cl (Å)	Ref.
$YCl_2P(O)$	CH_3	2.032(9)	212
$YCl_2P(O)$	CH_2Cl	2.008(4)	213
$YCl_2P(S)$	CH_2Cl	2.028(5)	214
$YCl_2P(O)$	$CH_2=CH$	2.016(4)	215
$Y_2ClP(O)$	$CH_2=CH$	2.029(4)	215
$YCl_2P(O)$	C_6H_5	2.025(5)	216

Table 2-42. ALKYLIDENEPHOSPHORANES

Compound	Bond lengths (Å)		Bond angles (deg)		Ref.
	P=C	P—C	∠C—P=C	∠CPC	
$(CH_3)_3P{=}CH_2$	1.640(6)	1.818(3)	116.5(0.6)	101.6(0.5)	217
$(CH_3)_3P{=}CHSiH_3$	1.653(11)	1.807(8)	115.0(1.3)	103.4(1.3)	218
$(CH_3)_3P{=}C{=}P(CH_3)_3$	1.594(3)	1.814(3)	115.4(0.6)	101.4(0.3)	219

about 25(14)°. The conformation is probably not a staggered one. This is in part substantiated by calculations on $H_3P{=}CH_2$,[220] although the difference in conformer energies and the barrier to rotation have been found to be exceedingly small (ie, <1.0 kJ mol^{-1}). A departure from linearity [ie, ∠PCP = 147.6(5)°] was observed for the allene bond system in $Me_3P{=}C{=}PMe_3$. A model with freely rotating $(CH_3)_3P$ groups provides the best fit to experiment.

Iminophosphoranes, $R_3P{=}N{-}X$. The only posphoranes that have been studied in the gas phase are shown in Table 2-43. The presence of *t*-Bu groups lengthens the P=N and P—C bonds and decreases the CPN angle. The methyl groups in the first phosphorane are in a staggered arrangement with respect to the P...Si axis, and the $(CH_3)_3P$ group is rotated by about 30(10)° with respect to the PNSi plane from the trans arrangement of the C—P=N—Si bonds. In the second molecule, steric interactions cause a 18.5(14)° rotation of the $(CH_3)_3C$ groups from the staggered conformation.

Phosphoric Acid Esters, Amides and Their Derivatives. The most interesting structural characteristics of these compounds are the different bond lengths, the PXC (X = O, S, N) valence angle, and the rotational isomerism about the P—X bond (Table 2-44). Methoxyl and amino substituents elongate the P=O, P=S, and P—Cl bonds. The P—O and P—N bond lengths are substantially smaller than the sums of the corresponding covalent radii. Phosphorus in its turn substantially increases the valence angles at O, S, and N.

A mixture of conformers occurs in $(CH_3O)_3PO$. The predominant conformer (70%) is characterized by a near trans, trans, trans arrangement of the $O{=}P{-}O{-}CH_3$ bonds, and the second conformer has a gauche, gauche, gauche conformation. Molecules with a single top about the P—O bond are mostly gauche conformers, except in the case of $CH_3OP(O)FCH_3$. Their P—S analogs are more likely to possess trans conformations. An equilibrium of two conformers of C_s symmetry with 62° and 131° angles of rotation about the P—O bond from the cis structure was observed in $(CH_3O)_2P(S)Cl$. The O—C and

Table 2-43. IMINOPHOSPHORANES

Compound	Bond lengths (Å)		Bond angles (deg)		Ref.
	P=N	P—C	∠CPN	∠PNX	
$(CH_3)_3P{=}N{-}Si(CH_3)_3$	1.542(5)	1.804(3)	115.7(11)	114.6(11)	221
$[(CH_3)_3C]_3P{=}N{-}H$	1.652(11)	1.913(6)	109.6(7)	—	210

Table 2-44. ESTERS AND AMIDES OF PHOSPHORIC ACID AND THEIR DERIVATIVES

| | Bond lengths (Å) | | | | |
Compound	$P{=}O$, $P{=}S$	$P{-}X^a$	$P{-}Cl$	$\angle PXC$ (deg)^a	Ref.
$(CH_3O)_3PO$	1.477(6)	1.580(2)		118.4(1.6)	222
$CH_3OP(O)Cl_2$	1.456(5)	1.575(4)	1.992(3)	117.8(2.6)	223
$CH_3OP(O)FCH_3$	1.488	1.546	—	115.8	224
$CH_3SP(S)Cl_2$	1.445(15)	2.097(15)	2.021(6)	117.4(8.0)	225
$CH_3OP(S)Cl_2$	1.922(6)	1.628(6)	2.013(4)	114.4(2.4)	226
$(CH_3O)_2P(S)Cl$	1.897(3)	1.580	2.033(4)	119.3	227
$(CH_3)_2NP(O)Cl_2$	1.47(1)	1.67(4)	2.033(8)	$116(2)_{avg}$	171
$O{=}C{=}NP(O)Cl_2$	1.455(10)	1.684(10)	2.006(3)	120.0(1.5)	228

^{a} X = O, S, N.

P—Cl bonds in the 131° conformer are nearly mutually cis. The conformation of $(CH_3)_2NP(O)Cl_2$ is nonsymmetric, and $O{=}C{=}NP(O)Cl_2{}^{228}$ exists as a cis conformer.

Bis Derivatives $CH_2(PCl_2O)_2$, $CH_2(PF_2S)_2$, and $O(PF_2S)_2$. The most interesting stereochemical parameters are given in Table 2-45. The bond lengths have their usual values. The valence angles at carbon and oxygen are, however, increased. Very likely, this results from an increase in polarity of the central bonds. All these molecules are characterized by a nearly staggered arrangement of bonds about the P—C and P—O axes.

Of the three possible conformers, the C_2 and C_s forms were found to occur in $CH_2(PCl_2O)_2$ in an equilibrium ratio of about 1.5:4; $CH_2(PF_2S)_2$ exists as a (roughly) 1:1 mixture of the C_2 and C_1 forms. The C_s conformer gives the best fit with experiment for $O(PF_2S)_2$. The occurence of the C_s form is somewhat unexpected, for at first sight it appears to embody strong steric and polar interactions. However, in fact, nonbonded distances correspond to the sums of the van der Waals radii, except in the case of the S...S distance, which is rather short, [3.364(37) Å]. This situation is similar to that for three-coordinated phosphorus derivatives (see above).

Cyclic Compounds. 1-Chloro-1-oxophosphathietane, $\overline{SCH_2P(O)ClCH_2},{}^{231}$ is the only four-membered ring-containing phosphorus compound whose structure has been studied by gas-phase electron diffraction. Interestingly, the compound possesses a nearly planar structure; the CPC–CSC dihedral angle is

Table 2-45. ACYCLIC BIS DERIVATIVES

Compound	$P{-}C(O)$ (Å)	$\angle PXP$ (deg)	Ref.
$CH_2(PCl_2O)_2$	1.814(11)	114.4(23)	229
$CH_2(PF_2S)_2$	1.807(7)	122.6(10)	230
$O(PF_2S)_2$	1.609(8)	130.9(35)	230

about 10°, and the P=O bond is in the axial position in the predominant (80%) conformer. Although the ring is strained rather severely [$\angle$ CPC = 86(2)°], the P—C [1.836)15) Å] and C—S [1.794(15) Å] bonds are not elongated noticeably. The structure cannot be discussed in more detail because of large experimental uncertainties.

1-Chloro-1-oxophosphacyclopentene-2, ClP(O)CH=CHCH$_2$CH$_2$,[232] and 1-chloro-1-oxophosphacyclopentene-3, ClP(O)CH$_2$CH=CHCH$_2$, contain five-membered rings that exist in envelope conformations. In the former molecule, the carbon atom of the second CH$_2$ group is displaced out of the P—CH=CH—C plane to produce a dihedral angle of about 23°. In the latter molecule phosphorus is displaced by a dihedral angle of about 13°. The P=O bond occupies the axial position. The P—C— [1.830(8) Å] and P—C [1.791(8) Å] bond lengths in the former molecule differ from each other by 0.04 Å. In both molecules, a tendency toward increasing P—Cl distances compared to OPCl$_3$ is observed [ie, 2.057(8) and 2.040(8) Å, respectively].

The five-membered rings in ethylenechlorophosphate, ClP(O)OCH$_2$CH$_2$O,[234] and in ethylenechlorothiophosphate, ClP(S)SCH$_2$CH$_2$S,[234] possess half-chair conformations. In pyrocatechyl chlorophosphate,[235] however, the ring is planar (**39**). In these three molecules

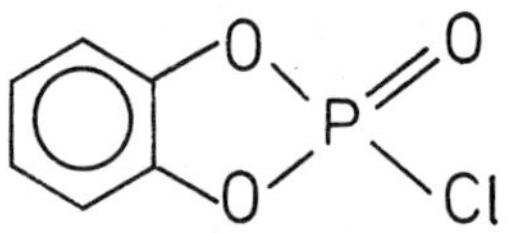

39

the P—Cl bonds are elongated compared to OPCl$_3$ (1.989 Å) to 2.057(10), 2.066(15), and 2.036(14) Å, respectively.

The (NP)$_3$ six-membered rings in inorganic "aromatic" structures of cyclophosphazenes (NPX$_2$)$_3$ (X = F, Cl) are planar, the P—N bonds have the same length [ie, 1.590(10 Å], and the PNP and NPN angles are also practically equal (120°). The XPX nonring angles are about 100°. The P—F [1.543(15) Å] and P—Cl [2.006(3) Å] bond lengths are not longer than in OPF$_3$ and OPCl$_3$.

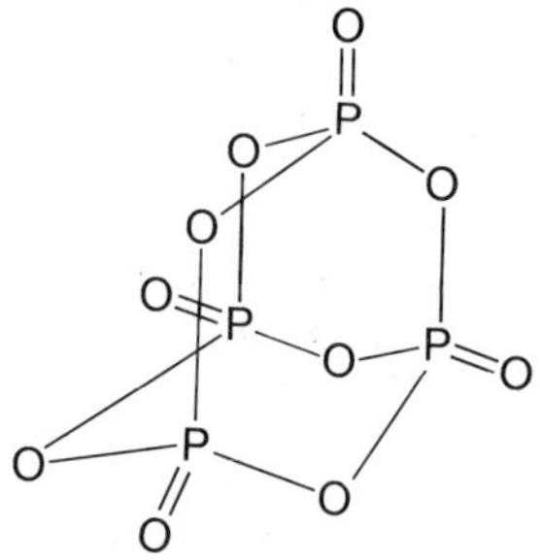

40

Table 2-46. DONOR-ACCEPTOR COMPLEXES

Compound	Bond lengths (Å)		$\angle$CPC (deg)	Ref.
	P—B	P—C		
$(CH_3)_3P \cdot BCl_3$	1.941(16)	1.800(4)	109.3(3)	236
$(CH_3)_3P \cdot BBr_3$	1.946(29)	1.804(4)	108.0(7)	237
$(CH_3)_3P \cdot BI_3$	1.947(11)	1.809(3)	106.0(5)	238

The P_4O_{10} molecule skeleton has an adamantane structure of T_d symmetry (**40**). The P=O [1.40(3) Å] and P—O [1.60(1) Å] bond lengths are close to those observed 'in acyclic esters of phosphoric acid. The POP valence angle [124.5(1)°] is smaller somewhat than in $O[P(S)F_2]_2$.

Four-Coordinated Phosphorus Atom $\diagdown\!\!\!\underset{\diagup}{P} \rightarrow$ **in Complexes.** Phosphines are active electron donors in coordination compounds. Boron and various metals act as acceptors. We shall briefly discuss phosphine geometry variations in complexes of the type $R_3P \cdot A$, where A is an acceptor such as BX_3 (X = Cl, Br, I) (Table 2-46). The P—B and P—C distances are constant along this series; the P—C bond is noticeably shorter and the CPC angle wider than in $P(CH_3)_3$. Similar effects are observed for P—C bonds in $(CH_3)_3P \cdot Ga(CH_3)_3$[239] and P—F bonds in $Pt(PF_3)_4$.[240]

Five-Coordinated Phosphorus Atom

Pentacoordination of a P atom is a new valence state of a Group V element that is not characteristic of nitrogen. Calculations of PH_5 and the experimental data on PF_5, PCl_5, and their derivatives indicate that phosphorus possesses a trigonal bipyramidal bond configuration, which results in axial and equatorial bonds being nonequivalent (Table 2-47). The P—F and P—Cl bond lengths differ from those observed in PF_3 and PCl_3, and also in OPF_3 and $OPCl_3$. Equatorial P—CH_3 bonds are shorter than in $P(CH_3)_3$ and $OP(CH_3)_3$; however, they undergo substantial elongation upon CF_3/CH_3 substitution (Table 2-48).

The $(P—C)_{eq}$ bond lengths in CH_3PF_4 and $C_6H_5PF_4$[243] are practically the same [1.796(10) Å]. The corresponding distance in $HC{\equiv}CPF_4$ is, however, significantly smaller [1.747(5) Å].[244]

Table 2-47. PHOSPHORUS PENTAHALIDES

PX_5	Bond lengths (Å)		Difference (Å)	Ref.
	$(P—X)_{eq}$	$(P—X)_{ax}$		
PF_5, r_g	1.534(4)	1.577(5)	0.043	LB
PCl_5, r_g	2.023(3)	2.127(3)	0.107(4)	241

Table 2-48. P—C Bond Lengths
in Phosphoranes

Compound	$(P—C)_{eq}$ (Å)	Ref.
CH_3PF_4	1.780(5)	LB
$(CH_3)_2PF_3$	1.798(5)	LB
$(CH_3)_3PF_2$	1.813(1)	LB
CF_3PF_4	1.881(8)	242
$(CF_3)_3PF_2$	1.888(4)	242

The $(P—N)_{eq}$ bonds in phosphoranes are somewhat shorter than those observed in aminophosphines; the corresponding distances vary only insignificantly over the series given in Table 2-49. A more detailed discussion is given in Chapter 4.

Table 2-49. P—N Bond Lengths
in Phosphoranes

Compound	$(P—N)_{eq}$ (Å)	Ref.
$(CH_3)_2NPF_4$	1.663(14)	67
$[(CH_3)_2N]_3PF_2$	1.647(5)	245
$(H_2N)_2PF_3$	1.648(13)	246
$(H_2N)_2PHF_2$	1.640(5)	247

Discussion

Bonds and Their Configurations

Structural studies of nitrogen and phosphorus derivatives have revealed more stereochemical differences than similarities between the two classes of compounds. Phosphorus valence states are more varied. Types and kinds of bonds formed by nitrogen and by phosphorus are essentially different. The bond type depends on the valence states of the atoms involved ($>$N—Cl, $>$P—Cl, $=$N—C$<$, $=$P—Cl$<$, etc); the kind of bond is determined by its environment.

Thus, the P—Cl bond length is affected substantially by the presence of other atoms attached to phosphorus, such as Cl, O, or N. The N—C and P—C bonds are different, however, for in $N(CF_3)_3$, the former is elongated compared to $N(CH_3)_3$, while the latter is longer in $P(CF_3)_3$ than in $P(CH_3)_3$. This result can be explained by different bond polarities in amines vis-à-vis phosphines.

A pyramidal bond configuration at a three-coordinated trivalent nitrogen atom is flattened compared to the corresponding bond pyramid at phosphorus. It is likely that for this reason nitrogen more readily changes its valence state upon transition to a planar bond configuration. Thus, the $\diagdown$N—C$\diagup$ bonds in organic acid amides and the $\diagdown$P—C$\diagup$ bonds in phosphides are substantially different. On the other hand, pyramidal bond configurations occur in ethylene imine $\left(\text{HN}\triangleleft\right)$ and its derivatives, which makes the stereochemistry of nitrogen in these compounds more akin to that of phosphorus.

Nitrogen bond lengths are subject to stronger variations because of conjugation of the lone electron pair on N with double bonds and vacant p_π–d_π atomic orbitals. Bonds formed by P with elements other than O and N do not differ significantly from the sums of the covalent radii.

Calculation of Internal Rotation Potential Function from Electron Diffraction Data

The study of nitrogen and phosphorus compounds contributed to the development of the electron diffraction technique. Thus the observation of a noticeable nonplanarity in aromatic nitro compounds with large (of the order of 20°) angles of rotation about the C—N bond led structural chemists to investigate the role of torsional vibrations in determining the average structure of a molecule. This nonplanarity in fact was shown to depend on torsional vibrations.[248] With molecules or molecular fragments that possess symmetry planes and undergo internal rotations that are asymmetric with respect to these planes, averaging over torsional vibrational states yields an effective configuration that lacks the initial symmetry plane.

Thus, a potential function that describes rotations about the C—N bond can be written in a parametric form:

$$V(\varphi) = \frac{V_0}{2}\,(1 - \cos 2\varphi) \tag{2-1}$$

where V_0 is the barrier to rotation, and φ is the rotation angle ($\varphi = 0°$ for a coplanar arrangement of the benzene ring and the NO_2 group in nitrobenzene). For the average configuration, this function affords a value for the root-mean-square angle $\langle\varphi^2\rangle^{1/2}$ close to the experimental value if V_0 is set at 14–17 kJ mol^{-1}. Conversely, if a planar equilibrium configuration is assumed, the barrier to rotation can be estimated from the experimental $\langle\varphi^2\rangle^{1/2}$ value.

With ortho derivatives, there are no reasons for assuming a planar equilibrium configuration. The electron diffraction data [for, eg, $o\text{-}NO_2C_6H_4Cl(Br)$] are indicative of a noticeable departure from planarity. Torsional vibration effects are difficult to determine in this case. No rotational potential function can be written in simple parametric form and then applied to this molecule.

The problem becomes even more general if other nitrogen and phosphorus derivatives are considered. Its solution can be obtained only by using new approaches—for example, by solving a Fredholm integral equation of the first kind. This equation appears in gas-phase electron diffraction as follows.

If internal rotation can be separated from small-amplitude frame vibrations, the general equation to describe the reduced scattering intensity component has the integral form:

$$sM(s) = \int_0^{2\pi} W(\varphi)sM(s, \varphi)\, d\varphi \qquad (2\text{-}2)$$

where $s = (4\pi/\lambda) \sin \vartheta/2$, λ is the wavelength, ϑ is the scattering angle, $W(\varphi)$ is the probability density for the angle φ, and $sM(s, \varphi)$ is the theoretical expression for the reduced intensity at a given φ value.

The classical Boltzmann distribution function can be used as a good approximation to $W(\varphi)$. We then have:

$$W(\varphi) = N \exp\left[\frac{-V(\varphi)}{kT}\right] \qquad (2\text{-}3)$$

where $V(\varphi)$ is the internal rotation potential function and N is the normalization factor.

The $sM(s, \varphi)$ function is in fact the reduced intensity for the molecular frame:

$$sM(s, \varphi) = \sum_{i \neq j} g_{ij}(s) \exp\left[-\frac{s^2 l_{frij}^2(\varphi)}{2}\right] \frac{\sin sr_{ij}(\varphi)}{r_{ij}(\varphi)} \qquad (2\text{-}4)$$

where l_{frij} is the vibrational amplitude for the pair of skeleton atoms i and j, r_{ij} is the internuclear distance [$r_{ij}(\varphi)$ and $l_{frij}(\varphi)$ depend on (φ)], and $g_{ij}(s)$ is the scattering function.

If r_{ij} values and valence angles in a molecule are known from a preliminary treatment of electron diffraction data on the level of a rougher approximation of small vibrational amplitudes, and if the force field for the frame is also known (and therefore, the $l_{frij}(\varphi)$ quantities can be calculated) then eq. (2-2) becomes a Fredholm equation of the first kind.

Special regularization techniques should, however, be employed to obtain a stable solution.[150,249] Accordingly, the sought function, $W(\varphi)$, was determined from the experimental data on $sM_{exp}(s)$, known r_{ij}, valence angle, and $l_{frii}(\varphi)$ values, by minimizing the functional:

$$\min\left\{\Phi_\alpha[W(\varphi)] = \int_{s_{min}}^{s_{max}} \left[\int_0^{2\pi} W(\varphi)sM_{theor}(s, \varphi)\, d\varphi \right.\right.$$
$$\left.\left. - sM_{exp}(s)\right]^2 ds + \alpha \int_0^{2\pi}\left[W^2(\varphi) + \left(\frac{dW(\varphi)}{d\varphi}\right)^2\right]d\varphi\right\}$$
$$(2\text{-}5)$$

where α is a numerical parameter ($\alpha > 0$). The solution to eq. (2-5) converges uniformly to the solution of eq. (2-2). The function $W(\varphi)$ is obtained as a set of W_i values that correspond to angles φ_i taken at intervals of $\Delta\varphi$. The parameter α is chosen such that the first term in eq. (2-5) corresponds to the uncertainty in $sM_{exp}(s)$.

This technique was applied to determine internal rotation potential functions for

$$p\text{-}(NO_2)_2C_6H_4, \; m\text{-}(NO_2)_2C_6H_4,$$
$$o\text{-}NO_2C_6H_4X \; (X = Cl, Br),$$
$$m\text{-}NO_2C_6H_4X \; (X = Cl, Br),$$

and also for $C_6H_5P(CH_3)_2$.

One-Top Examples

$m\text{-}NO_2C_6H_4X$ (X = Cl, Br). These molecules are of interest because the interaction between NO_2 and X is only weak, while the conformation-dependent contribution to the scattering is large. A broad rotation potential energy minimum occurs at $\varphi = 0°$. The parametric function $V(\varphi) = (V_0/2)(1 - \cos 2\varphi)$ predicts a narrower well. The barrier to rotation occurs at $\varphi = 90°$; its height is equal to 15.5(3.4) and 18.9(3.8) kJ mol^{-1} for X = Cl and Br, respectively. These quantities agree, within experimental error, with the barrier height of 12.2 kJ mol^{-1} found in a microwave study of $C_6H_5NO_2$.[250] It follows that the angle φ for the equilibrium conformation differs from that which characterizes the average conformation.

$o\text{-}NO_2C_6H_4X$ (X = Cl, Br). Narrow minima of $V(\varphi)$ curves correspond to $\varphi = 33°$ for X = Cl and $\varphi = 43°$ for X = Br, in agreement with the corresponding experimental φ values. Two maxima at $\varphi = 0°$ and $\varphi = 90°$ have different heights: $V(0°) = 4.2(1.3)$, $V(90°) = 15.1(3.4)$ for X = Cl, and $V(0°) = 17.7(4.2)$, $V(90°) = 8.4(2.5$ kJ mol^{-1} for X = Br. It is worthwhile mentioning that an inversion in relative maximum heights occurs on proceeding from X = Cl to X = Br.

$(CH_3)_2PC_6H_5$. The internal rotation potential curve passes through two minima in the range of 0 to 90°; the deeper one corresponds to $\varphi = 0°$, a bisector conformation. The difference in energy between the two conformers is 2.6(0.4) kJ mol^{-1}, and the barrier to transition from $\varphi = 0°$ to $\varphi = 90°$ amounts to 4.5(0.7) kJ mol^{-1}. The more stable conformation, therefore, does not meet the condition for conjugation (cf. dimethylaniline).

A Two-Top Example

$p\text{-}(NO_2)C_6H_4(NO_2)$. Calculations for this example are complicated. Their result is a two-dimensional internal rotation potential surface with a minimum corresponding to a planar configuration, whereas the average structure shows a certain departure from planarity. The surface maximum corresponds to twice the barrier height for separate NO_2 groups (ie, ~ 20 kJ mol^{-1}).

Conformations of Five-membered Rings Without Double Bonds

Unlike four-membered rings, the conformation of five-membered rings is determined by two parameters, the bond lengths being fixed. Available data show that the two most frequently occurring conformers in comparatively symmetric compounds that contain five-membered rings are the envelope and half-chair forms. Conformations are usually treated in a purely descriptive manner. An analysis of factors that determine the conformation, however, enables the principal forms to be identified. Stereochemical data on free molecules of phosphorus and arsenic compounds are very important in this respect.

Consider a ring with identical bonds. The geometry of chair and half-chair conformers is often described in terms of parameters that are less important than internal rotation angles. Thus, the envelope cyclopentane conformer **41** is characterized by the dihedral angle α made by the CCC and CCCC planes, and the half-chair conformer is characterized by the angle τ between the C—C bond normal to the C_2 axis and the plane of the other three carbon atoms (**42**).

C_s C_2

41   **42**

It can, however, be shown that both conformers are characterized by certain relations between internal rotation angles φ defined as follows:

$$\varphi_j = \varphi_m \cos\left(\frac{\Delta}{2} + \frac{j4\pi}{5}\right) \qquad j = 1, 2, 3, 4 \tag{2-6}$$

where φ_j is the dihedral angle for bond j, φ_m is the amplitude (the largest angle), and Δ is the phase of the pseudorotation angle ($\Delta = 0°$ for the half-chair and $36°$ for the envelope). Thus for the half-chair **43** we have:

$$\varphi_0 > \varphi_1 = \varphi_4 > \varphi_2 = \varphi_3 \qquad \text{and} \qquad \varphi_3 = 0°, \varphi_0 = \varphi_1 > \varphi_2 = \varphi_4$$

for the envelope **44**.

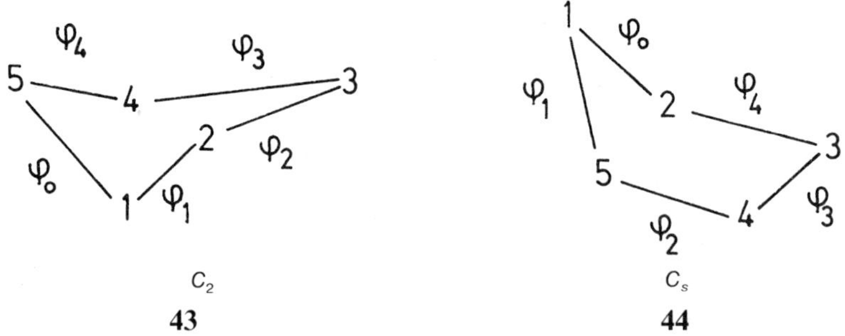

C_2 C_s

43   **44**

The same relations between the dihedral angles φ are preserved in rings that contain bonds of various lengths.

Nonplanar half-chair and envelope conformations are formed as a result of steric interactions in the initial planar structure, and different distributions of φ angles can be explained as follows. In a planar conformation, a substantial torsional momentum about each bond exists because of an eclipsed arrangement of adjacent bonds. The torsional momentum is determined by the height of the barrier to rotation about the corresponding bond. We lack such information about cycles. We can, however, suggest that the barriers to rotation should be proportional to the corresponding rotational barriers in acyclic compounds.

These qualitative considerations account for observed conformations for a number of cyclopentane derivatives, eg, $(CH_2)_4X$ (X = S, SiH_2, Se).[252] The barriers to rotation about the $-\overset{|}{\underset{|}{C}}-S$, $-\overset{|}{\underset{|}{C}}-Si-$ and $-\overset{|}{\underset{|}{C}}-Se$ bonds are noticeably lower than for the $-\overset{|}{\underset{|}{C}}-\overset{|}{\underset{|}{C}}-$ bond. The torsional momentum in the initial planar conformation will therefore be maximal for the C—C bond and minimal for the C—X bond. The distribution of dihedral angles in the rings under consideration should, therefore, correspond to a half-chair conformation, and half-chair conformers in fact are observed experimentally.

There exist, however, certain discrepancies. Thus, in tetrahydrofuran, a lower barrier to rotation about the C—O bond does not result in a half-chair conformation, and pseudorotation occurs in this compound as in cyclopentane itself. On the other hand, in pyrrolidine, $(CH_2)_4NH$,[116] which is intermediate between cyclopentane and tetrahydrofuran, an envelope conformation is observed. At present, a simple qualitative explanation to account for these findings is not forthcoming.

Further examination of structural data, however, reveals that the factors indicated above enable us to understand the patterns observed. Thus, the minima of pseudorotation potential in $(CH_2)_4CHCl$[253] and $(CH_2)_4CHCN$[254] correspond to envelope structures. In fact, torsional moments about the C—C—C(Cl)—C and C—C—C(CN)—C bonds should exceed all the other torsional moments in these molecules.

The same arguments, however, fail to explain the predominance of the envelope conformer in $(CH_2O)_2PCl$,[190] $(CH_2S)_2PCl$,[199] and $[CH_2N(CH_3)]_2PCl$,[197] since the highest barrier to rotation and therefore the largest torsional momentum should involve the C—C bond rather than the C—O and O—P bonds. An additional factor that favors the envelope conformation may be the so-called anomeric effect, which predicts the preferred conformation of the C—O—P—Cl bond system to be gauche. The competing effect of torsional momentum about the C—C bond clearly manifests itself in 4,4,5,5-tetramethyl-2-chloro-1,3,2-dioxaphospholane, $[(CH_3)_2CO]_2PCl$,[191] which possesses a half-chair conformation.

Since P=O and P=S multiple bonds also give rise to anomeric effects, half-chair conformations in the molecules $(CH_2O)_2P(O)Cl$[234] and $(CH_2S)_2P(S)Cl$[234] can be explained by following the same line of reasoning. The P—Cl, P=O, and P=S bonds all show about the same tendency to be oriented gauche with respect to the O—C and S—C bonds.

Until recently, it was not known whether the arrangement of bonds in envelope conformers (phospholanes in particular) was strictly eclipsed. The answer to this question was obtained via structural studies of a number of arsolanes, $(CH_2O)_2AsBr$, $(CH_2S)_2AsBr$, and $(CH_2O)_2AsCl$.[256] The five-membered rings in these compounds were found to be nonsymmetric, with carbon atoms displaced on the same side of the X_2As (X = O, S) plane and with different carbon–halogen distances (**45**). A repeated refinement of the $(CH_2O)_2PCl$[257] structure confirmed the suggestion of a decrease of ring symmetry from C_s.

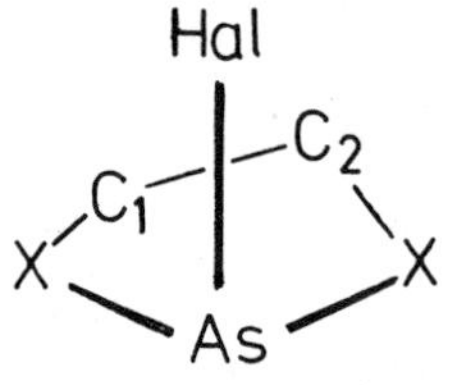

Hal

C₁ — C₂

X — As — X

45

Differences between interatomic distances in phosphorus derivatives were smaller than in arsenic compounds, and this is why they were not observed earlier. Further studies showed that as anomeric effects decrease, torsional momentum regains its importance as a factor determining conformation. Thus, 2-methyl-1,3,2-dithiarsolane, $(CH_2S)_2AsCH_3$,[258] and 1-chlorotetrahydroarsolane, $(CH_2CH_2)_2AsCl$,[258] occur as half-chair conformers.

The observed patterns enable us to estimate intermolecular interaction effects in crystal molecules that contain five-membered rings.[259]

The considerations discussed above allow us to predict ring conformations in yet unstudied molecules. Thus, eg, a half-chair conformation is predicted for $(CH_2CH_2)_2PH$, and an envelope form is predicted for $(CH_2CH_2)_2C(CH_3)H$.

Acknowledgments

We are grateful to Dr. V. A. Sipachev for translating our manuscript and to L. M. Shkolnikova for typing it.

References

1. Almenningen, A.; Nor, O.; Strand, T. G. *Acta Chem. Scand.* **1976**, *A30*, 567.
2. Typke, V.; Dakkouri, M.; Schumberger, F. *J. Mol. Struct.* **1980**, *62*, 111.
3. Fernholt, L.; Kveseth, K. *Acta Chem. Scand.* **1979**, *A33*, 335.
4. Almenningen, A.; Bak, B.; Jansen, P.; Strand, T. G. *Acta Chem. Scand.* **1973**, *27*, 1531.
5. Schei, S. H. *J. Mol. Struct.* **1983**, *98*, 141.
6. Traetteberg, M.; Bakken, P.; Seip, R. Annual Report of Norwegian ED group, **1982**, p. 45.
7. Khaikin, L. S.; Vilkov, L. V.; Andrutskaya, L. G.; Zenkin, A. A. *J. Mol. Struct.* **1975**, *29*, 171.

8. Hargittai, I.; Paul, I. C. "The chemistry of cyanates and their thio derivatives", In "The Chemistry of Functional Groups", Part 1, Patai, S., Ed; Wiley: New York, 1977, Chap. 2, pp. 69–128.

9. Vilkov, L. V.; Khaikin, L. S. *Top. Curr. Chem.* **1975**, *53*, 25–70.

10. Duckett, J. A.; Robiette, A. G.; Gerry, M.C.L. *J. Mol. Spectrosc.* **1981**, *90*, 374.

11. Cradock, S.; Huntley, C. M.; Durig, J. R. *J. Mol. Struct.* **1985**, *127*, 319.

12. Durig, J. R.; Sullivan, J. F.; Li, Y. S.; Mohamad, A. B. *J. Mol. Struct.* **1982**, *79*, 235.

13. Mürdoch, J. D.; Rankin, D. W. H.; Beagley, B. *J. Mol. Struct.* **1976**, *31*, 291.

14. Blair, P. D. *J. Mol. Struct.* **1983**, *97*, 147.

15. Cradock, S.; Laurenson, G. S.; Rankin, D.W.H. *J. Chem. Soc. Dalton Trans.* **1981**, 187.

16. Brunvoll, J.; Hargittai, I.; Seip, R. *J. Chem. Soc. Dalton Trans.* **1978**, 299.

17. Oberhammer, H.; Seppelt, K.; Mews, R. *J. Mol. Struct.* **1983**, *101*, 324.

18. Seppelt, H.; Oberhammer, H. *Inorg. Chem.* **1985**, *24*, 1227.

19. Christe, K. O.; Christen, D.; Oberhammer, H.; Schack, C. J. *Inorg. Chem.* **1984**, *23*, 4283.

20. Pearson, R.; Lovas, F. J. *J. Chem. Phys.* **1977**, *66*, 4149.

21. Christen, D.; Oberhammer, H.; Hammaker, R. M.; Shi-cheng Chang; Des Marteau, D. D. *J. Am. Chem. Soc.* **1982**, *104*, 6186.

22. Christen, D.; Kalcher, K. *J. Mol. Struct.* **1983**, *97*, 143.

23. Noakes, T. J.; Beagley, B.; Foord, A. *J. Mol. Struct.* **1976**, *35*, 115.

24. Naumov, V. A.; Litvinov, O. A.; Kibardin, A. M. *Zh. Strukt. Khim.* **1983**, *24*(2), 159.

25. Naumov, V. A.; Litvinov, O. A.; Kibardin, A. M. *Zh. Strukt. Khim.* **1984**, *25*(5), 35.

26. Traetteberg, M.; Hilmo, I.; Abraham, R. J.; Ljunggren, S. *J. Mol. Struct.* **1978**, *48*, 395.

27. Hargittai, I.; Seip, R. *Acta Chem. Scand.* **1976**, *A30*, 540.

28. Alderliesten, P.; Almenningen, A.; Strand, T. G. *Acta Chem. Scand.* **1975**, *B29*, 811.

29. Hagen, K.; Bondybey, V. V.; Hedberg, K. *J. Am. Chem. Soc.* **1977**, *99*, 1365.

30. Hagen, K.; Bondybey, V. V.; Hedberg, K. *J. Am. Chem. Soc.* **1978**, *100*, 7178.

31. Hargittai, I.; Schultz, G.; Naumov, V. A.; Kitaev, Y. P. *Dokl. Akad. Nauk SSSR* **1976**, *227*, 1131.

32. Bohn, R. K.; Bauer, S. H. *Inorg. Chem.* **1967**, *6*, 309.

33. Almenningen, A.; Anfinsen, I. M.; Haaland, A. *Acta Chem. Scand.* **1970**, *24*, 1230.

34. (a) Chang, C. H.; Porter, R. F.; Bauer, S. H. *J. Am. Chem. Soc.* **1970**, *92*, 5313. (b) Bürger, H.; Pawelke, G.; Oberhammer, H. *J. Mol. Struct.* **1982**, *84*, 49.

35. Traetteberg, M.; Hilmo, I.; Hagen, K. *J. Mol. Struct.* **1977**, *39*, 231.

36. Chiang, J. F.; Chiang, R. L.; Kratus, M. T. *J. Mol. Struct.* **1975**, *26*, 175.

37. Ilenchor, J. L.; Bauer, S. H. *J. Am. Chem. Soc.* **1967**, *89*, 5527.

38. Astrup, E. E.; Bouzga, A. M.; Ostoja Starzewski, K. A. *J. Mol. Struct.* **1979**, *51*, 51.

39. Oberhammer, H.; Strähle, J. *Z. Naturforsch.* **1975**, *30a*, 296.

40. Beagley, B.; Chantrell, S. J.; Kirby, R. G.; Schmidling, D. G. *J. Mol. Struct.* **1975**, *25*, 319.

41. Cradock, S.; Ebsworth, E.A.V.; Meikle, G. D.; Rankin, D.W.H. *J. Chem. Soc. Dalton Trans.* **1975**, 805.

42. Kuyper, J.; Isselmann, P. H.; Mijlhoff, F. C.; Spelbos, A.; Renes, G. H. *J. Mol. Struct.* **1975**, *29*, 247.

43. Oberhammer, H.; Kumar, R. C.; Knerr, G. D.; Shreeve, J. M. *Inorg. Chem.* **1981**, *20*, 3871.

44. Karl, R. R.; Bauer, S. H. *Inorg. Chem.* **1975**, *14*, 1859.

45. Des Marteau, D. D.; Eysel, H. H.; Oberhammer, H.; Günther, H. *Inorg. Chem.* **1982**, *21*, 1607.

46. Pauling, L. "The Nature of the Chemical Bond", 3rd ed. Cornell University Press; Ithaca, NY, **1963**.

47. Bunker, P. R.; Kraemer, W. P.; Špirko, V. *Can. J. Phys.* **1984**, *62*, 1801.

48. Beagley, B.; Medwid, A. R. *J. Mol. Struct.* **1977**, *38*, 229.

49. Bürger, H.; Niepel, H.; Pawelke, G.; Oberhammer, H. *J. Mol. Struct.* **1979**, *54*, 159.

50. Ter Brake, J.H.M.; Mom, V.; Mijlhoff, F. C. *J. Mol. Struct.* **1980**, *65*, 303.

51. Van des Does, H.; Mijlhoff, F. C.; Renes, G. H. *J. Mol. Struct.* **1981**, *74*, 153.

52. Sasaki, Y.; Kimura, K.; Kubo, N. *J. Chem. Phys.* **1959**, *31*, 477.

53. Brinker, W. D.; Huisman, P. A. G.; Mijlhoff, F. C. *J. Mol. Struct.* **1979**, *54*, 293.

54. Lister, D. G.; Tyler, J. K.; Hög, J. H.; Larsen, N. W. *J. Mol. Struct.* **1974**, *23*, 253.

55. Oberhammer, H.; Günther, H.; Bürger, H.; Heyder, F.; Pawelke, G. *J. Phys. Chem.* **1982**, *86*, 664.

56. Gundersen, G.; Mayo, R. A.; Rankin, D.W.H. *Acta Chem. Scand.* **1984**, *A38*, 579.

57. Gundersen, G.; Rankin, D.W.H.; Robertson, H. E. *J. Chem. Soc. Dalton Trans.* **1985**, 191.

58. Gundersen, G.; Rankin, D.W.H. *Acta Chem. Scand.* **1984**, *A38*, 647.

59. Fjeldberg, T. *J. Mol. Struct.* **1984**, *112*, 159.

60. Ebsworth, E.A.V.; Murray, E. K.; Rankin, D.W.H.; Robertson, H. E. *J. Chem. Soc. Dalton Trans.* **1981**, 1501.
61. Khaikin, L. S.; Belyakov, A. V.; Koptev, G. S.; Golubinskii, A. V.; Vilkov, L. V.; Girbasova, N. V.; Bogoradovskii, E. T.; Zavgordnii, V. S. *J. Mol. Struct.* **1980**, *66*, 191.
62. Fjeldberg, T. Studies on the structures of metal derivatives of low coordination numbers containing bulky-carbon-, nitrogen- or oxygen-centred ligands. Department of Chemistry. University of Trondheim, AVH, Norway, May **1985**.
63. Zaripov, N. M.; Naumov, V. A.; Tuzova, L. L.; *Phosphorus* **1974**, *4*, 179.
64. Arnold, D. E. J.; Rankin, D.W.H.; Todd, M. R. *J. Chem. Soc. Dalton Trans.* **1979**, 1290.
65. Hedberg, E.; Hedberg, L.; Hedberg, K. *J. Am. Chem. Soc.* **1974**, *95*, 4417.
66. Laurenson, G. S.; Rankin, D.W.H. *J. Chem. Soc. Dalton Trans.* **1981**, 1047, 425; *J. Mol. Struct.* **1979**, *54*, 111.
67. Khaikin, L. S.; Andrutskaya, L. G.; Vilkov, L. V. Ninth Hungarian Diffraction Conference, Pécs, **1978**, p. 23.
68. Hargittai, I. "The Structure of Volatile Sulfur Compounds". Akadémiai Kiadó; Budapest; Reidel: Dordrecht, 1985.
69. Almenningen, A.; Gundersen, G.; Mangerud, M.; Seip, R. *Acta Chem. Scand.* **1981**, *A35*, 341.
70. Rankin, D.W.H.; Todd, M. R.; Riddell, F. G. *J. Mol. Struct.* **1981**, *71*, 171.
71. Glidewell, C.; Marsden, C. J.; Robiette, A. G.; Sheldrick, G. M. *J. Chem. Soc. Dalton Trans.* **1972**, 1735.
72. Antipin, M. Y.; Struchkov, Y. T.; Shishkov, I. F.; Elfimova, T. L.; Vilkov, L. V.; Bredikhin, A. A.; Vereshchagin, A. N.; Ignatov, S. M.; Rudchenko, V. F.; Kostyanovskii, R. G. *Izv. Akad. Nauk SSSR Khim.* **1986**, 2235.
73. Kohata, K.; Fukuyama, T.; Kuchitsu, K. *J. Chem. Phys.* **1982**, *86*, 602.
74. Cardillo, M. J.; Bauer, S. H. *Inorg. Chem.* **1969**, *8*, 2086.
75. Gilbert, M. M.; Gundersen, G.; Hedberg, K. *J. Chem. Phys.* **1972**, *56*, 1691.
76. Nakata, M.; Takeo, J.; Matsumura, C.; Yamanouchi, K.; Kuchitsu, K.; Fukuyama, T. *Chem. Phys. Lett.* **1981**, *83*, 246.
77. Naumov, V. A.; Litvinov, O. A.; Geise, H. J.; Dillen, J. *J. Mol. Struct.* **1983**, *99*, 303.
78. Bartell, L. S.; Higgenbotham, H. K. *Inorg. Chem.* **1965**, *4*, 1346.
79. Glidewell, C.; Rankin, D.W.H.; Robiette, A. G.; Sheldrick, G. M. *J. Chem. Soc. A* **1970**, 318.
80. Naumov, V. A.; Litvinov, O. A.; Kitaev, Y. P. *Dokl. Akad. Nauk SSSR* **1981**, 256, 1158.
81. Kitano, M.; Kuchitsu, K. *Bull. Chem. Soc. Japan* **1974**, *47*: (a) 67, (b) 631, (c) **1973**, *46*, 3048.
82. Kitano, M.; Fukuyama, T.; Kuchitsu, K. *Bull. Chem. Soc. Japan* **1973**, *46*, 384.
83. (a) Samdal, S.; Seip, R. J. *J. Mol. Struct.* **1979**, *52*, 195. (b) Samdal, S.; Seip, R. J. *J. Mol. Struct.* **1980**, *62*, 131. (c) Saebo, S.; Klewe, B.; Samdal, S. *Chem. Phys. Lett.* **1983**, *97*, 499.
84. Gallaher, K. L.; Bauer, S. H. *J. Chem. Soc. Faraday Trans. 2* **1975**, *71*, 1423.
85. Hargittai, M.; Samdal, S.; Seip, R. *J. Mol. Struct.* **1981**, *71*, 147.
86. Fernholt, L.; Samdal, S.; Seip, R. *J. Mol. Struct.* **1981**, *72*, 217.
87. Khaikin, L. S.; Andrutskaya, L. G.; Vilkov, L. V. Ninth Hungarian Diffraction Conference, Pécs, **1978**, p. 21.
88. Sadova, N. I.; Vilkov, L. V. *Usp. Khim.* **1982**, *51*, 153.
89. Cox, A. P. *J. Mol. Struct.* **1983**, 97, 61.
90. Shishkov, I. F.; Sadova, N. I.; Vilkov, L. V.; Pankrushev, Y. A. *Zh. Strukt. Khim.* **1983**, *24*(2), 25.
91. Shishkov, I. F.; Sadova, N. I.; Vilkov, L. V.; Pankrushev, Y. A. *Zh. Strukt. Khim.* **1983**, *24*(3), 173.
92. Shishkov, I. F.; Sadova, N. I.; Novikov, V. P.; Vilkov, L. V. *Zh. Strukt. Khim.* **1984**, *25*(2), 98.
93. Penionzhkevich, N. P.; Sadova, N. I.; Popik, N. I.; Vilkov, L. V.; Pankrushev, Y. A. *Zh. Strukt. Khim.* **1979**, *20*(4), 603.
94. Batyukhnova, O. G.; Sadova, N. I.; Vilkov, L. V.; Pankrushev, Y. A. *J. Mol. Struct.* **1983**, *97*, 153.
95. Batyukhnova, O. G.; Sadova, N. I.; Vilkov, L. V.; Pankrushev, Y. A. *Zh. Strukt. Khim.* **1985**, 26(5), 175.
96. Penionzhkevich, N. P.; Sadova, N. I.; Vilkov, L. V. *Zh. Strukt. Khim.* **1976**, *17*(4), 753.
97. Sadova, N. I.; Popik, M. V.; Batyukhnova, O. G.; Sokolkov, S. V.; Pankrushev, Y. A.; Vilkov, L. V. "IV. Vsesoyuznoe Soveshchanie po Organicheskoi Kristallokhimii", Zvenigorod, 1984, p. 191 (Fourth USSR Conference on Organic Crystallochemistry).
98. Almenningen, A.; Brunvoll, J.; Popik, M. V.; Vilkov, L. V.; Samdal, S.; Sokolkov, S. V. *J. Mol. Struct.* **1984**, *118*, 37.

99. Batyukhnova, O. G.; Sadova, N. I.; Vilkov, L. V.; Pankrushev, Y. A. *Zh. Strukt. Khim.* **1984**, *25*(3), 166.·

100. Sadova, N. I.; Penionzhkevich, N. P.; Vilkov, L. V.; *Zh. Strukt. Khim.* **1976**, *17*(6), 1122.

101. Penionzhkevich, N. P.; Sadova, N. I.; Vilkov, L. V.; *Zh. Strukt. Khim.* **1979**, *20*(3), 527.

102. Almenningen, A.; Brunvoll, J.; Popik, M. V.; Sokolkov, S. V.; Vilkov, L. V.; Samdal, S. *J. Mol. Struct.* **1985**, *127*, 85.

103. Schultz, G.; Hargittai, L.; Kapovits, I.; Kucsman, A. *J. Chem. Soc. Faraday Trans. 2* **1984**, *80*, 1273.

104. Zaripov, N. M.; Golubinskii, A. V.; Sokolkov, S. V.; Vilkov, L. V.; Mannafov, T. G. *Dokl. Akad. Nauk. SSSR* **1984**, *278*, 664.

105. Batyukhnova, O. G.; Sadova, N. I.; Vilkov, L. V.; Ivshin, V. P.; Pankrushev, Y. A. *Zh. Strukt. Khim.* **1984**, *25*(6), 47.

106. McClelland, B. W.; Gundersen, G.; Hedberg, K.; *J. Chem. Phys.* **1972**, *56*, 4541.

107. McClelland, B. W.; Hedberg, L.; Hedberg, K.; Hagen, K. *J. Am. Chem. Soc.* **1983**, *105*, 3789.

108. Cox, A. P.; Waring, S. *Trans. Faraday Soc.* **1971**, *67*, 3441.

109. Kulikov, V. A.; Ugarov, V. V.; Rambidi, N. G. *Zh. Strukt. Khim.* **1981**, *22*(5), 183.

110. Kulikov, V. A.; Ugarov, V. V.; Rambidi, N. G. *Zh. Strukt. Khim.* **1981** (a) *22*(2), 196, (b) *22*(3), 168, (c) *22*(3), 166.

111. Shibata, S.; Iijima, K. *J. Mol. Struct.* **1984**, *117*, 45.

112. Touseev, N. I.; Golubinskii, A. V.; Zasorin, E. Z.; Spiridonov, V. P. "Sixth Austin Symposium on Gas-Phase Molecular Structure, Austin, TX, March 1976, No. A34.

113. Mastryukov, V. S.; Dorofeeva, O. V.; Vilkov, L. V.; Golubinskii, A. V. *J. Mol. Struct.* **1976**, *32*, 161.

114. Günther, H.; Schrem, G.; Oberhammer, H. *J. Mol. Spectrosc.* **1984**, *104*, 152.

115. Gebhardt, K. F.; Oberhammer, H.; Zeil, W. *J. Chem. Soc. Faraday Trans. 2* **1980**, *76*, 1293.

116. Pfafferott, G.; Oberhammer, H.; Boggs, J. E.; Caminiti, W. *J. Am. Chem. Soc.* **1985**, *107*, 2305.

117. Pfafferott, G.; Oberhammer, H.; Boggs, J. E. *J. Am. Chem. Soc.* **1985**, *107*, 2309.

118. Rademacher, P. *J. Mol. Struct.* **1975**, *28*, 97.

119. Gundersen, G.; Rankin, D.W.H. *Acta Chem. Scand.* **1983**, *A37*, 865.

120. Schei, H.; Shen, Q.; Hilderbrandt, R. L. *J. Mol. Struct.* **1980**, *65*, 297.

121. Hargittai, M.; Hargittai, I. "The Molecular Geometries of Coordination Compounds in the Vapour Phase". Akadémiai Kiadó: Budapest; Elsevier: Amsterdam, 1975.

122. Plato, V.; Hartford, W. D.; Hedberg, K. *J. Chem. Phys.* **1970**, *53*, 3488.

123. Lavigne, J.; Pépin, C.; Cabanna, A. *J. Mol. Spectrosc.* **1984**, *104*, 49.

124. Kroto, H. W.; Nixon, J. F.; Simmons, N. P. C. *J. Mol. Spectrosc.* **1980**, *82*, 185.

125. Kroto, H. W.; Nixon, J. F.; Simmons, N. P. C. *J. Mol. Spectrosc.* **1979**, *77*, 270.

126. Oberhammer, H.; Becker, G.; Gresser, G. *J. Mol. Struct.* **1981**, *75*, 283.

127. Nguyen, M. T.; Ruelle, P. *J. Chem. Soc. Faraday Trans. 2* **1984**, *80*, 1225.

128. Khaikin, L. S.; Grikina, O. G.; Vilkov, L. V.; Boggs, J. E. *Zh. Strukt. Khim.* **1987**, *28*(4), 167.

129. Vilkov, L. V.; Khaikin, L. S.; Vasil'ev, A. F.; Ignatova, N. P.; Mel'nikov, N. N.; Negrebetskii, V. V.; Shvetsov-Shilovskii, N. I. *Dokl. Akad. Nauk SSSR* **1971**, *197*, 1081.

130. Khaikin, L. S.; Vilkov, L. V.; Boggs, J. E. "Eleventh Austin Symposium on Gas-Phase Molecular Structure", Austin, TX, 1986, S16.

131. Andrianov, V. G.; Struchkov, Y. T.; Shvetsov-Shilovskii, N. I.; Ignatova, N. P.; Bobkova, R. G.; Mel'nikov, N. N. *Dokl. Akad. Nauk SSSR* **1973**, *211*, 1101.

132. Hogel, J.; Schnidpeter, A.; Sheldrick, W. S. *Chem. Ber.* **1983**, *116*, 549.

133. Boggs, J. E.; Niu, Z.; Khaikin, L. S. *J. Mol. Struct. (Theochem.)* **1985**, *124*, 155.

134. Ashe, A. J., III. *J. Am. Chem. Soc.* **1971**, *93*, 3293.

135. Wong, T. C.; Bartell, L. S. *J. Chem. Phys.* **1974**, *61*, 2840.

136. Hedberg, K.; Iwasaki, M. *J. Chem. Phys.* **1962**, *36*, 589.

137. Cowley, A. H.; Kilduff, J. E.; Ebsworth, E.A.V.; Rankin, D.W.H.; Robertson, H. E.; Seip, R. *J. Chem. Soc. Dalton Trans.* **1984**, 689.

138. Beagley, B.; Medvid, A. R. *J. Mol. Struct.* **1977**, *38*, 229.

139. Oberhammer, H.; Schmutzler, R.; Stelzer, O. *Inorg. Chem.* **1978**, *17*, 1254.

140. Marsden, C. J.; Bartell, L. S. *Inorg. Chem.* **1976**, *15*, 2713.

141. Rankin, D.W.H.; Robertson, H. E.; Karsch, H. H. *J. Mol. Struct.* **1981**, *77*, 121.

142. Naumov, V. A.; Tuzova, L. L.; Zaripov, N. M. *Zh. Strukt. Khim.* **1977**, *18*, 67.

143. Naumov, V. A.; Kataeva, O. N. *Zh. Strukt. Khim.* **1983**, *24*, 160.

144. Naumov, V. A.; Ziatdinova, R. N.; Romanov, G. V.; Pudovik, A. N. *Dokl. Akad. Nauk SSSR* **1980**, *253*, 167.

145. Naumov, V. A.; Kataeva, O. N. *Zh. Strukt. Khim.* **1983**, *24*, 96.
146. Novikov, V. P.; Kolomeets, V. I.; Golubinskii, A. V.; Vilkov, L. V.; Raevskii, O. A.; Novikova, Z. S.; Kurkin, A. N. *Zh. Strukt. Khim.* **1986**, *27*, 39.
147. Tuzova, L. L.; Naumov, V. A. *Zh. Strukt. Khim.* **1979**, *20*, 923.
148. Fernholt, L.; Stølevik, R. *Acta Chem. Scand.* **1975**. *A29*, 651.
149. Naumov, V. A.; Kataeva, O. N. *Zh. Strukt. Khim.* **1984**, *25*, 140.
150. Novikov, V. P.; Kolomeets, V. I.; Syshchikov, Y. N.; Vilkov, L. V.; Yarkov, A. V.; Tsvetkov, E. N.; Raevskii, O. A. *Zh. Strukt. Khim.* **1984**, *25*, 27.
151. Purt, A. W.; Rankin, D.W.H.; Stelzer, O. *J. Chem. Soc. Dalton Trans.* **1977**, 1752.
152. Naumov, V. A.; Zaripov, N. M.; Gulyaeva, N. A. *Zh. Strukt. Khim.* **1972**, *13*, 917.
153. Fong, G. D.; Kuczkowski, R. L. *Inorg. Chem.* **1981**, *20*, 2342.
154. Naumov, V. A.; Nesterov, V. Y. *Zh. Strukt. Khim.* **1984**, *25*, 137.
155. Shidulin, S. A.; Naumov, V. A. *Zh. Strukt. Khim.* **1979**, *20*, 728.
156. Khaikin, L. S.; Andrutskaya, L. G.; Grikina, O. E.; Vilkov, L. V.; El'natanov, Y. I.; Kostyanovskii, R. G. *J. Mol. Struct.* **1977**, *37*, 237.
157. Holywell, G. C.; Rankin, D.W.H. *J. Mol. Struct.* **1971**, *9*, 11.
158. Naumov, V. A.; Nesterov, V. Y.; Novikov, V. P.; Katsyuba, S. A. *Zh. Strukt. Khim.* **1984**, *25*, 129.
159. Boggs, J. E.; Khaikin, L. S. Eleventh Austin Symposium on Gas-Phase Molecular Structure, Austin, TX, March 1986.
160. Golding, E. G.; Jones, C. E.; Schwendeman, R. H. *Inorg. Chem.* **1974**, *13*, 178.
161. Zaripov, N. M. *Zh. Strukt. Khim.* **1982**, *23*, 142.
162. Yow, H. Y.; Rudolph, R. W.; Bartell, L. S. *J. Mol. Struct.* **1975**, *28*, 205.
163. Zaripov, N. M.; Naumov, V. A.; Tuzova, L. L. *Dokl. Akad. Nauk SSSR* **1974**, *218*, 1132.
164. Khaikin, L. S.; Vilkov, L. V. *Zh. Strukt. Khim.* **1969**, *10*, 722.
165. Arnold, D.E.J.; Gundersen, G.; Rankin, D.W.H. *J. Chem. Soc. Dalton Trans.* **1983**, 1989.
166. Ebsworth, E.A.V.; Macdonald, E. K.; Rankin, D.W.H. *Monatsh. Chem.* **1980**, *111*, 221.
167. Naumov, V. A.; Kataeva, O. N.; Sinyashin, O. G. *Zh. Strukt. Khim.* **1984**, *25*, 79.
168. Tuzova, L. L.; Naumov, V. A.; Galiakberov, R. M.; Ofitserov, E. H.; Pudovik, A. N. *Dokl. Akad. Nauk SSSR* **1981**, *256*, 891.
169. Vilkov, L. V.; Khaikin, L. S.; Evdokimov, V. V. *Zh. Strukt. Khim.* **1972**, *13*, 7.
170. Zaripov, N. M.; Naumov, V. A.; Tuzova, L. L. *Phosphorus* **1974**, *4*, 79.
171. Vilkov, L. V.; Khaikin, L. S. *Dokl. Akad. Nauk SSSR* **1966**, *168*, 810.
172. Holywell, G. C.; Rankin, D.W.H.; Beagley, B.; Freeman, J. M. *J. chem. Soc. A* **1971**, 785.
173. Laurenson, G. S.; Rankin, D.W.H. *J. Chem. Soc. Dalton Trans.* **1981**, 425.
174. Huntley, C. M.; Laurenson, G. S.; Rankin, D.W.H. *J. Chem. Soc. Dalton Trans.* **1980**, 954.
175. Hedberg, E.; Hedberg, K. *J. Am. Chem. Soc.* **1974**, *96*, 4417.
176. Laurenson, G. S.; Rankin, D.W.H. *J. Chem. Soc. Dalton Trans.* **1981**, 1047.
177. Arnold, D. E. J.; Rankin, D.W.H.; Todd, M. R. *J. Chem. Soc. Dalton Trans.* **1979**, 1290.
178. Rankin, D. W. H.; Cyvin, S. J. *J. Chem. Soc. Dalton Trans.* **1972**, 1277.
179. Cradock, S.; Laurenson, G. S.; Rankin, D.W.H. *J. Chem. Soc. Dalton Trans.* **1981**, 187.
180. Rankin, D.W.H. *J. Chem. Soc. Dalton Trans.* **1972**, 869.
181. Laurenson, G. S.; Rankin, D.W.H. *J. Mol. Struct.* **1979**, *43*, 111.
182. Beagley, B.; Conrad, A. R.; Freeman, J. M.; Monaghan, J. J.; Norton, B. G.; Holywell, G. C. *J. Mol. Struct.* **1972**, *11*, 371.
183. Durig, J. R.; Carreira, L. A.; Odom, J. D. *J. Am. Chem. Soc.* **1974**, *96*, 2688.
184. Hodges, H. L.; Su, L. S.; Bartell, L. S. *Inorg. Chem.* **1975**, *14*, 599.
185. Ebsworth, E. A. V.; Hutchison, D. J.; Rankin, D.W.H. *J. Chem. Res. Synop.* **1980**, 393.
186. McAdam, A.; Beagley, B.; Hewitt, T. G. *Trans. Faraday Soc.* **1970**, *66*, 2732.
187. Glidewell, C.; Pinder, P. M.; Robiette, A. G.; Sheldrick, G. M. *J. Chem. Soc. Dalton Trans.* **1972**, 1402.
188. Demuth, R.; Oberhammer, H. *Z. Naturforsch.* **1973**, *28a*, 1862.
189. Beagley, B.; Medwid, A. R. *J. Mol. Struct.* **1977**, *38*, 239.
190. Naumov, V. A.; Zaripov, N. M.; Dashevskii, V. G. *Dokl. Akad. Nauk SSSR* **1969**, *188*, 1062.
191. Naumov, V. A.; Zaripov, N. M. *Zh. Strukt. Khim.* **1970**, *11*, 1108.
192. Tuzova, L. L.; Naumov, V. A. *Dokl. Akad. Nauk SSSR* **1978**, *243*, 666.
193. Arbuzov, B. A.; Naumov, V. A.; Shaidulin, S. A.; Mukmenev, E. T. *Dokl. Akad. Nauk SSSR* **1972**, *204*, 859.
194. Naumov, V. A.; Pudovik, M. A. *Dokl. Akad. Nauk SSSR* **1972**, *203*, 351.
195. Naumov, V. A.; Bezzubov, V. M.; Pudovik, M. A. *Zh. Strukt. Khim.* **1975**, *16*, 3.

196. Khaikin, L. S.; Grikina, O. E.; Vilkov, L. V. *Zh. Strukt. Khim.* **1983**, *82*, 115.

197. Naumov, V. A.; F. Gulyaeva, N. A.; Pudovik, M. A. *Dokl. Akad. Nauk SSSR* **1972**, *203*, 590.

198. Khaikin, L. S.; Abrabenkov, A. V.; Smirnov, V. V.; Vilkov, L. V. Eleventh Austin Symposium on Gas-Phase Molecular Structure, Austin, TX, March 1986, S15.

199. Schultz, G.; Hargittai, I.; Matin, J.; Robert, J. B. *Tetrahedron* **1974**, *30*, 2365.

200. Petrov, K. P.; Ugarov, V. V.; Rambidi, N. G. *Zh. Strukt. Khim.* **1981**, *22*, 158.

201. Sen, K. D. *J. Chem. Phys.* **1981**, *75*, 1041.

202. Petrov, K. P.; Kolesnikov, A. U.; Ugarov, V. V.; Rambidi, N. G. *Zh. Strukt. Khim.* **1980**, *21*, 198.

203. Smith, J. G. *Mol. Phys.* **1976**, *32*, 621.

204. Karakida, K.; Kuchitsu, K. *Inorg. Chim. Acta* **1976**, *16*, 29.

205. Acha, L.; Cromie, E. R.; Rankin, D.W.H. *J. Mol. Struct.* **1981**, *73*, 111.

206. Boyd, A. S. F.; Laurenson, G. S.; Rankin, D.W.H. *J. Mol. Struct.* **1981**, *71*, 217.

207. Jacob, E. J.; Danielson, D. D.; Samdal, S. *J. Mol. Struct.* **1980**, *62*, 143.

208. Schmidt, M. W.; Gordon, M. S. *J. Am. Chem. Soc.* **1985**, *107*, 1922.

209. Vilkins, C. J.; Hagen, K.; Hedberg, L.; Shen, Q.; Hedberg, K. *J. Am. Chem. Soc.* **1975**, *97*, 6352.

210. Rankin, D.W.H.; Robertson, H. E.; Seip, R.; Schmidbaur, H.; Blaschke, G. *J. Chem. Soc. Dalton Trans.* **1985**, 827.

211. Jacob, E. J.; Samdal, S. *J. Am. Chem. Soc.* **1977**, *99*, 5656.

212. Naumov, V. A.; Semashko, V. N. *Zh. Strukt. Khim.* **1971**, *12*, 317.

213. Vajda, E.; Kolonits, M.; Hargittai, I. *J. Mol. Struct.* **1976**, *35*, 235.

214. Khaikin, L. S.; Vilkov, L. V.; Vasil'ev, A. F.; Mel'nikov, N. N.; Tulyakova, T. F.; Anashkin, M. G. *Dokl. Akad. Nauk SSSR* **1972**, *203*, 1090.

215. Naumov, V. A.; Shaidulin, S. A. *Zh. Strukt. Khim.* **1976**, *17*, 304.

216. Vilkov, L. V.; Sadova, N. I.; Zil'berg, I. Y. *Zh. Strukt. Khim.* **1967**, *8*, 528.

217. Ebsworth, E.A.V.; Fraser, T. E.; Rankin, D.W.H. *Chem. Ber.* **1977**, *110*, 3491.

218. Ebsworth, E.A.V.; Rankin, D.W.H.; Zimmer-Gasser, B.; Schmidbaur, H. *Chem. Ber.* **1980**, *113*, 1637.

219. Ebsworth, E.A.V.; Fraser, T. E.; Rankin, D.W.H.; Gasser, O.; Schmidbaur, H. *Chem. Ber.* **1977**, *110*, 3508.

220. Eades, R. A.; Gassman, P. G.; Dixon, D. A. *J. Am. Chem. Soc.* **1981**, *103*, 1066.

221. Astrup, E. E.; Bouzga, A. M.; Astoja-Starzewski, K. A. *J. Mol. Struct.* **1979**, *51*, 51.

222. Oberhammer, H. *Z. Naturforsch.* **1973**, *28a*, 1140.

223. Oberhammer, H. *J. Mol. Struct.* **1975**, *29*, 370.

224. Zeil, W.; Krats, H.; Haase, J.; Oberhammer, H. *Z. Naturforsch.* **1973**, *28a*, 1717.

225. Naumov, V. A.; Bezzubov, V. M. *Dokl. Akad. Nauk SSSR* **1976**, *228*, 888.

226. Bezzubov, V. M.; Naumov, V. A. *Zh. Strukt. Khim.* **1976**, *17*, 98.

227. Oberhammer, H. *J. Mol. Struct.* **1975**, *29*, 375.

228. Naumov, V. A.; Semashko, V. N.; Shatrukov, L. F. *Dokl. Akad. Nauk SSSR* **1973**, *209*, 118.

229. Novikov, V. P.; Kolomeets, V. I.; Vilkov, L. V.; Yarkov, A. V.; Raevskii, O. A.; Kurkin, A. N.; Novikova, Z. S. *Zh. Strukt. Khim.* **1986**, *27*, 45.

230. Rankin, D.W.H.; Todd, M. R. *J. Chem. Soc. Dalton Trans.* **1982**, 2079.

231. Naumov, V. A.; Semashko, V. N. *Dokl. Akad. Nauk SSSR* **1971**, *200*, 882.

232. Naumov, V. A.; Semashko, V. N. *Dokl. Akad. Nauk SSSR* **1970**, *193*, 348.

233. Naumov, V. A.; Semashko, V. N. *Zh. Strukt. Khim.* **1970**, *11*, 979.

234. Naumov, V. A.; Semashko, V. N.; Zav'yalov, A. P.; Cherkasov, R. A.; Grishina, L. N. *Zh. Strukt. Khim.* **1973**, *14*, 787.

235. Naumov, V. A.; Shaidulin, S. A. *Zh. Strukt. Khim.* **1974**, *15*, 133.

236. Iijima, K.; Shibata, S. *Bull. Chem. Soc. Japan* **1979**, *52*, 3204.

237. Iijima, K.; Koshimizu, E.; Shibata, S. *Bull. Chem. Soc. Japan* **1981**, *54*, 2255.

238. Iijima, K.; Koshimizu, E.; Shibata, S. *Bull. Chem. Soc. Japan* **1982**, *55*, 2551.

239. Golubinskaya, L. M.; Golubinskii, A. V.; Mastryukov, V. S.; Vilkov, L. V.; Bregadze, V. I. *J. Organomet. Chem.* **1976**, 117.

240. Pitz, C. L.; Bartell, L. S. *J. Mol. Struct.* **1976**, *31*, 73.

241. McClelland, B. W.; Hedberg, L.; Hedberg, K. *J. Mol. Struct.* **1983**, *99*, 309.

242. Oberhammer, H.; Grobe, J.; LeVan, D. *Inorg. Chem.* **1982**, *21*, 275.

243. Dittebrandt, C.; Oberhammer, H. *J. Mol. Struct.* **1980**, *63*, 227.

244. Oberhammer, H. *J. Mol. Struct.* **1979**, *53*, 139.

245. Oberhammer, H. *J. Chem. Soc. Dalton Trans.* **1976**, 1454.

246. Marsden, C. J.; Hedberg, K. K.; Shreeve, J. M.; Gupta, K. D. *Inorg. Chem.* **1984**, *23*, 3659.

247. Arnold, D.E.J.; Rankin, D.W.H.; Robinet, G. *J. Chem. Soc. Dalton Trans.* **1977**, 585.
248. Vilkov, L. V.; Penionzhkevich, N. P.; Brunvoll, J.; Hargittai, I. *J. Mol. Struct.* **1978**, *43*, 109.
249. Syshchikov, Y. N.; Vilkov, L. V.; Yagola, A. G. *Vest. Mosk. Univ. Ser. Khim.* **1983**, *24*, 541.
250. Høg, J. H.; Nygaard, L.; Sørensen, G. O. *J. Mol. Struct.* **1971**, *7*, 111.
251. Vilkov, L. V.; Mastryukov, V. S.; Dorofeeva, O. V.; Zaripov, N. M. *Zh. Strukt. Khim.* **1985**, *26*, 51.
252. Bastiansen, O.; Kveseth, K.; Møllendal, H. *Top. Curr. Chem.* **1979**, *81*, 101.
253. Hilderbrandt, R. L.; Shen, Q. *J. Phys. Chem.* **1982**, *86*, 587.
254. Hilderbrandt, R. L.; Leavitt, H.; Shen, Q. *J. Mol. Struct.* **1984**, *116*, 29.
255. Kirby, A. J. "The Anomeric Effect and Related Stereoelectronic Effects at Oxygen". Springer-Verlag: New York, 1982.
256. Zaripov, N. M.; Galiakberov, R. M.; Golubinskii, A. V.; Vilkov, L. V.; Chadaeva, N. A. *Zh. Strukt. Khim.* **1983**, *24*, 87.
257. Zaripov, N. M.; Galiakberov, R. M. *Zh. Strukt. Khim.* **1984**, *25*, 143.
258. Zaripov, N. M.; Galiakberov, R. M.; Golubinskii, A. V.; Vilkov, L. V.; Chadaeva, N. A. *Zh. Strukt. Khim.* **1984**, *25*, 25.
259. Newton, M. G.; Brown, H. C.; Finder, C. J. *J. Chem. Soc. Chem. Commun.* **1974**, 455.

3

OXYGEN AND
SULFUR COMPOUNDS

Victor A. Naumov

ARBUZOV INSTITUTE OF ORGANIC AND PHYSICAL CHEMISTRY
ACADEMY OF SCIENCES OF THE USSR
KAZAN, USSR

CONTENTS

INTRODUCTION   94
ONE- AND TWO-COORDINATED COMPOUNDS   94
  Alcohols   94
  Thioalcohols (Thiols)   99
  Ethers and Sulfides   101
  Hypohalides and Related Compounds   110
  Esters and Their Analogs   111
  Aldehydes   113
  Ketones   116
  Carboxylic Acids   117
  Acyl Halides   120
  Peroxides and Their Analogs   122
THREE-COORDINATED SULFUR COMPOUNDS   123
  Sulfoxides, $R_2S{=}O$ 123
  Compounds with S$=$N and S$-$N Bonds   123
  Sulfites   124
FOUR-COORDINATED SULFUR COMPOUNDS   125
  Sulfones   125
  Amides and Imides   128
  Sulfates and Halosulfuric Acid Esters   130
FIVE- AND SIX-COORDINATED SULFUR COMPOUNDS   131

CYCLIC COMPOUNDS OF OXYGEN AND SULFUR   133
    Three-Membered Ring Compounds   133
    Four-Membered Ring Compounds   135
    Five-Membered Ring Compounds   136
    Six-Membered Ring Compounds   138
    Ring Compounds Having Seven or More Members   140
REFERENCES   141

Introduction

Oxygen- and sulfur-containing compounds are of great importance in organic chemistry. One- and two-coordinated molecules are characteristic of oxygen, whereas sulfur may also form molecules with larger coordination numbers. Alcohols, ethers, aldehydes, ketones, carboxylic acids, etc, have been studied widely by gas-phase electron diffraction (ED) and by microwave (MW) spectroscopy. The ED data are considered according to the coordination number of the central atom and the class of the compound. Wherever necessary, MW results are also included.

One- and Two-Coordinated Compounds

Alcohols

The geometrical parameters of alcohols are given in Table 3-1. Because of the low scattering power of hydrogen, the accuracy with which such important parameters as C—O—H valence angles and O—H bond lengths are determined is low. More reliable are the results of MW investigations. The C—O bond length in alcohols is in the 1.40–1.43 Å range, the average value being 1.415(6) Å.* This value is smaller than the sum of the Pauling radii and is in good agreement with the value calculated by application of the Schomaker–Stevenson equation (1.42 Å).

Alcohols are of interest from the conformational view-point. Two conformers (**1**) have been found for ethanol[2] by MW spectroscopy. Valence angles and bond

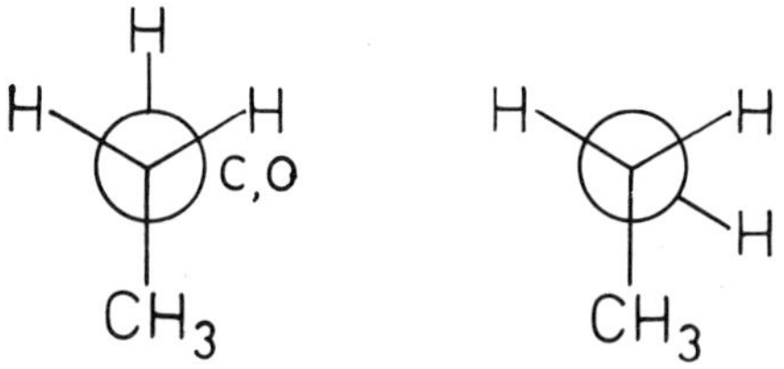

1

* Throughout this chapter, numbers in parentheses are the limits of error of the last digits.

Table 3-1. GEOMETRICAL PARAMETERS OF ALCOHOLS

Compound	Bond lengths (Å)		$\angle$ C—O—H (deg)	Ref.
	O—H	C—O		
CH_3OH	0.945(3)	1.421(3)	108.5(5)	1
C_2H_5OH	0.9710(6)	1.4310(32)	105.7(2)	2
$(CH_3)_2NCH_2CH_2OH$	1.235	—	100.5	3
FCH_2CH_2OH	0.966(9)	1.418(6)	—	4
$ClCH_2CH_2OH$	1.00	1.616(20)	—	5
$HOCH_2CH_2CH_2OH$	1.04(7)	1.410(6)	109(3)	6
	0.98(6)		97(3)	
$C(O)HOCH_2CH_2OH$	1.18(3)	1.412(7)	—	7
$C(O)HCH_2OH$	1.071–1.097	1.401–1.406	100.0–101.5	8
$(CF_3)_3COH$	0.96^a	1.41(2)	108.5^a	9
cyc-$(C_3H_5)CH_2OH$	0.9451	1.426	116.7	10
$C_6H_5CH_2OH$	1.034(11)	1.415(3)	105.3	11

[a] Assumed.

lengths are practically the same in both forms. The C—C bond, 1,515(34) Å, is shorter than that in ethane.

The conformational problem is more complicated in substituted alcohols XCH_2CH_2OH [X = F, Cl, N(CH$_3$)$_2$, OH] because it is necessary to consider the rotation about the C—O and C—C bonds. Gauche structure has been suggested for 2-dimethylaminoethanol,[3] $(CH_3)_2NCH_2CH_2OH$ (**2**), which favors O—H...N hydrogen bonding. As a result the O—H bond is extremely long ($\sim$ 1.235 Å). Another example of intramolecular hydrogen bonding can be

observed in the ED investigation[6] of 1,3-propanediol. Hydrogen bonding results in the quasi-cyclic structure **3**. The following torsional angles define the geometry of the cycle: O_1—C—C—C 68(3)°, O_2—C—C—C 61(5)°,

$C—C—O_1—H$ 46(5)°, and $C—C—O_2—H$ 180° (assumed). There is an appreciable difference between bond lengths $O_1—H$ (1.04 Å) and $O_2—H$ (0.98 Å) as well as between valence angles $C—O_1—H$ (97°) and $C—O_2—H$ (109°), since one of the hydroxyl hydrogens takes part in hydrogen bonding. The O...O distance is 2.70 Å.

ED investigations[4,5] have shown the gauche conformer to be the most favorable for FCH_2CH_2OH and $ClCH_2CH_2OH$ in the gas phase (**4**). At 156°C FCH_2CH_2OH exists mainly in the gauche form (90%). The energy difference

between the two forms is approximately 9.6 kJ mol^{-1}. For 2-chloroethanol, experimental data have been obtained at 37°C and at 200°C. At higher temperature the percentage of the trans conformer in the mixture is not large (15–20%), and at 37°C less than 10% is present in this form. The torsional angle O—C—C—Cl at 37°C is 60.3(3.9)°, increasing at higher temperature up to 70.2(2.0)°.

The above results are in good agreement with an MW study[12] of 1-fluoro-2-propanol, $CH_3CHOHCH_2F$ (**5**), for which conformer I predominates.

Intramolecular hydrogen bonding determines the dominant conformer in glycol monoformate,[7] $HOCH_2CH_2OCHO$. The conformation with torsional angles $\tau_i = 0°$ is given in Figure 3-1. When the bond is observed from the atom with a lower index to the atom with a higher index, positive rotation appears in the counter-clockwise direction. The $C_2—O_3$ and $C_1—O_4$ bond lengths are not identical, being 1.412(7) and 1.444(9) Å, respectively. The $C_1—C_2$ and C=O bond lengths are quite common, 1.525(4) and 1.203(3) Å. The mixture contains

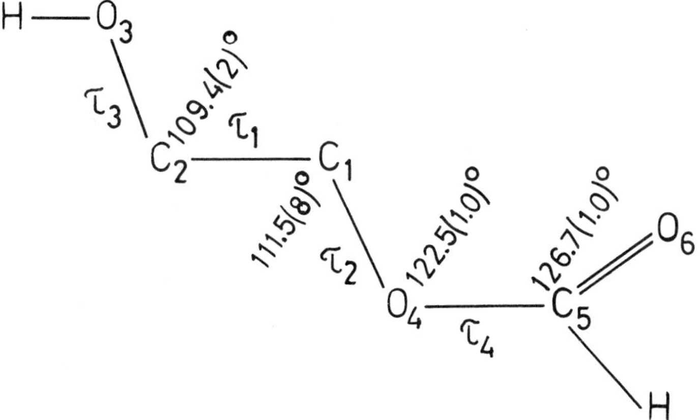

Figure 3-1. Conformation of glycol monoformate with $\tau_i = 0$.

58% gauche I conformer and the remainder gauche II with the following torsional angles:

	τ_1	τ_2	τ_3	τ_4
gauche I:	108.8(1.5)°	271.6(1.0)°	161°	8(3)°
gauche II:	124.7(3.0)°	299.5(1.5)°	170°	8(3)°

The important difference between the two conformers is in the hydrogen bonding. In gauche I the OH group is associated with the carbonyl oxygen (O_3...O_6 distance 2.785 Å), while in gauche II it is associated with the ether oxygen O_4 (O_3...O_4 distance 2.806 Å). In the infrared spectrum the O—H stretching band is shifted to 3430 cm^{-1} relative to that of glycol (3670 cm^{-1}). The C=O symmetric band is split into two peaks that have been assigned to two gauche conformers.

Microwave studies[13,14] have revealed the role of the NH_2 group in the stabilization of the predominant conformers. Thus, compound **6**, 1-amino-2-propanol,[13] $CH_3CHOHCH_2NH_2$, displays a unique conformation in the gas phase with $\tau(OCCN)$ 55.4°, $\tau(NCCC)$ 0°, $\tau(CCOH)$ 28.3°, and $\tau(CCNH)$ −101.8° and 20.5°. Conformer I is more stable than the other two

6

conformers by 4 kJ mol^{-1}. 2-Amino-1-propanol,[14] $CH_3CHNH_2CH_2OH$ has been observed to consist of a mixture of two conformers, the former with τ(NCCO) 54(2)° and τ(CCCO) 180(2)° being more stable than the latter with τ(NCCO) 61.5(2.0)° and τ(CCCO) 60(1)° by 2.1 kJ mol^{-1}.

In perfluoro-*tert*-butyl alcohol the bond length r(C—C) [1.566(9) Å] appeared to be longer than expected. The contrary is true for r(C—O), ie 1.41(2) Å, whereas the authors expected the 1.45–1.47 Å value.[9] The CF_3 groups are twisted from the symmetrical position by 23.0° and vibrate with a root-mean-square amplitude of 14.2°. The torsional vibration potential function possesses a flat minimum.

In compounds **7**, **8**, and **9**, respectively—cyclopropylcarbinol,[10] [CH_2—CH_2—CH—CH_2OH], allyl alcohol[15] [CH_2=CH—CH_2—OH], and 2,3-butadien-1-ol[16] [CH_2=C=CH—CH_2OH]—the molecular conformations are determined by the interaction of hydroxyl hydrogens with the electrons of the nearest strained cyclic bond in the first compound and with the unsaturated bonds in the two latter.

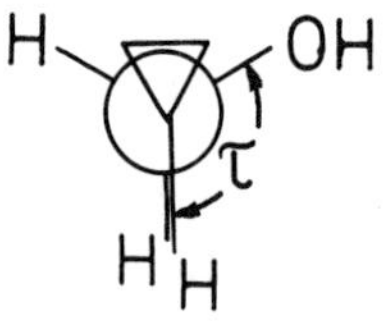

τ = 124.2°

7   **8** 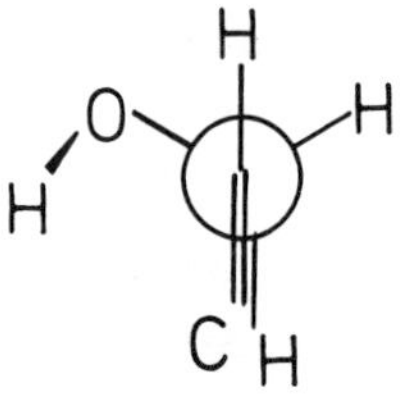   **9**

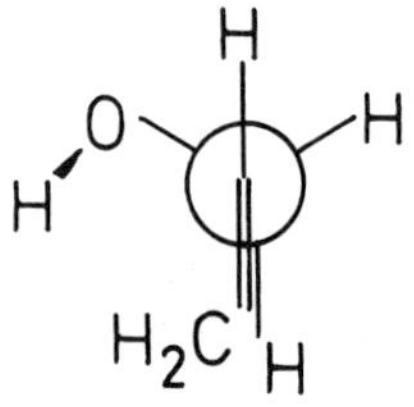

In **10**, benzyl alcohol,[11] oxygen is beyond the phenyl ring plane. The most stable conformer possesses the hydroxyl hydrogen in the anti position relative to the CH_2 group (endo conformer). The following geometrical parameters have

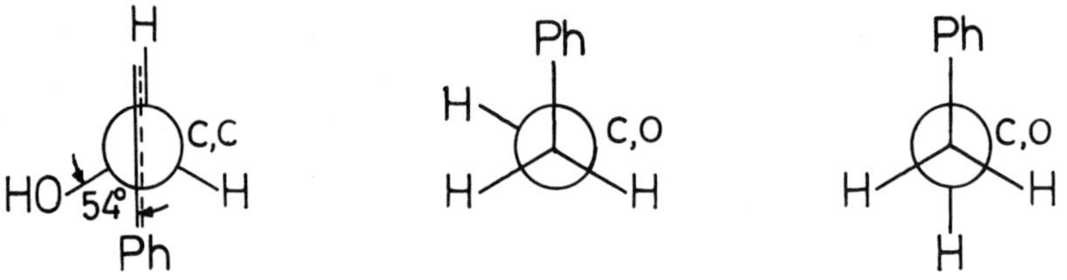

τ = 54°   endo   exo

10

been reported for benzyl alcohol: C—C, 1.394(1) Å; C—CH_2, 1.514(2) Å; C—O, 1.415(3) Å; O—H, 1.034(11) Å; $\angle$C—C—O, 114.5(5)°; $\angle$C—O—H, 105.3°; $\angle H_2C$—C—C, 118.0(8)°. The shrinkage of the H_2C—C—C valence angle may be explained in terms of the interaction between hydroxyl hydrogen and π-electrons of the phenyl ring.

A microwave study of *syn*-vinyl alcohol[17] has shown the C—O bond length to be shorter (1.372 Å) than that in saturated alcohols (Table 3-1). The other molecular parameters are common: C=C, 1.326 Å; O—H, 0.960 Å, $\angle$C=C—O, 126.2°; $\angle$C—O—H, 108.3°. The O—H bond adopts the syn configuration.

It is worthwhile to consider compounds that contain the N—O—H fragment. Molecules of pyruvaldehyde-1-oxime[18] adopt syn–trans configuration with a planar skeleton (**11**). The bond lengths and valence angles are as follows: C—CH$_3$, 1.513(6) Å; N=C, 1.260(4) Å; C=O, 1.212(3) Å; N—O, 1.382(4) Å; C—C, 1.479(4) Å; O—H, 0.962(15) Å; $\angle$C—C—C, 113.4(3)°; $\angle$C—C=O, 118.4(3)°; $\angle$C—C=N, 120.5(5)°; $\angle$C=N—O, 112.3(4)°; $\angle$N—O—H, 104.3(3.5)°. Rankin and coworkers studied the structure of *N*-methylhydroxyl-amine,[19] CH$_3$NHOH. Nitrogen adopts a pyramidal configuration with the mean valence angle 106.2°. As a result, the N—O bond is considerably longer: 1.477(2) Å.

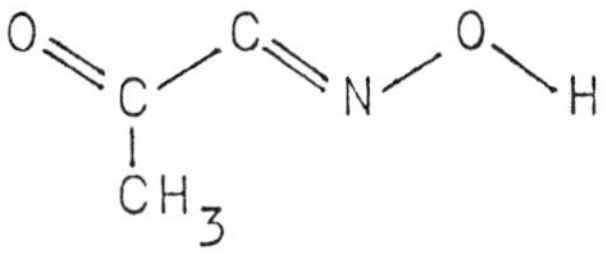

11

Thioalcohols (Thiols)

The simplest thiols, CH$_3$SH[20] and CF$_3$SH,[21] have been studied by MW spectroscopy. The S—H bond length is identical in these molecules (1.33 Å), while for C—S the bond length differs. In CH$_3$SH its value is 1.819 Å and in CF$_3$SH it is shorter, 1.800(5) Å. The C—S—H valence angle is 100.3(2)°.

In 1,2-ethanedithiol (**12**) two conformers have been found in the gas phase.[22]

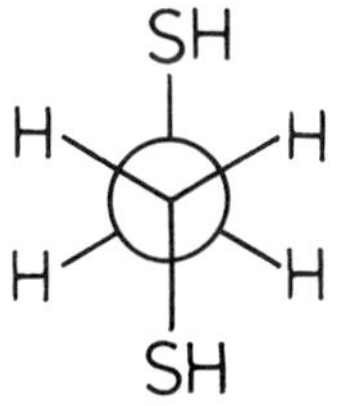

trans 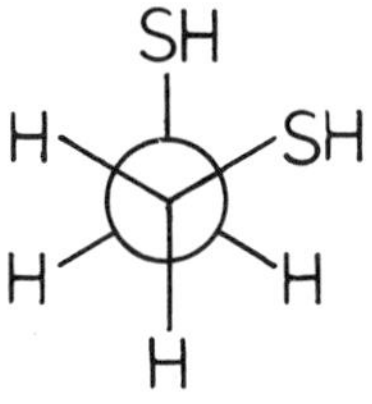   gauche

12

The trans form is the dominant (62%) one being more stable than the gauche conformer by 3.3 kJ mol^{-1}. The principal geometrical parameters are ordinary: S—H, 1.40(2) Å; S—C, 1.819(2) Å; C—C, 1.53(2) Å; $\angle$C—C—S, 112.9(9)°; $\angle$C—S—H, 90.5(3.2)°; $\angle$S—C—H, 107.0(2.2)°. The S—C—C—S torsional

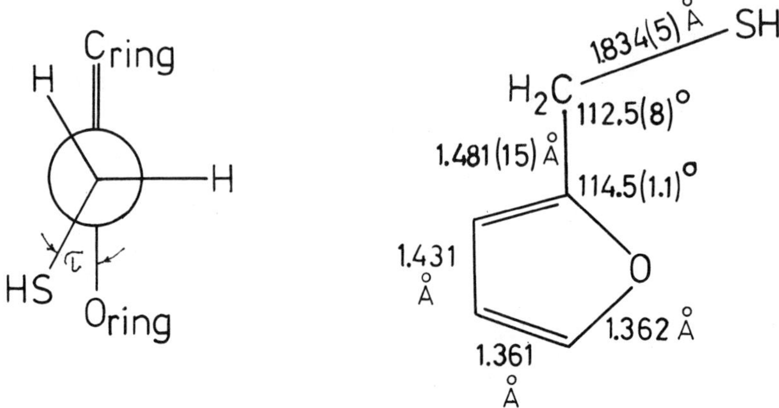

Figure 3-2. Conformation and the geometrical parameters of 2-furanmethanethiol.[24]

angle in the gauche form is 73.7(3.5)°. Similar parameters were obtained[23] by MW spectroscopy for ethanethiol: S—C, 1.820(5) Å; S—H, 1.322(6) Å; ∠C—S—H, 96.2(4)°; ∠C—C—S, 108.6(3)°; The molecule adopts a trans conformation with the H—S—C—C torsional angle 180°.

2-Furanmethanethiol[24] (Figure 3-2) provides an example of the unsaturated thiols. Sulfur is beyond the plane of the heteronuclear ring, the torsional angle τ(S—C—C—O) being 39(4)°. Such configuration favors S—H...O intramolecular hydrogen bonding.

As has been shown,[25] the OH and SH groups do not distort the benzene ring. In p-dihydroxy- and p-dithiolbenzenes the C—C$_{ipso}$—C valence angle is practically normal, the C_{2h} symmetry being assumed (Table 3-2). In agreement with the carbon valence state the ⋮C—O and ⋮C—S bonds are shorter than the corresponding —C—O and —C—S bonds.

Table 3-2. GEOMETRICAL PARAMETERS
OF p-HX—C$_6$H$_4$—XH MOLECULES

Parameter	X	
	O	S
C—C, Å	1.397(3)	1.400(3)
C—X, Å	1.384(4)	1.776(4)
∠C$_2$—C$_1$—C$_6$, deg	120.7(2)	120.2(2)
∠C$_2$—C$_1$—X, deg	124.4(6)	123.1(6)
∠C—X—H, deg	109.8(1.3)	96.2(1.3)

Ethers and Sulfides

The combined electron diffraction—microwave study[26] of dimethyl ether with the aid of *ab initio* geometry optimization has shown that this compound possesses C_{2v} symmetry and that each of the methyl groups is staggered with the opposite C—O bond (**13**). It has also been demonstrated that the methyl group symmetry axis does not coincide with the C—O bond by 3.7 (1.7)°. In methyl

ethyl ether[27] a gauche conformation with C—O—C—C torsional angle 84(6)° is preferable. It is important to point out that the O—CH$_3$ and O—CH$_2$ bonds differ by 0.009 Å. The C—O—C valence angle (Table 3-3) in most of the ethers is enlarged. For example, in di-*n*-propyl ether[42] it attains the value of 116.1(3.6)°. The C—O bonds in this molecule are shorter [1.404(6) Å] than in dimethyl ether. The mixture of 26% *aaaa* conformer and of 74% g^+aag^+ conformer

Table 3-3. GEOMETRICAL PARAMETERS OF CH$_3$—O—X ETHERS

Compound	Bond lengths (Å)		∠C—O—X (deg)	Ref.
	O—CH$_3$	O—X		
CH$_3$OCH$_3$	1.415(1)	1.415(1)	111.8(2)	26
CH$_3$OC$_2$H$_5$	1.413(9)	1.422(7)	111.9(5)	27
CH$_3$OCH$_2$Cl	1.414	1.368	113.1	28
CH$_3$OCHCl$_2$	1.405(9)	1.383(9)	118.6(2.1)	29
CH$_3$OCH=CH$_2$	1.427(7)	1.359(15)	116.8(1.8)	30
CH$_3$OC(CH$_3$)=CH$_2$	1.416(5)	1.353(5)	116.0(1.1)	31
CH$_3$OCH=C=CH$_2$	1.427(7)	1.375(7)	115.0(1.2)	32
CH$_3$OCH$_2$CH$_2$OCH$_3$	1.410(3)	1.410(3)	110.7(1.4)	33
CH$_3$OCH$_2$OCH$_3$	1.432(4)	1.382(4)	114.6(5)	34
CH$_3$OC(CH$_3$)$_2$OCH$_3$	1.423(6)	1.423(6)	114.0(1.4)	35
CH(OCH$_3$)$_3$	1.418(6)	1.382(6)	114.3(6)	36
C(CH$_3$)(OCH$_3$)$_3$	1.431(6)	1.398(6)	114.0(4)	37
C(OCH$_3$)$_4$	1.422(5)	1.395(6)	113.9(4)	38
CH$_3$OC$_6$H$_5$	1.423(15)	1.361(15)	120.0(2.0)	39
p-CH$_3$OC$_6$H$_4$CHO	1.432(10)	1.360(12)	120.8(1.7)	40
(cyclohexenyl)—OCH$_3$	1.421(6)	1.364(6)	119.7(2.5)	41

provides the best agreement with experimental data. The sequence of torsional angles (*a* and *g*) and the corresponding bonds is as follows:

$$C_1{-}C_2, \quad C_1{-}O, \quad O{-}C_1', \quad C_1'{-}C_2'$$

In gauche conformation the $C_3{-}C_2{-}C_1{-}O$ torsional angle is 71.8°.

Dimethyl ether may form stable complexes, eg, with BF_3. In the $(CH_3)_2OBF_3$ complex,[43] lengthening of the C—O bond up to 1.425(10) Å and decrease of the C—O—C valence angle to 108.4(3.1)° have been observed. The BF_3 fragment adopts a pyramidal configuration with relatively long B—F bonds [1.358(7) Å]. The angle between the B—O bond (1.75 Å) and the C—O—C plane is 40(8)°.

The substitution of hydrogen atoms in dimethyl ether by fluorine strongly affects the molecular geometry. Thus, the C—O bonds in CF_3OCF_3[44] become significantly shorter [1.369(4) Å], whereas the C—O—C valence angle increases up to 119.1(8)°. However, in corresponding sulfides the picture is not the same. In $(CH_3)_2S$,[45] the C—S bond equals 1.805(2) Å and the C—S—C angle is 99.1(4)°, whereas in $(CF_3)_2S$[46] the C—S bond length is 1.819(2) Å and the C—S—C angle is 97.3(8)°.

The anomeric effect in methyl chloromethyl ether,[28] $ClCH_2OCH_3$, results in contraction of the $O{-}CH_2$ bond (Table 3-3) and in lengthening of the C—Cl bond (1.813 Å). The molecule adopts a gauche conformation with torsion angle C—O—C—Cl 74.3°. The corresponding sulfide, $ClCH_2SCH_3$,[47] exists in the same form with τ(C—S—C—Cl) 49°. However, no C—Cl bond lengthening has been observed, 1.766(5) Å, though the $S{-}CH_2$ and $S{-}CH_3$ bonds differ, being equal to 1.785(5) and 1.827(5) Å, respectively.

Gauche orientation of the C—O and C—Cl bonds has also been observed in **14** (Cl_2CHOCH_3)[29] and in **15** $[(ClCH_2)_2O]$, the Cl—C—O—C torsional angles being 60.6(1.6) and 69.6(1.6)°, respectively. The overall symmetry of the $(ClCH_2)_2O$ molecule is C_2. The main geometrical parameters are as follows: C—O, 1.393(3) Å; C—Cl, 1.800(3) Å; ∠C—O—C, 114.2(1.5)°. The largest C—O—C valence angle among the saturated ethers and their derivatives has been observed in CH_3OCHCl_2 (Table 3-3). This is likely to be connected with steric interactions [C...Cl nonbonded distance is 3.114(12) Å]. Steric factors stabilize the gauche-trans conformer, but it is the anomeric effect that dominates. As a result the gauche- configuration is more stable.

14

15

It is of interest to consider the methoxymethane series, $CH_{4-n}(OCH_3)_n$ ($n = 1, 2, 3, 4$) (Figure 3-3). In dimethoxymethane[34] the $C—O—C_0—O—C$ chain adopts the gauche-gauche conformation with the $O—C_0—O$ valence angle 114.3(7)°. In the gauche-gauche chain of trimethoxymethane this angle is 115.0(1.0)°, whereas in the gauche-trans chains ($CH_3—O_3—C_0—O_2—CH_3$ and $CH_3—O_1—C_0—O_2—CH_3$) these angles are 6° smaller. This is likely to be the result of the anomeric effect.[49] *Ab initio* calculations on the $HO—CH_2—OH$ molecule have shown the gauche-gauche conformation to be more stable than the gauche-trans and trans-trans forms by 17 and 50 kJ mol^{-1}, respectively. The $C—O$ bonds in the gauche-gauche conformer are 0.016 Å shorter than in methanol, whereas in the gauche-trans form the difference amounts to 0.009 and 0.041 Å, respectively. The $C—O$ bonds in the methoxymethane series differ by 0.050 Å for $n = 2$, 0.036 Å, for $n = 3$ and 0.027 Å for $n = 4$. In $CH(OCH_3)_3$[36] and $C(OCH_3)_4$,[38] the terminal bond lengths are similar to those in dimethyl ether.

In $(CH_3O)_2C(CH_3)_2$[35] and $(CH_3O)_3CCH_3$,[37] molecular conformations are identical to the structures of di- and trimethoxymethanes. In the $CH_3—O—C_0—O—CH_3$ gauche-gauche chain of the former compound the $O—C_0—O$ angle amounts to 117.4(2)° and the $C—O—C—O$ torsional angle is 52.0(1.2)°. The $O—C_0—O$ angle in the gauche-gauche chain of the latter compound is 110.8(9)°, and in the gauche-trans chains it is 106.7(9)° and 108.5(9)°, respectively. One of the $CH_3—C_0—O—CH_3$ torsional angles amounts to 109(2)°, the other two torsion angles being 59(2)°. In the $C—C_0—O—C$ gauche-trans chain $\angle C—C_0—O$ is 106.6(9)°, and in the gauche-gauche chains it is 111.1(9) and 113.1(9)°, respectively.

Several investigations are concerned with the structures of unsaturated ethers: $CH_3OCH{=}CH_2$,[30] $CH_3OC(CH_3){=}CH_2$,[31] and $CH_3OCH{=}C{=}CH_2$,[32] methyl vinyl ether having been studied three times. The last combined ED/MW study[30] with *ab initio* calculations has shown the molecule to exist in cis conformation; only a small fraction exists in the skew form. A similar structure has been observed for methyl isopropenyl and methyl allenyl ethers. Table 3-4 indicates that the geometrical parameters of unsaturated ethers do not differ much (the parameters of vinyl alcohol are also included for comparison). Unlike methyl vinyl ether the corresponding sulfide, $CH_3SCH{=}CH_2$,[50] exists as a mixture of 38% cis and 62% gauche conformer in the gas phase, the latter having the $C—S—C{=}C$ torsional angle 135.8(6.5)°. The bond lengths and valence angles of methyl vinyl sulfide are as follows: $S—CH_3$, 1.795(8) Å; $S—CH$, 1.759(8) Å; $\angle C—S—C$, 102.1(5)°.

Anisole, $CH_3OC_6H_5$[39] and p-methoxybenzaldehyde[40] have planar molecular skeletons with the $C—O—C$ valence angle 120° (Table 3-3). Due to its internal rotation about the $C—O$ bond by 40°, anisole is no longer planar at higher temperatures (200°C). A nonplanar structure has also been found for its analogs thioanisoles and selenoanisoles,[51,52] with $\varphi_{X—C}$ of 45.3(2.8)° and 40(13)°, respectively. The $C—X—C$ valence angles amount to 105.6(7)° for the sulfur and 99.6(2.5)° for the selene derivative. The $C—O$ bonds in anisole differ by 0.06 Å; in thioanisole this difference is 0.054 Å (the $C—S$ bonds are 1.749 and

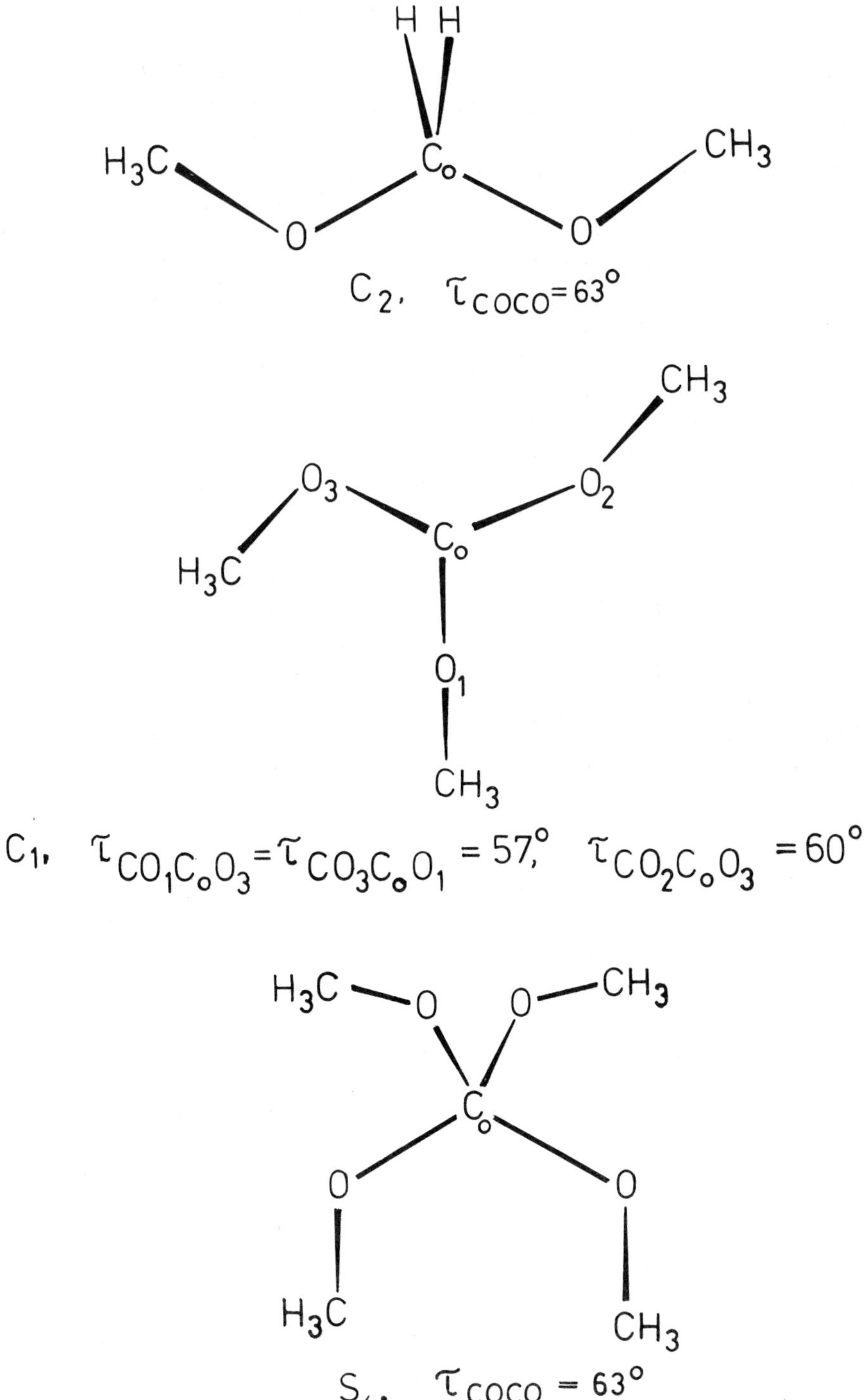

Figure 3-3. Conformation and geometrical parameters of the $CH_2(OCH_3)_2$, $CH(OCH_3)_3$, and $C(OCH_3)_4$ molecules.

Table 3-4. GEOMETRICAL PARAMETERS OF UNSATURATED ETHERS

Compound	Bond lengths (Å)			Bond angles (deg)	
	$O—CH_3$	$O—C$	$C=C$	$\angle C—O—C$	$\angle C=C—O$
$CH_3OCH=CH_2$	1.427(7)	1.359(15)	1.337(20)	116.8(1.8)	127.3(1.8)
$CH_3OC(CH_3)=CH_2$	1.416(5)	1.353(5)	1.330(7)	116.0(1.1)	125.8(7)
$CH_3OCH=C=CH_2$	1.427(8)	1.375(7)	1.318(4)	115.0(1.2)	125.3(1.2)
$CH_2=CHOH$	—	1.372	1.326	—	126.2

1.803 Å), and in selenoanisole 0.045 Å (C—Se bonds amount to 1.912 and 1.957 Å). Anisole is likely the only example of a planar structure in the gas phase at 55°C in this group of compounds.

In silyl phenyl ether[53] the torsional angle about the C—O bond is 68(3)°. The principal molecular parameters are as follows: C—O, 1.357(9) Å; Si—O, 1.648(7) Å; and $\angle$ C—O—Si 121(1)°.

The structures of diphenyl ether[54] $[(C_6H_5)_2O]$, and diphenyl sulfide[55] $[(C_6H_5)_2S]$ have been determined by gas-phase ED. In the oxide the benzene rings are twisted by 29° and 70° from the planar configuration. The C—O—C valence angle is 117.5(2.5)°, and C—O and C—C bond distances are 1.385(30) and 1.398(7) Å, respectively. If oxygen lone electron pairs are assumed to occupy p orbitals, the observed conformation favors p-π interaction. For the sulfide molecule two models with C_2 symmetry and torsional angles of 43 and 56° fit the experiment best. Di-2-pyridyl sulfide[56] has been shown to have C_1 symmetry with torsional angles about the C—S bonds 23 and 67°.

Silyl ethers that have been widely studied are listed in Table 3-5. In saturated methyl silyl ethers, except for the case of CH_3OSiF_3 the Si—O bond length varies between 1.632 and 1.640 Å, the average value being 1.638(4) Å. The C—O average distance is 1.421(3) Å; both average values are approximately equal to those in $(CH_3)_2O$ and $(SiH_3)_2O$. The opening of the C—O—Si angle up to 121–124° in saturated methyl silyl ethers is likely to be caused by the interaction between the oxygen lone electron pairs and the vacant silicon d orbitals.

The molecular structure of $(CH_3)_nSi(OCH_3)_{4-n}(n = 0, 1, 2, 3)$ series is identical to that of the corresponding methane derivatives mentioned above. 1,1-Dimethoxysilacyclohexane[71] **(16)** is the structural analog of

16

Table 3-5. GEOMETRICAL PARAMETERS OF ETHERS WITH Si—O AND Ge—O BONDS

Compound	Bond lengths (Å)		$\angle$C—O—Si (Si—O—Si) (deg)	Ref.
	Si—O, (Ge—O)	O—C		
CH_3OSiH_3	1.640(3)	1.418(9)	120.6(1.0)	57
CH_3OSiF_3	1.580(ass.)	1.392(15)	131.4(3.2)	58
$CH_3OSi(CH_3)_3$	1.639(4)	1.423(5)	122.5(6)	59
$(CH_3O)_2Si(CH_3)_2$	1.641(3)	1.418(3)	123.4(1.1)	60
$(CH_3O)_3SiCH_3$	1.632(4)	1.425(4)	123.6(5)	61
$Si(OCH_3)_4$	1.614(1)	1.416(2)	122.3(3)	62
$CH_2{=}CHOSi(CH_3)_3$	1.663(5)	1.355(6)	133(4)	63
$(SiH_3)_2O$	1.634(2)	—	144.1(8)	64
$CH_3SiH_2OSiH_2CH_3$	1.642(3)	—	143.0(6)	65
$H(CH_3)_2SiOSiH(CH_3)_2$	1.635(2)	—	148.4(9)	65
$(CH_3)_3SiOSi(CH_3)_3$	1.631(3)	—	148(3)	66
$F_3SiOSiF_3$	1.580(25)	—	155.7(2.0)	67
$Cl(CH_3)_2SiOSi(CH_3)_2Cl$	1.624(2)	—	154.0(1.5)	68
$Cl_3SiOSiCl_3$	1.592(10)	—	146(4)	58
$(H_3Ge)_2O$	1.766(4)	—	126.5(3)	69
$[(CH_3)_3Ge]_2O$	1.770(10)	—	141(1)	70

$(CH_3)_2Si(OCH_3)_2$. The H_3C—O—Si—O—CH_3 gauche-gauche chain is defined by the C—O—Si—O torsional angle, 60.0°. The two molecules have similar geometrical parameters. For 1,1-dimethoxysilacyclohexane they are as follows: Si—O, 1.632(2) Å; Si—C, 1.866(2) Å; C—O, 1.408(3) Å; $\angle$C—O—Si, 122.3(2)°; $\angle$O—Si—O, 123.0(4)°; $\angle$C—Si—O, 105.6(2)°. A considerable difference is only observed in the C—Si—C and O—Si—O valence angles, which are 114.5(1.6) and 110.1°, respectively, in $(CH_3)_2Si(OCH_3)_2$. Most probably this is due to the strain of the cycle, since the endocyclic angles are far from normal values for carbon.

The lengthening of the Si—O bond and the opening of the Si—O—C valence angle as compared to $CH_3OSi(CH_3)_3$ (Table 3-5) have been observed in the unsaturated ether $(CH_3)_3SiOCH{=}CH_2$.[63] In the gas phase this compound exists as a mixture of two conformers: trans and nearly cis with the Si—O—C=C torsional angle in the range of 0-40°. The presence of a mixture of two conformers has been confirmed in the liquid phase by spectroscopic methods,[72] the trans form being more stable, whereas in methyl vinyl ether the cis form is the dominant one.

In structure **17**, $[Cl(CH_3)_2Si]_2O$,[68] the $Cl(CH_3)_2Si$ groups are twisted by 78(6) and 141(19)° from the position in which Si—O and Si—Cl bonds are trans.

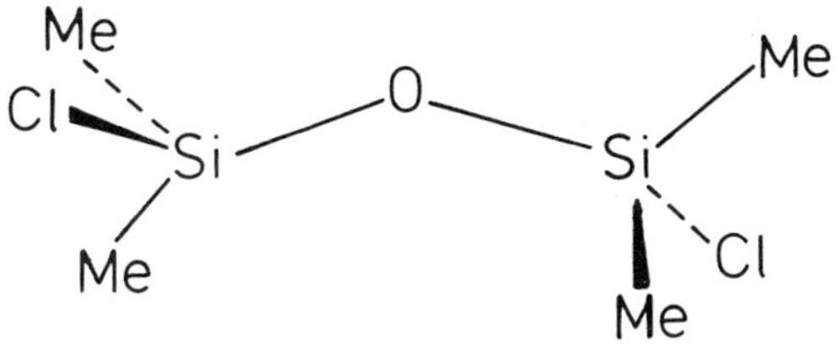

17

The $(CH_3H_2Si)_2O^{65}$ molecule (**18**) has a similar structure. The dominant conformer (64%) has the methyl groups twisted by 58(8) and 124(4)° from the trans configuration of the C—Si—O—Si fragment. As determined, the dimethylsilyl groups in $[(CH_3)_2HSi]_2O^{65}$ (**19**) were rotated by 41(4) and 101(8)° from the cis configuration of the H—Si—O—Si fragments.

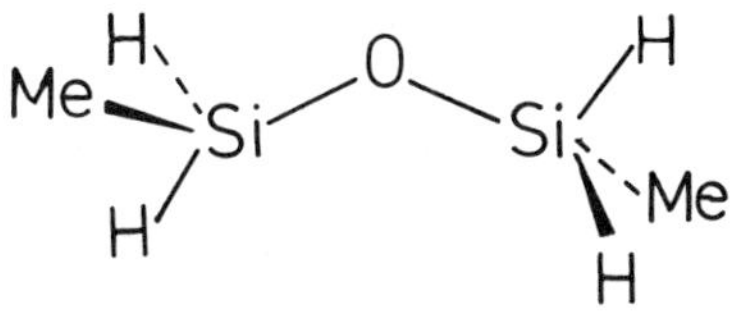

18 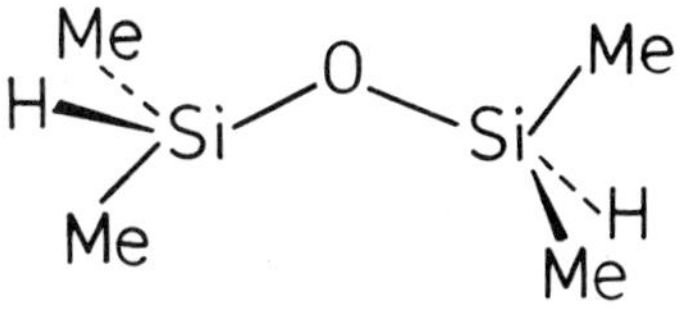   **19**

The principal geometrical parameters of sulfides are presented in Table 3-6. Some of them have been discussed earlier. In the equilibrium mixture of methyl ethyl sulfide the gauche conformer with τ(C—S—C—C) 66(9)° predominates (75%). In sulfides containing unsaturated groups (—CH=CH—,—CH=C=CH—,—C≡C—) the C—S—C valence angle is a little larger than in dimethyl sulfide, but in methyl allenyl sulfide this angle is even smaller. In 1,1-bis(methylthio)ethylene[75] a mixture of three conformers fits the experimental data best (**20**). The gauche orientation of the methyl groups is

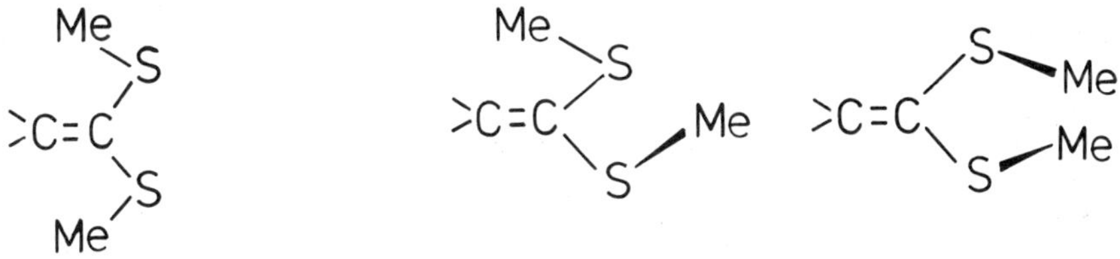

cis-cis   cis-gauche   gauche-gauche

20

defined by the C=C—S—C angle, 122(3)°. The composition of the mixture has not yet been determined, but the content of cis–cis and gauche–gauche forms must be the same.

Table 3-6. GEOMETRICAL PARAMETERS OF SULFIDES

Compound	C—S (Å)	∠C—S—C (deg)	Other parameters (Å or deg)	Ref.
$(CH_3)_2S$	1.805(2)	99.1(4)		45
$CH_3SC_2H_5$				
S—CH_3	1.806(27)	97.1(1.1)	C—C, 1.536(8)	73
S—CH_2	1.818(27)		∠C—C—S, 114.0(5)	
CH_3SCH_2Cl				
S—CH_3	1.827(5)	99.9(1.2)	C—Cl, 1.766	47
S—CH_2	1.785(5)		∠S—C—Cl, 114.6(8)	
$(CF_3)_2S$	1.819(3)	97.3(8)	C—F, 1.330(2)	46
			∠F—C—F, 108.1(2)	
$CH_3SCH{=}CH_2$				
S—CH_3	1.795(8)	102.1(5)	C=C, 1.343(1)	50
S—CH	1.759(8)		∠C=C—S, 127.5(7)	
$CH_3SCH{=}C{=}CH_2$				
S—CH_3	1.800(10)	98.1(8)	$H_2C{=}C$, 1.282	74
S—CH	1.745(10)		C=CH, 1.327	
			∠S—C—C, 125.4	
$(CH_3S)_2C{=}CH_2$				
S—CH_3	1.815(5)	102.1(2.5)	∠S—C—C, 122.4(1.0)	75
S—C=C	1.767(5)			
$CH_3SC{\equiv}CSCH_3$				
S—CH_3	1.806(2)	102.7(2)		76
$C_6H_5SCH_3$				
S—CH_3	1.803(4)	105.6(7)		51
S—Ph	1.749(4)			
$C_6H_5SCH{=}CH_2$	1.749(5)	109.2(1.8)		77
$(C_6H_5)_2S$	1.771(5)	103.7(1.3)		55
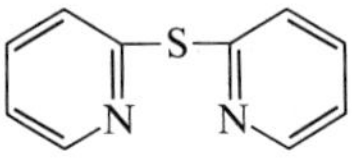 di-2-pyridyl sulfide	1.786(4)	104.4(1.7)		56
CH_3SPF_2				
S—CH_3	1.822(5)	102.0(1.2)	P—S, 2.085	78
			P—F, 1.589	
			∠S—P—F, 101.2	
$(PF_2)_2S$	2.132(4)	91.3(1.1)	P—F, 1.572	78
			∠S—P—F, 100.2(4)	
$(SiH_3)_2S$	2.136(2)	97.4(7)		79
$(GeH_3)_2S$	2.209(4)	98.4(1)		69
F_2PSGeH_3				
S—P	2.115(8)	99.0(6)	∠F—P—F, 97.0	80
S—Ge	2.256(4)		∠S—P—F, 99.9	

Methyl allenyl sulfide[74] prefers a structure with cis orientation of the CH_3 group and the unsaturated fragment. Approximately 28-37% of the gauche conformer may be present in the mixture.

As mentioned above, dimethyl ether forms stable complexes. Dimethyl sulfide may form a complex with trimethylaluminum[81] (**21**). The geometry of $(CH_3)_2S$ does not differ much from that of the free molecule: C—S, 1.817(5) Å; $\angle$ C—S—C, 99(3)°. The AlC_3 fragment is slightly pyramidal with the C—Al—C angle 118.9(3)°. The angle between the Al—S bond and the C—S—C plane is 31(5)°. The molecular symmetry is C_{2v}. Sulfides containing

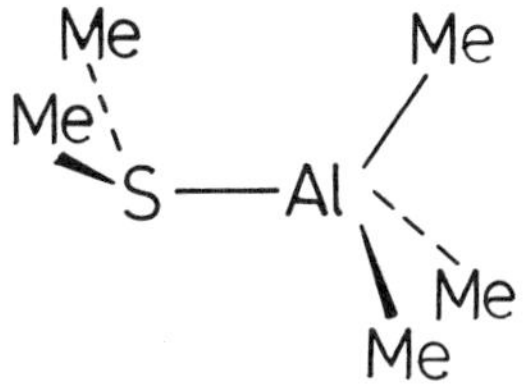

21

C—S, Si—S, and Ge—S bonds are included in Table 3-6. The $S(PF_2)_2$ molecule has an approximate C_{2v} symmetry, and the mean amplitude of vibration about the S—P bond is 22(2)°. Three models of CH_3SPF_2[78] provide agreement with the experimental data (**22**). The authors prefer the model with $\tau = 106°$: ie, with the P—F and C—S bonds mutually eclipsed. The F_2PSGeH_3[80] molecule adopts the distorted cis conformation with $\tau = 162°$ that is identical to the third model of compound **22**.

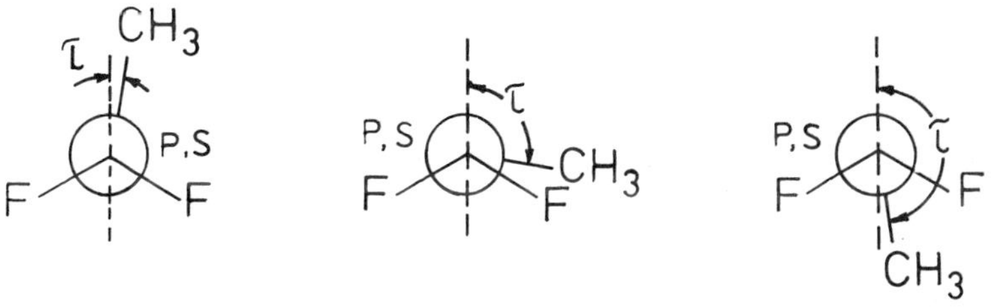

I $\tau = 19°$ II $\tau = 106°$ III $\tau = 171°$

22

Relatively few structural studies of selenides and tellurides have been reported: $Se(CF_3)_2$,[82] $Se(SiH_3)_2$,[83] $Se(GeH_3)_2$,[84] and $Te(CH_3)_2$.[85] The Se—C, Se—Si, and Se—Ge bonds are 1.978(9), 2.273(6), and 2.344(3) Å, respectively, and valence angles at the central atom are 95-96°. In $Te(CH_3)_2$ the Te-C bond length is 2.142 Å, and the C—Te—C valence angle is 94(2)°. It can be seen that in the series XMe_2 (X = O, S, Te) the C—X—C valence angle regularly decreases from 112° to 94°. Unfortunately, there are no corresponding data for dimethyl selenide. The C—Se—C valence angle is likely to be 94°. In H_2Se[86] the H—Se—H angle is 91°.

Hypohalides and Related Compounds

The oxygen compounds of the hypohalide class are less stable than sulfenyl halides. The simplest member of this class is trifluoromethylhypofluorite[87] (**23**). This molecule possesses C_s symmetry with the trifluoromethyl group tilted from the C—O bond by δ 4.1(8)°:

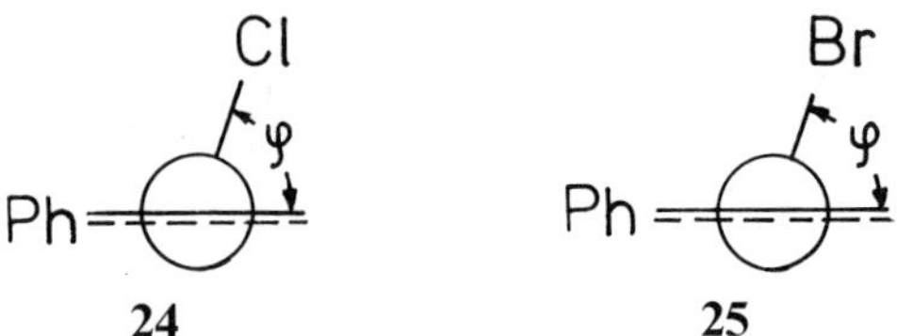

23

All the sulfenyl halides studied (CH_3SCl,[88] CF_3SF,[46] CF_3SCl,[47] CCl_3SCl[89]) possess similar structures with the principal parameters given in Table 3-7.

Electron diffraction studies[90,91] of phenyl sulfenyl chloride (**24**) and phenyl selenyl bromide (**25**) have shown these molecules to possess a nonplanar structure with torsional angles φ of 71.8(3.4) and 68.4(2.4)°, respectively:

24   **25**

The shortening of the C—S bond to 1.764(12) Å and the increase of the C—S—Cl angle up to 102.6(1.6)° in C_6H_5SCl as compared with CH_3SCl may be related to differences in carbon hybridizations in these compounds. The S—Cl bond length in PhSCl (see Table 3-8) is nearly the same as in methyl sulfenyl chloride.

Substitution of ortho hydrogen atoms in the phenyl rings of C_6H_5SCl and C_6H_5SeCl by a nitro group produces changes in the conformations about the

Table 3-7. GEOMETRICAL PARAMETERS OF SULFENYL HALIDES

Compound[a]	Bond lengths (Å)		Angles (deg)	
	C—S	S—X	∠C—S—X	δ
CF_3SF	1.805(9)	1.605(3)	97.1(2)	5.1(3)
CF_3SCl	1.824(6)	2.015(3)	98.9(4)	4.8(6)
CH_3SCl	1.787(3)	2.037(2)	99.4	3.7(2.0)

[a] Data for CCl_3SCl are less accurate and are not listed.

Table 3-8. GEOMETRICAL PARAMETERS OF PHENYL SULFENYL CHLORIDE, PHENYL SELENYL BROMIDE, AND THEIR ORTHO-NITRO DERIVATIVES

Parameter	o-NO$_2$C$_6$H$_4$SCl[a]	o-NO$_2$C$_6$H$_4$SeBr[b]	PhSCl[c]	PhSeBr[d]
C—C, Å	1.397(3)	1.396(6)		
C—N, Å	1.459(9)	1.466(24)	—	—
N=O, Å	1.229(3)	1.229(8)	—	—
C—S (Se), Å	1.776(5)	1.917(11)	1.764(12)	1.899(6)
S—Cl (Se—Br), Å	2.046(4)	2.354(3)	2.051(6)	2.325(3)
∠C—S—Cl (CSeBr), deg	102.3(1.4)	98.5(7)	102.6(1.6)	99.8(1.2)
∠C—N=O, deg	115.5 119.4(2.8)	117.6(6)$_{avg}$	—	—
∠N—C—CS (Se), deg	118.5(2.9)	118.5(6)	—	—
∠S(Se)—C—C(N), deg	118.5(1.6)	117.8(5)	—	—
O...S(Se), Å	2.41(1)	2.36(3)	—	—

[a] Reference 92.
[b] Reference 93.
[c] Reference 90.
[d] Reference 91.

C—S and C—Se bonds. In nitroderivatives, the halogen atom and the nitro group are in the aryl ring plane (**26**). The existing O ... S(Se) interaction in both molecules leads to a decrease in the exocyclic angles at C(S, Se) and C(N) carbon atoms (Table 3-8).

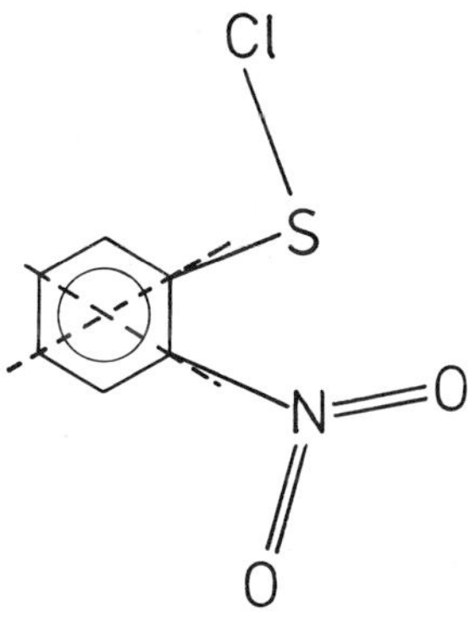

Cl

S

N=O

O

26

Esters and Their Analogs

Most of the esters listed in Table 3-9 have been studied by gas-phase ED. Both the ED[94] and MW[101] studies determined that methyl formate has a planar skeleton with cis-oriented C—O and C=O bonds. Methyl chloroformate, CH$_3$OC(O)Cl,[96] has a similar structure, but at 20°C about 5% of the trans conformer may also be present. The amount of trans conformer increases to

Table 3-9. GEOMETRICAL PARAMETERS OF R—O—C(O)H, R—O—C(S)H ESTERS AND THEIR ANALOGS

	Bond lengths (Å)			Bond angles (deg)		
	$C=O$		$>C-O$	$\angle O-C=O$	$\angle C-O-C$	
Compound	(C=S)	=C—O	(Si—O)	(O—C=S)	(Si—O—C)	Ref.
CH$_3$OCHO	1.206(5)	1.341(7)	1.455(5)	126.8(1.6)	114.3(1.6)	94
CH$_3$OCHO	1.202(2)	1.340(2)	1.435(3)	125.4(5)	115.9(5)	95
CH$_3$OC(O)Cl	1.190(4)	1.325(6)	1.443(7)	128.1(6)	114.4(1.7)	96
CH$_3$OC(S)H	1.612(3)	1.369(3)	1.369(3)	126.6(5)	115.5(6)	97
SiH$_3$OC(O)H	1.209(7)	1.351(6)	1.695(3)	123.5(5)	116.8(5)	98
SiH$_3$OC(O)CH$_3$	1.214(4)	1.358(4)	1.685(3)	121.7(8)	116.5(3)	99
SiH$_3$OC(S)CH$_3$	1.615(8)	1.345(7)	1.717(6)	127(2)	118(2)	100

10–20% at 200°C. The presence of chlorine in methyl chloroformate shortens the C=O bond and increases the O—C=O valence angle up to 128°.

Small deviations from planarity have been observed in CH$_3$OC(S)H [τ_{C-O} torsion 15.8(2.5)°] in SiH$_3$OCHO (21°), SiH$_3$OC(O)CH$_3$ (10°), and SiH$_3$OC(S)CH$_3$ (10°). For the two latter compounds a minor degree of nonplanarity also has been proved by X-ray crystallography.[99,100]

Among the thioformates, the structure of CH$_3$C(O)SGeH$_3$ has been determined.[102] The molecule adopts the planar configuration of C=O and S—Ge bonds, and the O—C—S—Ge torsional angle is only 5°. The molecular parameters are as follows: C—C, 1.493(10) Å; C=O, 1.224(8) Å; C—S, 1.765(7) Å; Ge—S, 2.233(4) Å; $\angle$C—C=O, 116.4(1.3)°; $\angle$O=C—S, 124.1(1.0)°; $\angle$C—S—Ge, 96.7(4)°.

In methyl esters of carboxylic and thiocarboxylic acids the heavy-atom skeleton is planar, with cis orientation of methyl groups and double bonds (**27**).

27

Shortening of the C=S bond in (CH$_3$S)$_2$C=S should be noted (Table 3-10). The remaining geometrical parameters are common. Thus, esters are characterized by a planar molecular structure with cis-oriented double (C=O or C=S) and single C—X (X = O, S) bonds.

In connection with the results discussed above, it is useful to consider the structure of anhydrides of organic acids. In the simplest member of this class,

Table 3-10. Geometrical Parameters of $(CH_3X)_2C=X$ ($X = O, S$) Esters

Compound	Bond lengths (Å)			Bond angles (deg)		Ref.
	C=X	C—X	X—CH$_3$	∠X=C—X	∠C—X—C	
$(CH_3O)_2C=O$	1.203(9)	1.343(10)	1.423(10)	122.8	114.5(4)	103
$(CH_3S)_2C=O$	1.206(4)	1.777(3)	1.802(4)	124.9(2)	99.3(4)	104
$(CH_3S)_2C=S$	1.634(5)	1.752(4)	1.800(6)	126.2(2)	103.3(4)	105

formic anhydride, one of the carbonyl groups is cis and the other is trans to the C—O bonds with a slight deviation from the C—O—C plane (Table 3-11). The O...H distance is 2.5 Å.

In formic acetic anhydride, HC(O)OC(O)CH$_3$,[107] the formyl and acetyl groups are twisted respectively by 21 and 46° from the C—O—C plane; the dihedral angle between the O=CH—O and O=C(CH$_3$)—O planes is 126°. Another conformation has been observed in acetic anhydride. Acetyl groups are rotated by 46° from the position with cis orientation of the double bond. Hexafluoroacetic anhydride, CF$_3$(O)COC(O)CF$_3$,[109] has a similar structure (cf. Table 3-11).

Aldehydes

The principal parameters of the aldehydes studied are presented in Table 3-12. The general property of all the aldehydes is the syn-eclipsed orientation of the single bond and the double bond. Thus, in acetaldehyde[111] the C=O and methyl C—H bonds are in the cis position, in propionaldehyde[113] **(28)** the dominant conformer is characterized by the syn-eclipsed configuration of the

Table 3-11. Geometrical Parameters of X—C(O)—O—C(O)—X′ Anhydrides

Parameter	X = X′ = H[a]	X = H, X′ = CH$_3$[b]	X = X′ = CH$_3$[c]	X = X′ = CF$_3$[d]
C—O, Å	1.384(3)	1.397(1)	1.405(1)	1.360(19)
C=O, Å	1.188(2)	1.195(1)	1.183(1)	1.203(10)
∠C—O—C, deg	115(2)	113.2(6)	115.8(8)	118.5(2.6)
∠O—C(CH$_3$)=O, deg	—	122.4(7)	121.7(2)	120.5(1.9)
∠O—C(H)=O, deg	123.4(5)	118.0(6)	—	—
∠O—C—C, deg	—	112.1(1.2)	108.3(2)	122.6(1.1)
τ(C—O—C—H), deg	17(6)	45.6(1.9)	—	—
τ(C—O—C=O), deg	26(9)	20.8(2.0)	48.5(8)	20.3(3.5)

[a] Reference 106.
[b] Reference 107.
[c] Reference 108.
[d] Reference 109.

Table 3-12. GEOMETRICAL PARAMETERS OF ALDEHYDES

Compound	C=O (Å)	$\angle$C—C=O (deg)	Other parameters (Å or deg)	Ref.
HC(O)H	1.209(3)	125.8 $\angle$HCO	C—H, 1.097	110
$CH_3C(O)H$	1.207(4)	124.2(5)	C—C, 1.512	111
$ClCH_2C(O)H$	1.206(3)	123.3(3)	C—C, 1.521; C—Cl, 1.782 $\angle$C—C—Cl, 110.4	112
$CH_3CH_2C(O)H$	1.209(4)	124.5(3)[a] 125.1[b]	C—CH, 1.515 C—CH_3, 1.521,[a] 1.569[b] $\angle$C—C—C, 113.8,[a] 110.2[b]	113
$(CH_3)_2CHC(O)H$	1.206(2)	123.3(1.9)	C—C_{avg}, 1.528	114
$CH_2CHC(O)H$	1.209(1)	124.7(2)	C—C, 1.481; C=C, 1.340	115
$H_2NC(O)H$	1.212(3)	125.0 $\angle$HCO	N—C, 1.368	116
▷—C(O)H	1.216(2)	122.0(1.8)	C—C_{avg}, 1.507	117
(furyl)—C(O)H	1.212(4)	122.7(8)	$\angle$OC—C—O, 117.6	118
$o\text{-}ClC_6H_4C(O)H$	1.195(7)	120.0(2.5)	C—CO, 1.466, C—C_{ring}, 1.396	119
$m\text{-}ClC_6H_4C(O)H$	1.220(5)		C—CO, 1.512; C—Cl, 1.732 C—C_{ring}, 1.399	120
H(O)C—C(O)H	1.208(2)	121.6(3)		121
H(O)CCH=CHC(O)H	1.207(2)	123.6(5)	C—C, 1.479; C=C, 1.336 $\angle$C—C=C, 122.7	122
$p\text{-}H(O)C\text{-}C_6H_4\text{-}OH$	1.208(3)	123.8(3)	C—CO, 1.487 C—C_{ring}, 1.401	123

[a] Syn conformer.
[b] Skew conformer.

C=O and C—CH_3 bonds. In the remaining conformer (19%) the C=O and C—H bonds are in eclipsed form with the O=C—C—C torsional angle 123.7°.

28

29

In isobutylaldehyde, $(CH_3)_2CHCHO$[114] (**29**) trans conformer has been reported to be the major component (90%). In chloroacetaldehyde, $ClCH_2CHO$[112] (**30**), and acrolein, CH_2=CHCHO[115] (**31**), the dominant conformers are characterized by the trans configuration of the C—Cl and C=O bonds in the former compound and the C=O and C=C bonds in the latter.

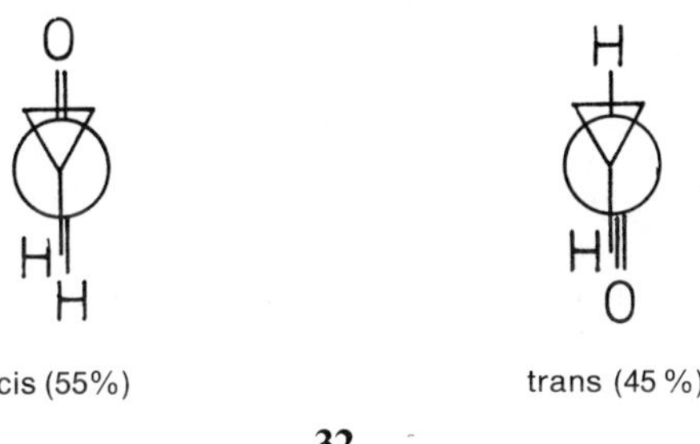

30   **31**

On the potential curve of chloroacetaldehyde there is a wide minimum with a small maximum at 180°. The O...Cl nonbonded distance (3.868 Å) corresponds to the effective torsional angle O=C—C—Cl approximately 170°. The trans configuration of double bonds also has been observed in fumaric aldehyde, HC(O)CH=CHC(O)H, but in 2-formylfuran the C=O and C=C bonds adopt a cis configuration, and the trans conformer constitutes only 31%. The greater stability of the cis conformer apparently is connected with the O...H interaction between the hydrogen atom of the aldehyde fragment and the furan oxygen. Similar effects are also observed in cyclopropylcarboxaldehyde[117] (**32**).

cis (55%)   trans (45%)

32

In the above-mentioned aldehydes the C=O bond length varies from 1.206 to 1.209 Å (Table 3-12). The shortest bonds 1.188(4) and 1.1717(13) Å have been observed in HC(O)F[124] and F_2CO,[125] respectively. The NH_2 group with its strong electron-donating ability does not change the C=O bond length. In HC(O)NH_2[116] it amounts to 1.212(3) Å.

Several studies have dealt with aromatic aldehydes. In *o*-chloro-benzaldehyde[119] (**33**) the trans conformer has been reported to be the major component due to possible H...Cl interaction. Sixty four percent of the trans

cis (20%)   trans (80%)

33

conformer is present[120] in the meta derivative. There is a significant difference between the C=O and C—CO bonds in these molecules.

Ketones

In acetone,[126] as well as in acetaldehyde, the C=O and C—H bonds are syn eclipsed. The dominant conformer (95%) of methyl ethyl ketone[127] possesses eclipsed C=O and C—CH$_3$ bonds. The C—C—C—C torsional angle in the gauche conformer is 70°. The structure of this ketone is similar to that of propionaldehyde.

Acetylacetone may exist in two tautomeric forms that are in equilibrium (**34**).

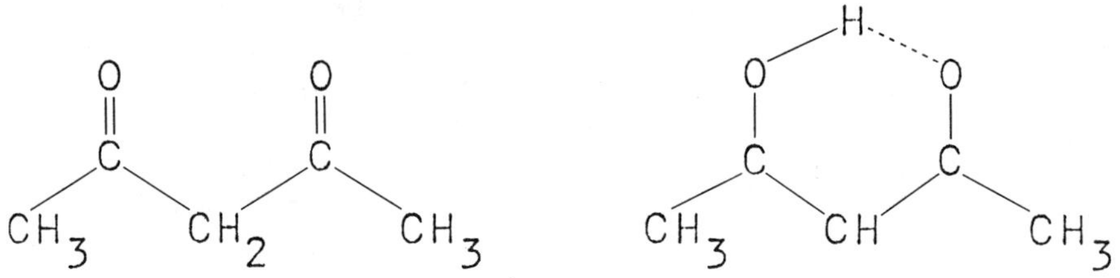

keto form   enol form

34

As has been shown by an electron diffraction study,[128] 66(5)% of the enol form is present at 105°C. The geometrical parameters of the keto form are listed in Table 3-13. Those of the enol form are somewhat different: C=O, 1.315(7) Å; HC—C, 1.416(10) Å; C—CH$_3$, 1.497(10) Å; ∠C—C—CH$_3$, 120.0°, ∠O=C—CH, 120.0°, ∠C—CH—C 118.0°. The extremely long C=O bond and large C—CH—C angle are characteristic of the enol form. At room temperature[129] the enol form is the major component (97%). This form is also more stable for fluoroderivatives of acetylacetone, CH$_3$C(O)CH$_2$C(O)CF$_3$,[129] and CF$_3$C(O)CH$_2$C(O)CF$_3$,[109] which have similar structure. The C—CH$_2$ and C—CF$_2$ end fragments contain torsional angles of about 40°. In biacetyl, CH$_3$C(O)C(O)CH$_3$,[130] the C=O bonds are mutually trans, and the O=C—C—H torsional angles are 0°.

Diacetamide,[133] the structural analog of acetylacetone, possesses its C=O bonds in trans configuration (**35**). The heavy-atom skeleton is not planar; the

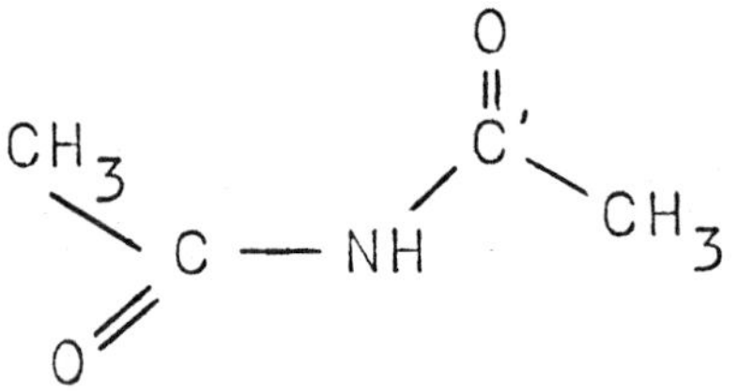

35

Table 3-13. GEOMETRICAL PARAMETERS OF KETONES

Compound	C=O (Å)	∠C—C=O (deg)	Other parameters (Å or deg)	Ref.
$CH_3C(O)CH_3$	1.210(3)	122.0(3)	C—C, 1.517	126
$CH_3C(O)CH_2CH_3$	1.218(1)	121.3(7)[a] 122.5(9)[b]	C—CO, 1.518 C—CH$_3$, 1.531	127
$CH_3C(O)CH_2C(O)CH_3$	1.225(10)	120.0(1.8)[a] 115.0(1.8)[b]	C—C, 1.540 ∠OC—C—CO, 114.0	128
$CF_3C(O)CH_2C(O)CH_3$	1.270(8)	123.6(1.7)	C—CH$_3$, 1.511 C—CF$_3$, 1.536 C—CO, 1.416	129
$CF_3C(O)CH_2C(O)CF_3$	1.259(18)	126.4(1.3)	C—CH, 1.407 C—CF, 1.546 ∠OC—C—CO, 115.2	109
$CH_3C(O)CH=CHC(O)CH_3$	1.212(1)	121.0(3)	C—C, 1.489 C=C, 1.331 ∠C=C—C, 123.1	122
$CH_3C(O)C(O)CH_3$	1.214(1)	120.3(7)	OC—CO, 1.507 C—CH, 1.527 ∠C—C—C, 116.3	150
$CH_3C(O)CF_3$	1.207(6)	122.0(1.1)[b] 116.8(7)[c]	C—CH, 1.482 C—CF, 1.562	129
$CF_3C(O)CF_3$	1.246(14)	119.3(4)	C—C, 1.549	131
$CCl_3C(O)CCl_3$	1.184(13)[d] 1.200(14)[e]	118.9(7)[d] 118.3(9)[e]	C—C, 1.574[d] C—C, 1.595[e]	132

[a] ∠O=C—CH$_2$.
[b] ∠O=C—CH$_3$.
[c] ∠O=C—CF$_3$.
[d] $T_1 = 65°C$.
[e] $T_2 = 300°C$.

C—N—C=O torsional angle is 12.8(16.0)° and the C′—N—C=O torsion angle is 156.7(8.6)°. The geometrical parameters are as follows: C—N, 1.402(2) Å; C=O, 1.210(2) Å; C—C, 1.518(3) Å; ∠C—N—C, 129.2(1.5)°; ∠O=C—C, 123.3(1.5)°; ∠N—C′=O 123.7(1.5)°; ∠N—C=O, 118.8(1.8)°. The dihedral angle between acetyl groups is 36.1°.

It should be noted that the C=O bond is elongated in fluoroderivatives by about 0.05 Å compared to that in acetone (Table 3-13).

Carboxylic Acids

Under normal conditions carboxylic acids form dimers due to intermolecular hydrogen bonding. The monomer ⇌ dimer equilibrium shifts to the left as the temperature increases. In formic, acetic, and other acids studied by gas-phase electron diffraction, the influence of hydrogen bonding on the geometrical parameters of the dimers is obvious. The C—O bonds in dimers are extremely shortened, whereas smaller changes occur in double bonds (see Table 3-14). At 175°C formic acid begins to decompose.[137] The energy of dimer dissociation

Table 3-14. Geometrical Parameters of Carboxylic Acids

Parameter	Formic acid[a]		Acetic acid[b]		Propionic acid[c]	
	HCOOH	$(HCOOH)_2$	CH_3COOH	$(CH_3COOH)_2$	CH_3CH_2COOH	$(CH_3CH_2COOH)_2$
$C{=}O$, Å	1.217(3)	1.220(3)	1.214(3)	1.231(3)	1.211(3)	1.232(6)
$C{-}O$, Å	1.361(3)	1.323(3)	1.364(3)	1.334(4)	1.367(4)	1.329(8)
$O{-}H$, Å	0.984(29)	1.036(17)	—	—	—	—
$C{-}CH_2$, Å	—	—	—	—	1.518(10)	1.518(15)
$C{-}CH_3$, Å	—	—	1.520	1.506(5)	1.543(10)	1.547(15)
$O{-}H{...}O$, Å	—	2.703(7)	—	2.68(1)	—	2.711(14)
$\angle O{-}C{=}O$, deg	123.4(5)	126.2(5)	122.8(5)	123.4(5)	122.1(8)	123.7(1.7)
$\angle C{-}C{-}O$, deg	—	—	110.6(6)	113.0(8)	111.2(8)	113.4(1.6)

[a] Reference 134.
[b] Reference 135.
[c] Reference 136.

amounts to 51.9(4.2) kJ mol^{-1} for HCOOH and 49.0(3.8) kJ mol^{-1} for HCOOD. Bastiansen and co-workers[138] analyzed the geometrical parameters of two isotopic species of formic acid. The C—O bond, 1.332(3) Å, in $(HCOOD)_2$ is somewhat longer than the corresponding C—O bond in $(HCOOH)_2$. The other parameters are the same within their respective limits of error. The monomer of acetic acid prefers the same conformation as that of acetaldehyde. In propionic acid[136] 50% of the molecules contain the C—CH$_3$ bond cis to the double bond, 10% of the molecules adopt the trans conformation, and the remainder adopt a nonplanar structure.

In the haloderivatives of acetic acid (Table 3-15), question of rotational isomerism about the C—C bond must be considered. In CF$_3$COOH[142] the CF$_3$ fragment is twisted by 17.3(9)° from the eclipsed position of the C—F and C=O bonds. Monofluoroacetic acid[139] at 112°C exists as a mixture of 64% cis and 36% trans forms, with the O—H bond directed toward the fluorine atom in the latter. The trans form favors O—H...F hydrogen bonding. The cis configuration is more stable than the trans form by 2.5(1.3) kJ mol^{-1}. As determined, monochloroacetic acid[140] is a mixture of three conformers (**36**) with τ(O=C—C—Cl) 10° (56%), 79° (30%), and 131° (14%):

$$\tau = 10° \qquad \tau = 79° \qquad \tau = 131°$$

36

Table 3-15. GEOMETRICAL PARAMETERS OF THE HALODERIVATIVES OF ACETIC ACID

Parameter	XCH$_2$COOH		F$_2$CHCOOH[c]	F$_3$CCOOH[d]
	X = F[a]	X = Cl[b]		
C=O, Å	1.200(4)	1.223(4)	1.212(4)	1.195(3)
C—O, Å	1.344(10)	1.352(5)	1.345(9)	1.353(14)
C—C, Å	1.534(8)	1.508(6)	1.517(6)	1.546(5)
C—F(Cl), Å	1.387(10)	1.778(5)	1.354(7)	1.325(3)
O—H, Å	0.971(7)	0.97	0.96(2)	0.96
∠C—C=O, deg	126.1(7)	126.1(5)	123.9(1.0)	126.8(8)
∠C—C—O, deg	108.8(4)	110.6(4)	110.6(1.0)	111.1(9)
∠C—C—F(Cl), deg	109.2(1.0)	112.5(4)	108.7(7)	109.5(3)
∠C—O—H, deg	105.9(5)	105.8(1.1)	107[e]	107[e]

[a] Reference 139.
[b] Reference 140.
[c] Reference 141.
[d] Reference 142.
[e] Assumed.

Difluoroacetic acid[141] exists as a mixture of two conformers with H—C—C=O torsion angles equal to 82.5°(74%) and 18° (26%). Both forms favor O—H...F hydrogen bonding.

Acyl Halides

Acetyl halides, $CH_3C(O)X$ (Table 3-16), have a planar skeleton with cis orientation of the C=O and C—H bonds. In these compounds the C=O bond is shorter than that in acetic acid, and C—X bonds are elongated as compared with the =C—X distance in haloethylenes.

Propenoyl chloride, CH_2=CHC(O)Cl (Table 3-17) exists as a mixture of two conformers (**37**).

cis   trans
37

In the more stable trans conformer the C—Cl bond is elongated [1.804(3) Å] in comparison with the cis form, in which it is 1.772(4) Å. The C—C—Cl valence angle amounts to 116.3(8)° in the trans form and 111.8(1.5)° in the cis form.

Propionyl chloride[145] as well as the chlorine and bromine derivatives[147,149] of acetyl chloride exist as a mixture of two conformers. In the more stable conformer of the first compound the C=O and C—CH$_3$ bonds are eclipsed. The relative amount of this conformer decreases from 77% to 65% and then to 53% as the temperature increases from 20 to 115 and finally to 215°C, respectively. The gauche conformer has the C—C—C=O torsional angle 120.0(7.7)°. The same trend is observed in ClCH$_2$C(O)Cl. Seventy percent of the cis conformer with τ(Cl—C—C=O) = 0° is present at room temperature, 67% at 110°C, and 57% at 215°C. In the gauche conformer the Cl—C—C=O

Table 3-16. GEOMETRICAL PARAMETERS OF ACETYL HALIDES, $CH_3C(O)X$

	Bond lengths (Å)			Bond angles (deg)			
X	C=O	C—X	C—C	∠C—C=O	∠C—C—X	∠O=C—X	Ref.
F	1.186(2)	1.364(3)	1.507(4)	130.5	109.8(1.2)	119.7(1.0)	143
Cl	1.185(3)	1.796(2)	1.505(3)	127.2	111.6(6)	121.2(6)	144
Br	1.181(3)	1.974(3)	1.516(3)	126.7	111.0(1.5)	122.3(1.5)	144
I	1.192(21)	2.20(2)	1.512	126.1	111.3(3.3)	123(3)	143

Table 3-17. GEOMETRICAL PARAMETERS OF ACYL HALIDES AND THEIR DERIVATIVES

| Compound | Bond lengths (Å) | | Bond angles (deg) | | Ref. |
	C=O	=C—X	∠C—C=O	∠C—C—X[a]	
$CH_3CH_2C(O)Cl$	1.184(5)	1.800(6)	127.1(6)	112.0(3)	145
CH_2=CHC(O)Cl	1.192(2)	1.804(3)[b]	125.2(2)	116.3(8)[b]	146
		1.772(4)[c]		111.8(1.5)[c]	
$ClCH_2C(O)Cl$	1.182(4)	1.777(3)	126.9(7)	110.7(1.0)	147
$Cl_2CHC(O)Cl$	1.189(3)	1.752(9)	123.3(1.3)	113.9(1.8)	148
$BrCH_2C(O)Cl$	1.188(9)	1.789(11)	127.6(1.3)	111.3(1.1)	149
$Br_2CHC(O)Br$	1.175(13)	1.987(20)	129.4(1.7)	110.7(1.5)	149
$CF_3C(O)F$	1.158(7)	1.324(2)	129(2)	109.5(5)	150
$CF_3C(O)Cl$	1.18(2)	1.750(5)	126(3)	109(3)	151

[a] X = F, Cl, Br.
[b] Trans conformer.
[c] Cis conformer.

torsional angles are 116.4(7.7)°, 120.2(9.6)°, and 121.5(6.5)° at the temperatures mentioned above. In $Cl_2CHC(O)Cl^{148}$ the form with τ(H—C—C=O) 0° is the major component (72%). In the gauche conformer the torsional angle amounts to 138.2(5) and 136.2(10.6)°, respectively, at 20 and 119°C. The conformations of $BrCH_2C(O)Cl$ and $BrCH_2C(O)Br$ resemble those of $ClCH_2C(O)Cl$, but the cis form is less stable (47% in the first compound and 62% in the latter).

Cis conformers predominate (85 and 80%, respectively) in cyclopropylcarbonyl chloride,[152] ▷—C(O)Cl, as well as in cyclopropyl methyl ketone,[152] ▷—C(O)CH₃ (**38**).

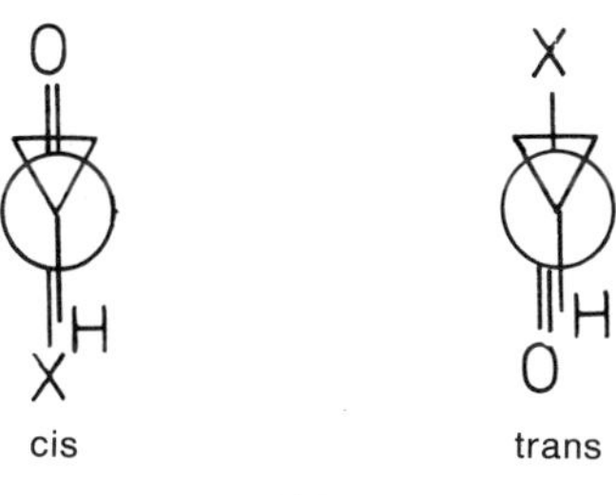

38

At 0, 80, and 190°C oxalyl chloride exists as a mixture of trans and gauche conformers.[153] As the temperature increases, the relative amount of trans conformer decreases (68, 51, and 42%, respectively). In the gauche conformer the O=C—C—O angle amounts to 122–128°. In oxalyl chloride the C—Cl bond is significantly shorter than that in acyl halides. The C=O bond is shortened to 1.182(2) Å in comparison with oxalic acid,[154] in which the geometrical parameters are as follows: C=O, 1.208(1) Å; C—O, 1.339(2) Å; ∠O=C—C, 125.0(2)°. In oxalyl chloride the O=C—C valence angle amounts to 124.3(3)°. Oxalic acid displays a unique trans conformation, which is stabilized by intramolecular O—H...O hydrogen bonding.

Peroxides and Their Analogs

The results of studies concerned in the structure of highly reactive peroxide compounds have only recently appeared. Two sets of parameters were obtained for CF_3OOF (Table 3-18). The main ambiguity concerns the O—O bond length. *Ab initio* calculations for CH_3OOCH_3 and CF_3OOCF_3 were carried out by Gase and Boggs.[159] The agreement between the experimental and theoretical parameters can be checked. The latter are as follows:

Compound	O—O (Å)	C—O (Å)	∠C—O—O (deg)	∠C—O—O—C (deg)
CH_3OOCH_3	1.411	1.422	105.4	115.5
CF_3OOCF_3	1.408	1.368	105.6	123.3

In $(CH_3)_3XOOX(CH_3)_3$ (X = C, Si),[158] an increase of torsional angles is observed. Several studies deal with peroxide derivatives: disulfides $(CH_3S)_2$,[160] $(CF_3S)_2$,[161] and diselenides $(CH_3Se)_2$,[162] and $(CF_3Se)_2$.[163] In the methyl derivatives $(CH_3X)_2$ (X = S, Se), the C—X—X—C torsional angle amounts to 85.3 and 87.5°, respectively; in $(CF_3S)_2$ it is 104.4°. The torsion angle in $(CF_3Se)_2$ is 84.5°.

Table 3-18. GEOMETRICAL PARAMETERS OF PEROXIDES

Compound	Bond lengths (Å)		Angles (deg)		Ref
	C—O	O—O	∠C—O—O	τ(O—O)	
CF_3OOH	1.376(10)	1.447(8)	107.6(8)	95	155
CF_3OOF					
I	1.419(24)	1.366(33)	108.2(1.2)	97(6)	155
II	1.415(15)	1.454(18)	105.7(1.5)	100(3)	155
CF_3OOCl	1.372(22)	1.447(15)	108.1(4.0)	93(7)	155
CH_3OOCH_3	1.420(7)	1.457(12)	105.2(5)	119(10)	156
CF_3OOCF_3	1.399(9)	1.419(20)	107.2(1.2)	123(4)	157
$(CH_3)_3COOC(CH_3)_3$	1.460	1.480(ass.)	103.9(1.2)	166(2)	158
$(CH_3)_3SiOOSi(CH_3)_3$	[a]	1.481(8)	[b]	144	158

[a] Si—O, 1.681 Å
[b] ∠Si—O—O, 106.6(1.4)°.

Three-Coordinated Sulfur Compounds

A great variety of organic compounds containing three-coordinated sulfur, sulfoxides, sulfites, sulfimines, etc have been studied by gas-phase ED and MW spectroscopy. By way of contrast, oxygen does not form molecules in which the coordination number of oxygen is larger than two.

Sulfoxides, $R_2S{=}O$

The study[164] of dimethyl sulfoxide, $(CH_3)_2S{=}O$, by microwave spectroscopy has resulted in the determination of the pyramidal configuration of sulfur: $\angle C{-}S{-}C$, 96.3(1)°; $\angle C{-}S{=}O$, 106.6(1)°. The valence angles at sulfur decrease in $(CF_3)_2S{=}O$[165]: $\angle C{-}S{-}C$, 94.2(8)°, $\angle C{-}S{=}O$, 104.5(1.1)°. The C—S bond in bis(trifluoromethyl)sulfoxide is longer [1.885(4) Å] and the S=O bond is shorter [1.469(4) Å] than the corresponding bonds in dimethyl sulfoxide [1.799(5) and 1.485(6) Å, respectively]. The CF_3 group is tilted from the C—S bond by 3.6(5)°.

In diphenyl sulfoxide, $(C_6H_5)_2S{=}O$,[166] the benzene rings are nearly perpendicular to the C—S—C plane with torsional angles about the C—S bonds of $\pm 84.0°$. The overall molecular symmetry is C_s. The geometrical parameters are C—S, 1.804(6) Å; S=O, 1.489(5) Å; $\angle C{-}S{-}C$, 93.9°; $\angle C{-}S{=}O$, 107.6°; they do not differ much from those in dimethyl sulfoxide.

In cyclic ethylene sulfoxide, $CH_2CH_2S{=}O$,[167] and trimethylene sulfoxide, $CH_2CH_2CH_2S{=}O$,[168] the C—S bonds are elongated up to 1.822(3) and 1.836(3) Å, respectively. Trimethylene sulfoxide has a puckered ring with a dihedral angle between the C—C—C and C—S—C planes equal to 34.9(6)°. The exocyclic S=O bond in trimethylene sulfoxide occupies the equatorial position.

Compounds with $S{=}N$ and $S{-}N$ Bonds

The $CF_3N{=}SF_2$[169] (**39**), $ClN{=}SF_2$[170] (**40**), and $ClN{=}S(CF_3)_2$[165] (**41**) molecules have staggered structures. The N—Y (Y = Cl, CF_3) bond is syn-eclipsed with the X—S—X (X = F, CF_3) angle bisector (**39–41**). The C_s

Table 3-19. GEOMETRICAL PARAMETERS OF $X_2S{=}N{-}Y$ MOLECULES

Parameter	$F_2S{=}NCF_3$[a]	$F_2S{=}NCl$[b]	$(CF_3)_2S{=}NCl$[c]
S—X, Å	1.583(4)	1.596(2)	1.878(6)
S=N, Å	1.447(6)	1.476(3)	1.434(8)
N—Y, Å	1.469(10)	1.723(4)	1.676(8)
∠X—S—X, deg	81.8(1.6)	89.8(2)	99.4(9)
∠X—S=N, deg	112.6(5)	111.2(1)	108.8(2.5)
∠S=N—Y, deg	130.4(7)	120.0(2)	138.2(3.8)

[a] Reference 169.
[b] Reference 170.
[c] Reference 165.

symmetry of the $CF_3N{=}SF_2$ molecule is slightly distorted, since the N—C bond deviates from the bisector plane by 7.9(1.5)° (**39**). The nature of the substituents in the S=N group affects the molecular parameters (Table 3-19).

In the $[(CH_3)_2N]_2S{=}O$ diamide[171] three possible models fit the experimental data (**42**). The authors prefer the first conformation with $\tau = 97°$. The nitrogen configuration in this molecule is nearly planar: ∠S—N—C, 116.1(5)°; ∠C—N—C, 113.9(1.5)°. The N—S—N and O=S—N valence angles are 96.9(1.2) and 105.5(8)°, respectively. These values are similar to the C—S—C and O=S—C angles in dimethyl sulfoxide.

42

Sulfites

Sulfites are esters of sulforous acid. Five- and six-membered ring sulfites are listed in Table 3-20. Ethylene sulfite and its selenium analog contain planar heteronuclear rings. In cis-trans and trans-trans isomers of 1,2-dimethyl ethylene sulfite the five-membered rings possess C_2 symmetry (half-chair). The C—C—C—C torsional angle amounts to 88° in the cis-trans isomer and 44° in the trans-trans isomer.

The five-membered ring of o-phenylene sulfite exists as an envelope with an axial S=O bond. The presence of the phenylene group leads to the lengthening of the S—O bond as compared with the S—O bond in ethylene sulfites (Table 3-20).

Table 3-20. GEOMETRICAL PARAMETERS OF SULFITES

Compound	Bond lengths (Å)		Bond angles (deg)			Ref.
	S=O	S—O	$\angle$ O—S—O	$\angle$ O=S—O	$\angle$ S—O—C	
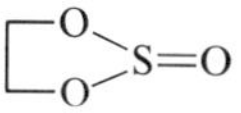 (ethylene sulfite)	1.439(15)	1.629(10)	102.0(3.0)	104.5(3.0)	108.8	172
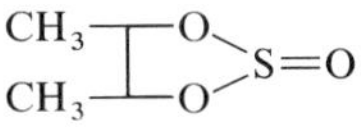 (dimethyl sulfite)						
trans–trans	1.432(24)	1.625(6)	93.5(1.0)	107.3(2.1)	112.3(1.0)	173
cis–trans	1.432(24)	1.624(6)	93.5(1.0)	107.0(2.1)	112.2(1.7)	173
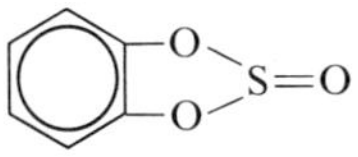 (catechol cyclic sulfite)	1.424(6)	1.672(4)	92.9(3)	107.9(4)	109.6(4)	174
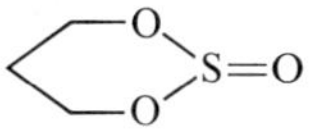 (trimethylene sulfite)	1.45(2)	1.62(1)	99(2)	113.5(3.0)	113(2)	175
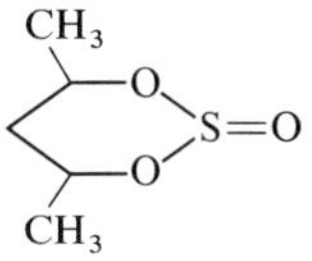 (2,4-pentanediol cyclic sulfite)	1.480(22)	1.622(9)	100.7(2.0)	116.7(1.4)	114.1(1.0)	176

Four-Coordinated Sulfur Compounds

The structure of four-coordinated sulfur compounds has been extensively studied. Most of the investigations of sulfones, amido- and iminosulfones (Table 3-21), etc, have been carried out in the Budapest laboratory.

Sulfones

The geometrical parameters of chlorosulfuric acid derivatives are included in Table 3-22. The S—Cl bond length in these compounds varies between 2.01 Å in SO_2Cl_2 and 2.06 Å in $(CH_3)_2NSO_2Cl$, depending on the electronegativity of the substituent. The S=O bond length is in the range 1.415–1.424 Å, the average value being 1.419(2) Å. The O=S=O and O=S—X valence angles are constant.

The conformations of sulfones have been discussed in details in monographs,[202,203] therefore only the results of recent studies will be considered further.

The $(CX_3)_2SO_2$ (X = F, Cl) molecules (**43**) have their CX_3 groups twisted about the C—S bonds in the opposite directions by 14° in the fluorine

Table 3-21. X—SO$_2$—Y Compounds Studied

Y	F	Cl	CH$_3$	CF$_3$	CCl$_3$	CH$_2$=CH	C$_6$H$_5$	(CH$_3$)$_2$N	OCH$_3$
F	177								
Cl		180							
CH$_3$	178	181	191				199		
CH$_2$Br			192						
CH$_2$=CH		182	193		198				
CH$_2$=CBr			192						
CF$_3$		183		195					
CCl$_3$		184			197				
C$_6$H$_5$		185					166		
C$_6$F$_5$		186							
(CH$_3$)$_2$N		187	194					200	
N=C=O		188							
OCH$_3$	179	189							201
OH				196					
(thiophene-2-yl)		190							

derivative[195] and by 12° in the chlorine analog.[197] The CX$_3$ fragments are tilted by 2.1° (X = F) and 4.9° (X = Cl). In the fluorine derivative, the C—S bond length [1.858(5) Å] and the C—S—C valence angle [102.2(8)°] are significantly smaller than those in the chlorine analog. At the same time, lengthening of the C—S bond in comparison with dimethyl sulfone has been observed in all molecules that contain CF$_3$ and CCl$_3$ groups: CF$_3$SO$_2$Cl, 1.856(6) Å; CCl$_3$SO$_2$Cl, 1.865(30) Å; CF$_3$SO$_2$OH, 1.833(5) Å; and in HC(SO$_2$F)$_3$.[204]

43

The results of a study of divinyl sulfone indicate tht the S—C and C=C, as well as the S—C and S=O bonds are mutually eclipsed in the predominant conformer. This conclusion agrees with the results of the investigations of methyl vinyl sulfone (**44**) and α-bromovinyl methyl sulfone (**45**). It has been

Table 3-22. GEOMETRICAL PARAMETERS OF CHLOROSULFURIC ACID DERIVATIVES, $X{-}SO_2Cl$

X	Bond lengths (Å)			Bond angles (deg)			
	S—Cl	S—X	S=O	$\angle$O=S=O	$\angle$O=S—Cl	$\angle$O=S—X	$\angle$X—O—Cl
Cl	2.012(4)	2.012(4)	1.418(3)	123.5(2)	108.0(1)	108.0(1)	100.3(2)
CH_3	2.046(4)	1.763(5)	1.424(3)	120.8(2.4)	107.1(7)		101.0(1.5)
$CH{=}CH_2$	2.035(5)	1.744(5)	1.420(6)	122.0(1.0)	106.4(5)	109.8(4)	100.2(6)
CF_3	2.015(5)	1.856(6)	1.415(7)	122.4(1.0)	108.3(3)	108.3(7)	98.7(4)
CCl_3	2.020(5)	1.865(30)	1.420(3)	121.5(9)	109.2(6)	108.3(7)	97.9(8)
C_6H_5	2.047(8)	1.764(9)	1.417(12)	122.5(3.6)	105.5(1.8)	110.0(2.5)	100.9(2.0)
$N(CH_3)_2$	2.064(5)	1.618(5)	1.421(4)	122.7(2.3)	105.8(4)	108.8(1.4)	103.0(5)
$O{=}C{=}N$	2.019(3)	1.656(4)	1.417(3)	122.8(2.3)	107.8(7)	108.3(2.2)	98(3)
OCH_3	2.023(4)	1.562(4)	1.419(3)	122.2(1.5)	106.4(6)	108.7(8)	102.8(1.4)
C_6F_5	2.027(5)	1.798(6)	1.415(3)	123.6(1.0)	105.3(2)	108.2(6)	104.8(8)
(thiophene) S	2.041(4)	1.73(2)	1.420(3)	123(2)	101(1)	109.7(1.5)	100.6–101.2

shown unambiguously that the latter compound adopts a conformation with eclipsed S=O and C=C bonds. Two models of methyl vinyl sulfone with C_s and C_1 symmetry provide sufficient agreement with the experimental data. Their geometrical parameters are given in Table 3-23.

In such compounds as $ClSO_2C_6H_5$[185] and $ClSO_2C_6F_5$[186] the S=O bond and the benzene ring plane are close to the eclipsed conformation, with τ being 75.3 and 61.8°, respectively (**46**). In diphenyl sulfone[166] and in 2-thiophene sulfonyl chloride,[190] the rings are perpendicular to the C—S—C and Cl—S—C planes, respectively. The C—S bonds in chlorine derivatives $ClSO_2Ar$ (Ar = C_6H_5, C_6F_5) differ by 0.03 Å. The situation in $ClSO_2C_6H_5$ is the same as in diphenyl sulfone [1.772(5) Å] and methyl phenyl sulfone [1.771(12) Å]. The C—S bond is significantly shortened in 2-thiophene sulfonyl chloride, 1.73(2) Å, and in $ClSO_2CH=CH_2$, 1.744(5) Å. However, in divinyl sulfone, methyl vinyl sulfone, and bromovinyl sulfone the C—S bond is 1.769(4), 1.774(8), and 1.763(6) Å, respectively. This fact reveals the influence of the chlorine atom on the S—C bond lengths in $ClSO_2R$ (R-alkenyl) sulfones.

Amides and Imides

The compounds with S—N and S=N bonds that have been studied include amides $(CH_3)_2NSO_2X$ [X = Cl, CH_3, $N(CH_3)_2$][187,194,200] and H_2NSO_2F,[204] and imides $(CH_3)_2S(O)(=NH)$[205] and $(CH_3)_2S(=NH)_2$.[206] The $(CH_3)_2NSO_2Cl$ molecule shows only one conformation of C_s symmetry, and $(CH_3)_2N]_2SO_2$ favors the trans form (80%). The same conformer has been

Table 3-23. Geometrical Parameters of Methane Sulfonyl Acid Derivatives, CH_3SO_2R

R	Bond lengths (Å)			Bond angles (deg)			
	$S-CH_3$	$S-C(R)$	$S=O$	$\angle O=S=O$	$\angle O=S-CH_3$	$\angle O-S-C(R)$	$\angle C(R)-S-CH_3$
CH_3	1.771(4)	1.771(4)	1.435(3)	119.7(1.1)	108.3	108.3	102.6(9)
CH_2Br	1.784(5)	1.784(5)	1.437(3)	116.8(1.2)	111.5	105.7(7)	104.5(1.2)
$CH=CH_2$	1.776(8)	1.774(8)	1.429(4)	120.0(1.5)	106.9	110.0(1.5)	103.2(1.6)
$CBr=CH_2$	1.773(6)	1.763(6)	1.438(4)	121.6(2.6)	109.1	105.6(1.7)	104.4(2.5)
C_6H_5	1.785(12)	1.771(12)	1.445(3)	118.4(6)	108.6	108.2(4)	103.9(7)

Table 3-24. Geometrical Parameters of Sulfones with S—N Bonds

Parameter	$ClSO_2N(CH_3)_2$[a]	$CH_3SO_2N(CH_3)_2$[b]	$[(CH_3)_2N]_2SO_2$[c]	FSO_2NH_2
S—N, Å	1.618(5)	1.645(10)	1.655(3)	1.61(3)
S=O, Å	1.421(4)	1.431(5)	1.436(10)	1.412(3)
C—S, Å	—	1.771(12)	—	—
∠C—N—C, deg	114.6(2.2)	122.4(2.1)	118.0(3.2)	—
∠O=S=O, deg	122.7(2.3)	118.0(9)	119.7(4)	123.4(2.3)
∠N—S=O, deg	108.8(1.4)	106.8(9)	107.9	109.3(1.7)
∠S—N—C, deg	115.8(7)	114.9(2.1)	115.2(1.1)	—
∠X—S—N, deg[e]	103.0(5)	106.1(1.2)	110.5(1.3)	99(6)

[a] Reference 187.
[b] Reference 194.
[c] Reference 200.
[d] Reference 204.
[e] X = F, Cl, C, N.

found in the crystalline phase. Later $CH_3SO_2N(CH_3)_2$[194] **(47)** was shown to favor conformations with C_s symmetry. About 20% of another asymmetric C_1 conformer with $\tau = 117(27)°$ may be present:

47

The geometrical parameters of amidosulfones are given in Table 3-24.

Sulfates and Halosulfuric Acid Esters

Only three molecules of this class have been investigated: FSO_2OCH_3,[179] $ClSO_2OCH_3$,[189] and $(CH_3O)_2SO_2$[201] (Table 3-25). Although the conclusions concerning the structure of FSO_2OCH_3 are not obvious, microwave spectroscopic study of this molecule has revealed the presence of a small amount of the gauche conformer, whereas electron diffraction data indicate the reverse.[179] As for $ClSO_2OCH_3$, the gauche form with $\tau(Cl—S—O—C)$ 74(4)° has been shown unambiguously to be the major component (89%). The conformational equilibrium has also been observed in dimethyl sulfate. The ratio of gauche–gauche to trans–trans forms is 2:1.

In addition let us consider the results of the electron diffraction study of the CF_3SO_2OH[196] molecule: S=O, 1.417(1) Å; S—O, 1.557(2) Å; C—S, 1.832(3) Å; ∠O=S=O, 122.0(9)°; ∠O—S=O, 109.9(5)°; ∠F—C—F,

Table 3-25. GEOMETRICAL PARAMETERS OF XSO_2OCH_3 MOLECULES[a]

Parameter	CH_3OSO_2F[b]	CH_3OSO_2Cl[c]	$(CH_3O)_2SO_2$[d]
S=O, Å	1.410(2)	1.412(3)	1.419(4)
S—X, Å	1.545(6)	2.023(3)	1.567(3)
S—O, Å	1.558(7)	1.562(4)	1.567(3)
∠O=S=O, deg	124.4(7)	122.2(1.5)	122.3(8)
∠O=S—X, deg	106.8(5)	106.4(6)	108.7(4)
∠O=S—O, deg	109.5(6)	108.7(8)	108.7(4)
∠O—S—X, deg	96.8(6)	102.8(1.4)	98.2(8)
∠S—O—C, deg	116.5(7)	114.4(1.1)	115.2(7)

[a] X = F, Cl, OCH_3.
[b] Reference 179.
[c] Reference 189.
[d] Reference 201.

108.6(2); ∠C—S—O, 102.3(1.6)°; ∠C—S=O, 105.3(8)°; ∠S—C—F, 110.3(2)°. The molecule has C_1 symmetry; the F—C—S—O torsional angle is 10.5(1.3)°.

Five- and Six-Coordinated Sulfur Compounds

The structure of five-coordinated sulfur compounds resembles that of phosphoranes. The main feature is the trigonal bipyramidal configuration of the sulfur bonds. A series of such compounds (**48–50**) has been studied.

The S=O and S=N bonds are in the equatorial plane, and the axial S—F_{ax} bonds are longer than the equatorial S—F_{eq} bonds. The CF_3 groups, which are less electronegative than the fluorine atoms (4.0 and 3.2), adopt an equatorial position in the $(CF_3)_2F_2S=O$ molecule. In the $XN=SF_4$ compounds (X = F, CH_3) the difference between the two axial bonds is 0.08 and 0.1 Å, respectively (Table 3-26). The lengths of the shorter bonds (S—F_{ax}) appeared to be even smaller than those of the equatorial S-F_{eq} bonds. These results are in agreement with the results of *ab initio* calculations[209] for the $HN=SF_4$ molecule: S=N, 1.461 Å; S—F_{eq}, 1.550 Å; S—F_{ax}, 1.611 and 1.569 Å; ∠F_{ax}—S—F_{ax}, 166.3°; ∠F_{eq}—S—F_{eq}, 104.9°; ∠N=S—F_{ax}, 96.9°. In the $XN=SF_4$ series the substituents H, F, and CH_3 are in the axial plane. There is less difference between axial bonds in $H_2C=SF_4$, as was shown by ED and X-ray analysis as well as by *ab*

Table 3-26. GEOMETRICAL PARAMETERS OF COMPOUNDS WITH FIVE-COORDINATED SULFUR

Parameter	$O{=}SF_4$[a]	$O{=}S(CF_3)_2F_2$[b]	$FN{=}SF_4$[c]	$CH_3N{=}SF_4$[c]
S=O, Å	1.409(9)	1.422(7)	—	—
S=N, Å	—	—	1.520(9)	1.400(6)
S—F_{eq}, Å	1.539(3)	—	1.564(5)	1.567(4)
S—F_{ax}, Å	1.596(3)	1.641(4)	1.615(7)	1.643(4)
S—F'_{ax}, Å	—	—	1.535(12)	1.546(7)
$\angle F_{eq}$—S—F_{eq}, deg	112.8(4)	97.8(8) $\angle$CSC	99.8(3)	102.6(1.6)
$\angle$O(N)=S—F_{ax}, deg	97.7(2)		96.9(4)	98.4(4)
$\angle$N=S—F'_{ax}, deg			90.6(5)	94.6(4)
$\angle F_{ax}$—S—F'_{ax}, deg	164.6(2)	173.1(6)	172.5(7)	167.0(6)

[a] Reference 207.
[b] Reference 203.
[c] Reference 209.

initio calculations[210] (Table 3-27). The C=S bond is in the equatorial plane due to its low polarity.

All the compounds with six-coordinated sulfur that have been studied are inorganic: SF_6,[211] SF_5Cl,[212] F_5XOXF_5 (X=S, Se, Te),[213] SF_5NF_2,[214] XF_5NCO (X = S, Se, Te),[215] and also $(ClNSF_4)_2$.[216] The SF_6 molecule adopts an octahedral configuration with sulfur in the center (S—F, 1.561 Å). Distortions arise when fluorine is partially substituted by other atoms or groups (**51**). Thus, the SF_5Cl molecule has a C_{4v} symmetry. The axial bonds differ from

X
F F
S
F F
F

X = Cl, NCO, NF$_2$, OSF$_5$

51

Table 3-27. GEOMETRICAL PARAMETERS OF $H_2C{=}SF_4$ BY DIFFERENT METHODS[a]

| Parameter | Method | | |
	ED	X-ray	*Ab initio*
S—F_{ax}, Å	1.595(15)	1.593	1.591
S—F_{eq}, Å	1.575(15)	1.561	1.562
S=C, Å	1.550(20)	1.553	1.542
$\angle F_{ax}$—S—F_{ax}, deg	170.0(20)	170.4	169.9
$\angle F_{eq}$—S—F_{eq}, deg	97.0(2.0)	96.4	98.8

[a] *Source*: Reference 210.

Table 3-28. GEOMETRICAL PARAMETERS OF COMPOUNDS
WITH SIX-COORDINATED SULFUR.

Compound	Bond lengths (Å)			
	$S-F_{ax}$	$S-F_{eq}$	$S-X$	Ref.
SF_5Cl	1.588(9)	1.566(3)	2.047(3)	212
SF_5NF_2	1.560(7)	1.546(2)	1.696(5)	214
F_5SOSF_5	1.572(34)	1.558(8)	1.586(11)	213
SF_5NCO	1.567(2)	1.567(2)	1.668(6)	215

the equatorial bonds, as indicated in Table 3-28. In isocyanates, XF_5NCO, the NCO chain is not linear, and the oxygen atom is beyond the basic plane. The isocyanate $N=C=O$ angle is 170, 173, and 176° for X = S, Se, and Te, respectively. In the F_5XOXF_5 molecules the deviations from linearity are even larger: $\angle X-O-X$ amounts to 143, 142, and 146° for X = S, Se, Te.

In the $(F_4SNCl)_2$ molecule the four-membered ring is planar. Chlorine atoms are mutually trans (**52**). The $N-S-N$ endocyclic angle is 80.7(6)°. The $S-F_{ax}$ bond [1.590(6) Å] is much longer than the equatorial [1.545(5) Å], and the deviation of the $F_{ax}-S-F_{ax}$ angle from 180° amounts to 3.6°.

52

Cyclic Compounds of Oxygen and Sulfur

There is much in common between the stereochemistry of oxygen and sulfur cyclic organic compounds and their carbocyclic analogs. However, some differences can be observed that are due to the substitution of one or several methylene groups by oxygen or sulfur.

Three-Membered Ring Compounds

Oxirane and thiirane are the simplest three-membered ring molecules with oxygen and sulfur. As determined by microwave spectroscopy, the $C-C$ bond in oxirane[217] [1.466(2) Å] is shorter than that in cyclopropane by 0.03 Å, and

Table 3-29. GEOMETRICAL PARAMETERS FOR
OXIRANE DERIVATIVES

| | Bond lengths (Å) | | |
Compound	C—O	C—C	Ref.
 (diepoxybutane)	1.439(4)	1.463(5)	219
 (oxirane–CH₂Cl)	1.437(2)	1.489(7)	220
 (oxirane–CH₂Br)	1.435(5)	1.472(15)	220
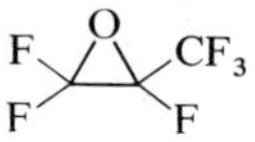 (F₂C–O–CFCF₃ epoxide)	1.410(8)	1.467(7)	221

the C—O bonds are longer than those in dimethyl ether. As for thiirane,[218] its molecular parameters are nearly the same as those of cyclopropane and dimethyl sulfide. The C—C bond length is 1.484(3) Å and the C—S bond length amounts to 1.815(3) Å.

Similar parameters have been determined for oxirane derivatives (Table 3-29). The C—C cyclic bonds in *d, l*-1, 2, 3, 4-diepoxybutane[219] are nearly trans to each other, with a deviation of 10°, and the C—O bonds adopt a gauche conformation (**53**). Three conformations are possible for chloro- and bromo-

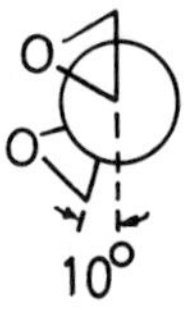

53

methyl oxiranes[220] (**54**). As determined by gas-phase ED these molecules adopt the first conformation with the C—X and C—O bonds mutually trans. However, a mixture of conformers I and II was observed in the microwave study.[222]

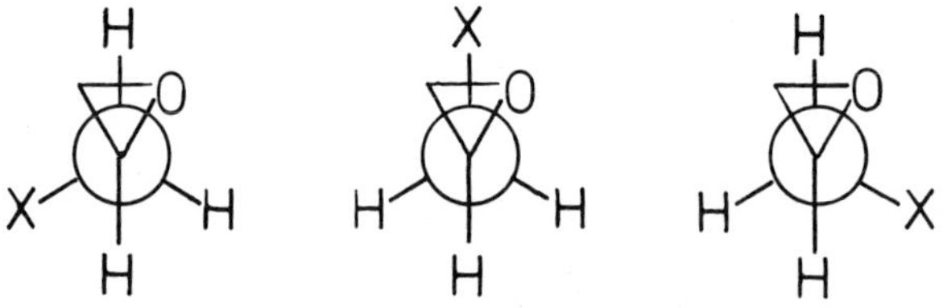

I   II   III

54

The parameters of the oxirane ring[223] in cyclopentene, cyclohexene, and cycloheptene oxides are usual. The condensed three-membered ring exerts considerable influence on the molecular conformations. Cyclopentene oxide and cyclohexene oxide are observed in the boat and half-chair forms, respectively. Cycloheptene oxide was found to exist as a 66:34 conformational mixture of two chair forms (**55**).

55

Four-Membered Ring Compounds

Several studies deal with heteronuclear four-membered ring compounds containing oxygen and sulfur. Oxetane, the simplest among them, has a planar structure.[224] Thietane[225] has a puckered ring with a dihedral angle between the C—C—C and C—S—C planes of 26(2)°. The geometrical parameters of oxetane and thietane differ from those of dimethyl oxide and dimethyl sulfide.

Compound	C—O(S) (Å)	C—C (Å)	∠C—O(S)—C (deg)	∠C—C—C (deg)
$(CH_2)_3O$	1.448(5)	1.546(5)	91.9(1.0)	84.6(1.0)
$(CH_2)_3S$	1.847(6)	1.549(3)	76.8(3)	

The lengthening of the C—O and C—S bonds observed in these compounds is a general characteristic of four-membered rings. The oxetane ring is also planar in 2-methyloxetane[226] and in 3,3-bis(azidomethyl)oxetane,[227] which have similar parameters: C—O, 1.448(5) Å; ∠C—C—C, 85.5°; and C—O 1.458(7), Å; ∠C—C—C, 87.3°, respectively.

In contrast to thietane, the four-membered ring in perfluoro-1,3-dithiacyclobutane[228] is planar, and the C—S bonds are considerably shorter: 1.820(2) Å. Substitution of the methylene group in thietane by a dimethylsilyl group results in an increase in the dihedral angle that is equal to 13(3)° in 3,3-dimethyl-3-silathietane.[229] The C—S—C valence angle amounts to 89.1(3.0)° and the C—S bond length [1.854(10) Å] is identical to that in thietane.

The unsaturated ring in 3,4-bistrifluoromethyl-1,2-dithiacyclobut-3-ene, $CF_3C=C(CF_3)—S—S$, is planar.[230] The overall molecular symmetry is C_2. The geometrical parameters are as follows: C=C, 1.40(3) Å; C—S, 1.73(1) Å; S—S, 2.05 Å; ∠C=C—S, 100.8(6)°. The S—S bond is somewhat longer (2.05 Å) than that in dimethyldisulfide (2.02 Å).[203]

Five-Membered Ring Compounds

The structures of a great variety of five-membered ring compounds have been determined. Oxacyclopentane exhibits nearly free pseudorotation, as determined by molecular mechanics calculations.[231] The results of an ED study[232] indicate that pseudorotation occurs as well when one of the hydrogen atoms in oxacyclopentane is substituted by bromine. In oxacyclopentane the C—O bond length is 1.428(1) Å, and the C—C bond length is 1.536(1) Å. In the 3-bromoderivative[233] the C—O and C—C bond lengths are 1.428(5) and 1.527(8) Å, respectively. The substitution of one of the methylene groups in cyclopentane by heavier atoms like sulfur or selenium results in stabilization of one of the conformers. Thus, tetrahydrothiophene[234] and tetrahydroselenophene[235] are observed in half-chair C_2 forms with the following geometrical parameters:

Compound	C—X (Å)	C—C (Å)	∠C—X—C (deg)	∠C—C—X (deg)	C—C—C (deg)
$(CH_2)_4S$	1.838(2)	1.536(2)	93.4(5)	106.1(4)	105.0(5)
$(CH_2)_4Se$	1.975(3)	1.537(4)	89.1(5)	105.8(3)	106.0(7)

There is no pseudorotation in trisubstituted cyclopentanes either. Thus, 1,2,4-trioxacyclopentane[236] has C_2 symmetry with the following main parameters: C—O, 1.414(2) Å; ∠C—O—C, 105.9(1.1)°; O—O, 1.487(6) Å; ∠C—O—O, 99.2(7)°; ∠O—C—O, 105.3(8)°. They are close to the corresponding values of ethers and peroxides (see Table 3-3 and 3-18).

Among the unsaturated oxacyclopentenes, the structure of oxacyclopent-3-ene is known.[237] Its dynamic model is characterized by the following geometrical parameters: C—O, 1.440(2) Å; C—C, 1.502(3) Å; C=C, 1.347(4) Å; ∠C—O—C, 108.3(5)°; ∠C—C—O, 106.5(4)°; ∠C—C=C, 109.3(2)°.

The five-membered heteronuclear rings with two unsaturated bonds are planar, eg, furan, thiophene, selenophene, and their derivatives. The C—O and C—C bonds of furan,[238] 2-chlorofuran,[239] 2-bromofuran,[239] and 3-bromofuran[240] are shorter: 1.36 and 1.39–1.43 Å, respectively. The C=C bonds are longer: 1.35–1.39 Å. The C—S and C=C bonds in thiophene[241] are 1.716(4) and 1.366(4) Å, respectively. In contrast to thiophene, these values differ in 2-chloro- and 2-bromothiophenes.[242] Minimal difference is observed in the chloroderivative: S—C(Cl), 1.722(4) Å; S—CH, 1.717(5) Å; HC=CH, 1.378(9) Å; ClC=CH, 1.389(11) Å. In 2-bromothiophene the difference is considerable: S—CBr, 1.749(8) Å; S—CH, 1.668(9) Å; HC=CH, 1.338(16) Å; HC=CBr, 1.384(15) Å. In the series of furan < thiophene < selenophene the C—X—C (X = O, S, Se) valence angle decreases from 106.5(1)° to 91.8(3)° and 87.7(1)°. The molecular structure of selenophene has been determined by the MW method[243]: C—Se, 1.855(1) Å; C—C, 1.433(3) Å; C=C, 1.370(1) Å; ∠C—Se—C, 87.7(1)°.

Table 3-30. GEOMETRICAL PARAMETERS OF SUCCINIC
ANHYDRIDE AND ITS DERIVATIVES[a]

| | X | | |
Parameter	H	F	CH_3
C=O, Å	1.190(2)	1.187(2)	1.197(3)
C—O, Å	1.389(3)	1.387(3)	1.396(4)
∠C—O—C, deg	109.5(5)	112.5(9)	109.6(5)
∠C—C—O, deg	110.5(4)	107.0(7)	109.2(4)
∠C—C—C, deg	103.8(5)	100.7(3)	100.9(4)

[a] *Sources*: H: reference 244, F: reference 245, CH_3: reference 246.

A planar ring structure was found in succinic anhydride,[244] **56** with X = H.

56

However, substituting fluorines or methyl groups for the hydrogens, destroys
the planarity of the ring. The F...F and CH_3...CH_3 nonbonded interactions
result in twisting of the ring about the C—C bond by 30(1)° in the
tetrafluoroderivative[245] and by 33(2)° in the tetramethyl derivative.[246] As a
result the endocyclic angles also change (Table 3-30). The C=O and C—O
bonds are identical to those in acyclic anhydrides (see Table 3-11). The
heteronuclear rings of maleic anhydride[247] and its dichloro-derivative[248]
XC=CX—C(O)—O—C(O) are planar. The geometrical parameters of both
molecules are similar (Table 3-31).

Table 3-31. GEOMETRICAL PARAMETERS OF
MALEIC ANHYDRIDE AND ITS
DICHLORO-DERIVATIVE[a]

| | X | |
Parameter	H	Cl
C=O, Å	1.195(3)	1.188(2)
C—O, Å	1.39(1)	1.389(3)
C=C, Å	1.33(3)	1.332(5)
C—C, Å	1.500(5)	1.495(3)
∠C—O—C, deg	107(1)	107.5
∠C—C=O, deg	129(2)	128.5(4)
∠C—C—O, deg	109(1)	108.4

[a] *Sources*: H: reference 247, Cl: reference 248.

Six-Membered Ring Compounds

Various six-membered ring compounds have been investigated by gas-phase ED. Cyclohexane is known to exist in a chair conformation with the C—C—C—C torsional angle 55°. The conformation remains the same after substitution of one methylene group by oxygen. The C—O—C—C, O—C—C—C, and C—C—C—C torsional angles in oxacyclohexane[249] are 59.9, 56.9, and 52.9°, respectively. The puckering of both rings is the same. The sum of the endocyclic angles equals 665° in oxacyclohexane and 668° in cyclohexane.

Puckering increases in thiacyclohexan-4-one[250] (**58**) in comparison with cyclohexanone[251] (**57**) as is obvious from inspection of their torsional angles.

The principal geometrical parameters of thiacyclohexan-4-one are as follows: ∠C—S—C, 97.0(1.9)°; ∠S—C—C, 113.2(4)°; ∠C—C—CO, 112.5(1.0)°; ∠C—CO—C, 118.9(1.1)°; C—S 1.804(3) Å. The C—S bond length and the C—S—C valence angle are identical to those of dimethyl sulfide.

1,3-Dioxane[252] and 1,3-dithiane[253] are reported to exist in chair forms, although a small amount (about 10%) of another form may be present in the latter compound. Puckering of the dithiane ring is greater than that of dioxane as can be seen from their parameters (Table 3-32). The C—X—C valence angles in both molecules are similar to those in oxides and sulfides. The C—S bond lengths in 1,3-dithiane are identical [1.812(3) Å]. The C—O bonds in 1,3-dioxane, however, are different: OC—O, 1.393(25) Å; CC—O, 1.439(39) Å.

Table 3-32. TORSIONAL AND BOND ANGLES OF 1,3-DIOXANE AND 1,3-DITHIANE[a]

	X	
Parameter	O	S
τ(XCXC), deg	58.9	61.4
τ(CXCC), deg	56.0	58.3
τ(XCCC), deg	57.4	64.1
∠C—X—C, deg	110.9(1.5)	98.1(7)
∠X—C—X, deg	115.0(3)	115.0(3)
∠C—C—X, deg	109.2(8)	114.9(4)
∠C—C—C, deg	107.7(1.1)	113.6(3.3)

[a] *Sources*: O: reference 252, S: reference 253.

Table 3-33. GEOMETRICAL PARAMETERS OF
1,4-DIOXANE AND 1,4-OXATHIANE[a]

Parameter	1,4-Dioxane	1,4-Oxathiane
C—O, Å	1.423(3)	1.418(4)
C—S, Å	—	1.826(4)
∠C—O—C, deg	112.4(5)	115.1(2.2)
∠C—S—C, deg	—	97.1(2.0)
∠C—C—O, deg	109.2(5)	113.2(1.7)
∠C—C—S, deg	—	111.4(1.0)

[a] *Sources*: 1,4-dioxane: reference 254, 1,4-oxathiane: reference 255.

Their average value (1.416 Å) is in good agreement with the bond length determined in the MW study (1.410 Å).

The molecules of 1,4-dioxane[254] and 1,4-oxathiane[255] adopt chair conformations with identical puckering. The principal geometrical parameters of these molecules are given in Table 3-33.

The main geometrical parameters of trioxane[256] and 2,4,6-trimethyltrioxane[257] are identical:

Compound	C—O (Å)	∠O—C—O (deg)	∠C—O—C (deg)
Trioxane	1.411(2)	111.0(7)	109.2(1.0)
2,4,6-Trimethyltrioxane	1.410(4)	110.7(7)	112.3(8)

Both molecules adopt a chair conformation, and the methyl groups in the trimethyl derivative are equatorial (**59**).

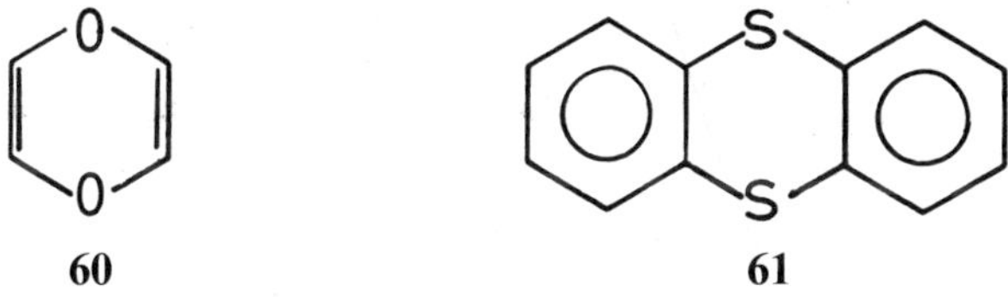

1,4-Dioxin[258] (**60**) thianthrene[259] (**61**) are examples of unsaturated six-membered heteronuclear ring compounds. 1,4-Dioxin has a planar structure

with the following parameters: C—O, 1.401(2) Å; C=C, 1.337(4) Å; ∠C—O—C, 109.8(5)°, ∠C=C—O, 124.2(5)°. For comparison, the C—O—C valence angle is 117° and the C—O bond length is 1.359 Å (see Table 3-3) in methyl vinyl ether. It should be noted that the six-membered ring in the carbon analog of 1,4-dioxin, 1,4-cyclohexadiene, is not planar. In thianthrene the dihedral angle between the two S—C—C—S planes is 131.4(3)°, and the C—S—C valence angle is 104.1°. The latter value is in good agreement with the C—S—C valence angles in diphenyl sulfide and in bis-2-pyridyl sulfide (Table 3-6). As a result, the conformation of thianthrene is not planar, since the C—S—C valence angle should be about 120° for the planar structure. The C—S bonds are shortened to 1.770(3) Å, in agreement with the carbon atom hybridization.

Ring Compounds Having Seven or More-Members

Oxepane[260] and 3,3,6,6-tetramethylthiacyclohept-4-ine[261] provide examples of seven-membered ring compounds that contain oxygen and sulfur. At 40°C oxepane exists in two twist–chair forms: $^3TC_4$ (53%) and $^2TC_3$ (47%) (**62**).

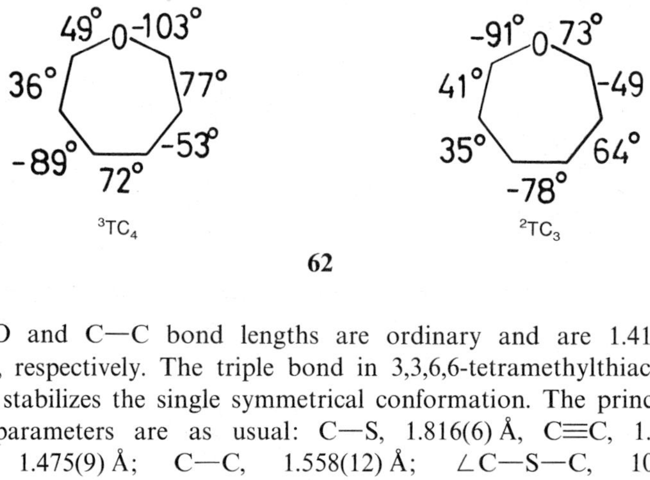

62

The C—O and C—C bond lengths are ordinary and are 1.419(5) and 1.531(3) Å, respectively. The triple bond in 3,3,6,6-tetramethylthiacyclohept-4-ine (**63**) stabilizes the single symmetrical conformation. The principal geometrical parameters are as usual: C—S, 1.816(6) Å, C≡C, 1.209(9) Å; ≡C—C, 1.475(9) Å; C—C, 1.558(12) Å; ∠C—S—C, 103.8(1.1)°; ∠S—C—C, 110.2(1.5)°; ∠C—C(Me)₂—C, 107.8(1.0)°; ∠C≡C—C, 145.8(7)°.

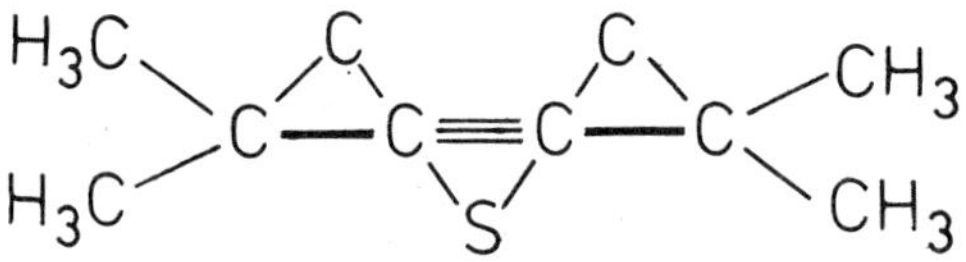

63

1,3,5,7-Tetraoxacyclooctane[262] provides an example of an eight-membered ring compound. A mixture of 32% crown form and 68% chair–boat form is

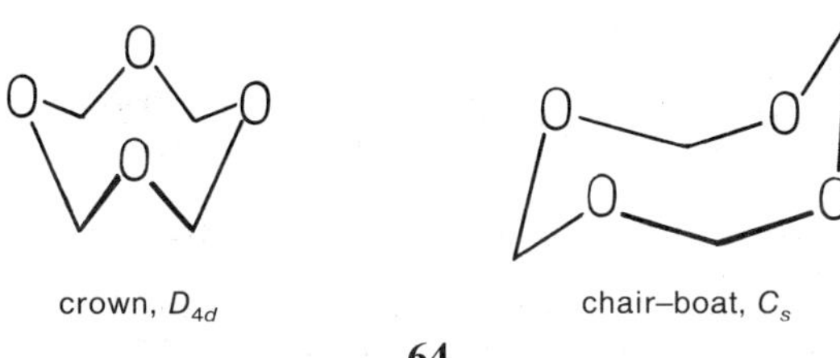

64

reported for this molecule in the gas phase (**64**). Both conformations are defined by identical valence angles:

Valence angle	D_{4d}	C_s
∠C—O—C, deg	114.6(2.0)	114.4(2.0)
∠O—C—O, deg	114.2(2.2)	114.3(2.1)

They are about 3–5° larger than those in 1,3,5-trioxane, and the C—O bond length, 1.404(1) Å, is identical to that in trioxane.

The maximal C—O bond length in the oxacycloalkane series is observed in oxacyclobutane (1.45 Å), this being the result of ring strain. In oxacyclopentane the C—O bond is shortened to 1.43 Å and in oxacyclohexane to 1.42 Å. The minimal value (1.404 Å) is observed in 1,3,5,7-tetraoxacyclooctane. The C—O—C valence angle in oxacyclohexanes is similar to that in dimethyl ether. It becomes larger, up to 115–116°, in the seven- and eight-membered rings.

References

1. Gelly, M.C.L.; Lees, R. M.; Winnewisser, G. *J. Mol. Spectrosc.* **1976**, *61*, 231.
2. Cuolt, J. P. Fourth Austin Symposium on Gas-Phase Molecular Structure, 1972, Austin, TX, p. 47.
3. Penn, R. E.; Birkenmeier, J. A. *J. Mol. Spectrosc.* **1976**, *62*, 416.
4. Almenningen, A.; Bastiansen, O.; Fernholt, L.; Hedberg, K. *Acta Chem. Scand.* **1971**, *25*, 1946.
5. Hagen, K.; Hedberg, K. *J. Am. Chem. Soc.* **1973**, *95*, 8263.
6. Kinneging, A. J.; Mom, V.; Mijlhoff, F. C.; Renes, G. H. *J. Mol. Struct.* **1982**, *82*, 271.
7. Bijen, J.M.J.M. *J. Mol. Struct.* **1973**, *17*, 69.
8. Marstokk, K.-M.; Møllendal, H. *J. Mol. Struct.* **1971**, *7*, 101.
9. Yokozeki, A.; Bauer, S. H. *J. Phys. Chem.* **1975**, *79*, 155.
10. Baumik, A.; Brooks, W.V.F.; Dass, S. C.; Sastry, K.V.L.N. *Can. J. Chem.* **1970**, *48*, 2949.
11. Traetteberg, M.; Østensen, H.; Seip, R. *Acta Chem. Scand.* **1980**, *34A*, 449.
12. Marstokk, K.-M.; Møllendall, H. *J. Mol. Structure* **1977**, *40*, 1.
13. Marstock, K.-M.; Møllendall, H. *J. Mol. Struct.* **1976**, *35*, 57.
14. Ellengsen, B. H.; Marstokk, K.-M.; Møllendall, H. *J. Mol. Struct.* **1978**, *48*, 9.
15. Murty, A. N.; Curl, R. F. *J. Chem. Phys.* **1967**, *46*, 4176.
16. Horn, A.; Marstokk, K.-M.; Møllendall, H.; Priebe, H. *Acta Chem. Scand.* **1973**, *37A*, 679.
17. Rodler, M.; Bauder, A. *J. Am. Chem. Soc.* **1984**, *106*, 4025.
18. Alderliesten, P.; Almenningen, A.; Strand, T. G. *Acta Chem. Scand.* **1975**, *29B*, 811.
19. Rankin, D.W.H.; Todd, M. R.; Riddell, F. G.; Turner, E. S. *J. Mol. Struct.* **1981**, *71*, 171.
20. Kojima, T. *J. Phys. Soc. Japan* **1960**, *15*, 1284.

21. Marsden, C. J. *J. Mol. Struct.* **1974**, *21*, 168.
22. Hargittai, I.; Schultz, G. *J. Chem. Soc. Chem. Commun.* **1972**, 323.
23. Hayashi, M.; Imaishi, H.; Kuvada, K. *Bull. Chem. Soc. Japan* **1974**, *47*, 2382.
24. Schultz, G.; Fellegvari, I.; Kolonits, M.; Kiss, A. I.; Pete, B.; Banki, J. *J. Mol. Struct.* **1978**, *50*, 325.
25. Domenicano, A.; Hargittai, I.; Portalone, G.; Schultz, G. Seventh European Crystallography Meeting, Jerusalem, 1982, p. 155.
26. Tamagava, K.; Takemura, M.; Konaka, S.; Kimura, M. *J. Mol. Struct.* **1984**, *125*, 131.
27. Oyanagi, K.; Kuchitsu, K. *Bull. Chem. Soc. Japan* **1978**, *51*, 2237.
28. Planie, M. C.; Toneman, L. H.; Dallinga, G. *Rec. Trav. Chim. Pays-Bas* **1965**, *84*, 232.
29. Astrup, E. E. *Acta Chem. Scand.* **1978**, *32A*, 115.
30. Pyckhout, W.; Van Nuffel, P.; Van Alsenoy, C.; Van den Enden, L.; Geise, H. J. *J. Mol. Struct.* **1983**, *102*, 333.
31. Schei, S. H. *Acta Chem. Scand.* **1983**, *37A*, 153.
32. Bijen, J.M.J.M.; Derissen, J. L. *J. Mol. Struct.* **1972**, *14*, 229.
33. Astrup, E. E. *Acta Chem. Scand.* **1979**, *33A*, 655.
34. Astrup, E. E. *Acta Chem. Scand.* **1973**, *27*, 3271.
35. Astrup, E. E.; Aomar, A. M. *Acta Chem. Scand.* **1975**, *29A*, 794.
36. Spelbos, A.; Mijlhoff, F. C.; Faber, D. H. *J. Mol. Struct.* **1977**, *41*, 47.
37. Spelbos, A.; Mijlhoff, F. C.; Renes, G. H. *J. Mol. Struct.* **1978**, *44*, 73.
38. Mijlhoff, F. C.; Geise, H. J.; Van Schaick, E.J.M. *J. Mol. Struct.* **1974**, *20*, 393.
39. Seip, H. M.; Seip, R. *Acta Chem. Scand.* **1973**, *27*, 4024.
40. Brunvoll, J.; Bohn, R. K.; Hargittai, I. *J. Mol. Struct.* **1985**, *129*, 81.
41. Lowrey, A. H.; George, C. F.; D'Antonio, P.; Karle, J. *J. Chem. Phys.* **1973**, *58*, 2840.
42. Astrup, E. E. *Acta Chem. Scand.* **1977**, *31A*, 125.
43. Iijima, K.; Yamada, T.; Shibata, S. *J. Mol. Struct.* **1981**, *77*, 271.
44. Lowrey, A. H.; George, C.; D'Antonio, P.; Karle, J. *J. Mol. Struct.* **1980**, *63*, 243.
45. Iijima, T.; Tsuchiya, S.; Kimura, M. *Bull. Chem. Soc. Japan* **1977**, *50*, 2564.
46. Oberhammer, H.; Gombler, W.; Willner, H. *J. Mol. Struct.* **1981**, *70*, 273.
47. Naumov, V. A.; Garaeva, R. N.; Geise, H. J.; Mijlhoff, F. C. *Dokl. Akad. Nauk SSSR* **1979**, *246*, 650.
48. Astrup, E. E.; Aomar, A. M. *Acta Chem. Scand.* **1976**, *30A*, 289.
49. Jeffrey, G. A.; Pople, J. A.; Rodom, L. *Carbohydr. Res.* **1972**, *25*, 117.
50. Samdal, S.; Seip, H. M.; Torgrimsen, T. *J. Mol. Struct.* **1979**, *57*, 105.
51. Zaripov, N. M. *Zh. Strukt. Khim.* **1976**, *17*, 741.
52. Zaripov, N. M.; Golubinskii, A. V.; Chmutova, G. A.; Vilkov, L. V. *J. Mol. Struct.* **1978**, *19*, 894.
53. Glidewell, C.; Rankin, D.W.H.; Robiette, A. G.; Sheldrick, G. M.; Beagley, B.; Freeman, J. M. *Trans. Faraday Soc.* **1969**, *65*, 2621.
54. Naumov, V. A.; Ziatdinova, R. N. *Zh. Strukt. Khim.* **1984**, *25*, 88.
55. Rozsondai, B.; Moore, J. H.; Gregory, D. C.; Hargittai, I. *Acta Chim. (Budapest)* **1977**, *94*, 321.
56. Rozsondai, B.; Hargittai, I.; Pappalardo, G. C. *Z. Naturforsch.* **1979**, *34a*, 752.
57. Glidewell, C.; Rankin, D.W.H.; Robiette, A. G.; Sheldrick, G. M. *J. Mol. Struct.* **1970**, *5*, 417.
58. Airey, W.; Glidewell, C.; Robiette, A. G.; Sheldrick, G. M. *J. Mol. Struct.* **1971**, *8*, 413.
59. Csákvári, B.; Wagner, Z.; Gömöry, P.; Hargittai, I.; Rozsondai, B.; Mijlhoff, F. C. *Acta Chim. (Budapest)* **1976**, *90*, 149.
60. Gergö, E.; Hargittai, I. Sixth International Symposium on Organosilicon Chemistry, Budapest, 1981, p. 139.
61. Gergö, E.; Hargittai, I.; Schultz, G. *J. Organomet. Chem.* **1976**, *112*, 29.
62. Boonstra, L. H.; Mijlhoff, F. C.; Renes, G.; Spelbos, A.; Hargittai, I. *J. Mol. Struct.* **1975**, *28*, 129.
63. Shen, Q. *J. Mol. Struct.* **1979**, *51*, 61.
64. Almenningen, A.; Bastiansen, O.; Ewing, V.; Hedberg, K.; Traetteberg, M. *Acta Chem. Scand.* **1963**, *17*, 2455.
65. Rankin, D.W.H.; Robertson, H. E. *J. Chem. Soc. Dalton Trans.* **1983**, 265.
66. Csákvári, B.; Wagner, Z.; Gömöry, P.; Mijlhoff, F. C.; Rozsondai, B.; Hargittai, I. *J. Organomet. Chem.* **1976**, *107*, 287.
67. Airey, W.; Glidewell, C.; Rankin, D.W.H.; Robiette, A. G.; Sheldrick, G. M. *Trans. Faraday Soc.* **1970**, *66*, 551.
68. Shen, Q. *J. Mol. Struct.* **1983**, *102*, 325.

69. Glidewell, C.; Rankin, D.W.H.; Robiette, A. G.; Sheldrick, G. M.; Beagley, B.; Gradock, S. *J. Chem. Soc. A* **1970**, 315.

70. Vilkov, L. V.; Tarasenko, N. A. *Zh. Strukt. Khim.* **1969**, *10*, 1102.

71. Carleer, R.; Van den Enden, L.; Geise, H. J.; Mijlhoff, F. C. *J. Mol. Struct.* **1978**, *50*, 345.

72. Dédier, J.; Marchand, A. *Spectrochim. Acta* **1982**, *38A*, 339.

73. Oyanagi, K.; Kuchitsu, K. *Bull. Chem. Soc. Japan* **1978**, *51*, 2243.

74. Derissen, J. L.; Bijen, J.M.J.M. *J. Mol. Struct.* **1973**, *16*, 289.

75. Jandal, P.; Seip, H. M.; Torgrimsen, T. *J. Mol. Struct.* **1976**, *32*, 369.

76. Beagley, B.; Ulbrecht, V.; Katsumata, S.; Lloyd, D. R.; Connor, J. A.; Hudson, G. A. *J. Chem. Soc. Faraday Trans. 2* **1977**, *73*, 1278.

77. Zaripov, N. M.; Mannafov, T. G. *Zh. Strukt. Khim.* **1975**, *16*, 832.

78. Arnold, D.E.J.; Gunderson, G.; Rankin, D.W.H.; Robertson, H. E. *J. Chem. Soc. Dalton Trans.* **1983**, 1989.

79. Almenningen, A.; Hedberg, K.; Seip, R. *Acta Chem. Scand.* **1963**, *17*, 2264.

80. Ebsworth, E.A.V.; Macdonald, E. K.; Rankin, D.W.H. *Monatsh. Chem.* **1980**, *111*, 221.

81. Fernholt, L.; Haaland, A.; Hargittai, M.; Seip, R.; Weidlein, J. *Acta Chem. Scand.* **1981**, *35A*, 529.

82. Marsden, C. J.; Sheldrick, G. M. *J. Mol. Struct.* **1971**, *10*, 405.

83. Almenningen, A.; Fernholt, L.; Seip, H. M. *Acta Chem. Scand.* **1968**, *22*, 51.

84. Murdoch, J. D.; Rankin, D.W.H.; Glidewell, C. *J. Mol. Struct.* **1971**, *9*, 17.

85. Blom, R.; Haaland, A.; Seip, R. *Acta Chem. Scand.* **1983**, *37A*, 595.

86. Oka, T.; Morino, Y. *J. Mol. Spectrosc.* **1962**, *8*, 300.

87. Diodati, F. P.; Bartell, L. S. *J. Mol. Struct.* **1971**, *8*, 395.

88. Guarnieri, A.; Charpentier, L.; Kück, B. *Z. Naturforsch.* **1973**, *28a*, 1721.

89. Alekseev, N. V.; Velichko, F. K. *Zh. Strukt. Khim.* **1967**, *8*, 8.

90. Zaripov, N. M.; Popik, M. V.; Vilkov, L. V. *Zh. Strukt. Khim.* **1980**, *21*, 38.

91. Zaripov, N. M.; Popik, M. V.; Vilkov, L. V.; Mannafov, T. G. *Zh. Strukt. Khim.* **1980**, *21*, 37.

92. Schultz, G.; Korovik, I.; Kucsman, A.; Hargittai, I. *J. Chem. Soc. Faraday Trans. 2* **1984**, *80*, 1273.

93. Zaripov, N. M.; Golubinskii, A. V.; Sokolkov, S. V.; Vilkov, L. V.; Mannafov, T. G. *Dokl. Akad. Nauk SSSR* **1984**, *278*, 664.

94. Shen, Q. *Acta Chem. Scand.* **1977**, *31A*, 795.

95. Gradock, S.; Rankin, D.W.H. *J. Mol. Struct.* **1980**, *69*, 145.

96. Shen, Q. *Acta Chem. Scand.* **1978**, *32A*, 245.

97. Rooij, J.; Mijlhoff, F. C.; Renes, G. *J. Mol. Struct.* **1975**, *25*, 169.

98. Bett, W.; Gradock, S.; Rankin, D.W.H. *J. Mol. Struct.* **1980**, *66*, 159.

99. Barrow, M. J.; Gradock, S.; Ebsworth, E.A.V.; Rankin, D.W.H. *J. Chem. Soc. Dalton Trans.* **1981**, 1988.

100. Barrow, M. J.; Ebsworth, E.A.V.; Huntley, C. M.; Rankin, D.W.H. *J. Chem. Soc. Dalton Trans.* **1982**, 1131.

101. Curl, R. F. *J. Chem. Phys.* **1959**, *30*, 1529.

102. Ebsworth, E.A.V.; Huntley, C. M.; Rankin, D.W.H. *J. Chem. Soc. Dalton Trans.* **1983**, 835.

103. Mijlhoff, F. C. *J. Mol. Struct.* **1977**, *36*, 334.

104. Auberg, E.; Samdal, S.; Seip, H. M. *J. Mol. Struct.* **1979**, *57*, 95.

105. Almenningen, A.; Fernholt, L.; Seip, H. M.; Henriksen, L. *Acta Chem. Scand.* **1974**, *28A*, 1037.

106. Boogaard, A.; Geise, H. J.; Miljhoff, F. C. *J. Mol. Struct.* **1972**, *13*, 53.

107. Vledder, H. J.; Mijlhoff, F. C.; Well, F. P.; Dofferhoff, G.M.T.; Leyte, J. C. *J. Mol. Struct.* **1971**, *9*, 25.

108. Vledder, H. J.; Mijlhoff, F. C.; Leyte, J. C.; Romers, C. *J. Mol. Struct.* **1971**, *7*, 421.

109. Andreassen, A. L.; Zebelman, D.; Bauer, S. H. *J. Am. Chem. Soc.* **1971**, *93*, 1148.

110. Kato, C.; Konaka, S.; Iijima, T.; Kimura, M. *Bull. Chem. Soc. Japan* **1969**, *42*, 2148.

111. Iijima, T.; Kimura, M. *Bull. Chem. Soc. Japan* **1969**, *42*, 2159.

112. Dyngeseth, S.; Schei, H.; Hagen, K. *J. Mol. Struct.* **1983**, *102*, 45.

113. Van Nuffel, P.; Van den Enden, L.; Van Alsenoy, C.; Geise, H. J. *J. Mol. Struct.* **1984**, *116*, 99.

114. Guillory, J. P.; Bartell, L. S. *J. Chem. Phys.* **1965**, *43*, 654.

115. Traetteberg, M. *Acta Chem. Scand.* **1970**, *24*, 373.

116. Kitano, M.; Kuchitsu, K. *Bull. Chem. Soc. Japan* **1974**, *47*, 67.

117. Bartell, L. S.; Guillory, J. P. *J. Chem. Phys.* **1965**, *43*, 647.

118. Schultz, G.; Fellegvari, I.; Kolonits, M.; Kiss, A. I.; Pete, B.; Banki, J. *J. Mol. Struct.* **1978**, *50*, 325.

119. Schäfer, L.; Samdal, S.; Hedberg, K. *J. Mol. Struct.* **1976**, *31*, 29.
120. Chiu, N. S.; Ewbank, J. D.; Askari, M.; Schäfer, L. *J. Mol. Struct.* **1979**, *54*, 185.
121. Kuchitsu, K.; Fukuyama, T.; Morino, Y. *J. Mol. Struct.* **1968**, *1*, 463.
122. Paulen, G.; Traetteberg, M. *Acta Chem. Scand.* **1974**, *28A*, 1155.
123. Domenicano, A.; Hargittai, I.; Portalone, G.; Schultz, G. Fifteenth Meeting of the Italian Crystallography Association, Monteporsio Catone, 1984, O2.
124. Huisman, P.A.G.; Klebe, K. J.; Mijlhoff, F. C.; Renes, G. H. *J. Mol. Struct.* **1979**, *57*, 71.
125. Nakata, M.; Kohata, K.; Fukuyama, T.; Kuchitsu, K.; Wilkins, C. J. *J. Mol. Struct.* **1980**, *68*, 271.
126. Iijima, T. *Bull. Chem. Soc. Japan* **1972**, *45*, 3526.
127. Abe, M.; Kuchitsu, K.; Shimanouchi, T. *J. Mol. Struct.* **1969**, *4*, 245.
128. Lowrey, A. H.; George, C.; D'Antonio, P.; Karle, J. *J. Am. Chem. Soc.* **1971**, *93*, 6399.
129. Andreassen, A. L.; Bauer, S. H. *J. Mol. Struct.* **1972**, *12*, 381.
130. Hagen, K.; Hedberg, K. *J. Am. Chem. Soc.* **1973**, *95*, 8266.
131. Hilderbrandt, R. L.; Andreassen, A. L.; Bauer, S. H. *J. Chem. Phys.* **1970**, *74*, 1586.
132. Andersen, P.; Astrup, E. E.; Borgan, A. *Acta Chem. Scand.* **1974**, *28A*, 239.
133. Gallaher, K. L.; Bauer, S. H. *J. Chem. Soc. Faraday Trans. 2* **1975**, *71*, 1423.
134. Almenningen, A.; Bastiansen, O.; Motzfeldt, T. *Acta Chem. Scand.* **1969**, *23*, 2848.
135. Derissen, J. L. *J. Mol. Struct.* **1971**, *7*, 67.
136. Derissen, J. L. *J. Mol. Struct.* **1971**, *7*, 81.
137. Bonham, R. A.; Su, L. S. Second Austin Symposium on Gas-Phase Molecular Structure, Austin, TX, 1968, Paper No. M1.
138. Almenningen, A.; Bastiansen, O.; Motzfeldt, T. *Acta Chem. Scand.* **1970**, *24*, 747.
139. Eijck, B. P.; Plaats, G.; Roon, P. H. *J. Mol. Struct.* **1972**, *11*, 67.
140. Derissen, J. L.; Bijen, J.M.J.M. *J. Mol. Struct.* **1975**, *29*, 153.
141. Bijen, J.M.J.M.; Derissen, J. L. *J. Mol. Struct.* **1975**, *27*, 233.
142. Maagdenberg, A.A.J. *J. Mol. Struct.* **1977**, *41*, 61.
143. Tsuchiya, S. *J. Mol. Struct.* **1974**, *22*, 77.
144. Tsuchiya, S.; Iijima, T. *J. Mol. Struct.* **1972**, *13*, 327.
145. Dyngeseth, S.; Schei, S. H.; Hagen, K. *J. Mol. Struct.* **1984**, *116*, 257.
146. Hagen, K.; Hedberg, K. *J. Am. Chem. Soc.* **1984**, *106*, 6150.
147. Steinnes, O.; Shen, Q.; Hagen, K. *J. Mol. Struct.* **1980**, *64*, 217.
148. Shen, Q.; Hilderbrandt, R. L.; Hagen, K. *J. Mol. Struct.* **1981**, *71*, 161.
149. Steinnes, O.; Shen, Q.; Hagen, K. *J. Mol. Struct.* **1980**, *66*, 181.
150. Brake, J.H.M.; Driessen, R.A.J.; Mijlhoff, F. C.; Renes, G. H. Lowrey, A. H. *J. Mol. Struct.* **1982**, *81*, 277.
151. Boulet, G. A. Ph. D. thesis, University of Michigan, 1964, Abstr. V.25, 12562.
152. Bartell, L. S. *J. Phys. Chem.* **1965**, *69*, 3043.
153. Hagen, K.; Hedberg, K. *J. Am. Chem. Soc.* **1973**, *95*, 1003.
154. Nàhlovska, Z.; Nàhlovsky, B.; Strand, T. G. *Acta Chem. Scand.* **1970**, *24*, 2617.
155. Marsden, C. J.; Des Marteau, D. D.; Bartell, L. S. *Inorg. Chem.* **1977**, *16*, 2359.
156. Haas, B.; Oberhammer, H. *J. Am. Chem. Soc.* **1984**, *106*, 6146.
157. Marsden, C. J.; Bartell, L. S.; Diodati, F. P. *J. Mol. Struct.* **1977**, *39*, 253.
158. Käss, D.; Oberhammer, H.; Brandes, D.; Blaschette, A. *J. Mol. Struct.* **1977**, *40*, 65.
159. Gase, W.; Boggs, J. E. *J. Mol. Struct.* **1984**, *116*, 207.
160. Yokozeki, A.; Bauer, S. H. *J. Chem. Phys.* **1976**, *80*, 618.
161. Marsden, C. J.; Beagley, B. *J. Chem. Soc. Faraday Trans. 2* **1981**, *77*, 2213.
162. D'Antonio, P.; George, C.; Lowrey, A. H.; Karle, J. *J. Chem. Phys.* **1971**, *55*, 1071.
163. Marsden, C. J.; Sheldrick, G. M. *J. Mol. Struct.* **1971**, *10*, 419.
164. Dreizler, H.; Dendl, G. *Z. Naturforsch.* **1964**, *19a*, 512.
165. Oberhammer, H.; Kumar, R. C.; Knerr, G. D.; Shreeve, J. M. *Inorg. Chem.* **1981**, *20*, 3871.
166. Rozsondai, B.; Moore, J. H.; Gregory, D. C.; Hargittai, I. *J. Mol. Struct.* **1979**, *51*, 69.
167. Saito, S. *Bull. Chem. Soc. Japan* **1969**, *42*, 663.
168. Bevan, J. W.; Legon, A. C.; Millen, D. J. *Proc. R. Soc. (London)* **1977**, *354A*, 491.
169. Karl, R. R.; Bauer, S. H. *Inorg. Chem.* **1975**, *14*, 1859.
170. Haase, J.; Oberhammer, H.; Zeil, W.; Glemser, O.; Mews, R. *Z. Naturforsch.* **1970**, *25a*, 153.
171. Hargittai, I.; Vilkov, L. V. *Acta Chim. (Budapest)* **1970**, *63*, 143.
172. Arbuzov, B. A.; Naumov, V. A.; Zaripov, N. M.; Pronicheva, L. D. *Dokl. Akad. Nauk SSSR* **1970**, *195*, 1333.
173. Geise, H. J.; Van Laere, E. *Bull. Soc. Chim. Belg.* **1975**, *84*, 775.

174. Schultz, G.; Serke, I.; Kapovits, I. *J. Chem. Soc. Faraday Trans. 2* **1979**, *75*, 1612.
175. Naumov, V. A.; Zaripov, N. M.; Shatrukov, L. F. *Zh. Strukt. Khim.* **1970**, *11*, 579.
176. Mustoe, F. J.; Hencher, J. L. *Can. J. Chem.* **1972**, *50*, 3892.
177. Hagen, K.; Cross, V. R.; Hedberg, K. *J. Mol. Struct.* **1978**, *44*, 187.
178. Hargittai, I.; Hargittai, M. *J. Mol. Struct.* **1973**, *15*, 399.
179. Hargittai, I.; Seip, R.; Nair, K.P.R.; Britt, C. O.; Boggs, J. E.; Cyvin, B. N. *J. Mol. Struct.* **1977**, *39*, 1.
180. Hargittai, I. *Acta Chim. (Budapest)* **1969**, *60*, 231.
181. Hargittai, M.; Hargittai, I. *J. Chem. Phys.* **1973**, *59*, 2513.
182. Brunvoll, J.; Hargittai, I. *Acta Chim. (Budapest)* **1977**, *94*, 333.
183. Brunvoll, J.; Hargittai, I.; Kolonits, M. *Z. Naturforsch.* **1978**, *33a*, 1236.
184. Brunvoll, J.; Hargittai, I.; Seip, R. *Z. Naturforsch.* **1978**, *33a*, 222.
185. Brunvoll, J.; Hargittai, I. *J. Mol. Struct.* **1976**, *30*, 361.
186. Vajda, E.; Hargittai, I. *Z. Naturforsch.* **1983**, *38a*, 765.
187. Hargittai, I.; Brunvoll, J. *Acta Chem. Scand.* **1976**, *30A*, 634.
188. Brunvoll, J.; Hargittai, I.; Seip, R. *J. Chem. Soc. Dalton Trans.* **1978**, 299.
189. Hargittai, I.; Schultz, G.; Kolonits, M. *J. Chem. Soc. Dalton Trans.* **1977**, 1299.
190. Brunvoll, J.; Hargittai, I.; Szekely, T.; Pappalardo, G. C. *J. Mol. Struct.* **1980**, *66*, 173.
191. Hargittai, M.; Hargittai, I. *J. Mol. Struct.* **1974**, *20*, 283.
192. Naumov, V. A.; Ziatdinova, R. N. *Zh. Strukt. Khim.* **1983**, *24*, 48.
193. Naumov, V. A.; Ziatdinova, R. N.; Berdnikov, E. A. *Zh. Strukt. Khim.* **1981**, *22*, 89.
194. Naumov, V. A.; Garaeva, R. N.; Butenko, G. G. *Zh. Strukt. Khim.* **1979**, *20*, 1110.
195. Oberhammer, H.; Knerr, G. D.; Shreeve, J. M. *J. Mol. Struct.* **1982**, *82*, 143.
196. Schultz, G.; Hargittai, I.; Seip, R. *Z. Naturforsch.* **1981**, *36a*, 917.
197. Hargittai, I.; Vajda, E.; Nielsen, C. J.; Klaeboe, P.; Seip, R.; Brunvoll, J. *Acta Chem. Scand.* **1983**, *37A*, 341.
198. Hargittai, I.; Rozsondai, B.; Nagel, B.; Bulcke, P.; Robinet, G.; Labarre, J.-F. *J. Chem. Soc. Dalton Trans.* **1978**, 861.
199. Brunvoll, J.; Exner, O.; Hargittai, I.; Kolonits, M.; Scharfenberg, P. *J. Mol. Struct.* **1984**, *117*, 317.
200. Hargittai, I.; Vajda, E.; Szöke, A. *J. Mol. Struct.* **1973**, *18*, 381.
201. Brunvoll, J.; Exner, O.; Hargittai, I. *J. Mol. Struct.* **1981**, *73*, 99.
202. Hargittai, I. "Sulphone Molecular Structures," Vol. 6, Lecture Notes in Chemistry, Springer-Verlag, Berlin, 1978.
203. Hargittai, I. "The Structure of Volatile Sulphur Compounds". Akadémiai Kiadó; Budapest, Reidel; Dordrecht, 1985.
204. Brunvoll, J.; Kolonits, M.; Bliefert, C.; Seppelt, K.; Hargittai, I. *J. Mol. Struct.* **1982**, *78*, 307.
205. Oberhammer, H.; Zeil, W. *Z. Naturforsch.* **1970**, *25a*, 845.
206. Oberhammer, H.; Zeil, W. *Z. Naturforsch.* **1969**, *25a*, 1612.
207. Hedberg, L.; Hedberg, K. *J. Phys. Chem.* **1982**, *86*, 598.
208. Oberhammer, H.; Shreeve, J. M.; Gard, G. L. *Inorg. Chem.* **1984**, *23*, 2820.
209. DesMarteau, D. D.; Eysel, H. H.; Oberhammer, H.; Günther, H. *Inorg. Chem.* **1982**, *21*, 1607.
210. Bock, H.; Boggs, J. E.; Kleeman, G.; Lentz, D.; Oberhammer, H.; Peters, E. M.; Seppelt, K.; Simon, A.; Solouki, B. *Angew. Chem.* **1979**, *91*, 1008.
211. Bartell, L. S.; Doun, S. K. *J. Mol. Struct.* **1978**, *43*, 245.
212. Marsden, C. J.; Bartell, L. S. *Inorg. Chem.* **1976**, *15*, 3004.
213. Oberhammer, H.; Seppelt, K. *Inorg. Chem.* **1978**, *17*, 1435.
214. Haase, J.; Oberhammer, H.; Zeil, W.; Glemser, O.; Mews, R. *Z. Naturforsch.* **1971**, *26a*, 1333.
215. Oberhammer, H.; Seppelt, K.; Mews, R. *J. Mol. Struct.* **1983**, *101*, 325.
216. Oberhammer, H.; Waterfeld, A.; Mews, R. *Inorg. Chem.* **1984**, *23*, 415.
217. Hirose, C. *Bull. Chem. Soc. Japan* **1974**, *47*, 1311.
218. Okije, K.; Hirose, C.; Lister, D. G.; Sheridan, J. *Chem. Phys. Lett.* **1974**, *24*, 111.
219. Smith, Z.; Kohl, D. A. *J. Chem. Phys.* **1972**, *57*, 5448.
220. Shen, Q. *Abstracts*, Tenth Austin Symposium on Gas-Phase Molecular Structure, Austin, TX, 1984, p. 62.
221. Beagley, B.; Pritchard, R. G.; Banks, R. E. *J. Fluor. Chem.* **1981**, *18*, 159.
222. Mohammadi, M. A.; Brooks, W.V.F. *J. Mol. Spectrosc.* **1979**, *78*, 89.
223. Traetteberg, M.; Sandnes, T. W.; Bakken, P. *J. Mol. Struct.* **1980**, *67*, 235.
224. Creswell, R. A. *Mol. Phys.* **1975**, *30*, 217.
225. Karakida, K.; Kuchitsu, K. *Bull. Chem. Soc. Japan* **1975**, *48*, 1691.

226. Schultz, G.; Bartok, M. *Z. Naturforsch.* **1979**, *34a*, 1130.
227. George, C.; Lowrey, A.; Karle, J. Tenth Austin Symposium on Gas-Phase Molecular Structure, Austin, TX, 1984, p. 39.
228. Smith, Z.; Seip, R. *Acta Chem. Scand.* **1976**, *30A*, 759. Hargittai, I. *J. Mol. Struct.* **1979**, *54*, 287.
229. Mastryukov, V. S.; Strelkov, S. A.; Golubinskii, A. V.; Vilkov, L. V.; Kirpichenko, S. V.; Suslova, E. N.; Voronkov, M. G. *Dokl. Akad. Nauk SSSR* **1983**, *271*, 384.
230. Hencher, J. L.; Shen, Q.; Tuck, D. G. *J. Am. Chem. Soc.* **1976**, *98*, 899.
231. Seip, H. M. *Acta Chem. Scand.* **1969**, *23*, 2741.
232. Geise, H. J.; Adams, W. J.; Bartell, L. S. *Tetrahedron* **1969**, *25*, 3045.
233. Smith, Z.; Seip, H. M.; Náhlovský, B.; Kohl, D. A. *Tetrahedron* **1975**, *29A*, 513.
234. Náhlovská, Z.; Náhlovský, B.; Seip, H. M. *Tetrahedron* **1969**, *23*, 3534.
235. Náhlovská, Z.; Náhlovský, B.; Seip, H. M. *Tetrahedron* **1970**, *24*, 1903.
236. Almenningen, A.; Kolsaker, P.; Seip, H. M.; Willadsen, T. *Tetrahedron* **1969**, *23*, 3398.
237. Tamagava, K.; Hilderbrandt, R. L. *J. Am. Chem. Soc.* **1984**, *106*, 20.
238. Bak, B.; Christensen, D.; Dixon, W. *J. Mol. Spectrosc.* **1962**, *9*, 124.
239. Shcherbak, G. A.; Sadova, N. I.; Vilkov, L. V.; Boiko, Y. A. *Zh. Strukt. Khim.* **1979**, *20*, 532.
240. Shcherbak, G. A.; Sadova, N. I.; Vilkov, L. V.; Boiko, Y. A. *Zh. Strukt. Khim.* **1979**, *20*, 530.
241. Harshbarger, W.; Bauer, S. H. *Acta Crystallogr.* **1970**, *26B*, 1010.
242. Karl, R. R.; Bauer, S. H. *Acta Crystallogr.* **1972**, *28B*, 2619.
243. Pozdeev, N. M.; Akulinin, O. B.; Shapkin, A. A.; Magdesieva, N. N. *Zh. Strukt. Khim.* **1970**, *11*, 869.
244. Brendhaugen, K.; Fikke, M. N.; Seip, H. M. *Acta Chem. Scand.* **1973**, *27*, 1101.
245. Fernholt, L.; Seip, H. M. *J. Mol. Struct.* **1978**, *49*, 333.
246. Almenningen, A.; Fernholt, L.; Rustad, S.; Seip, H. M. *J. Mol. Struct.* **1976**, *30*, 291.
247. Hilderbrandt, R. L.; Peixoto, E.M.A. *J. Mol. Struct.* **1972**, *12*, 31.
248. Hagen, K.; Hedberg, K. *J. Mol. Struct.* **1978**, *50*, 103.
249. Breed, H. E.; Gundersen, G.; Seip, R. *Acta Chem. Scand.* **1979**, *33A*, 225.
250. Seip, R.; Seip, H. M.; Smith, Z. *J. Mol. Struct.* **1976**, *32*, 279.
251. Dillen, J.; Geise, H. J. *J. Mol. Struct.* **1980**, *69*, 137.
252. Schultz, G.; Hargittai, I. *Acta Chim. (Budapest)* **1974**, *83*, 331.
253. Adams, W. J.; Bartell, L. S. *J. Mol. Struct.* **1977**, *37*, 261.
254. Davis, M. I.; Hassel, O. *Acta Chem. Scand.* **1963**, *17*, 1181.
255. Schultz, G.; Hargittai, I. *J. Mol. Struct.* **1972**, *14*, 353.
256. Clark, A. H.; Hewitt, T. G. *J. Mol. Struct.* **1971**, *9*, 33.
257. Astrup, E. E. *Acta Chem. Scand.* **1973**, *27*, 1345.
258. Peixoto, E.M.A.; Iga, I.; Mu-tao, L. *J. Chem. Res. Synop.* **1980**, 111.
259. Gallaher, K. L.; Bauer, S. H. *J. Chem. Soc. Faraday Trans. 2* **1975**, *71*, 1173.
260. Dillen, J.; Geise, H. J. *J. Mol. Struct.* **1980**, *64*, 239.
261. Haase, J.; Krebs, A. *Z. Naturforsch.* **1972**, *27a*, 624.
262. Astrup, E. E. *Acta Chem. Scand.* **1980**, *34A*, 85.

FLUORINE DERIVATIVES

Heinz Oberhammer

INSTITUT FÜR PHYSIKALISCHE UND THEORETISCHE CHEMIE
UNIVERSITÄT TÜBINGEN
TÜBINGEN, WEST GERMANY

CONTENTS

INTRODUCTION   147
HYDROGEN FLUORIDE   149
GROUP III COMPOUNDS   149
GROUP IV COMPOUNDS   151
    Fluorocarbons   151
    Silicon, Germanium, and Tin Compounds   160
GROUP V COMPOUNDS   165
    Nitrogen Fluorine Compounds   165
    Phosphorus Fluorine Compounds   168
GROUP VI COMPOUNDS   182
    Oxygen Fluorine Compounds   182
    Sulfur, Selenium, and Tellurium Compounds   184
GROUP VII COMPOUNDS   196
GROUP VIII COMPOUNDS   198
CONCLUSION   198
ACKNOWLEDGMENT   199
REFERENCES   199

Introduction

Fluorine is the most electronegative element, and it is well known that fluorine compounds exhibit behavior that differs in many respects from that of such analogous compounds as hydrogen or chlorine derivatives. In many cases, fluorination affects the stability and chemical properties of substances very strongly, and it is not surprising that such differences are reflected in their structural features. Many molecular species of main group fluorine compounds are gases or highly volatile liquids at room temperature and are therefore amenable to gas-phase structural studies. Although a number of fluorine derivatives have been studied by microwave spectroscopy (MW), complete and accurate determinations by this technique for compounds characterized by a larger number of geometric parameters are hampered by the absence of stable fluorine isotopes. Gas-phase electron diffraction (ED) is in most cases a more suitable method to obtain structural information on fluorinated molecules. The general experience that joint analysis of ED and MW data enhances the accuracy and reliability of geometric parameters applies to fluorine derivatives as well. Valuable information on stereochemistry or conformational properties is also obtainable from ^{19}F NMR spectroscopy. Since such spectra are usually recorded in the liquid phase and at a much longer characteristic time scale than ED, these differences must be considered when comparing results from both methods.

A comprehensive review of gas-phase structural data for fluorocarbons has been published by Yokozeki and Bauer.[1] These authors made an extensive survey of C—F bond lengths, the effect of fluorine/hydrogen substitution on C—C, C=C, and C≡C bonds, and the effect of CH_3/CF_3 substitution on adjacent bond lengths. These topics will be discussed in this chapter only insofar as new structural data have become available since 1975. Furthermore, fluorine-substituted benzene derivatives and metal fluorides are not considered here, since these compounds are discussed elsewhere in this volume (see Chapters 7 and 9). Except for the foregoing limitations, an attempt is made to give a nearly complete review of ED results for main-group fluorine compounds based on the sector–microphotometer method. The data have been taken from the literature published between about 1960 and summer 1985, and unpublished data of our laboratory are also included. In some cases MW results are consulted where they are helpful for comparison.

In general, the geometric structure of an isolated molecule is not of great interest, and only comparison between the structural parameters of similar or analogous compounds permits discussion of the various substitutents' effects on the bonding properties of a molecule. In this chapter special attention is directed toward such comparisons within families of compounds. Our main interest is in structural parameters of the nearest neighbors of fluorine, ie, adjacent bond lengths and bond angles. Attempts are made to rationalize these parameters or trends within an analogous group of compounds on the basis of simple models, such as polar effects or valence shell electron-pair repulsion (VSEPR) theory.[2] Isolated compounds, from which structural data of similar derivatives are not known, will be listed without further discussion. In general, the compounds are arranged according to their "central" atom, ie, the atom to which fluorine or a CF_3 group is bonded. This system is not followed strictly, however, when other arrangements are convenient for reasons of comparison. Different structural features of one compound may be discussed in different sections. Geometric parameters are in general cited in the representation ($r_a, r_g, r_\alpha^0, r_0,$ or r_s) and with the experimental uncertainties of the original literature. For definition of the various parameter types see reference 3 or Chapter 1 in volume A. Reported experimental uncertainties are in general two- or threefold standard deviations of the least-squares analysis and include in some cases estimated systematic errors. In some cases bond lengths have been converted to a common basis (r_g values) to allow direct comparison. Differences between the various types of geometric parameters are in most cases smaller or in the order of the experimental uncertainties.

Hydrogen Fluoride

Hydrogen fluoride is the only Group I compound discussed in this chapter. It is the most important starting material in preparative fluorine chemistry, and it is used as a versatile solvent for organic and inorganic compounds. Its high boiling point relative to HCl or HBr is a consequence of strong H...F bridges in the condensed phases. Even in the gas phase, HF exists as a mixture of monomers and mainly cyclic hexamers in a ratio of 4.2:1 (at 22°C)[4]. The ED value for the H—F bond distance in the monometer [0.933(13) Å]* is in good agreement with the MW value ($r_0 = 0.9232$ Å).[5] The cyclic hexamer forms a puckered ring with F—H...F = 2.525(3) Å [F—H = 0.973(9) and H...F = 1.552 Å] and F—F—F angles of 104°.

Group III Compounds

The parent boron compound, BF_3, has a planar configuration, and B—F distances of 1.3133(10) Å have been derived by joint analysis of ED intensities and high-resolution infrared data.[6] In general, substitution of fluorine by less electronegative ligands causes lengthening of the B—F bonds as is indicated by the structure of $(SiH_3)_2NBF_2$.[7] The heavy-atom skeleton is essentially planar with a small twist of 9(4)° between the BF_2 and NSi_2 planes, which may be a consequence of a low-frequency torsional vibration. The B—F and B—N bond lengths are 1.330(6) and 1.496(17) Å and angles FBF and SiNSi are 123.2(18) and 123.9(3)°, respectively. For SiF_3—SiF_2BF_2 a planar Si—Si—BF_2 skeleton has been reported with B—F = 1.309(3), B—Si = 2.008(5) Å and FBF = 118.8(7)°.[8a] In this compound the B—F bonds appear to be slightly shorter than in BF_3. Two novel boron compounds containing CF_3 groups, viz, $CF_3B[N(CH_3)_2]_2$ and $(CF_3)_2BN(CH_3)_2$, have been synthesized very recently and studied by ED.[8b] In both compounds the B—C bonds are substantially

---

* Throughout, numbers in parentheses are the limits of error of the last digits.

longer than those in analogous methyl compounds: 1.648(8) Å in $CF_3B[N(CH_3)_2]_2$ vs 1.586(3) Å in $CH_3B[NH(CH_3)]_2$[8c] and 1.623(5) Å in $(CF_3)_2BN(CH_3)_2$ vs 1.587(2) Å in $(CH_3)_2BNHCH_3$.[8c] $(CF_3)_2BN(CH_3)_2$ has a planar C_2BNC_2 skeleton with B—N = 1.422(9)Å and CBC = 122.2(10)°, and the dimethylamino groups in $CF_3B[N(CH_3)_2]_2$ are twisted by 24(3)° around the B—N bonds [B—N = 1.425(19) Å, NBN = 125(3)°].

Among the diboranes, the perfluorinated compound B_2F_4 is an exception insofar as it has a planar equilibrium configuration with a small twofold barrier to rotation around the B—B bond [1.8(7) kJ/mol^{-1}],[9] whereas the analogous chlorine[10] and bromine[11] derivatives have staggered conformations with the BX_2 planes perpendicular to each other. Tetrafluorodiborane has been studied at various temperatures ($-50, 22$, and 150°C), and the room temperature results are: B—F = 1.317(2) Å, B—B = 1.720(4) Å and FBF = 117.2(2)°. The structure reported for bis-(trifluorophosphine)diborane(4)[12] (**1**) has a planar PBBP skeleton, with the two PF_3 groups mutually trans and four terminal hydrogen atoms. A model with bridging hydrogens is excluded on the basis of the ED data and vibrational spectroscopy. The distance between the four-coordinated boron atoms [1.800(12) Å] is considerably longer than in B_2F_4, and the B—P bond lengths of 1.848(9) Å are close to that in $F_3P\cdot BH_3$ [1.836(7) Å from MW[13]]. Other skeletal parameters are: BBP = 100.2(15)°, BPF = 118.1(2)°, FPF = 99.6(2)°.

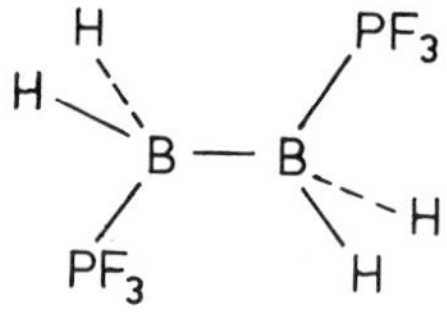

H PF_3

H B — B H

PF_3 H

1

Two donor–acceptor complexes of BF_3 have been studied in the gas phase: $BF_3\cdot N(CH_3)_3$[14–16] and $BF_3\cdot O(CH_3)_2$.[15,17] Reference 16 reports a joint analysis of ED and spectroscopic data for the trimethylamine complex, the results of which must be regarded as being more reliable than data obtained via the other investigations. The B—N distance of 1.674(4) Å is much longer than the B—N bond in $(SiH_3)_2NBF_2$ [1.496(17) Å] and the B—F [1.374(2) vs 1.3133(10) Å in BF_3] and N—C bonds [1.485(2) vs 1.462(2) Å in $N(CH_3)_3$][18] lengthen upon complex formation. The FBF angles decrease from 120° to 112.6(3)°. The study of the dimethyl ether complex is hampered by partial decomposition [57(4), 60(6), and 80(4)% at 16, 30, and 70°C, respectively], and the results for the geometric parameters depend on the sample temperature. The following parameters were derived from data obtained at 16°C[17]: B—O = 1.73(5) Å [cf B—O = 1.344(10) Å in BF_2OH from an MW study[19]], B—F = 1.361(8) Å, FBF = 117(2)°, C—O = 1.440(13) Å [vs 1.411(1) Å in $O(CH_3)_2$[20]]. The angle between the COC plane and the B—O bond is 40(8)°.

The only fluorinated cyclic boron compound studied in the gas phase is B-trifluoroborazine, $(FBNH)_3$,[21] a planar six-membered ring with B—N = 1.432(10), B—F = 1.361(10) Å, and NBN = 119(1)°. The ring parameters are equal to those in borazine[22] [B—N = 1.4355(21) Å and NBN = 117.7(12)°].

Group IV Compounds

Fluorocarbons

Methanes. The gradual shortening of C—F bond lengths in $CH_{4-n}F_n$ with increasing fluorination has been pointed out by Yokozeki and Bauer.[1] Since 1975 new results have been reported for the two extreme members of this series; CH_3F [C—F = 1.392(1) Å from a joint ED/MW analysis,[23] confirming a previous ED study,[24] which reported C—F = 1.392(5) Å], and CF_4[1.3203(4) Å].[25] This shortening of C—F bonds with increasing fluorination is commonly referred to as the "fluorine effect" and is observed for most X—F bonds. An explanation was formulated in terms of valence bond theory involving hyperconjugation.[26] An alternative explanation is based on polar contributions to the C—F bonds. The positive net charge of the carbon atom increases with increasing number of electronegative substituents, thereby enhancing the polar attraction between C and F. The results of semiempirical calculations indicate that covalent contributions to the C—F bonds depend little on other substituents.[27] The intermediate C—F bond length in CHFBrCl [1.348(5) Å][28] is in qualitative agreement with this simple concept.

Trifluoromethyl halides, CF_3X, with X = Cl, Br, I have been reinvestigated by combining ED and MW data.[29] When compounds with X = H and X = F are included in the comparison, the C—F bond lengths in the series decrease almost linearly with increasing electronegativity of X (Table 4-1), in agreement with the polarity concept. Except for X = H, the FCF angles increase with increasing electronegativity of X, thereby resulting in a remarkably constant value for the nonbonded F...F distances. Both observations made for these simple CF_3 compounds (ie, decreasing C—F bond lengths with increasing "effective" electronegativity of the adjacent atom and constancy of the F...F distances) apply also to larger compounds containing CF_3 groups. With few exceptions, F...F geminal distances of 2.155 ± 0.01 Å are reported.

Table 4-1. GEOMETRIC PARAMETERS OF CF_3 GROUPS IN
XCF_3 COMPOUNDS (r_α° VALUES)

Compound	Bond lengths (Å)		$\angle FCF$ (deg)	Ref.
	C—F	F...F		
HCF_3	1.334(3)	2.168(5)	108.7(3)	30
ICF_3	1.330(2)	2.153(2)	108.1(2)	29
$BrCF_3$	1.326(2)	2.156(3)	108.8(3)	29
$ClCF_3$	1.325(2)	2.151(3)	108.6(2)	29
FCF_3	1.318(0_4)	2.154(0_3)	109.47	25

Alkanes. Investigations or reinvestigations by electron diffraction of the entire series of fluorinated ethanes have been reported since 1975. The data now available (Table 4-2) allow the discussion of trends in C—F and C—C bond lengths within this series.[42]

1. The C—F bonds shorten with increasing fluorination at the carbon atom considered [from 1.399(4) Å in CH_2F—CH_3 to 1.340(2) Å in CF_3—CH_3]. This trend corresponds to the "fluorine effect" in the methanes.
2. A minor, but still detectable shortening of C—F bonds with increasing fluorination at the other carbon atom [eg, C—F bond lengths of the CF_3 group decrease from 1.340(2) Å in CF_3—CH_3 to 1.326(2) Å in CF_3—CF_3].
3. The C—C bond lengths *decrease* with increasing fluorination at one carbon atom from 1.533(2) Å in CH_3—CH_3 to 1.497(3) Å in CF_3—CH_3 and *increase* with further fluorination to 1.545(2) Å in CF_3—CF_3.

Table 4-2. C—C AND C—F BOND LENGTHS (r_g VALUES) IN
FLUORINATED ETHANES[a]

Compound	Bond lengths (Å)		Ref.
	C—C	C—F	
CH_3—CH_3	1.533(2)	—	31,32
CH_2F—CH_3	1.503(5)	1.399(4)	33
CHF_2—CH_3	1.499(5)	1.366(2)	34
CF_3—CH_3	1.497(3)	1.340(2)	35
CH_2F—CH_2F	1.506(4)	1.388(2)	36–39
CHF_2—CH_2F	1.502(5)	1.354(4) (CHF_2)	40
		1.388(8) (CH_2F)	
CHF_2—CHF_2	1.520(5)	1.351(2)	41
CF_3—CH_2F	1.503(4)	1.336(2) (CF_3)	42
		1.391(6) (CH_2F)	
CF_3—CHF_2	1.527(4)	1.329(2) (CF_3)	33
		1.349(2) (CHF_2)	
CF_3—CF_3	1.545(3)	1.326(2)	43

[a] Parameters listed are from the first reference.

The trends with partial fluorination on both carbon atoms are more subtle. It should be pointed out that four independent ED studies have been reported for 1,2-difluoroethane, three of which are in essential agreement with each other (values in Table 4-2), while the bond lengths derived in reference 39 are considerably longer [C—C = 1.535(2) and C—F = 1.394(1) Å]. In fluorinated ethanes the CCF angles vary between 108.2(3)° in CHF_2—CHF_2 and 112.3(4)° in CF_3—CH_3.

Within the 1,2-substituted ethanes, difluoroethane is an exception with respect to its conformation. While in all other known cases the preferred conformation is anti, in CH_2F—CH_2F the gauche conformation is lower in energy than the anti form. Studies at various temperatures result in $\Delta E =$ $-3.9(17)^{36}$ or $-7.4(21)^{37}$ kJ mol^{-1} and $\Delta S = 6.3(33)^{36}$ or $4.1(40)^{37}$ J K^{-1} mol^{-1} between trans and gauche conformations. The existence of two conformations has also been observed for 1,1,2-trifluoro-[40] and 1,1,2,2-tetrafluoroethane[41]; the anti conformations (hydrogen atoms in anti positions) are strongly preferred in both cases (92% at 265 K and 84% at 253 K, respectively). The prevailing conformation ($\sim$90% at 429 K) in 2-fluoroethanol[44] is gauche (fluorine with respect to oxygen) with C—C = 1.513(3), C—F = 1.400(5) Å, CCF = 107.7(14)°, and CCO = 112.2(19)°.

Propanes. The effect of fluorination in propane is analogous to that in ethane. In 2-fluoropropane[45] and 1,3-difluoropropane[46] the C—C bonds are shorter [1.514(4) and 1.515(3) Å] than in the parent compound [1.523(2) Å[47]], and the C—F bonds are long [1.405(5) and 1.392(2) Å]. The CCC angles [114.5(15)° and 112.9(8)°] are hardly affected by fluorination [112.4(12)° in propane]. The conformational composition of 1,3-difluoropropane has been studied in combination with microwave spectroscopy, resulting in 63(4)% GG, 27(2)% AG

AA

AG

GG

GG''

and 10(5)% GG'' (at 293 K). No AA conformer has been detected. In 1,3-dichlorohexafluoropropane[48] a different conformational composition has been determined: 53(6)% AA, 39(8)% AG, and 10%GG (293 K). Here the C—C bonds [1.560(6) Å] are longer than in CF_3—CF_3 and the mean C—F bond lengths are 1.337(4) Å.

Tertiary Butanes. The longest C—F bond [1.425(8) Å] for any gas-phase compound has been determined for $(CH_3)_3CF^{49}$ [C—C = 1.520(8) Å, CCC = 111.0(5)°]. The relevant geometric parameters of three perfluoro-*tert*-butyl derivatives and of perfluoroneopentane are listed in Table 4-3. Small variations are observed for the C—F bond lengths and FCF angles. The C—C

Table 4-3. PRINCIPAL PARAMETERS (r_g VALUES) OF PERFLUORO *tert*-BUTYL
COMPOUNDS AND PERFLUORONEOPENTANE

Parameters	$(CF_3)_3CH$[a]	$(CF_3)_3CI$[b]	$(CF_3)_3COH$[c]	$(CF_3)_4C$[d]
C—F, Å	1.336(2)	1.333(4)	1.335(4)	1.328(2)
∠F—C—F, deg	108.0(2)	107.9(6)	108.3(4)	108.1(2)
C—C, Å	1.539(3)	1.544(15)	1.566(9)	1.566(3)
∠C—C—C, deg	112.9(2)	111.5(17)	110.4(8)	109.47

[a] Reference 50.
[b] Reference 51.
[c] Reference 52.
[d] Reference 53.

bonds lengthen with increasing electronegativity of the substituent (H < I <
OH < CF_3), and the CCC angles decrease in the same direction. This lengthening of the C—C bonds can be rationalized either in terms of hybridization
effects (decrease in *s* character with decreasing CCC angle), polar effects
(increasing positive net charge at the tertiary carbon atom increases the polar
repulsion with the positively charged CF_3 carbon atoms), or a combination of
both effects. In perfluoroneopentane the CF_3 groups are twisted by 18.3(5)°
from the exact staggered positions, thereby reducing the overall symmetry from
T_d to T.

Alkenes. The structures of fluorinated ethenes (Table 4-4) have been studied
in two laboratories, at Cornell University (Cornell data) and at the University of
Leiden (Leiden data). While the Cornell data are based on ED intensities alone,
the Leiden data were derived by joint analyses of ED intensities and MW
rotational constants (except for trans CHF=CHF). The study of this series is

Table 4-4. C=C AND C—F BOND LENGTHS (Å) IN FLUORINATED ETHENES

Compound	C=C Cornell[a]	C=C Leiden[b]	C—F Cornell[a]	C—F Leiden[b]
$CH_2=CH_2$	1.337(2)[c]		—	—
$CHF=CH_2$	1.333(7)	1.330(18)	1.348(4)	1.351(15)
$CHF=CHF$ (trans)	1.329(4)	1.320(9)	1.344(2)	1.338(3)
$CHF=CHF$ (cis)	1.331(4)	1.330(11)	1.335(2)	1.342(5)
$CF_2=CH_2$	1.316(6)	1.340(6)	1.324(3)	1.315(3)
$CF_2=CHF$	1.309(6)	1.341(12)	1.336(2)[d]	1.316(11)[e]
				1.342(24)[f]
$CF_2=CF_2$	1.311(7)		1.319(2)	

[a] Reference 54.
[b] Reference 55–59.
[c] Reference 60.
[d] Mean value.
[e] For CF_2.
[f] For CHF.

hampered by large correlations between the closely spaced $C{=}C$ and $C{-}F$ bond lengths. The results reported by both laboratories are essentially in agreement for $CHF{=}CH_2$ and $CHF{=}CHF$ (cis and trans) but differ considerably for $CF_2 = CH_2$ and $CF_2{=}CHF$. Both data demonstrate shorter $C{-}F$ bonds in CF_2 groups than in CHF groups (the Cornell $C{-}F$ value for $CF_2{=}CHF$ is a mean value). This trend is analogous to the "fluorine effect" in the methanes, but much smaller for sp^2-hybridized carbon atoms. For the $C{=}C$ bonds the Cornell data indicate gradual shortening with increasing fluorination, despite large error limits, while the Leiden data show no systematic trend in the $C{=}C$ bond lengths. Within the experimental uncertainties the $C{=}C$ bond is essentially not affected by fluorination. Although, in general, structural information from joint ED/MW analyses (Leiden data) are more reliable, the trends in the $C{=}C$ force constants, MW results, and *ab initio* calculations favor the Cornell data. The $C{=}C$ force constants increase with increasing fluorination and correlate closely with this set of bond lengths. For $CF_2{=}CH_2$ r_s values of 1.315(5) and 1.323(5) Å for $C{=}C$ and $C{-}F$ were derived from rotational constants,[61] in perfect agreement with the Cornell data. Also in the case of $CF_2{=}CHF$ the MW r_0 value for $C{=}C$ $[1.315(20)$ Å$]$[62] favors the Cornell results, although the large error limit does not allow a definite conclusion. Furthermore, *ab initio* calculations[63] confirm the trend observed for the $C{=}C$ bonds lengths in the Cornell data, ie, shortening with double fluorination at one carbon atom. Comparison of $CF_2{=}CF_2$ and $CH_2{=}CH_2$ demonstrates that perfluorination has opposite effects on the $C{=}C$ double bond in ethene (shortening by about 0.03 Å) and on the $C{-}C$ single bond in ethane (lengthening by about 0.01 Å). The CCF bond angles of the two investigations are in close agreement. These angles are around $122°$ in CHF groups and $125°$ in CF_2 groups. For 1,2-difluoroethene the cis isomer is more stable than the trans form [by 4.5(5) kJ mol^{-1}].[64] This feature only recently has been reproduced correctly by *ab initio* calculations.[65]

The geometry of the CF_2 group is not affected by chlorine substitution in $CF_2{=}CCl_2$[66] $[C{-}F = 1.315(15), C{=}C = 1.345(25)$ Å, CCF $= 124.0(13)°$, and CCCl $= 120.5(5)°]$.

Propenes and But-2-enes. Table 4-5 compares skeletal parameters of fluorinated propenes and but-2-enes to those of the parent compound. In general, fluorination of one or two sp^3 carbon atoms results in shortening of double and single bonds, except in the case of two geminal CF_3 groups, where both bonds are lengthened. The exceptional behavior of $CH_2{=}C(CF_3)_2$ is also demonstrated by the $C{=}C{-}C$ bond angles, which are smaller by about $7°$ than in all other compounds. The torsional position of the CF_3 groups is generally eclipsed with respect to the $C{=}C$ bond, except for the case of *cis*-$CHCF_3{=}CHCF_3$, where the CF_3 groups are staggered, and in $CH_2 = C(CF_3)_2$, where the CF_3 groups occupy a position that is intermediate between staggered and eclipsed. Only one predominant conformer has been detected for *cis*-1-chloro-3-fluoropropene[71] **(2)** [fluorine close to trans with respect to the $C{=}C$ bond, $\tau(CCCF) = 149(4)°$] and 1,1,2-trichloro-3,3-difluoro-1-propene[72]

Table 4-5. SKELETAL PARAMETERS (r_g VALUES) FOR FLUORINATED PROPENES AND BUT-2-ENES

| | Bond lengths (Å) | | | |
Compound	C=C	C—C	∠C=C—C (deg)	Ref.
CH_2=$CHCH_3$	1.342(2)	1.506(3)	124.3(4)	67
CH_2=$CHCF_3$	1.318(8)	1.495(6)	125.8(11)	67
CF_2=$CFCF_3$	1.329(3)	1.513(3)	127.8(7)	68
$CHCF_3$=$CHCF_3$ (trans)	1.296(20)	1.481(5)	125.1(16)	69
$CHCF_3$=$CHCF_3$ (cis)	1.310(16)	1.492(5)	126.0(5)	69
CH_2=$C(CF_3)_2$	1.375(13)	1.533(6)	118.2(2)	70

(**3**) (hydrogen eclipsed to C=C bond). 2-Chloro-3-fluoro-1-propene[73] exists as a mixture of syn (**4a**) and gauche (**4b**) conformers in a ratio of 3:1.

The only gas-phase study of a fluorinated alkyne reported since 1975 is a joint ED/MW analysis of CF_3—C≡C—SiH_3,[74] where the carbon–carbon triple and single bonds are slightly longer [1.225(8) and 1.493(8) Å] than in CF_3—C≡C—CF_3 [1.201(3) and 1.474(4) Å, average values of references 75 and 76].

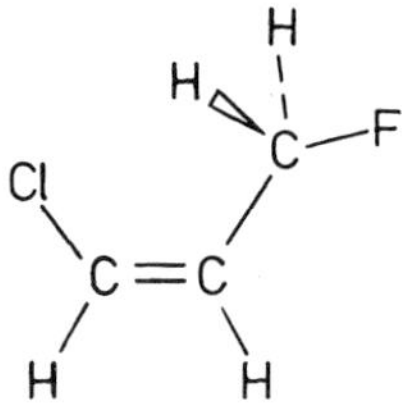

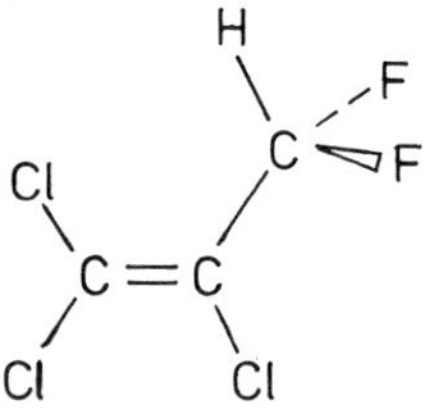

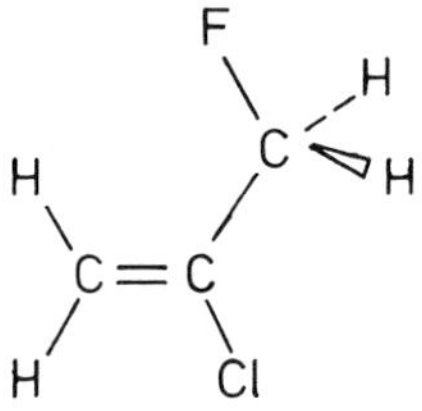

Compounds Containing C=O, C=S, and C=Se Groups. Precise structural data from joint ED/MW analyses have been reported for some carbonyl, thiocarbonyl, and selenocarbonyl compounds (Table 4-6). The geometry of the CF_2 groups in X=CF_2 (X = O, S, Se) is remarkably constant. Substitution of one fluorine by chlorine has analogous effects in carbonyl and thiocarbonyl compounds; that is, the C—F bond is lengthened by about 0.02 Å, no change occurs in the C=X bond lengths, and the XCF and XCCl bond angles are equal

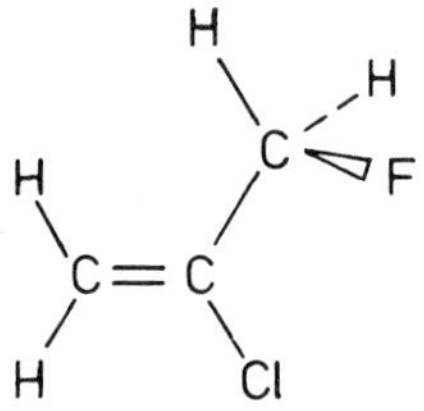

Table 4-6. Geometric Parameters of Carbonyls, Thiocarbonyls, and Selenocarbonyls, X=CFY (r_α° values)

Compound	Bond lengths (Å)			Bond angles (deg)		Ref.
	C=X	C—F	C—Y	∠XCF	∠XCY	
O=CF$_2$	1.1717(13)	1.3157(8)	—	126.15(4)	—	77
O=CFCl	1.173(2)	1.334(2)	1.725(2)	123.7(2)	127.5(3)	78
O=CFH	1.186(4)	1.345(3)	1.096(11)	122.3(2)	130(4)	79
S=CF$_2$	1.589(2)	1.316(1)	—	126.5(1)	—	80
S=CFCl	1.592(2)	1.336(3)	1.716(2)	123.8(3)	127.5(2)	81
Se=CF$_2$	1.743(3)	1.314(2)	—	126.3(2)	—	80

in OCFCl and SCFCl. No pronounced structural trends with fluorination are observed in mono,[82] di-,[83] and trifluoroacetic acid.[84] Two conformations have been detected for monofluoroacetic acid (trans with zigzag FCCOH chain and cis with a cyclic FCCOH chain) and for difluoroacetic acid with different rotational positions of the CHF$_2$ groups. A joint ED/MW analysis of 2-fluoroacetamide, NH$_2$C(O)CH$_2$F,[85] results in a planar heavy-atom skeleton with the C—F bond syn to the C—N bond.

Cyclic Compounds. The effects of perfluorination on carbon ring skeletons have been discussed by Yokozeki and Bauer.[1] Minor shortening of the C—C bonds occurs in cyclopropane, and the ring bonds in cyclobutane and cyclohexane become lengthened. In addition, simultaneous decrease of the puckering angles (flattening of the ring) occurs. These trends have been confirmed by recent studies of perfluoronitrosocyclobutane[86] [C—C = 1.573(5) Å, puckering angle 23.0(18)°, nitroso group in exo orientation] and *trans-* and *cis-*perfluorodecalin.[87] In spiropentane (**5**) the effect of fluorination on the ring

F$_2$C   CF$_2$
4   3   1
C
5   2
F$_2$C   CF$_2$

5

bond lengths is not uniform[88]: the C$_1$—C$_2$ bonds shorten from 1.519(3) Å in the parent compound[89] to 1.487(6) Å, and the C$_1$—C$_3$ bonds lengthen from 1.469(1) to 1.492(4) Å. Thus, the original difference in the C—C bond lengths of about 0.05 Å vanishes with fluorination. This result is in agreement with theoretical predictions for the effect of fluorination on adjacent and opposite C—C bonds in three-membered rings.[90]

The most interesting structural features of fluorinated biphenyls are the central bond length (C$_1$—C$_1'$) and the torsional angle between the two phenyl rings. Fluorination in the para position causes shortening of the central bond length from 1.509(4) in biphenyl[91] to 1.497(3) in 4-fluorobiphenyl[92] to

1.483(4) Å in 4,4′-difluorobiphenyl.[92] The torsional angle is not affected by substitution in the para position and is about 45° for the three compounds. In perfluorobiphenyl[93] the torsional angle increases to 70(2)°, and the central bond length [1.51(1) Å] is the same as in biphenyl. Both effects, ie, shortening of the central bond and increase of the torsional angle, are observed in 4-fluoro-2′,4′,6′-trimethylbiphenyl[94] [1.488(9) Å and 77.5(21)°].

The effect of fluorinating nitrogen-containing heterocycles is different from that in saturated carbon rings. In diazirine the $N=N$ bond shortens substantially and the $N-C$ bonds lengthen [$N=N$ = 1.228(5) and $N-C$ = 1.482(5) Å in diazirine[95] vs 1.293(9) and 1.426(4) Å in difluorodiazirine[96]]. The geometry of the C_3N_3 ring in triazine,[97] however, is not affected by fluorination [$C-N$ = 1.3381 Å and NCN = 126.8° in triazine vs 1.333(9) Å and 127(1)° in trifluorotriazine].[21]

An ED study of perfluoro(methyl)oxirane[98] revels that the $C-C$ ring bond length [1.467(7) Å] is approximately equal to that in oxirane [1.466(2) Å, r_s value from MW[99]], while the $C-O$ bonds shorten with fluorination [1.410(8) vs 1.431(2) Å in oxirane]. A microwave spectroscopy study of perfluorooxirane[100] revealed that shortening of all ring bonds occurs as a result of fluorination by 0.04 Å [$C-C$ = 1.426(4) and $C-O$ = 1.391(2) Å]. Fluorination of succinic anhydride[101] (6) causes lengthening of the $C-C$ bonds [C_2-C_3 = 1.528(5) and C_3-C_4 = 1.573(9) Å vs 1.510(4) and 1.535 Å in the parent compound][102] and, as an exception to the general trend, slight increase of the ring puckering is observed.

6

Three four-membered heterocycles that contain two sulfur or selenium atoms have been studied by ED: 1,2-bis(trifluoromethyl)dithiete[103] (7), tetrafluoro-1,3-dithietane[104,105] (8), and tetrafluoro-1,3-diselenetane[106] (9). The four-membered rings are planar in all three compounds. The S—S bond length in

7 is 2.05(1) Å. The bond lengths reported in two independent ED studies for 8 differ systematically by about 2%. On the basis of the constancy of geminal F…F distances in fluoromethanes (F…F = 2.15 Å, see above), the parameters of

reference 104 [S—C = 1.820(2), C—F = 1.344(2) Å, SCS = 97.3(2), FCF = 106.5(2)°] with F...F = 2.15 Å should be favored.[107] This F...F distance is unreasonably small (2.09 Å), as reported in reference 105. In the analogous selenium compound **9**, Se—C = 1.968(4) Å, C—F = 1.353(3) Å, SeCSe = 98.5(4)°, and FCF = 106.3(8)°. The S—C and Se—C bond lengths in these four-membered rings are equal to the corresponding values in the unstrained bis(trifluoromethyl) compounds (see section on chalcogen(II) compounds).

4-(Trifluoromethyl)-1,2,3,5-dithiadiazole[108] (**10**) is a stable radical in the gas phase, with a planar five-membered ring. The S—S bond in the 7π radical is longer [2.113(6) Å] than in the dimeric solid [2.087 Å] or in the 6π cation 2.086 Å), whereas the other ring parameters are nearly constant in all three forms.

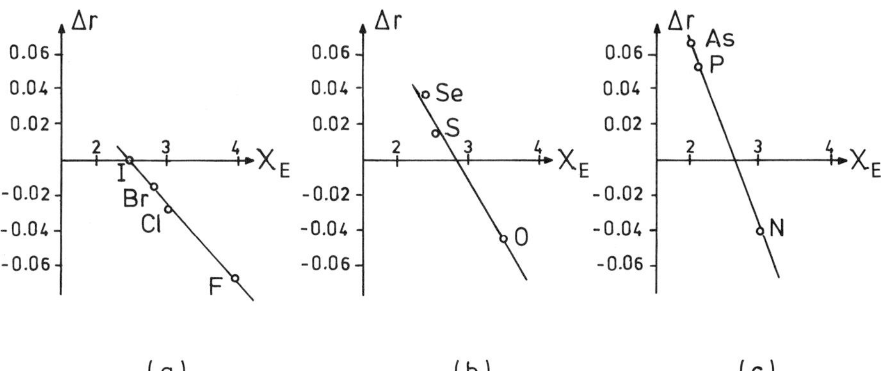

10

The CH$_3$/CF$_3$ Substituent Effect. It has been pointed out by Yokozeki and Bauer[1] that E—C distances in E(CH$_3$)$_n$ compounds, with $n = 1, 2, 3$ and E = main group element, decrease or increase with substitution of CH$_3$ by CF$_3$, depending on the electronegativity χ of the central atom E. If χ_E is larger than approximately 2.5 (that of carbon), the bonds shorten, and vice versa. Additional and more precise structural data have been reported since 1975, and this correlation between bond length variation and electronegativity can now be discussed in more detail. Plots of this substitution effect (Δr vs χ_E) for $n = 1, 2, 3$ (Figure 4-1) afford nearly linear correlations, and the slope increases with increasing n. From this trend of increasing slope with increasing number of

Figure 4-1. CH$_3$/CF$_3$ substitution effect on adjacent bond lengths. $\Delta r =$ (E—CH$_3$)–(E—CF$_3$) vs electronegativity χ_E: (a) ECH$_3$/ECF$_3$, (b) E(CH$_3$)$_2$/E(CF$_3$)$_2$, (c) E(CH$_3$)$_3$/E(CF$_3$)$_3$.

substituents we would expect an even larger variation of Δr with electronegativity for four-coordinated central atoms [$E(CH_3)_4$ vs $E(CF_3)_4$, E = C, Ge, Sn]. This expectation, however, is not confirmed by experimental data: $\Delta r = 0.025(4)$ Å for E = C ($\chi = 2.5$), $\Delta r = 0.049(6)$ Å for E = Ge, and $\Delta r = 0.058(6)$ Å for E = Sn ($\chi = 1.8$).

A simple qualitative explanation of the CH_3/CF_3 substitution effect, based on polar contributions to the E—C bonds, has been suggested.[109] Methyl carbon atoms carry small net charges, and polar effects are negligible in methyl compounds. Trifluoromethyl group carbon atoms, however, have high positive net charges that lead to attractive polar interactions and to bond shortening if the central atom carries a negative net charge (high electronegativity, eg, E = F, O, N), and vice versa (for, eg, E = Se, P, As). For compounds with $n = 4$, especially for perfluoroneopentane, steric interactions may be important in addition to these polar effects. This concept of polar contributions is applicable for nonuniformly substituted compounds as well. The effect of all substituents on the net charge of the central atom must be considered. This fact is demonstrated by the dependence of substituent effects on the oxidation state of the central atom. For S(II) compounds the lengthening of the S—C bonds upon fluorination is small ($\Delta r = +0.014$ Å), but for S(IV) and S(VI) compounds the effect is much larger ($\Delta r = +0.077$ Å for sulfoxides and $+0.087$ Å for sulfones). A similar observation is made for phosphorus compounds, where $\Delta r = +0.057$ Å in phosphines and $+0.083$ Å in phosphine oxides.

Replacement of CH_3 by CF_3 affects the bond angles at the central atom in different directions, depending on the "size" of the cental atom. For third-row or larger atoms, where the E—C bonds are long, the bond angles are slightly smaller in the CF_3 compounds: $\Delta\alpha = -1.8(9)°$ in the sulfides and $-1.6(9)°$ in the phosphines. This trend is in accord with predictions based on the VSEPR model. Since CF_3 groups have a higher electronegativity than CH_3, the interactions between the bonding electron pairs are reduced, thereby leading to an angle decrease. For second-row atoms, such as C, N, or O, where the E—C bonds are short, CH_3/CF_3 substitution leads to a substantial increase of CEC angles: $\Delta\alpha = +2.0(3)°$ in the isobutanes, $+7.0(8)°$ in the amines, and $+6.7(10)°$ in the ethers. In these cases repulsive interactions between nonbonded fluorine atoms override the interactions between bonding electron pairs, and these trends contradict the VSEPR predictions. Additional data and references are given in the respective sections.

Silicon, Germanium, and Tin Compounds

The "fluorine effect", ie, decreasing bond lengths with increasing fluorination, is less pronounced for silanes and germanes than it is for the methane series. While the C—F bonds shorten by 0.072(2) Å from CH_3F to CF_4, the corresponding effect is only 0.039(6) Å for the silanes [Si—F = 1.593(5) Å in SiH_3F[110] and 1.555(3) Å in SiF_4[111]; SiH_3F, SiH_2F_2,[112] and $SiHF_3$[113] have been studied by

Table 4-7. PRINCIPAL PARAMETERS FOR $(CH_3)_{4-n}SiF_n$, $n = 0\text{-}4$
(r_g VALUES)

Compound	Bond lengths (Å)		Bond angles (deg)		
	Si—F	Si—C	∠ FSiF	∠ CSiC	Ref.
$(CH_3)_4Si$	—	1.876(2)	—	109.47	116
$(CH_3)_3SiF$	1.600(1)	1.848(1)		111.5(2)	117
$(CH_3)_2SiF_2$	1.586(1)	1.836(2)	104.6(4)	116.7(6)	117
$(CH_3)SiF_3$	1.570(1)	1.828(4)	106.8(5)	—	117
SiF_4	1.555(3)	—	109.47	—	111

MW] and 0.06(3) Å for the germanes [1.732 Å in GeH_3F^{114} and 1.67(3) Å in $GeF_4{}^{115}$]. The latter value was derived by visual analysis. Similar effects are observed for the methylfluorosilanes $(CH_3)_{4-n}SiF_n$ (Table 4-7), where both Si—F and Si—C bonds shorten by about the same amount with increasing fluorination. This simultaneous shortening of both bonds can be rationalized in terms of increasing polar contributions to both bonds (the methyl carbon atom is assumed to carry a small negative net charge) and by contraction of the silicon valence shell ($3s$ and $3p$ orbitals). The importance of valence shell contraction in this series has been demonstrated by NMR data.[118] Hybridization effects may contribute additionally to bond shortening, as indicated by the increase of FSiF and CSiC angles with increasing fluorination. The short Si—F bonds in SiF_4 as compared to the Schomaker–Stevenson value of 1.71 Å given in reference 26 have often been attributed to π-bond contributions with d-orbital involvement. The Schomaker–Stevenson rule, however, is based on diatomic molecules and badly underestimates polar contributions in polyatomic compounds that contain several fluorine atoms. *Ab initio* calculations for $SiF_4{}^{119}$ demonstrate that the inclusion of d functions in the basis set has a minor effect on the Si—F bond lengths (1.59 Å without and 1.57 Å with d functions). A Mulliken population analysis indicates that the d functions to a large extent just compensate for an insufficient s,p basis set, while $(p - d)\pi$ contributions are negligible. The polarity of the Si—F bond, however, is much larger than can be accounted for by the Schomaker–Stevenson rule. Furthermore, $(p - d)\pi$ contributions would affect only the Si—F bond and could not explain the similar shortening of the Si—C bonds in this series. For the analogous germane series structural data are not complete. No structure for $(CH_3)_3GeF$ has been reported, and the Ge—F bond length in GeF_4 [1.672(3) Å] is a recent result from our laboratory. The available data demonstrate similar trends as observed for the silane series: Ge—F bonds shorten from 1.739(2) Å in $(CH_3)_2GeF_2{}^{120}$ to 1.714(2) Å in CH_3GeF_3,[120] and Ge—C bonds shorten from 1.945(3) Å in $(CH_3)_4Ge^{121}$ to 1.928(3) Å in $(CH_3)_2GeF_2$ and to 1.904(9) Å in CH_3GeF_3. In bis(*tert*-butyl)difluorosilane[117] the Si—F and Si—C bonds are longer than in dimethyldifluorosilane [1.606(4) vs 1.586(1) Å and 1.869(3) vs 1.836(2) Å]. Hybridization effects [FSiF = 97.7(8)

Table 4-8. Si—C and Ge—C Bond Lengths (Å) in Fluorinated Methyl Silanes and Germanes (r_g values, unless otherwise indicated)

Bond	CH_3XH_3	Ref.	CH_3XF_3	Ref.	CF_3XH_3	Ref.	CF_3XF_3	Ref.
Si—C	1.867(3)[a]	122	1.828(4)	117	1.925(3)	123	1.913(6)	124
Ge—C	1.945(5)[b]	125	1.904(1)	120	1.997(6)	126	1.970(9)	126

[a] r_0 value.
[b] r_s value.

and CSiC = 125.5(11)°] and steric repulsion between the two *tert*-butyl groups can rationalize these trends qualitatively.

Fluorination of methylsilane and germane has remarkable effects on the Si—C and Ge—C bond lengths (Table 4-8), which depend strongly on the site of fluorination (C or Si/Ge). Fluorination of Si/Ge causes *shortening* of the X—C bonds by about 0.04 Å, whereas fluorination of carbon causes *lengthening* by about 0.05 Å. This leads to a bond length difference of almost 0.1 Å between CH_3—XF_3 and CF_3—XH_3. Such an opposite effect of fluorination at C or Si has also been predicted for the Si=C bond on the basis of *ab initio* calculations.[127] The X—C bonds in the perfluorinated compounds CF_3—XF_3 are longer than in the parent compounds, but shorter than in CF_3—XH_3. The concept of polar contributions and valence shell contraction can explain these trends: fluorination of X leads to attractive polar interaction ($C^{-\delta}$—$X^{+\delta}$) and valence shell contraction at the X atom and thus to bond shortening. In the CF_3—XH_3 compounds the polar contribution is repulsive ($C^{+\delta}$—$X^{+\delta}$), and there is no orbital contraction at X. This leads to bond lengthening. In the perfluorinated compounds the polar repulsion increases further, but its effect is compensated in part by valence shell contraction, thereby leading to bond lengths that are intermediate between CH_3—XH_3 and CF_3—XH_3. Whereas the X—C bond lengths in methyl trifluoro compounds CH_3XF_3 increase substantially upon CH_3/CF_3 substitution (~0.08 Å), the X—F bond lengths decrease slightly [Si—F = 1.570(1) vs 1.556(2) Å and Ge—F = 1.714(3) vs 1.692(2) Å in the CH_3- and CF_3-substituted derivatives, respectively].

No further trifluoromethyl silanes in addition to the compounds listed in Table 4-8 have been synthesized so far, whereas several germane derivatives are known and have been studied by ED. A similar CH_3/CF_3 substitution effect as pointed out above for CH_3GeF_3 is observed between $(CH_3)_2GeF_2$[120] [Ge—C = 1.928(3), Ge—F = 1.739(2) Å, FGeF = 105.4(20)°, CGeC = 121.0(35)°] and $(CF_3)_2GeF_2$[128] [Ge—C = 2.001(5), Ge—F = 1.698(3) Å, FGeF = 109.3(34), CGeC = 117.4(17)°]. As expected from the foregoing arguments, the Ge—C bond length in CH_3GeCl_3 also increases with CH_3/CF_3 substitution [Ge—C = 1.929(6) Å in CH_3GeCl_3[129] vs 1.990(12) Å in CF_3GeCl_3],[126] but this effect is smaller than in the fluorogermanes. The bond

Table 4-9. Principal Structural Parameters of SiF_3 and SiH_3 Compounds

Compound	SiF₃ derivative			SiH₃ derivative		
	Si—E (Å)	α_E (deg)[a]	Ref.	Si—E (Å)	α_E (deg)[a]	Ref.
$(SiX_3)_2CH_2$	1.827(4)	117.7(4)	135	1.873(2)	114.4(2)	136
$(SiX_3)_2CCl_2$	1.871(8)	119.4(6)	135	—	—	
SiX_3—CH_2—CH=CH_2	1.837(10)	109.9(57)	137	1.875(4)	113.1(4)	138
$SiX_3N(CH_3)_2$	1.654(15)	119.7(16)	139	1.715(4)	120.0(4)	140
SiX_3NCO	1.648(10)	160.7(12)	141	1.703(4)	151.7(12)	142
$(SiX_3)_2O$	1.580(25)	155.7(20)	143	1.634(2)	144.1(9)	144
SiX_3OCH_3	1.58 (fixed)	131.4(32)	145	1.640(3)	120.6(10)	146
SiX_3PH_2	2.207(3)	90.2(33)	147	2.249(3)	91(3)	148
$SiX_3As(CH_3)_2$	2.334(9)	93.5(9)	149	2.355(1)[b]	—	

[a] Bond angle at E involving SiE bond.
[b] Si—As bond length in $As(SiH_3)_3$ (see reference 150).

length variation in tetramethyl compounds with substitution of CH_3 by CF_3 was discussed in the preceding section. The Ge—C bond length in $Ge(CF_3)_4$[130] is 1.994(5) Å [1.945(3) Å in $Ge(CH_3)_4$[131]] and Sn—C = 2.202(5) Å in $Sn(CF_3)_4$[132] [2.144(3) Å in $Sn(CH_3)_4$[133]]. Both fluorinated compounds have T_d symmetry, ie, CF_3 groups stagger the bonds around the central atom, with large-amplitude torsional vibrations. The only other tin compound containing fluorine that has been studied in the gas phase is tetrakis(trifluoropropynyl)tin, $Sn(C\equiv C—CF_3)_4$[134]: Sn—C = 2.070(7), C—C = 1.460(7) and C$\equiv$C = 1.215(6) Å.

ED studies for several compounds containing SiF_3 groups have been reported. Principal parameters are listed in Table 4-9 and are compared to those of the analogous silyl compounds, insofar as their structures have been determined. In addition to the compounds listed in Table 4-9, ED studies of cyclopropyl-[151] and 2,2-difluorocyclopropyltrifluorosilane[152] have been reported. The Si—F bond lengths (not given in Table 4-9) vary between 1.553(4) Å in SiF_3NCO and 1.600(3) Å in 2,2-difluorocyclopropyltrifluorosilane and decrease in general with increasing electronegativity of the atom to which SiF_3 is bonded. This correlation is similar to the observation made for the CF_3 groups, but the variations are larger. The FSiF angles are in all cases smaller than tetrahedral, ie, $107 \pm 2°$. More interesting than the geometric parameters of the SiF_3 groups themselves, is their effect upon the adjacent bond length and bond angle, as compared to that of the SiH_3 group. Whereas CH_3/CF_3 substitution leads to shortening or lengthening of C—E bonds, depending on whether the electronegativity of E is larger or smaller than that of carbon, the Si—E bonds shorten in all known examples with fluorination at Si. Since Si is more electropositive than all atoms E in Table 4-9, this observation is compatible with the concept of increased bond polarity in SiF_3 derivatives, which was suggested in the CH_3/CF_3 case. Due to large experimental uncertainties, the data in Table 4-9 do not show a strict correlation between bond shortening and electronegativity of E, but in general the effect increases with increasing electronegativity. In addition to these polar effects, the contraction of the Si valence shell with fluorination (which was discussed above) may enhance this substitution effect. The limited data available in the literature also indicate a general increase of the bond angle at the adjacent atom with SiH_3/SiF_3 substitution, thereby indicating larger steric requirements of the SiF_3 group as compared to SiH_3. These steric interactions, of course, increase with the observed bond shortening.

The differences in the Si—C bond lengths in disilylmethane and bis(trifluorosilyl) methane and in the allyl derivatives of about 0.04 Å correspond almost exactly to the observed difference between CH_3SiH_3 and CH_3SiF_3. The long Si—C bonds in $(SiF_3)_2CCl_2$ can be interpreted as being a consequence of chlorination, which renders the polar contribution to the Si—C bond repulsive $(Si^{+\delta} - C^{+\delta})$. Fluorination of silyldimethylamine leads to shortening of the Si—N bonds of about 0.06 Å. It does not affect the SiNC angles but increases the CNC angles from 111.1(12) to 120.3(30)°, thereby resulting in a planar heavy-atom skeleton in trifluorosilyl dimethylamine. The effect of fluorination on the Si—N bond in silylisocyanate is very similar to that in the amines. Both SiH_3NCO and SiF_3NCO are nonrigid molecules with large-amplitude SiNC bending vibrations, and the SiNC angles must be considered as being "effective" angles. For SiH_3NCO the ED intensities are reproduced best with a double-minimum bending potential that possesses minima at 159° and a barrier to linearity of $20 \, cm^{-1}$. These results are in good agreement with the results of an extensive MW study.[153] The effective SiNC angle increases with fluorination, and the ED analysis cannot discriminate between a double-minimum potential with low barrier and a flat single-minimum potential. A similar situation exists for the siloxanes, where the oxygen bond angles increase with fluorination by about 10° and the Si—O bonds shorten (this bond length has not been refined for SiF_3OCH_3). In the phosphorus and arsenic compounds the effect of fluorination on the Si—P and Si—As bonds is only about 0.02 Å, consistent with the low electronegativities of P and As. The bond angles in these compounds appear not to be affected.

Two fluorinated disilanes have been studied by ED: hexafluorodisilane[154,155] and $SiF_3—SiF_2BF_2$.[8] The results of two independent analyses for Si_2F_6, using either a rigid[154] or a nonrigid[155] molecular model, afford geometric parameters in mutual agreement within their error limits, which are listed with the first value from reference 154 and second value from reference 155: Si—Si = 2.319(6), 2.326(6) Å; Si—F = 1.569(2), 1.570(2) Å; FSiF = 108.7(3), 108.3(3)°. The Si—Si bond is only slightly shorter than in Si_2H_6 [2.332(3) Å].[156] The strongly repulsive contribution in Si_2F_6 $(Si^{+\delta} - Si^{+\delta})$ is apparently compensated by Si valence shell contraction. Bonding with d-orbital participation, $(p-d)\pi$ or $(d-d)\pi$, is unlikely, since the barrier to internal rotation in the fluorinated species is low $(2.0–2.8 \, kJ/mol^{-1})$[155]. A somewhat longer Si—Si bond has been reported for $SiF_3—SiF_2BF_2$ [2.361(3) Å].

Group V Compounds

Nitrogen Fluorine Compounds

A small number of fluorinated amines have been studied by ED and MW. An ED study of the difluoroamino radical NF_2[157] resulted in N—F = 1.363(8) Å and FNF = 102.5(9)° (r_a values), in fair agreement with a later MW report[158]: N—F = 1.3528(1) Å and FNF = 102.5(9)° (r_z values). Slightly longer N—F bonds have been determined by MW for NF_3 (r_0 values): N—F = 1.371(2) Å and FNF = 102.2(1)°. According to an MW study,[160] the N—F bonds increase in HNF_2 (r_z values): N—F = 1.400(3) Å, FNF = 102.9(3)°, and HNF = 99.8(3)°. This lengthening is not confirmed by an ED study [N—F = 1.367(4) Å],[161] the execution of which may have been hampered by impurities. Monofluoroamine, H_2NF, has been synthesized only recently, and an MW study, which is still in progress, indicates an even longer N—F bond [preliminary value 1.433(5) Å].[162] Thus, a "fluorine effect," which was pointed out for the fluorocarbons and silanes, seems to be present also in the fluoroamines. This conclusion is supported by the observation of an intermediate N—F bond length in $ClNF_2$[163]: N—F = 1.382(9) Å, FNF = 102.8(10)°. Although the N—F bonds vary considerably in these amines, the FNF angles are fairly constant (102–103°).

Some compounds that are composed of the highly electronegative elements nitrogen, oxygen, and fluorine (eg, nitrosylfluoride, FNO,[164] nitryl fluoride, FNO_2,[165] and trifluoroamine oxide, F_3NO[166]) have been studied by ED or by MW. The N—F bonds shorten with increasing coordination number from 1.512(5) Å in FNO to 1.467(15) Å in FNO_2 to 1.431(3) Å in F_3NO and in all three compounds are considerably longer than in NF_3. The NO bond lengths increase from 1.136(5) Å in FNO to 1.180(5) Å in FNO_2 and decrease again to 1.158(4) Å in F_3NO. The apparently slightly smaller FNF bond angles in F_3NO [100.8(12)°] relative to NF_3 [102.2(1)°] indicate a somewhat stronger steric effect of the N=O bond as compared with that of the nitrogen lone electron pair.

The series $(CF_3)_nNF_{3-n}$ demonstrates the effect of substituting fluorine by CF_3 groups on the skeletal parameters. Whereas the N—F bond lengths do not change with increasing number of CF_3 groups [1.371(2), 1.371(4),[167] and 1.378(12)[167] Å for n = 0, 1, 2], the N—C bonds shorten [1.476(5), 1.446(4), 1.426(6) Å[168] for n = 1, 2, 3]. This shortening has been rationalized quantitatively on the basis of polar contributions.[167] Furthermore, hybridization effects may be important, since the sum of the nitrogen bond angles increases from 306.6(1)° in NF_3 to 313.5(11)° in CF_3NF_2 [FNF = 105.3(5)°, CNF = 104.1(7)°] to 326.2(12)° in $(CF_3)_2NF$ [CNF = 104.8(7)°, CNC = 116.6(6)°] to 353.7(7)° in $(CF_3)_3N$ [CNC = 117.9(4)°]. While CH_3/CF_3 substitution in trimethylamine causes shortening of the N—C bonds by 0.032(7) Å [N—C = 1.458(2) Å in $(CH_3)_3N$],[18] a lengthening of 0.027(8) Å is observed between CH_3NF_2[169] and

CF_3NF_2. This reversed CH_3/CF_3 effect indicates a change of the nitrogen net charge from negative in $(CF_3)_3N$ to positive in CF_3NF_2.[109] The most striking effect between methyl- and trifluoromethyl-substituted amines is the increase in the CNC bond angles from 110.9(6)° in $(CH_3)_3N$ to 117.9(4)° in $(CF_3)_3N$. A still larger CNC bond angle has been reported for $(CF_3)_2NOH$[170]: CNC = 120.6(13)°, N—C = 1.435 Å (assumed).

Exceptional conformational properties are observed in tetrafluorohydrazine.[171,172] Both studies result in a mixture of two conformers with the trans conformer being more stable [71(8)% at 25°C][171] than the gauche form [dihedral angle 64.2(37)°].[172] The N—F bond lengths [1.373(4) Å, average value of references 171 and 172] are equal to those in NF_3, and the N—N bond is considerably longer than in hydrazine [1.490(6) Å in F_2N—NF_2 vs 1.449(2) Å in H_2N—NH_2].[173] Hydrazine exists only in the gauche conformation and possesses a dihedral angle of 91(2)°. Tetrakis(trifluoromethyl) hydrazine, $(CF_3)_2N$—$N(CF_3)_2$,[174] also exists only in the gauche form and possesses a dihedral angle of 88(4)°. Despite nonbonded repulsions between the CF_3 groups, the N—N bond [1.402(20) Å] is appreciably shorter than for the two hydrazines discussed above. The steric interactions are minimized by increasing the nitrogen bond angles [CNC = 121.2(15)° and NNC = 119.0(15)°], leading to an almost planar configuration at N.

Difluorodiazine, FN=NF, also exists as two conformers, the cis conformation being lower in energy[64] by about 12 kJ mol^{-1}, in contrast to HN=NH, where only the trans conformer has been observed.[175] The two conformers of FN=NF have been studied by separate ED experiments.[176] In the cis form the N=N bond appears to be slightly shorter than in the trans species [1.214(12) vs. 1.231(10) Å], and these bonds are shorter in both conformers than in the parent compound [1.252(2) Å in HN=NH].[175] The N—F bonds in the two conformers are equal within their error limits [1.410(9) and 1.396(8) Å for cis and trans, respectively], but the NNF bond angles differ appreciably, [NNF = 114.4(10)° for cis and 105.5(7)° for trans]. The results of an MW study for the cis conformer[177] are in essential agreement with the ED results.

Whereas methanimine, HN=CH_2, which is isoelectronic to HN=NH, is very unstable,[178] the perfluorinated compound, FN=CF_2 is sufficiently stable for a joint ED/MW study.[179] The N=C bond [1.274(6) Å in FN=CF_2 and 1.273 Å in HN=CH_2] is not affected by fluorination. The C—F bonds are extremely short [1.300(3) Å] when compared to such bonds in other =CF_2 groups such as O=CF_2 [1.3157(8) Å] or F_2C=CF_2 [1.319(2) Å]. The FCF angle in perfluoromethanimine is 112.5(2)°, and the CF_2 group is tilted toward the nitrogen lone electron pair by an angle of 3.7(2)°. The N—F bond [1.389(2) Å] is slightly shorter than the corresponding bonds in diazine, and the FNC angle [107.9(2)°] is appreciably smaller than the FNN angle in cis-FN=NF [114.4(10)°].

Several compounds containing the CF_3N=X moiety have been studied; the principal parameters are listed in Table 4-10. The parameters in the three N-trifluoromethyl methanimines are very similar; the C—N bond lengths are

Table 4-10. PRINCIPAL PARAMETERS OF $CF_3N=X$ MOIETIES

Compound	Bond lengths (Å)		$\angle CNX$ (deg)	Ref.
	C—N	N=C		
$CF_3N=CF_2$	1.419(13)	1.275(15)	117.3(18)	180
$CF_3N=CHF$ (trans)	1.411(7)	1.277(7)	117.3(9)	181
$CF_3N=CHCl$ (trans)	1.427(6)	1.277(11)	116.3(8)	181
$CF_3N=CO$	1.392(8)	1.233(6)	124.1(17)	180
$CF_3N=CS$	1.398(9)	1.210(5)	131.4(17)	180
$CF_3N=NCF_3$ (trans)	1.460(6)	1.235(10)	113.4(16)	69
$CF_3N=NCF_3$ (cis)	1.494(5)	1.229(8)	122.4(8)	182
$CF_3N=NN$	1.427(5)	1.254(5)	112.4(2)	183

about 1.42 Å, the N=C bonds are about 1.28 Å, and the CNC 117°. Both $CF_3N=CHF$ and $CF_3N=CHCl$ exist as mixtures of trans and cis isomers (F or Cl with respect to CF_3 group), the trans forms being more stable [94(2) % for F and 87(2) % for Cl, from room-temperature NMR data]. In the isocyanate and isothiocyanate (which contain sp-hybridized carbon instead of sp^2-hybridized carbon in the methanimines), the N=C bonds are considerably shorter. In addition the C—N single-bond lengths are slightly reduced, and the CNC angles are larger by 7° and by 14°. The NCO and NCS groups deviate appreciably from linearity, by 15° and by 10°, respectively. Both trans and cis conformers of hexafluoroazomethane, $CF_3N=NCF_3$, have been synthesized, and then have been studied by separate ED experiments. The increased steric strain in the cis form relative to the trans species is released by lengthening of the C—N bonds and by increasing the NNC angles by about 10°. Furthermore, the CF_3 groups are rotated from the eclipsed position (with respect to the N=N bond) in the trans isomer, becoming staggered (or nearly staggered) in the cis species.

Gas-phase structures of methyl analogs have been reported for four compounds listed in Table 4-10 and provide a basis for the discussion of the CH_3/CF_3 effect. The adjacent C—N bond lengths are, in part, appreciably shorter in the CF_3 derivatives vis à vis the corresponding C—N bonds in the methyl analogs: 1.452(4) Å in CH_3NCO,[184] 1.481(8) Å in CH_3NCS,[184] 1.478(5) Å in trans-$CH_3N=NCH_3$,[185] and 1.470(5) Å in CH_3NNN.[184] The effect of fluorination on the N=C bond length appears to be unreasonably large in methylisocyanate [N=C = 1.169(5) Å in CH_3NCO; this distance has possibly been interchanged with the C=O distance in the ED analysis] and is small in methylisothiocyanate [1.193(6) Å]. In the azido compound the $N_\alpha N_\beta$ bond length increases with fluorination from 1.217(4) to 1.254(5) Å, whereas the $N_\beta N_\omega$ bond is shortened concomitantly from 1.132(3) to 1.120(5) Å. The opposite effect of fluorination on the N_3 bond lengths and the shortening on the C—N bond has been attributed to an increased contribution of the resonance structure (see next page) in the CF_3 compound. The CNN angles in *trans*-$CH_3N=NCH_3$ [112.1(4)°] are not affected by fluorination, and this angle decreases slightly in CH_3NNN [116.8(3) vs 112.4(2)°]. The CNC angles in methyl isocyanate [140.3(4)°] and in methyl isothiocyanate [141.6(4)°], however, decrease by 16 and 10°, respectively. Methyl- and CF_3-substituted isocyanates and isothiocyanates exhibit nonrigid behavior under the conditions of microwave spectroscopy due to the existence of low barriers to internal rotation and CNC bending.[186,187]

$$M{\diagup} N^- {-} N^+ {\equiv} N$$

In nitrosomethane and nitromethane the effects of fluorination are different from those in the compounds listed in Table 4-10. Here fluorination leads to *lengthening* of the C—N bonds by about 0.05 and 0.07 Å, respectively (Table 4-11). This substituent effect and the extremely long C—N bonds in CF_3NO and CF_3NO_2 as compared to isocyanate indicate reversed polarity of the C—N bond, ie, $C^{+\delta} - N^{+\delta}$. The N—O bond lengths and CNO angles appear not to be affected by fluorination. Bis(trifluoromethyl) nitroxyl has a pyramidal structure with bond lengths and angles distinctly different from those of the two former CF_3 compounds. The geometric parameters of the $(CF_3)_2N$ moiety are very similar to those in $(CF_3)_2NOH$.[170]

Phosphorus Fluorine Compounds

Phosphorus compounds are arranged according to oxidation state and coordination number. A small number of arsenic compounds and one antimony derivative are included in this section.

Compounds with Oxidation State III. The only fluorinated P(III) compound with coordination number 2 studied by ED is the phosphaalkene $CF_3P=CF_2$. The following principal geometric parameters were derived[193]: $P—C = 1.901(4)$ Å, $P=C = 1.690(5)$ Å, and $CPC = 108.8(8)°$. The phosphorus–carbon

Table 4-11. PRINCIPAL PARAMETERS OF TRIFLUOROMETHYL-NITROGEN-OXYGEN COMPOUNDS AND METHYL DRIVATIVES

Compound	C—N (Å)		N=O (Å)		∠CNO (deg)	
	CF_3	CH_3	CF_3	CH_3	CF_3	CH_3
CX_3NO	1.546(8)[a]	1.49(3)[b]	1.197(5)[a]	1.22(3)[b]	113.2(13)[a]	112.6(10)[b]
CX_3NO_2	1.56(2)[c]	1.489(5)[d]	1.21(1)[c]	1.224(5)[d]	112(2)[c]	117.4(3)[d]
$(CX_3)_2NO$[e]	1.441(8)	—	1.26(3)	—	117.2(6)	—

[a] Reference 188.
[b] Reference 189.
[c] Reference 190.
[d] Reference 191.
[e] Reference 192.

single and double bonds are by about 0.06 Å longer than the corresponding Schomaker–Stevenson values (1.84 and 1.63 Å). The single bond length is in agreement with long $P-CF_3$ bonds in other compounds (see below). The phosphaalkene $CF_3P=CF_2$ is thermally unstable and dimerizes to a four-membered cyclic species (see below).

The structure of the central P(III) fluorine compound, PF_3, has some historical significance. A very early ED study[194] reported an FPF angle of 104°. This value is larger than the bond angles for PH_3 and the other P(III) halides and thus contradicts the VSEPR ideas. This discrepancy was explained by mesomeric structures with $P=F$ double bonds.[2] A later study based on ED and MW data corrected the original value for this angle to $FPF = 97.8(2)°$,[195] which agrees perfectly with the other halides (XPX decreasing in the sequence $I > Cl > Br > F$). Although the assumption of double-bond contributions originated from a wrong experimental result and is no longer required to explain the FPF angle, this concept is still widely used to rationalize the $P-F$ bond lengths [1.570(2) Å], which are considerably shorter than the Schomaker–Stevenson value of 1.67 Å. As pointed out in the silicon section, the Schomaker–Stevenson rule badly underestimates polar contributions in highly fluorinated compounds, and strongly attractive polar contributions can explain equally well the discrepancy between experimental and predicted values. Arsenic trifluoride has been studied by ED,[196] MW,[197] and joint ED/MW analysis[198,199] with essentially identical results: $As-F = 1.7102(5)$ Å[199] and $FAsF = 95.9(3)°$. The FXF angles in the Group V trifluorides, XF_3, decrease steadily with increasing size of the central atom, from $102.2(1)°$ for $X = N$, to $97.8(2)°$ for $X = P$ to $95.9(3)°$ for $X = As$.

Substitution of one fluorine atom in PF_3 by chlorine or by a cyano group has little or no effect on bond lengths and bond angles, as demonstrated by the results of an MW study for PF_2Cl[200] [$P-F = 1.571(5)$ Å, $FPF = 97.3(5)°$, $FPCl = 99.2(10)°$] and by the results of an ED study for PF_2CN[201] [$P-F = 1.569(3)$ Å, $FPF = 97.9(3)°$, $FPC = 98.3(3)°$]. Much larger effects are observed in the series $(t\text{-Bu})_nPF_{3-n}$.[202] The $P-F$ bond lengths increase from 1.570(2) Å in PF_3 to 1.590(4) Å in $t\text{-BuPF}_2$ to 1.620(7) Å in $(t\text{-Bu})_2PF$. The $P-C$ bonds lengthen by about 0.1 Å from 1.823(12) Å in $t\text{-BuPF}_2$ to 1.920(5) Å in $(t\text{-Bu})_3P$. These trends in the $P-F$ and $P-C$ bond lengths are similar to those observed in the methyl fluorosilanes, whereby steric interactions between the bulky t-Bu groups enhance the lengthening of the $P-C$ bonds. Within their experimental uncertainties, the FPF and FPC angles in this series are equal to the FPF angle in PF_3 and the CPC angles are larger by about 12° [113.8(19) in $(t\text{-Bu})_2PF$ and 109.9(7)° in $(t\text{-Bu})_3P$], indicating the operation of strong steric repulsions. Such steric interactions are expected to be small in tris(trifluoromethyl) phosphine. Nevertheless the $P-C$ bonds in this compound [1.904(7) Å][203] are appreciably longer than in the methyl derivative [1.848(3) Å].[18] This reversed CH_3/CF_3 substitution effect relative to the analogous amines has been pointed out previously and was rationalized by the different electronegativities of N and P. The effect on the bond angles is also reversed. The CPC angles

Table 4-12. Principal Parameters for $(PF_2)_nN(CH_3)_{3-n}$, $n = 1, 2, 3$

Compound	Bond lengths (Å)		Bond angles (deg)		
	P—F	P—N	∠FPF	∠NPF	Ref.
$PF_2N(CH_3)_2$	1.589(3)	1.648(8)	99(3)	97(4)	205
$(PF_2)_2NCH_3$	1.583(2)	1.680(6)	95.1(3)	99.6(3)	206
$(PF_2)_3N$	1.574(2)	1.712(4)	97.1(5)	99.0(4)	207

decrease slightly with fluorination from 98.8 to 97.2(7)°, whereas the CNC angles in the amines increase by 7.0(8)°. Comparison of $As(CF_3)_3$ [As—C = 2.053(19) Å from visual analysis[204]] and $As(CH_3)_3$ [As—C = 1.983(10) Å][18] indicates a similar substitution effect for the arsenic compounds.

A considerable number of compounds containing PF_2 groups have been studied by ED. Typical structural features of PF_2-substituted amines and trends with increasing number of PF_2 groups are demonstrated in Table 4-12. The P—F and P—N bond lengths vary in the opposite direction (ie, P—F bonds shorten and P—N bonds lengthen) with increasing number of PF_2 groups. The FPF angles are generally smaller ($\sim 96 \pm 2°$) than NPF angles ($99 \pm 2°$). The results of an ED study of $PF_2N(CH_3)_2$ suggest a pyramidal configuration at N, but a later MW investigation[208] demonstrated that the PNC_2 skeleton is planar. This discrepancy between ED and MW results possibly arises from a large-amplitude out-of-plane vibration. The PF_2 group staggers one of the N—C bonds with the N and P lone electron pairs mutually orthogonal. $(PF_2)_2NCH_3$ and $(PF_2)_3N$ also have planar skeletons with N and P lone electron pairs mutually orthogonal. The two P lone electron pairs in $(PF_2)_2NCH_3$ point toward one another [C_{2v} symmetry and PNP = 116.1 (8)°], and the symmetry of $(PF_2)_3N$ is C_{3h} (all P lone electron pairs pointing in the same direction). The principal structural parameters and trends in amino-difluorophosphines (PF_2NH_2,[205] PF_2NHCH_3,[209] $(PF_2)_2NH$,[210]) and in the difluorophosphino silylamines [$PF_2N(SiH_3)_2$,[211] and $(PF_2)_2NSiH_3$[211]] are very similar to those in the methyl series. The hydrogen atoms in PF_2NH_2 could not be located in the ED study, and the results of an MW investiation[212] suggest a planar configuration at N. The reported value for the PNC angle in PF_2NHCH_3 is very large [125.3(20)°], and the configuration at N is probably planar. The ED intensities can be reproduced with one single conformer with the P lone electron pair trans to the N—C bond. The $(PF_2)_2NH$ molecule exists as a mixture of two conformations, and the predominant form (72% at room temperature) has essentially C_{2v} symmetry, thereby corresponding to the structure of the analogous methyl compound. The PNP angle increases with replacement of CH_3 by H to 122.1(7)°. The P—N bonds in the silyl-substituted amines are longer than in the methyl analogs, [P—N = 1.680(4) and 1.691(4) Å in $PF_2N(SiH_3)_2$ and $(PF_2)_2NSiH_3$, respectively]. Within experimental uncertainties the P—N bond length in bis(difluorophosphino)germylamine[213]

[1.698(8) Å] is equal to that in the silyl compound, and the PNP angle is slightly smaller [114.0(8)° for germyl- and 117.6(7)° for silylamine].

Structure determinations for PF_2NCO and PF_2NCS are hampered by large correlations between geometric parameters in the ED analysis. Two ED studies[214,215] and a combined analysis of ED intensities and liquid crystal nuclear magnetic resonance (LCNMR) data[215] have been reported. For the isocyanate the ED and ED/LCNMR results are identical, whereas the parameters derived from the combined analysis differ appreciably from both ED studies in the case of the isothiocyanate. The geometries of the NPF_2 moieties of both compounds are very similar to those reported for the amines. The NCO and NCS groups are trans to the two fluorine atoms; ie, the P lone pair eclipses the $N{=}C$ bonds. The PNC angles [132.0(12)° in PF_2NCO[215] and 138.7(6)° in PF_2NCS[215]] are larger by about 7–8° than the nitrogen bond angles in the CF_3 analogs (see Table 4-10). A similar PNC bond angle [132.8(5)°] was reported for bis(difluorophosphine)carbodiimide, $PF_2N{=}C{=}NPF_2$.[216] The PF_2 groups are trans to the $N{=}C$ bonds, and the dihedral angle PNCNP is about 55°.

Beside these PF_2-substituted nitrogen compounds, structure investigations for PF_2-substituted ether, sulfides, and selenide have been reported. The parameters of greatest interest are the P—X bond lengths (X = O, S, Se) and the bond angles at the central atom X (Table 4-13). Bis(difluorophosphino)ether has been studied in two laboratories with quite different results. Arnold and Rankin[217] report the presence of only one single isomer, and Yow and co-workers[218] observe a mixture of four rotational isomers. It is impossible to resolve this discrepancy on the basis of ED data alone; additional information is required. The skeletal parameters of reference 217 (model B) and reference 218 are essentially equal within their uncertainties. The short P—O bond lengths relative to the Schomaker–Stevenson value of 1.71 Å, and the increased POP angle are discussed in the section on oxygen compounds. An MW investigation of PF_2OCH_3[221] reports a still shorter P—O bond [1.56(2) Å] and a bond angle [POC = 123.7(7)°] that is intermediate between $(PF_2)_2O$ and $(CH_3)_2O$. For bis(difluorophoshino)-sulfide and -selenide, the ED intensities are interpreted in

Table 4-13. PRINCIPAL PARAMETERS FOR
PF_2-SUBSTITUTED ETHER, SULFIDES, AND
SELENIDE (r_g VALUES)

Compound	P—X (Å)[a]	α_X (deg)[b]	Ref.
PF_2OPF_2	1.631(10)	135.2(18)	218
PF_2SPF_2	2.133(4)	91.3(11)	219
PF_2SCH_3	2.085(3)	102.0(12)	219
PF_2SGeH_3	2.116(8)	99.0(6)	220
PF_2SePF_2	2.273(5)	94.6(8)	219

[a] X = O, S, or Se.
[b] Bond angle at X atom.

terms of one conformer with C_{2v} symmetry in which the PF_2 groups point away from each other, ie, with the P lone electron pairs in the PXP plane and pointing toward each other. The PF_2 groups perform large-amplitude torsional motions. In the case of PF_2SCH_3 it is not possible to distinguish between three conformations that differ by the rotational position of the PF_2 group, whereas only one conformer appears to be present in PF_2SGeH_3. The sulfur bond angle increases by about $10°$ upon replacing one PF_2 group by CH_3, an opposite trend to that observed for the ethers (decrease of about $12°$). The P—S bond lengths vary substantially, from $2.133(4)$ Å in PF_2SPF_2 to $2.085(3)$ Å in PF_2SCH_3.

The only fluorinated cyclic P(III) compound whose structure has been studied in the gas phase is the dimer of $CF_3P{=}CF_2$ (**11**).[193] According to the result of NMR studies,[222] the trans isomer is formed predominantly upon

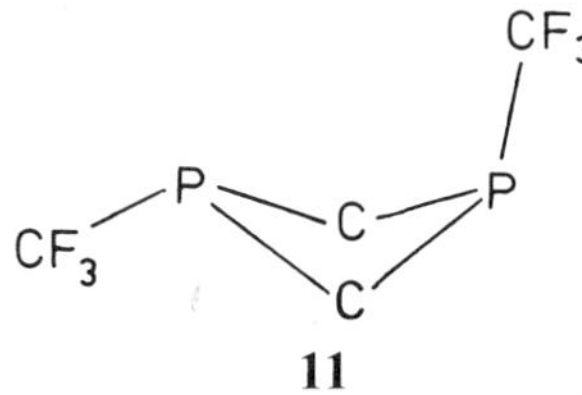

11

dimerization with small amounts of cis isomer. The cis isomer is thermally less stable and inverts slowly to the trans species at room temperature. The average P—C distance $[(P{-}C)_{avg} = 1.899(3)$ Å, $P{-}C_{endo} = 1.896(11)$ Å, and $P{-}C_{exo} = 1.906(21)$ Å] is equal to the P—C single bond in the monomer $[1.901(4)$ Å] and to those in $(CF_3)_2P$ $[1.904(7)$ Å]. The four-membered PCPC ring is strongly puckered [puckering angle 35.4 $(12)°$] and distorted to a rhombus $[CPC = 77.6(8)°$ and $PCP = 95.6(6)°]$. The exo CPC angle is $101.7(11)°$, slightly larger than the angle in tris(trifluoromethyl)phosphine $[97.2(7)°]$.

The conformational properties of fluorinated diphosphines differ from those of the hydrazines. The ED intensities of $F_2P{-}PF_2$[223] and $(CF_3)_2P{-}P(CF_3)_2$[223] were interpreted in terms of trans conformations that possess large-amplitude torsional vibrations around the P—P bonds. The corresponding hydrazines exist as a mixture of trans and gauche conformers $(F_2N{-}NF_2)$ or as a single gauche conformation $[(CF_3)_2N{-}N(CF_3)_2]$. Diphosphine itself, $H_2P{-}PH_2$, exists only in a gauche conformation,[224] just as hydrazine does. Fluorination of diphosphine leads to a marked lengthening of the P—P bond $[2.281(6)$ Å in P_2F_4 vs $2.218(4)$ Å in $P_2H_4]$.[225] The P—F bonds $[1.587()$ Å] are slightly longer than in PF_3, and the reported bond angles, $FPF = 99.1(4)°$ and $PPF = 95.4(3)°$, contradict those predicted by VSEPR theory (FPF < PPF). In the CF_3-substituted diphosphine, the P—P bond $[2.182(16)$ Å] is shorter than in P_2H_4; the P—C bonds $[1.914(4)$ Å] are similar to those in the phosphine $[1.904(7)$ Å], while the phosphorus valence angles are increased $[CPC = 103.8(8)°$ and $PPC = 106.7(7)°]$.

Compounds with Oxidation State V

Coordination Number Four. Two phosphoryl fluorides, a larger number of thiophosphoryl fluorides, and one selenophosphoryl fluoride have been studied by electron diffraction (Table 4-14). The P—F bonds shorten with increasing oxidation number [1.524(3) Å in O=PF_3 vs 1.570(2) Å in PF_3[195]], and the FPF angles increase slightly from 97.8(2) to 101.3(2)°. In the VSEPR formalism, this result is indicative of the reduced steric requirement of the P=O bond relative to the P lone electron pair. Different geometric parameters have been reported for O=PF_2CH_3 from MW, *ab initio*, and ED studies. Assuming the values of all bond lengths, unreasonable bond angles [FPF = 115.2(3)° and OPC = 105.8(10°)] have been derived from three rotational constants[231] that were based on a wrong assignment. The assignment was later corrected, and these values were revised to FPF = 99.2(10)° and OPC = 118.2(15)°,[232] which are in good agreement with the results of *ab initio* calculations[233] (FPF = 98.9°, OPC = 119.7°) and the ED investigation [FPF = 99.2(2)°, OPC = 117.8(8)°, FPC = 103.7(8)°, and P—C = 1.770(5) Å].

In the trifluoromethyl-substituted phosphoryl, O=$P(CF_3)_3$,[234] (not listed in Table 4-14) the P—C bonds are very long [1.895(6) Å]; this strong CH_3/CF_3 substituent effect in P(V) compounds [P—C = 1.810(2) Å in O=$P(CH_3)_3$[235]] has been pointed out earlier. The P=O bond lengths are not affected by fluorination [1.475(2) and 1.477(2) Å], but are considerably longer than in O=PF_3. The CPC angles are equal in both compounds [104.3(8) and 104.1(7)°], and they are larger than the FPF angles in O=PF_3, as predicted by VSEPR theory. In the thiophosphoryl fluorides, S=PF_2X where X = F, Cl, Br, and H, the P—F bond lengths increase steadily with decreasing electronegativity of X, while the other parameters remain nearly constant. Comparison of P—F distances and FPF angles along the series Y=PF_2H (Y = O, S, Se) indicates a minor increase of the P—F distances with decreasing electronegativity of Y: P—F = 1.539(5) Å in O=PF_2H,[236] 1.552(3) Å in S=PF_2H and

Table 4-14. PRINCIPAL PARAMETERS OF PHOSPHORYL, THIOPHOSPHORYL, AND SELENOPHOSPHORYL FLUORIDES (r_g VALUES)

| Compound | Bond lengths (Å) | | ∠FPF (deg) | Ref. |
	P—F	P=X[a]		
O=PF_3	1.524(3)	1.436(6)	101.3(2)	226
O=PF_2CH_3	1.540(2)	1.444(4)	99.6(5)	227
S=PF_3	1.538(3)	1.866(5)	99.6(3)	228
S=PF_2Cl	1.536(2)	1.865(8)	100.5(8)	229
S=PF_2Br	1.544(3)	1.882(4)	98.3(10)	229
S=PF_2H	1.552(3)	1.877(3)	98.3(4)	229
Se=PF_2H	1.559(3)	2.027(4)	98.1(7)	230

[a] X = O, S, or Se.

1.559(3) Å in $Se{=}PF_2H$. The FPF angles decrease slightly in the same direction [FPF = 101.9(20) in $O{=}PF_2H$].

Structures for two difluorothiophosphoryl compounds, $CH_2(PF_2S)_2$ and $O(PF_2S)_2$, have been reported.[237] The parameters of the $S{=}PF_2$ groups are very similar to those of $S{=}PF_3$, PCP = 122.6(10)° and POP = 130.9(35)°. The latter angle is smaller by about 5° than that in $O(PF_2)_2$. For the methane derivative a roughly 1:1 mixture of gauche–gauche (C_s symmetry) and gauche–trans conformers (C_1 symmetry) has been observed, and for the ether derivative only the gauche–gauche form has been observed (gauche/trans implies a dihedral angle PXPS of 60/180°, X = C or O).

For the phosphonitrilic fluoride trimer $(PNF_2)_3$, ED analysis results in an essentially regular hexagon (D_{3h} symmetry), with P—N = 1.586(13) Å and P—F = 1.548(13) Å (model A of reference 238).

Phosphoranes. Stereochemical, structural, and dynamic properties of phosphoranes have attracted considerable interest for many years. A large number of stereochemical and dynamic studies by NMR spectroscopy have been reported in the liquid phase. In addition, structure investigations in the solid phase have been reported. Excellent reviews on these results have been published by Homes[239] and by Luckenbach.[240] Only a relatively small number of structural studies have been performed in the gas phase, almost exclusively by ED. The results of all gas- and solid-phase studies confirm the preference of the trigonal bipyramidal skeleton over the square pyramidal structure in phosphoranes that contain monodentate ligands. In trigonal bipyramidal structures, axial and equatorial ligands can exchange their positions via pseudorotational processes (Berry or turnstile type). If the barrier to such an exchange is low and these processes are fast relative to the NMR time scale, this spectroscopic technique does not allow a distinction to be made between axial and equatorial positions. In such cases NMR spectroscopy and ED, the latter of which has a much shorter characteristic time ($\sim 10^{-17}$ s), may result in different stereochemical conclusions.

This influence of the experimental technique is demonstrated for the parent compound of this series, PF_5, where pseudorotation cannot be stopped on the NMR time scale. Even at very low temperatures one single fluorine signal is observed, indicating the presence of five equivalent fluorine atoms. The barrier to pseudorotation is estimated[241] to be 14 kJ mol^{-1}, in good agreement with the results of *ab initio* calculations[242] [16(2) kJ mol^{-1}]. The ED study, however, confirms a trigonal bipyramidal structure with distinctly different axial and equatorial P—F distances[243]: $P{-}F_{eq} = 1.534(4)$ Å and $P{-}F_{ax} = 1.577(5)$ Å. The same situation exists for AsF_5[244] [$As{-}F_{eq} = 1.656(4)$ Å, $As{-}F_{ax} = 1.711(5)$ Å]. The difference between axial and equatorial bond lengths in PF_5 and AsF_5 is the same within experimental uncertainties [$\Delta(As{-}F) = 0.055(7)$; $\Delta(P{-}F) = 0.043(7)$ Å]).

In heterosubstituted phosphoranes the electronegativity (or polarity) rule is generally applied to predict the preference of ligands for axial or equatorial positions. The more electronegative substituents, starting with F > CF_3 > Cl, prefer axial positions. This rule is based on VSEPR theory[2] and was confirmed experimentally.[245] It was later slightly modified in the apicophilicity series,[246] where the positions of CF_3 and Cl are interchanged; ie, a higher preference for axial positions is assigned to Cl than to CF_3 despite the lower electronegativity of Cl. In both these empirical rules fluorine has the highest preference for axial positions.

Fluoro-chloro phosphoranes, Cl_nPF_{5-n}, which are often used as model compounds, have been scrutinized by various methods, such as NMR, vibrational, and nuclear quadrupole resonance (NQR) spectroscopy (see reference 239 and references cited therein). The stereochemistry of this series has been determined unambiguously by these methods; ie, fluorine always prefers axial positions. ED studies for Cl_2PF_3[247] and Cl_3PF_2[248] have been reported only recently; simultaneously with these studies, investigations of the entire series $n = 1$–4 have been performed in our laboratory[249] (Table 4-15). The results cited in references 247 and 248 essentially agree with the values in Table 4-15. All bond lengths increase with decreasing number of fluorine atoms; the axial bonds lengthen about twice as much as the equatorial bonds. These trends confirm the results of extended Hückel theory (EHT) calculations *without* phosphorus 3d functions,[251] whereas such calculations *with* 3d functions predict similar lengthening for equatorial and axial bonds. The distortions of the trigonal bipyramid are surprisingly small, less than 1° for axial–equatorial bond angles and only about 2° for equatorial angles. The Cl_2AsF_3 molecule[252] shows the same trends in bond lengths relative to AsF_5 as the phosphorane analog [$As—F_{eq} = 1.663(7)$ Å, $As—F_{ax} = 1.723(4)$ Å, $As—Cl_{eq} = 2.096(3)$ Å] with small angular distortions [$F_{eq}AsF_{ax} = 89.2(2)°$, $Cl_{eq}AsCl_{eq} = 120.6(5)°$].

Table 4-15. GEOMETRIC PARAMETERS OF THE FLUOROCHLOROPHOSPHORANE SERIES PCl_nF_{5-n}, $n = 0$–5 (r_g VALUES)

Parameter	PF_5	$PClF_4$	PCl_2F_3	PCl_3F_2	PCl_4F	PCl_5
P—F_{eq}, Å	1.534(4)	1.535(3)	1.538(7)	—	—	—
P—F_{ax}, Å	1.577(5)	1.581(4)	1.593(4)	1.596(2)	1.597(4)	—
P—Cl_{eq}, Å	—	2.000(3)	2.002(3)	2.005(3)	2.011(3)	2.023(3)
P—Cl_{ax}, Å	—	—	—	—	2.107(6)	2.127(3)
$X_{eq}PX_{eq}$, deg	120[a]	117.8(7)[a]	121.8(4)[b]	120[b]	120.0(1)[b]	120[b]
$X_{eq}PY_{ax}$, deg	90[c]	90.3(4)[c]	90.0(3)[c]	90[d]	90.9(2)[e]	90[e]
Ref.	243	249	249	249	249	250

[a] $\angle F_{eq}PF_{eq}$.
[b] $\angle Cl_{eq}PCl_{eq}$.
[c] $\angle F_{eq}PF_{ax}$.
[d] $\angle Cl_{eq}PF_{ax}$.
[e] $\angle Cl_{eq}PCl_{ax}$.

Although both empirical rules, the electronegativity rule and apicophilicity series, predict an equatorial position of the CF_3 group in trifluorometyl-tetrafluorophosphorane, controversial results for CF_3PF_4 have been reported in the literature: $^{19}F(CF_3)$ NMR chemical shifts were interpreted in terms of the axial conformer[245] (CF_3 group axial). This interpretation was confirmed by MW,[253] where the typical spectrum of a symmetric rotor (C_3 or higher symmetry) was observed. Gas-phase infrared spectra,[254] however, indicate lower symmetry, C_s or C_1, ie, where the CF_3 group occupies equatorial position. This conclusion is confirmed via interpretation of later ^{19}F and ^{13}C NMR spectra.[246,255] The interpretation of NMR spectra of this compound is not straightforward, since all fluorine atoms bonded to phosphorus appear equivalent due to pseudorotation, even at very low temperatures. The NMR interpretations are based on "typical" chemical shifts and coupling constants for the CF_3 group, which lead to controversial results. Semiempirical MO calculations in the CNDO/2 approximation strongly favor the equatorial conformer by about 96 kJ mol^{-1}.[256] ED analysis[257] of CF_3PF_4 suggests that this compound exists as a mixture of both conformers in the ratio equatorial/axial of about 60:40. The presence of two conformers is confirmed by matrix infrared spectroscopy[258] and by the results of *ab initio* calculations[259] that predict a very small energy difference between the two conformations ($\Delta E = 1.2$ kJ mol^{-1}). This perfect agreement with the ED experiment [$\Delta G = 0.8(8)$ kJ mol^{-1}] should be regarded as fortuitous. The presence of a mixture of two conformers is also compatible with the infrared gas and MW spectra. The superposition of vibrational spectra for the equatorial (C_1 or C_s symmetry) and axial conformers (C_{3v} symmetry) is qualitatively identical to the spectrum of the low-symmetry species alone. The splitting of some vibrations is not detectable in the gas-phase spectrum. In the MW spectrum the rotational transitions for the equatorial conformer are expected to be much weaker than those for the axial conformer due to a smaller dipole moment, lower symmetry, and low-frequency CF_3 torsions in the former. Hence, these rotational transitions have not been observed for the equational conformer.

Controversial interpretations of ^{19}F and ^{13}C NMR spectra have also been reported for $(CF_3)_2PF_3$. The ^{19}F (CF_3) chemical shift has been interpreted as providing evidence that both CF_3 groups occupy axial positions.[245] The corresponding $^{13}CF_3$ NMR chemical shifts have been interpreted in terms of equatorial CF_3 groups.[255] The three fluorines bonded to phosphorus appear to be equivalent at all temperatures, either because of degeneracy or due to pseudorotation. The ED experiment[257] confirms the interpretation of the ^{19}F chemical shifts, ie, both CF_3 groups occupy axial positions. Small contributions ($<10\%$) of other conformers cannot be excluded. For $(CF_3)_3PF_2$,^{19}F,[245] ^{13}C,[255] NMR spectra, and ED[257] are in agreement with respect to the stereochemistry of this phosphorane, ie, equatorial positions of all three CF_3 groups. The conformations of these CF_3-substituted phosphoranes, where three CF_3 groups occupy equatorial positions, two CF_3 groups prefer axial positions, and the CF_3 group in CF_3PF_4 occupies an equatorial or axial position with

Table 4-16. Bond Lengths (r_g values) for $(CF_3)_n PF_{5-n}$, $n = 0$–3

Bond	$>\!\!P\!\!-$	$>\!\!P\!\!-CF_3$	CF_3 $>\!\!P\!\!-$	CF_3 $>\!\!P\!\!-$ CF_3	CF_3 $>\!\!P\!\!-CF_3$ CF_3
P—F$_{eq}$, Å	1.534(4)	1.538(5)	1.538(5)[a]	1.560(3)	
P—F$_{ax}$, Å	1.577(5)	1.574(7)	1.574(7)[a]	—	1.601(4)
P—C$_{eq}$, Å	—	1.882(8)	—	—	1.889(4)
P—C$_{ax}$, Å	—	—	1.90(2)	1.885(6)	—

[a] Assumed to be equal for equatorial and axial conformers.

nearly equal probability, indicate that other effects, such as steric interactions between ligands, in addition to their electronegativity, determine the stereochemistry. It appears that in $(CF_3)_2 PF_3$, the only example thus far studied that contradicts both empirical rules, these other effects override the electronegativity difference between F and CF_3 groups. The geometric parameters of this series are collected in Table 4-16. The P—F bond lengths in $CF_3 PF_4$ are equal to those in PF_5, whereas P—F$_{eq}$ and P—F$_{ax}$ increase slightly with further F/CF$_3$ substitution. The surprising observation is that P—C bond lengths in all three compounds are equal within their experimental uncertainties, irrespective of equatorial or axial position. This conclusion is also indicated by the results of *ab initio* calculations for $CF_3 PF_4$,[259] where P—C$_{eq}$ and P—C$_{ax}$ differ by less than 0.01 Å. The absolute values for P—C bond lengths, however, are poorly reproduced by these calculations (1.82 and 1.83 Å with 4-21 G* basis set). This essential equivalency of equatorial and axial bond lengths in trifluoromethyl phosphoranes may indicate that coupling constants $^1J_{PC}$ and $^2J_{PF}$ are not "typical" for equatorial or axial position of a CF_3 group in this series.

A similar problem with the interpretation of coupling constants exists for the trifluoromethylchlorophosphoranes $(CF_3)_n PCl_{5-n}$ with $n = 1, 2, 3$. Muetterties and co-workers[245] concluded from ^{19}F chemical shifts in $(CF_3)PCl_4$ and $(CF_3)_2 PCl_3$ that CF_3 groups occupy axial positions, in agreement with the electronegativity rule. This interpretation is supported by infrared (gas) and Raman (liquid) spectra[260,261] and by NQR spectra (solid).[262] Cavell and associates, however, conclude from NMR coupling constants that CF_3 groups occupy equatorial positions in the trifluoromethylchlorophosphoranes.[246] In addition to the coupling constants, the equivalence of all three CF_3 groups in $(CF_3)_3 PCl_2$ at temperatures as low as $-160°C$ is considered as further evidence for equatorial CF_3 groups. The ED analysis clearly demonstrates axial positions of CF_3 groups in $CF_3 PCl_4$,[263] $(CF_3)_2 PCl_3$,[264] and two axial and one equatorial CF_3 groups in $(CF_3)_3 PCl_2$.[264] Thus, application of the electronegativity rule predicts the correct sequence for axial preferences ($CF_3 >$ Cl) in contrast to the apicophilicity series (Cl $> CF_3$). The vibrational spectra [infrared (gas) and Raman (liquid)] demonstrate that the conformation does not change with the

Table 4-17. PRINCIPAL GEOMETRIC PARAMETERS FOR $(CF_3)_nPCl_{5-n}$, $n = 0-3$ (r_g VALUES)

Bond	PCl_4	CF_3PCl_3	$(CF_3)_2PCl_2$	$(CF_3)_3PCl_2$
$P-Cl_{eq}$, Å	2.023(3)	2.023(3)	2.038(2)	2.055(6)
$P-Cl_{ax}$, Å	2.127(3)	2.084(9)	—	—
$P-C$, Å	—	1.980(10)	1.951(11)	$1.944(5)^a$
$\angle Cl_{eq}PCl_{eq}$, deg	120	119.8(1)	120	133.0(17)
$\angle X_{ax}PY_{eq}$, deg	90^b	$92.4(2)^b$	90^c	$95.5(19)^d$
Ref.	250	263	264	264

^{a} Mean values; $P-C_{eq} = 1.939(31)$ Å; $P-C_{ax} = 1.947(14)$ Å.
^{b} $\angle Cl_{ax}PCl_{eq}$.
^{c} $\angle C_{ax}PCl_{eq}$.
^{d} $\angle C_{ax}PC_{eq}$.

physical state (gas or liquid) in CF_3PCl_4 and $(CF_3)_2PCl_3$. Such a change is thus unlikely in $(CF_3)_3PCl_2$. The principal geometric parameters are summarized in Table 4-17. The equatorial P—Cl distances increase slightly with increasing number of CF_3 groups. The axial P—Cl bond, however, shortens by about 0.04 Å with substitution of a CF_3 group for the opposite chlorine. The P—C bonds are extremely long and shorten with increasing CF_3 substitution. Within the large experimental uncertainties equatorial and axial P—C bonds in $(CF_3)_3PCl_2$ are indistinguishable. In this phosphorane, the trigonal bipyramid is strongly distorted in the direction of the transition state for a Berry-type pseudorotation. This result renders a low barrier to ligand exchange plausible and explains the equivalency of the three CF_3 groups in the NMR time scale even at very low temperatures.

The ground-state conformations of the methylfluorophosphoranes, $(CH_3)_nPF_{5-n}$, $n = 1, 2, 3$, have been determined unambiguously by NMR[245] and vibrational spectroscopy.[265,266] As expected from the large electronegativity difference between F and CH_3, the methyl groups occupy equatorial positions only. The results of the ED analyses are listed in Table 4-18. Whereas the equatorial bond lengths P—F and P—C increase only slightly (by ~0.02–0.03 Å) with increasing methyl substitution, the axial P—F bonds lengthen by more than 0.1 Å between PF_5 and $(CH_3)_3PF_2$. The methylfluorophosphorane series, in which the substituents differ strongly in their electronegativities, offers suitable examples for discussing bonding properties in phosphoranes. Three different models have been suggested.

1. Purely covalent bonding with participation of d orbitals, three sp^2 hybrids for the equatorial bonds, and two pd hybrids for the axial bonds.[269] This model results in weaker axial than equatorial bonds. In PF_5, which contains five highly

Table 4-18. Principal Geometric Parameters (r_g values) of Methyl Fluorophosphoranes $(CH_3)_nPF_{5-n}$, $n = 0$–3

Compound	Bond lengths (Å)			Bond angles (deg)		Ref.
	$P—F_{eq}$	$P—F_{ax}$	$P—C_{eq}$	$\angle F_{eq}PF_{ax}$	$\angle X_{eq}PX_{eq}{}^a$	
PF_5	1.534(4)	1.577(5)	—	90	120	243
CH_3PF_4	1.543(4)	1.612(4)	1.780(5)	89.1(4)	118.9(13)	267
$(CH_3)_2PF_3$	1.553(6)	1.643(3)	1.798(4)	89.9(3)	124.0(12)	267
$(CH_3)_3PF_2$	—	1.685(1)	1.813(1)	—	120	268

^{a} X = F or C.

electronegative substituents, the d oribitals are contracted and participate effectively in bonding. With the introduction of more electropositive substituents, the d orbitals expand and the axial bonds weaken. Since the $3s$ and $3p$ orbitals also expand, but to lesser extent, the equatorial bond lengths increase less.

2. Rundle[270] suggested an electron-deficient model without d orbitals that contains three-center four-electron bonds in axial directions. This model implies an axial bond order of one-half with polar contributions, leading also to weaker axial bonds. Substituting the equatorial fluorines by methyl groups decreases the positive phosphorus net charge and, thus, the polar contributions to the axial bonds, which lead to lengthening of these bonds. An argument similar to that used in model 1 can rationalize the slight increase of the equatorial bond lengths. Both models explain qualitatively the longer axial bonds and the trends that occur when equatorial fluorines are replaced by more electropositive substituents. The actual bonding situation is probably somewhere in between these two models.

3. VSEPR theory[2] predicts correctly that the most electronegative fluorines prefer axial positions and that the axial bond lengths in PF_5, with three nearest neighbors at 90°, are longer. This model also accounts for (a) the observed increase of bond lengths if equatorial fluorines are replaced by substituents that contain larger bonding electron pairs, such as methyl, and (b) angle deformations. It is surprising, however, that this minimization of overall repulsions in $(CH_3)_2PF_3$ has a relatively large effect on the axial bond lengths (which increase by 0.066 Å relative to PF_5), while the angular distortion in axial direction is negligible [$F_{eq}PF_{ax} = 89.9(3)°$].

The equatorial P—C bond lengths in methyl-, phenyl-,[271] and ethinyltetrafluorophosphorane[272] do not follow the expected trend. Instead of gradual bond shortening due to increased s character in the carbon bonding orbital, a longer P—C bond is observed in the phenyl compound [1.796(9) Å] relative to the methyl derivative [1.780(5) Å]. This bond length is 1.747(5) Å in the ethynyl phosphorane. The orientation of the phenyl ring is nearly perpendicular to the equatorial plane [angle between equatorial plane and ring plane is

72.9(36)°]. This sterically unfavorable orientation of the phenyl ring can be explained by two effects: (a) π bonding between the phenyl π system and the equatorial phosphorus orbitals or (b) formation of F...H bridges between the axial fluorines and the nearest phenyl hydrogens. These distances are only 2.18 Å. The long P—C bond is not consistent with explanation a. The geometries of the PF_4 groups in these three compounds are very similar except for slightly shorter P—F_{ax} bonds in the ethinyl derivative [1.599(3) Å vs 1.616(4) and 1.612(4) Å in phenyl- and methyl tetrafluorophosphorane].

In phosphoranes with equatorial P—N bonds the formation of π bonds has been predicted theoretically[273] and confirmed by the results of dynamic NMR studies[273] [barrier to internal rotation in $(NH_2)_2PF_3 = 46.0$ kJ mol^{-1}]. Theory predicts that π-donor substituents (eg, amino groups) are oriented with the donor orbitals in the equatorial plane to maximize π overlap with P. Results of ED studies for amino- and dimethylamino-substituted phosphoranes are consistent with these predictions (Table 4-19). In all four compounds the configuration at nitrogen is planar (or nearly planar) and the amino (dimethylamino) groups are oriented perpendicular (or nearly perpendicular) to the equatorial plane. In $(NH_2)_2PHF_2$ the result of CNDO/2 calculations concerning the orientation (90°) has been used as a constraint in the ED analysis. The largest deviation from perpendicular orientation is observed in $(NMe_2)_3PF_2$ where $\tau = 70.1(27)°$. This is an effective value due to large-amplitude torsional vibrations. The P—N bond lengths are shorter by about 0.10–0.13 Å than the Schomaker–Stevenson value of 1.77 Å. This difference can be explained in terms of π bonding and polar contributions. The axial P—F bonds lengthen with increasing number of electropositive substituents in the equatorial plane (NH_2, NMe_2, or H). The equatorial angles deviate in some cases appreciably from the ideal 120°: NPN = 127.4(10)° in $(NH_2)_2PF_3$ and $F_{eq}PF_{eq} = 112.9(21)°$ in NMe_2PF_4.

In cyclic phosphoranes ring strain is an additional factor that influences the stereochemistry. Ring strain may work with or against the electronegativity effect in this regard. In trifluorophosphoranes the stereochemistry of the ground-state conformation depends strongly on ring size. In $(NMePF_3)_2$ (**12**),

Table 4-19. PRINCIPAL GEOMETRIC PARAMETERS OF AMINO-SUBSTITUTED PHOSPHORANES

| | Bond lengths (Å) | | | | |
Compound	P—F_{eq}	P—F_{ax}	P—N_{eq}	τ (deg)^a	Ref.
NMe_2PF_4	1.563(9)	1.595(9)	1.642(12)	$\sim$90	274
$(NH_2)_2PF_3$	1.560(10)	1.619(7)	1.648(13)	$\sim$90	275
$(NH_2)_2PHF_2$	—	1.644(5)	1.641(5)	90^b	276
$(NMe_2)_3PF_2$	—	1.633(6)	1.675(5)	70.1(27)	277

^{a} Angle between amino/dimethylamino group and equatorial plane.
^{b} Assumed.

12

ring strain overrides the electronegativity effect, and the planar four-membered ring has equatorial and axial P—N bonds of 1.599 and 1.735 Å, respectively.[278] The trigonal bipyramid appears to be strongly distorted with NPN = 77.9° and $F_{eq}PF_{eq} = 103.9°$. Axial–equatorial binding has also been suggested for the five-membered ring in cyclotetramethylenetrifluorophosphorane (**13**), with the argument that a CPC angle near 90° would cause much less strain in the PC_4 ring than would an angle near 120° for diequatorial binding.[256] Spectra obtained from ^{19}F NMR analysis[245] show equivalent fluorine atoms at room temperature, thereby indicating fast permutational exchange between axial–equatorial and diequatorial binding. Below $-70°C$, however, two fluorine signals with a ratio of $F_{ax}/F_{eq} = 2:1$ are observed, thereby demonstrating diequatorial binding of the ring in the ground state. This conclusion is confirmed by the results of an ED study[279] that indicate a half-chair (twist) conformation of the ring with C_2 overall symmetry. Thus, in five-membered rings with methylene groups bonded to phosphorus the electronegativity effect overrides ring strain. This strain, however, leads to a lowering of the barrier to pseudorotation (~ 28 kJ mol^{-1})[245] relative to the strain-free $(CH_3)_2PF_3$, in which no positional exchange is observed at room temperature. The dynamic behavior of **13** is characterized by two types of pseudorotation, one for the phosphorus skeleton and the other for the five-membered ring. The energy barrier of the latter type is estimated to be lower than the barrier of the former. In the six-membered ring of cyclopentamethylenetrifluorophosphorane (**14**), the CPC angle is closer to 120 than to 90°, and ring strain enhances the electronegativity effect for diequatorial binding. The ^{19}F NMR spectrum consists of two fluorine signals ($F_{ax}/F_{eq} = 2:1$) at temperatures up to 100°C,[245] thereby indicating a high barrier to pseudorotation. The ED analysis[279] results

13   **14**

Table 4-20. PRINCIPAL GEOMETRIC PARAMETERS OF DIMETHYL-,
CYCLOTETRAMETHYLENE-, AND CYCLOPENTAMETHYLENETRIFLUOROPHOSPHORANE
(r_g VALUES)

Compound	Bond lengths (Å)			Bond angles (deg)		Ref.
	$P\text{—}F_{eq}$	$P\text{—}F_{ax}$	$P\text{—}C$	$\angle CPC$	$\angle F_{eq}PF_{ax}$	
$(CH_3)_2PF_3$	1.553(6)	1.643(3)	1.798(4)	124.0(12)	89.9(3)	267
$(CH_2)_4PF_3$	1.547(8)	1.635(3)	1.788(4)	103.5(8)	87.2(4)	279
$(CH_2)_5PF_3$	1.562(6)	1.637(3)	1.789(4)	115.7(8)	87.1(2)	279

in a chair conformation of the ring, which is much flatter at phosphorus [flap angle 23.8(12)°] than at the opposite carbon [flap angle 56.8(21)°, equal to the flap angles in cyclohexane, 55.1(7)°].[280] Table 4-20 compares the principal geometric parameters of the two cyclic phosphoranes with those of $(CH_3)_2PF_3$. Whereas the bond lengths in these three compounds are very similar, a drastic change in the CPC angles is observed. In the six-membered ring this angle decreases by about 10° relative to the strain-free compound, and it decreases again by 10° to 103.5(8)° in the five-membered ring.

Group VI Compounds

Oxygen Fluorine Compounds

Three oxides containing CF_3 groups have been studied by electron diffraction: CF_3OF,[281] CF_3OCl,[282] and CF_3OCF_3.[283] The O—F bond in the hypofluorite [1.421(6) Å] is slightly longer than in oxygen difluoride [1.4087(4) Å, r_s value].[284] The O—C bond lengths in CF_3OCl and CF_3OCF_3 are equal within their error limits [1.365(7) and 1.369(4) Å, respectively] and shorter than the O—C bond in CF_3OF [1.395(6) Å]. The oxygen bond angles vary significantly among these three compounds: 104.9(6)° in CF_3OF, 112.9(5)° in CF_3OCl, and 119.1(8)° in CF_3OCF_3. The CH_3/CF_3 substituent effect for dimethylether, ie, shortening of the O—C bonds by 0.047(5) Å [from 1.416(2)[285] to 1.369(4) Å] and increase of the COC angle by 6.7(10)° [from 112.4(6) to 119.1(8)°] has been pointed out previously. A similar substitution effect for the O—C bond lengths is observed between CH_3OCl [1.389(28) Å][286] and CF_3OCl [1.365(7) Å], whereas the bond angles are equal in both hypochlorites.

Oxygen bond angles in compounds of the type XOX, with X being fluorine or a fluorine-containing group vary by more than 50° (Table 4-21). With increasing valence angle, simultaneous bond shortening relative to Schomaker–Stevenson values is observed. In the classical example for a large oxygen valence angle, ie, disiloxane, this large bond angle in combination with the short O—Si bonds has been rationalized by partial delocalization of the oxygen lone pairs

Table 4-21. Oxygen Valence Angles and O—X Bond Lengths in
XOX Compounds[a]

Compound	$\angle$ XOX (deg)	$(O-X)_{expt}$ (Å)	$(O-X)_{SS}$ (Å)[b]	Δ (Å)[c]
F—O—F	103.32(5)	1.4087(4)	1.42	−0.01
CF_3—O—CF_3	119.1(8)	1.369(4)	1.43	−0.06
FO_2S—O—SO_2F	123.6(5)	1.611(5)	1.70	−0.09
F_2SP—O—PSF_2	130.9(35)	1.609(8)	1.73	−0.12
PF_2—O—PF_2	135.2(18)	1.631(10)	1.73	−0.10
SF_5—O—SF_5	142.5(16)	1.585(11)	1.70	−0.11
SeF_5—O—SeF_5	142.4(19)	1.697(13)	1.82	−0.12
TeF_5—O—TeF_5	145.5(21)	1.832(12)	2.00	−0.17
SiF_3—O—SiF_3	155.7(20)	1.580(25)	1.77	−0.19

[a] See respective sections for references.
[b] Schomaker–Stevenson value.
[c] Difference between $(O-X)_{expt}$ and $(O-X)_{SS}$.

and by $(p-d)\pi$ back-bonding.[287] This concept can in principle be applied also to the corresponding P, S, Se, and Te compounds. An alternative explanation for the short O—X bonds involves strong polar contributions, $O^{-\delta} - X^{+\delta}$. The large oxygen valence angles would then be a consequence of electrostatic and steric repulsions between the substituents, which increase with bond shortening. Both concepts explain qualitatively the observed structural features, except for CF_3OCF_3, where the $(p-d)\pi$ concept cannot be applied. Furthermore, *ab initio* calulations for disiloxane show that $(p-d)\pi$ contributions are negligible, whereas the polar contributions are much larger than might be expected on the basis of electronegativity differences.[288] Thus, the concept of polar contributions, which are expected to be even stronger for the compounds in Table 4-21 than for disiloxane, is favored by the author as a simple model for rationalizing these structural features.

Difluorine dioxide has a very unusual peroxide structure[289] that contains an extremely short O—O bond [1.217(5) Å], similar to that in O_2, long O—F bonds [1.575(5) Å vs 1.4087(4) Å in OF_2], and an FOOF dihedral angle of less than 90° [87.5(5)°]. Several bonding models have been suggested,[289] but so far these structural features could not be reproduced fully by *ab initio* methods.[290,291] Trifluoromethyl-substituted peroxides, CF_3OOX, where X = H,[292] Cl,[292] F,[292] and CF_3,[293] have O—O bond lengths that are more similar to that in hydrogen peroxide (1.464 Å),[294] that is, 1.447(8) Å for X = H, 1.447(15) Å for X = Cl, 1.419(20) Å for X = CF_3, and 1.366(33) Å for X = F. Although the experimental uncertainties are large, the gross trend is toward shortening of O—O with increasing electronegativity of X. The O—C bonds [between 1.372(22) Å in CF_3OOCl and 1.419(24) Å in CF_3OOF] are in general slightly shorter than are the corresponding bonds in CH_3OOCH_3 [1.422(7) Å][295] or CH_3OOH [1.437 Å].[296] The dihedral angles in the CF_3-substituted peroxides lie between 90 and 100° for X = H, Cl, and F, and

123.3(40)° in CF_3OOCF_3. This latter value is perfectly reproduced by *ab initio* calculations.[297] Within their experimental uncertainties, the oxygen valence angles lie between 108 and 110° in all peroxides, including FOOF.

Sulfur, Selenium, and Tellurium Compounds

Compounds with Oxidation State II. The parent compound in this category, sulfur difluoride, is a highly unstable species with a reasonable lifetime in the gas phase only at very low pressures. An MW study[298] gave in S—F = 1.59208(8) Å and FSF = 98.197(11)° (r_z values). Trifluoromethyl-substituted sulfides are more stable and are amenable to ED studies. Within the series CF_3SX, where X = F,[299] CF_3,[299] Cl,[299] Br,[300] and H,[301] the reported S—C bond lengths for the most electronegative and electropositive substituents F and H are shorter [1.805(3) and 1.800(5) Å], than for the intermediate members (close to 1.82 Å). The S—F bond in CF_3SF [1.605(3) Å] is only slightly longer than are the corresponding bonds in SF_2. In contrast to the analogous series of oxygen compounds, the sulfur valence angle shows little variation within this series [between 97.1(7)° in CF_3SF and 99.4(3)° in CF_3SBr].

The chalcogen bond angles in $(CF_3)_2X$ with X = S, Se,[302] and Te[303] decrease with increasing size of the central atom [97.3(8), 95.5(20), and 90.2(14)°] and are slightly smaller than in the analogous methyl compounds. Methyl/trifluoromethyl substitution causes minor lengthening of the X—C bonds by about 0.02 Å [S—C = 1.820(3), Se—C = 1.980(9), Te—C = 2.160(7) Å in the CF_3 compounds; see references 304–306 for data relating to the corresponding methyl compounds]. This result is consistent with the fact that the electronegativities of S, Se, and Te are close to that of carbon. In CF_3SeCN the two Se—C bond lengths differ appreciably [Se—C(F_3) = 1.984(20) Å and Se—C(N) = 1.854(16) Å], thereby reflecting the difference in hybridization of the two carbon atoms; also, CSeC = 92.2(20)°.

Three compounds that contain trifluoromethylthio groups have been studied by ED. Tetrakis(trifluoromethylthio)methane[307] (**15**) possesses S_4 symmetry with the CS_4 skeleton flattened along the S_4 axis [the SCS angles including the S_4 axis are 116.6(6)°] and C—S bond lengths (mean value of central and

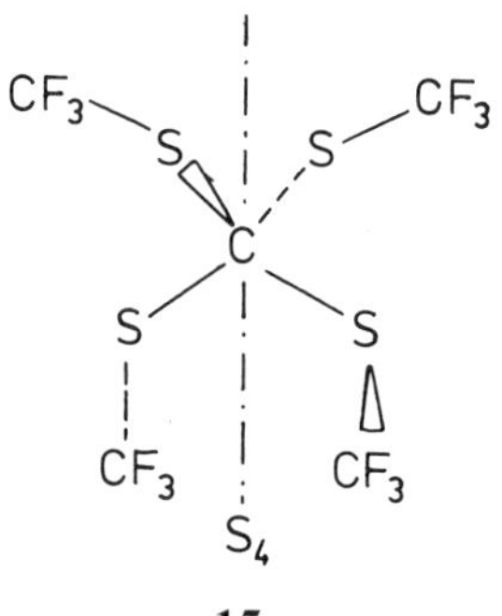

15

peripheral bonds) of 1.828(3) Å. The trifluoromethylthio groups are rotated by 78(2)° from the eclipsed positions. In the electron spin resonance (ESR) spectrum of hexakis(trifluoromethylthio) ethane, $(CF_3S)_3C$—$C(SCF_3)_3$, homolytic cleavage of the central C—C bond was observed at room temperature,[308] and this instability is reflected in an extremely long C—C bond length of 1.624(37) Å.[309] The experimental ED intensities are reproduced with C_1 symmetry for the two $C(SCF_3)_3$ groups [(C—S)$_{mean}$ = 1.832(4) Å and SCS = 113.9(3)°] and C_i overall symmetry. Tris(trifluoromethylthio)amine, $N(SCF_3)_3$,[310] has a planar or nearly planar NS_3 skeleton with two CF_3 groups oriented upward and one CF_3 group oriented downward (perpendicular to the NS_3 plane). The N—S and S—C bond lengths are 1.705(5) and 1.825(6) Å, respectively, with NSC angles of 100.1(10)°. The analogous selenium compound, $N(SeCF_3)_3$,[311] shows the same structural features with N—Se = 1.841(3) Å, Se—C = 1.973(5) Å, and NSeC = 104.3(9)°.

Disulfur difluoride is the only S_2X_2 compound that exists in two isomeric forms, a nonplanar chain species FSSF and a branched form S=SF$_2$. Both isomers have recently been reinvestigated by joint ED/MW analysis.[312] These results confirm an earlier MW study,[313] but with considerably reduced experimental uncertainties. The more stable isomer, S=SF$_2$ will be discussed in the section that deals with compounds of oxidation state IV. Disulfur difluoride shows structural features that are similar to those of FOOF, ie, a short S—S bond [1.890(2) Å], equal to that in S_2 (1.889 Å[314]), long S—F bonds [1.635(2) Å], SSF = 108.3(2)°, and an FSSF dihedral angle of 87.7(4)°. In the CF_3-substituted disulfide, CF_3S—SCF_3, the S—S bond is appreciably longer [2.030(5) Å],[315] similar to the corresponding bond in CH_3S—SCH_3 [2.039(4) Å],[316] and the S—C bonds [1.835(5) Å] are slightly longer than in $(CF_3)_2S$ [1.820(3) Å]. The SSC angle of 101.6(6)° is smaller than the SSF angle in FSSF, and the dihedral angle increases to 104.4(40)° upon F/CF_3 substitution. For the analogous diselenide, CF_3Se—$SeCF_3$, the following geometric parameters have been reported[317]: Se—Se = 2.292(10) Å, Se—C = 2.018(20) Å, SeSeC = 98.0(5)°, and CSeSeC = 84.5(30)°.

Compounds with Oxidation State IV. Chalcogen(IV) compounds can have coordination numbers three or four. Three-coordinated compounds that contain one double bond possess pyramidal structures. Principal paremeters for three-coordinated S(IV) compounds with S=O, S=S, and S=N are listed in Table 4-22. The ED parameters for O=SF$_2$[318] agree essentially with those obtained via an MW study.[321] The S=O and S=S bonds are shorter than the corresponding bonds in SO (1.493 Å)[314] or S_2 (1.898 Å),[314] and the S—F bonds are slightly shorter (O=SF$_2$) or longer (S=SF$_2$) than the corresponding bonds in SF$_2$. The FSF angles are in all cases considerably smaller than this angle in SF$_2$ [98.197(11)°], whereas the angle reported for F_2S=NCF$_3$ appears to be unreasonably small. The X=SF angles increase along the sequence O < S < N. The S=N bond length decreases simultaneously with increasing nitrogen valence angle [SNCl = 120.0(2)°, SNC = 130.4(7)°]. The sulfur and nitrogen lone electron pairs in the imino compounds are mutually cis. Replacement of

Table 4-22. Principal Geometric Parameters (r_g values)
for Three-Coordinated S(IV) Compounds

Compound	Bond lengths (Å)		Bond angles (deg)		
	S=X	S—F	∠FSF	∠XSF	Ref.
$F_2S=O$	1.421(3)	1.584(3)	92.2(3)	106.2(2)	318
$F_2S=S$	1.858(2)	1.610(2)	91.7(3)	108.1(2)	312
$F_2S=NCl$	1.477(3)	1.597(2)	89.8(2)	111.2(1)	319
$F_2S=NCF_3$	1.447(6)	1.583(4)	81.1(16)	112.6(5)	320

fluorines by CF_3 groups in $F_2S=O$ and in $F_2S=NCl$ has rather different structural effects in both compounds. The S=O bond in $(CF_3)_2S=O$ $[1.470(4)\,\text{Å}]^{322}$ is longer by about 0.05 Å than in $F_2S=O$, whereas the S=N bond in $(CF_3)_2S=NCl$ $[1.435(8)\,\text{Å}]^{322}$ is shorter relative to the analogous fluorine compound. This shortening and other structural changes that occur upon F/CF_3 substitution [ie, N—Cl shortens from 1.724(4) to 1.677(8) Å and SNCl increases from 120.0(2) to 138.2(38)°] are consistent with partial delocalization of the N lone electron pair with consequent double- and triple-bond contributions, respectively to the N—Cl and S=N bonds. It is not obvious, however, why such delocalization should occur in $(CF_3)_2S=NCl$ and not in $F_2S=NCl$. In the sulfoxide the sulfur valence angles are smaller by about 4–5° [CSC = 94.2(8), CSO = 104.5(11)°] than in the imido compound [CSC = 99.4(9), CSN = 108.8(25)°]. Comparison of the CF_3-substituted derivatives with $F_2S=O$ and $F_2S=NCl$ shows that the CSC angles are larger than the FSF angles, whereas the CSX angles are smaller than the corresponding FSX angles (X = O or N). It has been pointed out previously that the CH_3/CF_3 substituent effect is much larger for S(IV) than for S(II) compounds. In dimethylsulfoxide the S—C bonds lengthen by about 0.08 Å upon fluorination [1.808(4) Å in $(CH_3)_2SO^{323}$ vs 1.887(4) Å in $(CF_3)_2SO$]. Similarly, long S—C bonds have been observed for $(CF_3)_2SNCl$ [1.880(6) Å].

Four-coordinated chalcogen(IV) compounds possess trigonal bipyramidal structures in which the lone electron pair is equatorial. The parent compound in this category, SF_4, has been studied by MW^{324} and by $ED.^{325}$ The two investigations are in perfect agreement with respect to the bond lengths [S—F_{eq} = 1.542(5) and S—F_{ax} = 1.643(5) Å]^{325} and differ by less than 2° for the bond angles [$F_{eq}SF_{eq}$ = 103.8(6)°, $F_{eq}SF_{ax}$ = 89.0(7)°].^{325} The difference between axial and equatorial bond lengths in SF_4 [Δ(S—F) = 0.101(7) Å] is more than twice as large as that in PF_5 [Δ(P—F) = 0.043(7) Å]. A similarly large difference between axial and equatorial bond lengths has been observed in an MW analysis of SeF_4 [Se—F_{eq} = 1.682(4), Se—F_{ax} = 1.771(4) Å]^{326} with slightly smaller bond angles [$F_{eq}SeF_{eq}$ = 100.6(7), $F_{eq}SeF_{ax}$ = 86.6(3)°].

The stereochemistry of $(CF_3)_2SF_2$ is in accord with expectations based on application of the electronegativity rule, ie, the less electronegative CF_3 groups occupy the equatorial positions. Comparison with the isoelectronic phosphorane, $(CF_3)_2PF_3$, where the CF_3 groups occupy the axial positions of the trigonal bipyramid, demonstrates that formal replacement of one equatorial fluorine in the phosphorane by a lone pair $[(CF_3)_2SF_2]$ or double bond $[(CF_3)_2S(O)F_2$, see below] has a marked effect on the stereochemistry. The axial S—F bonds in the CF_3-substituted compound [1.683(3) Å] are longer by 0.04 Å than in SF_4, and the S—C bonds [1.889(4) Å] are similar to those in other S(IV) compounds. The equatorial [CSC = 97.3(8)°] and axial bond angles [CSF_{ax} = 88.0(3)°] decrease upon F/CF_3 substitution. This trend differs from the observations made for three-coordinated S(IV) compounds, for which the CSC angles are larger than the FSF angles. Some structural trends that are observed in the series $(CF_3)_2XF_2$ with X = S, Se,[327] and Te[303] are rather unexpected, and no explanation can be given. All three compounds have the same stereochemistry, ie, equatorial CF_3 groups. The axial bond angles CXF_{ax} decrease with increasing size of the central atom [88.0(3), 84.4(10), and 83.2(12)°], whereas the reported values for the equatorial angles vary irregularly [CSC = 97.3(8)°, CSeC = 118.7 (117)°, CTeC = 85(2)°]. A regular decrease of the CXC angles was observed for the two-coordinated $(CF_3)_2X$ compounds (see above). The X—C bonds in the four-coordinated compounds are longer by about 0.07 and 0.04 Å for X = S and Se, but shorter by 0.02 Å for X = Te, relative to those distances in $(CF_3)_2X$.

A mixed S(IV)–S(II) compound, SF_3SF (**16**), is formed by an unusual dimerization process of SF_2.[328] This rapid dimerization is responsible for the low stability of SF_2 in the gas phase and for its nonexistence in the condensed phase. A combination of ED and MW data and support from *ab initio* calculations were required to obtain an unambiguous structure for this small but highly unsymmetric molecule.[329] The basic structure corresponds to the predictions of the VSEPR theory, ie, a trigonal bipyramid with the lone electron pair and the electropositive SF group in equatorial positions. The SF_3 group deviates markedly from C_s symmetry, with the two axial bond lengths [S—F_{ax} = 1.624(6) Å and S—F'_{ax} = 1.722(8) Å] differing by 0.10 Å and the equatorial–axial angles [S′SF_{ax} = 92.2(6)° and S′SF'_{ax} = 76.0(10)°] differing by 16°. The S′F bond is oriented axially [F_{eq}SS′F = 95(4)°]. This sterically unfavorable orientation is stabilized by S—S′π bonding in the equatorial plane. According to the results of *ab initio* calculations, this π bonding contributes about 20% to the overall S—S′ bond strength and is due predominantly to

16   **17**

$(p - p)\pi$ overlap, with small contributions from $(d - d)$ and $(p - d)\pi$ overlap. This strong π bonding is consistent with the high barrier to internal rotation of the SF group, which is evident from temperature-dependent NMR spectra, and with the large value for the S—S′ torsional force constant. On the other hand, the S—S′ distance [2.040(5) Å] is longer by 0.15 Å than the corresponding bond distance in FSSF and is equal to the "single" bond value in CH_3SSCH_3 [2.039(4) Å].[316] The existence of strong polar repulsions in this S(IV)—S(II) bond may rationalize the discrepancy between long bond length and strong π bonding. The remaining geometric parameters for SF_3SF are as expected: $S-F_{eq} = 1.569(8)$ Å, $S'-F = 1.602(5)$ Å, $S'SF_{eq} = 104.9(14)°$, $SS'F = 105.9(10)°$.

An electronically similar system that possibly contains π bonding in the equatorial plane is dimethylaminotrifluorosulfurane[330] (**17**). The configuration at nitrogen is planar within the experimental uncertainties [sum of nitrogen valence angles is 357(3)° with SNC = 120.9(10) and CNC = 114.9(23)°], and the CNC plane is nearly perpendicular to the equatorial plane of the SF_3 group [$\tau = 78(4)°$]. Chen and Hoffmann[331] have pointed out in their theoretical study of a model compound, H_2NSH_3, that dative π bonding between the nitrogen lone electron pair and the equatorial sulfur orbitals favor the perpendicular conformation ($\tau = 90°$) with the nitrogen lone electron pair in the equatorial plane, whereas repulsion between the N and S lone electron pairs favors the conformation with the amino group oriented in the equatorial plane ($\tau = 0°$). The experimental result for $(CH_3)_2NSF_3$ indicates that the π bonding interaction dominates, a result that is similar to the situation in the aminophosphoranes discussed earlier. The results of an X-ray study for $[(CH_3)_2N]_2SF_2$[332] suggest that the dimethylamino groups adopt intermediate orientations ($\tau = 48.2°$) in which the configuration about the nitrogen atoms is pyramidal. It is not obvious whether this difference in bonding properties between $(CH_3)_2NSF_3$ and $[CH_3)_2N]_2SF_2$ is due to the presence of a second dimethylamino group or due to intermolecular interactions in the crystal. Whereas the S—N bond lengths are equal within their uncertainties [1.639(13) Å in **17** and 1.648(2) Å in the bis(dimethylamino)sulfurane], the $S-F_{ax}$ bond lengths differ markedly [1.670(7) and 1.770(2) Å]. The S—N bonds are considerably shorter than the Schomaker–Stevenson value of 1.74 Å or the experimental value in the S(IV) compound $[(CH_3)_2N]_2SO$ [1.693(4) Å],[333], results that are consistent with the existence of π bonding in both sulfuranes.

Compounds with Oxidation State VI. Two four-coordinated S(VI) compounds with a triple bond are known: $N\equiv SF_3$ and the recently synthesized $CF_3-C\equiv SF_3$.[334] When NSF_3 was studied by MW,[335] the following geometric parameters were reported: $S\equiv N = 1.416(3)$ Å, $S-F = 1.552(3)$ Å, $FSF = 94.03(30)°$. These S—F bonds are markedly shorter than those in the S(IV) derivative $N\equiv SF$ [$S\equiv N = 1.448(2)$, $S-F = 1.643(2)$ Å, $NSF = 116.91(8)°$].[336] An ED study for this compound based on visual analysis[337] resulted in an incorrect structural assignment, $F-N\equiv S$ with $F-N = 1.42(3)$ Å, $N\equiv S = 1.62(3)$ Å, and $FNS = 122(3)°$.

The compound CF_3—$C\equiv SF_3$ appears to have unexpected structural features that are not fully confirmed. *Ab initio* calculations in the HF approximation[338,339] result in a linear C—$C\equiv S$ chain, as one would ordinarily expect for an *sp*-hybridized central carbon atom and extremely short $C\equiv S$ (1.404 and 1.414 Å) and C—C (1.435 and 1.469 Å) bond lengths. These calculations also predict a linear structure for $HC\equiv SF_3$. The results of a low-temperature X-ray study[340] essentially confirm the short bond lengths ($C\equiv S$ = 1.420 Å and C—C = 1.439 Å). However, the results of the X-ray study indicate the presence of a slightly bent C—$C\equiv S$ chain (171.5°). This bending can be rationalized by a small contribution of the mesomeric structure **18**, but,

$$
\overset{\ominus}{C} = \overset{\oplus}{SF_3}
$$

$$CF_3$$

18

equally well, this result can be ascribed to intermolecular interactions in the crystal. An ED analysis[341] based on a rigid model results in a larger deviation from linearity, with CCS = 154.5(25)°. Two different nonrigid models were also used to fit the experimental intensities. Allowing free internal rotation of the CF_3 group leaves the quality of the fit unchanged relative to the rigid model. A model with a double-minimum potential for the CCS bending motion leads to solutions with different values for the barrier height at the linear configuration and the equilibrium angle. Barrier heights between 2 and 25 kJ mol^{-1} and equilibrium angles between 145 and 155° fit the experimental data. These two parameters, however, are strongly correlated in such a way as to yield average values for the CCS angle of 155(2)°. The equilibrium angle is 155° for high barriers (rigid model) and increases with decreasing barrier. A flat single-minimum potential does not fit the experimental intensities. The remaining geometric parameters for the various nonrigid models agree with the rigid-model values within the error limits. Although the ED data do not give any conclusive information on the dynamic behavior of this compound (CF_3 torsion and CCS bending potential), all possible solutions agree in producing an effective CCS angle of about 155°. Excluding the CCS angle, the geometric parameters of the ED analysis agree with those of the X-ray study: $C\equiv S$ = 1.434(14) Å. C—C = 1.45(3) Å, S—F = 1.561(3) Å, and FSF = 93.2(9)°. *Ab initio* calculations for the model compound H—$C\equiv SF_3$[341] also result in a bent configuration with a carbon valence angle of about 140°, if electron correlation is included at the level of MP2 [Møller–Plesset] approximation. The results of further test calculations for $HC\equiv SH_3$ indicate that bending of the skeleton is even greater if more advanced levels of theory are used, so it is unlikely that the MP2 calculations on $HC\equiv SF_3$ have overestimated the influence of correlation.

Four-coordinated chalcogen(VI) compounds with two double bonds have pseudotetrahedral structures. Principal parameters for some sulfones are listed in Table 4-23. Sulfonyl fluoride, for which the ED values in Table 4-23 are in

Table 4-23. PRINCIPAL GEOMETRIC PARAMETERS (r_g VALUES)
FOR FOUR-COORDINATED S(VI) COMPOUNDS WITH TWO
DOUBLE BONDS

Compound	Bond lengths (Å)		Bond angles (deg)		Ref.
	SF/SC	S=O	—S—	=S=	
F_2SO_2	1.531(2)	1.398(2)	96.7(11)	122.6(12)	342
$F_2S(O)NCl$	1.549(3)	1.395(3)	92.6(8)	117.4(31)	343
$(CF_3)_2SO_2$	1.860(5)	1.425(4)	102.2(8)	122.9(26)	344
CF_3ClSO_2	1.857(6)	1.416(7)	98.7(4)	122.4(10)	345
$CF_3(OH)SO_2$	1.833(5)	1.418(2)	102.3(23)	122.0(13)	346
CH_3FSO_2	1.563(4)[a]	1.411(3)	98.2(15)	123.1(15)	347
CH_3OFSO_2	1.546(6)	1.411(2)	96.8(6)	124.4(7)	348
$F(NH_2)SO_2$	1.56(2)	1.412(3)	99(6)	123.4(23)	349

[a] The S—C distance for this compound is 1.761(6) Å.

perfect agreement with an earlier MW study,[350] shows the typical structural features of sulfones. The S—F (S—C) and S=O bonds are shorter by about 0.05 and 0.03 Å relative to the corresponding bonds in S(IV) compounds, and FSF angles are larger by about 3–4°. In all cases OSO angles are close to 123°. Similar trends with increasing oxidation number are observed for the Se—F distances and FSeF angles in the analogous selenium compounds [Se—F = 1.686(2) Å, FSeF = 94.1(5)° in F_2SeO_2,[342] vs 1.730(2) Å and 92.22(5)° in F_2SeO[351]]. The Se=O bond lengths do not change with increasing oxidation number [1.575(2) Å in F_2SeO_2 vs 1.576(4) Å in F_2SeO]. In $F_2S(O)NCl$ the OSN angle is smaller than OSO angles in the sulfones, and the S=N bond length [1.485(7) Å] is very similar to that in the S(IV) compound F_2SNCl. The chlorine atom in $F_2S(O)NCl$ is trans to O. The S—C bond lengths in CF_3-substituted sulfones show a similar large CH_3/CF_3 substitution effect [S—C = 1.773(4) Å in $(CH_3)_2SO_2$[352]] as in the S(IV) compounds.

Gas-phase structures of three more compounds that contain sulfonyl fluoride groups, SO_2F, have been reported. In the polysulfuryl fluorides $O(SO_2F)_2$[353] (**19**) and $SO_2(OSO_2F)_2$[353] (**20**), the SO_2F groups have equivalent geometries (Table 4-24). The SOS bridging angles are 124° in both compounds and

19

20

Table 4-24. GEOMETRICAL PARAMETERS[a] FOR
POLYSULFURYL FLUORIDES $O(SO_2F)_2$ AND
$SO_2(OSO_2F)_2$

Parameter	$O(SO_2F)_2$	$SO_2(OSO_2F)_2$
S=O, Å	1.398(2)	1.402(3)
S—F, Å	1.525(5)	1.525(12)
S—O, Å	1.611(5)	1.613(6)
∠O=S=O, deg	126.8(24)	128.6(28)
∠F—S—O, deg	102.4(36)	101.3(30)

[a] Reference 353.

O—S—O in **20** is 97.8(20)°. Slightly different geometric parameters have been reported for the SO_2F groups in $HC(SO_2F)_3$[349]: S=O = 1.416(3) Å, S—F = 1.558(4) Å, and OSO = 123.0(5)°. The S—C bond lengths are 1.831(5) Å and the central SCS angles 111.1(3)°.

Five-coordinated S(VI) compounds have distorted trigonal bipyramidal structures with the double bond in the equatorial plane. The central compound in this category, thionyl tetrafluoride, has presented a puzzling problem for nearly two decades, which has been solved only recently by a combined ED/MW analysis.[354] Various structures have been suggested that differ mainly with respect to the bond angles. The original model[355] had $F_{eq}SF_{ax}$ angles slightly larger than 90°. Based on the same experimental intensities, two more models with $F_{eq}SF_{ax}$ less than 90°, as predicted by VSEPR theory, that differ in the $F_{eq}SF_{eq}$ angles were proposed later.[356] An independent ED study demonstrated that in fact four different models fit the experimental intensities equally well,[357] but the "wrong" model was favored on the basis of VSEPR arguments. Two independent MW studies resulting in different sets of rotational constants[358,359] further increased the confusion. It was demonstrated later that the set of constants given in reference 358 is based on an incorrect assignment in the MW spectra. The results of the combined analysis which favor one of the initially rejected models of reference 357, were confirmed by *ab initio* calculations.[360]

Table 4-25 compares structural parameters of some trigonal bipyramidal sulfur compounds that contain a lone electron pair (SF_4), an S=O, an S=C, or an S=N double bond in an equatorial position. These "isoelectronic" compounds show remarkable structural differences. The axial S—F bonds shorten with replacement of the S lone pair in SF_4 by the S=O double bond, whereas the equatorial bonds are not affected. The ED results for methylenesulfur tetrafluoride are in good agreement with the corresponding results of a low-temperature X-ray study[363] and *ab initio* calculations.[360] In this compound the equatorial S—F bonds are longer than in thionyl tetrafluoride, thereby reducing the difference between axial and equatorial bond lengths to 0.020 (21) Å. The combined ED/MW analysis for FN = SF_4 results in a structure with the N—F bond oriented in the axial direction and a strongly distorted SF_4 group. The axial S—F bond that is trans to the N—F bond (or cis to the nitrogen lone

Table 4-25. STRUCTURAL PARAMETERS (r_g VALUES) OF THE TRIGONAL
BIPYRAMIDAL TETRAFLUOROSULFUR COMPOUNDS SF_4, $O{=}SF_4$, $H_2C{=}SF_4$,
AND $FN{=}SF_4$

Parameter	$O{-}S{<}$	$O{=}S{<}$	$H_2C{=}S{<}$	$FN{=}S{<}$
$S{=}X$, Å	—	1.411(3)	1.550(20)	1.521(9)
$S{-}F_{eq}$, Å	1.542(5)	1.540(3)	1.575(15)	1.567(5)
$S{-}F_{ax}$, Å	1.643(5)	1.597(3)	1.595(15)	1.617(7)[a]
				1.537(12)[b]
$\angle F_{eq}SF_{eq}$, deg	103.8(6)	112.8(4)	97.0(20)	99.8(3)
$\angle F_{eq}SF_{ax}$, deg	89.0(7)	85.7(1)	86.7(6)	85.6(3)[a]
				89.6(3)[b]
Ref.	325	354	361	362

[a] Cis to N—F bond.
[b] Trans to N—F bond.

electron pair) is shorter by 0.080 (14) Å than the opposite axial bond and is even
shorter than the equatorial bonds. The $F_{eq}SF_{ax}$ angles differ by 4°. The largest
variations in the compounds listed in Table 4-25 occur in the $F_{eq}SF_{eq}$ angles,
which can be interpreted as reflecting the operation of different steric requirements between the S lone electron pair and $S{=}X$ double bonds. In the
pseudotetrahedral molecules PF_3 and $O{=}PF_3$ the lone electron pair appears to
be somewhat more repulsive than the $P{=}O$ double bond [$FPF = 97.8(2)°$ in
PF_3 and $101.2(3)°$ in $O{=}PF_3$]. In the less symmetric trigonal bipyramidal
compounds a direction dependence of the repulsion effect is apparent.[364] The
$S{=}O$ double bond exerts stronger repulsion in the axial direction and reduced
repulsion in the equatorial direction relative to the lone pair, leading to smaller
$F_{eq}SF_{ax}$ and larger $F_{eq}SF_{eq}$ angles. The $S{=}C$ double bond in the methylene
compound has a steric effect remarkably different from that of the $S{=}O$ bond.
The $S{=}C$ double bond exerts stronger repulsion in the equatorial direction
than does a lone electron pair or an $S{=}O$ bond, and it exerts intermediate
repulsion in the axial direction. This difference between $S{=}O$ and $S{=}C$ double
bonds has been explained by differences in electron distributions between these
π bonds.[360] Whereas the $S{=}O$ π bond has a nearly cylindrical shape with
similar populations of the π orbitals in axial and equatorial direction, the $S{=}C$
π bond is located almost exclusively in the equatorial plane. This is expected to
result from the axial orientation of the CH_2 group. The geometry of the SF_4
group in methylenesulfur tetrafluoride (bond angles and S—F distances) is close
to that of SF_6 (see below), as has been observed for $CF_3C{\equiv}SF_3$. The sulfur
bond angles in $FN{=}SF_4$ indicate a similar shape of the $S{=}N$ π bond and the
$S{=}C$ bond. The reduced symmetry of the NF group relative to CH_2 is reflected
in strong distortions of the SF_4 group from C_{2v} symmetry. The surprisingly
large difference between the two axial S—F bonds [0.080(14) Å] is in part

reproduced by *ab initio* calculations for HN=SF$_4$,[365] where a difference of 0.042 Å is predicted. The SNF angle is 117.6(12)° and the N—F bond [1.357(8) Å] is shorter than such bonds in FN=CF$_2$ [1.389(2) Å] or *cis*- and *trans*-FNNF [1.410(9) and 1.396(8) Å]. A similar strong distortion of the SF$_4$ group is observed for CH$_3$N=SF$_4$,[365] where the methyl group also points in the axial direction: S—F$_{eq}$ = 1.567, S—F$_{ax(cis)}$ = 1.643(4), S—F$_{ax(trans)}$ = 1.546(7) Å, F$_{eq}$SF$_{eq}$ = 102.6(2), F$_{eq}$SF$_{ax(cis)}$ = 84.5(4) and F$_{eq}$SF$_{ax(trans)}$ = 87.1(4)°. The S=N bond in *N*-methyliminosulfur tetrafluoride [1.480(6) Å] is shorter than in the analogous fluorine compound, and the nitrogen bond angle [127.2(11)°] is larger by about 10°.

The marked variation in bond angles between SF$_4$ and O=SF$_4$, which has been attributed to different steric effects of the lone pair and the S=O double bond, is not observed for the CF$_3$-substituted analogs (CF$_3$)$_2$SF$_2$ and (CF$_3$)$_2$S(O)F$_2$.[366] As mentioned above, the two CF$_3$ groups occupy equatorial positions in both compounds, and within experimental uncertainties the S—C distances and sulfur bond angles in the latter compound [S—C = 1.892(5) Å, CSC = 97.8(8)°, CSF$_{ax}$ = 87.7(3)°] are equal to those in (CF$_3$)$_2$SF$_2$. The axial bonds in the S(VI) derivative are shorter [1.643(4) Å vs 1.683(3) Å in (CF$_3$)$_2$SF$_2$] and the S=O bond [1.423(7) Å] is slightly longer than in O=SF$_4$.

The parent six-coordinated chalcogen(VI) compounds, SF$_6$, SeF$_6$, and TeF$_6$ have regular octahedral structures with bond lengths (r_g values) of 1.5635(4),[367] 1.685(2),[368] and 1.816(2) Å[369]. Hexafluorides of sulfur and tellurium have been studied at various temperatures, and the values above refer to data obtained at room temperature. In monosubstituted SF$_6$ derivatives, XSF$_5$, the question arises whether the S—F bond trans to X (axial) is longer or shorter than the cis bonds (equatorial). The ClSF$_5$ molecule has been studied in great detail by joint ED/MW analyses,[370,371] but this question could not be answered definitely, except to say that the difference between the two bonds is smaller than experimental error limits [S—F$_{eq}$ = 1.570(3), S—F$_{ax}$ = 1.571(6), and S—Cl = 2.055(1) Å]. The results of *ab initio* calculations depend on the basis set that is employed in the calculation.[371] The F$_{ax}$SF$_{eq}$ angle is slightly smaller than 90° [89.6(1)°], as predicted by the VSEPR theory. A more definite answer to the question above is obtained in the case of CF$_3$SF$_5$,[372] where the difference between the two types of bond length is slightly larger than their combined uncertainties [S—F$_{eq}$ = 1.573(2), Å, S—F$_{ax}$ = 1.563(7) Å]. The observation of lengthened cis bonds (S—F$_{eq}$) relative to the trans bond (S—F$_{ax}$) is in agreement with VSEPR ideas and the molecular orbital analysis of the "trans influence."[373] The S—C bond length [1.889(8) Å] is consistent with the long S—CF$_3$ bonds in other S(VI) and S(IV) derivatives.

Structural parameters for some compounds containing SF$_5$ groups are collected in Table 4-26. Either the cis and trans S—F distances have been constrained to be equal in these ED analyses, or their differences are smaller than the combined experimental uncertainties. The mean values are similar to those in SF$_6$ except for the S—F distances in the hypofluorite. The S—C bonds in the methane derivative are very *long*, whereas the S—N and S—O bonds are

Table 4-26. PRINCIPAL GEOMETRIC PARAMETERS OF
SOME SF_5 DERIVATIVES

| | Bond lengths (Å) | | | |
Compound	S—F$_{mean}$	S—X	∠SXS (deg)	Ref.
$(SF_5)_2CF_2$	1.563(3)	1.910(7)	124.3(7)	374
$(SF_5)_2NH$	1.568(3)	1.680(7)	134.8(10)	375
$(SF_5)_2NF$	1.556(4)	1.686(5)	138.3(10)[a]	375
SF_5NF_2	1.551(4)	1.698(10)	110.9(10)[b]	376
$(SF_5)_2O$	1.563(4)	1.588(11)	142.5(16)	377
SF_5OF	1.53	1.64	108[c]	378

[a] Planar configuration at nitrogen.
[b] SNF angle.
[c] SOF angle.

short relative to the corresponding Schomaker–Stevenson values (1.81, 1.74, and 1.70 Å for S—C, S—N, and S—O). The carbon, nitrogen, and oxygen bond angles are extremely large compared with standard values, which in all three cases approach the tetrahedral. These structural features can be rationalized qualitatively by polar contributions to the S—X bonds, in combination with steric repulsions between the SF_5 groups, as pointed out above for XOX compounds. The polar contribution is strongly attractive for the S—O bonds ($S^{+\delta} - O^{-\delta}$), reduced, but still attractive for the S—N bonds and strongly repulsive for the S—C bonds ($S^{+\delta} - C^{+\delta}$). The steric repulsion between the SF_5 groups is evident from the shortest F...F contacts, which are 2.40, 2.44, and 2.55 Å in the oxide, amide, and methane derivatives, respectively. They are considerably shorter than the F...F van der Waals distance of 2.70 Å. The alternative rationalization for these structural features, ie, partial delocalization of the O or N lone electron pairs with formation of $(p - d)\pi$ bonds, would not account for the "normal" bond angles in the monosubstituted compounds SF_5OF and SF_5NF_2 [FNF = 98.1(8)°].

The bis(pentafluoroselenium)- and tellurium oxides,[377] $(SeF_5)_2O$ and $(TeF_5)_2O$, have structural features that are similar to those of the sulfur analog (see oxygen section). The Se—O [1.699(13) Å] and Te—O [1.834(12) Å] bonds are shorter than the corresponding Schomaker–Stevenson values, and the oxygen bond angles are about 144° in both compounds. The Se—F and Te—F distances correspond to those in SeF_6 and TeF_6.

Exceptional chemical and structural properties of a selenium derivative relative to the analogous sulfur or tellurium compounds are found in the NCO-substituted pentafluoro compounds. In the original study the ED intensities for all three derivatives were interpreted in terms of isocyanate structures,[379] F_5XNCO (X = S, Se, Te). Although the experimental intensities for the Se compound could be reproduced almost equally well with the cyanate structure, this possibility was considered unlikely on the basis of bond lengths in the NCO group, which are comparable for all three compounds, and on the basis of the

Se—Y bond length (Se—N for isocyanate or Se—O for cyanate). The only marked differences that occur in this series involve the N bond angles, which are very similar in the S and Te compounds (124.9(12) and 126.5(24)°], but much smaller in the Se derivative [116.9(8)°]. Subsequent NMR studies[380] (^{14}N, ^{15}N, ^{19}F, ^{77}Se) demonstrate that the original choice between the possible structural models was incorrect; the Se compound exists in the cyanate form, $F_5Se—O—C\equiv N$. This structure is also in agreement with the vibrational spectrum and the chemical behavior (explosive) of the Se derivative, which differ from those of the S and Te analogs. The X—F (X = S, Se, Te) bond lengths are very similar to those in the XF_6 compounds [S—N = 1.668(6) Å, Te—N = 1.859(21) Å, and Se—O = 1.794(6) Å]. The latter distance is longer by 0.10 Å than that in $(SeF_5)_2O$, and this difference possibly can be explained by strongly different oxygen valence angles [SeOSe = 142.4(9)° vs SeOC = 116.9(8)°] and by different polar contributions. The NCO groups have equal bond lengths in the S and Te compounds [N=C = 1.24(1) Å C=O = 1.18(1) Å] and in the cyanate group [O—C = 1.257(10) Å and $C\equiv N$ = 1.181(10) Å].

Whereas thionyl tetrafluoride is a stable species for which dimerization has not been observed, the analogous selenium and tellurium compounds cannot be isolated in their monomeric forms, and only the cyclic dimers (**21**) are stable compounds. A possible explanation for this different monomer–dimer stability of S and Se or Te compounds is the higher stability of S=O vis-à-vis Se=O or Te=O double bonds relative to the respective single bonds. The four-membered rings[381] are planar with Se—O = 1.779(7) Å and Te—O = 1.918(9) Å. Both bond distances are longer by about 0.08 Å than those in $(XF_5)_2O$. The oxygen bond angles [SeOSe = 97.5(5)° and TeOTe = 99.5(6)°] are about 45° smaller than in the unstrained oxides. The axial X—F bonds (perpendicular to the ring plane and cis to both X—O bonds) are longer than the equatorial bonds [Se—F_{eq} = 1.668(12) Å, Se—F_{ax} = 1.698(10) Å; Te—F_{eq} = 1.802(11) Å, Te—F_{ax} = 1.848(11) Å]. A similar monomer–dimer behavior is observed for S(VI) compounds that contain S=C and S=N double bonds. The $H_2C=SF_4$ molecule is characterized by high thermal stability, and its dimeric form has not been observed so far. On the other hand, its fluorinated species has not been observed in its monomeric form but is perfectly stable as a cyclic dimer (**22**). The cyclic dimer can be synthesized by fluorination of tetrafluoro-1,3-dithietane.[382] Several $RN=SF_4$ compounds are known to exist in their monomeric form (R = F, CH_3, CF_3, SF_5), whereas the N-chloro derivative is stable only in its

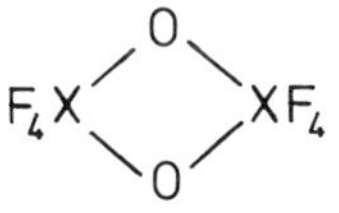

21 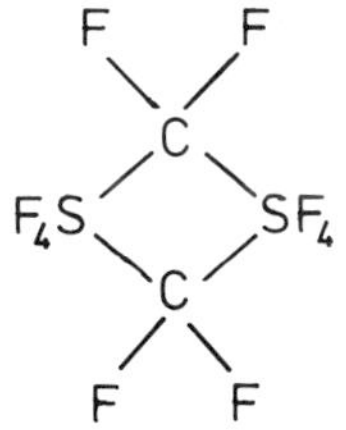　　　　**22** 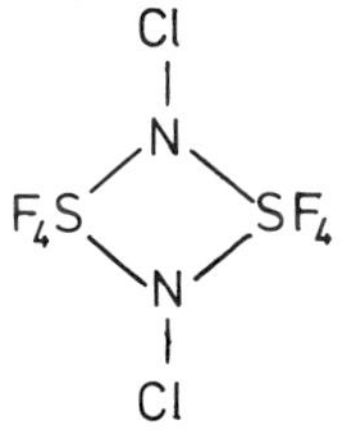　　　　**23**

dimeric form (**23**).[383] Both four-membered rings **22**[374] and **23**[384] are planar with S—C = 1.887(4) Å, SCS = 96.2(3)°, and S—N = 1.735(4) Å, SNS = 99.3(6)°. In both compounds, the axial S—F bonds are longer than the equatorial bonds, S—F_{eq} = 1.573(6) Å S—F_{ax} = 1.591(6) Å in **22**, and S—F_{eq} = 1.545(5) Å, S—F_{ax} = 1.591(6) Å in **23**.

Group VII Compounds

Among the interhalogen compounds the fluorides are by far the largest group. For fluorine itself, ED studies based on visual analysis only have been performed, resulting in F—F distances of 1.45(5)[385] and 1.433(10) Å,[386] respectively. Highly accurate MW results have been reported for the diatomic species ClF [r_e = 1.628341(4) Å][387] and BrF [r_e = 1.758987(4) Å].[387] The largest variety of oxidation states and coordination number exists for chlorine-fluorine-oxygen compounds (Table 4-27). The data for ClF and $ClFO_2$ are MW results, and the ED parameters for ClF_3 and ClF_5 are in good agreement with earlier MW studies.[393,394] The gross structures of these compounds obey the basic rules of the VSEPR theory. The Cl—F bond lengths can best be discussed if the bonds are separated into two types, mainly covalent and mainly semi-ionic three-center/four electron (3c–4e) bonds. In ClF_3 the equatorial bond is considered to be mainly covalent, and the axial bonds are 3c–4e bonds. However, in ClF_5, the bond types are reversed, and the axial bond is mainly covalent. Covalent bonds are shorter than 3c–4e bonds. Within each type the bonds shorten with increasing oxidation state and lengthen with increasing oxygen substitution.

Table 4-27. Geometric Parameters of Chlorine-Fluorine-Oxygen Compounds

Parameter	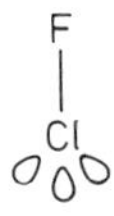 ClF	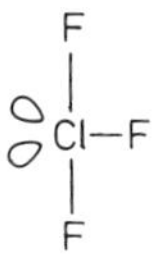 ClF₃	 ClF₅	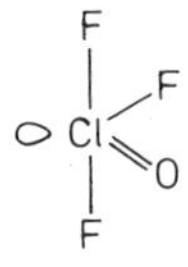 ClF₃O	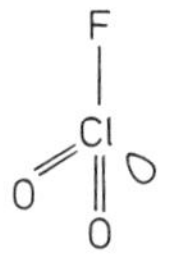 ClFO₂	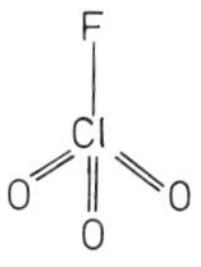 ClFO₃
Cl—F, Å	1.628	1.584(12)	1.669(15)	1.603(4)	1.697(4)	1.619(4)
Cl—F_{ax} Å		1.703(14)	1.571(14)	1.713(3)		
Cl=O, Å				1.405(3)	1.418(3)	1.404(2)
∠$F_{eq}ClF_{ax}$, deg		87(2)	86.0(15)	87.9(12)		
∠OClF, deg				108.9(9)[a] 94.7(20)[b]	101.7(1)	100.8(8)
∠OClO, deg					115.2(1)	116.6(5)
Ref.	387	388	389	390	391	392

[a] $OClF_{eq}$.
[b] $OClF_{ax}$.

The BrF_3,[388] BrF_5,[395] and $BrFO_3$[396] molecules have the same basic structures as do their chlorine analogs, and similar trends for the bond lengths are observed. The ED results for BrF_3[388] are in perfect agreement with those obtained via a corresponding MW study,[397] and the BrF_5 data were derived by a joint ED/MW analysis. The mainly covalent $Br-F$ bond lengths decrease from 1.759 Å in BrF to $Br-F_{eq} = 1.728(15)$ Å in BrF_3 to $Br-F_{ax} = 1.699(6)$ Å in BrF_5 and the mainly 3c–4e bonds from $Br-F_{ax} = 1.809(17)$ Å in BrF_3 to $Br-F_{eq} = 1.768(1)$ Å in BrF_5. The $F_{eq}BrF_{ax}$ angles are equal in BrF_3 $(85(2)°)$ and in BrF_5 $[85.1(4)°]$. For $BrFO_3$ the following parameters have been determined: $Br-F = 1.708(3)$ Å, $Br{=}O = 1.582(1)$ Å, and $OBrO = 114.9(3)°$.

Iodine-fluorine compounds are stable only in oxidation states V (IF_5) and VII (IF_7, IOF_5). According to the straightforward VSEPR theory, IF_5[398] and IOF_5[399] are closely analogous; accordingly, their structures should be very similar. Both compounds have been studied by joint ED/MW analyses. The angular distortions of the tetragonal bipyramids in which lone pairs and oxygen occupy axial positions are the same in both compounds $[F_{eq}IF_{ax} = 82.1(6)°$ in IF_5 and $82.0(3)°$ in $IOF_5]$. This result can be interpreted in terms of the VSEPR picture as reflecting equal steric requirements of the iodine lone electron pair and the $I{=}O$ double bond. The mean $I-F$ distances decrease with increasing oxidation number $[1.860(2)$ Å in IF_5 vs $1.826(1)$ Å in $IOF_5]$. The relative lengths observed for the $I-F$ bonds, however, are surprising. In IF_5, the equatorial bonds are longer than the axial bond ($I-F_{eq} = 1.868(4)$ Å and $I-F_{ax} = 1.834(10)$ Å], which corresponds to the observation made for ClF_5 and BrF_5. The differences between equatorial and axial bond distances decrease with increasing size of the central atom $[0.098(21)$ Å in ClF_5, $0.069(7)$ Å in BrF_5, and $0.034(11)$ Å in $IF_5]$. In IOF_5, however, the relative lengths of equatorial and axial bonds are reversed, the axial bond $[1.863(4)$ Å] being longer than the equatorial bonds $[1.817(2)$ Å]. Bartell and colleagues[399] rationalize these different structural features of IF_5 and IOF_5 by differentiating between a "primary" effect of deformation and "secondary relaxation" effect in the VSEPR picture. According to their arguments the "primary" effect is responsible for angle deformation and longer equatorial bonds relative to the axial bond in IF_5, whereas the "secondary relaxation" effect, which operates at the distorted structure, is responsible for the longer axial bond relative to the equatorial bonds in IOF_5. It is puzzling, however, that the steric effects of lone electron pair and $I{=}O$ bond are equal for the angular distortions and reversed for the bond lengths.

Iodine heptafluoride, IF_7,[400] has a pentagonal bipyramidal structure with axial bonds $[1.786(7)$ Å] shorter than equatorial bonds $[1.858(4)$ Å]. Repulsions between the crowded equatorial bonds lead to distortion from D_{5h} symmetry, similar to that found in five-membered rings (cyclopentane), with an average equatorial puckering of $7.5°$ and an axial bend of $4.5°$. ED intensities do not allow a distinction to be made between static deformation or a dynamic pseudorotational model.

Group VIII Compounds

Only five inert gas fluorides have been studied in the gas phase, and four of these are xenon derivatives. The only nonxenon fluoride, KrF_2, has a linear structure, and the values reported for the Kr—F distances from an ED study[401] $[r_g = 1.889(10)\,Å]$ and a high-resolution infrared investigation[402] $[r_0 = 1.875(2)\,Å]$ are essentially in agreement. The conformations of the xenon fluorides, XeF_2,[403] XeF_4,[404] and XeF_6[405] follow qualitatively the predictions of VSEPR theory. The difluoride XeF_2 is an AB_2E_3 type compound (E = lone electron pair) with a linear configuration, XeF_4 is an AB_4E_2 type compound with a square planar configuration, and XeF_6 is an AB_6E type compound with a distorted octahedral configuration. In its instantaneous configuration the lone electron pair occupies a position in the center of one face of the octahedron, so that the octahedron becomes distorted to C_{3v} symmetry (distortion in the T_{1u} bending mode). The angular displacements of the fluorine atoms range from 5-$10°$. It appears that the barriers between equivalent C_{3v} configurations are low, and the lone electron pair rapidly fluctuates between the eight possible positions. This results in nonrigid behavior of XeF_6 with a two-dimensional pseudorotation. The cause of distortion has been ascribed to the pseudo-Jahn–Teller effect. This interpretation of the ED intensities[405] confirms results obtained via earlier ED studies,[401,404,406,407] which also found distorted octahedral configurations for this compound. In addition, this interpretation is compatible with infrared, Raman, electric field deflection, and calorimetric data.[408] The Xe—F bond lengths decrease with increasing oxidation number from $1.9773(15)\,Å$ in XeF_2 to $1.94(1)\,Å$ in XeF_4 to $1.890(5)\,Å$ in XeF_6. The latter value is a mean value of Xe—F bonds, which are not exactly equivalent in the distorted configuration.

Xenon oxyfluoride, $XeOF_4$,[409] has a tetragonal bipyramidal structure in which the xenon lone electron pair and the Xe=O double bond occupy axial positions. The Xe—F bond lengths $[1.902(2)\,Å]$ are very similar to those in XeF_6 and Xe=O $= 1.708(7)\,Å$. The OXeF angle of $91.6(2)°$ indicates the operation of a slightly larger steric effect of the Xe=O double bond relative to the lone pair.

Conclusion

Geometric structures provide some insight into the bonding properties of compounds. More important for the elucidation of these properties is the examination of structural changes that occur upon various substitutions. As demonstrated in this chapter, fluorine causes quite drastic substituent effects in conformation, bond lengths, or angles, relative to hydrogen or other substituents. Some of these structural features of fluorine-containing compounds can be rationalized qualitatively with simple concepts, such as VSEPR theory or polar effects. A deeper insight into bonding properties and a qualitative "understanding" of geometries in terms of electronic structure may be gained from molecular orbital calculations. The prime condition for such calculations is that they reproduce correctly experimental structures and—even more important—substituent effects. Experience teaches that reproduction of experimental structures for compounds containing highly electronegative atoms, such as F, O, or N, is exceedingly difficult and is a challenge for high-quality *ab initio* methods. Thus, experimental methods will continue to be the preferred source of structural information for fluorinated compounds. The development of concepts that allow for more quantitative predictions of geometric structures and substituent effects requires studies of many more such compounds or, preferably, of families of related compounds. A prerequisite for the successful development of such improved concepts is collaboration among synthetic, structural, and theoretical chemists.

Acknowledgment

The author is indebted to Dr. Colin J. Marsden, University of Melbourne, for critical reading of the manuscript and for helpful suggestions to improve this chapter.

References

1. Yokozeki, A.; Bauer, S. H. *Top. Cur. Chem.* **1975**, *75*, 71.
2. Gillespie, R. J.; Nyholm, R. S. *Qt. Rev.* **1957**, *11*, 339; *Angew. Chem. Int. Ed. Engl.* **1967**, *6*, 819; "Molecular Geometry". Van Nostrand Reinhold: New York, 1972.
3. Kuchitsu, K.; Cyvin, S. J. In "Molecular Structures and Vibrations", Cyvin, S. J., Ed.; Elsevier: Amsterdam, 1972, Chap. 12.
4. Janzen, J.; Bartell, L. S. *J. Chem. Phys.* **1969**, *50*, 3611.
5. de Lucia, F. C.; Helminger, P.; Gordy, W. *Phys. Rev. A* **1971**, *3*, 1849.
6. Kuchitsu, K.; Konaka, S. *J. Chem. Phys.* **1966**, *45*, 4342.
7. Robiette, A. G.; Sheldrick, G. M.; Sheldrick, W. S. *J. Mol. Struct.* **1970**, *5*, 423.
8. (a) Chiang, C. H.; Porter, R. F.; Bauer, S. H. *J. Phys. Chem.* **1970**, *74*, 1363. (b) Hauser, R.; Oberhammer, H.; Bürger, H; Pawelke, G. *J. Chem. Soc. Dalton Trans.* **1987**, 1839. (c) Almenningen, A.; Gundersen, G.; Mangerud, M.; Seip, R. *Acta Chem. Scand.* **1981**, *A35*, 341.
9. Danielson, D. D.; Patton, J. V.; Hedberg, K. *J. Am. Chem. Soc.* **1977**, *99*, 6484.
10. Ryan, R. R.; Hedberg, K. *J. Chem. Phys.* **1969**, *50*, 4986.
11. Danielson, D. D.; Hedberg, K. Sixth Austin Symposium on Gas-Phase Molecular Structure, Austin, Texas 1976, Paper No. MM7.
12. Lory, E. R.; Porter, R. F.; Bauer, S. H. *Inorg. Chem.* **1971**, *10*, 1072.
13. Kuczkowski, R. L.; Lide, D. R. *J. Chem. Phys.* **1967**, *46*, 357.
14. Hargittai, M.; Hargittai, I. *J. Mol. Struct.* **1977**, *39*, 79.
15. Shibata, S.; Iijima, K. *Chem. Lett.* **1977**, 29.
16. Iijima, K.; Shibata, S. *Bull. Chem. Soc. Japan* **1979**, *52*, 711.
17. Iijima, K.; Yamada, T.; Shibata S. *J. Mol. Struct.* **1981**, *77*, 271.
18. Beagley, B.; Medwid, A. R. *J. Mol. Struct.* **1977**, *38*, 229.
19. Takeo, H.; Curl, R. F. *J. Chem. Phys.* **1972**, *56*, 4314.

20. Tamagawa, K.; Takemura, M.; Konaka, S.; Kimura, M. *J. Mol. Struct.* **1984**, *125*, 131.
21. Bauer, S. H.; Katada, K.; Kimura, K. In "Structural Chemistry and Molecular Biology". Freeman: New York, 1968, p. 653.
22. Harshbarger, W.; Lee, G.; Porter, R. F.; Bauer, S. H. *Inorg. Chem.* **1969**, *8*, 1683.
23. Oyanagi, K.; Kuchitsu, K. Seventh Austin Symposium on Gas-Phase Molecular Structure, Austin, TX, 1978, Paper No. A15.
24. Thornton, C. G. *Diss. Abstr.* **1953**, *14*, 604.
25. Fink, M.; Schmiedekamp, C. W.; Gregory, D. *J. Chem. Phys.* **1979**, *71*, 5238.
26. Pauling, L. "The Nature of the Chemical Bond," 2nd ed. Cornell University Press: Ithaca, NY, 1960.
27. Oberhammer, H. *J. Mol. Struct.* **1975**, *28*, 349.
28. Jacob, E. J. *J. Mol. Struct.* **1979**, *52*, 63.
29. Typke, V.; Dakkouri, M.; Oberhammer, H. *J. Mol. Struct.* **1978**, *44*, 85.
30. Gosh, S. N.; Trambarulo, R.; Gordy, W. *J. Chem. Phys.* **1952**, *20*, 605.
31. Iijima, T. *Bull. Chem. Soc. Japan* **1973**, *46*, 2311.
32. Kuchitsu, K. *J. Chem. Phys.* **1968**, *49*, 4456.
33. Beagley, B.; Jones, M. O.; Yavari, P. *J. Mol. Struct.* **1981**, *71*, 203.
34. Beagley, B.; Jones, M. O.; Houldsworth, N. *J. Mol. Struct.* **1980**, *62*, 105.
35. Beagley, B.; Jones, M. O.; Zanjanchi, M. A. *J. Mol. Struct.* **1979**, *56*, 215.
36. Fernholt, L.; Kveseth, K. *Acta Chem. Scand.* **1980**, *A34*, 163.
37. Friesen, D.; Hedberg, K. *J. Am. Chem. Soc.* **1980**, *102*, 3987.
38. Brunvoll, J. Ph. D. thesis, Trondheim, Norway, 1962.
39. Van Schaick, E.J.M.; Geise, H. J.; Mijlhoff, F. C.; Renes, G. *J. Mol. Struct.* **1973**, *16*, 23.
40. Beagley, B.; Brown, D. E. *J. Mol. Struct.* **1979**, *54*, 175.
41. Brown, D. E.; Beagley, B. *J. Mol. Struct.* **1977**, *38*, 167.
42. Al-Aidah, G.N.D.; Beagley, B.; Jones, M. O. *J. Mol. Struct.* **1980**, *65*, 271.
43. Gallaher, K. L.; Yokozeki, A.; Bauer, S. H. *J. Phys. Chem.* **1974**, *78*, 2389.
44. Hagen, K.; Hedberg, K. *J. Am. Chem. Soc.* **1973**, *95*, 8263.
45. Kakubari, H.; Iijima, T.; Kimura, M. *Bull. Chem. Soc. Japan* **1975**, *48*, 1984.
46. Klaeboe, P.; Powell, D. L.; Stolevik, R.; Vorren, O. *Acta Chem. Scand.* **1982**, *A36*, 471.
47. Iijima, T. *Bull. Chem. Soc. Japan* **1972**, *45*, 1291.
48. Fernholt, L.; Seip, R.; Stolevik, R. *Acta Chem. Scand.* **1978**, *A32*, 225.
49. Haas, B.; Haase, J.; Zeil, W. *Z. Naturforsch.* **1967**, *22a*, 1646.
50. Stolevik, R.; Thom, E. *Acta Chem. Scand.* **1971**, *25*, 3205.
51. Yokozeki, A.; Bauer, S. H. *J. Phys. Chem.* **1976**, *80*, 73.
52. Yokozeki, A.; Bauer, S. H. *J. Phys. Chem.* **1975**, *79*, 155.
53. Oberhammer, H. *J. Chem. Phys.* **1978**, *69*, 468.
54. Carlos, J. L.; Karl, R. R.; Bauer, S. H. *J. Chem. Soc. Faraday Trans. 2*, **1974**, 177.
55. Huisman, P.A.G.; Mijlhoff, F. C.; Renes, G. H. *J. Mol. Struct.* **1979**, *51*, 191.
56. Van Schaick, E.J.M.; Mijlhoff, F. C.; Renes, G. H.; Geise, H. J. *J. Mol. Struct.* **1974**, *21*, 17.
57. Spelbos, A.; Huisman, P.A.G.; Mijlhoff, F. C.; Renes, G. H. *J. Mol. Struct.* **1978**, *44*, 159.
58. Mijlhoff, F. C.; Renes, G. H.; Kohata, K.; Oyanagi, K.; Kuchitsu, K. *J. Mol. Struct.* **1977**, *39*, 241.
59. Mom, V.; Huisman, P.A.G.; Mijlhoff, F. C.; Renes, G. H. *J. Mol. Struct.* **1980**, *62*, 95.
60. Kuchitsu, K. *J. Chem. Phys.* **1966**, *44*, 906.
61. Laurie, V. W.; Pence, D. T. *J. Chem. Phys.* **1963**, *38*, 2693.
62. Bhaumik, A.; Brooks, W.V.F.; Dass, S. C. *J. Mol. Struct.* **1973**, *16*, 29.
63. Dixon, D. Seventh Winter Fluorine Conference, Orlando, FL, 1985, Paper No. 63.
64. Craig, N. C.; Piper, L. G.; Wheeler, V. L. *J. Chem. Phys.* **1971**, *75*, 1453.
65. Cremer, D. *Chem. Phys. Lett.* **1981**, *81*, 481.
66. Lowrey, A. H.; D'Antonio, P.; George, C. *J. Chem. Phys.* **1976**, *64*, 2884.
67. Tokue, I.; Fukuyama, T.; Kuchitsu, K. *J. Mol. Struct.* **1973**, *17*, 207.
68. Lowrey, A. H.; George, C.; D'Antonio, P.; Karle J. *J. Mol. Struct.* **1979**, *53*, 189.
69. Bürger, H.; Pawelke, G.; Oberhammer, H. *J. Mol. Struct.* **1982**, *84*, 49.
70. Hilderbrandt, R. L.; Andreassen, A. L.; Bauer, S. H. *J. Phys. Chem.* **1970**, *74*, 1586.
71. Schei, S. H.; Hagen, K. *J. Mol. Struct.* **1984**, *116*, 249.
72. Schei, S. H.; Seip, R. *Acta Chem. Scand.* **1984**, *A38*, 345.
73. Samdal, S.; Seip, H. M.; Torgrimsen, T. *J. Mol. Struct.* **1977**, *42*, 153.
74. Anderson, D.W.W.; Cradock, S.; Ebsworth, E.A.V.; Green, A. R.; Rankin, D.W.H.; Robiette, A. G. *J. Organomet. Chem.* **1984**, *271*, 235.

75. Chiang, C. H.; Andreassen, A.L.; Bauer, S. H. *J. Org. Chem.* **1971**, *36*, 920.
76. Kveseth, K.; Seip, H. M.; Stolevik, R. *Acta Chem. Scand.* **1971**, *25*, 2975.
77. Nakata, M.; Kohata, K.; Fukuyama, T.; Kuchitsu, K., Wilkins, C. J. *J. Mol. Struct.* **1980**, *68*, 271.
78. Oberhammer, H. *J. Chem. Phys.* **1980**, *79*, 4310.
79. Huisman, P.A.G.; Klebe, K. J.; Mijlhoff, F. C.; Renes, G. H. *J. Mol. Struct.* **1979**, *57*, 71.
80. Christen, D.; Oberhammer, H.; Zeil, W.; Haas A.; Darmadi, A. *J. Mol. Struct.* **1980**, *66*, 203.
81. Hamm, R.; Kohrmann, J.; Günther, H.; Zeil, W. *Z. Naturforsch.* **1976**, *31a*, 594.
82. Van Eijck, B. P.; Van der Plaats, G.; Van Roon, P. H. *J. Mol. Struct.* **1972**, *11*, 67.
83. Bijen, J.M.J.M.; Derissen, J. L. *J. Mol. Struct.* **1975**, *27*, 233.
84. Maagdenberg, A.A.J. *J. Mol. Struct.* **1977**, *41*, 61.
85. Samdal, S.; Seip, R. *J. Mol. Struct.* **1979**, *52*, 195.
86. Marsden, H. M.; Oberhammer, H.; Shreeve, J. M. *Inorg. Chem.* **1985**, *24*, 4756.
87. Mack, H. G. Diploma thesis, Universität Tübingen, 1984.
88. Dolbier, W. R., Jr.; Sellers, S. F.; Smart, B. E.; Oberhammer, H. *J. Mol. Struct.* **1983**, *101*, 193.
89. Dallinga, G.; Van der Draii, R. K.; Toneman, L. H. *Rec. Trav. Chim. Pays-Bas* **1968**, *87*, 897.
90. Skancke, A.; Flood, E.; Boggs, J. E. *J. Mol. Struct.* **1977**, *40*, 263.
91. Almenningen, A.; Bastiansen, O.; Fernholt, L.; Samdal, S.; Cyvin, B. N.; Cyvin, S. J. *J. Mol. Struct.* **1985**, *128*, 59.
92. Almenningen, A.; Bastiansen, O.; Gundersen S.; Samdal, S.; Skancke A. *J. Mol. Struct.* **1985**, *128*, 95.
93. Almenningen, A.; Hartmann, A. O.; Seip, H. M. ; *Acta Chem. Scand.* **1968**, *22*, 1013.
94. Zeitz, K.; Oberhammer, H.; Häfelinger, G. *Z. Naturforsch.* **1977**, *32b*, 420.
95. Pierce, L.; Doyns, Sister V. *J. Am. Chem. Soc.* **1962**, *84*, 2651.
96. Hencher, J. L.; Bauer, S. H. *J. Am. Chem. Soc.* **1967**, *89*, 5527.
97. Lancaster, J. E.; Stoicheff, B. P. *Can. J. Phys.* **1956**, *34*, 1016.
98. Beagley, B.; Pritchard, R. G.; Banks, R. E. *J. Fluorine Chem.* **1981**, *18*, 159.
99. Hirose, C. *Bull. Chem. Soc. Japan* **1974**, *47*, 1311.
100. Agopovich, J. W ; Alexander, J.; Gillies, C. W.; Raw, T. T. *J. Am. Chem. Soc.* **1984**, *106*, 2250.
101. Almenningen, A.; Fernholt, L.; Seip, H. M. *J. Mol. Struct.* **1978**, *49*, 333.
102. Brendhaugen, K.; Kolderup Fikke, M.; Seip, H. M. *Acta Chem. Scand.* **1973**, *27*, 1101.
103. Hencher, J. L.; Shen, Q.; Tuck, D. G. *J. Am. Chem. Soc.* **1976**, *98*, 899.
104. Smith, Z.; Seip, R. *Acta Chem. Scand.* **1976**, *A30*, 759.
105. Chiang, J. F.; Lu, K. C. *J. Phys. Chem.* **1977**, *81*, 1682.
106. Wehrung, T.; Oberhammer, H.; Haas, A.; Koch. B.; Welcman N. *J. Mol. Struct.* **1976**, *35*, 253.
107. Hargittai, I. *J. Mol. Struct.* **1979**, *54*, 287.
108. Höfs, H. U.; Bats, J. W.; Gleiter, R.; Hartmann, G.; Mews, R.; Eckert-Maksic, M.; Oberhammer, H.; Sheldrick, G. M. *Chem. Ber.* **1985**, *118*, 3781.
109. Oberhammer, H. *J. Fluorine Chem.* **1983**, *23*, 147.
110. Robiette, A. G.; Cartwright, G. J.; Hoy, A. R.; Mills, I. M. *Mol. Phys.* **1971**, *20*, 541.
111. Average value from two independent studies: Beagley, B.; Brown, D. P.; Freeman, J. M. *J. Mol. Struct.* **1973**, *18*, 337; Hagen, K.; Hedberg, K. *J. Chem. Phys.* **1973**, *59*, 1549.
112. Laurie, V. W. *J. Chem. Phys.* **1957**, *26*, 1359.
113. Hoy, A. R.; Bertram, M.; Mills, I. M. *J. Mol. Spectr.* **1973**, *46*, 429.
114. Kuchitsu, K. In "Landolt-Börnstein", New Series, Group II, Vol. 7, Springer-Verlag: Berlin, 1976, p. 58.
115. Gaunt, A. D.; Mackle, H.; Sutton, L. E. *J. Chem. Soc. Faraday Trans.* **1951**, *47*, 943.
116. Beagley, B.; Monaghan, J. J.; Hewitt, T. G. *J. Mol. Struct.* **1971**, *8*, 401.
117. Rempfer, B.; Oberhammer, H.; Auner, N. *Acta Chem. Scand. A*, in press.
118. Roelandt, F. F. *J. Organomet. Chem.* **1975**, *94*, 377.
119. Oberhammer, H. Seventh Winter Fluorine Conference, Orlando, FL, 1985.
120. Drake, J. E.; Hennings, R. T.; Hencher, J. L.; Mustoe, F. J.; Shen, Q. *J. Chem. Soc. Dalton Trans.* **1976**, 294.
121. Hencher, J. L.; Mustoe, F. J. *Can. J. Chem.* **1975**, *53*, 3542.
122. Kilb, R. W.; Pierce, L. *J. Chem. Phys.* **1957**, *27*, 108.
123. Beckers, H.; Bürger, H.; Eujen, R.; Rempfer, B.; Oberhammer, H.; *J. Mol. Struct.* **1986**, *140*, 281.
124. Rempfer, B.; Pfafferott, G.; Oberhammer, H.; Bürger, H.; Eujen, J.; Boggs, J. E. *Rev. Chim. minerale* **1986**, *23*, 551.
125. Laurie, V. W. *J. Chem. Phys.* **1959**, *30*, 1210.

126. Wehrlein, U. Ph. D. thesis, Universität Tübingen, 1982.
127. Apeloig, Y.; Karni, M. *J. Am. Chem. Soc.* **1984**, *106*, 6676.
128. Krauss, H. Diploma thesis, Universität Tübingen, 1981.
129. Drake, J. E.; Hencher, J. L.; Shen, Q. *Can. J. Chem.* **1977**, *55*, 1104.
130. Oberhammer, H.; Eujen, R. *J. Mol. Struct.* **1979**, *51*, 211.
131. Hencher, J. L.; Mustoe, F. J. *Can. J. Chem.* **1975**, *53*, 3542.
132. Eujen, R.; Bürger, H.; Oberhammer, H. *J. Mol. Struct.* **1981**, *71*, 109.
133. Nagashima, M.; Fujii, H.; Kimura, M. *Bull. Chem. Soc. Japan* **1973**, *46*, 3708.
134. Novikov, V. P.; Khaikin, L. S.; Vilkov, L. V. *J. Mol. Struct.* **1977**, *42*, 139.
135. Vajda, E.; Kolonits, M.; Fritz, G.; Thomas, J.; Sattler, E. *J. Mol. Struct.* **1984**, *117*, 329.
136. Almenningen, A.; Seip, H. M.; Seip, R. *Acta Chem. Scand.* **1970**, *24*, 1697.
137. Kuznetsova, T. M.; Alekseev, N. V.; Veniaminov, N. N. *Zh. Strukt. Khim.* (*Engl.*) **1979**, *20*, 281.
138. Beagley, B.; Foord, A.; Moutran, R.; Roszondai, B. *J. Mol. Struct.* **1977**, *42*, 117.
139. Airey, W.; Glidewell, C.; Robiette, A. G.; Sheldrick, G. M.; Freeman, J. M. *J. Mol. Struct.* **1971**, *8*, 423.
140. Glidewell, C.; Rankin, D.W.H.; Robiette, A. G.; Sheldrick, G. M. *J. Mol. Struct.* **1970**, *6*, 231.
141. Airey, W.; Glidewell, C.; Robiette, A. G.; Sheldrick, G. M. *J. Mol. Struct.* **1971**, *8*, 435.
142. Glidwell, C.; Robiette, A. G. Sheldrick, G. M. *Chem. Phys. Lett.* **1972**, *16*, 526.
143. Airey, W.; Glidewell, C.; Rankin, D.W.H.; Robiette, A. G.; Sheldrick, G. M.; Cruickshank, D.W.J. *J. Chem. Soc. Trans. Faraday Soc.* **1970**, *66*, 551.
144. Almenningen, A.; Bastiansen, O.; Ewing, V.; Hedberg, K.; Traetteberg, M. *Acta Chem. Scand.* **1963**, *17*, 2455.
145. Airey, W.; Glidewell, C.; Robiette, A. G.; Sheldrick, G. M. *J. Mol. Struct.* **1971**, *8*, 413.
146. Glidewell, C.; Rankin, D.W.H.; Robiette, A. G.; Sheldrick, G. M. *J. Mol. Struct.* **1970**, *5*, 417.
147. Demuth, R.; Oberhammer, H. *Z. Naturforsch.* **1973**, *28a*, 1862.
148. Glidewell, C.; Pinder, P. M.; Robiette, A. G.; Sheldrick, G. M. *J. Chem. Soc. Dalton Trans.* **1972**, 1402.
149. Oberhammer, H. Demuth, R. *J. Chem. Soc. Dalton Trans.* **1976**, 1121.
150. Beagley, B.; Robiette, A. G.; Sheldrick, G. M. *J. Chem. Soc. A* **1968**, 3006.
151. Il'enko, T. M.; Veniaminov, N. N.; Alekseev, N. V.; Shcherbinin, V. V. *Zh. Strukt. Khim.* (*Engl.*) **1976**, *17*, 294.
152. Kuznetsova, T. M.; Alekseev, N. V.; Shcherbinin, V. V.; Veniaminov, N. N.; Ronova, I. A. *Zh. Strukt. Khim.* (*Engl.*) **1976**, *17*, 922.
153. Duckett, J. A.; Robiette, A. G.; Gerry, M.C.L. *J. Mol. Spectrosc.* **1981**, *90*, 374.
154. Rankin, D.W.H.; Robertson, A. *J. Mol. Struct.* **1975**, *27*, 438.
155. Oberhammer, H.; *J. Mol. Struct.* **1976**, *31*, 237.
156. Beagley, B.; Conrad, A. R.; Freeman, J. M.; Monaghan, J. J.; Norton, B. G.; Holywell, G. C. *J. Mol. Struct.* **1972**, *11*, 371.
157. Bohn, R. K.; Bauer, S. H. *Inorg. Chem.* **1967**, *6*, 304.
158. Brown, R. D.; Burden, F. R.; Godfrey, P. D.; Gillard, I. R. *J. Mol. Spectrosc.* **1974**, *52*, 301.
159. Conwan, M.; Gordy, W. *Bull. Am. Chem. Soc. Ser. 2*, **1960**, *5*, 241.
160. Lide, D. R. *J. Chem. Phys.* **1963**, *38*, 456.
161. Hersh, O. L. Ph. D. thesis, University of Michigan; *Diss. Abstr.* **1963**, *24*, 2286.
162. Christen, D.; Minkwitz, R. Private communication.
163. Vilkov, L. V.; Nazarenko, I. I. *Zh. Strukt. Khim.* (*Engl.*) **1967**, *8*, 297.
164. Buckton, K. S.; Legon, A. C.; Millen, D. J. *J. Chem. Soc. Trans. Faraday Soc.* **1969**, *65*, 1975.
165. Legon, A. C.; Millen, D. J. *J. Chem. Soc. A* **1968**, 1736.
166. Plato, V.; Hartford, W. D.; Hedberg, K. *J. Chem. Phys.* **1970**, *53*, 3488.
167. Oberhammer, H.; Günther, H.; Bürger, H.; Heyder, F.; Pawelke, G. *J. Phys. Chem.* **1982**, *86*, 644.
168. Bürger, H.; Niepel, H.; Pawelke, G.; Oberhammer, H. *J. Mol. Struct.* **1979**, *54*, 159.
169. Pierce, L.; Hayes, R. G.; Beecher, J. F. *J. Chem. Phys.* **1967**, *46*, 4352.
170. Glidewell, C.; Marsden, C. J.; Robiette, A. G.; Sheldrick, G. M. *J. Chem. Soc. Dalton Trans.* **1972**, 1735.
171. Cardillo, M. J.; Bauer, S. H. *Inorg. Chem.* **1969**, *8*, 2086.
172. Gilbert, M. M.; Gundersen, G.; Hedberg, K. *J. Chem. Phys.* **1972**, *56*, 1691.
173. Kohata, K.; Fukuyama, T.; Kuchitsu, K. *J. Phys. Chem.* **1982**, *86*, 602.
174. Bartell, L. S.; Higginbotham, H. K. *Inorg. Chem.* 1965, *4*, 1346.
175. Carlotti, M.; Johns, J.W.C.; Trombetti, A. *Can. J. Phys.* **1974**, *52*, 340.
176. Bohn, R. K.; Bauer, S. H. *Inorg. Chem.* **1967**, *6*, 309.

177. Kuczkowski, R.; Wilson, E. B. *J. Chem. Phys.* **1963**, *39*, 1030.

178. Pearson, R.; Lovas, F. J. *J. Chem. Phys.* **1977**, *66*, 4149.

179. Christen, D.; Oberhammer, H.; Hammaker, R. M.; Chang, S. C.; Des Marteau, D. D. *J. Am. Chem. Soc.* **1982**, *104*, 6196.

180. Steger, B.; Oberhammer, H. Unpublished data.

181. Lentz, D.; Oberhammer, H. *Inorg. Chem.* **1985**, *24*, 4665.

182. Bürger, H.; Pawelke, G.; Oberhammer, H. *J. Mol. Struct.* **1985**, *128*, 283.

183. Christe, K. O.; Christen, D.; Oberhammer, H.; Schack, C. J. *Inorg. Chem.* **1984**, *23*, 4283.

184. Anderson, D.W.W.; Rankin, D.W.H.; Robertson, A. *J. Mol. Struct.* **1972**, *14*, 385.

185. Average value from two independent ED studies: Chiang, C. H.; Porter, R. F.; Bauer, S. H. *J. Am. Chem. Soc.* **1970**, *92*, 5313; Almenningen, A.; Anfinsen, I. M.; Haaland, A. *Acta Chem. Scand.* **1970**, *24*, 1230.

186. Koput, J. *J. Mol. Spectrosc.* **1984**, *106*, 12.

187. Steger, B.; Christen, D.; Oberhammer, H. In progress.

188. Bauer, S. H.; Andreassen, A. L. *J. Phys. Chem.* **1972**, *76*, 3099.

189. Coffey, D.; Britt, C. O.; Boggs, J. E. *J. Chem. Phys.* **1968**, *49*, 591.

190. Karle, I. L.; Karle, J. *J. Chem. Phys.* **1962**, *36*, 1969.

191. Cox, A. P.; Waring, S. *J. Chem. Soc. Faraday Trans.* **1972**, *68*, 1060.

192. Glidewell, C.; Rankin, D.W.H.; Robiette, A. G.; Sheldrick, G. M.; Williamson, S. M. *J. Chem. Soc. A* **1971**, 478.

193. Steger, B.; Oberhammer, H.; Grobe, J. Unpublished data.

194. Pauling, L.; Brockway, L. O. *J. Am. Chem. Soc.* **1935**, *57*, 2684.

195. Morino, Y.; Kuchitsu, K.; Moritani, T. *Inorg. Chem.* **1969**, *8*, 867.

196. Clippard, F. B.; Bartell, L. S. *Inorg. Chem.* **1970**, *9*, 85.

197. Chikaraishi, T.; Hirota, E. *Bull. Chem. Soc. Japan* **1973**, *46*, 2314.

198. Konaka, S.; Kimura, M. *Bull. Chem. Soc. Japan* **1970**, *43*, 1693.

199. Konaka, S. *Bull. Chem. Soc. Japan* **1970**, *43*, 1693.

200. Brittain, A. H.; Smith, J. E.; Schwendeman, R. H. *Inorg. Chem.* **1972**, *11*, 39.

201. Holywell, G. C.; Rankin, D.W.H. *J. Mol. Struct.* **1971**, *9*, 11.

202. Oberhammer, H.; Schmutzler, R.; Stelzer, O. *Inorg. Chem.* **1978**, *17*, 1254.

203. Marsden, C. J.; Bartell, L. S. *Inorg. Chem.* **1976**, *15*, 2713.

204. Bowen, H.J.M. *Trans. Faraday Soc.* **1954**, *50*, 463.

205. Holywell, G. C.; Rankin, D.W.H.; Beagley, B.; Freeman, J. M. *J. Chem. Soc. A* **1971**, 785.

206. Hedberg, E.; Hedberg, L.; Hedberg, K. *J. Am. Chem. Soc.* **1974**, *96*, 4417.

207. Arnold, D.E.J.; Rankin, D.W.H.; Todd, M. R.; Seip, R. *J. Chem. Soc. Dalton Trans.* **1979**, 1290.

208. Forti, P.; Damiani, D.; Favero, P. G. *J. Am. Chem. Soc.* **1973**, *95*, 756.

209. Laurenson, G. S.; Rankin, D.W.H. *J. Mol. Struct.* **1979**, *54*, 111.

210. Huntley, C. M.; Laurenson, G. S.; Rankin, D.W.H. *J. Chem. Soc. Dalton Trans.* **1980**, 954.

211. Laurenson, G. S.; Rankin, D.W.H. *J. Chem. Soc. Dalton Trans.* **1981**, 425.

212. Brittain, A. H.; Smith, J. E.; Lee, P. L.; Cohn, K.; Schwendeman, R. H. *J. Am. Chem. Soc.* **1971**, *93*, 6772.

213. Laurenson, G. S.; Rankin, D.W.H. *J. Chem. Soc. Dalton Trans.* **1981**, 1047.

214. Rankin, D.W.H.; Cyvin, S. J. *J. Chem. Soc. Dalton Trans.* **1972**, 1277.

215. Blair, P. D. *J. Mol. Struct.* **1983**, *97*, 147.

216. Rankin, D.W.H. *J. Chem. Soc. Dalton Trans.* **1972**, 869.

217. Arnold, D.E.J.; Rankin, D.W.H. *J. Fluorine Chem.* **1972/73**, *2*, 405.

218. Yow, H. Y.; Rudolph, R. W.; Bartell L. S. *J. Mol. Struct.* **1975**, *28*, 205.

219. Arnold, D.E.J.; Gundersen, G.; Rankin, D.W.H.; Robertson, H. E. *J. Chem. Soc. Dalton Trans.* **1983**, 1989.

220. Ebsworth, E.A.V.; Macdonald, E. K.; Rankin, D.W.H. *Monatsh. Chem.* **1980**, *111*, 221.

221. Codding, E. G.; Jones, C. E.; Schwendeman, R. H. *Inorg. Chem.* **1974**, *13*, 178.

222. Burg, A. B. *Inorg. Chem.* **1983**, *22*, 2573. Grobe, J.; Le Van, D. *Angew. Chem. Int. Ed. Engl.* **1984**, *23*, 710.

223. Hodges, H. L.; Su, L. S.; Bartell, L. S. *Inorg. Chem.* **1975**, *14*, 599.

224. Durig, J. R.; Carreira, L. A.; Odom, J. D. *J. Am. Chem. Soc.* **1974**, *96*, 2688.

225. Beagley, B.; Conrad, A. R.; Freeman, J. M.; Monaghan, J. J.; Norton, B. G.; Holywell, G. C. *J. Mol. Struct.* **1972**, *11*, 371.

226. Moritani, T.; Kuchitsu, K.; Morino, Y. *Inorg. Chem.* **1971**, *10*, 344.

227. von Carlowitz, S. Ph. D. thesis, Universität Tübingen, 1984.

228. Karakida, K.; Kuchitsu, K. *Inorg. Chem.* **1976**, *16*, 29.

229. Acha, L. Cromie, E. R.; Rankin, D.W.H. *J. Mol. Struct.* **1981**, *73*, 111.
230. Boyd, A.S.F.; Laurenson, G. S.; Rankin, D.W.H. *J. Mol. Struct.* **1981**, *71*, 217.
231. Durig, J. R.; Kalasinsky, K. S.; Kalasinsky, V. F. *J. Mol. Struct.* **1978**, *34*, 9.
232. Durig, J. R. *J. Molo. Struct.* **1982**, *78*, 247.
233. von Carlowitz, S.; Zeil, W.; Pulay, P.; Boggs, J. E. *J. Mol. Struct. Theochem.* **1982**, *4*, 113.
234. Marsden, C. J. Ninth Austin Symposium on Gas-Phase Molecular Structure, Austin, TX, 1982, Paper No. A7.
235. Wilkins, C. J.; Hagen, K.; Hedberg, L.; Shen, Q.; Hedberg, K. *J. Am. Chem. Soc.* **1975**, *97*, 6352.
236. Centofani, L. F.; Kuczkowski, R. L. *Inorg. Chem.* **1968**, *7*, 2582.
237. Rankin, D.W.H.; Todd, M. R.; Fild, M. *J. Chem. Soc. Dalton Trans.* **1982**, 2079.
238. Davis, M. I.; Paul, J. W., Jr. *J. Mol. Struct.* **1971**, *9*, 478.
239. Homes, R. R. "Pentacoordinated Phosphorus", Vols. I and II, ACS Monographs 175 and 176. American Chemical Society: Washington DC, 1980.
240. Luckenbach, R. "Dynamic Stereochemistry of Pentacoordineted Phosphorus and Related Elements". Thieme; Stuttgart, 1973.
241. Spiridonov, V. P.; Ischenko, A. A.; Ivashkevich, L. S. *J. Mol. Struct.* **1981**, *72*, 153.
242. Marsden, C. J. *J. Chem. Soc. Chem. Commun.* **1984**, 401.
243. Hansen, K. W.; Bartell, L. S. *Inorg. Chem.* **1965**, *4*, 1775.
244. Clippard, F. B.; Bartell, L. S. *Inorg. Chem.* **1970**, *9*, 805.
245. Muetterties, E. L.; Mahler, W.; Schmutzler, R. *Inorg. Chem.* **1963**, *2*, 613.
246. Cavell, R. G.; Poulin, D. D.; The, K. I.; Tomlinson, A. J. *J. Chem. Soc. Chem. Commun.* **1974**, 19.
247. French, R. J.; Hedberg, K.; Shreeve, J. M.; Gupta, K. D. *Inorg. Chem.* **1985**, *24*, 2774.
248. Hedberg, K. Tenth Austin Symposium on Gas-Phase Molecular Structure, Austin, TX 1984, Paper No. TA 6.
249. Macho, C.; Minkwitz, R.; Rohmann, J.; Steger, B.; Wölfel, V.; Oberhammer, H. *Inorg. Chem.* **1986**, *25*, 2828.
250. McClelland, B. W.; Hedberg, L.; Hedberg, K. *J. Mol. Struct.* **1983**, *99*, 309.
251. van der Vorn, P. C.; Drago, R. S. *J. Am. Chem. Soc.* **1966**, *88*, 3255.
252. Minkwitz, R.; Prenzel, H.; Schardey, A.; Oberhammer, H. *Inorg. Chem.* **1987**, *26*, 2730.
253. Cohen, E. A.; Cornwell, D. C. *Inorg. Chem.* **1968**, *7*, 398.
254. Griffiths, J. E. *J. Chem. Phys.* **1968**, *49*, 1307.
255. Cavell, R. G.; Gibson, J. A.; The, K. I. *J. Am. Chem. Soc.* **1977**, *99*, 7841.
256. Gillespie, P. ; Hoffmann, P. ; Klusacek, H.; Marquarding, D.; Pfohl, S.; Ramirez, F.; Tsolis, E. A.; Ugi, I. *Angew. Chem. Int. Ed. Engl.* **1971**, *10*, 687.
257. Oberhammer, H.; Grobe, J.; Le Van, D. *Inorg. Chem.* **1982**, *21*, 275.
258. Willner, H. Private communication.
259. von Carlowitz, M.; Boggs, J. E.; Oberhammer, H. Unpublished data.
260. Griffiths, J. E. *J. Chem. Phys.* **1964**, *41*, 3510.
261. Griffiths, J. E.; Beach, A. L. *J. Chem. Phys.* **1966**, *44*, 2686.
262. Holmes, R. R.; Carter, R. P., Jr.; Peterson, G. E. *Inorg. Chem.* **1964**, *3*, 1748.
263. Oberhammer, H.; Minkwitz, R.; Rohmann, J.; Wölfel, V. *J. Mol. Stuct.* **1986**, *147*, 185.
264. Oberhammer, H.; Grobe, J. *Z. Naturforsch.* **1975**, *30b*, 506.
265. Downs, A. J.; Schmutzler, R. *Spectrochim. Acta* **1965**, *21*, 1927.
266. Downs, A. J.; Schmutzler, R. *Spectrochim. Acta* **1967**, *23A*, 681.
267. Bartell, L. S.; Hansen, K. W. *Inorg. Chem.* **1965**, *4*, 1777.
268. Yow, H.; Bartell, L. S. *J. Mol. Struct.* **1973**, *15*, 209.
269. Craig, D. P.; Macoll, A.; Nyholm, R. S.; Orgel, L. E.; Sutton, L. E. *J. Chem. Soc.* **1954**, 332.
270. Rundle, R. E. *Rec. Chem. Prog.* **1962**, *23*, 195.
271. Dittebrandt, C.; Oberhammer, H. *J. Mol. Struct.* **1980**, *63*, 227.
272. Oberhammer, H. *J. Mol. Struct.* **1979**, *53*, 139.
273. Hoffmann, R.; Howell, J. M.; Muetterties, E. L. *J. Am. Chem. Soc.* **1972**, *94*, 3047.
274. Khaikin, L. S.; Andrutskaya, L. G.; Vilkov, L. V. Ninth Hungarian Diffraction Conference, Pécs, 1978.
275. Marsden, C. J.; Hedberg, K.; Shreeve, J. M.; Gupta, K. D. *Inorg. Chem.* **1984**, *23*, 3659.
276. Arnold, D.E.J.; Rankin, D.W.H.; Robinet, G. *J. Chem. Soc. Dalton Trans.* **1977**, 585.
277. Oberhammer, H.; Schmutzler, R. *J. Chem. Soc. Dalton Trans.* **1976**, 1454.
278. Almenningen, A.; Andersen, B.; Astrup, E. E. *Acta Chem. Scand.* **1969**, *23*, 2179.
279. Oberhammer, H.; Schmutzler, R. Unpublished data.
280. Ewbank, J. D.; Kirsch, G.; Schäfer, L. *J. Mol. Struct.* **1976**, *31*, 39.
281. Diodati, F. P.; Bartell, L. S. *J. Mol. Struct.* **1971**, *8*, 395.

282. Oberhammer, H.; Mehmood, T.; Shreeve, J. M. *J. Mol. Struct.* **1984**, *117*, 311.
283. Lowrey, A. H.; George, C.; D'Antonio, P.; Karle, J. *J. Mol. Struct.* **1980**, *63*, 243.
284. Morino, Y.; Saito, S. *J. Mol. Spectrosc.* **1966**, *19*, 435.
285. Tamagawa, K.; Takemura, M.; Konaka, S.; Kimura, M. *J. Mol. Struct.* **1984**, *125*, 131.
286. Ridgen, J. S.; Butcher, S. S. *J. Chem. Phys.* **1964**, *40*, 2109.
287. Cruickshank, D.W.J. *J. Chem. Soc.* **1961**, 5486, Bürger, H. *Angew. Chem. Int. Ed. Engl.* **1973**, *12*, 474.
288. Oberhammer, H.; Boggs, J. E. *J. Am. Chem. Soc.* **1980**, *102*, 7241.
289. Jackson, R. H. *J. Chem. Soc.* **1962**, 4585.
290. Lucchese, R. R.; Schaefer, H. F., III; Rodwell, W. R.; Radom, L. *J. Chem. Phys.* **1978**, *68*, 2507.
291. Ahlrichs, R.; Taylor, P. R. *Chem. Phys.* **1982**, *72*, 287.
292. Marsden, C. J.; Des Marteau, D.D.; Bartell, L. S. *Inorg. Chem.* **1977**, *16*, 2359.
293. Marsden, C. J.; Bartell, L. S.; Diodati, F. P. *J. Mol. Struct.* **1977**, *39*, 253.
294. Khachkuruzov, G. A.; Przhevalskii, I. N. *Opt. Spektrosk.* **1974**, *36*, 172, Cremer, D.; Christen, D. *J. Mol. Struct.* **1979**, *74*, 480.
295. Haas, B.; Oberhammer, H. *J. Am. Chem. Soc.* **1984**, *106*, 6146.
296. Tyblewski, M.; Meyer, R.; Bauder, A. Eighth Colloquium on High-Resolution Molecular Spectroscopy, Tours, France, 1983.
297. Gase, W.; Boggs, J. E. *J. Mol. Struct.* **1984**, *116*, 207.
298. Kirchhoff, W. H.; Johnson, D. R.; Powell, F. X. *J. Mol. Spectrosc.* **1973**, *48*, 157.
299. Oberhammer, H.; Gombler, W.; Willner, H. *J. Mol. Struct.* **1981**, *70*, 273.
300. Minkwitz, R.; Lekies, R.; Gadünz, A.; Oberhammer, H. *Z. anorg. allg. Chem.* **1985**, *531*, 31.
301. Marsden, C. J. *J. Mol. Struct.* **1973**, *21*, 168.
302. Marsden, C. J.; Sheldrick, G. M. *J. Mol. Struct.* **1971**, *10*, 405.
303. Ramme, K. J. Ph. D. thesis, Universität Tübingen, 1984.
304. Iijima, T.; Tsuchiya, S. ; Kimura, M. *Bull. Chem. Soc. Japan* **1977**, *50*, 2564.
305. Beecher, J. F. *J. Mol. Spectr.* **1966**, *21*, 414.
306. Blow, R.; Haaland, A.; Seip, R. *Acta Chem. Scand.* **1983**, *A37*, 595.
307. Oberhammer, H.; Haas, A.; Schlosser, K. Unpublished data.
308. Haas, A.; Schlosser, K. *Tetrahedron Lett.* **1976**, 4631.
309. Oberhammer, H.; Haas, A.; Schlosser, K. *J. Chem. Soc. Dalton Trans.* **1979**, 1075,
310. Marsden, C. J.; Bartell, L. S. *J. Chem. Soc. Dalton Trans.* **1977**, 1582.
311. von Carlowitz, M. Diploma thesis, Universität Tübingen, 1979.
312. Marsden, C. J.; Oberhammer, H.; Willner, H. Unpublished data.
313. Kuczkowski, R. L. *J. Am. Chem. Soc.* **1964**, *86*, 3617.
314. Herzberg, G. "Molecular Spectra and Molecular Structure", Vol. I. Van Nostrand Reinhold: New York, 1950.
315. Marsden, C. J.; Beagley, B. *J. Chem. Soc. Faraday Trans. 2* **1981**, *77*, 2213.
316. Kuhler, M.; Charpentier, L.; Sutter, D.; Dreizler, H. *Z. Naturforsch.* **1974**, *29a*, 1335.
317. Marsden, C. J.; Sheldrick, G. M. *J. Mol. Struct.* **1971**, *10*, 419.
318. Hargittai, I.; Mijlhoff, F. C. *J. Mol. Struct.* **1973**, *16*, 69.
319. Haase, J.; Oberhammer, H.; Zeil, W.; Glemser, O.; Mews, R. *Z. Naturforsch.* **1970**, *25a*, 153.
320. Karl, R. R., Jr.; Bauer, S. H. *Inorg. Chem.* **1975**, *14*, 1859.
321. Lucas, N.J.D.; Smith, J. G. *J. Mol. Spectrosc.* **1972**, *43*, 327.
322. Oberhammer, H.; Kumar, R. C.; Knerr, G. D.; Shreeve, J. M. *Inorg. Chem.* **1981**, *20*, 3871.
323. Typke, V. *Z. Naturforsch.* **1978**, *33a*, 842.
324. Tolles, W. M.; Gwinn, W. D. *J. Chem. Phys.* **1962**, *36*, 1119.
325. Kimura, K. Bauer, S. H. *J. Chem. Phys.* **1963**, *39*, 3172.
326. Bowater, I. C.; Brown, R. D.; Burden, F. R. *J. Mol. Spectrosc.* **1968**, *28*, 454.
327. Baxter, P. L.; Downs, A. J.; Forster, A. M.; Goode, M. J.; Rankin, D.W.H.; Robertson, H. E. *J. Chem. Soc. Dalton Trans.* **1985**, 941.
328. Seel, F.; Budenz, R.; Gombler, W. *Chem. Ber.* **1970**, *103*, 1701. Gombler, W.; Haas, A.; Willner, H. *Z. Anorg. Allg. Chem.* **1980**, *462*, 57.
329. von Carlowitz, M.; Oberhammer, H.; Willner, H.; Boggs, J. E. *J. Mol. Struct.* **1983**, *100*, 161.
330. Oberhammer, H.; Mews, R. Unpublished data.
331. Chen. M.M.L.; Hoffmann, R. *J. Am. Chem. Soc.* **1976**, *98*, 1647.
332. Cowley, A. H.; Riley, P. E.; Szobota, J. S.; Walker, M. L. *J. Am. Chem. Soc.* **1979**, *101*, 5620.
333. Hargittai, I.; Vilkov, L. V. *Acta Chim. Acad. Sci. Hung.* **1970**, *63*, 143.
334. Pötter, B.; Seppelt, K. *Angew. Chem. Int. Ed. Engl.* **1984**, *23*, 150.
335. Kirchhoff, W. H.; Wilson, E. B. *J. Am. Chem. Soc.* **1962**, *84*, 334.

336. Cook, R. L.; Kirchhoff, W. H. *J. Chem. Phys.* **1967**, *47*, 4521.
337. Rogowski, R. *Z. Phys. Chem.* **1961**, *27*, 277.
338. Boggs, J. E. *Inorg. Chem.* **1984**, *23*, 3577.
339. Dixon, D. A.; Smart, B. E.; *J. Am. Chem. Soc.* **1986**, *108*, 2688.
340. Pötter, B.; Seppelt, K.; Simon, A.; Peters, E. M.; Hettich, B. *J. Am. Chem. Soc.* **1985**, *107*, 980.
341. Christen, D.; Mack, H. G.; Marsden, C. J.; Oberhammer, H.; Schatte, G.; Seppelt, K.; Willner, H. *J. Am. Chem. Soc.* **1987**, *109*, 4009.
342. Hagen, K.; Cross, V. R.; Hedberg, K. *J. Mol. Struct.* **1978**, *44*, 187.
343. Oberhammer, H.; Glemser, O.; Klüver, H. *Z. Naturforsch.* **1974**, *29a*, 901.
344. Oberhammer, H.; Knerr, G. D.; Shreeve, J. M. *J. Mol. Struct.* **1982**, *82*, 143.
345. Brunvoll, J.; Hargittai, I.; Kolonits, M. *Z. Naturforsch.* **1978**, *33a*, 1236.
346. Schultz, G.; Hargittai, I.; Seip, R. *Z. Naturforsch.* **1981**, *36a*, 917.
347. Hargittai, I.; Hargittai, M. *J. Mol. Struct.* **1973**, *15*, 399.
348. Hargittai, I.; Seip, R.; Nair, K.P.R.; Britt, C. O.; Boggs, J. E.; Cyvin, B. N. *J. Mol. Struct.* **1977**, *39*, 1.
349. Brunvoll, J.; Kolonits, M.; Bliefert, C.; Seppelt, K.; Hargittai, I. *J. Mol. Struct.* **1982**, *78*, 307.
350. Lide, D. R.; Mann, D. E.; Fristrom, R. M. *J. Chem. Phys.* **1957**, *26*, 734.
351. Dowater, I. C.; Brown, R. D.; Burden, F. R. *J. Mol. Spectrosc.* **1968**, *28*, 461.
352. Hargittai, M.; Hargittai, I. *J. Mol. Struct.* **1972**, *20*, , 283.
353. Hencher, J. L.; Bauer, S. H. *Can. J. Chem.* **1973**, *51*, 2047.
354. Hedberg, L.; Hedberg, K. *J. Chem. Phys.* **1982**, *86*, 598.
355. Kimura, K.; Bauer, S. H. *J. Chem. Phys.* **1963**, *39*, 3172.
356. Hencher, J. L.; Cruickshank, D.W.J.; Bauer, S. H. *J. Chem. Phys.* **1968**, *48*, 518.
357. Gundersen, G.; Hedberg, K. *J. Chem. Phys.* **1969**, *51*, 2500.
358. Murty, K.S.R.; Mohanty, A. K. *Indian J. Phys.* **1971**, *45*, 535.
359. Graybeal, J. D. Sixth Austin Symposium on Gas-Phase Molecular Structure, Austin, TX, 1976, p. 53.
360. Oberhammer, H.; Boggs, J. E. *J. Mol. Struct.* **1979**, *56*, 107.
361. Bock, H.; Boogs, J. E.; Kleemann, G.; Lentz, D.; Oberhammer, H.; Peters, E. M.; Seppelt, K.; Simon, A.; Solouki, B. *Angew. Chem. Int. Ed. Engl.* **1979**, *18*, 944.
362. Des Marteau, D. D.; Eysel, H. H.; Oberhammer, H.; Günther, H. *Inorg. Chem.,* **1982**, *21*, 1607.
363. Simon, A.; Peters, E. M.; Lentz, D.; Seppelt, K. *Z. Anorg. Allg. Chem.* **1980**, *468*, 7.
364. Christe, K. O.; Oberhammer, H. *Inorg. Chem.* **1981**, *20*, 296.
365. Günther, H.; Oberhammer, H.; Mews, R.; Stahl, I. *Inorg. Chem.* **1982**, *21*, 1872.
366. Oberhammer, H.; Shreeve, J. M.; Gard, G. L. *Inorg. Chem.* **1984**, *23*, 2820.
367. Kelley, M. H.; Fink, M. *J. Chem. Phys.* **1982**, *77*, 1813.
368. Bartell, L. S.; Jin, A.; *J. Mol. Struct.* **1984**, *118*, 47.
369. Gundersen, G.; Hedberg, K.; Strand, T. G. *J. Chem. Phys.* **1978**, *68*, 3548.
370. Marsden, C. J.; Bartell, L. S. *Inorg. Chem.* **1976**, *15*, 3004.
371. Bartell, L. S.; Doun, S.; Marsden, C. J. *J. Mol. Struct.* **1981**, *75*, 271.
372. Marsden, C. J.; Christen, D.; Oberhammer, H. *J. Mol. Struct.* **1985**, *131*, 299.
373. Shustorovich, E. M.; Buslaev, Y. M. *Inorg. Chem.* **1976**, *15*, 1142.
374. Gupta, D. G.; Mews, R.; Waterfeld, A.; Shreeve, J. M.; Oberhammer, H. *Inorg. Chem.* **1986**, *25*, 275.
375. Waterfeld, A.; Oberhammer, H.; Mews, R. *Angew. Chem. Int. Ed. Engl.* **1982**, *21*, 355; *Angew. Chem. Suppl.* **1982**, 834.
376. Haase, J.; Oberhammer, H.; Zeil, W.; Glemser, O.; Mews, R. *Z. Naturforsch.* **1971**, *26a*, 1333.
377. Oberhammer, H.; Seppelt, K. *Inorg. Chem.* **1978**, *17*, 1435.
378. Crawford, R.; Dudley, F. B.; Hedberg, K. *J. Am. Chem. Soc.* **1959**, *81*, 5287.
379. Oberhammer, H.; Seppelt, K.; Mews, R. *J. Mol. Struct.* **1983**, *101*, 325.
380. Seppelt, K.; Oberhammer, H. *Inorg. Chem.* **1985**, *24*, 1227.
381. Oberhammer, H.; Seppelt, K. *Inorg. Chem.* **1979**, *18*, 2226.
382. Abe, T.; Shreeve, J. M. *J. Fluorine Chem.* **1973/74**, *3*, 17.
383. Waterfeld, A.; Mews, R. *Angew. Chem. Int. Ed. Engl.* **1981**, *20*, 1017.
384. Oberhammer, H.; Waterfeld, A.; Mews, R. *Inorg. Chem.* **1984**, *23*, 415.
385. Brockway, L. O. *J. Am. Chem. Soc.* **1938**, *60*, 1348.
386. Rogers, M. T.; Schomaker, V.; Stevenson, D. P. *J. Am. Chem. Soc.* **1941**, *63*, 2610.
387. Willis Jr., R. E.; Clark, W. W., III. *J. Chem. Phys.* **1980**, *72*, 4946.
388. Ischenko, A. A.; Miakshin, I. N.; Romanov, G. V.; Spiridonov, V. P.; Sukhoverkhov, V. F. *Dokl. Akad. Nauk SSSR (Engl.)* **1982**, *267*, 994.
389. Altman, A. B.; Miakshin, I. N.; Sukhoverkhov, V. F.; Romanov, G. V.; Spiridonov, V. P. *Dokl. Akad. Nauk SSSR (Engl.)* **1978**, *241*, 333.
390. Oberhammer, H.; Christe, K. O. *Inorg. Chem.* **1982**, *21*, 273.

391. Parent, C. R.; Gerry, M.C.L. *J. Mol. Spectrosc.* **1974**, *49*, 343.
392. Clark, A. H.; Beagley, B. ; Cruickshank, D.W.J.; Hewitt, T. G. *J. Chem. Soc. A* **1970**, 872.
393. Smith, D. F. *J. Chem. Phys.* **1953**, *21*, 609.
394. Goulet, P.; Juvek, R.; Chanussot, J. *J. Phys.* **1976**, *37*, 495.
395. Heenan, R. K.; Robiette, A. G. *J. Mol. Struct.* **1979**, *54*, 135.
396. Appelman, E. H.; Beagley, B.; Cruickshank, D.W.J.; Foord, A.; Rustad, S.; Ulbrecht, V. *J. Mol. Struct.* **1976**, *35*, 139.
397. Magnuson, D. W. *J. Chem. Phys.* **1957**, *27*, 223.
398. Heenan, R. K.; Robiette, A. G. *J. Mol. Struct.* **1979**, *55*, 191.
399. Bartell, L. S.; Clippard, F. B.; Jacob, E. J. *Inorg. Chem.* **1976**, *15*, 3009.
400. Adams, W. J.; Thompson, H. B.; Bartell, L. S. *J. Chem. Phys.* **1970**, *53*, 4040.
401. Harshbarger, W.; Bohn, R. K.; Bauer, S. H. *J. Am. Chem. Soc.* **1967**, *89*, 6466.
402. Murchison, C.; Reichman, S.; Anderson, D.; Overend, J.; Schreiner, F. *J. Am. Chem. Soc.* **1968**, *90*, 5690.
403. Reichman, S.; Schreiner, F. *J. Chem. Phys.* **1969**, *51*, 2355.
404. Bohn, R. K.; Katada, K.; Martinez, J. V.; Bauer, S. H. In "Noble Gas Compounds", Hyman, H., Ed.; University of Chicago Press: Chicago, 1963.
405. Gavin, R. M., Jr.; Bartell, L. S. *J. Chem. Phys.* **1968**, *48*, 2460. Bartell, L. S.; Gavin, R. M., Jr., *J. Chem. Phys.* **1968**, *48*, 2466.
406. Bartell, L. S.; Gavin, R. M., Jr.; Thompson, H. B.; Chernick, C. L. *J. Chem. Phys.* **1965**, *43*, 2547.
407. Hedberg, K.; Peterson, S. H.; Ryan, R. R.; Weinstock, B. *J. Chem. Phys.* **1966**, *44*, 1726.
408. Pitzer, K. S.; Bernstein, L. S. *J. Chem. Phys.* **1975**, *63*, 3849.
409. Jacob, E. J.; Thompson, H. B.; Bartell, L. S. *J. Mol. Struct.* **1971**, *8*, 383.

5

SATURATED

ORGANIC MOLECULES

Lawrence K. Montgomery

DEPARTMENT OF CHEMISTRY
INDIANA UNIVERSITY
BLOOMINGTON, INDIANA

CONTENTS

INTRODUCTION 210
NORMAL ALKANES 213
    Methane and Methane-d_4 213
    Ethane and Ethane-d_6 214
    Propane 215
    n-Butane 215
    n-Pentane, n-Hexane, and n-Heptane 217
    n-Hexadecane 218
BRANCHED ALKANES 221
    Isobutane 221
    Neopentane 222
    Hexamethylethane 224
    Tetramethylethane 225
    Tri-*tert*-butylmethane 227
    Di-*tert*-butylmethane 229
CYCLOALKANES 230
    Cyclopropane 230
    Cyclobutane 231
    Cyclopentane 232
    Cyclohexane 233
    Cycloheptane, Cyclooctane, and Cyclodecane 235
ACKNOWLEDGMENT 237
REFERENCES 237

Introduction

In addition to being of great practical significance, saturated hydrocarbons are an extremely important class of molecules from the structural point of view. Since they are nonpolar, do not enter into hydrogen bonding, and have been subjected to practically every conceivable experimental measurement, they are attractive candidates for the testing and development of all levels of theory and empirical relationships. Structural comparisons and conformational analyses of most aliphatic organic compounds are linked in one way or another to classic hydrocarbon prototypes. When the structure of a molecule, such as 1,2-dichloroethane, is ascertained experimentally or theoretically, hydrocarbon comparisons are inevitable. How does the C—C bond length compare with that in ethane? Does the ratio of gauche to anti conformations differ from that in *n*-butane? Because of the long-standing interest in the structures of saturated hydrocarbons, many pioneering electron diffraction studies have been carried out on alkanes an cycloalkanes. Consequently, the results presented in this chapter provide an unusually rich introduction to the potential and limitations of the electron diffraction technique.

Electron diffraction and microwave spectroscopy are the two most general methods of vapor-phase structure determination. Neither is particularly well-suited to handling large molecules, although electron diffraction is capable of yielding partial structures for many molecules that are too complex for spectroscopic methods. This chapter discusses hydrocarbons that range in size from methane to *n*-hexadecane ($C_{16}H_{34}$). In the former case, a highly precise C—H bond distance has been reported. It was even possible, with the aid of spectroscopic data, to obtain a good estimate of the equilibrium r_e (C—H) bond length. On the other hand, only average bond lengths, bond angles, and conformational preferences have been determined for *n*-hexadecane. In general, there is a steady decrease in the amount of detail that can be derived from an electron diffraction analysis as the size of the molecule is increased. This is a natural consequence of the electron diffraction experiment, which provides, ideally, the average interatomic distances and mean square amplitudes of vibration for all atom pairs in the molecule. In actuality, the resolution of internuclear distances is limited by thermal vibrations to about 0.1 Å. This means that in large molecules it is frequently not possible to differentiate among bond lengths, bond angles, and nonbonded distances that are very similar. Under such circumstances, it may be necessary to settle for average values of structural parameters. This is disappointing, but in many instances, this is the only way that any structural information can be acquired.

There are a number of pitfalls in reading the electron diffraction literature and in utilizing eletron diffraction results. This is particularly true for the nonspecialist. Several common problem areas are discussed briefly below, mainly because they are crucial to understanding the material that follows.

The enormous menu of bond distance parameters is a significant obstacle to comparing and interrelating bond lengths. There is no easy way around the problem. It is necessary to understand the various definitions of bond lengths, and it is preferable to have some knowledge of the manner in which they are interconverted. Several good reviews are available for this purpose,[1-3] see also Chapter 1 in volume A. The distances employed in this chapter are:

1. r_e, the distance between equilibrium nuclear positions (molecule in hypothetical vibrationless state)
2. r_a, the center of gravity of the electron diffraction $P(r)/r$ distribution function (the parameter most directly related to electron diffraction data)
3. r_g, the average value of the internuclear distance for a given temperature ($r_g = r_a + l_{ij}^2/r_e$, where l_{ij} is the amplitude of vibration)
4. r_z, the distance between the mean positions of atoms in the ground vibrational state, derived from spectroscopic data
5. r_α, the distance between the mean positions of atoms at a particular temperature (calculated from r_g)
6. r_α^0, the value of r_α extrapolated to 0 K (represents the same physical quantity as r_z but is derived from electron diffraction data)
7. r_{avg}, the same physical quantity as r_z and r_α^0, but derived from simultaneous refinement of spectroscopic and electron diffraction data
8. r_0, the parameter that reproduces the ground-state rotational constants A_0, B_0, and C_0

Electron diffraction bond lengths are most commonly reported as r_g or r_a representations. Unfortunately, the choice of parameter is not always clearly specified in the literature. Of the two alternatives, r_g is used exclusively here. Amplitudes of vibration are provided for most bonded distances so that interconversions can be made if desired. The physical significance of r_g is particularly simple, since it is the internuclear distance averaged over molecular vibrations. Another virtue of the r_g representation is that there is a straightforward approximate relationship between r_g and r_e:

$$r_e \sim r_g - \left(\frac{3a}{2}\right)l_{ij}^2 \tag{5-1}$$

where a is a Morse-like parameter representing the stretching anharmonicity of the bond. Among other applications, this relationship is useful in converting r_g distances to r_e distances for comparisons with *ab initio* molecular orbital calculations. The major drawback to carrying out the $r_e - r_g$ interconversion is the uncertainty (10-50%) in available a values, but this can be taken into account when establishing the error limits. Alternatively, *differences* in r_g bond distances can be compared with the corresponding *differences* in r_e bond distances. For similar bonds (eg, C—C comparisons), the term $(3a/2)l_{ij}^2$ is nearly constant. Such comparisons do not work nearly as well for most other parameters (eg, r_z).

An annoying feature of the r_g structure is that it does not obey Euclidean geometry. For example, two times the $r_g(C{=}S)$ bond distance in carbon disulfide is greater than the $r_g(S...S)$ nonbonded distance. Similar geometrical discrepancies are seen in nonlinear systems. Both effects are vibrational in origin and are called Bastiansen–Morino shrinkage effects.[1-3] The most commonly encountered structural consequence of shrinkage is the alteration of bond angles. In contrast, r_α structures are geometrically consistent (as are r_α^0, r_z, and r_{avg}). A disadvantage of the r_α representation is that it is more complicated conceptually; thus, it is less appealing as a basic structural parameter. A common practice is to report bond distances in r_g and calculate bond angles from $r_\alpha(\theta_\alpha)$. This corrects the angles for shrinkage. In the tables that follow in this chapter, bond angles labeled θ_α or θ_{avg} have been corrected systematically for shrinkage. In all other cases, it has been noted whether shrinkage corrections were applied. A variety of procedures exist for making approximate shrinkage corrections.

A second problem that arises when interpreting electron diffraction results involves identifying the assumptions that have been made and understanding how they influence the reported structural parameters. In principle, these assumptions should be carefully delineated. This is not done in many instances, although the situation has improved considerably in recent years. The assumptions associated with geometrical constraints deserve particular scrutiny. They are introduced in an effort to circumvent the problems of resolution of internuclear distances mentioned above. Overall symmetry and local symmetry (eg, local C_{3v} symmetry for methyl groups) are assumed in most electron diffraction analyses. When symmetry conditions are not stated, they can usually, but not always, be discerned from the parameters that are varied in the least-squares analysis (independent parameters). Symmetry is *rarely* determined in an electron diffraction experiment. In addition to symmetry and simple geometrical constraints, bond lengths, bond angles, and torsional angles are commonly fixed at values obtained from other experimental methods (microwave spectroscopy, liquid crystal NMR spectroscopy), molecular mechanics, or *ab initio* molecular orbital calculations. When a parameter from an electron diffraction study is being considered for comparison purposes, it is important to ascertain whether it was obtained directly from the least-squares procedure or was assigned or determined in part by assumption. An indirect method of applying constraints is to include rotational constants obtained from infrared, Raman, and microwave spectroscopy in the least-squares procedure (combined study). The structure obtained in this manner is listed as r_{avg} in this chapter, although geometries of this type are commonly referred to as r_z in the literature. In the tables that follow, "ED + SP" designates combined electron diffraction analysis utilizing infrared and Raman spectroscopy; "ED + MW" denotes the incorporation of microwave data.

A final aspect of any electron diffraction study that must be carefully considered involves uncertainties in the structural parameters. Unfortunately, there exists no universally accepted method of error evaluation. Although least-squares analyses are now most commonly carried out on some form of the molecular intensity function, a number of studies discussed in this chapter utilized the radial distribution function. If properly executed, both methods yield the same structural parameters; however, the estimated uncertainties are obviously somewhat different. This is not a major concern. Systematic errors are generally more difficult to assess. As a rough rule, this type of error in the basic electron diffraction experiment should be 0.5σ or less, if not discussed explicitly. In extensively constrained analyses, systematic errors can be quite large; even worse, they are frequently not considered. Finally, the discussion of errors in combined studies is usually sketchy. Sizable uncertainties can be introduced by the force fields that are employed for making the $B_z - B_0$ and other vibrational corrections[1-3]; the weighting of the electron diffraction data relative to the spectroscopic data is rarely cited, much less discussed. In the tables below, no effort was made to convert the reported uncertainties to a standard system. Nevertheless, an attempt has been made to specify the errors as precisely as the original literature permits.

References are provided with the title of each subsection. Most pertinent electron diffraction studies since the 1950s are cited. In general, only papers deemed to best represent the present status of structural results are discussed in detail.

Normal Alkanes

Methane and Methane-d_4[4,5]

In 1961 Bartell, Kuchitsu, and deNeui published a study of methane and methane-d_4 that represented an electron diffraction milestone in several ways. At the time good equilibrium structures were available for the other first-row hydrides, but not for methane. In addition to determining an r_e structure for both molecules, this work provided the first electron diffraction evidence for isotope effects on r_g bond lengths and mean square amplitudes of vibration, and it effectively treated the asymmetry of stretching vibrations. To arrive at an r_e structure, it was necessary to construct a plausible model force field to estimate the vibrational corrections needed to convert the r_g structures to equilibrium structures. Although the force field was reevaluated two decades later,[5] the resulting structural parameters differed little from the 1961 values (~ 0.002 Å). The original r_g results and the 1978 r_e bond lengths for CH_4 and CD_4 are given in Table 5-1. In a more recent, unpublished electron diffraction study, Fink[6] observed an equilibrium distance of 1.088_7 Å. Using a purely spectroscopic approach, Gray and Robiette[7] have refined the anharmonic force field of methane to fit the spectroscopic data for six isotopic species. They obtained a value for the equilibrium bond length [r_e(C—H) $= 1.0858(10)$ Å] that is in excellent agreement with the corresponding electron diffraction results. A similar, but less elaborate, spectroscopic investigation by Hirota[8] yielded r_e(C—H) $= 1.0870(7)$ Å. It is interesting to note that r_0(C—H), calculated directly from B_0 for CH_4 [$1.0939940(54)$ Å], is about 0.007 Å longer than the

Table 5-1. METHANE AND METHANE-d$_4$ STRUCTURAL PARAMETERS[a]

Parameter	r_g[b,c]	l_{ij}[b,c]	r_e[d,e]
C—H, Å	1.1068(10)	0.075(2)	1.0862(24)
H...H, Å	1.811(7)	0.120(6)	
C—D, Å	1.1027(10)	0.066(2)	1.0875(26)
D...D, Å	1.805(8)	0.105(6)	

[a] Although not explicitly stated, T_d symmetry assumed.
[b] Results of Bartell, Kuchitsu, and deNeui.[4]
[c] Random and systematic errors considered. Consult reference 4.
[d] Results of Bartell and Kuchitsu.[5]
[e] Uncertainties 2σ.

mean r_e value. One final point is worth mentioning: the difference in the mean internuclear C—D and C—H distances is only a few thousandths of an angstrom [r_g(C—D) = 1.1027(10) Å vs r_g(C—H) = 1.1068(10) Å], but the slightly smaller size of the deuterium is sufficient to provide significant effects on chemical reaction rates.[9]

Ethane and Ethane-d$_6$[10–12]

Electron diffraction structures for ethane and ethane-d$_6$ are given in Table 5-2. Also shown are the results of a combined study utilizing rotational constants obtained from high-resolution infrared and Raman spectroscopy.

Table 5-2. ETHANE AND ETHANE-d$_6$ STRUCTURAL PARAMETERS[a]

Parameter	ED results[b,c]		ED + SP results[d,e]	
	r_g, θ_g[f]	l_{ij}	r_g	r_{avg}, θ_{avg}
C—C(d$_0$), Å	1.5340(11)	0.0496(10)	1.5326(20)	1.5319(20)
C—C(d$_6$), Å	1.5323(11)	0.0517(10)	1.5310(20)	1.5300(20)
C—H, Å	1.1122(12)	0.0760(10)	1.1108(20)	1.0957(20)
C—D, Å	1.1071(12)	0.0668(10)	1.1053(20)	1.0941(20)
∠CCH, deg	111.0(0.2)			111.5(0.3)
∠HCH, deg				[107.4(0.3)][g]
∠CCD, deg	111.0(0.2)			111.4(0.3)
∠DCD, deg				[107.4(0.3)][g]

[a] Symmetry not explicitly stated, presumably D_{3h}.
[b] Results of Bartell and Higginbotham.[10]
[c] Standard errors, radial distribution curve. Consult reference 10 for details.
[d] Theoretical results of Kuchitsu.[11] High-resolution infrared and Raman rotational constants (SP) for CH$_3$CH$_3$ and CD$_3$CD$_3$ combined with ED data.
[e] Uncertainties 1σ plus systematic errors.
[f] Not corrected for shrinkage.
[g] Dependent variable.

Table 5-3. PROPANE STRUCTURAL PARAMETERS[a]

Parameter	ED[b,c] r_g, θ_α	ED + MW[d–f] r_{avg}, θ_{avg}
C—C, Å	1.5323(30)	1.5337(31)
C—H(CH$_3$), Å	1.1073(51)	1.0976(22)
C—H(CH$_2$), Å		1.0962(22)
C...C, Å	2.544(15)	
C...H, Å[g]	2.194(25)	
∠CCC, deg	[112.4(1.2)][h]	112.0(0.2)
∠HCH(CH$_3$), deg	[107(0.3)][h]	107.9(0.2)
∠HCH(CH$_2$), deg	[106.1][i]	107.8(0.2)
∠CCH, deg	[111.4][j]	

[a] Results of Iijima.[13]
[b] Major assumptions: overall C_{2v} symmetry and local C_{3v} methyl group symmetry. Methyl and methylene group C—H distances assumed to be equivalent. Mean amplitudes of vibration constrained to calculated values.
[c] Uncertainties approximately 3σ.
[d] Includes rotational constants of $CH_3CH_2CH_3$.
[e] Assumptions not precisely stated.
[f] Errors subjectively assessed at 0.2%.
[g] Distance between central carbon and methyl hydrogens.
[h] Dependent variable.
[i] Value fixed from reference 14.
[j] Calculated from reported r_α distances.

The parameter of greatest interest is the C—C bond length [r_g(C—C) = 1.5326(20) Å], since ethane is the simplest organic molecule that contains this structural unit. The primary r_g(C—H) − r_g(C—D) isotope effect of 0.0050(6) Å is similar to that in methane. The secondary isotope effect on the C—C distance is small or nonexistent [Δr_g = 0.0016(7) Å, Δr_{avg} = 0.0019(20) Å].

Propane[13]

Electron diffraction and combined electron diffraction–microwave studies of Iijima[13] are shown in Table 5-3. The C—C distances in propane [r_g = 1.5323(30) Å] and ethane [r_g = 1.5326(20) Å] are the same within experimental error, as are the C—H distances [r_g = 1.1073(51) Å vs r_g = 1.1108(20) Å] and methyl group HCH angles [θ_{avg} = 107.9(0.2)° vs θ_{avg} = 107.4(0.3)°]. The bond angle comparison would be more reliable, however, if rotational constants were available for more isotopic species.

n-Butane[15–18]

The structure and conformational analysis of n-butane represents one of the cornerstones of conformational theory.[19] n-Butane exists in two conformers, the trans or anti form (**1**) and the gauche form (**2**). The structural details and relative energetics of these species offer a prototypical system for discussing the conformational analyses of a large number of more complicated acyclic and cyclic systems.[19]

CH_3 CH_3

1   **2**

Several studies of *n*-butane have been published since 1959.[15–18] The most recent by Heenan and Bartell[15] is outlined in Table 5-4. The scattered intensities of the trans and gauche conformers differ mainly in terms of the contributions by the longer nonbonded distances. Accordingly, many of the structural parameters must be determined as average values. In obtaining the results in Table 5-4, a number of assumptions were made. A completely staggered conformation with C_{2h} symmetry was assumed for the trans isomer, whereas C_2 symmetry was assigned to the gauche form. Perfect staggering was not assumed, because nonbonded interactions between the two methyl groups produce a modest opening of the $C_1C_2C_3C_4$ torsional angle ($\tau > 60°$) and a slight twisting of the methyl groups. Since preliminary studies showed that the twist angle of the gauche methyl groups could not be determined with precision, this parameter was fixed at $-2.5°$, a value suggested by molecular mechanics calculations (MUB-2).[22] Parallel and perpendicular amplitudes of vibration were calculated for both the trans and gauche conformers from the general valence hydrocarbon force field of Schachtschneider and Snyder,[20] which was modified with the correct torsional force constant.[21] The perpendicular amplitudes were employed

Table 5-4. *n*-BUTANE STRUCTURAL PARAMETERS[a,b]

Parameter	$r_g, \theta_\alpha{}^c$	Parameter	$l_{ij}{}^c$
C—C, Å	1.531(2)	C—C, Å	0.054(3)
C—H, Å	1.119(2)	C—H, Å	0.088(3)
∠CCC, deg	113.3(0.4)	1,3-C...C, Å	0.068(6)
∠CCH, deg	110.7(0.4)	1,3-C...H, Å	0.111(4)
∠τ gauche, dihedral, deg	72.4(4.8)	1,4-C...C (gauche), Å	[0.185][d]
trans conformer	64(7)%	1,4-C...C (trans), Å	[0.072][d]

[a] Results of Heenan and Bartell.[15]
[b] Numerous assumptions: see discussion in text and in reference 15. Most parameters are mean values.
[c] Uncertainties 3σ and do not include any estimates for systematic errors.
[d] Fixed at calculated values.

in the shrinkage corrections. Morse asymmetry constants for all internuclear distances were set equal to 2.0 Å^{-1}. The gauche and trans isomers were required to have identical internal coordinates, except for the torsional angles and the shrinkages applied to the nonbonded distances. All the C—C and C—H bond lengths and the CCH bond angles were made equivalent. Five independent structural parameters were refined in addition to the conformer composition: r(C—C), r(C—H), $\angle$CCC, $\angle$CCH, and the gauche dihedral angle. Additionally, four amplitudes of vibration were allowed to vary independently. These were the bonded C—C and C—H and the geminal C...C and C...H amplitudes. Since the gauche and trans 1,4-C...C l_{ij}'s were strongly correlated with the trans mole fraction, they were held at their calculated values, as were all amplitudes not listed in Table 5-4. Although the information in the electron diffraction data is not sufficient to obtain independent values for the trans and gauche CCC angles, it was found that the remaining parameters refine to essentially the same values if the difference between the CCC angles is fixed at 1.26° (value predicted by the MUB-2 force field).[22]

Most of the molecular parameters in Table 5-4 are in reasonable agreement with average values for other n-alkanes (see Table 5-7). A notable exception is the CCC angle [113.3(0.4)°], which is slightly larger than the C_3 and C_5-C_7 entries in Table 5-7. Shrinkage corrections were not applied in the C_5-C_7 alkane investigations.[23] The corrections would increase the CCC angles by about 0.5°. The nonbonded asymmetry constant of 2.0 Å^{-1} adds an additional 0.3°. A more recent study of n-hexadecane,[24] which takes shrinkage into account, reports an average CCC angle of 114.6(0.6)°.

Two additional features of the n-butane analysis deserve comment. Although the error limit on the gauche dihedral angle is quite large (4.8°), the angle is unquestionably greater (72.4°) than the unperturbed staggered value (60°). Also, the percentage of trans conformer [64(7)%] is in agreement with a recent spectroscopic estimate of 68%.[25]

n-Pentane, n-Hexane, and n-Heptane[23,26]

Bonham, Bartell, and Kohl[23] have reported structural parameters for n-pentane, n-hexane, and n-heptane (Table 5-5). As was the case for n-butane, it was possible to extract mean values only for the C—C and C—H bond distances and the CCC and CCH bond angles. In addition to the large number of internuclear distances in these molecules, the analyses are complicated by an increased number of possible conformations. Three gauche–trans isomers were considered for n-pentane: the trans, trans (TT, **3**), the trans, gauche (TG, **4**), and the gauche, gauche (GG, **5**) forms. Six conformers (TTT, TTG, TGT, TGG, GTG, GGG) were utilized in analyzing the n-hexane data, while 10 were considered in the case of n-heptane. The gauche and anti CCCC torsional angles in all conformers were fixed at 60 and 180°, respectively. Conformers containing adjacent gauche^{+}–gauche^{-} interactions, such as **6**, were assumed to be sterically

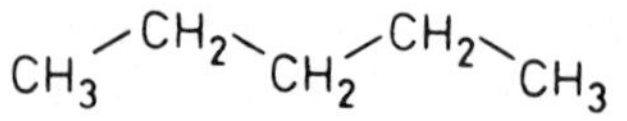

$$CH_3 \diagup \overset{CH_2}{} \diagdown \overset{CH_2}{} \diagup \overset{CH_2}{} \diagdown CH_3$$

3 (TT)

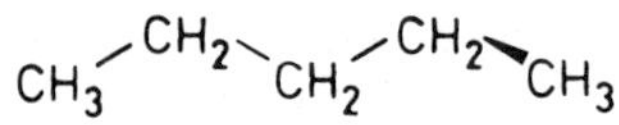

$$CH_3 \diagup \overset{CH_2}{} \diagdown \overset{CH_2}{} \diagup \overset{CH_2}{} CH_3$$

4 (TG)

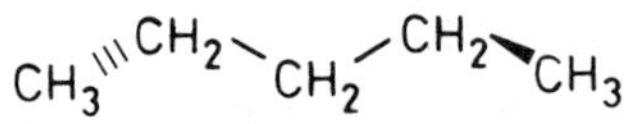

$$CH_3 \overset{CH_2}{} \diagdown \overset{CH_2}{} \diagup \overset{CH_2}{} CH_3$$

5 (GG)

$$CH_3 \overset{CH_2}{} \diagdown \overset{CH_2}{} \diagup \overset{CH_2}{} CH_3$$

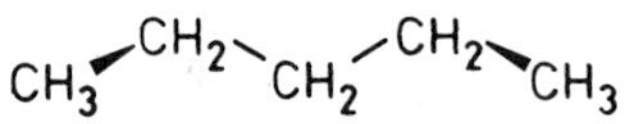

6 (G^+G^-)

unreasonable and were assigned zero weights. The relative weights of all other conformations were obtained by assuming that each gauche arrangement increases the conformer free energy by 2.93 kJ mol^{-1} (ΔG°),[26] a number estimated from the nonbonded peaks at distances greater than 2.7 Å. Although considerable uncertainty exists in the estimated value of ΔG°, it was shown that the experimental parameters in Table 5-5 are relatively insensitive to the value assumed. Shrinkage corrections were not applied in the bond angle calculations. A complementary paper[26] discusses the derivation of the gauche–trans ΔG°, shrinkage effects, the average gauche angle, the magnitude of $\angle$ CCC(gauche)–$\angle$ CCC(trans), and the distortion from planarity of trans segments when they are adjacent to gauche segments.

n-Hexadecane (C$_{16}$H$_{34}$)[24]

X-Ray crystallographic studies of both long- and short-chain normal alkanes show that the individual molecules are constrained to all-trans arrangements by crystal packing considerations.[27] It is extremely important, therefore, to explore the structures and conformations of free hydrocarbons in the vapor phase. The early papers on *n*-butane through *n*-heptane[17,18,23] represent the pioneering efforts in this direction. Fitzwater and Bartell[24] have carried out an electron diffraction study of *n*-hexadecane in an attempt to obtain limiting mean structural parameters for *n*-alkanes in the vapor phase.

n-Hexadecane has 1225 internuclear distances in a single conformation and can exist in well over a million distinct conformations. Obviously, electron

Table 5-5. *n*-PENTANE *n*-HEXANE, AND
n-HEPTANE STRUCTURAL PARAMETERS[a,b]

Parameter	$r_g, \theta_g{}^c$	l_{ij}
n-Pentane		
C—C, Å	1.531(2)	0.0575(25)
C—H, Å	1.118(4)	0.083(4)
1,3-C...H, Å	2.188(4)	0.116(5)
1,3-C...C, Å	2.552(3)	0.0800(35)
∠CCC, deg	[112.9(0.2)][d]	
∠CCH, deg	[110.4(0.3)][d]	
n-Hexane		
C—C, Å	1.533(3)	0.0575(25)
C—H, Å	1.118(6)	0.080(5)
1,3-C...H, Å	2.178(5)	0.123(5)
1,3-C...C, Å	2.541(4)	0.0800(35)
∠CCC, deg	[111.9(0.4)][d]	
∠CCH, deg	[109.5(0.5)][d]	
n-Heptane		
C—C, Å	1.534(3)	0.051(3)
C—H, Å	1.121(7)	0.078(7)
1,3-C...H, Å	2.185(4)	0.104(5)
1,3-C...C, Å	2.553(3)	0.0760(35)
∠CCC, deg	[112.6(0.3)][d]	
∠CCH, deg	[109.8(0.5)][d]	

[a] Results of Bonham, Bartell, and Kohl.[23] See also
Bartell and Kohl.[26]
[b] Numerous assumptions: see discussion in text and in
reference 23. All parameters mean values. Standard
errors.
[c] Shrinkage corrections not applied to bond angles.
[d] Dependent variable.

diffraction data analysis is not possible without introducing major simplifications. A statistical model was formulated that utilized mean bond lengths, angles, amplitudes, and a mean gauche–trans free energy difference ($\Delta G°$). Five independent geometrical parameters were used to specify the structure: the average CC and CH bond lengths, the average CCC and HCH (local C_{2v} symmetry assumed) bond angles, and the average gauche dihedral angle. All trans dihedral angles were fixed at 180°, although torsional and other shrinkage effects were applied. The shrinkage corrections were estimated by rough extrapolations from related values determined for *n*-butane. Morse asymmetry constants of 2.0 Å^{-1} were adopted for all distances. As with the C_5–C_7 alkanes, the relative percentage of various conformations were calculated from the number of gauche sites. The excess free energy of a given conformation with *n* gauche sites was assumed to be equal to *n* times the $\Delta G°$ value for a single gauche site. The best value of $\Delta G°$ (1.15 ± 1.46 kJ mol^{-1}) was taken to be that yielding the minimum value of $\sigma(I)$ for the intensity function. Conformations

Table 5-6. n-HEXADECANE STRUCTURAL
PARAMETERS[a–c]

Parameter	r_g, θ^d	l_{ij}
C—C, Å	1.542(4)	0.057(5)
C—H, Å	1.130(8)	0.087(5)
∠CCC, deg	114.6(0.6)	
∠CCH, deg	$[110.4(1.1)]^e$	
∠HCH, deg	99.5(4.5)	
∠gauche dihedral, deg	65(10)	

[a] Results of Fitzwater and Bartell.[24].
[b] Numerous assumptions: see discussion in text and in reference 24.
[c] Quoted uncertainties 3σ. Uncertainties arising from the shrinkages adopted or other assumptions not included.
[d] Shrinkage corrections estimated by rough extrapolations from related values determined for n-butane.
[e] Dependent variable.

with gauche$^+$–gauche$^-$ sites were excluded, although it was carefully pointed out that this is an approximation (estimated strain energy ≥ 12.5 kJ mol^{-1}). This assumption would be of dubious value when applied to chains much longer than C_{16}; in such cases, one segment of a chain might fold and pack against another.

Many of the longer internuclear distances in n-hexadecane give rise to very broad peaks in the radial distribution function and may, therefore, be neglected in an analysis that is concerned with structural features that lie in the 1–4 Å range. Accordingly, a model was adopted that accounted explicitly for all C...C distances through 1,7 and all C...H and H...H distances through 1,5. Only the mean C—C, C—H, 1,3-C...C and C...H, and 1,4-$gauche$ C...C amplitudes were varied in the least-squares analysis (molecular intensity function), along with the previously specified geometrical parameters. All other amplitudes were assigned plausible values.

The mean C—C and C—H bond distances for n-hexadecane are longer than those for the C_2–C_7 normal alkanes (Tables 5-6 and 5-7). Fitzwater and Bartell argue convincingly that it is unlikely that the observed differences result from experimental errors, particularly in the case of the C—C bonds. The increase in the C—H bond length is less certain, because the experimental uncertainties are fairly large. Carbon–hydrogen distances and their associated angles are difficult to determine precisely by electron diffraction. Hydrogen atoms are the weakest scatterers of electrons and have some of the largest vibrational effects. The increase may, nevertheless, reflect the dominance of CH_2 groups in n-hexadecane.

The mean C—C bond in n-hexadecane is approximately 0.01 Å longer than the C—C bonds of the other n-alkanes in Table 5-7. The value of 1.542 Å must correspond very nearly to the methylene–methylene distance in the interior of

Table 5-7. SUMMARY OF n-ALKANE STRUCTURAL PARAMETERS[a]

Alkane, n	Bond distances, r_g (Å)		Bond angles (deg)[b]	
	C—C	C—H	∠CCC	∠CCH
1	—	1.107(3)	—	—
2	1.533(6)	1.111(6)	—	111.5(0.9)
3	1.532(3)	1.107(5)	112.4(1.2)	111.4
4	1.531(2)	1.119(2)	113.3(0.4)	110.7(0.4)
5	1.531(6)	1.118(12)	112.9(0.6)[c]	110.4(0.9)[c]
6	1.533(9)	1.118(18)	111.9(1.2)[c]	109.5(1.5)[c]
7	1.534(9)	1.121(21)	112.6(0.9)[c]	109.8(1.5)[c]
16	1.542(4)	1.130(8)	114.6(0.6)	110.4(1.1)

[a] Uncertainties adjusted to approximately 3σ.
[b] Corrected for shrinkage effects except where noted.
[c] Not corrected for shrinkage.
Source: Data from Tables 5-1 to 5-6.

hydrocarbons. The results of molecular mechanics calculations that employ the MUB-2 force field[22] support this conclusion. They further imply that the lengthening is due to 1,3-C...H repulsions and to an increased concentration of gauche units, which tend to be longer than their trans counterparts.

The mean CCC angle for n-hexadecane of 114.6(0.6)° is substantially larger than all the entries in Table 5-7 with the exception of n-butane. The lack of vibrational corrections in the C_5–C_7 alkanes is very likely responsible for a large portion of this difference (see n-butane). The uncertainties in the CCH angles in Table 5-7 make comparisons difficult, although the CCH angles are surprisingly constant in the 110–111° range.

Branched Alkanes

Isobutane (2-Methylpropane)[28]

Hilderbrandt and Wieser[28] undertook an electron diffraction study of isobutane to probe the tertiary C—H bond length. Valence shell electron-pair repulsion (VSEPR) theory[29] suggests that the C—H bond distances in hydrocarbons should depend on the number of geminal C...H interactions; a systematic lengthening is expected along the series ethane (methyl hydrogens), propane (methylene hydrogens), and isobutane (tertiary hydrogens). It is impossible to determine the length of the tertiary C—H bond in isobutane from electron diffraction data alone, because the distance is too close to the methyl C—H bond lengths to be resolved. Fortunately, rotational constants were available for a number of isotopic species, permitting a combined electron diffraction-microwave spectroscopic determination. The results based on the electron diffraction data alone and the combined study are given in Table 5-8.

Table 5-8. ISOBUTANE STRUCTURAL PARAMETERS[a–c]

Parameter	ED		ED + MW[d]	
	r_g, θ_α	l_{ij}	r_g, θ_α	l_{ij}
C—H (methyl), Å	1.114(2)	0.075(2)	1.113(2)	0.075(2)
C—H (tertiary), Å			1.119(5)	e
C—C, Å	1.534(1)	0.048(1)	1.534(1)	0.048(1)
∠CCH (tertiary), deg	108.1(0.2)		108.1(0.2)	
∠CCH (methyl), deg	111.7(0.4)		111.4(0.2)	
∠CCC, deg	[110.9(0.2)][f]		[110.9(0.2)][f]	

[a] Results of Hilderbrandt and Wieser.[28]

[b] Major assumptions: a completely staggered C_{3v} model with local C_{3v} methyl group symmetry. Consult reference 28 for vibrational corrections. Many amplitudes constrained to calculated values.

[c] Uncertainties 3σ.

[d] Includes three rotational constants[30] for the on-axis-substituted species [$(CH_3)_3CH$, $(CH_3)_3CD$, and $(CH_3)_3{}^{13}CH$]. A third model is reported in which nine rotational constants are included. The only structural parameter to change significantly was C—H (tertiary) = 1.122(6) Å.

[e] Not reported.

[f] Dependent variable.

The average C—H value [$r_g = 1.114(2)$ Å] obtained from the electron diffraction analysis is similar to the corresponding average values for the *n*-alkanes in Table 5-7. Since the amplitude of vibration for this distance is not unusually large [$l_{ij} = 0.075(2)$ Å], whatever splitting is present must be less than a mean amplitude. In the combined analysis in Table 5-8, the tertiary and methyl C—H distances were varied independently, and the rotational constants[30] for three on-axis-substituted species [$(CH_3)_3CH$, $(CH_3)_3CD$, and $(CH_3)_3{}^{13}CH$] were included. A second combined analysis that used nine rotational constants yielded a tertiary C—H distance of $r_g = 1.122(6)$ Å. Hilderbrandt and Wieser's best estimate of the splitting of methyl and tertiary C—H distances was $\Delta r_g = 0.008_7(6_1)$ at the 99% confidence level. Whether systematic errors were included in this estimate is not clear. Once again it is obvious that it is very difficult to measure the difference between two C—H bond distances by electron diffraction, even if microwave data are included. It should also be noted that rotational constants, while very useful in combined analyses, are the least sensitive to hydrogen positions.

Neopentane (2,2-Dimethylpropane)[31,32]

Neopentane is the final compound in a structural series (methane, ethane, propane, isobutane, summarized in Table 5-9) in which the hydrogens of methane are successively replaced by methyl groups. The results of the two most

Table 5-9. COMPARISON OF METHYL-SUBSTITUTED METHANES[a]

Compound	Bond distances r_g (Å)		
	C—C	C—H[b]	$\angle$ CCC(deg)[b]
Methane		1.107(3)	
Ethane	1.533(6)	1.111(6)	
Propane	1.532(3)	1.107(5)	112.4(1.2)
Isobutane	1.534(1)	1.114(2)	110.9(0.2)
Neopentane	1.537(3)	1.114(8)	109.5

[a] Uncertainties adjusted to approximately 3σ.
[b] Average C—H value.
[c] Corrected for shrinkage.
Source: Tables 5-1 to 5-3, 5-8, 5-10.

recent electron diffraction studies[31,32] of neopentane are shown in Table 5-10. The 1977 study by Bartell and Bradford[31] incorporates vibrational corrections and has a smaller uncertainty in the C—C distance. Both studies assumed T_d symmetry, allowing the structures to be analyzed in terms of three independent structural parameters.

The number of geminal C...C interactions increases along sequence propane, isobutane, neopentane. Any C—C bond lengthening due to this effect is quite small (Table 5-9). Propane and isobutane can relieve unfavorable 1,3-nonbonded interactions by CCC bond angle bending. Because of symmetry constraints, neopentane does not have this option.

Table 5-10. NEOPENTANE STRUCTURAL PARAMETERS: T_d SYMMETRY ASSUMED

Parameter	ED (1977)[a,b]		ED (1969)[c,d]	
	r_g, θ^e	l_{ij}	r_g, θ^f	l_{ij}
C—C, Å	1.537(3)	0.051(4)	1.541(6)	0.058(9)
C—H, Å	1.114(8)	0.079(8)	1.128(9)	0.097(12)
$\angle$ CCH, deg	112.2(2.8)		110.0(1.2)	

[a] Results of Bartell and Bradford.[31]
[b] Uncertainties are 3σ and include both random and systematic errors. Bond angle uncertainty does not include any allowance for the uncertainties in shrinkages used.
[c] Results of Beagley, Brown, and Monaghan.[32]
[d] Uncertainties 3σ.
[e] Shrinkage corrections approximated.
[f] No shrinkage corrections.

Hexamethylethane (2,2,3,3-Tetramethylbutane)[33]

The steric interactions in hexamethylethane are sufficiently severe to permit signific̣nt structural deformations to be discerned, even when it is necessary to obtain many of the structural parameters as average values. The structural results of Bartell and Boates[33] are summarized in Table 5-11.

If the carbon skeleton is assumed to have D_3 symmetry and the methyl groups are assigned local C_{3v} symmetry, the structure of hexamethylethane may be described with seven parameters. The parameters chosen for refinement were:

1. $\bar{r}$, the average C—C bond length
2. Δr, the difference between the central and C—CH$_3$ bond lengths
3. r(C—H), the average C—H bond length
4. $\angle$ CCC, the angle between the central C—C and terminal C—CH$_3$ bonds
5. $\angle$ CCH, the average CCH bond angle
6. τ_C, the torsional displacement of one tertiary butyl group with respect to the other, taken as zero in the D_{3d} staggered form
7. τ_M, the torsional displacement about the C—CH$_3$ bond, taken as zero when the C—H bonds are in a staggered conformation.

The two torsional parameters, τ_C and τ_M, are determined only marginally by the data and were not varied simultaneously with the remaining parameters. Estimates of τ_C and τ_M were made by following the standard deviation between the experimental and theoretical radial distribution curves at various imposed parameter values. This procedure indicated a significant distortion from D_{3d} symmetry ($\tau_C > 0$), but no deviation from perfect methyl staggering. Accordingly, τ_C and τ_M were fixed at 5° and 0° in subsequent least-squares investigations on the intensity function. The deformation from a completely staggered D_{3h} symmetry is a reasonable consequence of methyl group interactions. In a

Table 5-11. Hexamethylethane Structural Parameters[a–d]

Independent parameters	r_g, θ^e	Key distances	r_g	l_{ij}
$\bar{r}$, Å	1.547(2)	C—CH$_3$, Å	1.542(2)	0.055(3)
Δr, Å	0.041(6)	C—C, Å	1.583(10)	0.056
C—H, Å	1.113(4)	C—H, Å	1.113(4)	0.082(4)
$\angle$ CCC, deg	111.0(0.3)			
$\angle$ CCH, deg	111.5(1.4)			
τ_C, deg	5(4)[f]			
τ_M, deg	0(7)[f]			

[a] Results of Bartell and Boates[33] (run II).
[b] Structural parameter definitions in text.
[c] Numerous assumptions: see discussion in text and reference 33; D_3 symmetry for the carbon skeleton and C_{3v} methyl group symmetry assumed.
[d] Uncertainties 2σ, including scale factor and intensity errors.
[e] Shrinkage corrections were estimated crudely.
[f] Parameters not varied in least-squares analysis. See text.

structure with full D_{3h} symmetry, six of the methyl hydrogens on one end of the molecule are projected directly at six on the other end that are only about 2 Å away. Increasing either τ_C or τ_M opens the H...H distances. Molecular mechanics calculations (MUB-1[34] and MM2[35]) indicate that both the central and the terminal bonds are twisted, with the central torsional angle being somewhat larger (13–15°) than the assigned value.

Both the C—C bonds in hexamethylethane are longer than those in neopentane [$r_g = 1.537(3)$ Å]. The central bond is stretched by 0.045(10) Å and the C—CH$_3$ bonds by a smaller extent [0.005(2) Å]. These deformations presumably result from severe intermethyl repulsions. In keeping with this point of view, the $\angle$CCC is 1.5° larger than that expected for an undistorted tetrahedral, quaternary carbon.

Tetramethylethane (2,3-Dimethylbutane)[33]

Tetramethylethane is a moderately congested hydrocarbon that presents some unique conformational and structural problems. There are two stable conformers, the trans (**7**) and the gauche (**8**) forms. In structures **7** and **8** the rear circle in the Newman notation represents carbon atom 3. Methyl groups 5 and 6 are attached to the second and third carbon atoms, respectively. The results of an electron diffraction study by Bartell and Boates[33] are provided in Table 5-12.

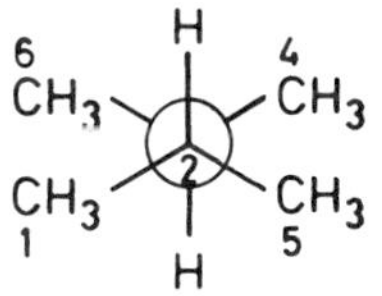

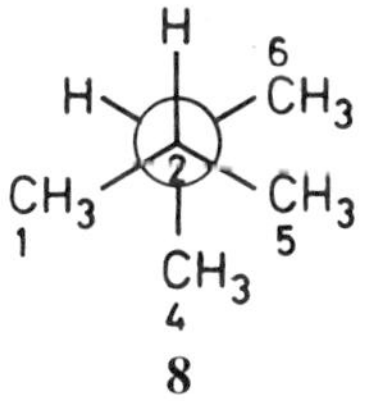

A number of well-chosen assumptions were necessary to make the data analysis tractable. Both isomers were assumed to have the same CH$_3$—C and C—H bond lengths, identical CCH bond angles, and identical terminal CCC angles ($\angle C_1C_2C_5 = \angle C_4C_3C_6$). The central C—C bond was assumed to be 0.002 Å longer in the gauche isomer. Symmetries of C_{2h} and C_2 were ascribed to the trans and gauche conformers, respectively. Two different internal angles ($\angle C_2C_3C_6$ and $\angle C_2C_3C_4$) were permitted in the gauche form. The CCH (tertiary) angles were adjusted to make the tertiary hydrogens equidistant from each of their carbon nearest neighbors.

The least-squares refinement was performed utilizing the following ten parameters:

1. The percent of gauche conformer
2. $\bar{r}$, the average C—C length
3. Δr, the difference between the central C—C and terminal C—CH$_3$ bonds in the trans isomer

Table 5-12. Tetramethylethane Structural Parameters[a–d]

Independent parameters	r_g, θ^e	Key distances	r_g	l_{ij}
$\bar{r}$, Å	1.5398(20)	C—H(methyl), Å	[1.113][f]	0.085(7)
Δr, Å	[0.005(7)][g]	**Trans isomer**[h]		
C—H, Å	1.115(4)	C—CH$_3$, Å	1.539	0.055(3)
$\angle$CCC(trans), deg	111.1(1.4)	C—C, Å	1.544	0.056
$\angle$CCC(gauche), deg	113.6(0.9)	**Gauche isomer**[h]		
$\Delta\angle$CCC(gauche), deg	[2(2)][g]	C—CH$_3$, Å	[1.539][i]	[0.055][i]
$\angle$C$_1$C$_2$C$_5$, deg	[110.1(0.8)][g]	C—C, Å	[1.546][i]	[0.056][i]
$\angle$CCH, deg	[110.5][f]			
τ (gauche), deg	[65(5)][g]			
$\angle$CCC(avg), deg	111.3(0.4)			
$\angle$CCC(inner, avg), deg	112.0(0.8)			

[a] Results of Bartell and Boates.[33]
[b] Structural parameter definitions in the text.
[c] Numerous assumptions: see discussion in text and in reference 33; C_{2h} and C_2 symmetries were assumed for the trans and gauche conformers, respectively.
[d] Estimated uncertainties 2σ.
[e] Shrinkage corrections were estimated crudely.
[f] Assumed r_g(C—H) same as in hexamethylethane.
[g] Parameters not varied in least-squares analysis. See text.
[h] Results determined at fixed composition with 60% gauche.
[i] Tied to trans values.

4. r(C—H), the bond length in the methyl groups (tertiary C—H bond assumed to be 0.02 Å longer)
5. $\angle$CCC(trans), the inner CCC angle in the trans isomer
6. $\angle$CCC(gauche), the inner C$_2$C$_3$C$_4$ bond angle in the gauche isomer
7. $\Delta\angle$CCC(gauche), the amount by which $\angle$C$_2$C$_3$C$_4$ exceeds $\angle$C$_2$C$_3$C$_6$
8. $\angle$C$_1$C$_2$C$_5$, the average CCC terminal bond angle
9. $\angle$CCH, the average CCH bond angle
10. τ(gauche), the torsional displacement about the central C—C bond in the gauche isomer, taken as zero in the eclipsed C_{2v} conformation

It was impossible to obtain convergence when all nine structural parameters were varied simultaneously in the least-squares refinement. Therefore, the four most subtle parameters, Δr, $\angle$C$_1$C$_2$C$_5$, τ(gauche), and $\Delta\angle$CCC(gauche) were fixed at various values while the other parameters were refined. Estimates of the four fixed parameters were obtained by a procedure similar to that employed in the analysis of hexamethylethane (see above).

Approximately 60% of tetramethylethane exists in the gauche form, which is within experimental error of the distribution expected if the free energy difference between the gauche and anti conformations is zero. This is consistent with spectroscopic investigations but violates the extremely useful rule of qualitative conformational analysis that asserts that gauche methyl interactions are additive.[19] Structure **7** has two such interactions; conformer **8** has three. Consequently, the anti conformer **7** should be more stable. Several factors very likely are responsible for the fact that the energy difference between the gauche and anti conformations is unexpectedly small. The terminal CCC angles ($\angle C_1 C_2 C_5$) are larger than the tetrahedral value. This forces the methyl groups into one another in the anti form and raises the energy of the conformation. At the same time, the gauche conformer can evade some of the methyl–methyl repulsion by undergoing a small rotation about the central C—C bond.

The mean C—C bond length in tetramethylethane is 1.540(2) Å, and the difference between the terminal and central C—C bonds in the anti conformer (Δr) is only 0.005(7) Å. Evidently the steric stretching displacements in tetramethylethane are much smaller than in hexamethylethane. This result is not surprising, since the high symmetry of hexamethylethane provides very few alternatives for removing unfavorable steric interactions.

Tri-tert-butylmethane
(3-tert-Butyl-2,2,4,4-tetramethylpentane)[36,37]

The steric crowding in tri-*tert*-butylmethane is unusually severe. If a C_{3v} model of tri-*tert*-butylmethane is constructed with normal bond lengths, bond angles, and torsional angles, the unrealistically strained molecule contains six pairs of hydrogens that are within 0.3 Å of their normal van der Waals contact distances. The manner in which the molecule distorts to relieve unavoidable gauche$^+$–gauche$^-$ interactions is of considerable significance in understanding highly strained molecules.

If it is assumed that tri-*tert*-butylmethane has overall C_3 symmetry, the radial distribution function is composed of 275 internuclear distances grouped into seven peaks, not all of which are well resolved. Thirty-nine geometrical parameters are needed to specify the most general model of this symmetry. The additional assumption of local C_3 symmetries in the *tert*-butyl groups and equivalence of all C—H bond lengths reduces the number of required parameters to eight. Burgi and Bartell[36] have carried out a conventional electron diffraction analysis of tri-*tert*-butylmethane data by employing the assumptions outlined above (Table 5-13). The independent parameters they selected were as follows (subscripts: m = methyl, t = tertiary, q = quaternary):

1. $r(C—H)$, the average C—H bond length
2. $r(C_q—C_m)$, the average C—CH$_3$ bond length
3. Δr, the difference between the central C$_t$—C$_q$ and terminal C$_q$—C$_m$ bonds
4. $\angle C_q C_t H_t$, the CCH bond angle involving the tertiary hydrogen
5. $\angle C_t C_q C_m$, the average CCC angle formed by the methyl groups and the tertiary carbon

Table 5-13. TRI-*tert*-BUTYLMETHANE STRUCTURAL
PARAMETERS[a–d]

Parameter	r_g, θ^e	l_{ij} (assigned)
C—H, Å	1.111(3)	[0.080]
C_q—C_m, Å	1.548(2)	[0.055]
Δr, Å	0.063(5)	
$\angle C_q C_t H_t$, deg	101.6(0.4)	
$\angle C_t C_q C_m$, deg	113.0(0.2)	
$\angle C_q C_m H_m$, deg	114.2(1.0)	
$\Delta\tau H_t C_t C_q C_m$, deg	10.8(0.5)	
$\Delta\tau$(methyl), deg	18.0(6.0)	
C_t—C_q, Å	$[1.611(5)]^f$	
$\angle C_q C_t C_q$, deg	$[116.0(0.4)]^f$	
$\angle C_m C_q C_m$, deg	$[105.7(0.2)]^f$	

[a] Results of Burgi and Bartell.[36]
[b] Parameter definitions in text.
[c] Numerous assumptions: see discussion in text and in reference 36. Overall C_3 symmetry and local C_3 symmetry for the *tert*-butyl groups and methyl groups assumed.
[d] Uncertainties 1σ and do not include most systematic errors.
[e] No shrinkage corrections.
[f] Dependent variable.

6. $\angle C_q C_m H_m$, the average *tert*-butyl CCH angle
7. $\Delta\tau(H_t C_t C_q C_m)$, the average torsional displacement of the *tert*-butyl groups about $C_t C_q$, taken as zero in the staggered conformation
8. $\Delta\tau$(methyl), a similar average torsional displacement of the methyl groups, taken as zero in the staggered conformation

The structural parameters in Table 5-13 clearly reveal the influence of the unfavorable gauche interactions. The three central C_t—C_q bond lengths [1.611(5) Å] are the longest reported (electron diffraction) for an acyclic hydrocarbon. Moreover, the nine *tert* butyl C—C bonds are longer [1.548(2) Å] than those in neopentane (1.537 Å). The average C—C value for the molecule as a whole is 1.564 Å. The $C_q C_t H_t$ angle [101.6(0.4)°] is 6.5° less than the corresponding angle in isobutane; the related $\angle C_q C_t C_q$ is 116.0(0.4)°. The extreme congestion is also reflected in the $C_m C_q C_m$ and $C_q C_m H_m$ angles of the *tert*-butyl groups [105.7(0.2) and 114.2(1.0)°, respectively].

In a complementary paper,[37] Bartell and Burgi relaxed the local C_{3v} symmetry constraint on the carbon skeleton of the *tert*-butyl groups and introduced four additional angle parameters. They point out in so doing that extreme caution must be exercised when increasing the number of independent parameters in a system as complex as tri-*tert*-butylmethane. In scanning the multidimensional space of structural parameters, it is possible to find several different structures that give fits of comparably good quality. In fact, six slightly different sets of structural parameters were found to correspond to least-squares minima. The noteworthy aspect of the study was that the authors employed vibrational spectroscopy and molecular mechanics to critique and to extend the electron diffraction results. The details of this important contribution are too complex to discuss at length here.

Di-tert-butylmethane(2,2,4,4-tetramethylpentane)[31]

While less strained than tri-*tert*-butylmethane, di-*tert*-butylmethane should also be significantly distorted sterically. It is of interest to determine the manner in which the *tert*-butyl groups arrange themselves about the less encumbered methane center. The electron diffraction results of Bartell and Bradford[31] are provided in Table 5-14.

Overall C_2 symmetry was assumed in the analysis. Local C_3 symmetry was attributed to the *tert*-butyl and methyl groups. The weakly scattering methylene group (HCH) was placed on the angle bisector of the central $C_qC_sC_q$ angle, and the HCH angle was constrained to 105°. The least-squares analysis was completed utilizing nine parameters:

1. $r(C-C)$, the average C—C bond length
2. Δ, the difference between the central C_s-C_q and *tert*-butyl C—C bond lengths
3. $r(C-H)$, the average C—H bond length
4. $\angle C_qC_sC_q$, the central CCC bond angle
5. $\angle C_sC_qC_m$, the average *tert*-butyl CCC bond angle
6. $\angle C_qC_mH_m$, the average methyl CCH bond angle
7. $\tau(t\text{-Bu})$, the torsional displacement of the *tert*-butyl groups, taken as zero in the staggered conformation
8. $\tau(CH_3)$, the torsional displacement of the methyl groups, taken as zero in the staggered conformation
9. $\varepsilon(\text{tilt})$, the tilt of the *tert*-butyl groups about an axis through their central carbon atom perpendicular to the plane defined by $C_qC_sC_q$; a positive tilt increases the clearances between the two *tert*-butyl groups

The average C—C bond length of di-*tert*-butylmethane is 1.545(5) Å. This is longer than the C—C bond length in neopentane (1.537 Å), but much less than the average C—C value for tri-*tert*-butylmethane (1.564 Å). Several attempts to determine Δ, the splitting between the central C_s-C_q and C_q-CH_3 bonds, were unsuccessful. The standard deviation for Δ was twice as large as the parameter. Apparently there is a splitting, but it is much less than in tri-*tert*-butylmethane. The most dramatic deformation is at the central $C_qC_sC_q$ angle, which is opened to 128.0(6.0)°. The two *tert* butyl groups also respond to the inescapable gauch$^+$-gauche$^-$ interactions by undergoing torsional displacements of 15.1(6.0)° and by tilting away from each other by 3.1(2.3)°.

Table 5-14. DI-*tert*-BUTYLMETHANE
STRUCTURAL PARAMETERS[a-d]

Parameter	r_g, θ^e	l_{ij}
C—H, Å	1.122(15)	0.087(15)
C—C, Å	1.545(5)	0.058(5)
$\angle C_q C_s C_q$, deg	128.0(6.0)	
$\angle C_s C_q C_m$, deg	110.6(2.8)	
$\angle C_q C_m H_m$, deg	112.1(3.0)	
$\tau(t\text{-Bu})$, deg	15.1(6.0)	
$\tau(CH_3)$, deg	13.2(7.0)^f	
$\varepsilon(\text{tilt})$, deg	3.1(2.3)^g	

[a] Results of Bartell and Bradford.[31]
[b] Parameter definitions in text.
[c] Numerous assumptions: see text and reference 31.
Overall C_2 symmetry assumed, also local *tert*-butyl
and methyl group C_3 symmetry.
[d] Uncertainties are 3σ and include both random and
systematic errors. Bond angle uncertainties do not
include any allowance for the uncertainties in
shrinkages used.
[e] Shrinkage corrections estimated.
[f] Uncertainty subjectively doubled.
[g] Parameter strongly correlated with $\angle C_q C_s C_q$.

Cycloalkanes

Cyclopropane[38,39]

Cyclopropane is a highly strained cyclic hydrocarbon. A recent joint electron
diffraction–spectroscopic analysis by Yamamoto and associates[38] has provided
r_{avg} and r_e structures in addition to the usual r_g representation. The combined
study utilized the rotational constants for the ground states of C_3H_6 and C_3D_6
and the v_{11} fundamental state (C—C stretching mode). The inclusion of the v_{11}
state was a novel aspect of the analysis. Since the differences in the nuclear
positions caused by isotopic substitution and by vibrational excitation originate
from the anharmonicity of the internuclear potential, they can be used as a
source of experimental information on vibrational anharmonicity.

Assuming D_{3h} symmetry, three parameters define the molecular geometry of
cyclopropane: $r(C—C)$, $r(C—H)$, and $\angle HCH$. The r_g bond lengths in Table
5-15 agree within experimental error with a 1964 electron diffraction study (r_a)
by Bastiansen, Fritsch, and Hedberg.[39] The C—C bond distance in cyclopropane (1.514 Å) is about 0.02 Å shorter than in strain-free cyclohexane (1.536 Å).
The shortening has been attributed to "bent" C—C bonds in the three-
membered ring.[40,41] The C—H bonds of cyclopropane (1.099 Å) are also
shorter than those in cyclohexane (1.121 Å). They are very close, in fact, to the
C—H bond distance in ethylene (1.103 Å).[42] This is reasonable, considering

Table 5-15. CYCLOPROPANE STRUCTURAL PARAMETERS[a–c]

Parameter	r_g	l_{ij}	r_{avg}, θ_{avg}	r_e, θ_e
C—H, Å	1.0991(20)	0.081(4)	1.0840(20)	1.083(5)
C—C, Å	1.5139(12)	0.054(2)	1.1527(12)	1.501(4)
∠HCH, deg			114.5(0.9)	114.5(0.9)

[a] Results of Yamamoto et al.[38]
[b] Assumption: D_{3h} symmetry: see text and reference 38 for details of combined electron diffraction–spectroscopic analysis.
[c] Uncertainties listed only as estimated limits of error. Systematic errors considered in making the vibrational interconversions.

that it has been proposed that the hybridization of the carbon atoms in cyclopropane is similar to that in ethylene.[40] One final point worth noting is that the bond-stretching anharmonicity constant obtained experimentally in this work $[a(C—C) = 3.2(1.0)\,\text{Å}^{-1}]$ is about 50% larger than the value generally assumed for C—C bonds $[a(C—C) \sim 2\,\text{Å}^{-1}]$. Cyclopropane is far from typical, however, and the uncertainty is quite large.

Cyclobutane[43–46]

Although it may seem surprising, a full-length electron diffraction study of cyclobutane has not appeared in the literature since 1952.[45] While it is relatively straightforward to obtain the bonded distances for cyclobutane, the more subtle features of the molecular structure are difficult to define because of the operation of puckering and rocking phenomena. Cyclobutane is not planar (D_{4h}); it is folded diagonally across the four-membered ring (D_{2h}). Moreover, the quasi-axial hydrogens in the 1,3 positions are tilted slightly toward one another. The barrier for inverting the cyclobutane ring between degenerate, puckered conformations is low (~ 6.3 kJ mol^{-1}).[64]

The results of a recent preliminary study of cyclobutane by Egawa and co-workers[43] are summarized in Table 5-16. This study involves a joint electron

Table 5-16. CYCLOBUTANE STRUCTURAL PARAMETERS[a]

Parameter	r_g	r_{avg}, θ_{avg}
C—H, Å	1.109(3)	1.093(3)
C—C, Å	1.554(1)	1.552(1)
∠HCH, deg		106.4(1.3)
θ(dihedral), deg		27.9(1.6)
β(rocks), deg		6.2(1.2)

[a] Preliminary results of Egawa et al.[43]

diffraction–spectroscopic analysis that utilizes one rotational constant obtained from the v_{14} (CH$_2$-scissoring, B_{1g}) and v_{16} (CH$_2$-rocking, A_{2u}) band spectra in the infrared region. The C—C [$r_g = 1.554(1)$ Å] and C—H [$r_g = 1.109(3)$ Å] bond lengths agree well with those that appeared in a preliminary report by Almenningen, Bastiansen, and Skancke in 1961.[46] The C—C bond length is slightly longer than that in cyclohexane (1.536 Å). This bond lengthening generally has been agreed to result from unfavorable transannular 1,3-non-bonded repulsions.[45] In Table 5-16, θ is the dihedral angle for ring folding [27.9(1.6)°], and β is the rocking angle between the bisectors of HCH and CCC angles that share a common carbon atom [6.2(1.2)°].

Cyclopentane[46,47]

Thermodynamic and spectroscopic evidence suggests that cyclopentane is puckered.[47] Furthermore, the deformation is dynamic; the puckering displacements pseudorotate in an essentially free manner (very small ten-fold barrier) around the five-membered ring. Adams, Geise, and Bartell[47] have published the most recent electron diffraction study of cyclopentane (Table 5-17).

In this work, a comparison of the experimental radial distribution function with the calculated distribution function for a planar (D_{5h}) model of cyclopentane clearly showed the ring to be nonplanar. A pseudorotation description was therefore adopted. Puckering displacements were described relative to the plane of the unpuckered D_{5h} conformation in terms of two independent parameters, q and ϕ, the pseudoradial puckering amplitude and the pseudorotational angle, respectively. The structural parameters shown in Table 5-17 were varied in the least-squares analysis. Local C_{2v} symmetry was assumed at each methylene

Table 5-17. CYCLOPENTANE STRUCTURAL PARAMETERS[a,b]

Parameter	r_g, θ^c	l_{ij}
C—H, Å	1.1135(15)	0.0815(21)
C—C, Å	1.5460(12)	0.0535(11)
1,3-C...H, Å	2.2140(21)	
1,3-C...C, Å	2.4440(15)[d]	
∠CCH, deg	[111.7(0.2)][e]	
$\langle q \rangle$, Å[f]	[0.427(15)][e]	

[a] Results of Adams, Geise, and Bartell.[47]
[b] Assumptions: see text and reference 47.
[c] Shrinkage corrections not applied except where noted.
[d] Standard errors, including both random and systematic errors.
[e] Dependent variable.
[f] Mean radial puckering coordinate of the pseudorotation derived by assuming a C...C shrinkage of 0.002 Å.

group, and all CCH angles were taken to be equal. Envelope (**9**, C_s symmetry, $\phi = 0°$) and half-chair (**10**, C_2 symmetry, $\phi_1 = 5°$ and $\phi_2 = 9°$) models gave equally excellent fits to the data. The experimental diffraction intensities were sufficiently sensitive to the puckering displacement, nevertheless, that the mean value, $\langle q \rangle$, could be determined by comparing the mean C—C bonded and C...C nonbonded distances.

$$\textbf{9}\,(C_s) \qquad\qquad\qquad \textbf{10}\,(C_2)$$

The C—C bond length in cyclopentane [1.5460(12) Å] is longer than that in cyclohexane (1.536 Å) or *n*-alkanes of comparable molecular weight (1.531 Å, Table 5-7). The lengthening is certainly ascribable in part to differences in 1,3-C...C nonbonded interactions; the mean 1,3-C...C distance in cyclopentane (2.444 Å) is about 0.1 Å shorter than in the model compounds mentioned above. In addition to the $\langle q \rangle$ reported in Table 5-17 ($\langle q \rangle = 0.427$ Å), a mean equilibrium value, q_e, was also estimated ($q_e = 0.438$ Å).

Cyclohexane[48–54]

Six-membered carbocyclic rings occur commonly in nature, because they are strain-free or essentially so. The conformational behavior of cyclohexane and its derivatives have been studied in great detail.[19] Sachse[55] first recognized in 1890 that cyclohexane might exist free of bond angle strain in two conformational forms, the "rigid" form and the "flexible" form. Subsequent experimental and theoretical studies have elaborated upon Sachse's analysis, and it is now generally agreed that there are two stable conformations of cyclohexane, the chair form (**11**, original rigid form) and the twist–boat or skew–boat form (**12**, one of the flexible conformations).[19] The twist–boat conformer lies approximately 21–25 kJ mol^{-1} higher in energy than the chair conformer and is not

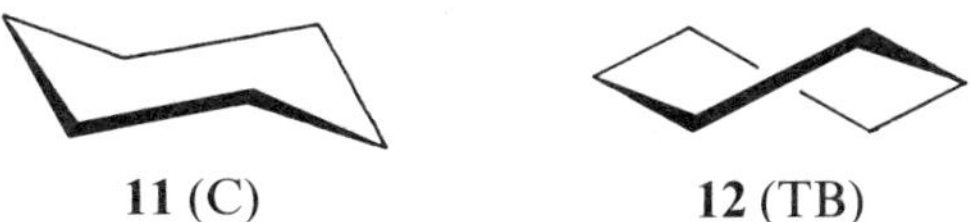

$$\textbf{11}\,(C) \qquad\qquad\qquad \textbf{12}\,(TB)$$

populated to a significant extent at room temperature.[19] Electron diffraction played a key role in identifying the chair conformation as the dominant conformer. It is also noteworthy in this regard that Odd Hassel shared the Nobel Prize in chemistry in 1969 for his experimental contributions to conformational analysis in general and his work on cyclohexane and related molecules in particular.

In 1973 a joint publication[48] of independent studies of cyclohexane by groups in Japan and Norway provided the best structural description to date of the chair conformation. There was an alarming lack of agreement among several earlier electron diffraction studies of cyclohexane.[50–54] Assuming D_{3d} symmetry, the geometry of the chair conformer can be represented in terms of five independent structural parameters: the average C—H distance, the C—C bond length, the CCC bond angle, the HCH bond angle, and the torsional angle τ for the C—C bonds. Although the two structure determinations were carried out independently, both groups used the same corrections for thermal vibrations. There was excellent agreement between the parallel investigations, and the average results are reported in Table 5-18.

The C—C bond in cyclohexane (1.536 Å) is probably shorter than the average C—C bond length in *n*-hexadecane (1.542 Å), although the outcome of the comparison depends on the assigned uncertainties. The C—C bonds in cyclohexane might reasonably be shorter, since they are free from certain nonbonded repulsions that exist in gauche sites in *n*-alkanes. Cyclohexane's CCC bond angle (111.4°) is significantly smaller than the average CCC angle in *n*-hexadecane (114.6°). Since the CCC angle is greater than the tetrahedral value ($\angle$HCH is less), however, the torsional angles are less than 60° ($\tau = 54.9°$). The major effect of the deviation of the carbons from perfect tetrahedral geometry is to flatten the ring somewhat. This tilts the axial hydrogens outward a few degrees, causing them to deviate from a parallel alignment.

At least one attempt has been made to detect the twist–boat conformation of cyclohexane at elevated temperatures. Montgomery and Wilson[56,57] collected electron diffraction data for cyclohexane at four different temperatures, ranging from 700 to 900°C. Combined molecular mechanics–statistical thermodynamics calculations indicated that the twist–boat conformer should be present to the extent of 15-23% at these temperatures.[57] Efforts to analyze the data as chair–twist–boat mixtures were unsuccessful. Among other problems, the diffraction data were insensitive to the molar composition. The uncertainties (3σ) were comparable to the twist–boat percentages.

Following a logic similar to that proposed by Allinger and Freiberg[58] in the classic stereochemical study of *trans*-1,3-di-*tert*-butylcyclohexane, Schubert,

Table 5-18. Cyclohexane Structural Parameters[a–c]

Parameter	r_g, θ_α	l_{ij}
C—H, Å	1.121(4)	0.051(2)
C—C, Å	1.536(2)	0.079(5)
$\angle$CCC, deg	111.4(0.2)	
$\angle$HCH, deg	107.5(1.5)	
τ, deg	54.9(0.4)	

[a] Average results of Bastiansen et al.[48] See also Ewbank, Kirsch, and Schäfer.[49]
[b] Assumptions: D_{3d} symmetry, see text.
[c] Estimated uncertainties include systematic errors.

Southern, and Schäfer[59] have obtained electron diffraction evidence that
cis-1,4-di-*tert*-butylcyclohexane exists in both chair and twist–boat forms, with a
probable composition of about one-third chair and two-thirds twist–boat forms.
A novel, combined electron diffraction–molecular mechanics approach was used
in the analysis. No structural parameters were obtained.

Cycloheptane,[60] Cyclooctane,[61,62] and Cyclodecane[63]

Cycloalkanes of a medium ring size present formidable problems to the electron
diffractionist. Experimental and molecular mechanics investigations indicate
that each cycloalkane can exist in several different conformations.[64] As the
number of carbons increases, it even becomes difficult to compile a list of
possible conformations systematically. One quantity the electron diffraction
experiment can obtain without auxiliary assumptions is average structural
parameters.

Average structural results for cycloheptane,[60] cyclooctane,[61] and cyclodecane[63] are shown in Table 5-19. The most striking feature of the results is the
similarity of structural parameters among the three systems. The C—C bond
lengths are particularly interesting (1.540–1.545 Å). The average C—C bond
length in *n*-hexadecane falls in the middle of this range (1.542 Å, see *n*-
hexadecane discussion). The CCC bond angles in cyclooctane and cyclodecane
are larger than the average CCC value for *n*-alkanes (again, see *n*-hexadecane).
This modest increase probably reflects the efforts of the two cycloalkanes to
remove unfavorable nonbonded interactions imposed by ring formation.

To obtain detailed structures for medium-sized rings is a much more difficult
task. The 1973 study of cyclodecane by Hilderbrandt, Wieser, and
Montgomery[63] illustrates the problems involved. Prior X-ray crystallographic
investigations of cyclodecane derivatives showed that the boat–chair–boat
(BCB, **13**) conformation is strongly preferred.[63] Consequently, the electron

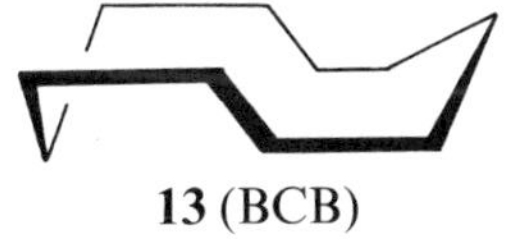

13 (BCB)

diffraction data were refined in terms of a simplified BCB model of C_{2h}
symmetry. The fit was excellent by all the usual statistical criteria. Even so, it
was conceivable that the agreement with the BCB model was fortuitous. A
number of the cyclodecane conformations are very similar in structure. To
explore this possibility, an alternative analytical approach was adopted in which
complete molecular mechanics structures and selected thermodynamic properties were calculated for 12 plausible conformations of cyclodecane. Theoretical
radial distribution curves were generated for the four conformations of lowest
energy. Least-squares analysis of the radial distribution curve utilizing the four
theoretical curves indicated that the BCB conformation represented only half

Table 5-19. AVERAGE CYCLOALKANE STRUCTURAL PARAMETERS[a,b]

Parameter	Cycloheptane[c]	Cyclooctane[d]	Cyclodecane[e]
C—H, Å	1.118(3)	1.116(2)	1.115(3)
l(C—H), Å	0.082(4)	0.081(2)	0.078(2)
C—C, Å	1.540(1)	1.540(1)	1.545(3)
l(C—C), Å	0.058(2)	0.051(1)	0.054(2)
∠CCC, deg	—	116.4	116.1(1.1)
∠CCH, deg	—	108.6	108.7(1.0)

[a] Bond lengths r_g values, bond angles r_α representations.
[b] Uncertainties 3σ.
[c] Results of Dillen and Geise.[60]
[d] Results of Dorofeeva et al.[61]
[e] Results of Hilderbrandt, Wieser, and Montgomery.[63]

the conformers present. The other main component was the twist–boat–chair ($\sim 35\%$). Complicating the picture further, several new cyclodecane conformations have been proposed since this work was published.[64]

In comparing the two methods of analysis, it is clear that extreme caution must be exercised in dealing with such systems. Furthermore, it is unlikely that electron diffraction alone can provide reliable structural information without the aid of auxiliary techniques. This is true even if the cycloalkane exists predominantly in a single conformation. Some other method must be employed to rule out the possibility of minor components. It also seems that if approaches assisted by molecular mechanics (such as the method outlined above) are to yield meaningful information about vapor-phase composition, the theoretical conformational models must be judiciously selected, and the systematic errors introduced by the application of this procedure must be carefully evaluated.

An elaborate conformational analysis of cyclooctane, based on molecular mechanics calculations, NMR spectroscopic results, and the X-ray crystal structures of cyclooctane derivatives reveals that there are numerous conformations of cyclooctane that fall into three families (boat–chair, chair–chair, and boat–boat).[64] A 1966 electron diffraction study of cyclooctane by the Oslo group[62] failed to model the data with a single conformation. Nine possible conformations were considered. More recently, Dorofeeva and co-workers[61] have reinvestigated cyclooctane by considering three single conformers and one conformational mixture. The authors conclude that the boat–chair conformer (BC, **14**) is the exclusive or strongly predominant form at 59°C in the vapor phase. Detailed structural parameters for the BC conformation were obtained. A word of caution is in order, however. A recent critical discussion of the absolute

14 (BC)

entropy of cyclooctane[65] points out that the entropy cannot be calculated satisfactorily from the BC conformation and the reported experimental frequencies of vibration. Agreement with the experimental entropy can be achieved by (a) introducing some unobserved low frequencies or (b) mixing several conformers with small energy differences.

Dillen and Geise[60] have investigated the structure of cycloheptane. The diffraction intensities were analyzed by assuming the existence of two conformations, the twist–chair and the chair. The twist–chair conformation accounted for 76(6)% of the conformers present. The validity of the analysis procedure is open to some question; it has been pointed out[64] that the chair form most likely corresponds (molecular mechanics calculations) to an energy maximum.

Acknowledgment

The author gratefully acknowledges the help of Cheryl Johnson and Martin P. Grendze in preparing this chapter.

References

1. Robiette, A. G. In "Molecular Structures by Diffraction Methods", Sim, G. A.; Sutton, L. E., Eds.; Specialist Periodical Report, The Chemical Society: London, 1973, Part 1, Chap. 4, pp. 160 197.
2. Kuchitsu, K. In "Diffraction Studies on Non-Crystalline Substances", Haigittai, I.; Orville-Thomas, W. J., Eds.; Elsevier: Amsterdam, 1981, pp. 63–116.
3. Kuchitsu, K.; Cyvin, S. J. In "Molecular Structures and Vibrations", Cyvin, S. J., Ed.; Elsevier: Amsterdam, 1972, Chap. 12, pp. 183–211.
4. Bartell, L. S.; Kuchitsu, K.; deNeui, R. J. *J. Chem. Phys.* **1961**, *35*, 1211.
5. Bartell, L. S.; Kuchitsu, K. *J. Chem. Phys.* **1978**, *68*, 1213.
6. Fink, M. Unpublished results cited in reference 5.
7. Gray, D. L.; Robiette, A. G. *Mol. Phys.* **1979**, *37*, 1901.
8. Hirota, E. *J. Mol. Spectrosc.* **1979**, *77*, 213.
9. See, eg, the elegant experiments of: Mislow, K.; Graeve, R.; Gordon, A. J.; Wahl, G. H., Jr. *J. Am. Chem. Soc.* **1963**, *85*, 1199.
10. Bartell, L. S.; Higginbotham, H. K. *J. Chem. Phys.* **1965**, *42*, 851.
11. Kuchitsu, K. *J. Chem. Phys.* **1968**, *49*, 4456.
12. Iijima, T. *Bull. Chem. Soc. Japan* **1973**, *46*, 2311.
13. Iijima, T. *Bull. Chem. Soc. Japan* **1972**, *45*, 1291.
14. Lide, D. R., Jr. *J. Chem. Phys.* **1960**, *33*, 1514.
15. Heenan, R. K.; Bartell, L. S. *J. Chem. Phys.* **1983**, *78*, 1270.
16. Bradford, W. F.; Fitzwater, S.; Bartell, L. S. *J. Mol. Struct.* **1977**, *38*, 185.
17. Bonham, R. A.; Bartell, L. S. *J. Am. Chem. Soc.* **1959**, *81*, 3491.
18. Kuchitsu, K. *Bull. Chem. Soc. Japan* **1959**, *32*, 748.
19. Eliel, E. L.; Allinger, N. L.; Angyal, S. J.; Morrison, G. A. "Conformational Analysis". Wiley-Interscience: New York, 1965.
20. Schachtschneider, J. H.; Snyder, R. G. *Spectrochim. Acta* **1963**, *19*, 117.
21. Hilderbrandt, R. L. *J. Mol. Spectrosc.* **1972**, *44*, 599.
22. Fitzwater, S.; Bartell, L. S. *J. Am. Chem. Soc.* **1976**, *98*, 5107
23. Bonham, R. A.; Bartell, L. S.; Kohl, D. A. *J. Am. Chem. Soc.* **1959**, *81*, 4765.

24. Fitzwater, S.; Bartell, L. S. *J. Am. Chem. Soc.* **1976**, *98*, 8338.
25. Compton, D. A. C.; Montero, S.; Murphy, W. F. *J. Phys. Chem.* **1980**, *84*, 3587.
26. Bartell, L. S.; Kohl, D. A. *J. Chem. Phys.* **1963**, *39*, 3097.
27. Kitaigorodsky, A. I. "Molecular Crystals and Molecules". Academic Press: New York, 1973, pp. 48-62.
28. Hilderbrandt, R. L.; Wieser, J. D. *J. Mol. Struct.* **1973**, *15*, 27.
29. Bartell, L. S. *J. Chem. Educ.* **1968**, *45*, 754.
30. Lide, D. R., Jr. *J. Chem. Phys.* **1960**, *33*, 1519.
31. Bartell, L. S.; Bradford, W. F. *J. Mol. Struct.* **1977**, *37*, 113.
32. Beagley, B.; Brown, D. P.; Monaghan, J. J. *J. Mol. Struct.* **1969**, *4*, 233.
33. Bartell, L. S.; Boates, T. L. *J. Mol. Struct.* **1976**, *32*, 379.
34. Jacob, E. J.; Thompson, H. B.; Bartell, L. S. *J. Chem. Phys.* **1967**, *47*, 3736.
35. Allinger, N. L. *J. Am. Chem. Soc.* **1977**, *99*, 8127.
36. Burgi, H. B.; Bartell, L. S. *J. Am. Chem. Soc.* **1972**, *94*, 5236.
37. Bartell, L. S.; Burgi, H. B. *J. Am. Chem. Soc.* **1972**, *94*, 5239.
38. Yamamoto, S.; Nakata, M.; Fukuyama, T.; Kuchitsu, K. *J. Phys. Chem.* **1985**, *89*, 3298.
39. Bastiansen, O.; Fritsch, F. N.; Hedberg, K. *Acta Crystallogr.* **1964**, *17*, 538.
40. Coulson, C. A.; Moffitt, W. E. *Phil. Mag.* **1949**, *40*, 1.
41. Stevens, R. M.; Switkes, E.; Laws, E. A.; Lipscomb, W. N. *J. Am. Chem. Soc.* **1971**, *93*, 2603.
42. Kuchitsu K. *J. Chem. Phys.* **1966**, *44*, 906.
43. Egawa, T.; Yamamoto, S.; Fukuyama, T.; Kuchitsu, K. *Abstracts*, Eleventh Austin Symposium on Gas-Phase Molecular Structure, Austin, TX, 1986, p. 93.
44. Takabayashi, F.; Kambara, H.; Kuchitsu, K. *Abstracts*, Seventh Austin Symposium on Gas-Phase Molecular Structure, Austin, TX, 1978, p. 63.
45. Dunitz, J. D.; Schomaker, V. *J. Chem. Phys.* **1952**, *20*, 1703.
46. Almenningen, A.; Bastiansen, O.; Skancke, P. N. *Acta Chem. Scand.* **1961**, *15*, 711.
47. Adams, W. J.; Geise, H. J.; Bartell, L. S. *J. Am. Chem. Soc.* **1970**, *92*, 5013.
48. Bastiansen, O.; Fernholt, L.; Seip, H. M.; Kambara, H.; Kuchitsu, K. *J. Mol. Struct.* **1973**, *18*, 163.
49. Ewbank, J. D.; Kirsch, G.; Schäfer, L. *J. Mol. Struct.* **1976**, *31*, 39.
50. Pauling, L. Brockway, L. O. *J. Am. Chem. Soc.* **1937**, *59*, 1223.
51. Hassel, O.; Ottar, B. *Arch. Math. Naturvidensk.* **1942**, *345*(10).
52. Davis, M.; Hassel, O. *Acta Chem. Scand.* **1963**, *17*, 1181.
53. Alekseev, N. V.; Kitaigorodsky, A. I. *Zh. Strukt. Khim.* **1963**, *4*, 163.
54. Geise, H. J.; Buys, H. R.; Mijlhoff, F. C. *J. Mol. Struct.* **1971**, *9*, 447.
55. Sachse, H. *Chem. Ber.* **1890**, *23*, 1363.
56. Montgomery, L. K.; Wilson, C. A. Unpublished results.
57. Wilson, C. A. Ph.D. thesis, Indiana University, 1984.
58. Allinger, N. L.; Freiberg, L. A. *J. Am. Chem. Soc.* **1960**, *82*, 2393.
59. Schubert, W. K.; Southern, J. F.; Schäfer, L. *J. Mol. Struct.* **1973**, *16*, 403.
60. Dillen, J.; Geise, H. J. *J. Chem. Phys.* **1979**, *70*, 425.
61. Dorofeeva, O. V.; Mastryukov, V. S.; Allinger, N. L.; Almenningen, A. *J. Phys. Chem.* **1985**, *89*, 252.
62. Almenningen, A.; Bastiansen, O.; Jensen, H. *Acta Chem. Scand.* **1966**, *20*, 2689.
63. Hilderbrandt, R. L.; Wieser, J. D.; Montgomery, L. K. *J. Am. Chem. Soc.* **1973**, *95*, 8598.
64. Burkert, U.; Allinger, N. L. "Molecular Mechanics". American Chemical Society: Washington, DC, 1982, Chap. 4.
65. Dorofeeva, O. V.; Gurvich, L. V.; Mastryukov, V. S. *J. Mol. Struct.* **1985**, *129*, 165.

6

UNSATURATED ORGANIC MOLECULES

Marit Traetteberg

Department of Chemistry
College of Arts and Sciences
The University of Trondheim
Dragvoll, Norway

CONTENTS

INTRODUCTION 239
COMPOUNDS WITH C=C BONDS 242
   n-Alkenes 242
   Substituent Effects 245
   Sterically Strained C=C Bonds 253
   Conjugated Double Bonds 260
   Cross-Conjugated C=C Bonds 268
   Cumulated C=C Bonds 269
COMPOUNDS WITH C≡C BONDS 271
   Aliphatic Alkynes 271
   Cyclic Alkynes 271
INTRAMOLECULAR INTERACTIONS INVOLVING C=C OR C≡C BONDS 273
ACKNOWLEDGMENTS 275
REFERENCES 275

Introduction

In this chapter it is assumed that unsaturated organic compounds are compounds containing at least one C=C or C≡C bond. Aromatic compounds and compounds containing π bonds involving heteroatoms are therefore included only insofar as they are of interest because of their unsaturated carbon–carbon bonds.

An unsaturated carbon–carbon bond will always have a profound influence on the conformation of a molecule. For a C=C bond this is due to π bonding, which normally prohibits rotation around this type of bond, because of the accompanying increase in energy. The four substituents at a C=C bond are therefore mutually locked relative to each other. A C$\equiv$C bond has two π bonds, but because of their mutual interaction the energy is insensitive to rotation around the C$\equiv$C bond. The C$\equiv$C—C bonding angles of 180° do, however, give rise to a linear fragment of about 4.10 Å, which is of great importance for the overall conformation of a molecule, because this fragment is very much longer than the other fragments (bonds) in organic molecules, typically 1.50 Å.

Today it is well known that stereochemistry is of the utmost importance for molecular properties. The relationships between molecular properties and stereochemistry of C=C bonds are known and understood in some instances, but in most cases they still remain speculative. Among the earliest known examples are the *cis-* and *trans*-1,4-polymers of isoprene (CH_2=C(CH_3)CH=CH_2), named rubber and guttapercha, respectively. The former is an elastic polymer, while the latter is hornlike. Hydrogenation of the two polymers gives products that are indistinguishable.

Stereochemically well-defined C=C bonds play a vital part in many biological reactions, for example, in redox processes. It is therefore not surprising that unsaturated C=C bonds often are incorporated in pharmaceuticals.

The sterically crowded drug tamoxiphene (**1**) is, for example, widely used in the treatment of certain types of cancer. It seems very likely that the stereochemistry of the compound, in its interaction with a cancer cell, is essential for the

Et, C_6H_5 / H_5C_6, C_6H_4-p-(O-CH_2CH_2-NMe_2)

1

functioning of the drug. The tamoxiphene molecule is too large to be studied by gas-phase electron diffraction (GED), but the study of smaller model compounds might give insight that will eventually contribute to explaining the biochemical reactions taking place.

What are the possibilities of determining the structure of an unsaturated group accurately by GED?

The contribution from an intramolecular distance r_{ij} to the scattered electron intensities (see also Chapter 1 of Part A) is approximately proportional to:

$$\frac{Z_i Z_j}{r_{ij}} \exp\left(\frac{-l_{ij}^2 \cdot s^2}{2}\right) \cdot \sin(s \cdot r_{ij}) \qquad (6\text{-}1)$$

where Z_i represents the atomic number of atom i, s is proportional to the scattering angle, and l_{ij} represents the average deviation from the equilibrium or average distance, r_{ij}. A small value for l_{ij} implies that the distance r_{ij} is fairly rigid. In such cases the exponential term will decrease more slowly than otherwise.

Compared with saturated carbon–carbon bonds, C=C and C≡C bonds are short and rigid; ie, they have relatively small r_{ij} and l_{ij} values. A decrease in r_{ij} will give an increase in the amplitude of the sine wave, $\sin(s \cdot r_{ij})$, while a decrease in l_{ij} will reduce the damping of the wave. Therefore C=C and C≡C bonds should be ideally suited to be the subjects of GED studies.

However, in practice problems often arise because the molecules of interest may contain relatively few unsaturated bonds compared to the number of saturated ones. For example, if one wants to study the effect of steric crowding on a C=C bond, as in **2**, the saturated CC bonds outnumber the C=C bond by 12 to 1. Electron scattering from intramolecular distances between the isopropyl groups across the double bond do, however, give additional information on the C=C bond, and will thereby diminish the problems due to abundance.

2

If the four substituents are of the same kind, as in the foregoing example, it is normally possible to determine the structure to a high degree of precision. The greatest difficulties will arise if the four substituents are different, but with qualitatively similar scattering power, as for example with a mixture of primary, secondary, and tertiary alkyl groups. If, additionally, several conformers have comparable energies and appear simultaneously, the complexities may be so overwhelming that the molecular structure might not be solved by GED alone. In such cases the combined use of several methods (GED, vibrational spectroscopy, molecular mechanics, and/or *ab initio* calculations) may enable one to elucidate the molecular structure and conformational composition. The accuracy of the determined structural parameters will of course be less than is possible for smaller, conformationally homogeneous molecules.

There are several ways of defining an internuclear distance r_{ij} between two atoms i and j. To discuss small bond length differences in terms of physical effects such as electronegativity, conjugation, and sterical strain, the bond lengths being discussed must be defined in the same way.

In electron diffraction studies bond lengths are normally presented as r_g or r_a values, where r_g corresponds to the position of the center of gravity of the probability function $P(r)$ and is the average value of the internuclear distance, and r_a is the distance obtained directly from the electron diffraction data. Normally r_a is closer than r_g to the equilibrium internuclear distance r_e, which usually is not experimentally obtainable. The relationship between r_g and r_a is given to a good approximation by:

$$r_g \approx r_a + \frac{l^2}{r_a} \qquad (6\text{-}2)$$

In the present chapter all bond lengths are given as r_a values. If a paper presents results only as r_g data, the above formula is used to transfer them to an r_a basis.

Uncertainties are most often given as standard deviations or estimated error limits.

Compounds with C=C Bonds

The stereochemical characteristics of an alkene or alkene derivative are primarily the C=C bond length, the valence angles at the double bond, and the local conformational properties based on the =C—Y— dihedral angles, where Y represents a substituent atom. In addition intramolecular interactions between the π electrons of the C=C bond and a heteroatom or group of atoms in other parts of the molecule may influence the conformation. Relatively little is known about the latter subject, which will be treated separately in the last section of this chapter.

n-Alkenes

In *n*-alkenes the C=C bond may be considered to be only moderately influenced by its substituents. The double-bond moiety will therefore be expected to assume its "normal" geometry, corresponding to minimum steric energy.

Table 6-1 shows geometrical parameters for some *n*-alkenes and structurally related molecules that have been studied by GED.

The 1-alkenes are found to have C=C bond lengths indistinguishable from that in ethylene.[1] In alkenes with nonterminal double bonds there is a slight, but not significant, tendency toward an increase in the C=C bond length. This tendency vanishes, however, if the 2-butene structures, which have not been corrected for shrinkage effects, are omitted. Based on the available GED structures, one might therefore conclude that the C=C length in *n*-alkenes appear to be the same as that in ethylene.[1]

The C=C—C valence angle in 1-*n*-alkenes and nonterminal *trans* *n*-alkenes is typically found to be 3-4° larger than the C=C—H angle in ethylene.[1] If conformational strain is present, as for example in the less abundant syn conformer of 1-butene, there is an additional 2-3° increase in the C=C—C

Table 6-1. *n*-ALKENES STUDIED BY GED

Molecule	r_a (Å)		∠C=C—C (deg)	Shrinkage corrected	Ref.
	C=C	C—C			
	1.336(2)			+	1
	1.335(3)	1.502(4)	124.3(8)	+	2
	1.339(4)	1.500(2)	125.6(3), gauche 127.2(3), syn	+	3
	1.336(2)	1.505(2)	125(2)	+	4
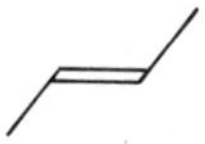	1.347(3)	1.508(2)	123.8(4)	−	5
	1.346(3)	1.506(2)	125.4(4)	−	5
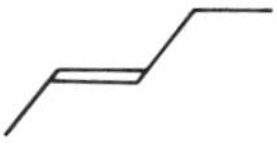	1.334(2)	1.484(1)	125.4(3)	+	6
	1.338(2)	1.490(1)	127.4(2)	+	7
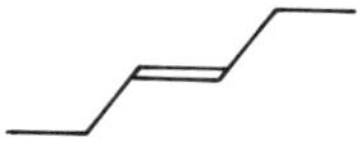	1.342(6)	1.506(2)	124.0(11), gauche 127.0(11), syn	+	8
	1.330(7)	1.505(2)	129.1(8)	+	8

angle. Steric crowding due to cis configuration at the double bond is also found to open the angle by 2–5° compared to the angle in the corresponding trans isomer. Syn conformers of *cis-n*-alkenes (alkyl groups larger than methyl) would exhibit configurational as well as conformational sterical strain. However, such conformers have not been observed by GED.

244   MARIT TRAETTEBERG

The C=C—H valence angles are normally determined with low precision in GED studies, due to the small scattering power of hydrogen, combined with the fact that most often there are many overlapping distances in the r region where the necessary information is stored. An exception is Kuchitsu's determination of the C=C—H angle in ethylene.[1]

The conformational preferences for some selected molecules from Table 6-1 are illustrated in Table 6-2. The most stable conformation of a saturated atom adjacent to an unsaturated C atom is generally found to have one of the former atom's bonds syn-eclipsed with the double bond. This is in accordance with a threefold rotational contribution to the torsional potential energy.

In alkenes with alkyl double-bond substituents larger than methyl, C—H as well as C—C bonds are observed to eclipse the double bond. The former

Table 6-2. Conformational Preferences in n-Alkenes, as Determined by GED

Molecule	Conformers (%)	θ, C=C—C— (deg)	Ref.
	H, H—⊖≡H, H	0	2
	CH₃, H—⊖≡H, H [83(10)] H, H—⊖≡CH₃, H [17(10)]	119.9(3) 0 (assumed)	3
	H, H—⊖≡H, H	0	5
	H, H—⊖≡, H	20.6(15.4)	5
	+ac, +ac (44) −ac, +ac (42) ac, sp (14)	120, 120 −120, 120 120, 0	8
	+ac, +ac (51) −ac, +ac (49)	120, 120 −120, 120	8

arrangement is, however, energetically favored, as is demonstrated in the study of 1-butene (83% gauche (*ac*); 17% syn (*sp*))[3]. The more abundant gauche conformer is also favored by entropy, and when this is corrected for, an enthalpy difference in favor of *gauche* of 2.22(1.75) kJ mol^{-1} is determined. The lower preference for C—C compared to C—H syn eclipsing at a C=C bond may primarily be ascribed to increased van der Waals repulsion. This is reflected by increased C=C—C angles for syn conformers as well (see Table 6-1).

When two alkyl groups are cis-substituted at the double bond, syn eclipsing at both =C—C bonds gives a nonbonded internuclear distance between the syn-oriented atoms that is smaller than the sum of their van der Waals radii. In *cis*-2-butene a deviation from syn eclipsing of about 20° is observed, but with an uncertainty in the determined =C—C— dihedral angle of the same order of magnitude.[5]

When two larger alkyl groups are substituted at the double bond, as in for example *cis*- or *trans*-3-hexene, the complexity of the structural problem is so greatly increased that one can no longer expect to accurately determine the =C—C— dihedral angles of the various conformers from GED data alone. In their study of these molecules van Hemelrijk and co-workers[8] very reasonably assumed the =C—C— dihedral angles to be ± 120 or 0°, corresponding to syn eclipsing of C—H or C—C bonds, respectively. In *trans*-3-hexene they found C—H syn eclipsing to the double bond to be clearly favored, in agreement with the results observed for 1-butene.[3] They found, however, 14% of an (*ac*, *sp*) conformer, with one C—C bond syn-eclipsed, while a possible (*sp*, *sp*) conformer was found to contribute less than 1%.

In the cis isomer no contribution from conformers with C—C bond syn eclipsing was observed.[8] This is hardly surprising, since C—C syn eclipsing in this case would have involved nonbonded internuclear distances of the same order of magnitude as bond distances.

Substituent Effects

The Effects of Halogen Substituents. Halogenated alkenes have been fairly extensively and systematically studied by GED. Table 6-3 shows geometrical parameters determined for halogen-substituted ethenes.

It is well known that a fluorine atom substituted at a carbon atom causes shortening of the other bonds in which the carbon atom takes part.[42] This is observed for C—Cl, C—Br, and C—C, as well as for C=C bonds. The C=C bond lengths shown in Table 6-3 are in accordance with these considerations. Fluorine derivatives are described elsewhere in this book (Chapter 4), so they are not further discussed here.

Chlorinated or brominated C=C bond lengths appear to be little affected by electronegative substituents. If, however, two halogen substituents are cis-oriented across the double bond, slightly increased C=C bond lengths are observed. With cis substitution at both sides of the double bond, this tendency

Table 6-3. HALOGEN-SUBSTITUTED ALKENES, (X = F, Cl, Br) AS STUDIED BY GED

Molecule	r_a C=C (Å)			∠C=C—X (deg)			Ref.		
	F	Cl	Br	F	Cl	Br	F	Cl	Br
	1.328(18) 1.332(1)	1.340(4)	1.347(8)	121.5(2) 121.0(2)	122.5(3)	122.8(8)	9 14	10	11
	1.339(6) 1.315(2)	1.333(5)	—	124.7(3) 125.3(3)	122.9(3)	—	12 14	13	
	1.328(4) 1.309(8)[a]	1.335(6)	—	119.3(2) 122.5(2)	120.7(3)	—	14 18	16	
	1.330(4) 1.328(11)	1.354(5) 1.345(6)	1.36(3)[b]	123.7(8) 122.0(2)	123.8(5) 123.8(2)	124(2)	14 21	160 15	17
	1.308(2) 1.340(12)	—	—	118.8(8)–125.4(3) 120(1)–124(1)	—	—	14 22		
	1.309(3)	1.354(3)	1.362(9)	123.8(2)	122.2(2)	122.4(3)	14	19	20

[a] Unusually large l values applied (0.56 and 0.57 Å)
[b] Unidentified, possibly r_a.

becomes more evident. The increase might merely be a result of increased sterical strain across the double bond, but since the effect is clearly more pronounced than for 1,2-*cis*-alkylated double bonds, it may reflect more complicated interactions between the orbitals of the halogen atom and the π bond.

The C=C—X valence angles presented in Table 6-3 are generally found to be somewhat smaller than comparable C=C—C angles. This shows that the ideal, or minimum energy, C=C—X angle is 1–2° smaller than the corresponding C=C—C angle.

Table 6-4 shows C=C—X and C=C—C angles for molecules containing $\mathrm{C{=}C{\diagdown}^{C}_{X}}$ fragments. If the last two entries (with more complicated stereochem-istry) are omitted, the average difference between observed C=C—C and

Table 6-4. 1.1-DISUBSTITUTED ETHENES, AS STUDIED BY GED

		Angles (deg)			
Molecule	r_a, C=C (Å)	∠C=C—C	∠C=C—X	Δ (deg)	Ref.
$H_2C=C(CH_2Cl)(Cl)$	1.334(4)	127.6(11)	122.2(10)	5.4	23
$H_2C=C(CH_2Cl)(Br)$	1.360(14)	127.0(10)	122.1(26)	4.9	24
$H_2C=C(CH_2Br)(Br)$	1.333(13)	124.2(17)	120.1(44)	4.1	25
$H_2C=C(CH_2F)(Cl)$	1.338(6)	125.6(6)	123.0(7)	2.6	26
$H_2C=C(CH_3)(Cl)$	1.338(3)	126.3(2)	119.0(3)	7.3	27
$H_2C=C(CH_3)(Br)$	1.343(4)	126.9(3)	118.3(5)	8.6	27
$Cl_2C=C(\triangle)(Cl)$	1.353	126.4(14)	117.6(9)	8.8	28
$Cl_2C=C(CF_2H)(Cl)$	1.365(18)	124.5(9)	120.1(10)	4.4	29

C=C—X angles is 5.5°. These differences are clearly larger than the "natural" differences of 1–2° (see above). The observed increases in C=C—C/C=C—X angle differences are in agreement with the relative sizes of the bending force constants of the angles involved. Representative values for the C=C—C, C=C—Cl, and C=C—Br binding force constants are for example given by Stølevik and Thingstad,[43] as 0.70, 0.130, and 0.120 mdyn Å $(rad)^{-2}$.

The simplest halogenated alkenes that show conformational isomerism are 3-substituted propenes. Table 6-5 presents observed conformational preferences for some 3-haloalkenes. 3-Chloro-[30] and 3-bromo-propene[33] do both clearly prefer a gauche conformation. Explanations may focus on expected increases in van der Waals energy and/or electrostatic repulsion between the electronegative halogen atom and the π electrons for syn compared to gauche conformers. However, since 3-halopropenes do adopt the syn conformation in measurable amounts, the syn position cannot be seriously hindered.

In the following discussion the terms syn and gauche refer to the relative orientation of a $C_1=C_2$ bond and a C_3—X bond.

If the hydrogen in 2 position of $CH_2=CH—CH_2X$ is substituted by a larger atom or group of atoms, the gauche conformer may be influenced by interactions between the substituents (X and Y in Table 6-5). 2-Methyl-substitution seems to further stabilize gauche relative to syn conformation.[31,34] This indicates that there are weak attractive forces between the halogen atom and the

Table 6-5. CONFORMATIONAL PREFERENCES IN $H_2C=C$ (with CH_2X and Y substituents) MOLECULES, AS STUDIED BY GED

X	Y	r_a, C=C (Å)	% syn	% gauche	θ, =C—C—X (deg)	Ref.
Cl	H	1.341(5)		82(9)	121(5)	30
Cl	CH_3	1.340(6)		87(7)	116(3)	31
Cl	CH_2Cl	1.331(10)	34(7)	66(7)	115(3)	32
Cl	Cl	1.334(4)	55(8)		109(4)	23
Cl	Br	1.360(14)	48(11)		109.5(35)	24
Br	H	1.335(7)		86(14)	117(5)	33
Br	CH_3	1.331(9)		96(8)	113(2)	34
Br	Br	1.333(13)	46(7)		112.0(43)	25
CN	H	1.326(7)	94(18)			37

[a] Syn and gauche refer to the C=C—C—X dihedral angle.

methyl group. However, it should be noted that the 2-methyl group causes the =C—C—X dihedral angle in 3-chloro- and 3-bromo-propene to decrease by 5(5) and 4(5)°, respectively, thereby increasing the methyl halogen distance. If the latter observation is considered separately, it might be interpreted in terms of repulsive sterical interactions between methyl and halogen. The simultaneous increase in gauche population does, however, refute an explanation in these terms, and rather suggests that the observed decrease in =C—C—X dihedral angle is a result of a balance between optimum van der Waals interaction and torsional energy.

If the 2-position in a 3-halopropene molecule is substituted with another halogen atom, syn becomes the energetically favored conformer.[23–25] This is hardly surprising, considering the expected electrostatic repulsion between the negative ends of the two C—X dipoles.

In 3-butenonitrile[37] the syn conformation is clearly favored. No complete explanation has been found as to which effects stabilize the eclipsing of the two unsaturated groups,[44] which in this case may be characterized as homoconjugated.

Table 6-6 shows structural and conformational results for halogenated alkenes that are additionally experiencing steric strain across the double bond. If the 3-halopropenes are cis-substituted in the 1-position, the added steric strain of the syn conformer is expected to increase the conformational energy. If the substituent is a halogen atom, the syn conformer will be further destabilized because of dipole–dipole repulsion. It is therefore not surprising that cis isomers of ClHC=CHCH$_2$X (X = F, Cl) are exclusively observed as gauche conformers by GED.[35,36]

The cis-1,4-dichloro-2-butene molecule is found to be predominantly in a form in which one of the C—Cl bonds eclipses the C=C bond, while the other C—Cl bond is rotated 114° away from the C=C bond, corresponding to a syn, gauche conformation.[38] It seems likely that the inherent favorable relative orientation of the C—Cl dipoles is a decisive factor in determining the conformational preference.

It will be noted that, except for two of the fluorinated alkenes, all C=C bonds of the sterically strained molecules in Table 6-6 are longer than the C=C bond in ethylene.[1] Eventual rehybridization of the unsaturated carbon atoms in accordance with enlarged C=C—C(X) angles would give the opposite effect, because of increased s character of the orbitals forming the C=C σ bond. It is therefore likely that the observed increases in C=C bond lengths are due to bond stretching because of steric strain across the C=C bond. The fact that increases in C=C bond lengths are more noticeable for cis-haloalkenes is probably connected to the differences in C=C—C and C=C—X bending force constants as discussed above.

The Effects of Other Heterosubstituents. Considerably less information is available from GED studies of alkenes substituted with heteroatoms other than halogen.

Table 6-6. EXAMPLES OF MOLECULES AFFECTED BY ELECTRONEGATIVE SUBSTITUENTS AND BY STERIC STRAIN

Molecule	r_a, C=C (Å)	Major conformer (%)	Angles (deg)			Ref.
			θ, C=C—C—X	$\angle$C=C—C	$\angle$C=C—X, cis	
Cl—CH=CH—CH$_2$Cl	1.339(5)	gauche(100)	124(5)	125.5(15)	124.6(16)	35
Cl—CH=CH—CH$_2$F	1.347(14)	gauche($\sim$100)	143(10)	125.4(8)	122.7(6)	36
ClH$_2$C—CH=CH—CH$_2$Cl	1.350(6)	syn, gauche	5(20), 114(13)	124.8(10)		38
F$_3$C—CH=CH—CF$_3$	1.309(16)		1.2(13)	126.0(5)		39
CH$_2$=C(CF$_3$)(CF$_3$)	1.371(2)		36.9(7)	118.2(3)		40
Cl(CHF$_2$)C=C(Cl)(Cl)	1.365(18)	syn(H)		124.5(9)	123.3(6)	29
F(F)C=C(CF$_3$)(F)	1.328(3)			127.8(7)	(123.9)	41

Table 6-7 shows GED structural results obtained for alkenes substituted by oxygen or sulfur. The C=C bond is apparently only modestly influenced by these substituents while the C=C—O and C=C—S angles are observed to be larger than most of the C=C—Y angles discussed above. This observation must, however, be correlated with the conformational preferences, which are generally syn for these molecules.

In the discussion of *n*-alkenes it was pointed out that syn conformers generally have C=C—C angles that are 2–3° larger than their gauche conformational isomers. The C=C—O(S) angles presented in Table 6-7 are in most cases similar to the C=C—C angles of alkene syn conformers shown in Table 6-1. If the C=C—O(S) angle opening is determined primarily by internal steric strain, the bending force constants for C=C—O and C=C—S would be expected to

Table 6-7. GED RESULTS FOR ALKENES SUBSTITUTED BY OXYGEN OR SULFUR

Molecule	r_a, C=C (Å)	∠C=C—O(S) (deg)	Preferred conformation	Ref.
(H₃C—O—CH=CH₂)	1.341 1.336(20)	127.7(14) 127.3(18)	73(2)% syn 99(5)% syn	45 46
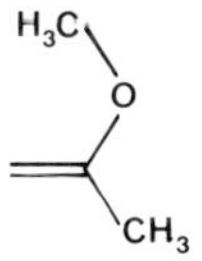 (H₃C—O—C(CH₃)=CH₂)	1.330(7)	125.8(7)[a]	≥90% syn	47
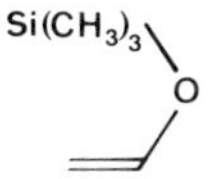 (Si(CH₃)₃—O—CH=CH₂)	1.340(ass.)	122.8(11)[b]	~65% syn[c]	48
(H₃C—S—CH=CH₂)	1.343(1) 1.342(7)	127.5(7) 127.0(15)	38(7)% syn 79–89% syn	45, 49 50
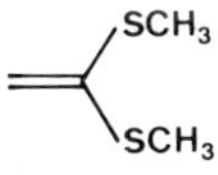 ((CH₃S)₂C=CH₂)	1.348(5)	122.4(10)	[d]	51

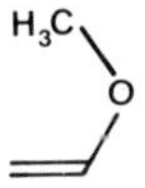

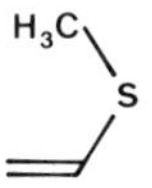

[a] ∠C=C—C: 123.9(8)°.
[b] ∠C—O—Si: 133.4(40)°.
[c] θ_{syn}: 41(14)°, 35% anti; θ_{anti}: 166(29)°.
[d] Mixture of syn, syn; syn, gauche, and gauche, gauche.

be larger and smaller, respectively, than k_{bend}(C=C—C), because of the differences in C—O, C—C, and C—S bond lengths. It is difficult to find reliable and comparable data for the actual bending force constants. The C=C—O and C=C—C bending force constants used by Schei in a recent study[47] [C=C—O: 1.27; C=C—C: 0.98 mdyn Å (rad)$^{-2}$] are, however, in accordance with this view. In his study of the two sulfur compounds in Table 6-7 Seip[49,51] used k_{bend}(C=C—S) of 0.70 mdyn Å (rad)$^{-2}$.

In n-alkenes, gauche conformation is clearly preferred over syn. 1-Butene was for example found to have 83(10)% of the molecules in gauche conformation at room temperature.[3] In oxygen-substituted alkenes the conformational preferences are very different, and there seems to be a remarkably strong preference for syn conformation. This is especially worth noticing for the sterically crowded CH_2=CH—O—Si (CH_3)$_3$ molecule, which is observed to exist in a conformational mixture of 65% of a somewhat distorted syn conformer [=C—O—: 41(14)°] and 35% anti [=C—O—: 166(29)°].[48] The steric problem of the syn conformer is eased by opening the C—O—Si angle to 133.4°.

The conformational preferences of the oxygen-substituted alkenes show these molecules to be strongly influenced by forces not present in alkenes of comparable size. It is logical to correlate these forces with the lone electron pairs of the oxygen atom. The preferred conformation corresponds to the lone-pair electrons being as far away as possible from the π electrons of the C=C bond (**3**).

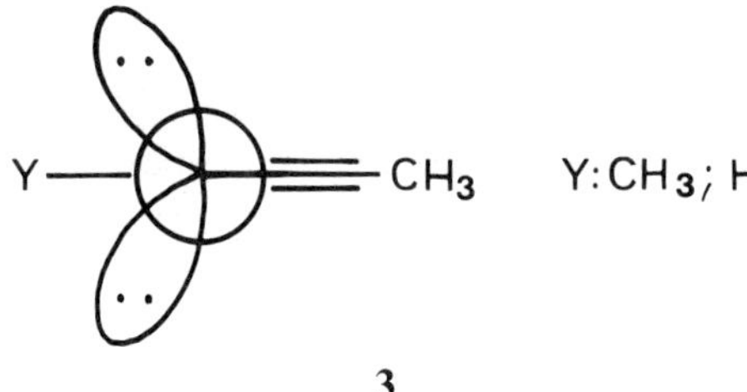

3

This so-called hetero effect is found to influence the conformational energy of sulfur-substituted alkenes as well, but to a much smaller extent, in accordance with the larger distances between the lone-pair electron densities and the π electrons in these molecules.

Table 6-8 gives GED structural results for some other heterosubstituted alkenes. In most cases the conformational situation is complicated. It is, however, of interest to note that while CH_2=CHSO$_2$CH$_3$ is observed to have 50% of the molecules with the CH_3 group syn-eclipsed to the C=C bond, the CH_3 group in CH_2=CBrSO$_2$CH$_3$ is only observed gauche to C=C. This is in accordance with an attractive interaction between CH_3 and Br (see Table 6-5). Repulsive interaction between the partly negative bromine and oxygen atoms will, however, give the same stereochemical result, and both effects are probably operative.

Table 6-8. GED RESULTS FOR SOME HETEROSUBSTITUTED ALKENES

Molecule	r_a, C=C (Å)	∠C=C—Z, (deg)	Preferred conformation	Ref.
CH₂=CH—SO₂Cl	1.357(18)	121.3(7)	[a]	52
CH₂=CH—SO₂CH₃	1.347(16)	118.4(21)	~50% syn (CH₃)[b]	53
CH₂=C(SO₂CH₃)(Br)	1.350(13)	120.9(28)[c]	gauche (CH₃)	54
(CH₂=CH)₂SO₂	1.332(4)	121.5(3)	[a]	55
CH₂=CH—PCl₂	1324(12)	124.4(21)	[d]	56
CH₂=CH—POCl₂	1.338(ass.)	117.5(27)	[a]	57
(CH₂=CH)₂P(O)Cl	1.346(9)	117.2(8)	[a]	57
H₃Si—CH=CH₂ (CH₂)	1.325(4)	125.6 (ass.)	gauche $[\theta = 102(1)°]$	58
F₃Si—CH=CH₂ (CH₂)	1333(18)	119.0(42)	gauche	59

[a] Mixture, not precisely defined.
[b] The other conformer, $\theta = 100°(CH_3)$.
[c] ∠C=C—Br: 123.0(27)°.
[d] Mixture with lone electron pairs on P anti [53(18)%] or syn relative to C=C bond.

Sterically Strained C=C Bonds

A C=C fragment may be forced to deviate from its normally optimum geometry because of steric strain. This will be the case if large groups are cis-substituted at the double bond, or if the substituents at one or both of the unsaturated carbon atoms force the angles at the double bond to deviate from their normal values.

Sterical Strain in 1,1-Disubstituted Alkenes. Table 6-9 shows GED results for alkenes with varying degrees of C=C—C angle strain. It is only in the most seriously strained molecule, bicyclopropylidene, that an unambiguous effect on the C=C bond length is observed.[64]

Bicyclopropylidene is ideally suited for a GED study because it is a nonflexible molecule with few geometrical parameters and several well-separated peaks (interatomic distances) in the radial distribution curve. This is reflected in the exceptionally low standard deviations of the geometrical parameters. The C=C bond shortening in bicyclopropylidene is probably primarily due to change in hybridization of the central carbon atoms, compared to regular trigonal sp^2 hybrid orbitals, which causes increased s character in the orbitals responsible for the σ bond between the two rings. The special bonding properties of the three-membered ring may also contribute to this effect.

The data in Table 6-9 indicate that it is easy to bend a C=C—C valence angle (note the C=C—C angle differences between the first two entries in Table 6-9) and that moderate bending (up to 15°) hardly influences the C=C bond length.

Table 6-9. GED RESULTS FOR ALKENES WITH VALENCE ANGLE STRAIN (1,1-DISUBSTITUTED OR EXOCYCLIC C=C BONDS)

Molecule	r_a, C=C (Å)	∠C=C—C (deg)	Ref.
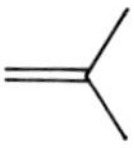	1.341(3)	122.2(2)	60
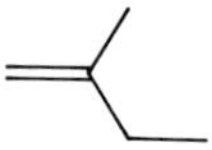	1.334 (ass.)	125.2, 122.5 (CH$_3$), syn[a] 124.9, 126.0 (CH$_3$), gauche[a]	61
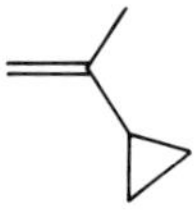	1.345(5)	121.2(25) (CH$_3$) 123.2(16)	62
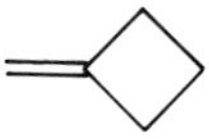	1.331(3)	134.1(6)	63
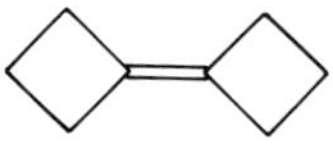	1.338(8)	133.3(8)	63
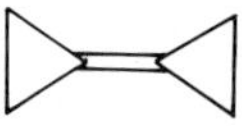	1.314(1)	148.1(1)	64

[a] Determined by a trial-and-error method.

Cyclic Alkenes. The rotational restrictions in cyclic compounds, compared to open chains of similar size, are in principle an advantage when the probabilities of getting reliable results from a GED study are considered. Gaseous 1-hexene is, for example, expected to appear as a mixture of several conformers, while cyclohexene probably will be conformationally homogeneous. On the other hand, it is generally more difficult to interpret the information from nonbonded distances in a cyclic molecule because these are normally restricted to a narrow r region, and the geometrical and vibrational parameters are often highly correlated. The nonbonded carbon–carbon distances in cyclohexene will, for example, be distributed in the region 2.5–3.0 Å, while the longest such distance in 1-hexene will be about 6.0 Å.

Correlations among independent parameters as well as ring flexibility increase with increasing ring size. It is therefore difficult to obtain reliable structural results from GED data alone for cycloalkenes with eight or more carbon atoms. An exception is *trans*-cyclooctene due to the restrictions induced by the configuration of the double bond.

When GED data are combined with other structural information (MW, NMR, etc), the possibilities of obtaining reliable structural results are promising. The combination of molecular mechanics calculations and GED has proved to be especially favorable in the study of many large ring compounds.[64–66]

Table 6-10 shows the available GED results for *cis*-cycloalkenes. All results were obtained more than 15 years ago, and the cyclobutene results were determined 30 years ago.[68] Improvements in the GED method would permit

Table 6-10. GED RESULTS FOR *cis*-CYCLOALKENES

| Molecule | r_a (Å) | | $\angle C{=}C{-}C$ (deg) | Ref. |
	C=C	C—C[a]		
	1.304(3)	1.519(2)	64.6[b]	67
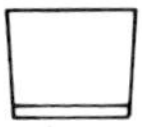	1.325(40)	⟨1.537⟩	94.0(8)	68
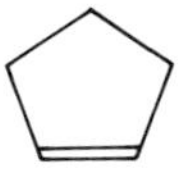	1.342(10)	1.519(30), 1.546(35)	111.0(12)	69
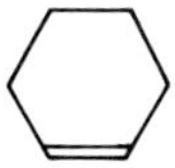	1.335(3)	1.504(6), 1.515(20), 1.550(40)	123.5[b]	70
	1.340(9)	1.503(9), 1.535(9)	124(2)	71
	1.334(4)	1.502(20), 1.515(20), 1.537(20)	123.4(8)	72

[a] These distances are arranged according to increasing distance from the C=C bond.
[b] Uncertainty not given.

Table 6-11. GED Results for Substituted *cis*-Cycloalkenes and for *cis,cis*-Cycloalkadienes with Isolated C=C Bonds

Molecule	r_a (Å) C=C	r_a (Å) C—C[a]	$\angle$C=C—C (deg)	Ref.
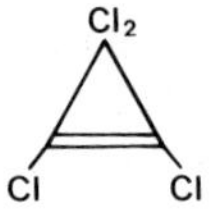 (Cl₂; Cl, Cl cyclopropene)	1.320(10)	1.479(11)	63.5	74
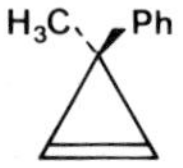 (H₃C, Ph cyclopropene)	1.311(4)	1.517(6)	64.4	75
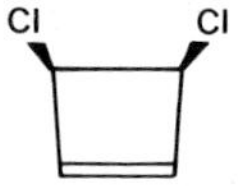 (Cl, Cl cyclobutene)	1.349(6)	1.505, 1.583		76
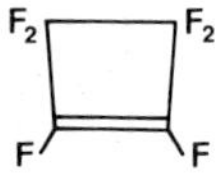 (F₂, F₂, F, F cyclobutene)	1.342(6)[b]	1.508(3)[b]	94.8(3)	77
 (F₂, F₂, F₂, F, F cyclopentene)	1.341[c]	1.507(28), 1.536(40)	113.6(12)	78
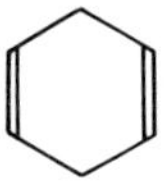 (cyclohexene)	1.347(10)	1.511(6)	123(1)	79
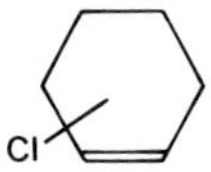 (Cl cyclohexene)				
1-Cl	1.335(6)	1.499(9), 1.532(10), 1.536[c]	123.9(8)	80
2-Cl	1.335[c]	1.500(10)–1.541[c]	123.2(10)	80
3-Cl	1.335[c]	1.506(7)–1.544[c]	123.3(15)	80
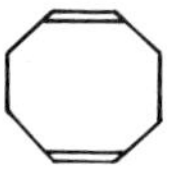 (cyclooctene)	1.340(3)	1.514(5), 1.536(9)	130.6(12)	81
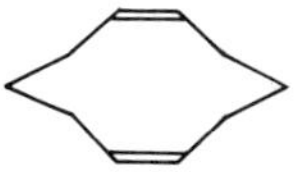 (cycloalkadiene)	1.326(4)	1.506(6), 1.534(6)	128.2(3)	82

[a] These distances are arranged according to increasing distance from the C=C bond.
[b] Unidentified, possibly r_a.
[c] Uncertainty not given.

more accurate determinations today, but apart from the cyclobutene data, which were obtained by the visual method, the results are probably reliable.

The C=C bond in cyclopropene[67] is observed to be about 0.03 Å shorter than that in ethylene[1] and about 0.20 Å shorter than the C—C bond in cyclopropane.[73] The difference between typical $\ce{>C=C<}$ and $\ce{-C-C-}$ bonds in unstrained hydrocarbons ($\sim$1.337 and 1.537 Å) is also about 0.20 Å. The short C=C bond in cyclopropene is therefore probably primarily due to the bonding properties of the three-membered ring.

Structural data determined for substituted *cis*-cycloalkenes and cycloalkadiens with isolated C=C bonds are presented in Table 6-11. Bicyclic and tricyclic alkenes are not included in this discussion.

The steric problems that are encountered when a trans C=C bond is part of a ring that is too small for the molecule to adopt normal structural parameters are illustrated by the stereochemistry of *trans*-cyclooctene.[83] The molecule is found to have C_2 symmetry. Some structural details are shown in Figure 6-1.

Distorted CCC valence angles are unavoidable, and the largest deviations are found at the $C_3(C_8)$ and $C_5(C_6)$ atoms. If the C=C π bond is assumed to be formed by p orbitals of trigonally hybridized (sp^2) carbon atoms, the observed C—C=C—C dihedral angle corresponds to an angle between the p-orbital axes of 44°, hence to a very weak C=C bond. This is not in agreement with the observed bond length of 1.332 Å.

The location of the hydrogens at the C=C bond shows, however, that the unsaturated carbon atoms are not hybridized in planar fashion. The structure corresponds to pyramidalization of the σ bonds at C_1 and C_2. The angle between the axes of the p-like orbitals forming the π bond is thereby reduced to about 20°. The geometry furthermore corresponds to increased s character of

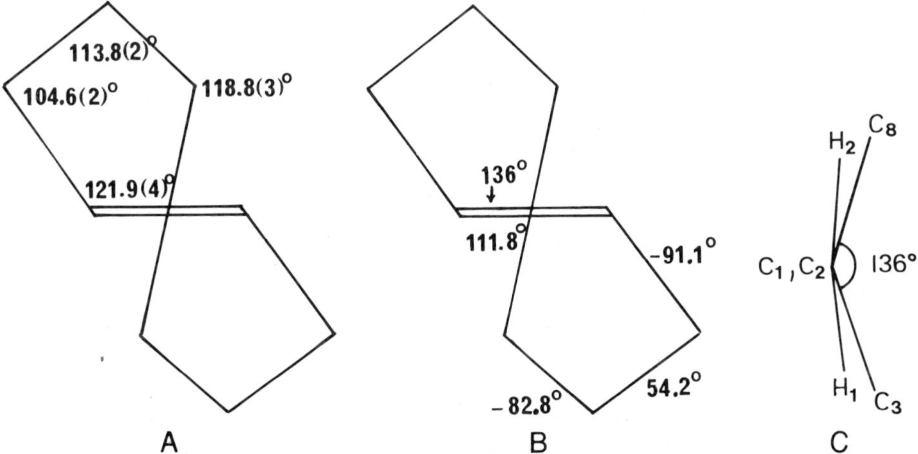

Figure 6-1. *trans*-Cyclooctene: (A) valence angles, (B) dihedral angles, and (C) projection of the dihedral angles at the C=C bond.

the orbitals forming the $C=C$ σ bond. The latter effect and the reduced *p*-orbital overlap are expected to have opposite influences on the $C=C$ bond length, which is observed to be normal or insignificantly shorter than normal.

The GED results for 1-methyl-*trans*-cyclooctene[84] confirm qualitatively those discussed above for the unsaturated fragment of *trans*-cyclooctene.

Steric Strain Across a $C=C$ Bond. Large cis substituents at a $C=C$ bond will not have enough space and will repel each other (**4**). Nonbonded repulsion

$$C=C$$

4

across the $C=C$ bond may be reduced by enlarging the $C=C-R$ angles and by stretching of or rotation around the $C=C$ bond. It is well known that valence angle bending normally costs less energy than bond stretching or twisting of π bonds.[85] The stereochemical consequences of *cis*-1,2-disubstitution at a $C_1=C_2$ bond are therefore expected to be noticeable primarily in enlarged $C=C-R$ angles.

If a $C=C$ bond is fully substituted by large groups, $C=C-R$ angle opening will be hampered by repulsions between the geminal substituents (**5**). In such cases the two other processes, $C=C$ bond stretching or π-bond twisting, might become competitive.

$$C=C$$

5

The structure of a sterically strained tetrasubstituted alkene molecule is intrinsically a complicated project for GED. The molecular scattering from the substituents will outweigh by far that from the double bond. Additional information on the $C=C$ bond will be available through the nonbonded internuclear distances across the double bond. This information will, however, often consist of a large number of closely spaced distances and may be difficult to interpret.

To minimize the problems encountered in a GED study of a sterically strained alkene, the following requirements should be fulfilled:

Identical substituents

Minimum conformational freedom of the substituents, to avoid a mixture of conformers.

Table 6-12 shows structural parameters for sterically strained alkenes studied by GED. In agreement with the general discussion above, it is only in the tetrasubstituted compounds that $C=C$ bonds are observed to be substantially longer than in ethylene.[1]

Table 6-12. STERICALLY STRAINED ALKENES STUDIED BY GED

Molecule	r_a (Å)		$\angle$ C=C—C(N) (deg)	Ref.
	C=C	=C—C		
	1.346(3)	1.506(2)	125.4(4)	5
	1.338(2)	1.490(1)	127.4(2)	7
	1.330(7)	1.505(2)	129.1(8)	8
	1.349(4)	1.511(7)	⟨124.5⟩	86
	1.344(4)	1.508(7)	⟨124.9⟩	86
	1.352(4)	1.509(2)	123.9(5)	60
	1.350(3)	1.509(2)	123.4(3)	87
	~1.40	1.54	126.5	88
	1.387(11)		123.7(3)	89

In accordance with expectations, the C=C—R valence angles of *cis*-1,2-disubstituted alkenes are found to be largest when none of the unsaturated carbon atoms are disubstituted.

A GED study of tetra-*t*-butylethene (**6**) would be of interest to elucidate the effects of steric strain across a C=C bond. The steric strain is expected to be very large, and the molecule fulfills the requirements set forth above. The steric strain is, however, so tremendous that no one has so far succeeded in synthesizing the compound, in spite of several attempts.[88] The last but one molecule in Table 6-12, which has the same substitution pattern and the same number of carbon atoms as tetra-*t*-butylethene, has, however, been studied,[88] and its $C=C$ bond is observed to be 0.06–0.07 Å longer than that in ethylene.[1] The severe strain across the $C=C$ bond causes other abnormalities as well, as for example the unusually long $=C-C$ bond of 1.54 Å.

6

The steric strain of the last molecule in Table 6-12 is also substantial, and the $C=C$ bond is found to be very long.[89] This molecule has, however, a cross-conjugated double bond, and since cross-conjugation lowers the π-bond order of a double bond,[90] the $C=C$ bond lengthening must be due to the combined effects of steric strain across the double bond and cross-conjugation.

Conjugated Double Bonds

Conjugated Aliphatic Polyenes. In GED studies of unsubstituted conjugated hydrocarbons the contribution from unsaturated bonds constitutes a major part of the molecular scattering. These molecules should therefore basically be well suited for GED studies. If there are more than two $C=C$ bonds in conjugation, the $C=C$ bond lengths are, however, expected to be slightly different. There will be mutual correlations involving $C=C$ bond lengths and the corresponding vibrational parameters, and this will complicate a GED study. The development of computer programs for normal coordinate analyses,[91–93] with which vibrational amplitudes and shrinkage corrections may be routinely calculated, has led to great improvements of the GED method. Moreover, such programs are especially advantageous in studies of conjugated polyenes because of the correlation problem and because shrinkage effects are especially large in these molecules. Most GED studies published during the past decade have been aided by normal coordinate analyses. Only few of those of earlier origin have had this advantage, and the resulting data are therefore often somewhat less accurate.

Table 6-13 shows structural and conformational results for conjugated polyenes that have been studied by GED. The preference for anti orientation of two $C=C$ bonds in conjugation is clearly demonstrated. When a conjugated polyene is substituted with *cis*-methyl groups (*cis* refers to the polyene chain) in relative positions 1 and 3, as in the molecules 5–7 of Table 6-13, steric repulsion between the methyl groups destabilizes the anti conformer. These molecules are all observed to assume gauche conformation.[98,99] These observations do not necessarily prove that *s*-gauche is normally the second rotational minimum in a conjugated polyene, because the alternative *s*-syn conformers of molecules 5–7 would all be more (5, 7) or less (6) destabilized by steric strain. The fact that all the dihedral ($=C-C=$) angles of the molecules discussed are close to 60°, does, however, support the idea of gauche being the second minimum when rotating around a $=\!C\!-\!C\!=$ bond of a conjugated polyene. The results obtained for isoprene support this view.[96]

Gauche conformation is also observed in some halosubstituted butadienes (**7–9**).

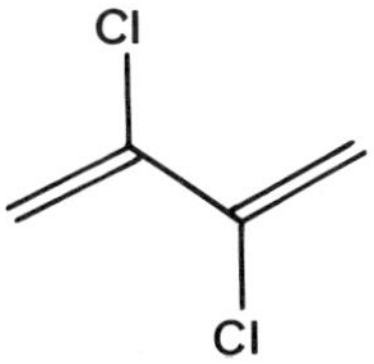

Reference 104:
72% anti
28% gauche
θ_g: 52.3(9.7)°

7

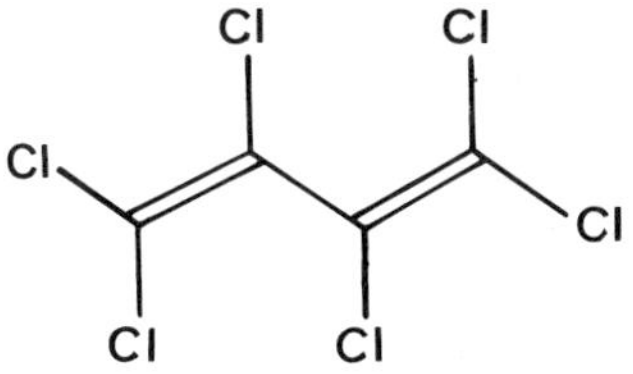

Reference 105:
100% gauche

θ_g: 78.1(1.1)°

8

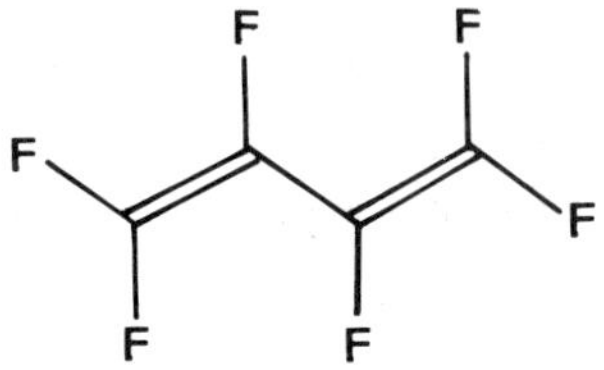

Reference 106:
100% gauche

θ_g: 47.4(8°)

9

GED has been used to study cis-[108] and trans-stilbene.[107] In the trans isomer the two phenyl rings and the C=C bond are practically coplanar, while the interaction between the phenyl rings in the cis isomer forces the rings to be rotated about 43° out of the plane of the double bond.[108]

In the nearly 20 years old studies of the cis[102] and trans[100] isomers of 1,3,5-hexatriene (molecules 10 and 8 of Table 6-13) the central C=C bond was found to be about 0.03 Å longer than the terminal ones. However, the GED studies were not combined with normal coordinate analyses, and shrinkage corrections were not carried out. It is not unlikely that the C=C bond differences have been overestimated in the two studies.

Conjugated Alicyclic Polyenes. Table 6-14 shows GED results for conjugated alicyclic polyenes. The C=C bonds in these molecules are forced to be approximately syn relative to one another, but this condition has no dramatic effect on the bond lengths of the conjugated systems.

As was pointed out above, the study of cyclic molecules by GED may be complicated. Most of the structures presented in Table 6-14 are probably basically correct, but 1,3-cyclooctadiene should be reinvestigated. In this study[119] no normal coordinate analysis was performed, and the study was carried out at a time when structural chemists did not normally consider the possibility of several conformers for cyclic molecules.

Notice the conformational differences between the boat-shaped 1,3,5-cycloheptatriene[117] and the essentially planar 2,4,6-cycloheptatrienone[118] (Table 6-14). The dipolar form of the carbonyl group allows six π electrons to be distributed over the seven cyclic carbon atoms, and thereby to achieve aromatic stabilization. The bond length distribution shows, however, that the conventional C=C bonds retain the greatest part of the double-bond character of a nonaromatic polyene system.

Table 6-13. Conjugated Aliphatic Polyenes Studied by GED

Molecule	r_a (Å) C=C	r_a (Å) C—C	∠C=C—C (deg)	Conformation	Ref.
1	1.340(2)	1.461(3)	123.6(3)	anti	94
	1.349(1)	1.467(2)	124.4(1)	anti	95
2	1.340(1)	1.463(2)	121.4(3)	95% anti[a]	
		1.512(2)	127.3(3)	5% gauche[a]	96
3	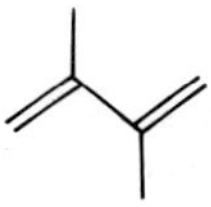 1.349(2)	1.491(2)	122.0(7)	anti	97
		1.504(2)			
4	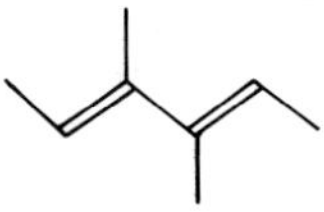 1.350(5)	1.473(9)	126.6(8)	anti	98
		1.521(5)			
5	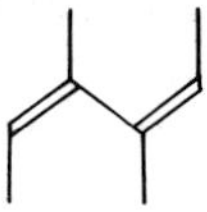 1.349(6)	1.479(10)	123.5(10)	gauche	98
		1.521(6)			
6	1.359(10)	1.460(20)	120.6(6)	gauche	98
		1.528(10)	123.3(4)		
7	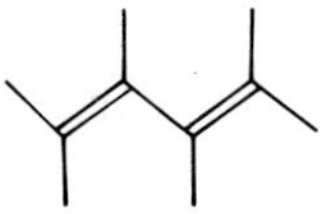 1.349(2)	1.487[b]	125.0(2)	gauche	99
		1.515(1)			
8	1.337(4)	1.458(4)	121.7(6)		100
	1.368(8)		124.4(6)	anti,anti	
9	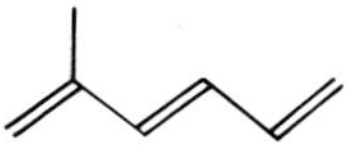 1.348(4)[c]	1.456(6)	119.1(C$_2$)[b]	anti,anti	101
		1.510(8)	⟨124.8⟩		
10	1.336(6)	1.462(4)	122.1(4)		102
	1.362(10)		125.9(4)	anti,anti	
11	 1.345(3)[c]	1.463(6)	123–128[b]	gauche,anti	103
		1.516(20)			

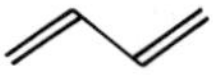

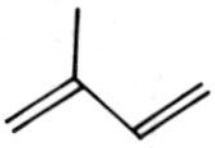

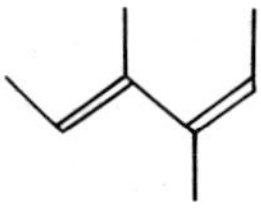

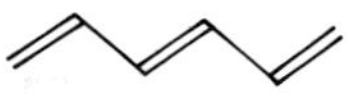

[a] Determined by a trial-and-error method.
[b] Uncertainty not given.
[c] The C=C bonds are assumed to be equal.

Table 6-14. Conjugated Alicyclic Polyenes, as Studied by GED

Molecule	r_a (Å)		Angles (deg)		Ref.
	C=C	C—C[a]	$\angle C_1\!=\!C_2\!-\!C_3$	θ[b]	
—SiH₃	1.389(13)		107.9(6)	0[c]	109
	1.348(2)	1.465(6) 1.519(3), 1.538(6)	120.3(2)	18.0(2)	112
	1.349(4)	1.466(14) 1.521(16), 1.532(20)	120.1(6)	18.3[d]	113
O / SiCl₂	(1.317)	1.481(9)	123.7(7)	8(6)	114
	1.35(1)	1.48(1) 1.54(1), 1.55(1)	129(2)	0[c]	115
	1.347(4)	1.450(12) 1.509(16), 1.522(16)	129.1(2)	0[c]	116
	1.354(5)	1.445(7) 1.504(7)	127.2[d]	41(2)	117
=O	1.362(12) 1.342(16)	1.429(12) 1.475(12)	125–133	0[a,e]	118
	1.347(5)	1.475(10) 1.509, 1.542	129(1)	38(2)	119
	1.340(2)	1.476(2)	126.1(5)	43(1)	120
	1.342(2)	1.486(9) 1.528, 1.573	134.7(7)	0[c]	183

[a] When several numbers are listed, they are given in the order C=C , C—C—, —C—C—.

[b] θ is the C=C—C=C dihedral angle.
[c] Fixed, lowest R factor.
[d] Uncertainty not given.
[e] Planar ring.

Conjugation Between C=C Bonds and Cyclopropyl Systems. In Walsh's description of the bonding in cyclopropane,[158] the carbon atoms are assumed to be sp^2-hybridized, with the hybrid orbitals directed toward the hydrogens and the center of the ring, while the p-orbital axis is normal to the plane of the other orbital axes. According to this description, the p orbital of each carbon atom should be able to interact with other π systems. Conjugation between cyclopropyl groups and C=C bonds has been observed in many molecules by different structural methods and has also been extensively studied theoretically.

Table 6-15 shows structural data for some molecules of this type studied by GED. The simplest molecule is vinylcyclopropane. A recent reinvestigation[122] of this compound confirmed the results obtained earlier,[121] in which anti was found to be the favored conformer and gauche a second contributing form.

A structural and conformational study of *trans*-1,2-divinylcyclopropane is intrinsically of considerably greater complexity than that of vinylcyclopropane, due to the increase in the number of possible conformers. The observed[123] conformational mixture of this molecule corresponds to the expected statistical

Table 6-15. MOLECULES WITH C=C BONDS IN CONJUGATION WITH CYCLOPROPYL GROUPS, AS STUDIED BY GED

Molecule	r_a (Å) C=C	r_a (Å) C—C[a]	$\angle$C=C—C (deg)	Conformation	Ref.
	1.334	1.475	126.2(14)	75(16)% anti 25(6)% gauche	121
	1.336(1)	1.470(2)	127.3(3)A 129.0G[b]	77(3)% anti 23(3)% gauche	122
trans	1.335(2)	1.474(4)	125.9(7)A 127.9G[b]	mixture[c]	123
	1.344(2)	1.473(5)	120.7(2)	[d]	124
	1.341(2)	1.509(3)[e]	109.2		125

[a] Bond between C=C and cyclopropyl.
[b] Uncertainty not given.
[c] 58% A/A, 29% A/−G$^+$ A/+G, 13% −G/+G.
[d] Two conformers (0.7:0.3).
[e] r(=C—C=) = 1.460(5) Å.

distribution, when each of the vinyl groups interacts with the cyclopropyl ring but there is no interaction between the two unsaturated substituents.

In 3-substituted cyclopropenes a π substituent has possibilities of interacting with the Walsh e_A orbital of the cyclopropane ring and with the π orbital of the cyclopropyl double bond. Interaction with the Walsh e_A orbital is expected to favor a bisected conformer, while interaction with the $C=C$ bond will favor a perpendicular conformer. GED data for 3-methyl-3-phenylcyclopropene[75] were in accordance with a bisected conformer having a C_3-C_{Ar} twist angle of 38°, or with a conformational mixture of 64% of the bisected conformer $[\theta(C_3-C_{Ar}) = 0°]$ and 36% of the perpendicular conformer $[\theta(C_3-C_{Ar}) = 90°]$.

Conjugation Between $C=C$ and $C=O$ Bonds: 10

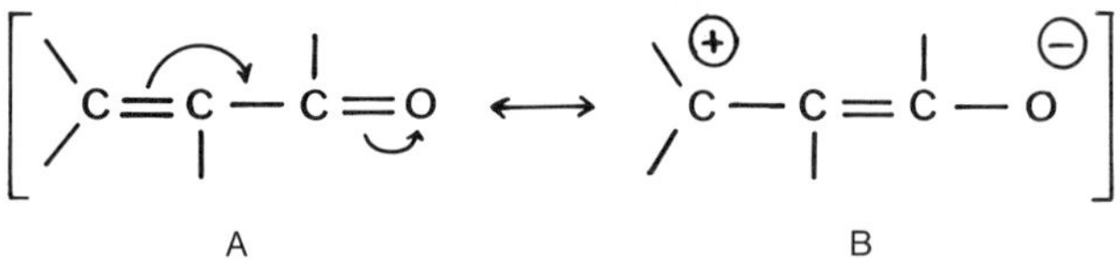

$$\left[\begin{array}{c} \diagdown \\ \diagup \end{array} C=C-C=O \quad \longleftrightarrow \quad \begin{array}{c} \diagdown \\ \diagup \end{array} \overset{\oplus}{C}-C=C-\overset{\ominus}{O} \right]$$

A   B

10

The dipolar form (**10B**) of an α,β-unsaturated molecule is expected to influence the structure to a greater extent than the dipolar forms affect the structure of a conjugated diene. The dipolar form might contribute to strengthening the $C-CO$ bond and to increasing the torsional barrier for rotation around the $C=C-C=O$ bond.

Table 6-16 shows GED results for α,β-unsaturated compounds that are potentially conformationally flexible. When these data are compared with those for flexible conjugated polyenes (Table 6-13) there are no obvious differences between the $C=C$ and $=C-C=$ bonds in the two series of molecules. There is, however, a striking difference in the conformational preferences. While anti is the preferred conformer for both types of molecule the second rotational minimum appears to be syn for a $C=C-C=O$ and gauche for a $C=C-C=C$ system.

The occurrence of anti and gauche conformers in $C=C-C=C$ systems may be interpreted in terms of twofold and threefold rotational contributions to the torsional energy. The observation of syn as the second stable conformer in $C=C-C=O$ systems indicates that a twofold rotational contribution is larger and/or that a threefold rotational contribution is smaller than in conjugated polyenes. The polarity of a $C=C-C=O$ fragment supports the idea of increased twofold rotational contribution. The conformational properties of the molecules shown in Tables 6-13 and 6-16 are not, however, directly comparable: all but two of the unsaturated carbonyl compounds with syn conformers are chlorine-substituted, while the polyenes are hydrocarbons. The conformational differences may therefore have a more complicated origin.

Table 6-17 shows GED results for nonflexible α,β-unsaturated compounds. The data appear to be quite normal and do not require special comment.

Table 6-16. MOLECULES WITH C=C BONDS IN CONJUGATION WITH CARBONYL GROUPS, AS STUDIED BY GED

Molecule	r_a (Å)		$\angle$C=C—CO (deg)	Conformation[a]	Ref.
	C=C	C—CO			
(acrolein)	1.332(12)	1.485(10)	120.7(5)	anti (100%)	126
	1.340(11)	1.481(13)	119.9(4)	anti (100%)	127
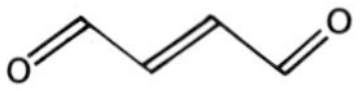 (OHC—CH=CH—CHO)	1.336(5)	1.479(2)	122.7(6)	anti,anti	128
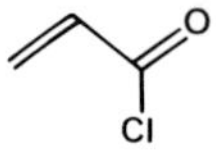 (acryloyl chloride)	1.339(2)	1.484(4)	123.4(7)	60(7)% anti[b] 40(7)% syn	129
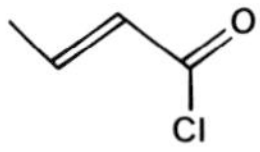 (crotonyl chloride)	1.349(7)	1.473(9)	126.8(14)	53(11)% anti 47(11)% syn	130
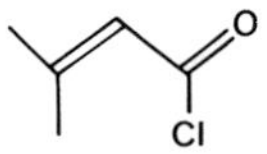 (3-methylcrotonoyl chloride)	1.333(7)	1.467(8)	127.1(8)	100% syn	130
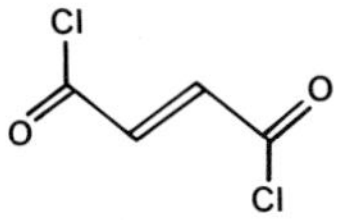 (fumaryl chloride)	1.334(5)	1.488(3)	125.2(6)	Mixture[c]	131
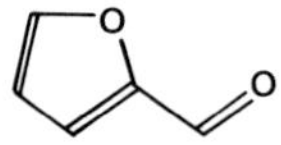 (furfural)		1.453(7)		31(9)% anti 69(9)% syn	159
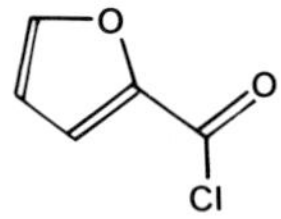 (2-furoyl chloride)	(1.37)	1.465(13)	131.6(9)	70(14)% anti 30(14)% syn	132
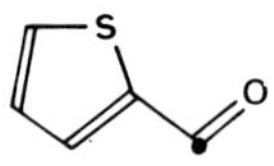 (2-thiophenecarboxaldehyde)	1.375(7)	1.466(16)	126.4(13)	80(8)% anti 20(8)% syn [θ:21.5(23.2)°]	133
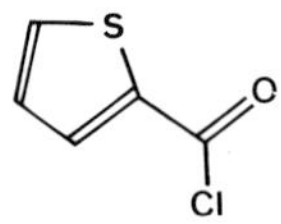 (2-thiophenecarbonyl chloride)	1.380(11)	1.475(20)	127.0(16)	59(11)% anti 41(11)% syn	134

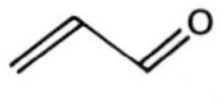

[a] The C=C—C=O dihedral angle θ is approximately as follows: syn, 0°; anti, 180°.
[b] Studied at three temperatures.
[c] 36% anti,anti, 42% anti,syn, 22% syn,syn.

Table 6-17. NONFLEXIBLE MOLECULES WITH C=C BONDS IN CONJUGATION WITH CARBONYL GROUPS, AS STUDIED BY GED

Molecule	r_a (Å)			$\angle$ C=C—CO (deg)	Ref.
	C=C	C—CO	C=O		
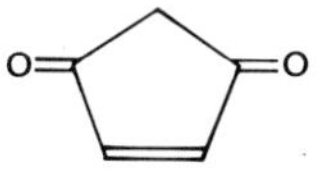 (4-cyclopentene-1,3-dione)	1.341(5)	1.493(5)	120.8(2)	110.4(3)	135
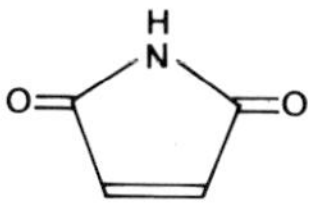 (maleimide, N—H)	1.342(5)	1.507(3)	120.4(2)	108.6(1)	136
 (maleic anhydride)	(1.33)	1.500(5)	119.5(3)	107.7(10)	137
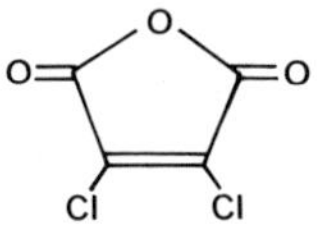 (dichloromaleic anhydride)	1.332(5)	1.495(3)	118.8(2)	107.9(2)	138
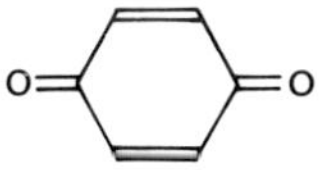 (p-benzoquinone)	1.344(3)	1.481(2)	122.5(2)	121.0(3)	139
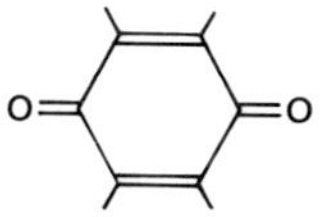 (dimethyl-p-benzoquinone)	1.352(8)	1.491(11)	122.9(8)	119.6(8)	140
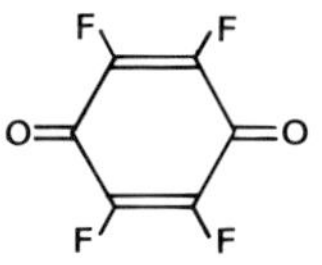 (tetrafluoro-p-benzoquinone)	1.339(12)	1.489(5)	121.1(6)	121.6(7)	140
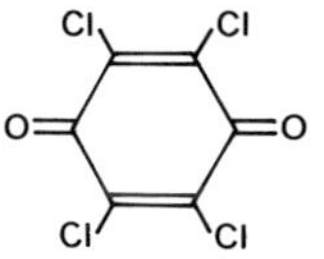 (tetrachloro-p-benzoquinone)	1.353(3)	1.492(3)	121.6(4)	122.0(7)	141

Cross-Conjugated C═C Bonds

Theoretical semiempirical calculations indicate that bond orders in cross-conjugated compounds differ systematically from those in similar through-conjugated ones.[90]

If one π system (eg, in an alkene or a polyene) is extended by linearly adding another π system, the effect will be increased delocalization throughout the system (**11**).

The effect of adding a cross-conjugated π system will be different in the two directions from the point of coupling (**12**). In the direction through the double bond there will be a delocalization (D) effect, while in the direction through the single bond the effect will be increased localization (L) (less p bond order to the formal single bonds and more to the formal double bonds).

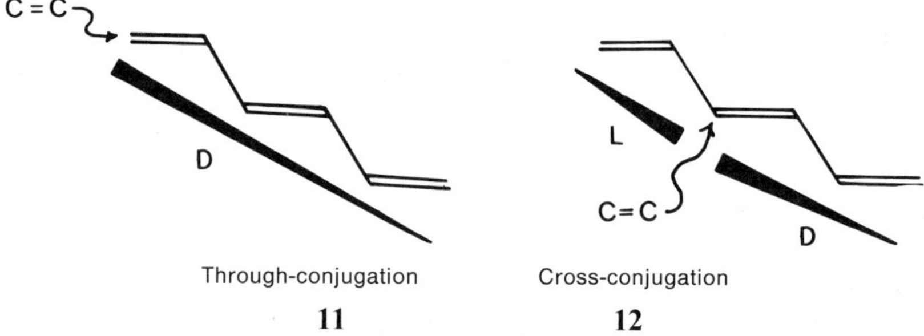

On the basis of these general ideas, the largest differences in bond lengths between cross-conjugated and through-conjugated compounds, which are otherwise comparable, will probably be in the single bond at the point of connection, where the localization effect should result in elongation of the C—C bond. The effect will probably be small.

If, however, a C═C bond is cross-conjugated at both carbon atoms (**13**), the effects from the cross-conjugating groups will reinforce one another in the doubly substituted C═C bond, while they will more or less cancel for the other bonds. In such cases the effect might be large enough to be significantly established experimentally.

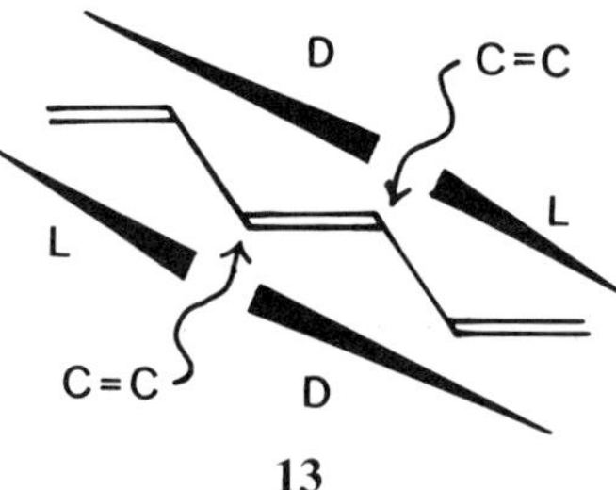

In Table 6-18, which shows results for cross-conjugated compounds that have been studied by GED, we see that in compound 2 the cross-conjugated C═C

Table 6-18. MOLECULES WITH CROSS-CONJUGATED C=C BONDS, AS STUDIED BY GED

Molecule	r_a (Å)			$\angle$ C=C—C(N) (deg)	Ref.
	C=C[a]	C=C	=C—C(N)		
1	1.357[b]	1.352(1)	1.478(6)	122.2(3)	142
2	1.382(8)	1.350(2)	1.472(5)	122.4(3)	142
3	1.347(10)	1.340(6)	1.476(8)	127[b]	143
4	1.357[b]		1.435[b]	121.1[b]	144
5	1.387(11)		1.397(4)	124.4(3)	145

[a] Cross-conjugated bond.
[b] Uncertainty not given.

bond is 0.03 Å longer than the other C=C bonds.[142] The other results shown in Table 6-18 are also in agreement with the description given above. In compound 5 the C=C bond experiences some steric strain from the methyl substituents. This effect might contribute to the lengthening of the C=C bond.

Cumulated C=C Bonds

Compounds with cumulated C=C bonds are generally well suited to GED studies because of the rigidity of a C=C=C fragment, which has relatively high scattering power.

A cumulated C=C bond is expected to be shorter than an isolated or conjugated C=C bond because the central carbon atom in a C=C=C fragment is sp-hybridized. The GED data for cumulated molecules (Table 6-19) show that a $C_{sp^2}=C_{sp}$ bond is about 0.02 Å shorter than the C=C bond in ethylene,[1] while a $C_{sp}=C_{sp}$ bond is 0.05–0.06 Å shorter.

The preference of unsaturated compounds with OR or SR substituents for the syn conformation (see Table 6-7) is confirmed also for cumulenes.

Table 6-19. Molecules With Cumulated C=C Bonds, as Studied by GED

Molecule	r_a (Å)		∠C=C—Y (deg)	Conformation[a]	Ref.
	C=C	=C—Y			
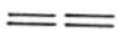 (structure)	1.312[b]				146
 (structure)	1.312(1)	1.468(4)	124.0(3)	anti	147
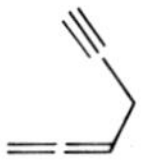 (structure)	1.307(1)	1.514(3)	124.4(6)	39(7)% syn 61(7)% gauche	148
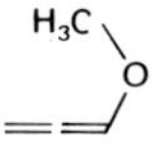 H₃C / O (structure)	1.318(4)	1.375(7)	125.3(12)	syn	149
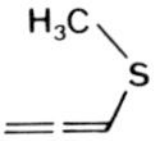 H₃C / S (structure)	1.282(10) 1.327(10)	1.745(10)	125.4(6)	Two conformers[c]	150
 (structure)	1.318(5) 1.283[b]				151
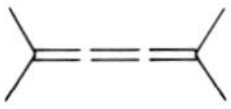 (structure)	1.330(3) 1.270(5)	1.514(3)	122.0(3)		152
 Cl (structure)	1.327(2) 1.268(4)	1.734(2)	121.9(3)		153
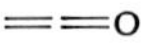 ==O (structure)	1.317(1)				154
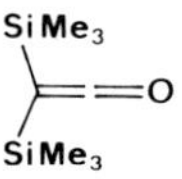 SiMe₃ / SiMe₃ (structure) ==O	1.318(13)		115.5 (+80, −40)	155	
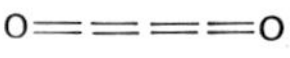 O=====O (structure)	1.294(2) 1.288(2)				156 157

[a] syn refers to θ(C=C—Z—) or θ(C=C—C=C) equal to 0.
[b] Uncertainty not given.
[c] 63–72% of conformer with θ = 6–19°; second conformer, θ = 45–75°.

Compounds with C≡C Bonds

Aliphatic Alkynes

A molecule with the general formula

$$A—C≡C—B$$

almost invariably has C≡C—A(B) valence angles of 180° and almost free rotation around the ≡C—A(B) bonds. The effects from substituents of different kinds on a —C≡C— fragment will therefore normally be noticeable, if at all, only in the C≡C bond length. Structural studies of substituted alkynes will therefore generally not give as much information on the nature of intramolecular forces as will those of alkene derivatives.

Table 6-20 shows GED C≡C bond lengths determined for a number of substituted alkynes. There is a slight tendency toward an increase in C≡C bond length when the substituents are electropositive, and a decrease with electronegative substituents. Not all data in Table 6-20 are in accordance with this tendency, however. If the C≡C bond in 2-butyne of 1.212(1) Å is used as a reference, most data in Table 6-20 are not more than 3 standard deviations away from this reference value.

Cyclic Alkynes

Because the two substituents of a C≡C bond are directed 180° away from each other, they will not mutually interfere. Sterically strained C≡C bonds will therefore be present primarily in medium-sized alicyclic compounds.

Table 6-20. GED DETERMINATIONS OF THE C≡C BOND IN MOLECULES OF THE TYPE
A—C≡C—B

A	B	r_a, C≡C (Å)	Ref.	A	B	r_a, C≡C (Å)	Ref.
H	 ▷ (cyclopropyl)	1.208(2)	161	H	C(CH$_3$)$_3$	1.210(5)	170
		1.212(2)	162	CH$_3$	CH$_3$	1.212(1)	171
H	PF$_4$	1.217(7)	163	CH$_3$	SiH$_3$	1.230(4)	172
H	SiMe$_3$	1.20(1)	164	Br	Cl	1.206(4)	173
H	SnMe$_3$	1.232(20)	165	Br	I	1.206(8)	173
H	SnIX$_2$[a]	1.225(6)	166	Cl	C≡N	1.205(3)	173
H	SnX$_3$[a]	1.227(8)	166	Cl	SiH$_3$	1.238(6)	174
H	(CH$_2$)$_2$C≡CH	1.220(1)	167	H	CHO	1.210(7)	175
H	(CH$_2$)$_2$C≡CBr	1.217(2)	168	CH$_2$=CH	CH$_2$=CH	1.220(2)	176
Br	(CH$_2$)$_2$C≡CBr	1.219(2)	169	SCH$_3$	SCH$_3$	1.211[b]	177

[a] X = C≡CH.
[b] Fixed.

So far only three sterically strained cyclic alkynes have been studied by GED. The geometrical parameters that have been determined for these molecules are shown in Table 6-21. The seven-membered[178] and nine-membered[181] rings were studied several years ago, based on GED data alone, while the GED study of cyclooctyne[180] was combined with normal coordinate analysis, molecular mechanics calculations and data from NMR coupling constants ($^1J_{CC}$).

The $C{\equiv}C{-}C$ bending force constant is known to be substantially smaller than those for $C{=}C{-}C$ and $C{-}C{-}C$ angles.[182] It is therefore not surprising that the $C{-}C{\equiv}C{-}C$ fragments of sterically strained cycloalkynes deviate from linearity. The data in Table 6-21 show that the $C{\equiv}C{-}C$ angles decrease with decreasing ring size.

The $C{\equiv}C$ triple bond of cyclooctyne is found to be twisted about 40° away from a planar cis conformation.[180] If it is energetically favourable for the $C{-}C{\equiv}C{-}C$ fragment in cyclooctyne to be twisted, this will probably be the case also in the seven- and nine-membered rings. In the GED studies of the latter

Table 6-21. GED Results (distances, Å; angles, deg) for Three Sterically Strained Cyclic Alkynes

	1^a	2^b	3^c
Bond lengths, r_a			
$C{\equiv}C$	1.209(9)	1.228(2)	1.212(4)
$C_2{-}C_3$	1.475(9)	1.457(2)	1.456(9)
$C_3{-}C_4$	1.558(12)	1.534(5)	1.513(40)
$C_4{-}C_5$		1.565(3)	1.538(47)
$C_5{-}C_6$			1.572(17)
Valence angles			
$C{\equiv}C{-}C$	145.8(7)	154.5(8)	160.2(107)
$C_2C_3C_4$	107.8(10)	108.3(3)	112.7(21)
$C_3C_4C_5$	110.2(15)	115.3(4)	124.1(15)
$C_4C_5C_6$		118.9(5)	113.3(24)
$C_5C_6C_7$			110.4(46)
Dihedral angles			
$C{-}C{\equiv}C{-}C$	0 (fixed)	40.2(104)	0 (fixed)
$C{\equiv}C{-}C{-}C$		−39.1(53)	60.5(77)
$C_2{-}C_3{-}C_4{-}C_5$		50.7(10)	−34.8(99)
$C_3{-}C_4{-}C_5{-}C_6$		−87.1(5)	58.4(109)
$C_4{-}C_5{-}C_6{-}C_7$		105.0(10)	−129.2(136)

[a] Data from reference 178.
[b] Data from reference 180.
[c] Data from reference 181.

molecules, however, the C—C≡C—C dihedral angles were assumed to be zero. It is difficult to know the extent to which the geometrical parameters that were determined are dependent on this assumption.

Cyclooctyne has been studied earlier with the assumption of a cis planar C—C≡C—C fragment,[179] and the resulting geometrical parameters differed substantially from those of the more recent study.[180] In the earlier study the C—C—C angles in the ring varied between 108 and 110°, while abnormally long C—C bonds were found. These results are unreasonable, since it is well known that valence angle bending is energetically more favourable than bond stretching. The GED results obtained for molecules 1 and 3 in Table 6-21 do, however, appear to be more reasonable than those discussed above for cyclooctyne.[179]

GED studies of medium-sized cyclic alkynes are hampered by the same difficulties as those described for cyclic alkenes earlier. When the GED studies of molecules 1 and 3 of Table 6-21 were carried out, the available literature described strained C—C≡C—C fragments as being planarly cis-distorted. The assumptions about the C≡C dihedral angle in these molecules were therefore not unreasonable.

These and other[15] experiences show that assumptions about molecular parameters should be made with the utmost care and that complicated GED studies should, as far as possible, be supplemented with structural information from other sources.

Intramolecular Interactions Involving C=C or C≡C Bonds

Intramolecular interactions involving π bonds have attracted the interest of chemists for some time.[184] Intramolecular OH...π hydrogen bonding has for example been studied by infrared,[185,186] microwave,[187–189] and NMR[190,191] spectroscopy, as well as by GED[192] and X-ray diffraction.[193]

Most of the molecules in which such intramolecular interactions have been observed have had fairly limited conformational freedom, because of either limited size of the molecules or partly cyclic molecular structure.

Some recent GED studies of relatively large, potentially flexible molecules[194–196] show, however, that intramolecular interactions between π systems (C=C or C≡C) and a hydroxyl group[195,196] or a metal atom (Zn)[194] are strong enough to keep most of these gas molecules in conformations that are favourable for the particular intramolecular interactions but are otherwise not favored.

Table 6-22 shows the most relevant conformational data for these molecules, which are all potentially flexible and with a variety of possible conformers. A large number of geometrical parameters are necessary to describe one single

Table 6-22. CONFORMATIONAL PREFERENCES IN SOME MOLECULES STABILIZED BY INTRAMOLECULAR INTERACTIONS INVOLVING C=C OR C≡C BONDS, GED RESULTS

Number of possible conformers[a]	5^b	14^b	14^c	5^c
Preferred conformation (A)	g, syn	g_+, g_-, ac_+ or g_+, g_-, syn	g_+, g_-, g_+ (C—O: g_-)	g_+, g_- (C—O: g_+)
%A (from GED)	80	≥80	50(20)	≥50
%A estimated (no interaction)	13	7	7	9
Estimated interaction energy (kJ mol^{-1})	−4.5	−6	(−4)	(−4)
Bond lengths (Å)				
=C$_1$...Zn(O),	3.15(6)	3.41(8)	3.53	3.19
=C$_2$...Zn(O)	3.44(3)	3.00(8)	2.97	2.82
Reference	194	194	195	196

[a] A mirror image pair is counted as one conformer.
[b] Involving a metal atom and one hydrocarbon chain.
[c] Does not include the OH hydrogen atom.

conformer, and in such cases it is not possible to determine accurately from GED data alone the geometry of each of several simultaneously occurring conformers. It is necessary to introduce assumptions of various kinds and/or structural information from other sources. Even then it will normally be possible to make a reliable structural determination only when one conformer dominates the conformational mixture.

In the two organometallic compounds,[194] experimental information on the conformational behavior of structurally related fragments (1-butene,[3] butane[197]) was used to estimate the conformational distribution of the hydrocarbon chains in the absence of π...metal interaction. Similar estimates were made for the unsaturated alcohols[195,196] on the basis of molecular mechanics calculations for all possible conformers.

Considerable more experimental and theoretical information about intramolecular interactions between π systems and other groups of atoms are necessary before these phenomena may be properly described and understood. GED studies have already contributed to the elucidation of these effects and will certainly give valuable information to this field in the future.

Acknowledgments

I thank Dr. Kolbjørn Hagen for permission to include unpublished structural results, for reading the manuscript, and for valuable advice. Pirkko Bakken prepared the formulas and illustrations, Kari Traetteberg typed the manuscript, and Dr. David Nicholson corrected the English of the manuscript. I express my sincere thanks and appreciation to them all.

References

1. Kuchitsu, K. *J. Chem. Phys.* **1966**, *44*, 906.
2. Tokue, I.; Fukuyama, T.; Kuchitsu, K. *J. Mol. Struct.* **1973**, *17*, 207.
3. van Hemelrijk, D.; van den Enden, L.; Geise, H. J.; Sellers, H. L. *J. Am. Chem. Soc.* **1980**, *102*, 2189.
4. ter Brake, J.H.M. *J. Mol. Struct.* **1984**, *118*, 73.
5. Almenningen, A.; Anfinsen, I. M.; Haaland, A. *Acta Chem. Scand.* **1970**, *24*, 43.
6. ter Brake, J.H.M.; Mijlhoff, F. C. *J. Mol. Struct.* **1981**, *77*, 253.
7. ter Brake, J.H.M. *J. Mol. Struct.* **1984**, *118*, 63.
8. van Hemelrijk, D.; van den Enden, L.; Geise, H. J. *J. Mol. Struct.* **1981**, *74*, 123.
9. Huisman, P.A.G.; Mijlhoff, F. C.; Renes, G. H. *J. Mol. Struct.* **1978**, *51*, 191.
10. Huisman, P.A.G.; Mijlhoff, F. C. *J. Mol. Struct.* **1979**, *54*, 145.
11. Huisman, P.A.G.; Mijlhoff, F. C. *J. Mol. Struct.* **1979**, *57*, 83.
12. Mijlhoff, F. C.; Renes, G. H.; Kohata, K.; Oyanagi, K.; Kuchitsu, K. *J. Mol. Struct.* **1977**, *39*, 241.
13. Nakata, M.; Kutchitsu, K. *J. Mol. Struct.* **1982**, *95*, 205.
14. Carlos, J. L.; Karl, R. R.; Bauer, S. H. *J. Chem. Soc. Faraday Trans. 2* **1974**, *70*, 177.
15. Hagen, K.; Stølevik, R. *J. Mol. Struct.* To be published.
16. Hagen, K.; Stølevik, R.; Thingstad, Ø. *J. Mol. Struct.* **1982**, *78*, 313.
17. Davis, M. I.; Kappler, H. A.; Cowan, D. J. *J. Phys. Chem.* **1964**, *68*, 2005.
18. Van Schaick, E.J.M.; Miljhoff, F. C.; Renes, G.; Geise, H. J. *J. Mol. Struct.* **1974**, *21*, 17.
19. Strand, T. G. *Acta Chem. Scand.* **1967**, *21*, 2111.
20. Strand, T. G. *Acta Chem. Scand.* **1967**, *21*, 1003.
21. Spelbos, A.; Huisman, P.A.G.; Mijlhoff, F. C.; Renes, G. *J. Mol. Struct.* **1978**, *44*, 159.
22. Mom, V.; Huisman, P.A.G.; Miljhoff, F. C.; Renes, G. H. *J. Mol. Struct.* **1980**, *62*, 95.
23. Trongmo, Ø.; Shen, Q.; Hagen, K.; Seip, R. *J. Mol. Struct.* **1981**, *71*, 185.
24. Søvik, O. I.; Schei, S. H.; Stølevik, R.; Hagen, K.; Shen, Q. *J. Mol. Struct.* **1984**, *116*, 239.
25. Søvik, O. I.; Trongmo, Ø.; Shen, Q.; Hagen, K.; Schei, S.; Stølevik, R. *J. Mol. Struct.* **1984**, *118*, 1.
26. Samdal, S.; Seip, H. M.; Torgrimsen, T. *J. Mol. Struct.* **1977**, *42*, 153.
27. Hilderbrandt, R. L.; Schei, S. H. To be published.
28. Schei, S. H.; de Meijere, A. To be published.
29. Schei, S. H.; Seip, R. *Acta Chem. Scand.* To be published.
30. Schei, S. H.; Shen, Q. To be published.
31. Schei, S. H. *Acta Chem. Scand.* **1983**, *A37*, 671.
32. Shen, Q. *J. Mol. Struct.* **1979**, *53*, 61.
33. Schei, S. H.; Shen, Q. *J. Mol. Struct.* **1982**, *81*, 269.
34. Schei, S. H. *J. Mol. Struct.* **1983**, *102*, 305.
35. Shen, Q. *J. Mol. Struct.* **1981**, *75*, 303.
36. Schei, S. H.; Hagen, K. *J. Mol. Struct.* **1984**, *116*, 249.
37. Schei, S. H. *J. Mol. Struct.* **1983**, *98*, 141.
38. Shen, Q. *J. Mol. Struct.* **1984**, *112*, 327.

39. Bürger, H.; Pawelke, G.; Oberhammer, H. *J. Mol. Struct.* **1982**, *84*, 49.
40. Hilderbrant, R. L.; Andreassen, A. L.; Bauer, S. H. *J. Chem. Phys.* **1970**, *74*, 1586.
41. Lowrey, A. H.; George, C.; Antonio, P. D.; Karle, J. *J. Mol. Struct.* **1979**, *53*, 189.
42. Vilkov, L. V.; Mastryukov, V. S.; Sadova, N. I. In "Determination of the Geometrical Structure of Free Molecules". Mir: Moscow, 1983.
43. Stølevik, R.; Thingstad, Ø. *J. Mol. Struct.* **1984**, *106*, 333.
44. Schei, S. H. "Structural and Conformational Studies of Some Halogenated Propenes, Butenes and Related Molecules". Dissertation, University of Trondheim, Norway, 1984.
45. Samdal, S.; Seip. H. M. *J. Mol. Struct.* **1975**, *28*, 193.
46. Pyckout, W. "Conformational Analysis and Structural Chemistry". Dissertation, University of Antwerp, Belgium, 1986.
47. Schei, S. H. *Acta Chem. Scand* **1983**, *A37*, 153.
48. Shen, Q. *J. Mol. Struct.* **1979**, *51*, 61.
49. Samdal, S.; Seip, H. M.; Torgrimsen, T. *J. Mol. Struct.* **1979**, *57*, 105.
50. Derissen, J. L.; Bijen, J.M.J.M. *J. Mol. Struct.* **1973**, *16*, 289.
51. Jondal, P.; Seip, H. M.; Torgrimsen, T. *J. Mol. Struct.* **1976**, *32*, 369.
52. Brunvoll, J.; Hargittai, I. *Acta Chim. Acad. Sci. Hung.* **1977**, *94*, 333.
53. Naumov, V. A.; Ziatdinova, R. N.; Berdnikov, E. A. *Zh. Strukt. Khim.* (*Engl.*) **1981**, *22*, 382.
54. Naumov, V. A.; Ziatdinova, R. N. *Zh. Strukt. Khim.* (*Engl.*) **1983**, *24*, 370.
55. Hargittai, I.; Rozsondai, B.; Nagel, B.; Bulcke, P.; Robinet, G.; Labarre, J.-F. *J. Chem. Soc. Dalton Trans.* **1978**, 861.
56. Naumov, V. A.; Nesterov, V. Y.; Aleksandrova, I. A. *Zh. Strukt. Khim.* (*Engl.*) **1984**, *25*, 137.
57. Naumov, V. A.; Shaidulin, S. A. *Zh. Strukt. Khim.* (*Engl.*) **1976**, *17*, 304.
58. Beagley, B.; Foord, A.; Moutran, R.; Rozsondai, B. *J. Mol. Struct.* **1977**, *42*, 117. *J. Mol. Struct.* **1979**, *51*, 156.
59. Kuznetsova, T. M.; Alekseev, N. V.; Veniaminov, N. N. *Zh. Strukt. Khim.* (*Engl.*) **1979**, *20*, 281.
60. Tokue, I.; Fukuyama, T.; Kuchitsu, K. *J. Mol. Struct.* **1974**, *23*, 33.
61. Shimanouchi, T.; Abe, Y.; Kutchitsu, K. *J. Mol. Struct.* **1968**, *2*, 82.
62. Konaka, S.; Suga, H.; Kimura, M. *J. Mol. Struct.* **1983**, *98*, 133.
63. Traetteberg, M.; Simon, A.; Peters, E. M.; de Meijere, A. *J. Mol. Struct.* **1984**, *118*, 333.
64. Traetteberg, M.; Bakken, P.; Almenningen, A. *J. Mol. Struct.* **1981**, *74*, 321.
65. Traetteberg, M.; Bakken, P.; Almenningen, A. *J. Mol. Struct.* **1981**, *70*, 287.
66. Traetteberg, M.; Lüttke, W.; Machinek, R.; Krebs, A.; Hohlt, H. J. *J. Mol. Struct.* **1985**, *128*, 217.
67. Chiang, J. F. *J. Chin. Chem. Soc.* **1970**, *17*, 65.
68. Coldish, E.; Hedberg, K.; Schomaker, V. *J. Am. Chem. Soc.* **1956**, *78*, 2714.
69. Davis, M. I.; Muecke, T. W. *J. Phys. Chem.* **1970**, *74*, 1104.
70. Chiang, J. F.; Bauer, S. H. *J. Am. Chem. Soc.* **1969**, *91*, 1898.
71. Naumov, V. A.; Dashevskii, V. G.; Zaripov, N. M. *Zh. Strukt. Khim.* (*Engl.*) **1970**, *11*, 736.
72. Geise, H. J.; Buys, H. R. *Rec. Trav. Chim. Pays-Bas* **1970**, *89*, 1147.
73. Yamamoto, S.; Nakata, M.; Fukayama, T.; Kuchitsu, K. *J. Phys. Chem.* **1985**, *89*, 3298.
74. Mair, H. J.; Bauer, S. H. *J. Phys. Chem.* **1971**, *75*, 1681.
75. Traetteberg, M.; de Meijere, A.; Domnin, I. N. *J. Mol. Struct.* **1985**, *128*, 207.
76. Bastiansen, O.; Derissen, J. L. *Acta Chem. Scand.* **1966**, *20*, 1089.
77. Chang, C. H.; Porter, R. F.; Bauer, S. H. *J. Mol. Struct.* **1971**, *7*, 89.
78. Chang, C. H.; Bauer, S. H. *J. Phys. Chem.* **1971**, *75*, 1685.
79. Oberhammer, H.; Bauer, S. H. *J. Am. Chem. Soc.* **1969**, *91*, 10.
80. Lu, K. C.; Chiang, R. L.; Chiang, J. F. *J. Mol. Struct.* **1980**, *64*, 229.
81. Hagen, K.; Hedberg, L.; Hedberg, K. *J. Phys. Chem.* **1982**, *86*, 117.
82. Almenningen, A.; Jacobsen, G. G.; Seip, H. M. *Acta Chem. Scand.* **1969**, *23*, 1495.
83. Traetteberg, M. *Acta Chem. Scand.* **1975**, *B29*, 29.
84. Traetteberg, M.; Bakken, P.; Almenningen, A. *J. Mol. Struct.* **1981**, *74*, 321.
85. Eliel, E. L.; Allinger, N. L.; Angyal, S. J.; Morrison, G. A. "Conformational Analysis". Wiley-Interscience: New York, 1966.
86. Geise, F. J.; Mijlhoff, F. C.; Renes, G.; Rummens, F.H.A. *J. Mol. Struct.* **1973**, *17*, 37.
87. Carlos, J. L.; Bauer, S. H. *J. Chem. Soc. Faraday Trans. 2* **1974**, *70*, 171.
88. Traetteberg, M.; Bakken, P.; Krebs, A. To be published.
89. Traetteberg, M. To be published.
90. Janssen, J.; Lüttke, W. *J. Mol. Struct.* **1972**, *81*, 73.
91. Gwinn, W. D. *J. Chem. Phys.* **1971**, *55*, 447.

92. Stølevik, R.; Seip, H. M.; Cyvin, S. J. *Chem. Phys. Lett.* **1972**, *15*, 263.
93. Hilderbrandt, R. L.; Wieser, J. D. *J. Chem. Phys.* **1966**, *42*, 4648.
94. Kuchitsu, K.; Fukuyama, T.; Morino, Y. *J. Mol. Struct.* **1968**, *1*, 463.
95. Kveseth, K.; Seip, R.; Kohl, D. A. *Acta Chem. Scand.* **1980**, *A34*, 31.
96. Traetteberg, M.; Paulen, G.; Cyvin, S. J.; Panchenko, Y. N.; Mochaklov, V. I. *J. Mol. Struct.* **1984**, *116*, 141.
97. Aten, C. F.; Hedberg, L.; Hedberg, K. *J. Am. Chem. Soc.* **1968**, *90*, 2463.
98. Traetteberg, M. *Acta Chem. Scand.* **1970**, *24*, 2295.
99. Traetteberg, M.; Sydnes, L. K. *Acta Chem. Scand.* **1977**, *B31*, 387.
100. Traetteberg, M. *Acta Chem. Scand.* **1968**, *22*, 628.
101. Traetteberg, M.; Paulen, G. *Acta Chem. Scand.* **1974**, *A28*, 1150.
102. Traetteberg, M. *Acta Chem. Scand.* **1968**, *22*, 2294.
103. Traetteberg, M.; Paulen, G. *Acta Chem. Scand.* **1974**, *A28*, 1.
104. Hagen, K.; Hedberg, K.; Neisess, J.; Gundersen, G. *J. Am. Chem. Soc.* **1985**, *107*, 341.
105. Gundersen, G. *J. Am. Chem. Soc* **1975**, *97*, 6342.
106. Chang, C. H.; Andreassen, A. L.; Bauer, S. H. *J. Org. Chem.* **1971**, *36*, 920.
107. Traetteberg, M.; Frantsen, E. B.; Mijlhoff, F. C.; Hoekstra, A. *J. Mol. Struct.* **1975**, *26*, 57.
108. Traetteberg, M.; Frantsen, E. B. *J. Mol. Struct.* **1975**, *26*, 69.
109. Bentham, J. E.; Rankin, D.W.H. *J. Organomet. Chem.* **1971**, *30*, C54.
110. Veniaminov, N. N.; Ustynyuk, Y. A.; Alekseev, N. V.; Ronova, I. A.; Struchkov, Y. T. *Zh. Strukt. Khim. (Engl.)* **1972**, *13*, 118.
111. Veniaminov, N. N.; Ustynyuk, Y. A.; Alekseev, N. V.; Ronova, I. A.; Struchkov, Y. T. *Zh. Strukt. Khim. (Engl.)* **1971**, *12*, 879.
112. Traetteberg, M. *Acta Chem. Scand.* **1968**, *22*, 2305.
113. Oberhammer, H.; Bauer, S. H. *J. Am. Chem. Soc.* **1969**, *91*, 10.
114. Nipan, M. E.; Sadova, N. I.; Golubinskii, A. V.; Vilkov, L. V.; Krasnova, T. L.; Chernyshev, E. A.; Labartkava, M. O. *Zh. Strukt. Khim. (Engl.)* **1984**, *25*, 701.
115. Chiang, J. F.; Bauer, S. H. *J. Am. Chem. Soc* **1966**, *88*, 420.
116. Hagen, K.; Traetteberg, M. *Acta Chem. Scand.* **1972**, *26*, 3643.
117. Traetteberg, M. *J. Am. Chem. Soc* **1964**, *86*, 4265.
118. Ogasawara, M.; Ijima, T.; Kimura, M. *Bull. Chem. Soc. Japan* **1972**, *45*, 3277.
119. Traetteberg, M. *Acta Chem. Scand.* **1970**, *24*, 2285.
120. Haugen, W.; Traetteberg, M. In "Selected Topics in Structure Chemistry". Universitetsforlaget: Oslo, 1967.
121. de Meijere, A.; Lüttke, W. *Tetrahedron* **1969**, *25*, 2047.
122. Traetteberg, M.; Lüttke, W. To be published.
123. Traetteberg, M.; Schrumpf, G. To be published.
124. Hagen, K.; Traetteberg, M. *Acta Chem. Scand.* **1972**, *26*, 3636.
125. Chiang, J. F.; Wilcox, C. F. *J. Am. Chem. Soc.* **1973**, *95*, 2885.
126. Kuchitsu, K.; Fukuyama, T.; Morino, Y. *J. Mol. Struct.* **1969**, *4*, 41.
127. Traetteberg, M. *Acta Chem. Scand.* **1970**, *24*, 373.
128. Paulen, G.; Traetteberg, M. *Acta Chem. Scand.* **1974**, *A28*, 1155.
129. Hagen, K.; Hedberg, K. *J. Am. Chem. Soc.* **1984**, *106*, 6150.
130. Nordtømme, T.; Hagen, K. *J. Mol. Struct.* **1985**, *128*, 127.
131. Hagen, K. *J. Mol. Struct.* **1985**, *128*, 139.
132. Hagen, K. *J. Mol. Struct.* **1985**, *130*, 255.
133. Braathen, G. O.; Kveseth, K.; Nielsen, C. J.; Hagen, K. To be published.
134. Hagen, K. To be published.
135. Hagen, K.; Hedberg, K. *J. Mol. Struct.* **1978**, *44*, 195.
136. Harsanyi, L.; Vajda, E.; Hargittai, I. *J. Mol. Struct.* **1985**, *129*, 315.
137. Hilderbrandt, R. L.; Peixoto, E.M.A. *J. Mol. Struct.* **1972**, *12*, 31.
138. Hagen, K.; Hedberg, K. *J. Mol. Struct.* **1978**, *50*, 103.
139. Hagen, K.; Hedberg, K. *J. Chem. Phys.* **1973**, *59*, 158.
140. Schei, H.; Hagen, K.; Traetteberg, M.; Seip, R. *J. Mol. Struct.* **1980**, *62*, 121.
141. Hagen, K.; Hedberg, K. *J. Mol. Struct.* **1978**, *49*, 351.
142. Traetteberg, M.; Bakken, P.; Almenningen, A.; Lüttke, W.; Janssen, J. *J. Mol. Struct.* **1982**, *81*, 87.
143. Chiang, J. F.; Bauer, S. H. *J. Am. Chem. Soc.* **1970**, *92*, 261.
144. Hope, H. *Acta Chem. Scand.* **1968**, *22*, 1057.
145. Traetteberg, M. To be published.

146. Almenningen, A.; Bastiansen, O.; Traetteberg, M. *Acta Chem. Scand.* **1959**, *13*, 1699.
147. Traetteberg, M.; Paulen, G.; Hopf, H. *Acta Chem. Scand.* **1973**, *27*, 2227.
148. Seip, R.; Bakken, P.; Traetteberg, M.; Hopf, H. *Acta Chem. Scand.* **1981**, *A35*, 365.
149. Bijen, J.M.J.M.; Derissen, J. L. *J. Mol. Struct.* **1972**, *14*, 229.
150. Derissen, J. L.; Bijen, J.M.J.M. *J. Mol. Struct.* **1973**, *16*, 289.
151. Almenningen, A.; Bastiansen, O.; Traetteberg, M. *Acta Chem. Scand.* **1961**, *15*, 1557.
152. Almenningen, A.; Gundersen, G.; Aanonsen, J. E. Norwegian Electron Diffraction Group, Annual Report, 1984.
153. Almenningen, A.; Gundersen, G.; Granberg, M.; Karlsson, F. *Acta Chem. Scand.* **1975**, *A29*, 545.
154. Kuchitsu, K.; Takabayashi, F. Ninth Hungarian Diffraction Conference, Pécs, 1978.
155. Rozsondai, B.; Hargittai, I. *Acta Chim. Acad. Sci. Hung.* **1976**, *90*, 157.
156. Almenningen, A.; Arnesen, S. P.; Bastiansen, O.; Seip, H. M.; Seip, R. *Chem. Phys. Lett.* **1968**, *1*, 569.
157. Tanimoto, M.; Kuchitsu, K.; Morino, Y. *Bull. Chem. Soc. Japan* **1970**, *43*, 2776.
158. Walsh, A. D. *Trans. Faraday Soc.* **1949**, *45*, 179.
159. Schultz, G.; Fellegvari, I.; Kolonits, M.; Kiss, A. I.; Pete, B.; Bánki, J. *J. Mol. Struct.* **1978**, *50*, 325.
160. Davis, M. I.; Hanson, H. P. *J. Phys. Chem.* **1965**, *69*, 4091.
161. Klein, A. W.; Schrumpf, G. *Acta Chem. Scand.* **1981**, *A35*, 431.
162. Tamagawa, K.; Hilderbrandt, R. L. *J. Phys. Chem.* **1983**, *87*, 3839.
163. Oberhammer, H. *J. Mol. Struct.* **1979**, *53*, 139.
164. Zeil, W.; Haase, J.; Dakkouri, M. *Discuss. Faraday Soc.* **1969**, *47*, 149.
165. Belyakov, A. V.; Khaikin, L. S.; Vilkov, L. V.; Apalkova, G. M.; Nikitin, V. S.; Bogoradovskii, E. T.; Zavgorodnii, V. S. Ninth Austin Symposium on Gas-Phase Molecular Structure, Austin, TX, 1982, Paper No. A24.
166. Khaikin, L. S.; Belyakov, A. V.; Vilkov, L. V.; Bogaradovskii, E. T.; Zavgorodnii, V. S. *J. Mol. Struct.* **1980**, *66*, 149.
167. Traetteberg, M.; Bakken, P.; Seip, R.; Cyvin, S. J.; Cyvin, B. N.; Hopf, H. *J. Mol. Struct.* **1979**, *51*, 77.
168. Gogstad, E.; Stølevik, R.; Traetteberg, M. *J. Mol. Struct.* **1984**, *116*, 295.
169. Traetteberg, M.; Bakken, P.; Seip, R.; Cyvin, S. J.; Cyvin, B. N.; Hopf, H. *J. Mol. Struct.* **1979**, *55*, 199.
170. Haase, J.; Zeil, W. *Z. Naturforsch.* **1969**, *24a*, 1844.
171. Tanimoto, M.; Kuchitsu, K.; Morino, Y. *Bull. Chem. Soc. Japan* **1969**, *42*, 2519.
172. Cradock, S.; Koprowski, J.; Rankin, D.W.H. *J. Mol. Struct.* **1981**, *77*, 113.
173. Almenningen, A.; Nor, O.; Strand, T. G. *Acta Chem. Scand.* **1976**, *A30*, 567.
174. Cradock, S.; Fraser, A.; Rankin, D.W.H. *J. Mol. Struct.* **1981**, *71*, 209.
175. Lugié, M.; Fukuyama, T.; Kuchitsu, L. *J. Mol. Struct.* **1972**, *14*, 333.
176. Almenningen, A.; Gogstad, E.; Hagen, K.; Schei, H.; Stølevik, R.; Thingstad, Ø.; Traetteberg, M. *J. Mol. Struct.* **1984**, *116*, 131.
177. Beagley, B.; Ulbrecht, V.; Katsumata, S.; Lloyd, D. R.; Connor, J. A.; Hudson, G. A. *J. Chem. Soc. Faraday Trans. 2* **1977**, *73*, 1278.
178. Haase, J.; Krebs, A. *Z. Naturforsch.* **1972**, *27a*, 624.
179. Haase, J.; Krebs, A. *Z. Naturforsch.* **1971**, *26a*, 1190.
180. Traetteberg, M.; Lüttke, W.; Machinek, R.; Krebs, A.; Hohlt, H. J. *J. Mol. Struct.* **1985**, *128*, 217.
181. Typke, V.; Haase, J.; Krebs, A. *J. Mol. Struct.* **1979**, *56*, 77.
182. Herzberg, G. "Molecular Spectra and Molecular Structure". Van Nostrand Reinhold: New York, 1945.
183. Montgomery, L. K.; Wilson, C. A.; Wieser, J. D. *J. Mol. Struct.* **1985**, *129*, 69.
184. Aror, H. S. *Top. Stereochem.* **1979**, *11*, 10.
185. Oki, M.; Iwamura, H. *Bull. Chem. Soc. Japan* **1966**, *39*, 470.
186. Joris, L.; Schleyer, P.v.R.; Gleiter, R. *J. Am. Chem. Soc.* **1968**, *90*, 327.
187. Marstokk, K.-M.; Møllendal, H. *Acta Chem. Scand.* **1981**, *A35*, 395.
188. Horn, A.; Marstokk, K.-M.; Møllendal, H.; Priebe, H. *Acta Chem. Scand.* **1983**, *A37*, 679.
189. Tyblewski, M.; Bauder, A. *J. Mol. Struct.* **1983**, *102*, 267.
190. Abraham, R. J.; Bakke, J. M. *Acta Chem. Scand.* **1983**, *A37*, 865.
191. Senda, Y.; Ishiyama, J.; Imaizumi, S. *J. Chem. Soc. Perkin Trans.* **1980**, *2*, 90.
192. Traetteberg, M.; Østensen, H. *Acta Chem. Scand.* **1979**, *A33*, 491.

193. Schweizer, W. B.; Dunitz, J. D.; Pfund, R. A.; Tombo, G.M.R.; Ganter, C. *Helv. Chim. Acta* **1981**, *64*, 2738.
194. Haaland, A.; Lehmkuhl, H.; Nehl, H. *Acta Chem. Scand.* **1984**, *A38*, 547.
195. Traetteberg, M.; Bakken, P.; Seip, R.; Lüttke, W.; Knieriem, B. *J. Mol. Struct.* **1985**, *128*, 191.
196. Traetteberg, M.; Bakken, P.; Seip, R.; Lüttke, W.; Knieriem, B. *J. Mol. Struct.* To be published.
197. Bradford, W. F.; Fitzwater, S.; Bartell, L. S. *J. Mol. Struct.* **1977**, *38*, 185.

7

SUBSTITUTED BENZENE DERIVATIVES

Aldo Domenicano

DEPARTMENT OF CHEMISTRY, CHEMICAL ENGINEERING
AND MATERIALS
UNIVERSITY OF L'AQUILA
L'AQUILA, ITALY
CNR INSTITUTE OF STRUCTURAL CHEMISTRY
MONTEROTONDO STAZIONE, ITALY

CONTENTS

INTRODUCTION   282
THE NATURE OF THE BENZENE RING DEFORMATIONS   283
   Ring Deformations in Monosubstituted Derivatives   283
   The Origin of the Geometrical Effects   287
   Ring Deformations in Polysubstituted Derivatives   287
   Ring Deformations as a Measure of Electronic Substituent Effects   289
THE ROLE OF ELECTRON DIFFRACTION STUDIES   291
   Electron Diffraction Studies of Monosubstituted Benzene Derivatives   292
   Electron Diffraction Studies of Symmetrically para-Disubstituted
      Benzene Derivatives   308
   Electron Diffraction Studies of Symmetrically Trisubstituted Benzene
      Derivatives   312
   On the Additivity of the Ring Deformations   316
   Intermolecular Interactions and Substituent Effects   318
ACKNOWLEDGMENT   321
REFERENCES   322

                    ALDO DOMENICANO

Introduction

Given a hydrocarbon framework, chemists are able to substitute a hydrogen atom with another atom or functional group. This has the effect of modifying the distribution of valence electrons within the framework, which gives rise in turn to changes in many chemical and physical properties: eg, reaction rates, thermodynamic quantities, spectroscopic properties, and—last but not least —molecular geometry.

Although the changes in reactivity and spectral parameters are often substantial, the geometrical distortions of the framework are generally small. Because they seldom exceed a few hundredths of an angstrom for C—C bond distances and some degrees for C—C—C bond angles, accurate experimental methods are required for their determination.

Also the length and orientation of the C—H bonds is, of course, affected by substitution. The experimental study of their structural variation is, however, a very difficult task, which is beyond the possibilities of some techniques of structural determination, like gas-phase electron diffraction and X-ray crystallography.

On a relative scale the changes in bond angles tend to be larger than those in bond distances. This is because the energy required to stretch a bond is greater than that required to open an angle. Another characteristic of bond angles is that often they are more accurately measured than bond distances. For these reasons bond angles are more suited than bond distances to providing information on substituent effects.

The effects of substitution on molecular geometry have been most extensively investigated with benzene derivatives. These fairly rigid molecules are well suited to structural studies, and the high symmetry of their skeleton is instrumental in revealing small distortional effects.

The first observation of a ring distortion in a benzene derivative was reported by Keidel and Bauer in an early (1956) electron diffraction study of the molecular structure of phenylsilane, $C_6H_5\text{-}SiH_3$.[1] These authors found that the agreement between the experimental and theoretical radial distributions of this molecule could be improved substantially by allowing the internal ring angles at the ipso and para positions to become 117.4 and 120.8°, respectively. Subsequent studies by X-ray crystallography have confirmed that an ipso angle of about 117° is typical of silicon substitution in benzene derivatives.[2]

Unfortunately this early observation, dating from a time when the least-squares method was not yet in use for the refinement of molecular parameters, had no impact on the electron diffraction community. It was only in the late 1970s that electron diffraction began to be applied systematically to determine the geometrical distortions of the benzene ring caused by functional groups.

In the meantime, three studies by microwave spectroscopy had provided detailed accounts of the structural effects of substitution in fluorobenzene,[3] cyanobenzene,[4] and aniline.[5] Moreover, the molecular structures of hundreds of substituted derivatives of benzene had been determined by X-ray crystallography accurately enough to show at least some of the effects of substitution on the ring geometry. Almost invariably, however, the determination of these effects was not the primary aim of the X-ray diffraction studies, and thus the small deviations of the ring geometry from the reference geometry of unsubstituted benzene were seldom noticed.

The first attempts to rationalize the observed ring deformations were published in 1966[6] and 1968.[3] More general conclusions were derived during the 1970s through the systematic analysis of many accurate results, mostly obtained by X-ray crystallography.[2,7–10] A full statistical treatment of the experimental results available for monosubstituted benzene rings was published in 1983.[11]

The study of geometrical substituent effects in benzene derivatives is today an active field of research. All techniques of structure determination—including molecular mechanics and *ab initio* molecular orbital (MO) calculations with geometry optimization—are being used to augment the structural information available for these molecules and to investigate the electronic effects of the substituents through their geometric effects. This chapter briefly reviews the present knowledge of the skeletal distortions of the benzene nucleus caused by substitution and discusses in some detail the contribution of electron diffraction studies.

The Nature of the Benzene Ring Deformations

Ring Deformations in Monosubstituted Derivatives

In a C_6H_5X molecule the distortion of the carbon skeleton of the benzene ring conforms generally to C_{2v} symmetry.[7] Further distortion to C_s symmetry is appreciable only with some strongly asymmetric functional groups, eg, X = OMe,[12,13] OH,[14,15] CH=CH$_2$.[16] When C_{2v} symmetry is retained the geometrical changes consist of [7,11]:

1. *Bond distance variation.* The *a* bonds (see Figure 7-1) are lengthened—or shortened—with respect to the value found in unsubstituted benzene. The *b* and *c* bonds seem to be scarcely affected;
2. *Angular variation.* The internal ring angles α and β (and, to a lesser extent, γ and δ) deviate from the ideal value of 120°, with $\Delta\beta$ being approximately equal to $-\Delta\alpha/2$.

The parameter that is most affected by the presence of the substituent is the angle α. Its values span a range of 13° (viz, 112–125°), depending on the electronic properties of X. Unlike C—C bond distances, it can be measured with a good level of accuracy, within 0.1° in favorable cases. All this makes α an interesting indicator of electronic substituent effects.[2,9,10]

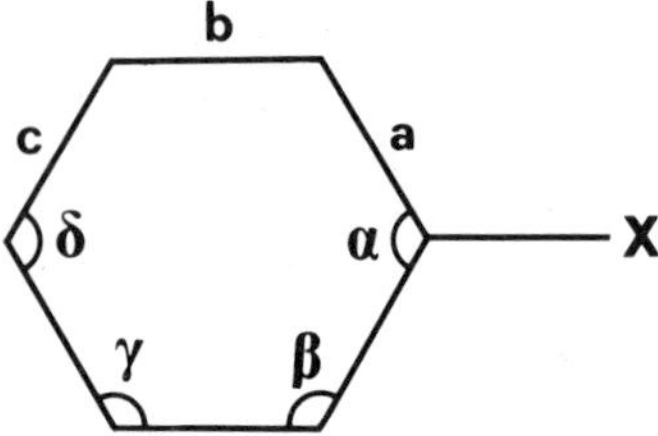

Figure 7-1. Labeling of the C—C bonds and C—C—C angles in monosubstituted benzene rings; C_{2v} symmetry has been assumed.

The angle α is particularly sensitive to the σ-inductive effect of the substituent. Within a row of the periodic table it increases linearly with the electronegativity of the substituent (Figure 7-2).[2,17] The regression lines in Figure 7-2 may be used to evaluate the electronegativity of functional groups from the observed ring distortion.[2,17,18] The increase in α is associated with a decrease in β, a small increase in δ, and a shortening in the a bonds.[11]

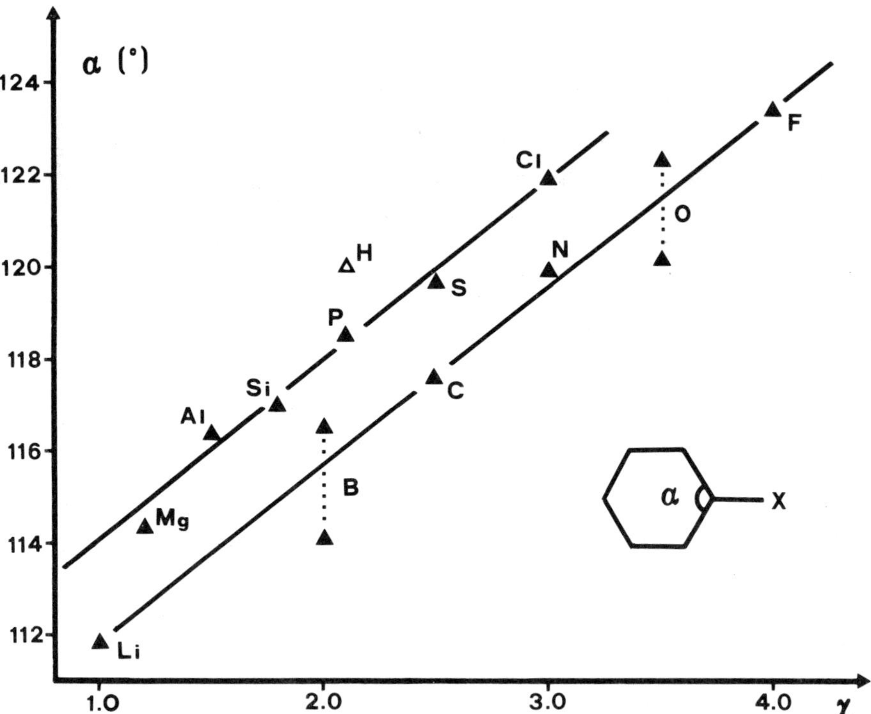

Figure 7-2. Empirical correlation between the angle α of a monosubstituted benzene ring and the electronegativity of the substituent, χ. (After reference 17.)

When a substituent is capable of sharing π electrons with the ring, the angle α decreases with increasing conjugation. This effect—which also involves other geometrical parameters of the ring—is not as large as the previous one, but certainly significant and well documented.[8] One of the best examples is provided by the molecular structure of diphenylaminotriphenylmethane, $Ph_3C\text{-}NPh_2$, as determined by low-temperature X-ray crystallography.[19] In this molecule the nitrogen atom has a planar coordination. The two phenyl groups bonded to N have different conformations in the crystal: the dihedral angles between the coordination plane of N and the planes of the rings are 75 and 12°, respectively. Both angles are essentially torsions about the N—C bonds. The different conformation strongly affects the extent of conjugation of the rings with the substituent, as witnessed by the different lengths of the two N—C bonds (Figure 7-3). It is seen that π donation from the substituent to the ring causes the angles α and δ to decrease by about 2°, and the angles β and γ to increase by about 1°. The angular changes are associated with a 0.01 Å lengthening of the a bonds.

A deeper understanding of the benzene ring deformations caused by substitution has been achieved through a statistical analysis of the geometrical variance of a large sample of monosubstituted benzene rings, mostly studied by X-ray crystallography.[11] The molecules of the sample were selected to cover a wide spectrum of substituent effects. Since accurate structural data were seldom available for derivatives with common functional groups, most molecules in the sample had large, relatively "unusual" molecular fragments as substituents. The results of the statistical analysis may be summarized as follows.

The angular variance of monosubstituted benzene rings with first-row substituents is fully described by two orthogonal components of distortion, involving angular changes in different ratios. The distortion that accounts for most of the variance is a concerted change of the internal angles α, β, and (to a minor extent) δ:

$$\Delta\alpha : \Delta\beta : \Delta\gamma : \Delta\delta = 1.00 : -0.69 : 0.06 : 0.26° \qquad (7\text{-}1)$$

This distortion is related to the electronegativity of the substituent. When bond distance variation is included in the model, a change of the a bond distances is also seen to be involved:

$$\Delta\alpha : \Delta a = 1.00° : -0.0027 \text{ Å} \qquad (7\text{-}2)$$

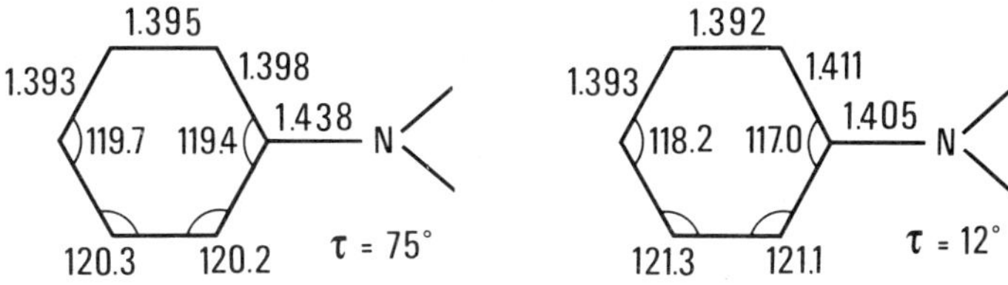

Figure 7-3. Effect of conjugation on the geometry of the benzene ring, as seen by comparing the geometries (Å, deg) of the two phenyl groups bonded to N in $Ph_3C\text{-}NPh_2$.[19] Least-squares standard deviations are about 0.001 Å for bond distances and 0.1° for bond angles; τ is the angle of twist between the coordination plane of N and the plane of the ring. (Adapted from reference 8.)

The second distortion involves mainly the internal angles γ and δ:

$$\Delta\alpha : \Delta\beta : \Delta\gamma : \Delta\delta = 0.08 : 0.19 : -0.73 : 1.00° \tag{7-3}$$

and appears to be controlled to a large extent by the π-donor/acceptor strength of the substituent.

Only the first distortion is of importance for benzene derivatives with second-row substituents.

Note how the separation of the two distortions *is not based on chemical assumptions*, but originates directly from the statistical treatment. Also note that increasing conjugation between the ring and the substituent, as in going from $\tau = 75°$ to $\tau = 12°$ in Figure 7-3, corresponds to a variation of the ring geometry that consists of a blend of the two distortions. This comes as no surprise, since the shortening of the N—C bond is associated with a decrease of the σ-electron-withdrawing properties of the substituent.

The existence of only two orthogonal components of angular distortion implies that the four different internal angles of a benzene ring of C_{2v} symmetry are related by two equations of constraint. One of these equations is purely geometrical (it expresses the condition of planarity):

$$\Delta\alpha + 2\Delta\beta + 2\Delta\gamma + \Delta\delta = 0 \tag{7-4}$$

The other is:

$$\Delta\beta = -0.591(7)\Delta\alpha - 0.301(15)° \tag{7-5}$$

for first-row substituents, and

$$\Delta\beta = -0.615(11)\Delta\alpha - 0.384(19)° \tag{7-6}$$

for second-row substituents. (The numbers in parentheses are the standard deviations of the coefficients, given as units in the last digit.) The correlation coefficients are -0.991 and -0.993 on 149 and 50 data points, respectively.

The physical meaning of eqns. (7-5) and (7-6) is that the angular distortion caused by a substituent at the ipso position of the ring is largely—though not entirely—relaxed by distortions of opposite sign at the ortho positions. The difference between the calculated coefficients for first- and second-row substituents is small and of marginal importance. Note that in both cases unsubstituted benzene ($\Delta\beta = \Delta\alpha = 0$) definitely lies off the regression line.

Only limited information about bond distance changes caused by substitution could be obtained from the statistical analysis of reference 11. A substantial amount of bond distance variance was found to arise from the serious systematic effects that influence the interatomic distances obtained by X-ray crystallography. A short description of these effects and of their treatment will be given later in this chapter.

The Origin of the Geometrical Effects

A qualitative explanation of the distortion of the carbon skeleton caused by substitution in the ipso region of the benzene ring has been given[3,6] in terms of hybridization effects. In *sp*-, *sp²*-, and *sp³*-hybridized atoms the *p* character concentrates in the hybrid orbitals that point toward more electronegative atoms.[20] This follows from the fact that *p* electrons are held more loosely than are *s* electrons, since they penetrate the inner core less than do *s* electrons, thus experiencing a smaller effective nuclear charge. In the case of a benzene derivative a shift of electron density from the ipso carbon to a σ-electron-withdrawing substituent is best accomplished through an increase in the *p* character of the hybrid orbital that points toward the substituent. This implies a decrease in the *p* character of the other two hybrid orbitals and leads, therefore, to a shortening of the *a* bonds and an increase of the angle α. The explanation is easily extended to the case of a substituent that is conjugated with the ring.[7]

A pictorial, but effective rationalization of the benzene ring distortion has been given[7] in terms of the valence-shell electron-pair repulsion (VSEPR) model[21]. According to this model a σ-electron-withdrawing substituent polarizes the C—X σ-bonding molecular orbital, so that the bonding electron pair takes up a decreasing amount of space in the valence shell of the ipso carbon and interacts less strongly with the two neighboring σ-bonding pairs. This causes the *a* bonds to become shorter and the angle α to increase with respect to the value α_0, expected for a substituent having the same electronegativity as the ring (Figure 7-4). Note that α_0 may differ from 120°, since the spatial requirements of a bonding electron pair are not necessarily the same for a C—X bond and a C—II bond. The reversed pattern of effects occurs with a σ-electron-releasing substituent.

If the substituent is conjugated with the ring, the electron density of the C—X σ-bonding molecular orbital will increase, due to the shorter interatomic distance. This will, in turn, increase the repulsive interactions between the electron pairs of the C—X and *a* bonds, so that the angle α becomes smaller and the *a* bonds longer, with respect to the values that would be found with the same substituent in the absence of conjugation.

Ring Deformations in Polysubstituted Derivatives

In most polysubstituted benzene derivatives the angular distortion of the ring may be interpreted—at least to a first approximation—as arising from the superposition of separate, independent contributions from each substituent.[7] Within this approximation the contribution of a substituent is expressed by a set of *angular parameters*, defined as the deviations of the ring angles from 120° caused by that substituent. If the distortion of the benzene ring caused by a substituent conforms to C_{2v} symmetry the angular parameters will be $\Delta\alpha$, $\Delta\beta$, $\Delta\gamma$, and $\Delta\delta$. Of course, these four parameters are not independent, due to the

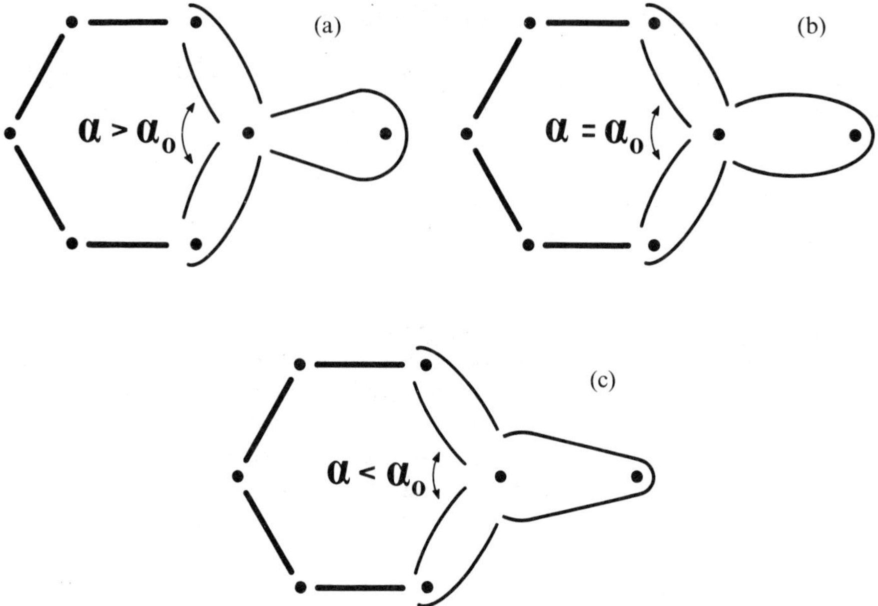

Figure 7-4. The effect of valence-shell electron-pair repulsions on the angle α of a monosubstituted benzene ring: (a) a σ-electron-withdrawing substituent, (b) a substituent having the same electronegativity as the ring, and (c) a σ-electron-releasing substituent. (After reference 7.)

exact geometrical constraint expressed by eq. (7-4), and, further, to the statistical constraint expressed by eq. (7-5) or (7-6).

Ideally, the angular parameters of common functional groups should be determined by direct measurement on monosubstituted benzene derivatives. Unfortunately, to date only few of these have been studied with the necessary accuracy. On the other hand, the ring geometries of a large number of polysubstituted derivatives with common functional groups have been determined accurately in the solid state. By using multiple regression it has proved possible to derive from these geometries the angular parameters of many functional groups.[10,22] The values obtained for some important groups are reported in Table 7-1.

The angular substituent parameters of Table 7-1 may be used to predict the geometries of other polysubstituted benzene rings. In many cases the predicted angles agree within a few tenths of a degree with the experimental results.

Small systematic deviations from additivity of angular distortions, amounting to no more than $1-2°$, have been observed with some benzene derivatives where cooperative interactions between substituents are known to occur. A reference case is that of p-nitroaniline.[23]

An analysis of bond distance changes in polysubstituted benzene rings, again based on the hypothesis that each substituent gives a separate contribution to

Table 7-1. ANGULAR SUBSTITUENT PARAMETERS (deg) FOR IMPORTANT FUNCTIONAL GROUPS IN THE SOLID STATE[a]

Substituent	$\Delta\alpha$	$\Delta\beta$	$\Delta\gamma$	$\Delta\delta$
F	3.4	−2.0	0.5	−0.4
NO_2	2.9	−1.9	0.3	0.4
Cl	1.9	−1.4	0.6	−0.2
NH_3^+	1.8	−1.2	0.4	−0.1
SO_2Me	1.6	−1.3	0.2	0.6
CN	1.1	−0.8	0.3	−0.1
OMe	0.2	−0.6	1.1	−1.1
OH	0.2	−0.4	0.6	−0.6
COOH	0.1	−0.2	0.1	0.2
NHAc	−0.1	−0.3	0.7	−0.6
N=NR	−0.1	−0.3	0.5	−0.4
COOR	−0.6	0.2	0.3	−0.3
COMe	−1.0	0.4	0.2	−0.3
NH_2	−1.2	0.2	1.0	−1.3
CH=CHR	−1.8	0.8	0.3	−0.4
Me	−1.9	1.0	0.4	−0.8
Ph	−2.3	1.0	0.6	−0.9
NMe_2	−2.4	0.6	1.4	−1.7

[a] For planar polyatomic substituents the values given refer to conformations where the dihedral angle between the plane of the ring and that of the substituent is never greater than 42°, and usually much less. Standard deviations are 0.1–0.2°.

Source: After reference 10.

the ring deformation, has also been carried out.[22]. Here, however, the rather small values of the effects and the lack of adequate corrections for systematic errors in most of the experimental data available make the results of the analysis less reliable than that with bond angles. It will certainly be useful to repeat the analysis when more accurate experimental information becomes available.

Ring Deformations as a Measure of Electronic Substituent Effects

The interpretation of the benzene ring distortion as due to an effect of the electronic interaction of the substituent with the ring makes it interesting to see how changes in the individual ring angles depend on inductive and resonance effects. Organic chemists are used to quantifying these effects in terms of two parameters derived from reactivity data. These are the inductive parameter σ_I (for which a single scale is in use) and the resonance parameter σ_R (for which several different scales have been proposed, to be used with different types of reactions).[24]

By regressing the angular substituent parameters of Table 7-1 against the corresponding σ_I and σ_R values, it is possible to ascertain the role that inductive and resonance effects have in determining the ring deformations. This procedure leads to the following results[25]:

1. The relatively large angular changes occurring at the ipso and ortho positions of the benzene ring ($\Delta\alpha$ and $\Delta\beta$, respectively) are controlled primarily by the inductive effects of the substituents.
2. The minute changes occurring at the meta position, $\Delta\gamma$, are controlled almost entirely by resonance effects.
3. The rather small changes occurring at the para position, $\Delta\delta$, are controlled by an approximately 1:1 blend of the two effects.

The quality of the fits that are obtained is exemplified in Figure 7-5, where the values of the angle α derived from Table 7-1 are plotted against the values of the inductive parameter σ_I. (This is derived in most cases from the pK_a values of substituted acetic acids in aqueous solution.[26])

The present findings are in keeping with eqs. (7-1) and (7-3), derived from statistical analysis of monosubstituted benzene rings.[11] Note that the two

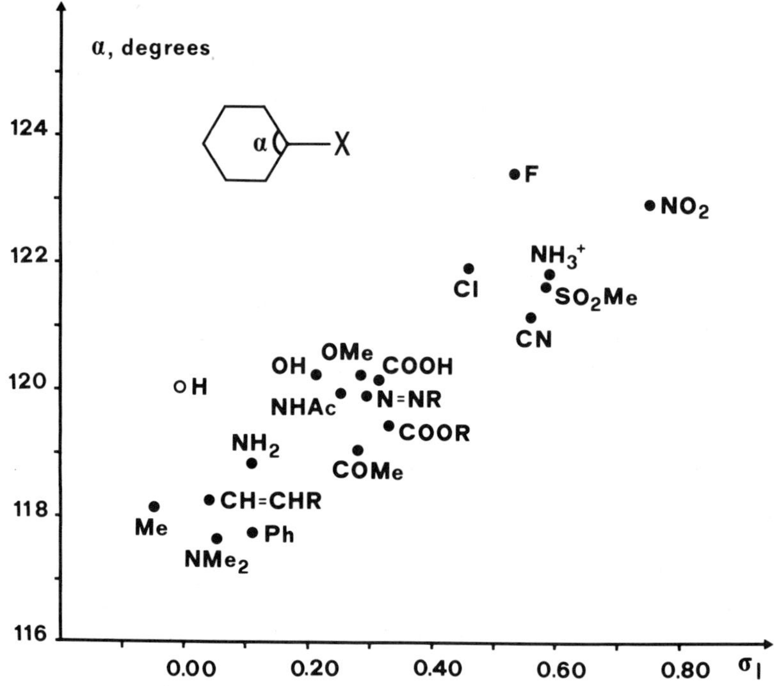

Figure 7-5. Plot of the angle α against the inductive parameter σ_I for monosubstituted benzene derivatives in the solid state. The values of α are derived from the $\Delta\alpha$ values of Table 7-1; those of σ_I are taken from reference 26. Also shown is the reference point of unsubstituted benzene. (Adapted from reference 10.)

approaches not only are different but are based on different sets of experimental data.

The angular substituent parameters may thus be used to investigate inductive and resonance effects in benzene derivatives.[10] These geometrical parameters are a valuable addition to traditional reactivity parameters, since they:

1. Measure the effect of the substituent not on a remote reaction center, but directly on the carbon skeleton of the benzene ring in the ground state.
2. Do not depend on the choice of suitable reaction series.
3. Are not disturbed by solvent effects (if measured in the free molecule), or, alternatively, refer to a fixed, well-known intermolecular environment (if measured in the crystal).
4. Can be related to the conformation of the substituent, since rotation about the $C-X$ bond may affect the extent of conjugation and hence the ring geometry.

The Role of Electron Diffraction Studies

During the last decade, gas-phase electron diffraction has been applied extensively to determine the structural effects of substitution in benzene derivatives. The main advantage of this technique is that it gives accurate information about the structure of free molecules. Other things being equal, the accuracy of the results increases with molecular symmetry. This makes electron diffraction complementary, rather than alternative, to microwave spectroscopy, the other important technique that is used for structural studies in the gaseous phase but cannot be applied to molecules that lack a permanent dipole moment.

The contribution of electron diffraction studies to the investigation of geometrical substituent effects in benzene derivatives has developed primarily along the following lines.

1. Determination of the ring geometry in the monosubstituted derivatives of benzene. Here the relatively low molecular symmetry makes it necessary to implement the electron diffraction data with information obtained by other techniques. It is of importance to select the information with care, and to use it properly in the analysis of the electron diffraction intensities. The purpose of these studies is to provide the basic reference data that are needed to achieve a better understanding of substituent effects.
2. Determination of the ring geometry in the symmetrically para-disubstituted derivatives of benzene, $1,4\text{-}C_6H_4X_2$, with the aim of investigating the nature of the ring deformations and their dependence on the electronic properties of the substituent.
3. Determination of the ring geometry in the symmetrically trisubstituted derivatives of benzene, $1,3,5\text{-}C_6H_3X_3$. Here the primary aim is to look at the effect of substitution on the length of the $C-C$ bonds, an effect that

cannot be determined accurately with the classes of molecules named previously.

4. Studies on the additivity of the structural (and electronic) effects of substitution in polysubstituted benzene rings, with the aim of establishing cases in which deviations from additivity occur, and the reasons for the deviations.

5. Studies on the dependence of the electronic properties of the substituents on intermolecular interactions in the solid state. This requires a comparison of the geometry of the benzene ring in the free molecule with that in the crystal molecule, as well as an analysis of the intermolecular interactions that occur in the crystal.

Electron Diffraction Studies of Monosubstituted Benzene Derivatives

Monosubstituted benzene derivatives with common functional groups have been studied repeatedly by various techniques of structure determination. In most cases, however, the accuracy of these studies has not been high enough to allow a reliable determination of the structural effects of substitution. In other cases, only partial information has been obtained about the geometry of the benzene ring. Disturbing differences between geometrical parameters obtained by different techniques are by no means uncommon.

For these reasons our knowledge of the benzene ring deformations caused by simple functional groups is far from being satisfactory. For several groups we know only the changes of the ring angles with respect to $120°$, as derived by regression from structural data on polysubstituted derivatives in the crystal.[10,22] Accurate, highly reliable experimental results obtained by direct study of the monosubstituted derivative in the gaseous phase are available only for a few groups. There is still serious uncertainty, or no information at all, as to the extent of the ring distortion caused in the isolated molecule by some common functional groups, like NO_2 and NMe_2.

Accurate studies of simple monosubstituted derivatives of benzene in the gaseous phase are strongly needed for a better undertanding of the structural effects of substitution. These studies provide the basic reference data that are required to ascertain the origin of the ring deformations. They serve to establish whether deviations from additivity of geometrical distortions occur in polysubstituted benzene rings. They are also needed to test the correctness of the geometries obtained from theoretical calculations. When the results from gas-phase studies are compared with those obtained by X-ray or neutron crystallography, the subtle effects that intermolecular interactions in the condensed phase may have on the electronic properties of the substituent may be revealed and interpreted.

Microwave spectroscopy and electron diffraction are the two principal experimental techniques employed to determine the structure of the free molecules. Using isotopic substitution, microwave spectroscopy can provide very accurate values for geometrical parameters, except when the atoms involved have small coordinates ($< \sim 0.4$ Å) in the inertial reference framework. This is unfortunately the case for atoms C1 (or C2 and C6) of many monosubstituted derivatives of benzene, since the center of gravity of these molecules is often close to C1, or to the line connecting C2 with C6 (see Figure 7-6a for the numbering of atoms). As a result, microwave spectroscopy may not be the method of choice for providing a reliable description of the geometrical distortion that occurs in the ipso region of the benzene ring.

On the other hand, the relatively low molecular symmetry makes the monosubstituted derivatives of benzene difficult subjects for electron diffraction studies. The carbon skeleton of a benzene ring of C_{2v} symmetry (see Figure 7-1) has, in general, three different C—C bond distances (a, b, c) and four different angles (α, β, γ, δ). These seven internal coordinates are linked by two equations of geometric constraint, expressing the conditions of planarity and ring closure:

$$\alpha + 2\beta + 2\gamma + \delta = 4\pi \tag{7-7}$$

$$a \sin\left(\frac{\alpha}{2}\right) + b \sin\left(\beta + \frac{\alpha}{2} - \pi\right) = c \sin\left(\frac{\delta}{2}\right) \tag{7-8}$$

Five independent parameters are therefore required to describe the geometry of the benzene nucleus in these systems.

The accurate determination of these five geometrical parameters solely by electron diffraction is a hopeless task. Particularly difficult is the measurement of the small differences in the lengths of the a, b, and c bonds, which seldom exceed 0.02 Å. It is clear that in the electron diffraction study of a monosubstituted benzene derivative full advantage must be taken of any reliable piece of information on the ring geometry that is available from studies by other techniques.

If a substitution geometry has been determined by microwave spectroscopy, the internal angle at the para position of the ring, δ, and the differences between the C—C bond distances that do not involve atoms with small coordinates may be taken from the microwave study and imposed as constraints in the least-squares refinement. But even if a substitution geometry is not available, the details of the ring deformation are expected to be more reliable when the refinement is based conjointly on the electron diffraction intensities and the experimental rotational constants taken from the microwave study.[27]

With some substituents the differences between the lengths of the ring C—C bonds are determined with reasonable accuracy by *ab initio* MO calculations with geometry optimization, although their absolute values may not be reliable. These differences may fruitfully be used as constraints in the analysis of the electron diffraction data.

Statistical analysis of the results of semiempirical MO calculations at the level of complete neglect of differential overlap, CNDO/2, carried out on about 20 monosubstituted derivatives of benzene, has shown that the b and c bond

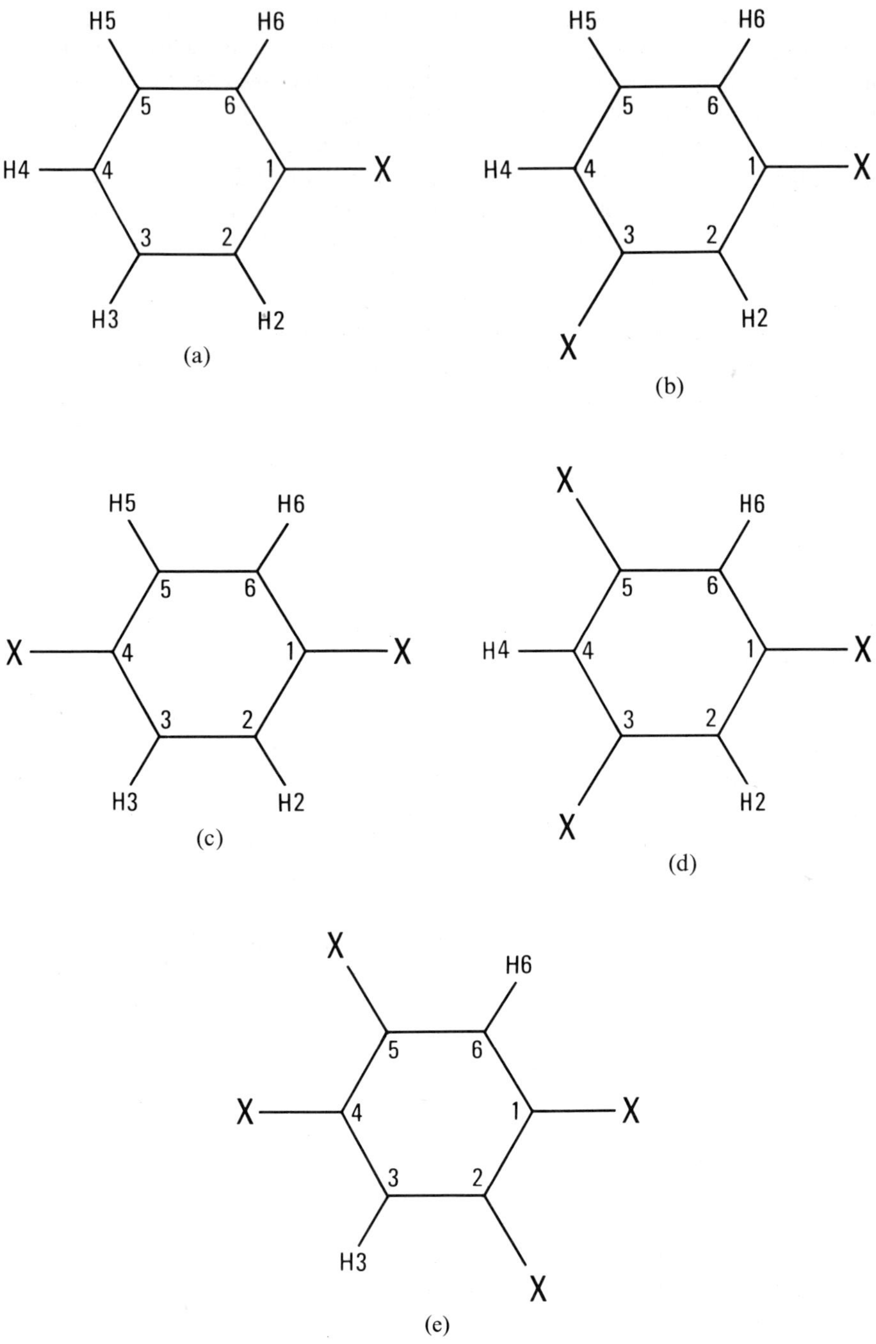

Figure 7-6. Numbering of the C and H atoms in substituted benzene rings: (a) monosubstituted, (b) 1,3-disubstituted, (c) 1,4-disubstituted, (d) 1,3,5-trisubstituted, and (e) 1,2,4,5-tetrasubstituted.

distances and the γ angle are the least affected by substitution.[28] It has been suggested[28] that these three variables may be fixed at the values found in unsubstituted benzene, which leaves only two independent ring parameters to be determined from the electron diffraction data.

Equations (7-5) and (7-6), relating the change of the angle β to the change of α in monosubstituted benzene rings with first- and second-row substituents, respectively, may also serve as constraints.[29] As we have seen, these equations were derived by regression from a large sample of monosubstituted benzene derivatives, mostly studied by X-ray crystallography.[11]

In the least-squares refinement of the molecular model not all the important amplitudes of vibration can be treated as independent variables. Amplitudes corresponding to interatomic distances that fall under the same peak of the radial distribution are often grouped together. Within each group, differences between amplitudes are either taken from spectroscopic calculations or fixed at reasonable values. The effect that the choice of these differences may have on the ring deformations should, of course, be thoroughly checked.

Several monosubstituted derivatives of benzene have been studied by gas-phase electron diffraction, in which the carbon skeleton of the ring has been allowed to distort from D_{6h} to C_{2v} symmetry. Important geometrical parameters for these molecules are reported in Table 7-2, where the geometrical constraints used in the refinement are also specified.

Some of the benzene ring geometries reported in Table 7-2 are in good agreement with the results obtained by other techniques of structure determination. In other cases, however, significant differences occur. Six important molecules that have been studied by several techniques will be discussed here in detail. Their geometries from electron diffraction studies are compared in Tables 7-3 to 7-8 with those obtained from other experimental and theoretical studies. When inspecting these tables it should be considered that a straightforward comparison of geometrical parameters obtained by different techniques may be attempted with the ring angles, but such a comparison is certainly inappropriate with bond distances. The variance in the values of $\langle r(C-C) \rangle$ and $r(C-X)$ reported in each table reflects prevalently the different physical meaning that a bond distance has in each technique of structure determination, rather than experimental error or possible geometrical effects caused by the different states of aggregation. *Differences* between bond distances are more easily transferable from one experiment to another than are the distances themselves, because some of the systematic effects cancel. For this reason the differences $a - b$ and $b - c$, rather than the individual values of a, b, and c, are reported in Tables 7-3 to 7-8.

Fluorobenzene (Table 7-3). The determination of the molecular structure of fluorobenzene has been the subject of a number of experimental and theoretical studies, whose results have been compared in detail.[30] The values of the internal ring angles from six different experiments[3, 30, 41–43] are seen to be in excellent agreement, the only exception being the angles α and β obtained by liquid crystal NMR spectroscopy in the ZLI 1167 mesophase.[43] The *ab initio* MO calculations[44–46] yield values of γ and δ that reproduce the experimental results

Table 7-2. IMPORTANT GEOMETRICAL PARAMETERS FOR MONOSUBSTITUTED BENZENE DERIVATIVES AS DETERMINED BY ELECTRON DIFFRACTION

Bond distances are given in Å, angles in degrees. Estimated total errors are given as error limits; σ (3σ) values from least-squares refinement are given in parentheses (square brackets) as units in the last digit. Geometrical parameters are labeled according to Figure 7-1; C_{2v} symmetry has been assumed throughout.

Substituent	α	β	γ	δ	$\langle r_g(C-C)\rangle^a$	$r_g(C-X)$	Constraints	Ref.
F	123.4 ± 0.2	118.0 ± 0.2	120.2 ± 0.3^b	120.2 ± 0.4^b	1.396 ± 0.003	1.356 ± 0.004	$\begin{cases} a - b = -0.012^c \\ b - c = -0.002^c \end{cases}$	30
Cl	$121.7[6]$	$118.8^{b,d}$	$120.1^{b,d}$	$120.4[15]$	$1.400[1]$	$1.737[5]$	$a = b = c$	31
Br	$121.5(4)$	$119.0(7)$	$120.0^{b,d}$	$120.5^{b,d}$	$1.396[3]$	$1.899[3]$	See note e	27
CN	121.9 ± 0.2	118.6^b	120.5^b	119.9^b	1.400 ± 0.003	1.438 ± 0.003	$\begin{cases} a = b = c \\ \Delta\beta = -0.591\Delta\alpha - 0.301^f \end{cases}$	29
SO_2Me^g	121.9 ± 1.3	118.9 ± 0.8	119.6 ± 0.7^b	121.2 ± 0.7^b	1.399 ± 0.003	1.771 ± 0.012	$a = b = c$	32
$NO_2{}^h$	$125.1[14]$	$115.7[11]$	$122.8^{b,d}$	$117.9^{b,d}$	$1.391[2]$	$1.478[13]$	$a = b = c$	33
OH	121.4 ± 0.2	118.9^b	120.6^b	119.7^b	1.398 ± 0.003	1.381 ± 0.004	$\begin{cases} a = b = c \\ \Delta\beta = -0.591\Delta\alpha - 0.301^f \end{cases}$	34
SPh^i	120.2 ± 0.6	119.7 ± 0.7	120.5^b	119.4^b	1.401 ± 0.003	1.772 ± 0.005^j	$a = b = c$	35
$CF_3{}^k$	119.7 ± 0.4	120.4 ± 0.4	119.4 ± 0.4^b	120.7 ± 0.6^b	1.397 ± 0.003	1.504 ± 0.004	$a = b = c$	18
Me	118.7 ± 0.4	120.4 ± 0.4	120.6 ± 0.5^b	119.4 ± 0.6^b	1.400 ± 0.003^l	1.507 ± 0.004	$b = c$	36, 37
Et^m	118.3 ± 1.0	121.1 ± 0.6^b	120.0	119.5 ± 0.3^b	1.399 ± 0.004^n	1.524 ± 0.009	$\begin{cases} b = c = 1.399\ (r_g) \\ \beta = 120 \end{cases}$	28, 38

Ph[o]	119.4(4)	119.9(4)	120.9(5)[b]	119.0(6)[b]	1.400[d]	1.509(4)[j]	See note p	39
Ph[o,q]	117.9(4)	121.3(4)	119.8(5)[b]	119.8(6)[b]	1.400[d]	1.491(4)[j]	See note r	39
p-F—C$_6$H$_4$[o]	118.4(6)	120.8(2)	120.7(6)[b]	118.5(7)[b]	1.398(1)[j]	1.499(3)[j]	See note s	40
p-Cl—C$_6$H$_4$[o,t]	117.4(6)	120.7(3)	123.0[b,d]	115.1[b,d]	1.399(1)[j]	1.482(4)[j]	See note s	40
SiH$_3$[u]	117.4	122.2[b,d]	118.7[b,d]	120.8	1.394 ± 0.005[j]	1.844 ± 0.005[j]	a = b = c	1

[a] Mean value of the C—C bond distances of the benzene ring.

[b] Dependent parameter.

[c] Taken from a study by microwave spectroscopy.[3]

[d] Calculated from the published ring parameters.

[e] The molecular geometry was determined by a conjoint analysis based on electron diffraction intensities and microwave rotational constants. The following values were obtained for the three independent C—C bonds of the benzene ring: $a = 1.394(3)$, $b = 1.396(5)$, $c = 1.394(7)$ Å (r_a values).

[f] From a statistical analysis of monosubstituted benzene rings.[11]

[g] The conformation of the molecule is such that the C—S—C plane is perpendicular to the plane of the ring.

[h] The twist angle of the substituent is found to be about 23°. The nonplanarity of the molecule may, however, be an effect of high-amplitude motions.

[i] At least two different conformations agree with the experimental data.

[j] From the r_a value given in the original paper.

[k] The experimental data agree with nearly free rotation of the CF$_3$ group about the C—CF$_3$ axis.

[l] The a bond distance and the difference $a - b$ were treated as independent variables. This led to $a = 1.408 ± 0.005$ Å, $b = c = 1.396 ± 0.004$ Å (r_g values).

[m] The experimental data indicate a perpendicular or highly twisted conformation.

[n] The a bond distance was treated as an independent variable; this led to $a = 1.400 ± 0.004$ Å (r_g value).

[o] The dihedral angle between the planes of the two rings is about 45°.

[p] The three different C—C bond distances of the two rings were treated as independent variables. This led to $a = 1.404(4)$, $b = 1.395(5)$, $c = 1.396(5)$ Å (r_a values). However, these values were kept fixed in the final least-squares refinements.

[q] Perdeuterated biphenyl.

[r] The three different C—C bond distances of the two rings were treated as independent variables. This led to $a = 1.403(6)$, $b = 1.396(8)$, $c = 1.398(13)$ Å (r_a values). However, these values were kept fixed in the final least-squares refinements.

[s] The 12 aromatic C—C bonds of the molecule were assumed to have equal length.

[t] The values of the ring angles given in the original paper are not consistent with eq (7-7).

[u] In this early work the molecular parameters were not refined by the least-squares method.

Table 7-3. INTERNAL RING ANGLES (deg) AND DISTANCE PARAMETERS (Å) IN FLUOROBENZENE[a]

Technique[b]	α	β	γ	δ	$\langle r(\text{C—C})\rangle$	$a - b$	$b - c$	$r(\text{C—F})$	Ref.
ED[c]	123.4	118.0	120.2	120.2	1.396	-0.012^d	-0.002^d	1.356	30
MW[e]	123.4	117.9	120.5	119.8	1.392	-0.012	-0.002	1.354	3
XD[f]	123.8	117.9	120.3	119.8	—	-0.017^g	—	1.364	41
XD[h]	123.3	118.0	120.3	120.0	—	-0.018^i	—	1.363	42
NMR (ZLI 1167)[j]	124.3	117.5	120.4	119.8	1.389	-0.015	-0.006	1.371	43
NMR (ZLI 1132)[j]	123.1	118.1	120.3	120.0	1.395	-0.016	0.004	1.355	43
MO (4-21G)	122.4	118.6	120.3	119.8	1.381	-0.009	-0.002	1.369	44
MO (6-31G)	123.0	118.2	120.3	120.0	1.385	-0.011	-0.001	1.376	45
MO (6-31G**)	122.4	118.4	120.5	119.8	1.383	-0.007	-0.001	1.331	46

[a] Experimental errors are not shown. They have been estimated in different ways by the various authors and can be found in the original papers. (Those on the geometrical parameters that have been determined by electron diffraction are given in Table 7-2). This table is not aimed at showing whether the geometrical parameters of the benzene ring obtained by different techniques are consistent within their estimated errors, but rather at ascertaining whether they can be determined reliably to within a few tenths of a degree (for bond angles) or a few thousandths of an angstrom (for bond distances). For this reason bond angles are presented with one decimal figure and bond distances with three, although in some cases one more figure appears in the original papers.

[b] The following abbreviations are used: ED, electron diffraction; MW, microwave spectroscopy; XD, X-ray crystallography; NMR (...), liquid–crystal nuclear magnetic resonance spectroscopy (with the nematic mesophase specified in parentheses); MO (...), *ab initio* molecular orbital calculations with geometry optimization (with the basis set specified in parentheses).

[c] Bond distances are r_g values.

[d] Assumed from the microwave study.[3]

[e] Substitution structure. The a coordinates of F and C2 are not substitution coordinates, however.

[f] From the molecular structure of p-fluorobenzoic acid,[41] after subtracting the known angular deformations caused by the COOH substituent.[10]

[g] The effect of the COOH substituent is probably very small and has been ignored.

[h] From the molecular structure of p-fluoroaniline hydrochloride,[42] after subtracting the known angular deformations caused by the NH_3^+ substituent.[10]

[i] The effect of the NH_3^+ substituent is probably very small and has been ignored.

[j] r_α structure. The c bond distance was assumed to be 1.3976 Å.

Source: Reference 30.

perfectly. They are less successful, however, in reproducing the large distortion of the ipso region of the ring caused by the F substituent. These difficulties are more likely to originate from the neglect of correlation energy than from basis set truncation.[46]

As regards the differences between bond distances, all the studies agree in indicating that the a bonds are shorter than the b bonds by about 0.01 Å, while the b and c bonds have nearly equal lengths. Note, however, the larger spread of values of these differences as compared with bond angles.

Chlorobenzene (Table 7-4). The values of the ipso angle determined by electron diffraction,[31] X-ray crystallography,[48] and liquid crystal NMR spectroscopy[49] are all in close agreement, $\alpha = 121.7–121.8°$. On the other hand, the substitution structure obtained by microwave spectroscopy[47] shows an almost undistorted benzene ring, with $\alpha = 120.2°$. It should be noted, however, that in chlorobenzene the a coordinates of C1, C2, and H2 are all rather small, which makes it impossible to determine accurately the geometry of the ipso region of the ring by the substitution method.[48, 50] Concerning the angles γ and δ, electron diffraction, microwave spectroscopy, and X-ray crystallography give similar results, while the values obtained by liquid crystal NMR spectroscopy differ considerably.

As Table 7-4 indicates, no reliable information on bond distance changes is available for chlorobenzene.

Cyanobenzene (Table 7-5). The molecular structure of cyanobenzene has also been investigated repeatedly. The values of the ring angles obtained by electron diffraction[29] agree with those from the microwave study.[4] The angular deformations of the benzene ring caused by the CN substituent have also been derived by regression from a large set of structural data, consisting prevalently of solid-state results on para-disubstituted derivatives.[10] These angular deformations may be considered—at least to a first approximation—as referring to cyanobenzene in a crystalline, highly polar environment. (Cyanobenzene itself has been studied by X-ray crystallography,[52] but the limited number of reflexions used in the analysis makes it impossible to determine accurately the structural effects of substitution.) The values of the angles γ and δ derived by regression are in agreement with the gas-phase results. The distortion of the ipso region, however, is somewhat less pronounced than that in the free molecule. This is probably an effect of the strong intermolecular interactions that involve the CN substituent in the crystals of all the substituted cyanobenzenes used in the regression. The nature of this effect will be treated in more detail in the last section of this chapter.

The values of α and β obtained by liquid crystal NMR spectroscopy[51] are close to the values derived by regression. This is not the case for the other ring angles, however.

The *ab initio* MO calculations[45, 46] give values of γ and δ that agree with the gas-phase and solid-state results, but they are unable to reproduce the angular distortion of the ipso region caused by the CN substituent. As in the case of fluorobenzene, this failure has been attributed to the neglect of correlation energy.[46]

Table 7-4. Internal Ring Angles (deg) and Distance Parameters (Å) in Chlorobenzene[a]

Technique[b]	α	β	γ	δ	$\langle r(\text{C—C})\rangle$	$a-b$	$b-c$	$r(\text{C—Cl})$	Ref.
ED[c]	121.7	118.8[d]	120.1[d]	120.4	1.400	0.000[e]	0.000[e]	1.737	31
MW[f]	120.2	119.8	120.2	119.8	1.394	0.013	−0.012	1.725	47
XD[g]	121.8	118.7	120.4	120.0	—	0.001[h]	—	1.736	48
NMR (ZLI 1167)[i]	121.8[d]	119.3	119.1	121.4[d]	1.401	−0.025	0.017	—	49

[a] See note a to Table 7-3 concerning experimental errors.
[b] See note b to Table 7-3 concerning abbreviations.
[c] Bond distances are r_g values.
[d] Calculated from the published ring parameters.
[e] Assumed.
[f] Substitution structure.
[g] From the molecular structure of p-chlorobenzoic acid,[48] after subtracting the known angular deformations caused by the COOH substituent.[10]
[h] The effect of the COOH substituent has been ignored.
[i] r_α structure. The c bond distance was assumed to be 1.3976 Å.

Table 7-5. INTERNAL RING ANGLES (deg) AND DISTANCE PARAMETERS (Å) IN CYANOBENZENE[a]

Technique[b]	α	β	γ	δ	$\langle r(\mathrm{C-C})\rangle_{\mathrm{ring}}$	$a - b$	$b - c$	$r(\mathrm{C-CN})$	Ref.
ED[c]	121.9	118.6	120.5	120.0	1.400	0.000[d]	0.000[d]	1.438	29
MW[e]	121.8	119.0	120.1	120.1	1.394	−0.008	−0.002	1.451	4
XD[f]	121.1	119.2	120.3	119.9	—	—	—	—	10
NMR (ZLI 1167)[g]	120.8	119.4	119.7	121.0	1.402	0.010	−0.002	1.434	51
MO (6-31G)	120.1	119.8	120.1	120.2	1.389	0.008	−0.003	1.435	45
MO (6-31G**)	120.4	119.6	120.1	120.2	1.387	0.007	−0.003	1.445	46

[a] See note a to Table 7-3 concerning experimental errors.
[b] See note b to Table 7-3 concerning abbreviations.
[c] Bond distances are r_g values.
[d] Assumed.
[e] Substitution structure. It should be considered, however, that the a coordinate of C2 is small ($\sim$0.15 Å) and has been calculated from the first moment equation.
[f] By regression from polysubstituted benzene rings.
[g] r_α structure. The H2...H6 distance was assumed to be 4.300 Å.
Source: Reference 29.

Concerning the differences between bond distances, the various studies consistently indicate that the b bonds are slightly shorter than the c bonds, while the indications on the relative lengths of the a and b bonds are contradictory.

Nitrobenzene (Table 7-6). The molecular structure of gaseous nitrobenzene has been determined by microwave spectroscopy[53] and electron diffraction.[33] Although the a coordinate of C1 is small (~ 0.16 Å), the value of the angle α produced by the microwave study is in excellent agreement with that obtained by electron diffraction. The two studies, however, give quite different results for the other angles. The value of δ obtained by electron diffraction, 117.9°, is too low in view of the high value of α.

An early low-temperature X-ray diffraction study of nitrobenzene[57] was not accurate enough to give reliable values for the benzene ring deformations caused by the NO_2 substituent. As with cyanobenzene, the ring deformations of the crystal molecule will be estimated here from the results obtained for many polysubstituted benzene derivatives studied in the solid state. Three separate regressions, based on different sets of structural data,[10,22,54] agree in indicating that the distortion of the ipso region is much less pronounced in the solid state (α = 122.7–122.9°) than in the gaseous phase (α = 125.0–125.1°). This difference is puzzling, since (a) a distortion even smaller than that occurring in the solid state is obtained via *ab initio* MO calculations using different basis sets (α = 122.2–122.4°),[45,56] and (b) in the case of p-dinitrobenzene the ipso angle has comparable values in the gaseous phase (122.9°)[58] and in the crystal (123.4°).[59]

Concerning the para region of the benzene ring, the values of the angles γ and δ obtained by regression from solid-state results and those produced via the MO calculations are all in substantial agreement with the results of the microwave study.

The molecular structure of nitrobenzene has also been investigated by liquid crystal NMR spectroscopy in two different mesophases.[55] The two experiments give consistent results, but the pattern of ring distortions emerging from this study is unusual, and suspiciously different from the pattern produced by any of the other methods of structure determination.

Needless to say, the molecular structure of nitrobenzene is still an open problem and calls for further, more accurate studies.

Phenol (Table 7-7). An experimental study by microwave spectroscopy[14] and three theoretical studies by *ab initio* MO calculations[15,45,60] have shown that in this molecule the asymmetry of the substituent gives rise to minor deviations of the ring geometry from the expected C_{2v} symmetry. The planarity of the ring is still preserved, but the axial symmetry is lost. In no case, however, are the deviations from C_{2v} symmetry found to exceed 0.006 Å for bond distances and 0.5° for bond angles.

The determination of these small, additional distortions is beyond the possibilities of gas-phase electron diffraction. Therefore the benzene ring was constrained to C_{2v} symmetry in the electron diffraction study.[34] The same symmetry was assumed in the two studies[10,22] in which the distortions of the

Table 7-6. INTERNAL RING ANGLES (deg) AND DISTANCE PARAMETERS (Å) IN NITROBENZENE[a]

Technique[b]	α	β	γ	δ	$\langle r(\text{C}-\text{C})\rangle$	$a - b$	$b - c$	$r(\text{C}-\text{NO}_2)$	Ref.
ED[c]	125.1	115.7	122.8[d]	117.9[d]	1.391	0.000[e]	0.000[e]	1.478	33
MW[f]	125.0	117.1	120.3	120.2	1.391	−0.028	0.007	1.492	53
XD[g]	122.9	118.1	120.3	120.4	—	—	—	—	10
XD[g]	122.9	118.1	120.0	120.8	—	−0.008[h]	0.004[h]	—	22
XD[g]	122.7	118.1	120.3	120.5	—	—	—	—	54
NMR (ZLI 1167)[i]	121.8	119.8	117.8	123.0	1.414	−0.024	0.012	1.44	55
NMR (PCH)[i]	121.1	120.2	117.6	123.3	1.416	−0.021	0.015	1.45	55
MO (STO-3G)	122.4	118.5	120.0	120.6	1.387	0.001	−0.003	1.506	56
MO (6-31G)	122.2	118.6	120.1	120.4	1.386	0.000	−0.004	1.449	45

[a] See note a to Table 7-3 concerning experimental errors.
[b] See note b to Table 7-3 concerning abbreviations.
[c] Bond distances are r_g values.
[d] Calculated from the published ring parameters.
[e] Assumed.
[f] Substitution structure. It should be considered, however. that the a coordinate of C1 is small (~ 0.16 Å) and has been calculated from the first moment equation.
[g] By regression from polysubstituted benzene rings.
[h] From the bond deformation parameters given in Table 6 of reference 22.
[i] r_α structure. The H2…H6 distance was assumed to be 4.300 Å.

Table 7-7. Important Bond Angles (deg) and Distance Parameters (Å) in Phenol[a,b]

Technique[c]	α	β	γ	δ	$\angle$ C2—C1—O[d]	$\langle r(\text{C—C})\rangle$	$a-b$	$b-c$	$r(\text{C—OH})$	Ref.
ED[e]	121.4	118.9	120.6	119.7	122.6	1.398	0.000[f]	0.000[f]	1.381	34
MW[g]	120.9	119.3	120.6	119.2	122.1	1.393	−0.002	−0.002	1.375	14
XD[h]	120.2	119.6	120.6	119.4	—	—	—	—	—	10
XD[h]	120.3	119.5	120.6	119.5	—	—	0.004[i]	0.001[i]	—	22
MO (STO-3G)	119.8	119.8	120.6	119.4	122.6	1.388	0.011	−0.003	1.395	15
MO (STO-3G)	120.0	119.8	120.2	120.0	122.8	1.388	0.009	−0.004	1.396	60
MO (6-31G)	120.7	119.4	120.5	119.4	122.5	1.387	−0.001	−0.002	1.378	45

[a] See note a to Table 7-3 concerning experimental errors.

[b] The geometrical parameters given in some of the original papers[14,15,45,60] have been made consistent with a benzene ring of C_{2v} symmetry by averaging appropriate bond distances and angles.

[c] See note b to Table 7-3 concerning abbreviations.

[d] The hydrogen atom of the hydroxyl group is cis to C2.

[e] Bond distances are r_g values.

[f] Assumed.

[g] Substitution structure. The a coordinates of C2 and C6 and the b coordinates of C1, C4, O, and H4 are not substitution coordinates, however.

[h] By regression from polysubstituted benzene rings.

[i] From the bond deformation parameters given in Table 6 of reference 22.

benzene ring caused by the OH substituent were derived by regression from the geometries of many polysubstituted derivatives in the solid state. To make Table 7-7 more readable, the results of all the structural studies on phenol have been made consistent with a benzene ring of C_{2v} symmetry, by averaging appropriate bond distances and angles whenever lower symmetry had been assumed.

The agreement of the results of the various studies is good for the angles γ and δ (except for one of the MO studies),[60] but is less satisfactory for the other geometrical parameters. As with cyanobenzene, the distortion of the ipso region derived by regression from solid-state results is somewhat less pronounced than that obtained from the gas-phase studies. This is probably an effect of the O—H...O hydrogen bonds that are formed in the crystal, as discussed in the last section of this chapter. Concerning the geometry obtained by *ab initio* MO calculations, the 6-31G level of approximation is clearly superior to STO-3G in reproducing the deformations of the benzene ring observed in the free molecule.

Toluene (Table 7-8). The structure of the free molecule has been determined by electron diffraction,[36,37] microwave spectroscopy,[61] and *ab initio* MO calculations with various basis sets.[45,62,63] The distortions of the benzene ring caused by the Me substituent have also been obtained by regression from solid-state results on polysubstituted derivatives.[10,22]

The results of the various studies are in good agreement, although the distortion of the ipso region in the solid state ($\alpha = 118.1-118.2°$) is slightly more pronounced than that obtained from gas-phase studies ($\alpha = 118.7-119.1°$). The MO calculations give an intermediate level of distortion, with $\alpha = 118.4-118.6°$.

As regards bond distances, there is a clear indication from Table 7-8 that the a bonds are slightly longer than the b bonds, while the b and c bonds have essentially equal length.

Analysis of Results of Studies on Monosubstituted Benzene Derivatives. From the six cases discussed above, it is clear that not all the geometrical parameters of Table 7-2 should be regarded as highly accurate, and that the reliability of the results of an electron diffraction study may differ considerably for the various parameters that define the geometry of a monosubstituted benzene ring. The angle α is probably well determined in the majority of the molecules considered. In all cases that have been studied in detail, this parameter appears to be substantially—if not totally—insensitive to the conditions of the analysis: ie, to the treatment of the variables that cannot be determined solely by electron diffraction, and thus must be assumed or constrained in the least-squares refinement. This may not be the case for the other ring angles, which are more sensitive to the geometrical constraints used. The effect that assuming different values for the difference $a - b$ has on the other geometrical parameters of the benzene ring has been investigated in some cases: the results obtained for fluorobenzene[30] are shown in Table 7-9.

In Figure 7-7 the values of the angle α for all the monosubstituted derivatives of benzene that have been studied in the gaseous phase, by either electron diffraction or microwave spectroscopy, are plotted against the values of the

Table 7-8. INTERNAL RING ANGLES (deg) AND DISTANCE PARAMETERS (Å) IN TOLUENE[a]

Technique[b]	α	β	γ	δ	$\langle r(\text{C--C})\rangle_{\text{ring}}$	$a - b$	$b - c$	$r(\text{C--Me})$	Ref.
ED[c]	118.7	120.4	120.6	119.4	1.400	0.012	0.000[d]	1.507	36, 37
MW[e]	119.1	120.5	120.3	119.4	1.396	0.002	0.000	1.513	61
RX[f]	118.1	121.0	120.4	119.2	—	—	—	—	10
RX[f]	118.2	121.0	120.2	119.5	—	0.006[g]	0.001[g]	—	22
MO (4-21G)[h]	118.6	120.8	120.2	119.5	1.385	0.004	0.001	1.519	62
MO (STO-3G)[h]	118.6	120.7	120.1	119.8	1.388	0.006	0.000	1.526	63
MO (4-31G)[h]	118.4	120.9	120.2	119.4	1.385	0.005	0.000	1.509	63
MO (6-31G)[i]	118.4	120.9	120.2	119.5	1.389	0.005	0.000	1.511	45

[a] See note a to Table 7-3 concerning experimental errors.

[b] See note b to Table 7-3 concerning abbreviations.

[c] Bond distances are r_g values.

[d] Assumed.

[e] Substitution structure. It should be considered, however, that the a coordinate of C2 is small (~ 0.24 Å) and has been calculated from the first moment equation.

[f] By regression from polysubstituted benzene rings.

[g] From the bond deformation parameters given in Table 6 of reference 22.

[h] One of the C—H bonds of the methyl group has been assumed to lie in a plane perpendicular to the plane of the ring.

[i] One of the C—H bonds of the methyl group has been assumed to lie in the ring plane. The geometrical parameters of the ring given in the original paper have been made consistent with a benzene ring of C_{2v} symmetry by averaging appropriate bond distances and angles.

Table 7-9. DEPENDENCE OF THE RING
GEOMETRY OF FLUOROBENZENE ON
THE VALUE ATTRIBUTED TO THE
DIFFERENCE $a - b$

Parameter	$a - b$, Å		
	-0.020^a	-0.010^a	0.000^a
a, Å (r_a)	1.381	1.387	1.393
$b = c$, Å^b (r_a)	1.401	1.397	1.393
r_a(C—F), Å	1.354	1.355	1.356
α, deg	123.4	123.5	123.5
β, deg	118.3	117.8	117.4
γ, deg^b	119.9	120.4	120.8
δ, deg^b	120.3	120.1	120.1
R factor	0.0246	0.0222	0.0216

^{a} Assumed.
^{b} Dependent parameter.
Source: Reference 30.

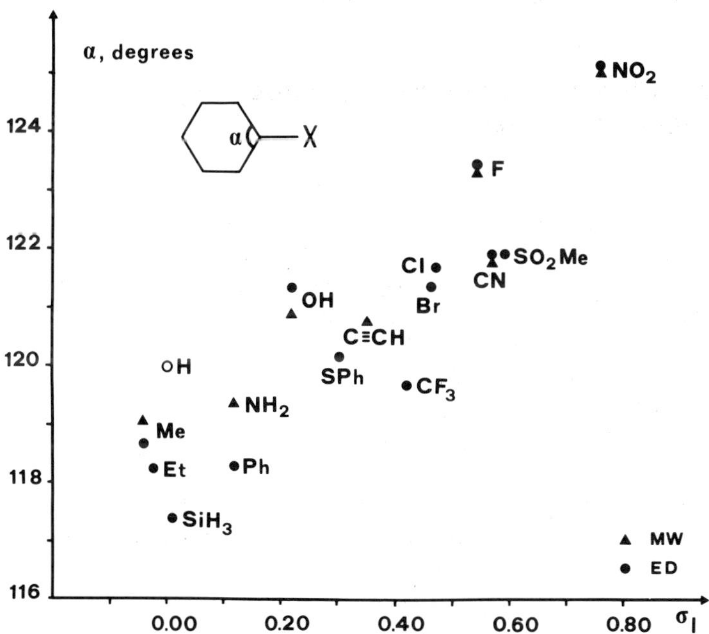

Figure 7-7. Plot of the angle α against the inductive parameter σ_I for monosubstituted benzene derivatives in the gaseous phase. The values of α obtained by electron diffraction are taken from Table 7-2 (that for the phenyl substituent is the average of four entries: biphenyl, perdeuterated biphenyl, p-fluorobiphenyl, p-chlorobiphenyl). The values of α obtained by microwave spectroscopy refer to the following molecules: fluorobenzene,[3] cyanobenzene,[4] aniline,[5] phenol,[14] nitrobenzene,[53] toluene,[61] phenylacetylene.[64] Also shown is the reference point of unsubstituted benzene.

inductive parameter σ_I. As expected, a correlation exists between α and σ_I ($r = 0.91$ on 20 data points), of the same general quality as that obtained with the values of α derived from solid-state measurements (Figure 7-5, $r = 0.91$ on 18 data points). This reinforces the view that the values of α reported in Table 7-2 are, in general, free from serious experimental errors.

Since the scatter of the data points in the plots of Figures 7-5 and 7-7 is far too large to arise solely from experimental errors in the values of α, other effects must be invoked. Resonance interactions between the ring and the substituent are expected to cause the angle α to decrease somewhat[8]: the decrease should be appreciable with the strongest π donors, eg, NH_2, OH, NMe_2, OMe, F. It should also be considered that the values of the inductive parameter σ_I may differ considerably when derived from reaction series other than the dissociation of substituted acetic acids, and even more so when obtained by entirely different methods, eg, by ^{19}F NMR. An extreme case is that of the NO_2 substituent, for which values of σ_I ranging from 0.57 to 0.76 have been reported.[26] Be that as it may, the main point is that an inherent difference exists between σ_I, a reactivity parameter that measures the polar effect of a substituent on a *remote* reaction center, and α, a ground-state structural parameter that measures the *local* effect of the substituent on the geometry of the benzene ring.

Electron Diffraction Studies of Symmetrically para-Disubstituted Benzene Derivatives

The para-disubstituted derivatives of benzene, $p\text{-}C_6H_4XY$, have been the subject of extensive structural research in the solid state, and, more recently, also in the gaseous phase. The information on the ring geometry of these molecules available by the end of 1978—originated almost entirely from solid-state measurements—was the major source of the data that were used[10] to derive the angular substituent parameters of Table 7-1.

A substantial proportion of the gas-phase studies has been devoted to the symmetrically para-disubstituted derivatives, $p\text{-}C_6H_4X_2$. These studies are aimed at determining accurately the structural effects of substitution in the free molecules, to allow an analysis of the nature of these effects.

The symmetrically para-disubstituted derivatives of benzene are not accessible to microwave studies, because they lack a permanent dipole moment. On the other hand, they are better suited for electron diffraction analysis than the monosubstituted or other disubstituted derivatives, because many of the atom–atom interactions double in the electron scattering from a $p\text{-}C_6H_4X_2$ molecule. At the same time some of the structural effects of substitution may superimpose and become more conspicuous. This is the case of any change occurring in the C1...C4 nonbonded distance, or in the length of the central C—C bonds of the ring, C2—C3 and C5—C6 (see Figure 7-6c for the numbering of atoms). But also the changes of the internal angles at the ipso and para positions of the ring may reinforce each other, since the angular parameters $\Delta\alpha$ and $\Delta\delta$ have equal sign with several substituents (see Table 7-1). Of course, additivity of the structural effects of substitution should not be taken for granted. But whenever deviations from addivity are observed, it is certainly of interest to understand why this is the case.

In all the electron diffraction studies of symmetrically para-disubstituted derivatives the benzene ring has been assumed to have D_{2h} symmetry. Strictly speaking this assumption is justified only if the substituent has axial symmetry. However, *ab initio* MO calculations carried out on *p*-dihydroxybenzene[65] and *p*-diformylbenzene[66] predict that these unsymmetrical substituents cause only relatively small deviations of the ring geometry from D_{2h} symmetry. This additional distortion cannot be determined by an electron diffraction study.

Three independent geometrical parameters are required to describe the carbon skeleton of a benzene ring with D_{2h} symmetry. These may be chosen in several ways; particularly convenient choice for interpreting the structural effects of substitution is the following.

1. The internal ring angle at the place of substitution, $\angle C2{-}C1{-}C6$
2. The C1...C4 nonbonded distance, $r(C1...C4)$
3. The difference between the lengths of the two nonequivalent C—C bonds of the ring, $\Delta(C{-}C) = r(C1{-}C2) - r(C2{-}C3)$

The first two parameters are measured accurately by electron diffraction, since their variation is associated with changes in several of the longest atom–atom separations.[36,67] Unfortunately the third parameter is poorly determined, since $r(C1{-}C2)$ and $r(C2{-}C3)$ have very close values.

Important geometrical parameters for the 12 symmetrically para-disubstituted derivatives that have been studied by electron diffraction are presented in Table 7-10. Inspection shows that a substantial part of the variance of the ring parameters is due to convariance, arising from a concerted change of $\angle C2{-}C1{-}C6$ and $r(C1...C4)$. The gradual decrease of $\angle C2{-}C1{-}C6$ from 123.5° (X = F) to 115.7° (X = SiMe$_3$) is accompanied by a simultaneous increase of $r_g(C1...C4)$ from 2.712 to 2.900 Å. The two geometrical parameters are linearly related (see Figure 7-8), with a correlation coefficient $r = -0.997$. The point corresponding to unsubstituted benzene [$\angle C2{-}C1{-}C6 = 120°$, $r_g(C1...C4) = 2.796$ Å[74]] lies on the regression line.

The existence of this linear relationship was first reported in 1980[75]; the very few data points from gas-phase studies available at that time were implemented with many results obtained in the solid state. This gave a less satisfactory correlation ($r = -0.97$ on 33 data points), since the C1...C4 distances determined by X-ray crystallography tend to be systematically shorter (by ~ 0.02 Å) with respect to the thermal-average internuclear separations r_g, obtained in the gaseous phase.

Although the individual C—C bond distances of the benzene ring are not measured accurately, their mean is very well determined by electron diffraction. The values of $\langle r_g(C{-}C) \rangle$ reported in Table 7-10 fall in a rather narrow range,

Table 7-10. IMPORTANT GEOMETRICAL PARAMETERS FOR SYMMETRICALLY para-DISUBSTITUTED BENZENE DERIVATIVES AS DETERMINED BY ELECTRON DIFFRACTION

Bond distances are given in angstroms, angles in degrees. Estimated total errors are given as error limits; σ (3σ) values from least-squares refinement are given in parentheses (square brackets) as units in the last digit. The C atoms of the benzene ring are numbered according to Figure 7-6c; D_{2h} symmetry has been assumed through-out.

Substituent	$\angle$ C2—C1—C6	r_g(C1...C4)	Δ(C—C)^a	$\langle r_g$(C—C)$\rangle^b$	r_g(C1—X)	Ref.
F	123.5 ± 0.1	2.712(2)	-0.012^c	1.392 ± 0.003	1.354 ± 0.004	67
NO$_2$^{d}	122.9[9]	2.720^e	0.000^c	1.392[2]	1.463[10]	58
CN	122.1 ± 0.2	2.745(4)	0.005(2)	1.397 ± 0.003	1.454 ± 0.005	68
Br	121.8 ± 0.2	2.755(3)	0.011(4)	1.399 ± 0.002	1.890 ± 0.003	69
NC	121.7 ± 0.2	2.760(3)	0.000^c	1.400 ± 0.003	1.388 ± 0.003	70
Cl	121.6 ± 0.2	2.749(3)	$-0.005(3)$	1.392 ± 0.003	1.730 ± 0.004	50
OH	120.7 ± 0.2	2.776(3)	0.009(3)	1.397 ± 0.003	1.384 ± 0.004	71
CHO^f	120.5 ± 0.4	2.785(5)	0.018(3)	1.402 ± 0.003	1.487 ± 0.004	66
SH	120.2 ± 0.2	2.792(3)	0.009(3)	1.400 ± 0.003	1.776 ± 0.004	71
NH$_2$^{g}	119.8 ± 0.2	2.795(5)	0.007(4)	1.397 ± 0.003	1.424 ± 0.005	72
Me	117.1 ± 0.3	2.856(2)	0.013(3)	1.400 ± 0.003	1.512 ± 0.003	36
SiMe$_3$^{h}	115.7 ± 0.6	2.900(8)	0.016(5)	1.408 ± 0.003	1.856 ± 0.007	73

^{a} Δ(C—C) = r(C1—C2) − r(C2—C3).

^{b} Mean value of the C—C bond distances of the benzene ring.

^{c} Assumed.

^{d} The twist angle of the substituent is found to be about 19°. The nonplanarity of the molecule may, however, be an effect of high-amplitude motions.

^{e} Calculated from the published ring parameters.

^{f} The substituent is coplanar with the benzene ring. The vapor consists of a mixture of the cis and trans conformers, in nearly 1:1 ratio.

^{g} The H—N—H plane makes an angle of 43 ± 4° with the C1...C4 axis.

^{h} The experimental data are consistent with either free rotation around the C1—SiMe$_3$ bond or a conformation deviating by about 15° from the eclipsed form.

1.392–1.408 Å. The correlation with $\angle$ C2—C1—C6 observed in 1982[73] is found to be rather poor ($r = -0.80$) when the present, more extensive set of experimental data is used.

The linear relationship of Figure 7-8 has a clear physical implication. It indicates that while some substituents—like F—are able to distort the benzene ring by *pushing* C1 and C4 toward the ring center, other substituents—like SiMe$_3$—cause the opposite type of distortion, by *pulling* C1 and C4 away from the ring center. The much smaller variance of Δ(C—C) and $\langle r_g$(C—C)$\rangle$, as compared with r_g(C1...C4), indicates that the relatively large distortion of the ring that takes place along the C1...C4 line is accompanied by only minor changes in the lengths of the C—C bonds.

A plot of $\angle$ C2—C1—C6 against the inductive parameter σ_I for the symmetrically para-disubstituted derivatives of Table 7-10 is shown in Figure 7-9. As with the angle α of the monosubstituted derivatives (see Figures 7-5 and 7-7), the distortion of $\angle$ C2—C1—C6 in these molecules is controlled primarily

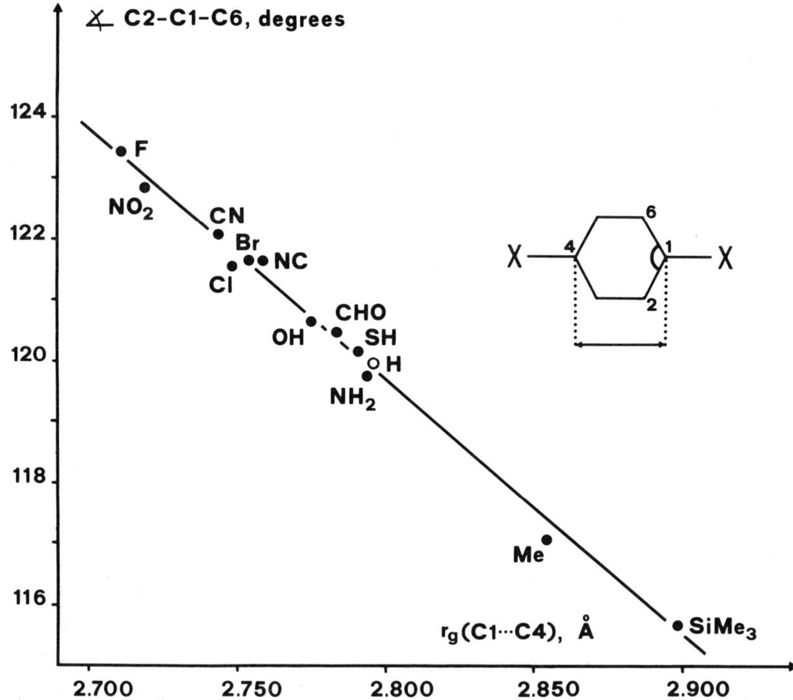

Figure 7-8. Correlation between the angle C2—C1—C6 and the distance r_g(C1...C4) in symmetrically para-disubstituted benzene rings in the gaseous phase. Also shown is the reference point of unsubstituted benzene. The equation of the least-squares line is $\angle$ C2—C1—C6 = $-41.86\ r_g$(C1...C4) $+ 236.94°$.

by the σ-electron-withdrawing or -releasing properties of the substituent. The correlation is better than indicated by the value of the correlation coefficient, $r = 0.92$, due to the appreciable deviation from linearity.

The angular distortion of the benzene ring in a symmetrically para-disubstituted derivative is determined by electron diffraction more accurately than in the corresponding monosubstituted species. The somewhat anomalous position of the data point for the F substituent in Figure 7-9 is certainly real and cannot be attributed to experimental error. The anomaly is not unique to the symmetrically para-disubstituted derivative in the gaseous phase: it is also seen in Figure 7-5, which refers to monosubstituted derivatives in the crystal. In Figure 7-7 the effect is probably obscured by the high value of α reported[33,53] for the NO₂ substituent.

A number of other disubstituted benzene derivatives with less symmetrical patterns of substitution have been studied by electron diffraction. In most of the recent studies the benzene ring has been allowed to distort from D_{6h} symmetry, which has led to the acquisition of further results on the geometrical effects of substitution. Among the molecules that have been studied are *p*-nitroaniline,[76] *p*-chloronitrobenzene,[77] *p*-bromonitrobenzene,[78] *p*-methoxybenzaldehyde,[79]

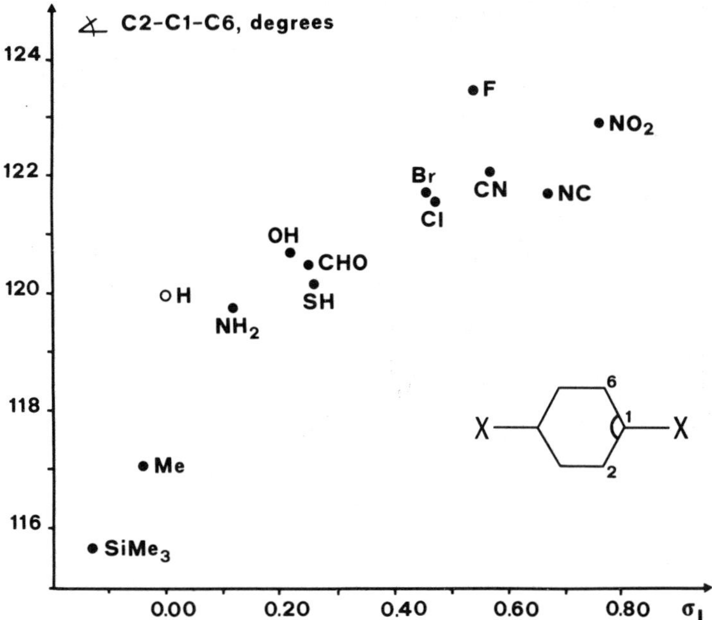

Figure 7-9. Plot of the angle C2—C1—C6 against the inductive parameter σ_I for symmetrically para-disubstituted benzene derivatives in the gaseous phase. Also shown is the reference point of unsubstituted benzene.

p-ethylbenzaldehyde,[80] p-isopropylbenzaldehyde,[80] p-fluorobiphenyl,[40] p-chlorobiphenyl,[40] p,p'-difluorobiphenyl,[40] p,p'-dichlorobiphenyl,[40] m-difluorobenzene,[81] m-dinitrobenzene,[82] m-chloronitrobenzene,[82] m,m'-dibromobiphenyl,[83] o-dinitrobenzene,[58] o-dicyanobenzene,[84] o-chloronitrobenzene.[85]

The benzene ring deformations obtained for some of these molecules should be viewed with caution, however, since the low molecular symmetry may have prevented their accurate determination. Nevertheless, the inclusion of the ring deformations in the analysis of electron diffraction data—even in the form of an assumed distortion based on literature data—is always recommended, as it may be instrumental in improving the accuracy of other geometrical parameters.

Electron Diffraction Studies of Symmetrically Trisubstituted Benzene Derivatives

The symmetrically trisubstituted derivatives of benzene, $1,3,5$-$C_6H_3X_3$, are particularly well suited for electron diffraction studies. In these molecules the carbon skeleton of the ring has D_{3h} symmetry, hence only two independent parameters are required to describe its geometry. The most obvious choice (see Figure 7-6d for the numering of atoms) is:

1. The internal ring angle at the place of substitution, $\angle C2-C1-C6$
2. The length of the six C—C bonds of the ring, $r(C-C)$

The equivalence of the six aromatic C—C bonds makes these molecules ideal subjects for determining accurately the effect of substitution on the C—C bond distances. This effect would otherwise be hard to determine by electron diffraction; and it appears from Tables 7-3 to 7-8 that accurate measurements of bond distances in monosubstituted benzene derivatives are difficult by any technique of structure determination. Note that the symmetrically trisubstituted derivatives of benzene do not possess a permanent dipole moment and are, therefore, inaccessible to microwave studies.

Only five $1,3,5\text{-}C_6H_3X_3$ molecules have been studied by electron diffraction, in which the carbon skeleton of the ring has been allowed to distort from D_{6h} to D_{3h} symmetry. Important geometrical parameters for these molecules are reported in Table 7-11. It is clear that the presence of the substituent gives rise to a variation of the ring angles *and* C—C bond distances. Figure 7-10 shows that $r_g(C-C)$ and $\angle C2-C1-C6$ are linearly related. From the slope of the regression line it is seen that an increase of $\angle C2-C1-C6$ by $1°$ implies a decrease or $r_g(C-C)$ by 0.0023 Å. This compares well with the ratio $1°: -0.0027$ Å [see eq. (7-2)], obtained by statistical analysis of monosubstituted benzene rings in the solid state.[11] It should be considered, however, that the linear relationship of Figure 7-10 is based on a limited number of data points and that the range of values of $r_g(C-C)$ is rather small; further experimental studies will be necessary to ascertain whether the relationship is of general validity.

Table 7-11. Important Geometrical Parameters for Symmetrically Trisubstituted Benzene Derivatives as Determined by Electron Diffraction

Bond distances are given in angstroms, angles in degrees. Estimated total errors are given as error limits; 3σ values from least-squares refinement are given in square brackets as units in the last digit. The C atoms of the benzene ring are numbered according to Figure 7-6d; D_{3h} symmetry has been assumed throughout.

Substituent	$r_g(C-C)$	$\angle C2-C1-C6$	$r_g(C1-X)$	Ref.
F	1.390 ± 0.002	123.7 ± 0.2	1.340 ± 0.003	86
$NO_2{}^a$	$1.388[2]$	$123.2[9]$	$1.475[5]$	58
Cl	1.392 ± 0.002	122.0 ± 0.2	1.734 ± 0.002	86
Br	$1.397[4]^b$	$120.4[8]$	$1.877[4]^b$	87
Me	1.401 ± 0.002	118.2 ± 0.2	1.509 ± 0.002	88

[a] The twist angle of the substituent is found to be $21°$. The nonplanarity of the molecule may, however, be an effect of high-amplitude motions.

[b] From the r_a value given in the original paper.

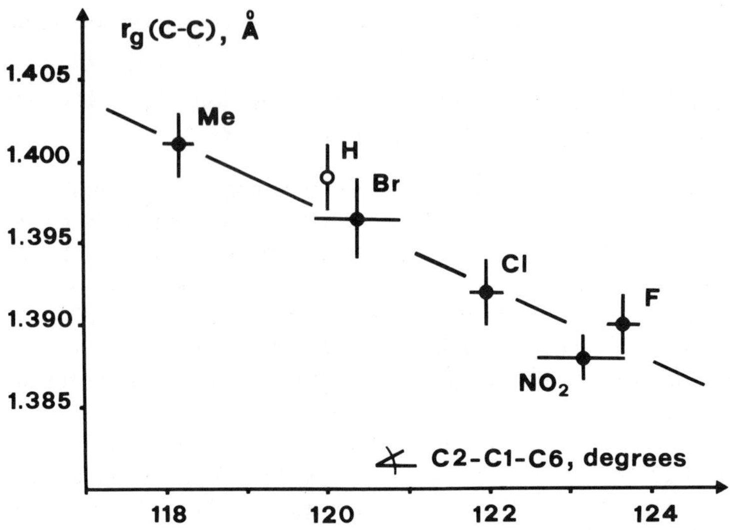

Figure 7-10. Correlation between the distance r_g(C—C) and the angle C2—C1—C6 in symmetrically trisubstituted benzene derivatives in the gaseous phase. Also shown is the reference point of unsubstituted benzene. The error bars correspond to either estimated total errors or 2σ values from least-squares refinement.

There is no doubt that the small variation of the aromatic C—C bond distances in the symmetrically trisubstituted benzene derivatives of Table 7-11 is a real chemical effect. A very similar variation has been observed by X-ray crystallography in some 3,5-disubstituted derivatives of benzoic acid,[89] as shown in Table 7-12.

The results presented in Table 7-12 require some comments. Comparing bond distances obtained by different techniques of structure determination is by no means straightforward because different physical meanings are involved.[91] The r_g bond distances produced by electron diffraction are thermal-average internuclear separations. The bond distances obtained by X-ray crystallography are separations between average atomic positions, but—unlike r_α distances—they refer to the centroids of electron density distribution rather than to the nuclei. Moreover, the motion of a molecule in a crystal has substantial contributions from librational rigid-body motions, which may seriously affect the observed geometry.

Thus, comparing bond distances obtained by the two techniques would ideally require[72]:

1. Changing the r_g values obtained by electron diffraction into r_α values
2. Using only high-θ reflexions in the refinement of the X-ray diffraction data, to eliminate the asphericities of the electron density distribution due to valence electrons, and to get essentially the positions of the nuclei
3. Correcting the atomic positions obtained in this way for the effects of librational motions.

Table 7-12. COMPARISON OF AROMATIC C—C BOND DISTANCES (Å)
OBTAINED BY X-RAY CRYSTALLOGRAPHY AND
GAS-PHASE ELECTRON DIFFRACTION

The distances obtained by X-ray crystallography refer to 3,5-disubstituted benzoic acids:

Each distance is the mean of four independent values, r(C2—C3), r(C3—C4), r(C4—C5), r(C5—C6). The distances obtained by electron diffraction refer to symmetrically trisubstituted derivatives.

	$\langle r(C-C)\rangle$ by X-ray crystallography[a]			r_g(C—C) by gas phase electron diffraction[b]
Substituent	Conventional refinement	High-θ refinement	High-θ refinement, corrected for libration	
NO$_2$	1.381	1.384	1.387	1.388[c]
Cl	1.383	1.389	1.394	1.392[d]
Me	1.390	1.395	1.401	1.401[e]

[a] References 89 and 90. Least-squares standard deviations are about 0.002 Å.
[b] Total errors and 3σ values are 0.002 Å.
[c] Reference 58.
[d] Reference 86.
[e] Reference 88.
Source: After reference 89.

Even so, it would be necessary to extrapolate the results to a common temperature, and to assume that the internal molecular motions are the same in the two phases.

Although these problems may appear formidable, there is now increasing evidence that the most serious effects can be treated, at least with relatively rigid aromatic molecules.[32,39,40,68,70,72] The results of Table 7-12 show that once the atomic coordinates from the high-θ refinement have been corrected for libration, the C—C bond distances obtained by X-ray crystallography agree surprisingly well with the values of r_g(C—C) obtained by electron diffraction. Note, however, that the r_g values (*not* r_α) from electron diffraction are being used in this comparison.

On the Additivity of the Ring Deformations

It has often been stated that the additivity of geometrical distortions in polysubstituted benzene rings works well to a first approximation.[10,22,40,83] Then the following problem arises: Do significant deviations from additivity occur? And, if so, do they convey any chemically relevant information about the nature of the ring–substituent interaction?

It is expected that deviations from additivity may occur whenever the interaction of a substituent with the ring is perturbed by the presence of other substituents. Significant deviations, interpreted in terms of cooperative interactions between π-donor and π-acceptor functional groups, have indeed been observed in at least two molecules studied by X-ray crystallography, p-nitroaniline[23] and 2,6-dinitroaniline.[92]

In some recent electron diffraction studies of polysubstituted benzene derivatives attempts have been made to verify whether the observed angular distortion of the ring could be reproduced by superimposing separate contributions from each substituent.[40,83,86,93] In most cases, however, the experimental bond angles were not of high accuracy, and the angular parameters used as predictors were those obtained by regression from solid-state results on polysubstituted benzene rings.[10,22] Even when more accurate experimental bond angles were available, and angular parameters from gas-phase studies were used as predictors, the comparison of experimental and predicted angles proved to be inconclusive.[86] It was recognized that more accurate sets of predictor parameters were necessary to ascertain whether significant deviations from additivity had occurred.

Ideally, the prediction of the ring angles in a polysubstituted benzene derivative should be based on highly accurate angular parameters obtained from gas-phase studies on monosubstituted benzene derivatives. The predictor parameters should be reliable to within 0.1–0.2°. It is clear from Tables 7-3 to 7-8 that within the monosubstituted derivatives of benzene considered there, this level of accuracy has only been reached for fluorobenzene and—possibly—cyanobenzene.

In the case of fluorobenzene, the results of various studies reported in Table 7-3 suggest that the following angular parameters should be used for the F substituent:

$$\Delta\alpha_F = 3.4° \qquad \Delta\beta_F = -2.0° \qquad \Delta\gamma_F = 0.3° \qquad \Delta\delta_F = 0.0°$$

Accurate gas-phase geometries have been determined for four fluoro-substituted benzene derivatives other than fluorobenzene: a comparison of the experimental and predicted ring angles for these molecules is presented in Table 7-13. The experimental and predicted values are seen to agree to within 0.3°, with the single exception of 1,2,4,5-tetrafluorobenzene,[93] for which the differences are 1–2°. The deviation from the additivity scheme that occurs in this molecule may originate from either a saturation phenomenon, or an ortho effect; an accurate gas-phase study of o-difluorobenzene would settle this point.

Table 7-13. Internal Ring Angles (deg) in Some Fluorosubstituted Benzene Derivatives in the Gaseous Phase: Experimental Versus Predicted Values

The predictions are based on the following predictor parameters, derived from the gas-phase molecular structure of fluorobenzene (see Table 7-3): $\Delta\alpha_F = 3.4°$, $\Delta\beta_F = -2.0°$, $\Delta\gamma_F = 0.3°$, $\Delta\delta_F = 0.0°$. Additivity of distortional effects has been assumed.

Molecule	Angle[a]	Experimental value[b]	Predicted value	Ref.
1,4-Difluorobenzene	C2—C1—C6	123.5 ± 0.1	123.4	67
	C1—C2—C3	$118.2_5{}^c$	118.3	
1,3-Difluorobenzene[d]	C2—C1—C6	123.9^c	123.7	94
	C1—C2—C3	116.0	116.0	
	C3—C4—C5	117.9^c	118.0	
	C4—C5—C6	120.4	120.6	
1,3,5-Trifluorobenzene	C2—C1—C6	123.7 ± 0.2	124.0	86
	C1—C2—C3	116.3^c	116.0	
1,2,4,5-Tetrafluoroben-zene	C2—C1—C6	120.6 ± 0.3	121.7	93
	C2—C3—C4	118.8^c	116.6	

[a] The numbering scheme of the various molecules is shown in Figure 7-6.
[b] Estimated total errors are given as error limits.
[c] Derived parameter.
[d] The experimental values are from a combined study by electron diffraction, microwave spectroscopy, and liquid- crystal NMR spectroscopy (model 3 of reference 94).

In the case of cyanobenzene, the structural results of Table 7-5 suggest the following angular parameters for the CN substituent in an isolated molecule:

$$\Delta\alpha_{CN} = 1.8° \qquad \Delta\beta_{CN} = -1.2° \qquad \Delta\gamma_{CN} = 0.3° \qquad \Delta\delta_{CN} = 0.0°$$

Comparison of the experimental and predicted values of the C2—C1—C6 angle in p-dicyanobenzene (experimental[68]: $122.1 \pm 0.2°$; predicted: $120° + \Delta\alpha_{CN} + \Delta\delta_{CN} = 121.8°$) indicates that the additivity scheme works well with this molecule.

Another substituent for which accurate angular parameters are available is NH_2. The complete substitution structure of aniline has been determined by microwave spectroscopy[5]; the values of the ring angles agree to within $0.3°$ with those obtained for the C_6H_5-NH- fragment by averaging the geometries of several C_6H_5-NHX molecules studied by X-ray crystallography.[8] The microwave study yields the following angular parameters:

$$\Delta\alpha_{NH_2} = -0.6° \qquad \Delta\beta_{NH_2} = 0.1° \qquad \Delta\gamma_{NH_2} = 0.7° \qquad \Delta\delta_{NH_2} = -1.1°$$

Comparison of the experimental and predicted values of the C2—C1—C6 angle in p-diaminobenzene (experimental[72]: $119.8 \pm 0.2°$; predicted: $120° + \Delta\alpha_{NH_2} + \Delta\delta_{NH_2} = 118.3°$) shows that in this molecule there is a remarkable deviation from the additivity scheme. The deviation is associated with a change in the $C-NH_2$ bond distance, which is 1.402 ± 0.002 Å in aniline (r_s value) and

1.424 ± 0.005 Å in p-diaminobenzene (r_g value). If one considers the geometrical changes that occur upon conjugation (see Figure 7-3), it may be concluded that π donation from the amino group to the ring is appreciably reduced when a second amino group is introduced in the para position.[72]

Intermolecular Interactions and Substituent Effects

The interaction of a substituent with the benzene ring should not be regarded as a mere intramolecular effect, unless the molecule is in the gaseous phase. Intermolecular forces in a condensed phase may influence the ring–substituent interaction. This is more likely to occur when the substituent is a π donor or a π acceptor, since π electrons are more easily polarized than σ electrons by interactions with adjacent molecules.

A change in the electronic interaction of the substituent with the ring is associated with a change of the ring geometry. The influence of intermolecular interactions on substituent effects may thus be investigated by comparing the geometry of the benzene ring in the free molecule with that in the crystal molecule.[95]

With respect to traditional solution studies, based on the measurement of suitable molecular parameters in different solvents, this approach has the advantage of providing results that pertain to a perfectly defined intermolecular environment. On the other hand, the changes of the ring geometry that occur in going from the free molecule to the crystal molecule are small and hard to measure accurately. A proper treatment of the various systematic effects inherent in each technique of structure determination is mandatory. It is, of course, preferable to compare bond angles rather than bond distances.

Four cases have been studied in detail; the molecules investigated are symmetrically para-disubstituted benzene derivatives, p-$C_6H_4X_2$ (X = CN, NC, NH_2, OH). As we have seen, the high symmetry of these molecules not only makes it possible to measure their gas-phase geometry more accurately than would be feasible with the corresponding monosubstituted derivatives, but also enhances the geometrical effects of substitution. In a monosubstituted derivative of benzene, increasing conjugation of the substituent with the ring causes the internal angles at the ipso and para positions to decrease simultaneously (see Figure 7-3). In a symmetrically para-disubstituted derivative these angular changes are superimposed, which makes the C2—C1—C6 angle more sensitive to changes in the extent of conjugation between the ring and the substituent.

In p-dicyanobenzene[68] the crystal structure is characterized by strong dipole- -dipole interactions between antiparallel cyano groups from neighboring molecules (Figure 7-11). This increases the contribution of polar canonical forms like **1** to the electronic structure of the crystal molecule. Indeed the C2—C1—C6

$$NC-\overset{+}{\underset{}{\bigcirc}}=C=\bar{N}$$

1

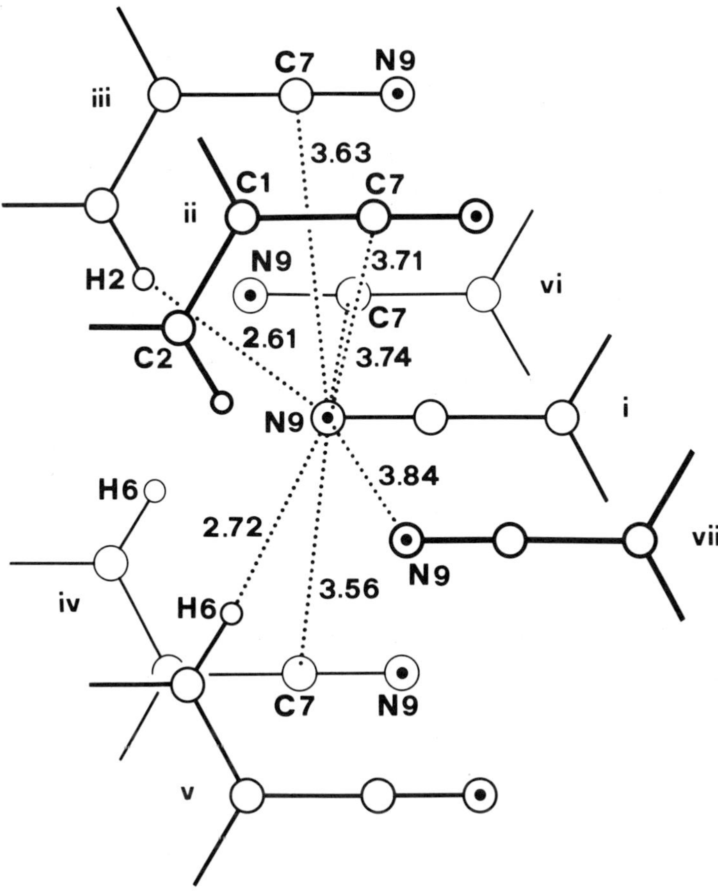

Figure 7-11. Projection of the crystal structure of *p*-dicyanobenzene, showing the network of dipole–dipole interactions between antiparallel cyano groups. (From reference 68.)

angle and the C1—X bond distance decrease slightly (by 0.5° and 0.012 Å, respectively) in going from the free molecule to the crystal molecule.

In crystalline *p*-diisocyanobenzene[70] the benzene ring is sandwiched between two isocyano groups from adjacent molecules (see Figure 7-12). The charge-transfer interaction requires the isocyano group to exist as $-N^+\equiv C^-$; this makes the contribution of canonical forms like **2** less important in the crystal molecule than in the free molecule. And this time the C2—C1—C6 angle and the C1—X bond distance are found to increase slightly (by 0.5° and 0.008 Å, respectively) in going from the free molecule to the crystal molecule. As in the

$$CN-\langle\hspace{-4pt}=\hspace{-4pt}\rangle=\overset{+}{N}=C$$

2

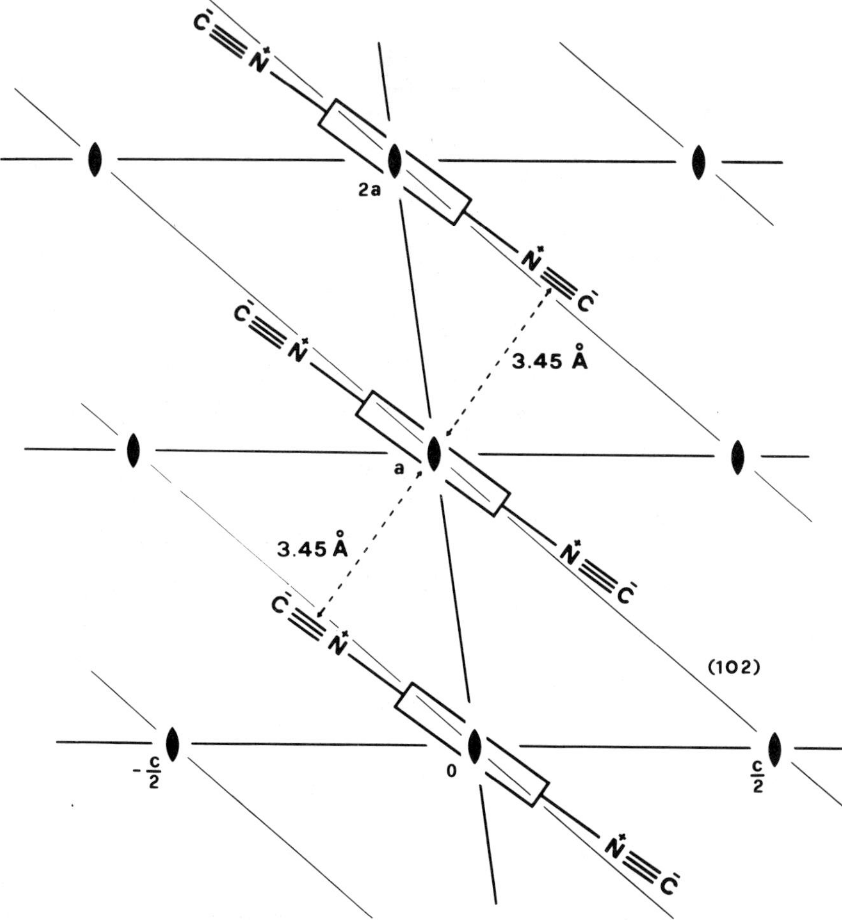

Figure 7-12. Projection of the crystal structure of *p*-diisocyanobenzene, showing the interaction of the benzene ring with two isocyano groups from adjacent molecules. (From reference 70.)

previous case, however, the effect is very small and only at the border of significance.

A highly significant variation of the ring geometry occurs in *p*-diaminobenzene.[72] In this molecule the C2—C1—C6 angle is found to be $119.8 \pm 0.2°$ in the gaseous phase and $117.9(1)°$ in the crystal. The crystal structure is based on a system of intermolecular N—H...N hydrogen bonds, whereby each amino group acts as a proton acceptor in a hydrogen bond and as a proton donor in another (Figure 7-13). Since the amino group is a good base but a very poor acid, a stronger hydrogen bond is formed when a partial positive charge resides on the nitrogen atom. This causes the contribution of polar

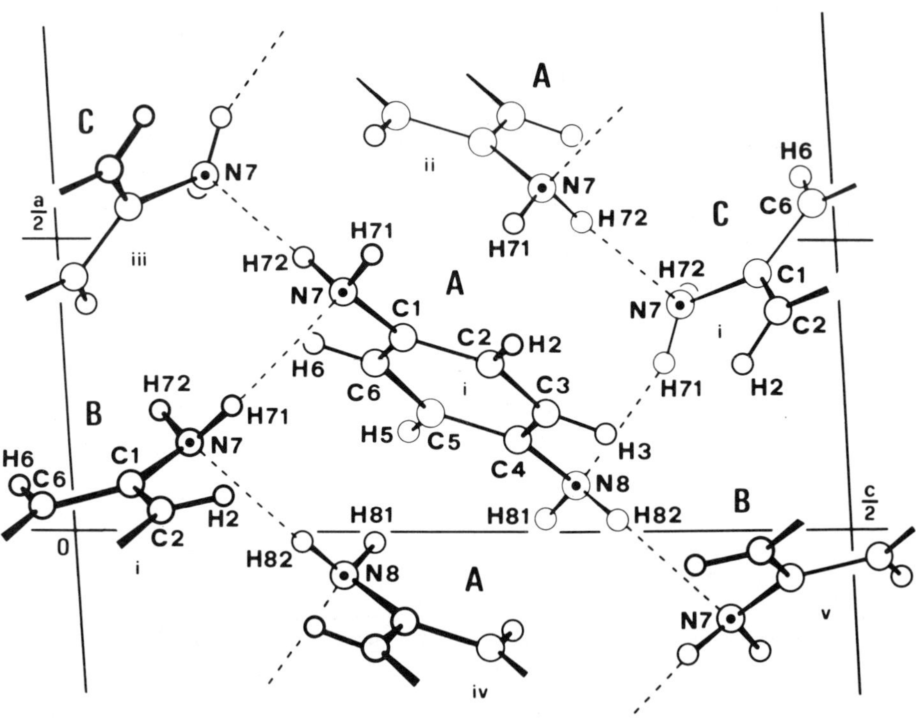

Figure 7-13. Projection of the crystal structure of *p*-diaminobenzene, showing the system of intermolecular hydrogen bonds. (From reference 72.)

canonical forms like **3** to be more important in the crystal molecule than in the free molecule.

$$H_2N - \langle - \rangle = \overset{+}{N}H_2$$

3

A similar effect is observed in *p*-dihydroxybenzene, where the C2—C1—C6 angle is $120.7 \pm 0.2°$ in the free molecule[71] and $119.7(1)°$ in the crystal molecule.[95] Here the dominant feature of the crystal structure is a system of O—H...O hydrogen bonds. Since the hydroxyl group is inherently more acidic than the amino group, the difference between gas-phase and solid-state geometries is less pronounced than in *p*-diaminobenzene.

Acknowledgment

I thank Mrs. Clara Marciante for help with the preparation of the figures.

References

1. Keidel, F. A.; Bauer, S. H. *J. Chem. Phys.* **1956**, *25*, 1218.
2. Domenicano, A.; Vaciago, A.; Coulson, C. A. *Acta Crystallogr.* **1975**, *B31*, 1630.
3. Nygaard, L.; Bojesen, I.; Pedersen, T.; Rastrup-Andersen, J. *J. Mol. Struct.* **1968**, *2*, 209.
4. Casado, J.; Nygaard, L.; Sørensen, G. O. *J. Mol. Struct.* **1971**, *8*, 211.
5. Lister, D. G.; Tyler, J. K.; Høg, J. H.; Larsen, N. W. *J. Mol. Struct.* **1974**, *23*, 253.
6. Carter, O. L.; McPhail, A. T.; Sim, G. A. *J. Chem. Soc. A* **1966**, 822.
7. Domenicano, A.; Vaciago, A.; Coulson, C. A. *Acta Crystallogr.* **1975**, *B31*, 221.
8. Domenicano, A.; Vaciago, A. *Acta Crystallogr.* **1979**, *B35*, 1382.
9. Domenicano, A.; Mazzeo, P.; Vaciago, A. *Tetrahedron Lett.* **1976**, 1029.
10. Domenicano, A.; Murray-Rust, P. *Tetrahedron Lett.* **1979**, 2283.
11. Domenicano, A.; Murray-Rust, P.; Vaciago, A. *Acta Crystallogr.* **1983**, *B39*, 457.
12. Di Rienzo, F.; Domenicano, A.; Portalone, G.; Vaciago, A. Second Yugoslav–Italian Crystallographic Conference, Dubrovnik, Yugoslavia, May 31–June 3, 1976; *Izvj. Jugoslav. Centr. Krist. (Zagreb)* **1976**, *11*, A-102.
13. Konschin, H. *J. Mol. Struct. (Theochem)* **1983**, *105*, 213.
14. Larsen, N. W. *J. Mol. Struct.* **1979**, *51*, 175.
15. Konschin, H. *J. Mol. Struct. (Theochem)* **1983**, *92*, 173.
16. Bock, C. W.; Trachtman, M.; George, P. *Chem. Phys.* **1985**, *93*, 431.
17. Domenicano, A.; Hargittai, I.; Schultz, G. *Abstracts*, Eighth Austin Symposium on Molecular Structure, Austin, TX, March 3–5, 1980; p. 17.
18. Schultz, G.; Hargittai, I.; Seip, R. *Z. Naturforsch.* **1981**, *36a*, 669.
19. Hoekstra, A.; Vos, A. *Acta Crystallogr.* **1975**, *B31*, 1716, 1722.
20. Walsh, A. D. *Discuss Faraday Soc.* **1947**, *2*, 18. Bent, H. A. *Chem. Rev.* **1961**, *61*, 275.
21. Gillespie, R. J. *J. Chem. Educ.* **1970**, *47*, 18. Gillespie, R. J. *Angew. Chem. Int. Ed. Engl.* **1967**, *6*, 819.
22. Norrestam, R.; Schepper, L. *Acta Chem. Scand.* **1981**, *A35*, 91.
23. Colapietro, M.; Domenicano, A.; Marciante, C.; Portalone, G. *Z. Naturforsch.* **1982**, *37b*, 1309.
24. Katritzky, A. R.; Topsom, R. D. *J. Chem. Educ.* **1971**, *48*, 427. Ehrenson, S.; Brownlee, R. T. C.; Taft, R. W. *Prog. Phys. Org. Chem.* **1973**, *10*, 1.
25. Domenicano, A.; Murray-Rust, P. To be published.
26. Exner, O. In "Correlation Analysis in Chemistry", Chapman, N. B.; Shorter, J. Eds.; Plenum Press: New York and London, 1978, Chap. 10, pp. 439–540.
27. Almenningen, A.; Brunvoll, J.; Popik, M. V.; Sokolkov, S. V.; Vilkov, L. V.; Samdal, S. *J. Mol. Struct.* **1985**, *127*, 85.
28. Scharfenberg, P.; Hargittai, I.; Rozsondai, B. *Abstracts*, Ninth Austin Symposium on Molecular Structure, Austin, TX, March 1–3, 1982; p. 83.
29. Portalone, G.; Domenicano, A.; Schultz, G.; Hargittai, I. *J. Mol. Struct.* **1987**, *160*, 97.
30. Portalone, G.; Schultz, G.; Domenicano, A.; Hargittai, I. *J. Mol. Struct.* **1984**, *118*, 53.
31. Penionzhkevich, N. P.; Sadova, N. I.; Vilkov, L. V. *Zh. Strukt. Khim.* **1979**, *20*, 527.
32. Brunvoll, J.; Colapietro, M.; Domenicano, A.; Marciante, C.; Portalone, G.; Hargittai, I. *Z. Naturforsch.* **1984**, *39b*, 607. Brunvoll, J.; Exner, O.; Hargittai, I.; Kolonits, M.; Scharfenberg, P. *J. Mol. Struct.* **1984**, *117*, 317.
33. Shishkov, I. F.; Sadova, N. I.; Novikov, V. P.; Vilkov, L. V. *Zh. Strukt. Khim.* **1984**, *25*(2), 98.
34. Portalone, G.; Domenicano, A.; Schultz, G.; Hargittai, I. *Abstracts*, Eighteenth Meeting of the Italian Crystallographic Association, Como; October 27–30, 1987; p. 54.
35. Rozsondai, B.; Schultz, G.; Hargittai, I. *J. Mol. Struct.* **1981**, *70*, 309. Rozsondai, B.; Moore, J. H.; Gregory, D. C.; Hargittai, I. *Acta Chim. (Budapest)* **1977**, *94*, 321.
36. Domenicano, A.; Schultz, G.; Kolonits, M.; Hargittai, I. *J. Mol. Struct.* **1979**, *53*, 197.
37. Seip, R.; Schultz, G.; Hargittai, I.; Szabó, Z. G. *Z. Naturforsch.* **1977**, *32a*, 1178.
38. Scharfenberg, P.; Rozsondai, B.; Hargittai, I. *Z. Naturforsch.* **1980**, *35a*, 431.
39. Almenningen, A.; Bastiansen, O.; Fernholt, L.; Cyvin, B. N.; Cyvin, S. J.; Samdal, S. *J. Mol. Struct.* **1985**, *128*, 59.
40. Almenningen, A.; Bastiansen, O.; Gundersen, S.; Samdal, S.; Skancke, A. *J. Mol. Struct.* **1985**, *128*, 95.
41. Colapietro, M.; Domenicano, A.; Pela Ceccarini, G. *Acta Crystallogr.* **1979**, *B35*, 890.
42. Colapietro, M.; Domenicano, A.; Marciante, C.; Portalone, G. *Acta Crystallogr.* **1981**, *B37*, 387.

43. Jokisaari, J.; Kuonanoja, J.; Pulkkinen, A.; Väänänen, T. *Mol. Phys.* **1981**, *44*, 197.
44. Boggs, J. E.; Pang, F.; Pulay, P. *J. Comput. Chem.* **1982**, *3*, 344.
45. Bock, C. W.; Trachtman, M.; George, P. *J. Mol. Struct. (Theochem)* **1985**, *122*, 155.
46. Bock, C. W.; Trachtman, M.; George, P. *J. Comput. Chem.* **1985**, *6*, 592.
47. Michel, F.; Nery, H.; Nosberger, P.; Roussy, G. *J. Mol. Struct.* **1976**, *30*, 409.
48. Colapietro, M.; Domenicano, A. *Acta Crystallogr.* **1982**, *B38*, 1953.
49. Diehl, P.; Jokisaari, J. *J. Mol. Struct.* **1979**, *53*, 55.
50. Schultz, G.; Hargittai, I.; Domenicano, A. *J. Mol. Struct.* **1980**, *68*, 281.
51. Diehl, P.; Amrein, J.; Veracini, C. A. *Org. Magn. Reson.* **1982**, *20*, 276.
52. Fauvet, G.; Massaux, M.; Chevalier, R. *Acta Crystallogr.* **1978**, *B34*, 1376.
53. Høg, J. H. A Study of Nitrobenzene, Thesis, University of Copenhagen, 1971.
54. Colapietro, M.; Domenicano, A.; Marciante, C.; Portalone, G. Thirteenth International Congress of Crystallography, Hamburg, Germany, August 9–18, 1984; *Acta Crystallogr.* **1984**, *A40*, C-98.
55. Catalano, D.; Forte, C.; Veracini, C. A. *J. Magn. Reson.* **1984**, *60*, 190.
56. Penner, G. H. *J. Mol. Struct. (Theochem)* **1986**, *137*, 121.
57. Trotter, J. *Acta Crystallogr.* **1959**, *12*, 884.
58. Penionzhkevich, N. P.; Sadova, N. I.; Popik, N. I.; Vilkov, L. V.; Pankrushev, Y. A. *Zh. Strukt. Khim.* **1979**, *20*, 603.
59. Di Rienzo, F.; Domenicano, A.; Riva di Sanseverino, L. *Acta Crystallogr.* **1980**, *B36*, 586.
60. Schaefer, T.; Wildman, T. A.; Sebastian, R. *J. Mol. Struct. (Theochem)* **1982**, *89*, 93.
61. Amir-Ebrahimi, V.; Choplin, A.; Demaison, J.; Roussy, G. *J. Mol. Spectrosc.* **1981**, *89*, 42.
62. Pang, F.; Boggs, J. E.; Pulay, P.; Fogarasi, G. *J. Mol. Struct.* **1980**, *66*, 281.
63. Gough, K. M.; Henry, B. R.; Wildman, T. A. *J. Mol. Struct. (Theochem)* **1985**, *124*, 71.
64. Cox, A. P.; Ewart, I. C.; Stigliani, W. M. *J. Chem. Soc. Faraday Trans. 2* **1975**, *71*, 504.
65. Konschin, H. *J. Mol. Struct. (Theochem)* **1984**, *110*, 267.
66. Bock, C. W.; Domenicano, A.; George, P.; Hargittai, I.; Portalone, G.; Schultz, G. *J. Phys. Chem.* **1987**, *91*, 6120.
67. Domenicano, A.; Schultz, G.; Hargittai, I. *J. Mol. Struct.* **1982**, *78*, 97.
68. Colapietro, M.; Domenicano, A.; Portalone, G.; Schultz, G.; Hargittai, I. *J. Mol. Struct.* **1984**, *112*, 141.
69. Schultz, G.; Kolonits, M.; Hargittai, I.; Portalone, G.; Domenicano, A. *J. Mol. Struct.* **1988**, in press.
70. Colapietro, M.; Domenicano, A.; Portalone, G.; Torrini, I.; Hargittai, I.; Schultz, G. *J. Mol. Struct.* **1984**, *125*, 19.
71. Domenicano, A.; Hargittai, I.; Portalone, G.; Schultz, G. *Abstracts*, Seventh European Crystallographic Meeting, Jerusalem, August 29–September 3, 1982; p. 155.
72. Colapietro, M.; Domenicano, A.; Portalone, G.; Schultz, G.; Hargittai, I. *J. Phys. Chem.* **1987**, *91*, 1728.
73. Rozsondai, B.; Zelei, B.; Hargittai, I. *J. Mol. Struct.* **1982**, *95*, 187.
74. Tamagawa, K.; Iijima, T.; Kimura, M. *J. Mol. Struct.* **1976**, *30*, 243. Schultz, G.; Kolonits, M.; Hargittai, I. Unpublished observations.
75. Colapietro, M.; Domenicano, A.; Hargittai, I.; Riva di Sanseverino, L.; Schultz, G. *Abstracts*, Sixth European Crystallographic Meeting, Barcelona; July 28–August 1, 1980; p. 22.
76. Sadova, N. I.; Penionzhkevich, N. P.; Vilkov, L. V. *Zh. Strukt. Khim.* **1976**, *17*, 1122.
77. Sadova, N. I.; Penionzhkevich, N. P.; Vilkov, L. V. *Zh. Strukt. Khim.* **1976**, *17*, 753.
78. Almenningen, A.; Brunvoll, J.; Popik, M. V.; Vilkov, L. V.; Samdal, S. *J. Mol. Struct.* **1984**, *118*, 37.
79. Brunvoll, J.; Bohn, R. K.; Hargittai, I. *J. Mol. Struct.* **1985**, *129*, 81.
80. Brunvoll, J.; Kolonits, M.; Bohn, R. K.; Hargittai, I. *J. Mol. Struct.* **1985**, *131*, 177.
81. Van Schaick, E. J. H.; Geise, H. J.; Mijlhoff, F. C.; Renes, G. *J. Mol. Struct.* **1973**, *16*, 389.
82. Batyukhnova, O. G.; Sadova, N. I.; Vilkov, L. V.; Pankrushev, Y. A. *J. Mol. Struct.* **1983**, *97*, 153.
83. Almenningen, A.; Bastiansen, O.; Fernholt, L.; Gundersen, S.; Kloster-Jensen, E.; Cyvin, B. N.; Cyvin, S. J.; Samdal, S.; Skancke, A. *J. Mol. Struct.* **1985**, *128*, 77.
84. Schultz, G.; Brunvoll, J.; Almenningen, A. *Acta Chem. Scand.* **1986**, *A40*, 77.
85. Batyukhnova, O. G.; Sadova, N. I.; Vilkov, L. V.; Pankrushev, Y. A. *Zh. Strukt. Khim.* **1985**, *26*(5), 175.
86. Almenningen, A.; Hargittai, I.; Brunvoll, J.; Domenicano, A.; Samdal, S. *J. Mol. Struct.* **1984**, *116*, 199.

87. Novikov, V. P.; Sokolkov, S. V.; Golubinskii, A. V.; Vilkov, L. V. *Zh. Strukt. Khim.* **1986**, *27*(1), 50.

88. Almenningen, A.; Hargittai, I.; Samdal, S.; Brunvoll, J.; Domenicano, A.; Lowrey, A. *J. Mol. Struct.* **1983**, *96*, 373.

89. Colapietro, M.; Domenicano, A.; Hargittai, I.; Portalone, G.; Samdal, S.; Schultz, G. *Abstracts,* Fifteenth Meeting of the Italian Crystallographic Association, Monteporzio Catone; October 2–5, 1984; p. 10.

90. Colapietro, M.; Domenicano, A.; Marciante, C.; Portalone, G. *Z. Naturforsch.* **1984**, *39b*, 1361.

91. Kuchitsu, K.; Cyvin, S. J. In "Molecular Structures and Vibrations", Cyvin, S. J., Ed.; Elsevier: Amsterdam, 1972, Chap. 12, pp. 183–211. Robiette, A. G. In "Molecular Structure by Diffraction Methods", Vol. 1, The Chemical Society: London, 1973, Chap. 4, pp. 160–197. Kuchitsu, K. In "Diffraction Studies on Non-Crystalline Substances", Hargittai, I.; Orville-Thomas, W. J., Eds.; Elsevier: Amsterdam, 1981, pp. 63–116.

92. Párkányi, L.; Kálmán, A. *J. Mol. Struct.* **1984**, *125*, 315.

93. Schei, S. H.; Almenningen, A.; Almlöf, J. *J. Mol. Struct.* **1984**, *112*, 301.

94. Den Otter, G. J.; Gerritsen, J.; MacLean, C. *J. Mol. Struct.* **1973**, *16*, 379.

95. Domenicano, A. *Abstracts,* Fifth Italian–Yugoslav Crystallographic Congress, Padua, Italy; June 3–6, 1986; Paper No. C-23.

ORGANOMETALLIC COMPOUNDS OF MAIN GROUP ELEMENTS

Arne Haaland

DEPARTMENT OF CHEMISTRY
UNIVERSITY OF OSLO
OSLO, NORWAY

CONTENTS

INTRODUCTION 326
GROUP 14 (Ge, Sn, Pb) 327
 Tetravalent Compounds 327
 Divalent Compounds 333
GROUP 13 (Al, Ge, In, Tl) 335
 Triorganyl Compounds 335
 Aluminum and Gallium Hydroborates 340
 Electron Donor-Acceptor Complexes 341
 Compounds of Composition R_2MX (M = Al and Ga, X = F, Cl, OR, SR, or NR$_2$) 346
GROUP 2 AND GROUP 12 353
 Dialkyl Compounds 353
 Electron Donor-Acceptor Complexes 357
GROUP 1 AND GROUP 11 359
PERIODIC TRENDS IN M—C BOND DISTANCES 359
CYCLOPENTADIENYL COMPOUNDS OF MAIN GROUP METALS 361
 Monocyclopentadienyl Compounds 361
 Dicyclopentadienyl Compounds 368
CONCLUDING REMARKS 377
REFERENCES 377

Introduction

The purpose of this chapter is to review the molecular structures of organometallic compounds determined by gas-phase electron diffraction (GED) over the past two or three decades. The intention has been to include all organic derivatives of main group metals in Groups 1, 2, 12, 13 (except boron[1]) and 14 (except silicon[1]). "Organic derivatives" is taken to mean compounds containing a direct bond between the metal (M) and at least one carbon atom. This chapter covers the gas-phase structures of about 120 compounds that are organometallic in this strict sense.

The hydrogen atom may be regarded as the zeroth member of the alkyl series C_nH_{2n+1}, and metal hydrides and alkyls are often chemically related. Comparison of their structures can bring valuable insights. This chapter therefore includes about a dozen compounds containing M—H bonds, though they are not organometallic in the strict sense.

The result of an investigation by GED is not an established fact set in the reinforced concrete of eternal certainty. The first source of error is the composition of the sample: in his index of structures of organometallic compounds studied by X-ray crystallography, Bruce lists three cases of "mistaken identity", where errors in the original chemical analysis have been detected and corrected in the literature.[2] Probably there are other cases of mistaken identity that have passed undetected till now. Since a GED investigation involves less information than an X-ray investigation, the probability that an error in the chemical analysis will pass undetected is considerably greater. When crystals are grown for X-ray diffraction, solvent molecules are rarely incorporated in the lattice. If they are incorporated, they usually appear in stoichiometric amounts and are easily detected by chemical analysis or in the course of the structure refinement. GED does not require a crystalline sample; often the sample is liquid or amorphous. In such cases there may be present significant amounts of solvent or other volatile impurities that if undetected, will lead to significant errors in the structure refinement.

Organometallic compounds are often unstable in air and react vigorously with traces of water. Since the electron diffraction experiment is conducted in high vacuum, air and moisture sensitivity should present no particular problems if one works with reasonable care. But the sample must be vaporized, which requires heating (or cooling) to a temperature where the vapor pressure is about 1—or more often 10—torr. Many GED investigations of organometallic compounds have had to be abandoned because the thermal stability of the sample was too low. In some cases thermal decomposition may have occurred without being detected.

Unless the molecule of interest is very small, the electron diffractionist must simplify the task by making assumptions: assumptions about the overall symmetry of the molecule, about the local symmetry of molecular fragments, about the relative magnitude or similarity of bond distances or valence angles.

When doing so, the electron diffractionist, like all chemists, operates within a framework of what is considered established knowledge, or at least highly probable, at the time. If the assumptions are unwarranted, the results of the structure refinement may be erroneous.

So, errors have been made in the past. And, unless electron diffractionists restrict themselves to well-charted waters, errors may occasionally be made in the future. In writing this chapter, I have felt free to point out some of the errors made in the past. Since two of those errors are my own, I feel I can do this without being suspected of arrogance.

What are the merits of gas-phase electron diffraction? What information can be obtained by GED that is not more easily or more reliably obtained by other methods (in particular by X-ray crystallography)? I shall return to these questions throughout and attempt to give more general answers in the concluding section.

The last review dealing specifically with the structures of organometallic compounds in the gas phase is now more than 10 years old.[3] In the meantime the subject has been covered, at least in part, in more general reviews published by M. and I. Hargittai[4] and by Vilkov, Mastryukov, and Sadova.[5] Organometallic compounds are also covered by the annual surveys on gas-phase molecular structures determined by electron diffraction published by the Royal Society of Chemistry in the Specialist Periodical Report on Spectroscopic Properties of Inorganic and Organometallic Compounds.

Group 14 (Ge, Sn, Pb)

Tetravalent Compounds

Table 8-1 lists the bond distances of some simple derivatives of tetravalent germanium, tin, and lead in the gas phase. In addition to results obtained by GED, the table contains bond distances determined by microwave spectroscopy (MW) and vibration–rotation spectroscopy (VRS).

Tetramethylgermane, -stannane, and -plumbane, $M(CH_3)_4$, (M = Ge, Sn, Pb) are all liquids at room temperature. Both melting points and boiling points increase with increasing atomic number of the metal: the melting points are $-80°C$ (Ge), $-55°C$ (Sn), and $-30°C$ (Pb). The normal boiling points are 43°C (Ge), 77°C (Sn), and 110°C (Pb). The compounds are monomeric in hydrocarbon solution. To grow and mount X-ray-quality crystals of compounds with very low melting points and high volatility is no trivial task—such compounds are more suited for gas-phase electron diffraction. GED data are consistent with tetrahedral molecular symmetry, T_d, and the bond distances are Ge—C, 1.945(3) Å[6]; Sn—C, 2.144(3) Å[11]; Pb—C, 2.238(3) Å[15]. The mean M—C bond dissociation energies decrease with increasing atomic number of the metal (and

Table 8-1. BOND DISTANCES IN ORGANIC DERIVATIVES OF TETRAVALENT GERMANIUM, TIN, AND LEAD IN THE GAS PHASE

Compound	Bond distances (Å)[a]		Method[b]	Ref.
	M—C	M—H		
$Ge(CH_3)_4$	1.945(3)	—	GED	6
$HGe(CH_3)_3$	1.947(6)	1.532(3)	MW	7
$H_2Ge(CH_3)_2$	1.950(10)	—	MW	8
H_3GeCH_3	1.945(5)	1.529(5)	MW	9
H_4Ge	—	1.525(1)	VRS	10
$Sn(CH_3)_4$	2.144(3)	—	GED	11
$HSn(CH_3)_3$	2.149(4)	1.71(7)	GED	12
$H_2Sn(CH_3)_2$	2.153(3)	1.680(15)	GED	12
H_3SnCH_3	2.143(2)	1.70(2)	MW	13
H_4Sn	—	1.711(1)	VRS	14
$Pb(CH_3)_4$	2.238(3)		GED	15

Compound	M—C	M—X	Method[b]	Ref.
$F_2Ge(CH_3)_2$	1.928(3)	1.739(2)	GED	16
F_3GeCH_3	1.904(9)	1.714(2)	GED	16
$ClGe(CH_3)_3$	1.940(3)	2.170(3)	MW	17
$Cl_2Ge(CH_3)_2$	1.926(4)	2.155(4)	GED	18
$Cl_2Ge(CH_3)_2$	1.928(6)	2.143(4)	GED	19
Cl_3GeCH_3	1.893(10)	2.132(4)	GED	18
Cl_4Ge	—	2.113(3)	GED	20
$BrGe(CH_3)_3$	1.936(7)	2.323(2)	MW	21
$Br_2Ge(CH_3)_2$	1.911(12)	2.303(2)	GED	22
Br_3GeCH_3	1.89(3)	2.276(2)	GED	22
Br_4Ge	—	2.272(3)	GED	23
$ClSn(CH_3)_3$	2.106(6)	2.351(2)	GED	12
$Cl_2Sn(CH_3)_2$	2.109(3)	2.327(2)	GED	24
$Cl_3Sn(CH_3)$	2.104(16)	2.304(3)	GED	12
Cl_4Sn		2.281(4)	GED	25
H_3GeCH_3	1.945(5)		MW	9
$H_3GeCH{=}CH_2$	1.926(12)		MW	26
$H_3GeC{\equiv}CH$	1.896(1)		MW	27

Compound	Sn—C≡	Sn—C<	Method[b]	Ref.
$(CH_3)_3SnC{\equiv}CH$	2.126(8)	2.141(3)	GED	28
$(CH_3)_3SnC{\equiv}CSn(CH_3)_3$	2.095(3)	2.127(2)	GED	29

Compound	Sn—C	Sn—I	Method[b]	Ref.
$ISn(C{\equiv}CH)_3$	2.060(6)	2.645(2)	GED	30

Compound	M—C		Method[b]	Ref.
$Sn(C{\equiv}CH)_4$	2.067(2)		GED	31
$Sn(C{\equiv}CCF_3)_4$	2.070(3)		GED	32
$Sn(CH{=}CH_2)_4$	2.116(2)		GED	33
$Sn(C_6H_5)_4$	2.168(3)		GED	34
$Ge(CF_3)_4$	1.989(2)		GED	35
$Sn(CF_3)_4$	2.201(2)		GED	36

[a] Estimated standard deviations in parentheses in units of the last digit.
[b] GED = gas-phase electron diffraction, MW = microwave spectroscopy, VRS = vibrational-rotational spectroscopy.

increasing bond distance): $\bar{D}(\text{Ge--Me}) = 258\,\text{kJmol}^{-1} > \bar{D}(\text{Sn--Me}) = 226\,\text{kJmol}^{-1} > \bar{D}(\text{Pb--Me}) = 161\,\text{kJmol}^{-1}$.[37]

In Table 8-1 we compare bond distances in the series H_nMMe_{4-n} (M = Ge and Sn). The Sn--C, Ge--C, and Ge--H bond distances are constant to within 0.01 Å in each series; Sn--H bond distances are too inaccurate to allow us to draw a similar conclusion.

In Table 8-1 we also compare bond distances in methyl/halogen derivatives X_nMMe_{4-n} (M = Ge and Sn). It is seen that replacement of electron-donating methyl groups by electron-withdrawing halogen atoms leads to progressive shortening of both M--C and M--X bond distances. Comparison of Ge--C bond distances in $Ge(CH_3)_4$ and F_3GeCH_3 or Cl_3GeCH_3 indicates that the bonding radius of Ge has decreased by 0.04-0.05 Å; comparison of M--X bond distances in the pairs $ClGe(CH_3)_3/Cl_4Ge$, $BrGe(CH_3)_3/Br_4Ge$, and $ClSn(CH_3)_3/Cl_4Sn$ indicates that the bonding radius of the metal atom has

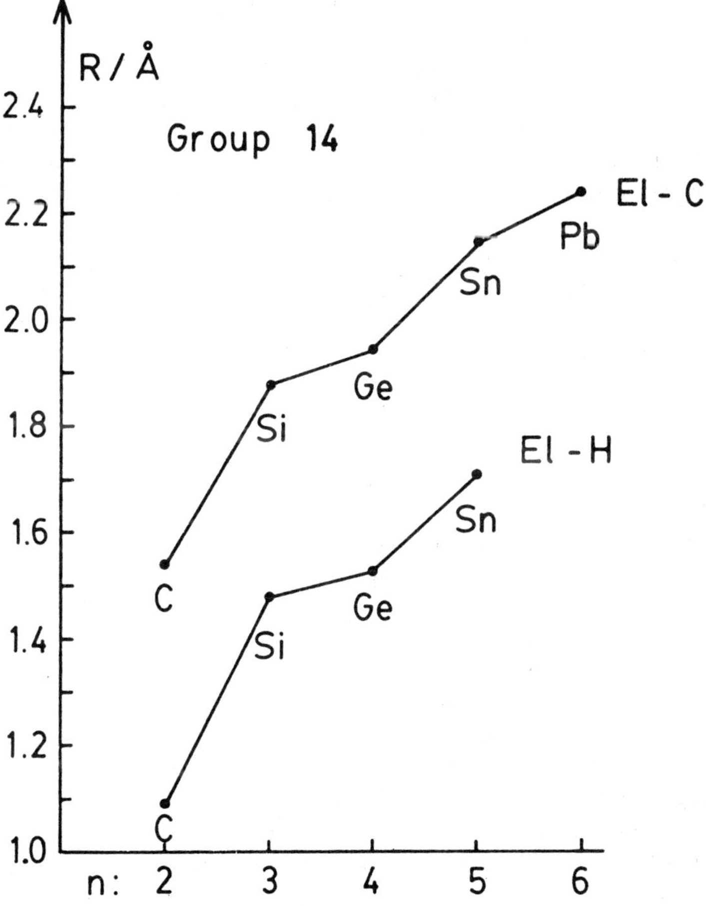

Figure 8-1. El--C bond distances of $El(CH_3)_4$ and El--H bond distances of ElH_4.

decreased by 0.05–0.07 Å. The apparent contraction of the metal atom by about 0.02 Å per halogen atom is easily rationalized in terms of an inductive effect: as the number of electronegative ligands increases, the positive charge on the metal atom increases and the atom contracts.

A vast number of C—C bond distances may be rationalized if it is assumed that the bonding radius of C depends on the configuration: tetrahedral (sp^3-hybridized) —C—, trigonal (sp^2-hybridized) —C or —C , or linear (sp-hybridized), —C≡. The bonding radius is commonly assumed to decrease by about 0.03 Å on going from tetrahedral to trigonal C, and to decrease by another 0.03 or 0.04 Å on going from trigonal to linear C. It does not follow that metal–carbon bond distances exhibit the same trend. However, the bond distances in Table 8-1 indicate that Ge—C and Sn—C bond distances vary with the configuration (hybridization) of the C atom in much the same way as C—C bond distances: thus the Ge—C bond distance in $H_3GeC\equiv CH$ is about 0.05 Å shorter than in H_3GeCH_3, and the Sn—C≡ bond distance in $(CH_3)_3SnC\equiv CSn(CH_3)_3$ is about 0.05 Å shorter than the Sn—C bond distance in $Sn(CH_3)_4$. The Sn—C≡ bond distance in $Sn(C\equiv CH)_4$ is about 0.08 Å shorter than in $Sn(CH_3)_4$: since the ethynyl radical is more electronegative than the methyl radical, it is not unreasonable to assume that an increasing number of ethynyl substituents leads to a small (say by 0.03 Å) contraction of the metal atom.

Only the Sn—C bond distance in tetraphenylstannane, $Sn(C_6H_5)_4$, fails to conform to the pattern: it is similar to, and may even be slightly longer than in $Sn(CH_3)_4$.[34]

Thermochemical results indicate that the four Sn—C bonds in $Sn(C_6H_5)_4$ are destabilized by 16 kJmol^{-1} each compared to the Sn—C (sp^2) bond in $(CH_3)_3SnCH=CH_2$.[37] According to Belyakov and co-workers: "It would be easy to suggest that the major reason for the elongation of the Sn—C bond in $Sn(C_6H_5)_4$ is steric repulsion [between the ligands]. However, the molecular geometry provides no argument for this suggestion."[34]

The M—C bond distances in the trifluoromethyl compounds $Ge(CF_3)_4$ and $Sn(CF_3)_4$ are 0.05 or 0.06 Å longer than in the hydrocarbon analogs $M(CH_3)_4$.[35,36] Thus, while introduction of electronegative ligands on the metal atom leads to a decrease of the bond distance, introduction of electronegative substituents on the carbon atom appears to have the opposite effect. Yokozeki and Bauer[38] have pointed out that for molecules of the type $E(CH_3)_n$ with $n = 1$, 2, or 3, the variation of the E—C bond distance on CH_3/CF_3 substitution is correlated with the electronegativity of the central atom E: if E is more electronegative than C, the E—CF$_3$ bonds are shorter than E—CH$_3$ bonds: if E is more electropositive than C, the E—CF$_3$ bonds are longer than the E—CH$_3$ bonds, and the elongation appears to increase with increasing electronegativity difference, $\chi_C - \chi_E$.

In the compounds in Table 8-1, the valence angles at the metal generally fall into the range 105–115°. In each molecule the smaller valence angles are those subtended by the most electronegative ligand pair. Thus in F_3GeCH_3 the angle $\angle FGeF = 105.9(9)°$ and $\angle CGeF = 113.2(6)°$.[16] The observation is in agreement with Bent's law[39] and may also be rationalized in terms of the valence-shell electron-pair repulsion (VSEPR) model[40]; the electron pairs that are polarized toward the ligand require less space. The largest deviations from tetrahedral angles are found in the dihalides $X_2Ge(CH_3)_2$ (X = F, Cl, Br), where $\angle CGeC$ appears to be about 120° and $\angle XGeX$ about 105°, though the error limits in each of the three molecules are rather wide.

1-Methyl-1-germaadamantane (**1**) has been investigated through GED by Shen and co-workers.[41] The adamantane frame constricts the three endocyclic $\angle CGeC$ angles to 101.9(5)°, but the average Ge—C bond distance, 1.954(3) Å, is indistinguishable from that of $Ge(CH_3)_4$. Cyclobutylgermane, $H_3Ge(c\text{-}C_4H_7)$ (**2**), has been investigated by Dakkouri.[42] The gas was found to consist of 77(3)% quasi-equitorial (**2a**) and 23(3)% quasi-axial (**2b**) conformers. The Ge—C bond distance is indistinguishable from that of H_3GeCH_3. The gas-phase structure of cyclopentadienylgermane, $H_3Ge(\eta^1\text{-}C_5H_5)$, will be discussed later.

$$\underset{\textbf{1}}{} \qquad \underset{\textbf{2a}}{} \qquad \underset{\textbf{2b}}{}$$

In the early 1960s Volpin and co-workers synthesized a series of compounds of empirical formula $H_2C_2GeX_2$ (X = CH_3, Cl, I).[43] Cryoscopic molecular weights in benzene and camphor indicated that they were monomeric at low concentrations, and on the basis of various physicochemical measurements they were assigned structures as 3-germacyclopropenes (germirenes) (**3**).[43]

A sample of $H_2C_2GeMe_2$ given to the electron diffraction group at Moscow State University did indeed yield an electron diffraction pattern consistent with the germirene structure.[44] A sample of $H_2C_2GeCl_2$ investigated later by the same group, however, yielded data consistent with a digermacyclohexadiene-like dimer (**4**).[44] A sample of $H_2C_2GeI_2$ yielded a diffraction pattern reminiscent of a germirene when evaporated at low temperature and a diffraction pattern reminiscent of a digermacyclohexadiene when evaporated at high temperature.[44]

Subsequent investigations by mass spectrometry[45] and determination of vapor density[44] have made it clear that all these compounds are dimeric in the gas phase, and the digermacyclohexadiene structures were confirmed by X-ray crystallography.[46] The electron diffraction patterns obtained for $H_2C_2GeMe_2$ and for $H_2C_2GeI_2$ at low temperatures remain unexplained. Very recently true

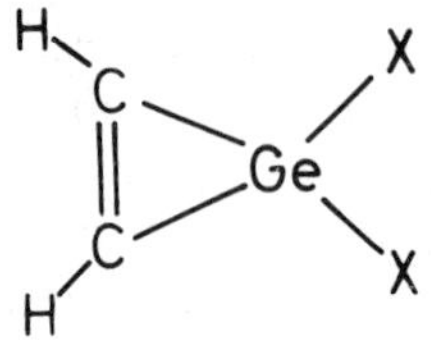

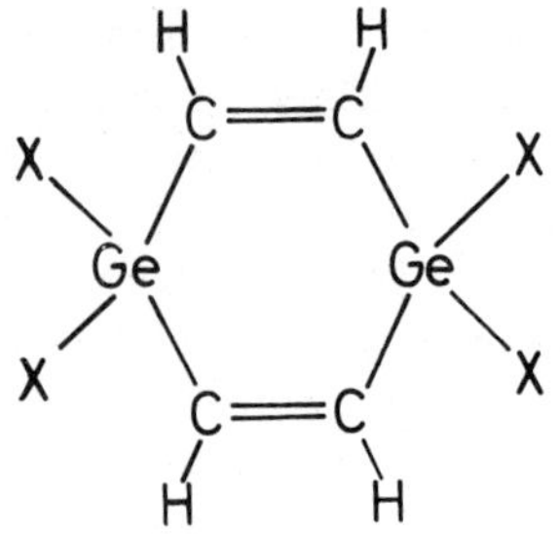

ARNE HAALAND

3

4

germirenes have in fact been synthesized,[47] and the structure of one was determined by X-ray diffraction.[48]

We now turn our attention to dinuclear and trinuclear species.

Table 8-2 compares bond distances and valence angles in the compounds $E(GeH_3)_2$ (E = element = O, S, Se) and $E(GeH_3)_3$ (E = N, P). The Ge—O bond distance in digermyloxane, $O(GeH_3)_2$, is 0.07 Å shorter than the Ge—O bond distance calculated from Pauling's covalent radii by the Schomaker-Stevenson rule.[56] The $\angle$ GeOGe angle is about 127°, though the VSEPR model predicts that it should be less than tetrahedral. These observations might be rationalized by evoking dative $p\pi$-$d\pi$ bonding between O and Ge, ie, that the lone electron pairs on O are delocalized into the vacant $4d$ orbitals on Ge.

Recently, we have shown that a slight modification of the Schomaker-Stevenson rule (hereafter referred to as MSS):

$$R(A—B) = r_A + r_b - c|\chi_A - \chi_B|^{1.4} \qquad c = 0.085 \text{ Å, (MSS)} \qquad (6-1)$$

Table 8-2. BOND DISTANCES, NONBONDED DISTANCES, AND VALENCE ANGLES FOR DIGERMYLOXANE, $O(GeH_3)_2$, TRIGERMYLAMINE, $N(GeH_3)_3$, AND RELATED COMPOUNDS STUDIED BY GED

	Bond distances (Å)			Nonbonded distance, M...M	
Compound	M—E	M—E$_{calc}$[a]	$\angle$ MEM (deg)	(Å)	Ref.
$O(GeH_3)_2$	1.766(4)	1.77	126.5(3)	3.15	49
$O(GeMe_3)_2$	1.770(10)	1.77	141.0(5)	3.34	50
$S(GeH_3)_2$	2.209(4)	2.20	98.9(1)	3.38	49
$Se(GeH_3)_2$	3.344(3)	2.33	94.6(5)	3.45	51
$N(GeH_3)_3$	1.836(5)	1.84	120.0	3.18	52
$P(GeH_3)_3$	2.308(3)	2.29	95.4(5)	3.41	53
$OCC(GeMe_3)_2$[b]	1.946(5)	1.96	127.6(13)	3.49	54
$O(SnMe_3)_2$	1.940(13)	1.93	141.0(5)	3.65	50
$N(SnMe_3)_3$	2.037(3)	2.00	120.0	3.53	55

[a] Calculated from the MSS rule.[57]
[b] $OCE(GeMe_3)_2$, E = C.

eliminates the discrepancy between observed and calculated bond distances in $O(GeH_3)_2$ and similar molecules.[57] On the other hand Bartell[58] and Glidewell[59] have argued that when two or more large atoms (like Ge) are bonded to a small atom (O or N), repulsion between the ligating atoms may override VSEPR and determine the angle subtended by the ligating atoms. On this basis Glidewell has assigned a hard-sphere radius to Ge limiting across-angle Ge...Ge contacts. The value selected was 1.58 Å, half the observed Ge...Ge nonbonded distance in $O(GeH_3)_2$.[58]

This is probably the place for the author to admit to a predilection for rationalizations in terms of steric rather than electronic effects when both seem to work: interatomic distances are clearly defined and easy to measure; atomic charges are neither.

The $\angle$ GeOGe angle in $O(GeMe_3)_2$ is 15° greater than in $O(GeH_3)_2$. It seems reasonable to regard the increase as due to repulsion between the ligating *groups* rather than between ligating atoms.

On going from $O(GeH_3)_2$ to $S(GeH_3)_2$ the $\angle$ GeEGe angle drops to 99°, ie, a value less than tetrahedral, as predicted by the VSEPR model. The S atom is so large (the S—Ge bond so long) that the Ge...Ge distance is greater than twice the Glidewell hard-sphere radius as long as the valence angle is greater than 90°. On going from $S(GeH_3)_2$ to $Se(GeH_3)_2$ the $\angle$ GeEGe angle drops by another 5°.

The coordination around N in $N(GeH_3)_3$ is trigonal planar (or very nearly so), corresponding to a $\angle$ GeNGe angle of 120°. The Ge—N bond distance is 0.04 Å smaller than calculated from the Pauling covalent radii and the Schomaker–Stevenson rule. The latter discrepancy is removed by the MSS rule. The wide angle may be rationalized in terms of dative π-bonding between N and Ge, but again we prefer to adopt a steric rationalization and note that a valence angle of 120° is just sufficient to make the Ge...Ge distance greater than twice the hard-sphere radius.

In the bis(trimethylgermyl)ketene, $(Me_3Ge)_2C=C=O$, the $\angle$ GeCGe angle is about 128°. The Ge...Ge distance is comfortably greater than the limiting value. The wide angle is presumably determined by repulsion between ligating groups rather than atoms.

On the basis of the Sn...Sn distance found in $O(SnMe_3)_2$, Glidewell has assigned to Sn a hard-sphere radius of 1.82 Å.[60] This radius is large enough to provide rationalization of the planar coordination of N in $N(SnMe_3)_3$.

Divalent Compounds

Compounds of composition R_2M, where M is Ge or Sn and R is alkyl or aryl, have been known for a long time. The *monomeric* metallenes (isoelectronic with methylene, CH_2) were, however, known only as short-lived intermediates that condense rapidly to metal–metal bonded rings or chains. Attempts to synthesize similar lead compounds failed, as the products disproportionated to R_6Pb_2 and metallic lead.

The first stable organometallic derivatives of divalent Ge, Sn, and Pb to be prepared were the dicyclopentadienyl compounds Cp_2M. These will be discussed in the section on cyclopentadienyl compounds of main group elements.

The synthesis of the first stable dialkylmetallene, R_2Sn with R = $CH(SiMe_3)_2$, was reported by Davidson and Lappert in 1973.[61a] Later Lappert and co-workers also prepared the Ge and Pb analogs.[61b,c] The metallenes R_2Ge and R_2Sn are monomeric in freezing cyclohexane and in the gas phase (by mass spectroscopy). Data for R_2Pb are lacking due to low solubility and thermal stability; R_2Ge and R_2Sn appear to be the only known stable (Group 14) metallenes. There can be little doubt that the stability of R_2Ge and R_2Sn is due to the extreme bulk of the ligands. In the solid phase, these metallenes form metal–metal bonded dimers.[61b,62] The structure of the monomers can thus be determined only by GED.[63] A molecular model is shown in Figure 8-2. The molecules are bent at the metal atom and the ligands oriented in such a fashion that interligand distances are maximized. Due to the large number of atoms (59) and the low symmetry (probably C_2), the complexity of the molecular structures is near the limit of what can be studied by GED, and the accuracy obtained is not very great. The bond distances are found to be 2.038(15) Å for Ge—C and 2.22(2) Å for Sn—C. The valence angles at the metal are 107(2) and 97(2)° for $\angle$CGeC and $\angle$CSnC, respectively.

Both R_2Ge and R_2Sn are diamagnetic, ie, have a singlet ground state.[61] Optimization of the molecular structures of Me_2Ge and Me_2Sn by *ab initio* molecular orbital calculations[63–65] yield the bond distances 2.02 and 2.20 Å for Ge—C and Sn—C, respectively, and the valence angles $\angle$CGeC = 97 or 98° and $\angle$CSnC = 96°. Both bond distances and the valence angle $\angle$CSnC are in

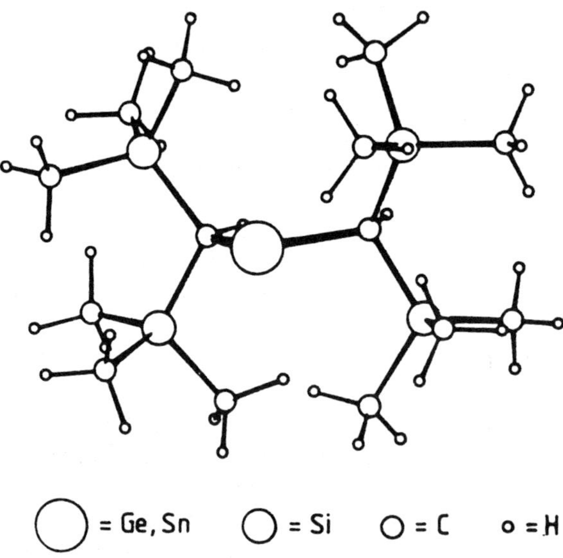

Figure 8-2. Molecular model of $M[CH(SiMe_3)_2]_2$.[63]

good agreement with the experimental values for MR_2. We assume that the difference between the angles $\angle CGeC$ in Me_2Ge and R_2Ge is due to interligand repulsion in the latter. The M—C bond distances are about 0.08 Å greater than in the tetravalent compounds $M(CH_3)_4$. The long M—C bonds and small $\angle CMC$ angles in MR_2 may be rationalized by assuming the lone electron pair to occupy nearly pure s orbitals on the metal and the M—C σ-bonding orbitals to be formed from nearly pure p orbitals on the metal atom. This view is in agreement with the orbital coefficients obtained in the *ab initio* calculations.

We have mentioned that R_2Ge and R_2Sn are dimeric in the crystalline phase. The dimetallenes R_4M_2 are isoelectronic with ethene and might be expected to have planar C_2MMC_2 centers and MM distances significantly shorter than M—M single bonds. In fact structure determinations by X-ray diffraction show that the C_2MMC_2 fragments are nonplanar and trans-folded, and that the M—M distances are intermediate between the values expected for single and double metal–metal bonds.[62]

Group 13 (Al, Ga, In, Tl)

Triorganyl Compounds

We will first consider compounds in which the metal atom is bonded to three organic ligands. As Table 8-3 indicates, the melting and boiling points of the trimethyl derivatives of the Group 13 elements form a very irregular sequence.

The crystal structures of trimethylboron and -gallium are unknown, but crystals presumably consist of weakly interacting monomer units. Trimethylaluminum forms a methyl-bridged dimer in the solid phase.[66] An early X-ray study of trimethylindium revealed that the compound is associated in the crystalline phase through formation of linear $In—CH_3...In$ bridges.[67] A more precise redetermination of this very interesting structure is long overdue. Solid

Table 8-3. MELTING POINTS AND BOILING POINTS (°C) FOR TRIMETHYL DERIVATIVES OF GROUP 13 ELEMENTS

Compound	Melting point	Boiling point
$B(CH_3)_3$	−160	−22
$Al(CH_3)_3$	15	126
$Ga(CH_3)_3$	−16	56
$In(CH_3)_3$	88	136
$Tl(CH_3)_3$	44	147[a]

[a] Extrapolated.

trimethylthallium is isostructural with $In(CH_3)_3$, but $CH_3...M$ contacts are significantly longer, indicating that the interaction is weaker.[68]

Trimethylboron, -gallium, -indium, and -thallium are monomeric in hydrocarbon solution and in the gas phase. Trimethylaluminum in hydrocarbon solution or in the gas phase forms a mixture of dimeric and monomeric species in a temperature and concentration/pressure-dependent equilibrium. The enthalpy of dissociation of one mole gaseous dimer into two moles of gaseous monomer is $\Delta H = 85.3 \pm 1.4\,kJ$.[69]

The molecular structure of monomeric trimethylaluminum (Figure 8-3B) has been determined by gas-phase electron diffraction with a nozzle temperature of about 215°C and a pushing pressure of about 20 torr.[70]

The coordination of the Al atom is trigonal planar ($\angle CAlC = 120°$). The barrier to internal rotation of the methyl groups appear to be negligible. Since a sixfold barrier is the simplest form compatible with the molecular geometry, this result is not unexpected. The effective molecular symmetry is thus D_{3h}.

The structures of the trimethyl derivatives of the other Group 13 elements are analogs to that of $Al(CH_3)_3$ and the bond distances obtained by GED are: $B—C = 1.578(1)$, $\mathring{A}$,[71] $Ga—C = 1.967(2)$ $\mathring{A}$,[72] $In—C = 2.161(3)$ $\mathring{A}$,[73] and $Tl—C = 2.206(3)\,\mathring{A}$.[73] Thus the $Ga—C$ bond distance is only marginally greater than the $Al—C$ bond distance. The similarity has been rationalized in terms of a "d-block contraction" to which we shall return. The mean $M—C$ bond dissociation energies decrease as the group is descended[37]:

$$\bar{D}(B—Me) = 374\,kJmol^{-1} \qquad \bar{D}(Al—Me) = 283\,kJmol^{-1},$$

$$\bar{D}(Ga—Me) = 256\,kJmol^{-1} \qquad \bar{D}(In—Me) = 169\,kJmol^{-1}$$

The bond distances and valence angles of the trimethylaluminum dimer given in Figure 8-3A are those obtained by GED with a nozzle temperature of about 60°C. The bond distances and the $\angle AlC_bAl$ valence angle are in excellent agreement with those obtained in the low-temperature X-ray study of Huffmann and Streib[66]: $Al—C_t = 1.949(2)$ and $1.956(2)\,\mathring{A}$, $Al—C_b = 2.129(2)$ and $2.152(3)\,\mathring{A}$, $\angle AlC_bAl = 75.7(1)°$ (t = terminal, b = bridge). The $\angle C_tAlC_t$ angle is, however, $123.2(1)°$ in the crystal, 6° greater than in the gas phase. The difference may be due to crystal packing forces. The hydrogen atoms are located in the crystal study: as expected, the three $C—H$ bonds in a bridging methyl group and the $C_b...C_b$ vector point toward the corners of a slightly distorted tetrahedron.

The terminal $Al—C$ bond distance is equal to the $Al—C$ bond distance found in monomeric $Al(CH_3)_3$ and may be assumed to correspond to normal, single $Al—C$ bonding. The bridging $Al—C_b$ bonds, on the other hand, are $0.18\,\mathring{A}$ longer than the terminal, indicating a fractional bond order. The $Al—Al$ distance is only $0.10\,\mathring{A}$ larger than the value calculated for a single bond by doubling Pauling's tetrahedral covalent radius[56] and $0.24\,\mathring{A}$ less than the $Al—Al$ distance in the metal: insofar as bond distances in "electron-deficient"

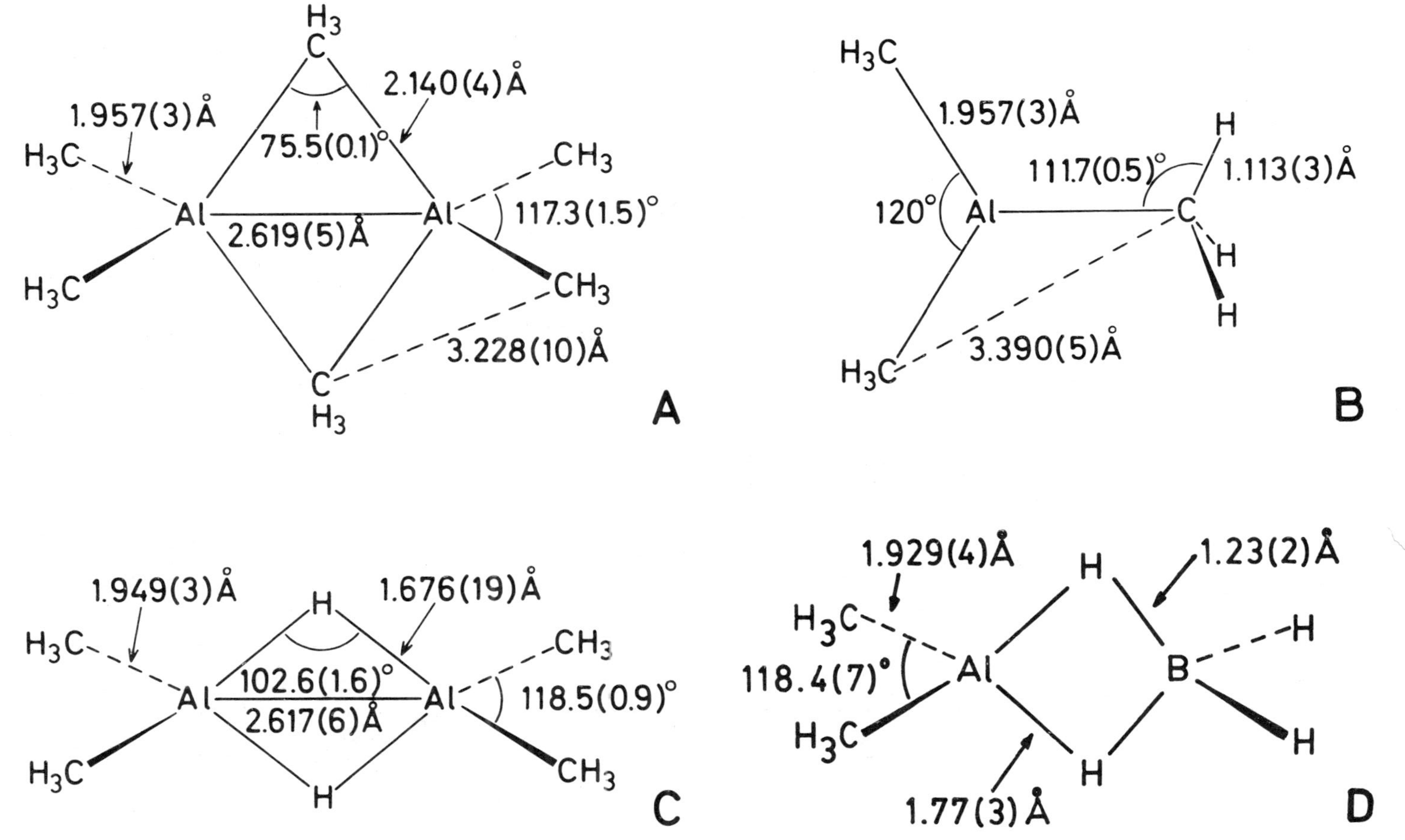

Figure 8-3. Molecular structures of monomeric (B) and dimeric (A) $Al(CH_3)_3$,[70] dimeric $(CH_3)_2AlH$[81] (C), and $(CH_3)_2AlBH_4$ (D).[96]

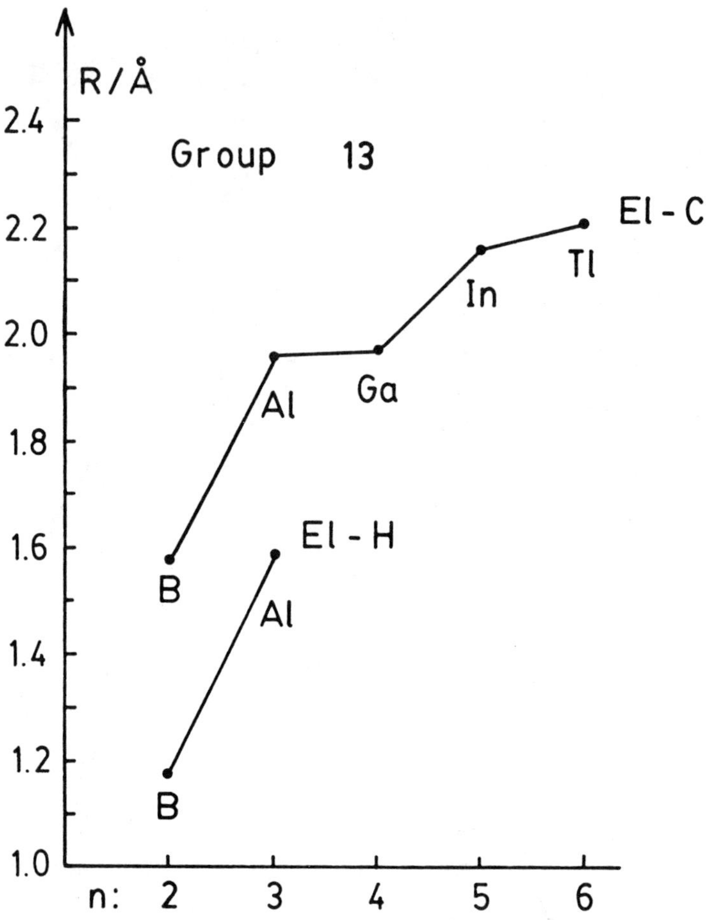

Figure 8-4. El—C bond distances in $El(CH_3)_3$ and El—H bond distances in the radical species $\cdot BH_2$,[74] and $\cdot AlH_2$.[75]

molecules of this kind give any information about bond strength, they indicate that Al—Al bonding is as important as Al—C_b bonding.

The ^{1}H-NMR spectrum of trimethylaluminum dimer in hydrocarbon solution at room temperature consists of one peak only, showing that exchange of terminal and bridging methyl groups is rapid on the NMR time scale. There is now general agreement that exchange is effected through dissociation into monomeric species followed by rapid recombination.[76]

The methyl groups in the trimethylaluminum dimer are closely packed, the eight $C_b...C_t$ distances being 3.23 Å as compared to a methyl group van der Waals diameter of 4.00 Å. If the methyl groups are replaced by more bulky substituents, the enthalpy of dissociation of the liquid dimer decreases from 81 kJmol^{-1} for trimethylaluminum[77] to 71 kJmol^{-1} for triethylaluminum[78] and 34 kJmol^{-1} for triisobutylaluminum.[79] The latter is only 40% associated as a pure liquid at 10°C.

Dimethylaluminum hydride is trimeric in hydrocarbon solution at room temperature.[80] Gas-phase molecular weight measurements show that the compound is dimeric at higher temperatures, but indicate the presence of significant amounts of trimeric species at lower temperatures.[80] The molecular structure of the dimer has been determined by GED with a nozzle temperature of about 170°C (Figure 8-3C). Attempts to determine the molecular structure of the trimer by recording GED data at much lower temperatures have failed till now.

The structure of $(Me_2AlH)_2$ is remarkably similar to that of the trimethylaluminum dimer: no significant differences are found between the Al—Al and Al—C$_t$ bond distances or the $\angle C_tAlC_t$ angles in the two compounds. The constancy of Al—Al distance is most easily rationalized if one assumes direct metal–metal bonding. The Al—H$_b$ bonds are about 0.12 Å longer than the (terminal) Al—H bonds in H_3AlNMe_3, 1.56(1) Å.[82]

A crystal structure of dimeric dimethyl(phenyl)aluminum have shown that the *phenyl* groups occupy the bridging positions.[83] The reason for this preference must obviously be electronic rather than steric. The plane of the phenyl rings (and the nodal plane of the π system) is perpendicular to the Al—Al vector. An elongation of the Al—Al bond indicates that the π system has been delocalized into a vacant Al—Al antibonding orbital. Similarly the X-ray structure of a dimeric dialkyl(1-alkenyl)aluminum reveals that the alkenyl group occupies the bridging position with the plane of the vinyl group perpendicular to the Al—Al vector.[84]

Trivinylgallium is dimeric in hydrocarbon solution.[85,86] In the hope of obtaining structural information on the dimer, we carried out an investigation by GED.[87] The gas was, however, found to contain monomeric species only. The Ga—C bond distance, 1.963(3) Å, is indistinguishable from that of $Ga(CH_3)_3$ and provides no evidence for partial Ga—C π bonding.

Dimethyl(phenylethynyl)aluminum, $Me_2AlC{\equiv}CPh$, is dimeric in benzene with the phenylethylene groups in the bridging position.[88,89] The slowness of the exchange of methyl groups with trimethylaluminum indicates that ethynyl bridges have particularly high thermodynamic stability.[89,90] In 1974 Stucky, Eisch, and co-workers published the structure of the diphenyl(phenylethynyl)-aluminum dimer as determined by X-ray crystallography.[91] Rather than being a symmetrically bridged dimer, the molecule was found to consist of two deformed monomer units joined by long bonds between the Al atom of one monomer unit and the C$_\alpha$ atom of the ethynyl group of the other.

We thought that the asymmetric structure of the molecule might be due to crystal packing forces, or to the presence of phenyl rings in conjugation with the $C{\equiv}C$ bonds or to a combination of both causes, and decided to determine the molecular structure of the dimethyl(1-propynyl)aluminum dimer by GED.[92]

The molecular structure as shown in Figure 8-5 turned out to be very similar to that of $(\mu\text{-}C{\equiv}C\text{—}Ph)_2(Ph_2Al)_2$. The monomer units are considerably distorted, in particular the Al—C$\equiv$ bonds are significantly longer than normal

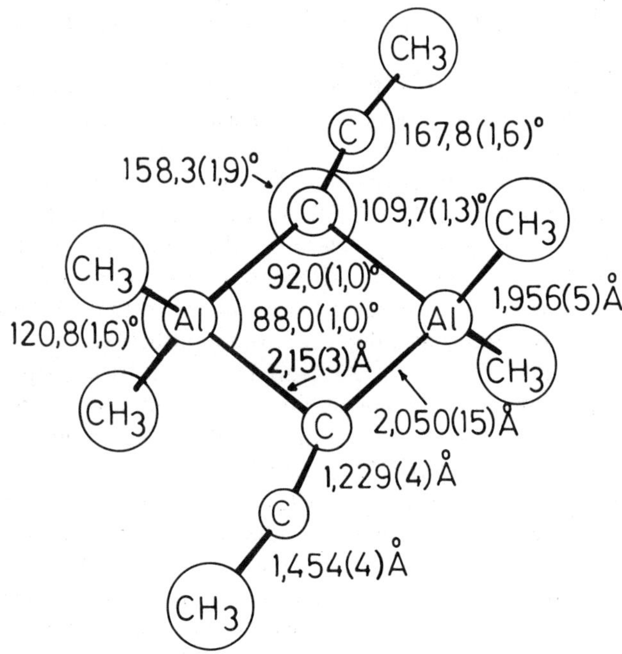

Figure 8-5. The molecular structure of dimeric $(CH_3)_2AlC{\equiv}CCH_3$.[92]

Al—C single bonds, and the $\angle$ Al—C$\equiv$C valence angle is about 160°. The two monomer units are joined by long Al—C' bonds [2.15(3) Å] that are roughly perpendicular to the approximate AlC_5 plane of the monomers.

The gas-phase structures of dimeric dimethyl(1-propyl)gallium and -indium[93] are similar to that of the Al analog, but the length of the M—C' bonds joining the monomer units [Ga—C' and In—C', 2.24(3) and 2.52(4) Å, respectively] indicate that the monomers are more loosely joined. The structures have been rationalized by assuming donation of CC π electrons from each unit into an empty p orbital on the metal atom of the other unit.

Since the solid-state structure of $(\mu{-}C{\equiv}CPh)_2(Me_2Ga)_2$[94] is very similar to the gas-phase structure of $(\mu{-}C{\equiv}CMe)_2(Me_2Ga)_2$, the latter probably retains its dimeric structure in the solid phase. 1-Propynyl(dimethyl)indium is, however, polymeric in the solid phase. Crystals consist of infinite stacks of monomer units with each In atom interacting with the C_α atoms of the units above and below. The two distances In...C'_α are about 0.40 Å greater than in the gas-phase dimer.[95]

Aluminum and Gallium Hydroborates

The whole series of methylaluminum tetrahydroborates, $Me_nAl(BH_4)_{3-n}$, $n = 0$, 1, 2, 3, is known, and the gas-phase molecular structures have been determined. In each compound Al and B are joined by double hydrogen bridges. No significant structural differences are found between the $Al(\mu\text{-}H_2)BH_2$ fragments in different molecules. The structure of Me_2AlBH_4[96] is shown in Figure 8-3D. The B—H_b distance is about 0.10 Å shorter than in diborane, the Al—H_b distance about 0.10 Å longer than in $(Me_2AlH)_2$, the Al—B distance is 0.07 Å shorter than the average of the B—B distance in diborane and Al—Al distance in $(Me_2AlH_2)_2$. Though there may be some Al—B bonding, the M—H_b distances clearly support a description as a borohydride. The Al—C bond distance is slightly but significantly shorter than in $Al(CH_3)_3$, as might be expected on the introduction of an electronegative ligand. The structure of Me_2GaBH_4 is very similar, except that all bond distances involving the metal (M—C, M—B, M—H_b) appear to be about 0.02 Å greater.[96] The structure of $MeAl(BH_4)_2$ is closely related to that of Me_2AlBH_4, the angle $\angle BAlB = 121.5(7)°$.[97] Rankin and co-workers have also determined the molecular structures of the dimethylaluminum octahydrotriborate (5) and the gallium analog.[99]

$$(CH_3)_2M \quad \underset{H \quad\quad H}{\overset{H \quad\quad H}{\diagup\,BH\,\diagdown}} \quad BH_2 \qquad M = Al, Ga$$

5

Electron Donor–Acceptor Complexes

Trimethylaluminum forms stable 1:1 complexes with electron donors like amines, phosphines, ethers, and thioethers. The enthalpies of dissociation in liquid hexane ($Me_3AlL = Me_3Al + L$) are:[100,101]

L	ΔH (kJmol^{-1})
NMe_3	125
PMe_3	88
OMe_2	85
SMe_2	45

Trimethylgallium forms similar complexes. The dissociation enthalpies appear to decrease in the same order, ie, $NMe_3 > PMe_3 > OMe_2 > SMe_2$, but are considerably smaller than for the Al analogs.

The molecular structure of the complex Me_3AlNMe_3 is shown in Figure 8-6C.[102] It is noteworthy that the Al—N bond distance is 0.10 Å *greater* than an Al—C bond distance and 0.30 Å greater than calculated by means of the MSS rule. The principal structure parameters of other electron donor–acceptor complexes are listed in Table 8-4. It is seen that replacement of the three methyl groups bonded to Al by three Cl atoms leads to a shortening of the Al—N bond

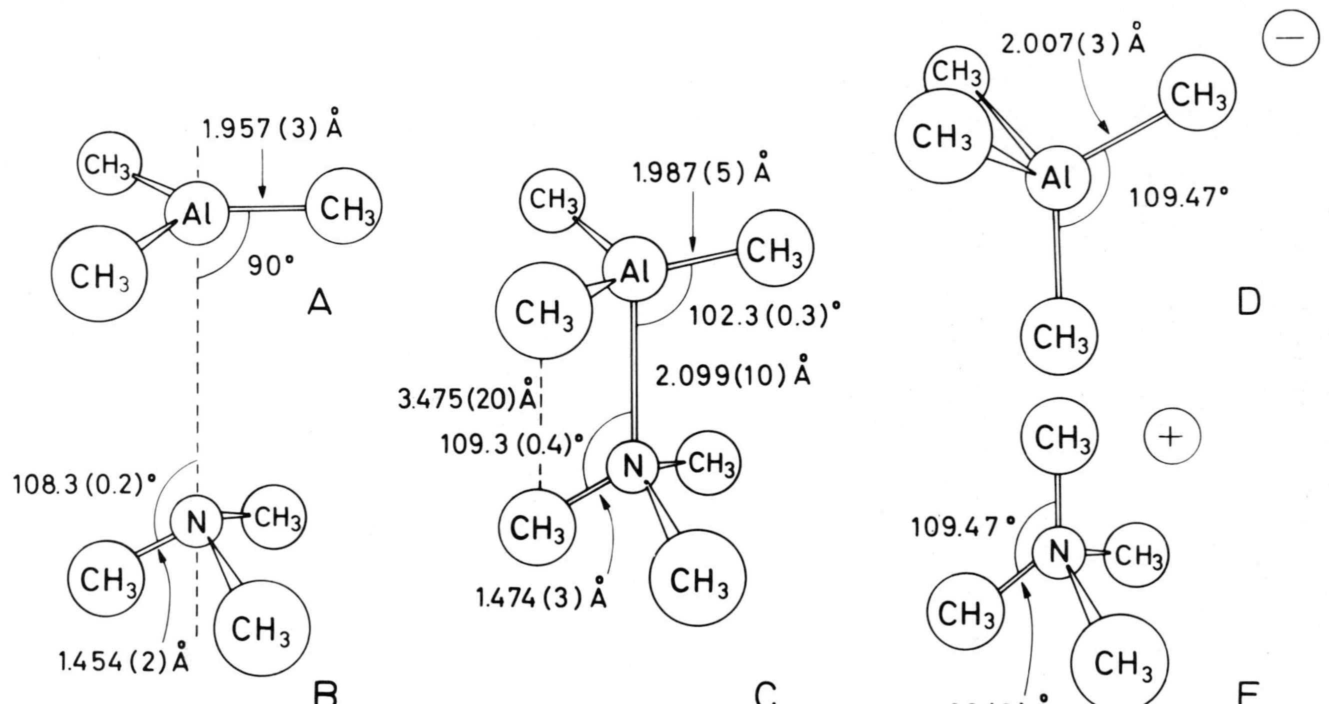

Figure 8-6. The molecular structure of (A) free $(CH_3)_3Al$,[70] (B) free $N(CH_3)_3$,[113] and (C) the complex $(CH_3)_3AlN(CH_3)_3$[102] all by GED. The idealized structures of the ions (D) $[Al(CH_3)_4]^-$ [bond distance from the solid-state structure of $NaAl(CH_3)_4$[115]] and (E) $[N(CH_3)_4]^+$ [bond distance from the solid-state structure of $(CH_3)_4NF·4H_2O$[116]]. (Reproduced with permission from reference 3.)

Table 8-4. STRUCTURE PARAMETERS OF ELECTRON DONOR–ACCEPTOR COMPLEXES OF Me_3M, H_3M, AND Cl_3M (M = Al OR Ga); X_3MDY_n [D(donor atom) = N, P, O, or S] DETERMINED BY GED UNLESS OTHERWISE INDICATED

Complex	Al—D (Å)	Al—X (Å)	∠DAlX (deg)	D—Y (Å)	∠MDY (deg)	Ref.
Me_3AlNMe_3	2.099(10)	1.987(5)	102.3(3)	1.474(3)	109.3(4)	102
H_3AlNMe_3	2.063(8)	1.560(11)	104.3(11)	1.476(3)	109.0(3)	82
Cl_3AlNMe_3	1.96(1)[a]	2.121(4)	104.9(7)	1.516(12)	112.6(15)	103
Me_3AlPMe_3	2.53(4)	1.973(3)	100.0(13)	1.822(3)	115.0(7)	104
Me_3AlOMe_2	2.014(14)	1.973(11)	98.7(15)	1.436(3)	122.6(5)	105
$Me_3AlO(CH_2)_3$	2.03(4)	1.965(7)	98.8(8)	1.465(2)	123(4)	106
Me_3AlSMe_2	2.55(2)	1.985(5)	96.0(10)	1.817(5)	124(2)	106
Me_3GaNMe_3	2.09(3)	1.992(6)	118(15)	1.484(6)	104(3)	107
H_3GaNMe_3	2.124(7)	1.50(15)	(116)	1.482(5)	109.1	108
H_3GaNMe_3[b]	2.111(2)	—	116	1.47	110	109
Me_3GaPMe_3	2.52(2)	1.997(9)	98(2)	1.84(1)	116(2)	107
Cl_3GaPMe_3[c]	2.35(3)	2.174(2)	109.3(6)	1.794(8)	107.6(3)	110
Me_3GaOMe_2	2.05(2)	1.965(4)	104(2)	1.431(5)	124(2)	111

[a] By X-ray crystallography.[112]
[b] By microwave spectroscopy.
[c] By X-ray crystallography.

by 0.14 Å. The dissociation enthalpy of Cl_3AlNMe_3 has been estimated as 200 ± 10 kJmol^{-1}, more than 50% higher than in Me_3AlNMe_3.[102] We suggest that electron donor–acceptor bonds differ from normal covalent bonds in:

1. Their low dissociation energies (cf the mean Al—C bond dissociation energy, 283 kJmol^{-1})
2. Their greater length
3. Their greater susceptibility to inductive effects

The structures of the other complexes in Table 8-3 are in agreement with these criteria.

In Figure 8-6 we compare the structure of the free acceptor with its structure in the complex. Trimethylaluminum is planar; the angle between the threefold symmetry axis and Al—C bonds is 90°, but in the complex it has increased to 102.3(3)°, corresponding to a ∠CAlC angle of 115.6°. The Al—C bonds are about 0.03 Å longer in the complex than in the free acceptor. The deformation of the acceptor may be rationalized in terms of the VSEPR model as being due to the introduction of a fourth electron pair on Al, or alternatively as due to steric repulsions between the donor and acceptor molecules.

The angle between the N—C bonds and the threefold symmetry axis in free NMe_3 is 108.3(2)°, corresponding to a ∠CNC angle of 110.6(2)°.[113] In the complex the angle between the threefold axis and the C—N bond has increased by 1° to 109.3(4)°. This difference hovers at the edge of statistical significance. The N—C bond increases by 0.02 Å on formation of the complex. Application of the VSEPR model leads to the prediction that N—C bonds should contract and

$\angle$CNC angles should open as the lone pair is partly removed. Steric repulsion between donor and acceptor molecules should lead to an increase in N—C bond distances and a closing of $\angle$CNC angles. Clearly the latter effect dominates.

Figure 8-7 compares the structure of the complex Me_3AlPMe_3 with the structures of the free acceptor and the free donor.[114] As predicted by the VSEPR model and from steric considerations, we again find that Al—C bond lengths increase and $\angle$CAlC angles decrease on formation of the complex. This behavior is indeed common to all the complexes in Table 8-4.

Comparison of the structure of the free donor, PMe_3, with its structure in the complex shows that the behavior of this ligand is opposite to that of NMe_3: on formation of the complex, the $\angle$CPC angle opens and the P—C bond distance decreases, as predicted by the VSEPR model; electron-pair repulsion appears to dominate when the atom (P) has the same size or is larger than its bonding partners, whereas steric repulsions dominate when the central atom (N) is small.

Figure 8-6 also compares the structure of the complex Me_3AlNMe_3 with the structure of the $[Al(CH_3)_4]^-$ ion (bond distance from the crystal structure of $NaAlMe_4$[115]) and with the structure of $[N(CH_3)_4]^+$ (bond distance from the crystal structure of $Me_4NF \cdot 4H_2O$[116]). We see that the structure of the acceptor in the complex is intermediate between the structure of the free acceptor and the acceptor fragment in the anion, which might be described as a complex between Me_3Al and a very strong donor, namely CH_3^-. Similarly the N—C bond distance of the donor in the complex is intermediate between the bond distances

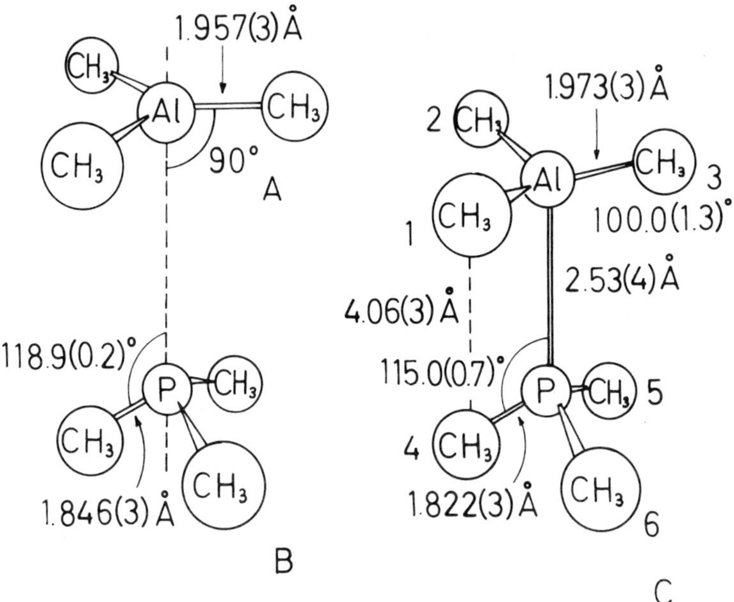

Figure 8-7. The molecular structures of (A) free $(CH_3)_3Al$, (B) free $P(CH_3)_3$,[114] and (C) the complex $(CH_3)_3AlP(CH_3)_3$.[104]

in free NMe_3 and the ion $N(CH_3)_4^+$, which might be described as a complex between NMe_3 and a very strong acceptor, namely CH_3^+.

We now turn our attention to the complexes with the ethers and thioethers. The molecular structure of the complex Me_3AlOMe_2 is shown in Figure 8-8.[105] The deformation of the donor on complex formation defies easy rationalization: the $\angle COC$ angle in the complex is larger than in free dimethylether,[117] $\angle COC = 111.8(2)°$ (as expected from the VSEPR model), but the O—C bond distance is greater than in free $OMe_2[1.417(1)\,Å]$[117] as expected from steric considerations.

The most interesting feature of this complex is perhaps the planarity—or near-planarity—of the oxygen atom. The sum of the three valence angles at O is $359.7°$. The angle between the Al—O bond and the OC_2 plane of the ether is $\varphi = 5(4)°$. The structure of Me_3AlOMe_2 is strikingly similar to the structure of the isoelectronic silylamide Me_3SiNMe_2, where $\angle SiNC = 121.4(5)°$, $\angle CNC = 117.1(10)°$, and the N atom is essentially planar.[118]

A rationalization of the essentially planar coordination of O in Me_3AlOMe_2 in terms of dative π bonding between O and Al would imply that the electronegative O atom donates electrons to an electropositive Al atom in both the σ and π systems. It is difficult to believe in such a model. Indeed, *ab initio* MO calculations on the model complex H_3AlOH_2 with double-zeta plus polarization (DZP) basis[119] give no indication for the formation of dative π

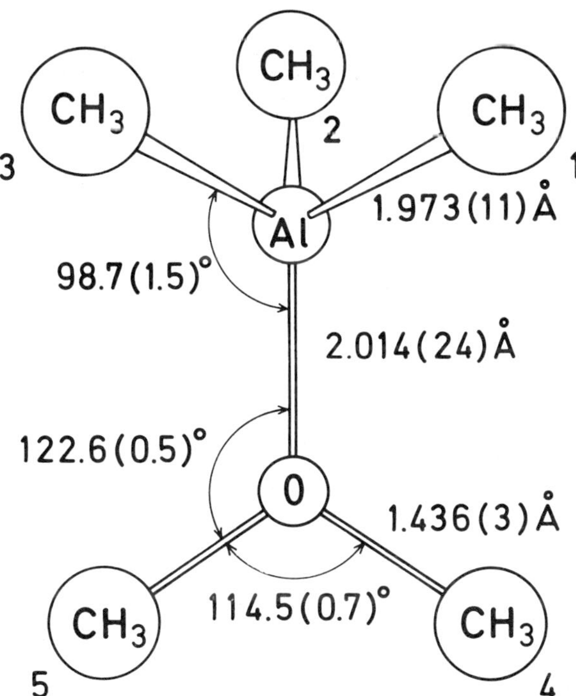

Figure 8-8. The molecular structure of the complex $(CH_3)_3AlO(CH_3)_2$.[105]

bonds between O and Al. The equilibrium value of φ **(6)** was calculated as 27°, the energy of a planar configuration is calculated to be only 0.8 kJmol^{-1} above the equilibrium conformation. Under the circumstances, we prefer to rationalize the planarity at O in Me$_3$AlOMe$_2$ as due to across angle repulsion between Al and C(O). The Al$\cdots$C distance is 3.04(2) Å, subtraction of a C hard-sphere radius of 1.25 Å yields an Al hard-sphere radius of 1.79 Å.

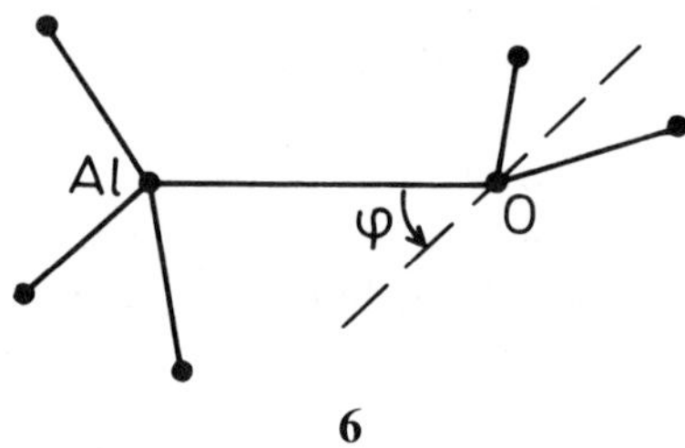

6

If this rationalization is valid, the coordination around O should deviate from planarity if dimethylether is replaced by a cyclic ether where $\angle$COC is constrained to a smaller value: a GED investigation of the complex of Me$_3$Al with oxetane,[106] O(CH$_2$)$_3$, shows that the Al—O and O—C bond distances and the $\angle$AlOC valence angle are indistinguishable from those of the dimethylether complex. The $\angle$COC angle is 95(1)°, $\varphi = 37(7)$°, and Al...C(O) = 3.07(3) Å.

The weakest complex of trimethylaluminum to be investigated is Me$_3$AlSMe$_2$.[106] It would be reasonable to expect the deformation of the donor to increase with increasing dissociation energy. Indeed the values listed in Table 8-4 show that the angular deformation as measured by $\angle$DAlC is only half as large in Me$_3$AlSMe$_2$ as in Me$_3$AlNMe$_3$. The angle between the Al—S bond and the SC$_2$ plane, $\varphi = 31(5)$° is intermediate between the values expected for trigonal and tetrahedral hybridization of the donor atom.

The structures of the gallium complexes listed in Table 8-4 have been determined with lower accuracy (Cl$_3$GaPMe$_3$, which has been studied by low-temperature X-ray diffraction,[110] is an exception). The structures are, however, very similar to those of the Al analogs. In particular, the O atom in Me$_3$GaOMe$_2$ appears to be near-planar.[111] NMe$_3$ forms a 2:1 complex with AlH$_3$ in addition to the 1:1 complex described in Table 8-4. X-Ray,[120] and GED[121] investigation of AlH$_3$(NMe$_3$)$_2$ shows Al to be pentacoordinated with N in the apical positions. The Al—N bond distance is 2.18(1) Å, 0.12 Å greater than in the 1:1 complex, and Al—H is 1.53(3) Å.

Compounds of Composition R$_2$MX (M = Al and Ga, X = F, Cl, OR, SR, or NR$_2$)

In all these compounds the metal atom is bonded to two C (or H) atoms and a potential donor atom, D, with one or more lone electron pairs. In solution and in the gas phase they form cyclic oligomers where the ring consists of alternating

M and D atoms. Oligomerization may be thought to be due to formation of electron donor–acceptor bonds between D and M. However, all investigations show that all in-ring M—D distances are equal, suggesting a rationalization in terms of resonance between covalent and donor–acceptor bonds. We shall return to this point below.

The molecular structure of $(Me_2AlF)_4$[122] is shown in Figure 8-9A. The most striking aspect of this structure is perhaps the presence of two-coordinated F with a very wide valence angle $\angle AlFAl = 146(3)°$. The Me_2AlF_2 fragments are considerably distorted from tetrahedral symmetry, with $\angle CAlC = 131(2)°$ and $\angle FAlF = 92.3(12)°$. As in the case of Me_2MX_2 (M = Ge or Sn, X = halogen), the distortion may be rationalized within the VSEPR model.

The valence angle at the bridging F atom in $(Me_2AlF)_4$ is remarkably similar to the valence angle at O in the isoelectronic cyclic siloxane $(Me_2SiO)_4$: $\angle SiOSi = 144.8(1.2)°$.[123] A rationalization of the wide valence angle at F in terms of dative π bonding between Al and F would again require the assumption that a very electronegative atom transfers charge to an electropositive atom in both the σ and π systems. We prefer to rationalize the wide valence angle in terms of across-angle repulsion between Al atoms. The observed Al...Al distance, 3.46 Å, suggests a hard-sphere radius of 1.73 Å as compared to an estimate of 1.79 Å from the Al...C(O) distance in Me_3AlOMe_2. In the following we shall use the intermediate value 1.75 Å.

The Al_4F_4 ring is nonplanar, hence presumably free from angle strain. The observed $\angle AlFAl$ and $\angle FAlF$ angles then suggest an answer to the question of why Me_2AlF is tetrameric rather than trimeric or dimeric: if a trimer with the same valence angles were formed, the Al_3F_3 ring would have to be nearly planar. This would imply a near-eclipsing of all Al—F bonds around the ring. We propose that a trimer would be destabilized by the resulting Pitzer strain.

The formation of dimers would of course imply introduction of considerable angle strain, while the formation of pentamers or higher associates would imply loss of entropy.

The trimer $[Me_2AlOMe]_3$, shown in Figure 8-9B, is trimeric in hydrocarbon solution and in the gas phase. The valence angles at the three-coordinated O atoms are very wide, $\angle AlOAl = 125.8(4)°$ and $\angle AlOC = 116.9(3)°$. Across-angle Al...Al and Al...C(O) distances are both 0.20 Å shorter than the sum of the hard-sphere radii. The O coordination is planar. Like the Me_2AlF_2 fragments in $(Me_2AlF)_4$, the Me_2AlO_2 fragments are considerably distorted from tetrahedral angles in the direction predicted on the basis of the VSEPR model.

The magnitude of the $\angle OAlO$ and $\angle AlOAl$ angles shows that formation of dimeric species must be accompanied by introduction of angle strain. On the other hand, distances between some of the methyl groups around the ring are shorter than the van der Waals diameter of a methyl group. Introduction of bulkier organyl groups is then expected to lead to increased van der Waals strain: such strain would be reduced by the formation of dimers.

Indeed, dimethylaluminum phenoxide is found to equilibrate to a mixture of dimers and trimers,[125] and dimethylaluminum t-butoxide is dimeric.[126] The

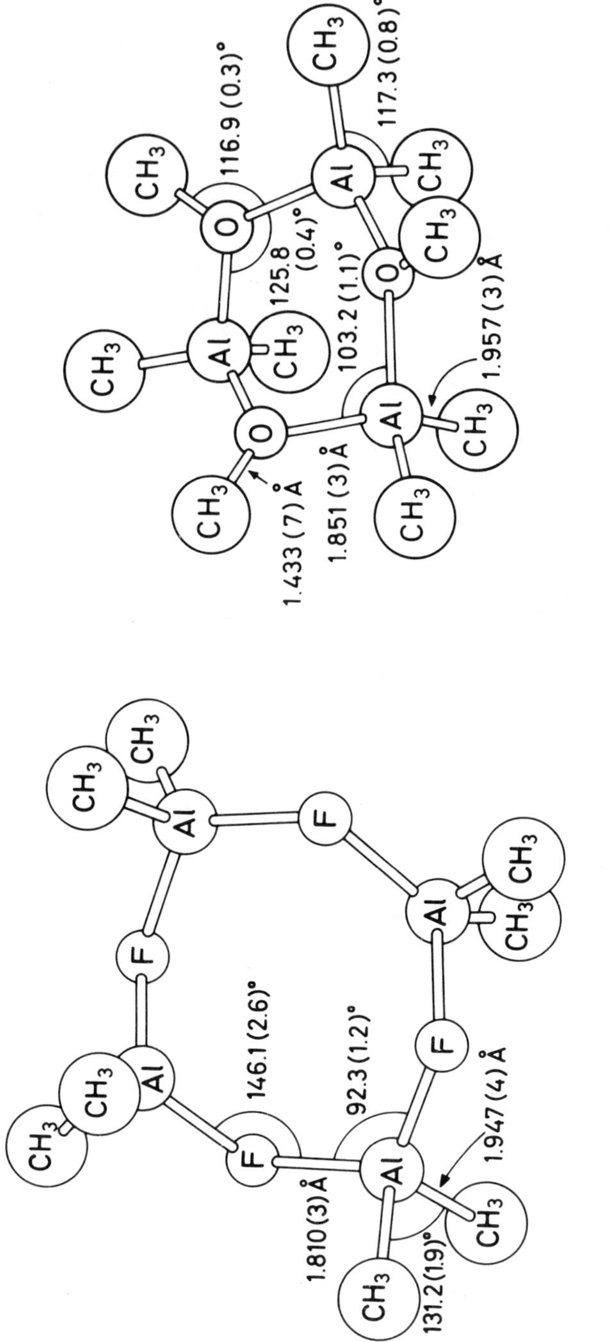

Figure 8-9. The molecular structures of (A) $[(CH_3)_2AlF]_4$[122] and (B) $[(CH_3)_2AlOCH_3]_3$.[124]

structure of this dimer is shown in Figure 8-10.[127] The $\angle$ AlOAl angle has been reduced by 28° to 98.1(7)° and $\angle$ OAlO by 21° to 81.9(7)°. The coordination around O is again planar.

Dialkylaluminum amides are trimeric or dimeric in solution and in the gas phase. One trimeric amide, H_2AlNMe_2, has been studied by both gas-phase electron diffraction[128] and by X-ray crystallography.[129] The results of the two investigations are in good agreement. The mean bond distances for Al—N and N—C are 1.950(2) and 1.497(5) Å, respectively, and the mean endocyclic valence angles $\angle$ AlNAl and $\angle$ NAlN are 114.9(1) and 108.8(1)°, respectively. Introducing bulkier substituents reduces the degree of association; Me_2AlNMe_2 is dimeric with a mean Al—N bond distance of 1.963(6) Å by X-ray diffraction.[130]

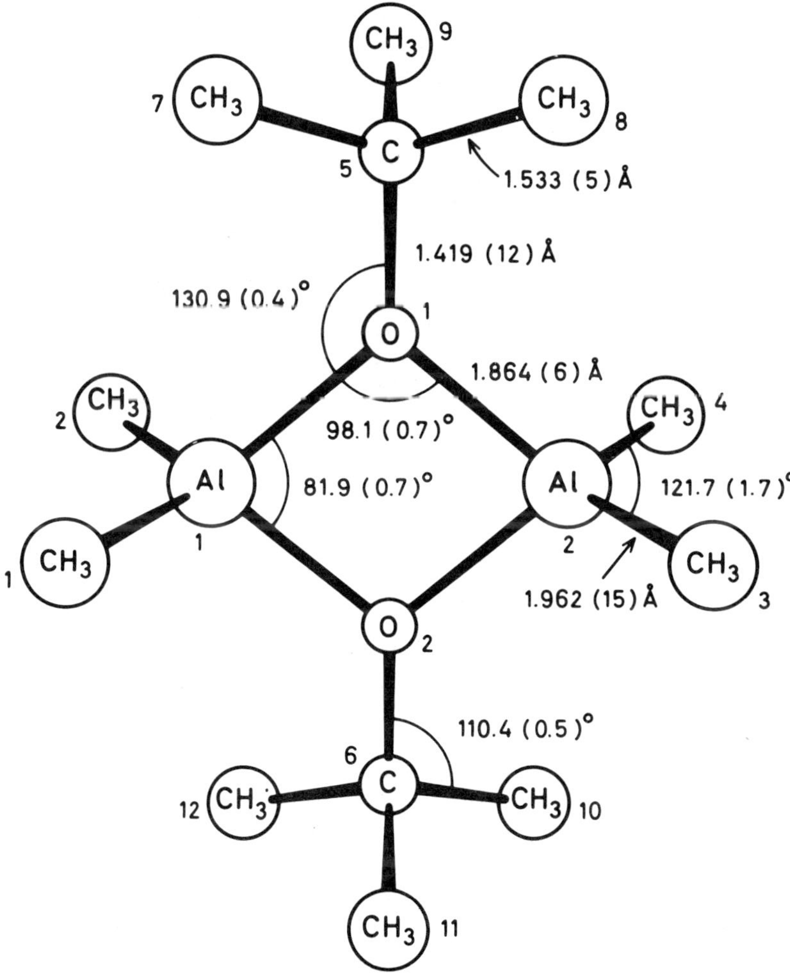

Figure 8-10. The molecular structure of $[(CH_3)_2AlOC(CH_3)_3]_2$.[127]

Before going on, we wish to comment on the Al—D bond distances in cyclic dialkylaluminum amides and alkoxides: these bond distances (Al—N = 1.95 or 1.96 Å, Al—O = 1.85–1.87 Å) are considerably *shorter* than the dative bond distances found in Me_3AlNMe_3 (2.10 Å), and Me_3AlOMe_2 ($\sim$2.01 A), and considerably *longer* than the single-bond distances calculated from the MSS rule (Al—N = 1.77 Å, Al—O = 1.72 Å). In the following we shall adopt the mean of the *terminal* Al—N bond distances of the cyclic dimer of $Al(NMe_2)_3$,[129] 1.81 Å, as a reference value for a single covalent Al—N bond distance and the mean Al—O *terminal* bond distance in trimeric aluminum triisopropoxide,[131] 1.71 Å, as a reference value for a single covalent Al—O bond distance. It is gratifying to note that the endocyclic Al—N bond distances in $(R_2AlNR'_2)_n$, $n = 2$ or 3, are very close to the average of our single covalent and dative bond distances $0.5(1.81 + 2.10)$ Å = 1.96 Å. Similarly the Al—O bond distance in $(R_2AlOR')_n$, $n = 2$ or 3, are very close to the average of our single covalent and dative bond distances, $0.5(1.71 + 2.01)$ Å = 1.86 Å.

In the case of Al—N bonds this discussion may be brought further: in solid aluminum nitride both Al and N are tetrahedrally coordinated. Each Al—N bond may be regarded as a resonance between three covalent and one dative bonds. The mean Al—N distance is 1.893 Å,[132] compared to a calculated value of Al—N = $(0.75 \times 1.81 + 0.25 \times 2.10)$ Å = 1.88 Å. The aluminum imines $PhAlNPh$[133] and $HAlN(i\text{-}C_3H_8)$[134] adopt tetrameric cubanelike structures: Each Al—N bond may be regarded as a resonance between two covalent and one dative bond. The mean Al—N bond distances are 1.914(5) and 1.913(2) Å, compared to a calculated value of $(0.67 \times 1.81 + 0.33 \times 2.10)$ Å = 1.91 Å. Thus a simple covalent/dative resonance model appears capable of rationalizing Al—N bond distances ranging from 1.80 to 2.10 Å.

Downs, Rankin, and their co-workers[137] have determined the gas-phase structure of H_2GaNMe_2. Even though the infrared spectrum of the gaseous compound had been interpreted as evidence for the predominance of monomeric species, the gas was found to consist of cyclic *dimers*. The bond distances were Ga—N = 2.027(4) Å (significantly longer than in the Al analogs) and N—C = 1.463(13) Å. The endocyclic valence angles are very close to 90°. It is not clear why this compound is dimeric when the Al analog is trimeric.

Both Me_2AlCl and Me_2AlSMe are dimeric in hydrocarbon solution and in the gas phase. The molecular structures are shown in Figure 8-12.[138,139] The Me_2AlD_2 fragments are considerably distorted from tetrahedral coordination in the direction predicted by the VSEPR model. The $\angle AlClAl$ valence angle is 90.6(5)°, considerably less than tetrahedral as predicted from VSEPR. On going from $(Me_2AlCl)_2$ to $(Me_2AlSMe)_2$ one lone pair on the bridging atom is replaced by a bonding pair, the $\angle AlDAl$ angle increases (by 4°), again as predicted by VSEPR; across-angle repulsion between ligating Al atoms appears to be significant only when the central atom is N, O, or F. A similar conclusion was reached when we compared the structures of $O(GeH_3)_2$ and $N(GeH_3)_3$ with the structures of $S(GeH_3)_2$ and $P(GeH_3)_3$.

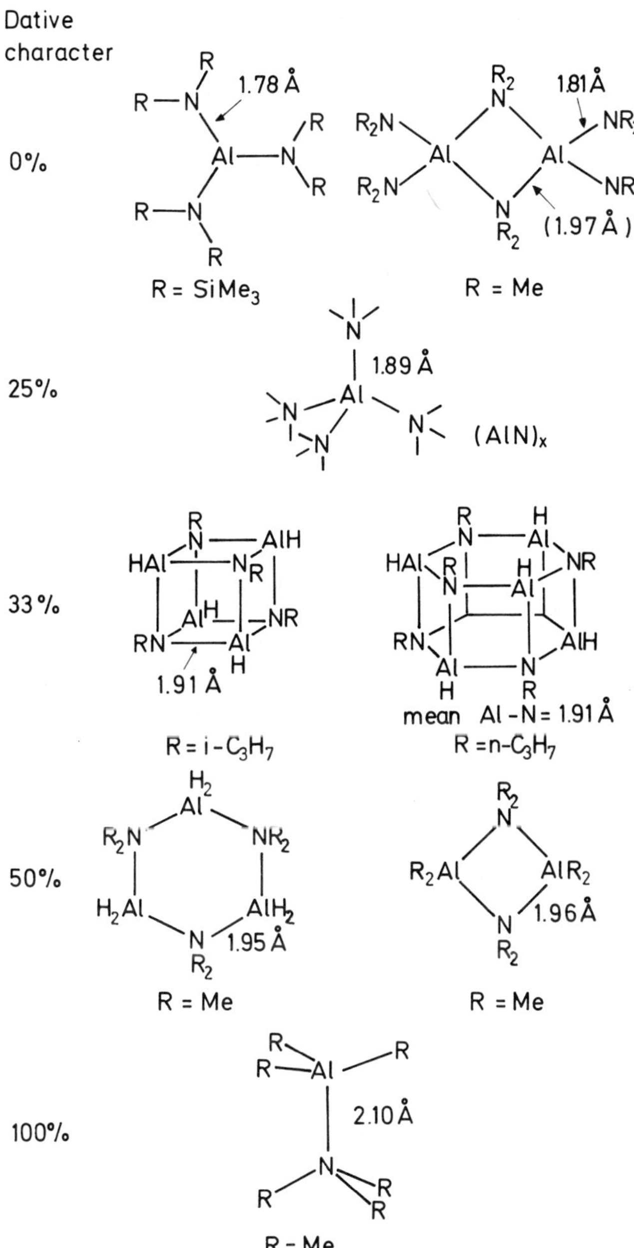

Figure 8-11. Al—N bond distances as function of dative character: Al[N(SiMe$_3$)$_2$]$_3$,[135] [Al(NMe$_2$)$_3$]$_2$,[129] (AlN)$_\infty$,[132] [HAlN(i-C$_3$H$_7$)]$_4$,[134] [HAlN(n-C$_3$H$_7$)]$_6$,[136] (H$_2$AlNMe$_2$)$_3$,[129] (Me$_2$AlNMe$_2$)$_2$,[130] and (Me)$_3$AlN(Me)$_3$.[102]

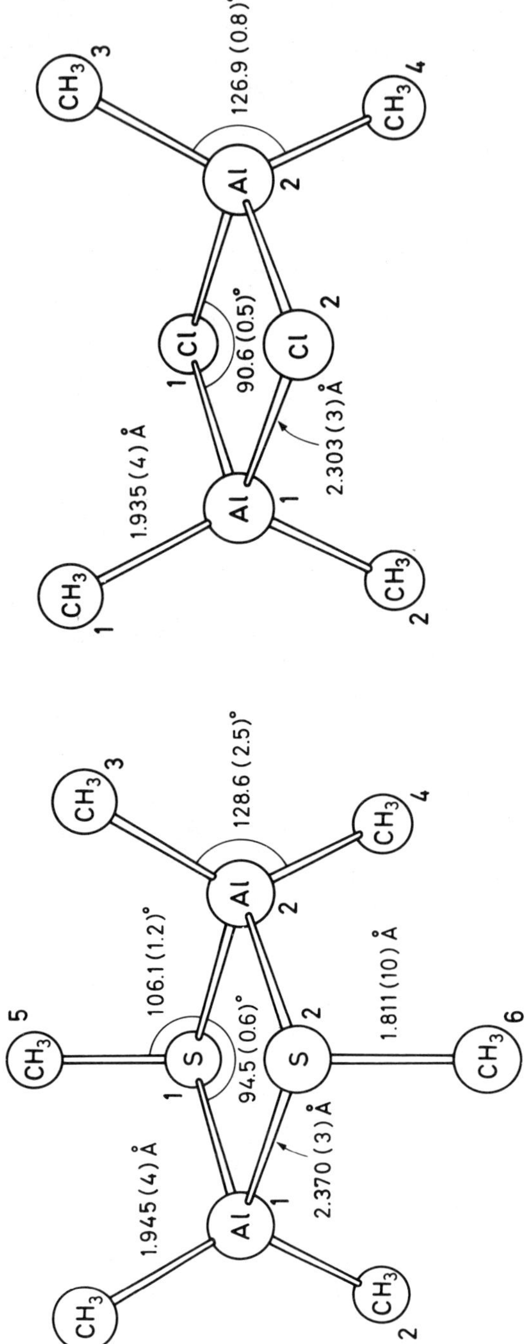

Figure 8-12. The molecular structures of $[(CH_3)_2AlSCH_3]_2$ (left)[139] and $[(CH_3)_2AlCl]_2$ (right).[138]

The endocyclic Al—Cl and Al—S bond distances are considerably longer than expected for covalent single bonds. The Al—S distance is about 0.18 Å shorter than in the donor–acceptor complex, Me_3AlSMe_2.

The complex Me_2AlSMe is polymeric in the solid state.[140]

Group 2 and Group 12

Dialkyl Compounds

Dimethylberyllium and dimethylmagnesium are colorless solids at room temperature. Neither solid has been observed to melt. Both have low volatility and very low solubility in noncoordinating solvents, though the beryllium compound is slightly more soluble and more volatile than the Mg analog.

The crystal structure of Me_2Be as determined in 1951 in an X-ray study by Snow and Rundle[141] is shown in Figure 8-13A. (A more accurate redetermination would be of great interest.) The crystal consists of infinite chains of Be atoms, each surrounded by four bridging methyl groups. The Be—C bridge bond distance is about 0.23 Å greater than the single Be—C bond distance in monomeric $BeMe_2$,[142] the Be—Be distance slightly less than twice the tetrahedral covalent radius: as in the case of the $AlMe_3$ dimer, interatomic distances indicate significant M...M bonding.

With a C...C distance of 3.11 Å, the methyl groups appear to be more closely packed than in Al_2Me_6. Replacement of the methyl groups by bulkier substituents leads to a breakdown of the polymer structure. Diethyl-, diisopropyl-, and di-*tert*-butylberyllium are all liquids at room temperature. Diethylberyllium and diisopropylberyllium are dimeric in hydrocarbon solution, di-*tert*-butylberyllium is monomeric.

Dimethylberyllium is sufficiently volatile to be studied by GED with a reservoir temperature of about 150°C.[142] The Be—C bond distance is indistinguishable from the Be—C bond distance in $Be(CMe_3)_2$[143] (see Figure 8-13B). Some geometrical parameters of gas-phase di-*tert*-butylberllium are indicated in Figure 8-13C.

The crystal structure of dimethylmagnesium has been determined by Weiss.[144] The crystal structure is like that of $BeMe_2$. The Mg—Mg distance is 2.73(2) Å, slightly less than twice the tetrahedral covalent radius; the Mg—C bridge bond distance is 2.24(3) Å as compared to a Mg—C single bond distance of 2.126(6) Å in monomeric bisneopentylmagnesium.[145] Since the M—C bond distances are greater, the Me groups are not nearly as densely packed as in $(BeMe_2)_2$.

The polymeric structure of $MgMe_2$ is not broken down as easily as the polymeric structure of $BeMe_2$: diethylmagnesium is still polymeric in the solid phase, while bisneopentylmagnesium, $Mg(CH_2CMe_3)_2$, is trimeric in hydrocarbon solution.[146] GED data collected with reservoir and nozzle temperatures of

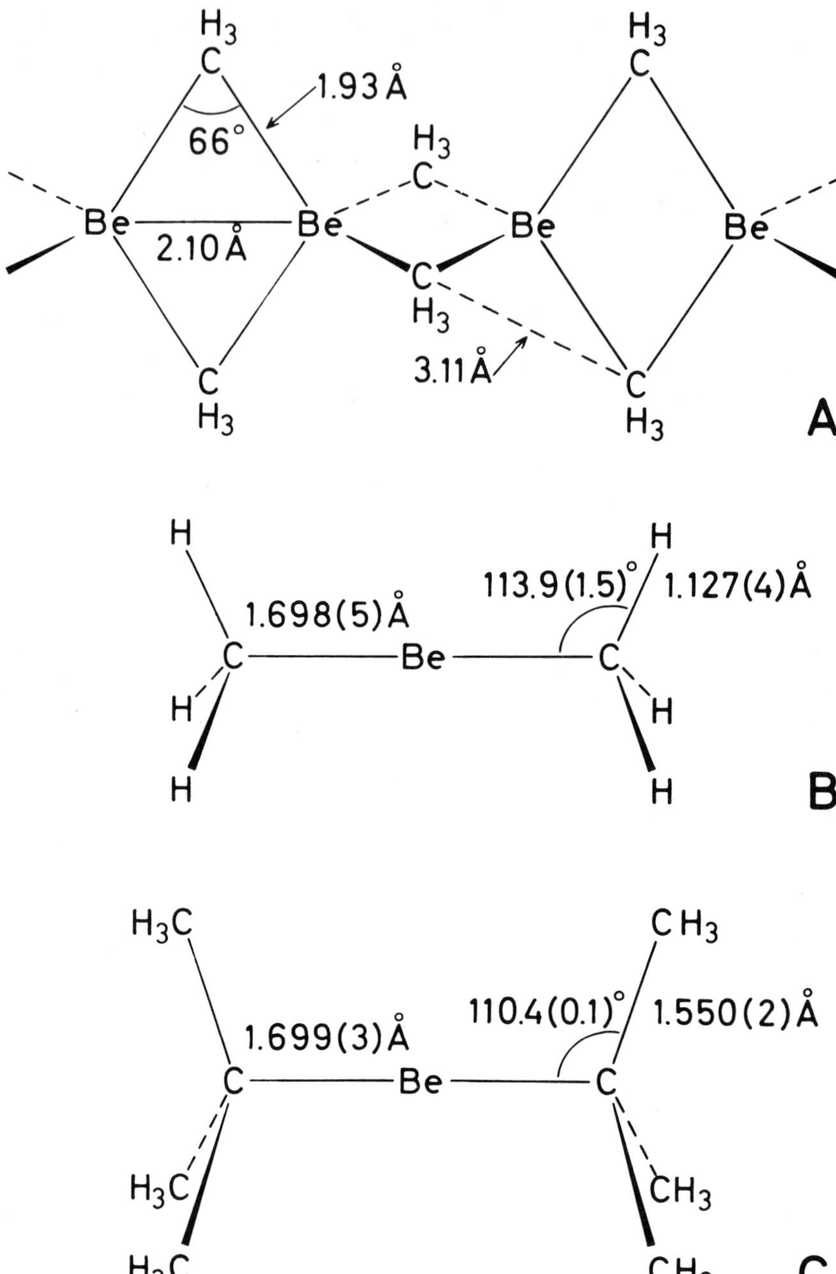

Figure 8-13. The structure of dimethylberyllium in the crystalline phase[141] (A) and in the gas phase[142] (B) and of di-*tert*-butylberyllium in the gas phase[143] (C). (Reproduced with permission from reference 3.)

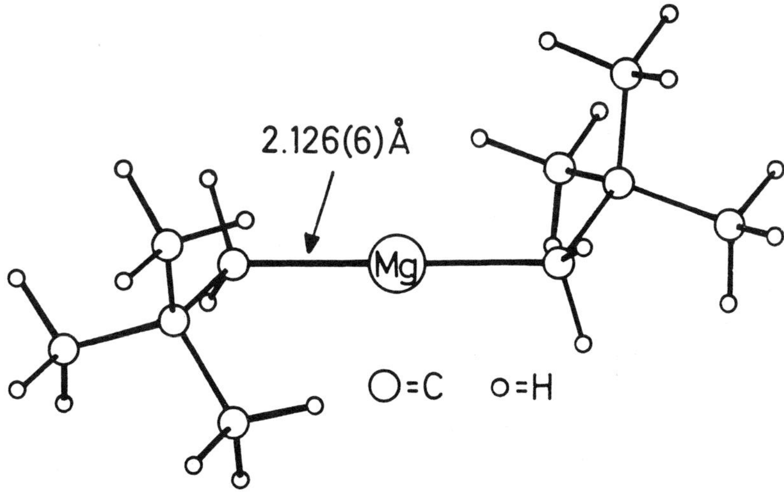

Figure 8-14. Molecular structure of bisneopentylmagnesium.[145]

about 160°C show that the gas consists of monomeric species, (Figure 8-14). The CMgC fragment is linear; the Mg—C bond distance is 2.126(6) Å. The Mg—C σ bond is not only very long, it is also very loose: the root-mean-square M—C vibrational amplitude l, which is approximately inversely proportional to the square root of the M—C stretching force constant, is 0.086(5) Å, as compared to M—C vibrational amplitudes in $Be(CMe_3)_2$, in monomeric $AlMe_3$, and in $ZnMe_2$, where l is 0.048(2), 0.066(1), and 0.050(2) Å, respectively. In the following we shall often use the vibrational amplitudes of bond distances as indicators of bond strength.

Dimethylzinc, -cadmium, and -mercury are mobile, volatile liquids at room temperature. All are monomeric in solution and in the gas phase. The M—C bond distances were first determined by rotational Raman spectroscopy in 1960[147] (see Table 8-5). Stoicheff and co-workers noted that the Hg—C bond distance is slightly, but significantly, shorter than the Cd—C bond distance and added that the origin of the anomaly was not understood. Indeed, an explanation in terms of relativistic effects emerged only 15 years later. (For reviews of relativistic effects on atomic "sizes" and bond distances, see references 157 and 158.)

The M—C bond distances of $ZnMe_2$[148] and $HgMe_2$[149] have also been determined by GED. The bond distances obtained are in good agreement with those of Stoicheff and co-workers (see Table 8-5). GED investigations of diethyl- and dipropylzinc show that the Zn—C bond distances are slightly, but significantly longer than in $ZnMe_2$.

The mean M—C bond dissociation enthalpies on the dimethyl derivatives of Be and Mg are not available. For the Group 12 metals they are as follows (kJ mol^{-1}): $\bar{D}$(Zn—C) = 186, $\bar{D}$(Cd—C) = 149, and $\bar{D}$(Hg—C) = 130.[37] As in

Table 8-5. BOND DISTANCES IN ORGANIC DERIVATIVES OF
TWO-COORDINATED BERYLLIUM, MAGNESIUM, ZINC, CADMIUM, AND
MERCURY IN THE GAS PHASE

Compound	Bond distances (Å)		Method[a]	Ref.
	M—C	M—X		
$Be(CH_3)_2$	1.698(5)		GED	142
$Be(CMe_3)_2$	1.699(3)		GED	143
$Mg(CH_2CMe_3)_2$	2.126(6)		GED	145
$Zn(CH_3)_2$	1.929(4)		RR	147
$Zn(CH_3)_2$	1.930(2)		GED	148
$Zn(C_2H_5)_2$	1.950(2)		GED	148
$Zn(C_3H_7)_2$	1.952(3)		GED	148
$Cd(CH_3)_2$	2.112(4)		RR	147
$Hg(CH_3)_2$	2.094(5)		RR	147
$Hg(CH_3)_2$	2.083(2)		GED	149
$ClHgCH_3$	2.055(2)	2.283	MW	150
Cl_2Hg		2.253(2)	GED	151
$BrHgCH_3$	2.072(2)	2.406(2)	MW	150
$IHgCH_3$	2.077(2)	2.571(2)	MW	150
I_2Hg		2.569(2)	GED	152
$Hg(CF_3)_2$	2.106(2)		GED	153
$Hg(CH_3)(CF_3)$	2.052(3)[b]		GED + MW	154
	2.116(2)[c]			
$HgPh_2$	2.092(5)		GED	155
$BrHgPh$	2.07(2)	2.435(4)	GED	156

[a] RR = rotational Raman spectroscopy; other abbreviations as in Table 8-1.
[b] $Hg—CH_3$
[c] $Hg—CF_3$

Groups 13 and 14, dissociation energies decrease as the group is descended. Though the Hg—C bond is shorter than the Cd—C bond, it is *not* stronger.

Comparison of Hg—C and Hg—Cl bond distances in the series Cl_nHgMe_{2-n}, $n = 0, 1, 2$, shows that the bond distances decrease by about 0.03 Å each time a methyl group is replaced by an electronegative Cl atom. Comparison of Hg—C bond distances in $HgMe_2$ and BrHgMe suggests a contraction by about 0.02 Å. A similar contraction has already been noted in the series X_nMMe_{4-n}, $n = 0, 1, 2, 3, 4$ (X = F, Cl, Br for M = Ge; X = Cl for M = Sn) and was rationalized as an inductive effect. The electronegativity of iodine is very similar to that of C; it is not clear from the bond distances in Table 8-5 whether the introduction of I leads to a detectable contraction.

Mercury also resembles germanium and tin in forming longer bonds to CF_3 than to CH_3.

The Hg—C bond distance in diphenylmercury, $HgPh_2$, is indistinguishable from that in $HgMe_2$. It appears that the bonding radii of sp^2- and sp^3-hybridized C are the same for bonding to Hg.

Electron Donor–Acceptor Complexes

Both dimethylberyllium and dimethylmagnesium dissolve readily in co-ordinating solvents like amines. The molecular structures of the 1:2 complexes $Me_2Be \cdot 2NC_7H_{13}$[159] (NC_7H_{13} = quinuclidin) and $Me_2Mg \cdot 2NC_7H_{13}$[160] have been determined by X-ray diffraction. The coordination around the metal atoms is distorted tetrahedral: $\angle CBeC = 118°$, $\angle CMgC = 129°$, $\angle NBeN = 111°$, and $\angle NMgN = 108°$. The M—C bond distances (Be—C = 1.83 Å, Mg—C = 2.19 Å) are significantly greater than in base-free $BeMe_2$ and $Mg(CH_2CMe_3)_2$. The M—N bond distances (Be—N = 1.91 Å, Mg—N = 2.24 Å) are about 0.30 Å longer than expected for single covalent bonds. These trends are similar to those observed for complexes of $AlMe_3$ and $GaMe_3$.

Dimethylzinc interacts with electron donors like amines and ethers, but even though there are no methyl bridges to break down, complexes are less stable than those of Me_2Be, Me_2Mg, and Me_3Al or, for that matter, Me_3Ga. The electron diffraction pattern of the gas evaporated from a sample of liquid "$Me_2Zn \cdot NMe_3$" (see reference 161) at 30°C indicated quantitative dissociation into Me_2Zn and NMe_3 in the gas phase.[162] Dimethylzinc and OMe_2 can be separated by fractional distillation.[163] Perhaps because of the weakness of the interactions, structural information on donor–acceptor complexes of diorgano-zinc compounds have been lacking.

However, donor–acceptor bonds that persist in the gas phase are formed if the donor atom is chained to the zinc atom through a methylene chain of suitable length, as in the compounds $Zn[(CH_2)_nOMe]_2$ $n = 3$ or 4. The structures of these compounds are shown in Figure 8-15.[164] Both chains are coiled back to allow O- to- Zn-coordination. The Zn—O distances are about 2.37 Å, about 0.40 Å greater than the Zn—O distances in solid ZnO. In the latter compound

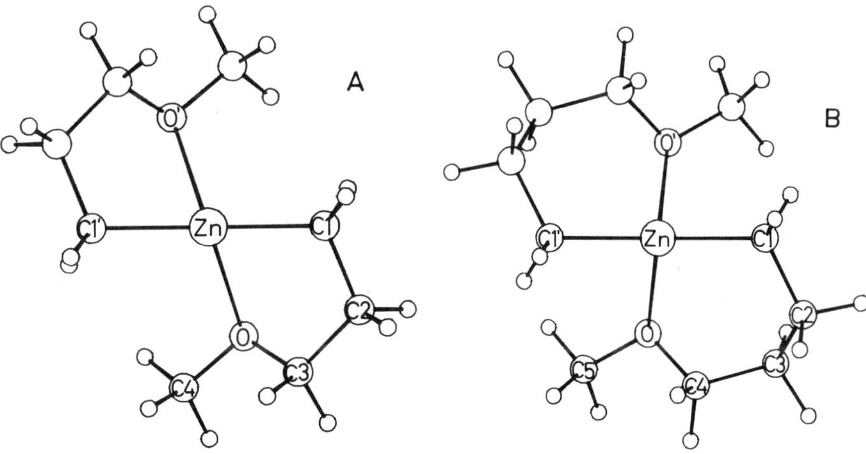

Figure 8-15. Molecular models of $Zn[(CH_2)_3OMe]_2$ (A) and $Zn[(CH_2)_4OMe]_2$ (B).[164]

both Zn and O are tetrahedrally coordinated, and each bond may be regarded as a 50/50 resonance of covalent and dative bonds.

Very recently we have completed structure determinations of $Zn[(CH_2)_3SMe]_2$ and $Zn[(CH_2)_3NMe_2]_2$.[165] Both compounds contain four-coordinated Zn: Zn—S = 2.73(1) Å, which is 0.40 Å greater than the Zn—S bond distance in solid ZnS, and Zn—N = 2.39(2) Å, which is more than 0.50 Å greater than the Zn—N bond distance in $Zn[N(SiMe_3)_2]_2$.[166] The weakness of the donor–acceptor interactions is also apparent from the large vibrational amplitudes of the Zn—D bonds: l(Zn—O) = 0.16(3) Å, l(Zn—S) = 0.13(2) Å, and l(Zn—N) = 0.17(3) Å.

While many transition metals form stable complexes with ethenes [eg, $(CO)_4FeC_2H_4$[167]] no stable ethene complex of a main group metal has ever been isolated. After Oliver and co-workers[168] had presented evidence for intramolecular M/CC double-bond interactions in dipent-4-enylzinc, we decided to investigate dibut-3-enylzinc and dipent-4-enylzinc, $Zn[(CH_2)_nCH{=}CH_2]_2$, n = 2 and 3.[169] About 80% of the hydrocarbon chains in the former were found to be in the coiled-back gauche–syn conformation shown in Figure 8-16A. In this way the C=C bonds are brought into the

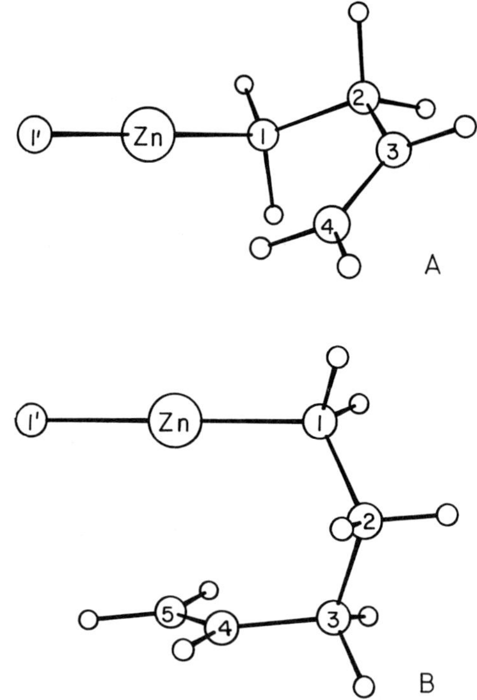

Figure 8-16. Molecular fragments of $Zn[(CH_2)_2CH{=}CH_2]_2$ (A) and $Zn[(CH_2)_3CH{=}CH_2]_2$ (B), showing the predominant coiled-back conformations of the hydrocarbon chains. Nonbonded distance Zn...C(4) = 3.15(6) and 3.00(8) Å in (A) and (B), respectively.[169]

vicinity of the metal atom; $Zn...C(4) = 3.15(6)$ Å. We assume the preponderance of this conformation to be due to M/CC double-bond interactions and estimate the interaction energy to be about $-\Delta G° = 4.5\,kJ\,mol^{-1}$. In gaseous dipent-4-enylzinc *at least* 80% of the chains are in the coiled-back gauche$_+$, gauche$_-$, anticlinal$_+$ conformation shown in Figure 8-16B, or in a coiled-back gauche$_+$, gauche$_-$, syn conformation. The distance from Zn to the nearest of the olefinic C atoms is $Zn...C(4) = 3.00(8)$ Å, and the estimated interaction energy is *at least* $-\Delta G° = 6\,kJ\,mol^{-1}$. These results are in good agreement with subsequent *ab initio* MO calculations on the model complex $H_2Zn—C_2H_4$.[170] The dissociation energy D_e of a complex where Zn is located above the midpoint of the $C{=}C$ bond was calculated as $5.0\,kJ\,mol^{-1}$ and $r_e(Zn—C) = 3.40$ Å. The Zn/CC double-bond interactions are so weak that one may prefer to describe the compounds as containing solvated, two-coordinated rather than four-coordinated Zn atoms.

Group 1 and Group 11

Very few organometallic compounds of the Group 1 or 11 metals are stable enough thermally and volatile enough to be investigated by GED. Very recently, however, Fjeldberg and co-workers[171] succeeded in recording GED data for an alkyllithium compound, $LiCH(SiMe_3)_2$. The compound is monomeric in the gas phase with $Li—C = 2.03(6)$ Å. Though very inaccurate, this value is in agreement with the bond distance $Li—C = 2.00$ Å obtained by *ab initio* calculations on LiMe.[172]

In the solid phase $LiCH(SiMe_3)_2$ is linear polymer with each Li atom surrounded by two bridging alkyl groups with $\angle C_bLiC_b$ and $\angle LiC_bLi$ both equal to about 150° and $Li—C_b$ bond distances ranging from 2.14 to 2.27 Å.[173]

Even more recently we have succeeded in recording the GED pattern of $MeAuPMe_3$, and structure refinement is in progress.[174]

Periodic Trends in M—C Bond Distances

Figure 8-17 plots the M—C bond distances in monomeric gaseous MMe_2 (M = Be, Zn, Cd, Hg) and in $Mg(CH_2CMe_3)_2$. Also included are bond distances in the diatomic, radical species $\cdot MH$ (M = Be, Mg, Ca, Zn, Cd, Hg,[175]: on going from $\cdot BeH$ to $\cdot MgH$ and on to $\cdot CaH$, the M—H bond distance increases in the way expected when a group is descended. Between Ca and Zn the nuclear charge increases by 10 units, while 10 electrons are added to the $3d$ orbitals. The d electrons do not shield the nucleus very effectively—the

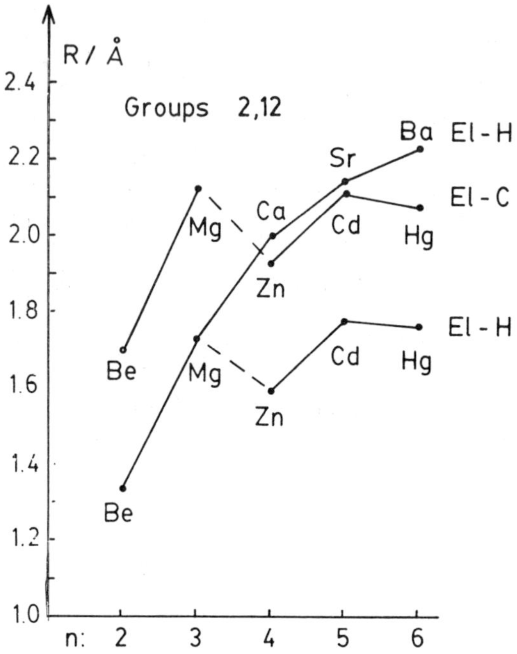

Figure 8-17. El—C bond distances in $El(CH_3)_2$ and in $Mg(CH_2CMe_3)_2$; El—H bond distances in the radical diatomic species $\cdot ElH$.[175]

effective nuclear charge increases and the atoms contract. The Zn—H distance is consequently 0.40 Å shorter than the Ca—H bond distance. We have termed this the "d-block contraction".[176] The M—C bond distance increases in the normal way on going from Be to Mg. Extrapolation suggests a Ca—C single-bond distance around 2.40 Å; on going from Ca to Zn, M—C decreases to 1.93 Å (0.20 Å shorter than Mg—C).

We now see why there was a break at Al in the M—C bond distance curve for the Group 13 elements (see Figure 8-4): The effect of the d-block contraction lingers on in Ga. It is also perceptible in Ge in the plot of M—C distances in the Group 14 elements (see Figure 8-1).

In copper, the element preceding zinc, d electrons are still part of the valence shell. In zinc they are just beneath the skin. As the periodic table is followed further to the right, the $3d$ electrons gradually sink into the atomic core. As they do so, they gradually become more efficient in shielding the nuclear charge; the d-block contraction is gradually eliminated.

In Figure 8-18 we show the variation of main group element to C bond distances across the second to sixth periods. The curves have been discussed elsewhere.[176]

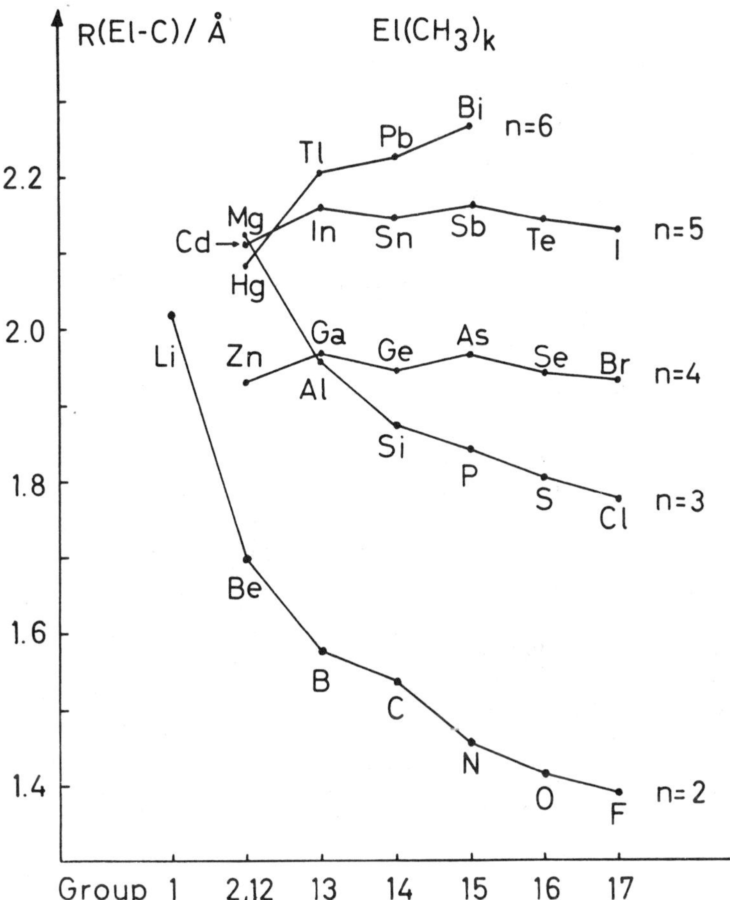

Figure 8-18. Periodic variation of El—C bond distances in simple methyl compounds El(CH$_3$)$_k$.[176]

Cyclopentadienyl Compounds of Main Group Metals

Monocyclopentadienyl Compounds

The molecular structure of germylcyclopentadiene (5-germylcyclopenta-1,3-diene, Figure 8-19), has been determined both by low-temperature X-ray crystallography and by gas-phase electron diffraction.[177] The cyclopentadienyl ring is planar, and the C(1)—C(2) and the C(2)—C(3) bond distances are not significantly different from those of gaseous cyclopenta-1,3-diene itself[178] and correspond to localized C=C and C—C bonds, respectively. The C(1)—C(5) bond, 1.474(13) Å by X-ray diffraction and 1.478(13) Å by GED, may be slightly

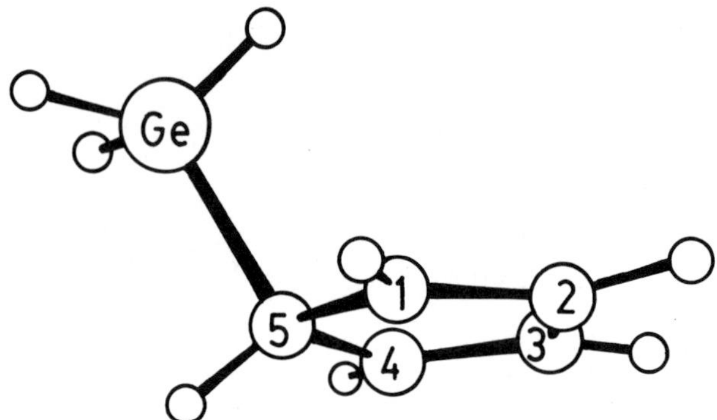

Figure 8-19. The molecular structure of $H_3Ge(\sigma\text{-}C_5H_5)$.[177]

shortened compared to cyclopentadiene, 1.506(1) Å. The Ge—C bond is slightly, but significantly longer (0.02 Å) than in $GeMe_4$ or H_3GeMe, and the angle between the Ge—C bond and the C_5 ring plane, $\angle C_5$, C—Ge, is 60.4° by X-ray and 64.1(7)° by gas-phase electron diffraction, about 7 or 11° greater than the angle $\angle C_5$, C—H in cyclopentadiene. Taken together, the short C(1)—C(5) bond distance, the long Ge—C bond distance, and the fact that Ge has moved toward a position perpendicularly above C(5), may be taken as indications for a somewhat delocalized bonding mode within the GeC(4)C(5)C(1) fragment. Nevertheless we shall regard the structure of $H_3GeC_5H_5$ as the prototype for metal–cyclopentadienyl σ bonding, M—σCp.

The ^1H NMR spectrum of $H_3GeC_5H_5$ at 35°C consists of a sharp peak at $\delta = 6.67$ corresponding to $3H(GeH_3)$ and a broad peak at $\delta = 4.44$ corresponding to $5H(Cp)$.[179] This demonstrates the occurrence of a rapid rearrangement rendering all five ring protons equivalent. Such rearrangement is common for σ-bonded metal cyclopentadienyls, and the accumulated evidence points to a mechanism involving a series of jumps of the metal atom to one of the two neighboring ring carbon atoms. We shall see below that the structure of $Me_2AlC_5H_5$ provides a possible model for the transition state in such a rearrangement. The activation energies ($kJ\,mol^{-1}$) for the migration of the Me_3M groups in Me_3MCp are 54 for M = Si, 38 for M = Ge, and 33 for M = Sn.[180,181]

Methyl(cyclopentadienyl)beryllium, MeBeCp, provides a prototype for a *pentahapto-* or π-bonded Cp ring. The molecular structure is shown in Figure 8-20.[182] The C_5H_5 ring has D_{5h} symmetry, the C—C bond distance is typical for rings π-bonded to a main group metal. The Be metal resides on the fivefold symmetry axis $h = 1.497(4)$ Å above the ring plane. The Be—C(Me) bond distance is indistinguishable from that of $BeMe_2$. If the Me group acts as a one-electron ligand and the Cp ring as a five-electron ligand, the Be atom is surrounded by an electron octet. One may assume the Be atom to be approximately (sp)-hybridized. One such hybrid points toward the methyl group and interacts with an sp^3 atomic orbital on the C atom; the other points toward the

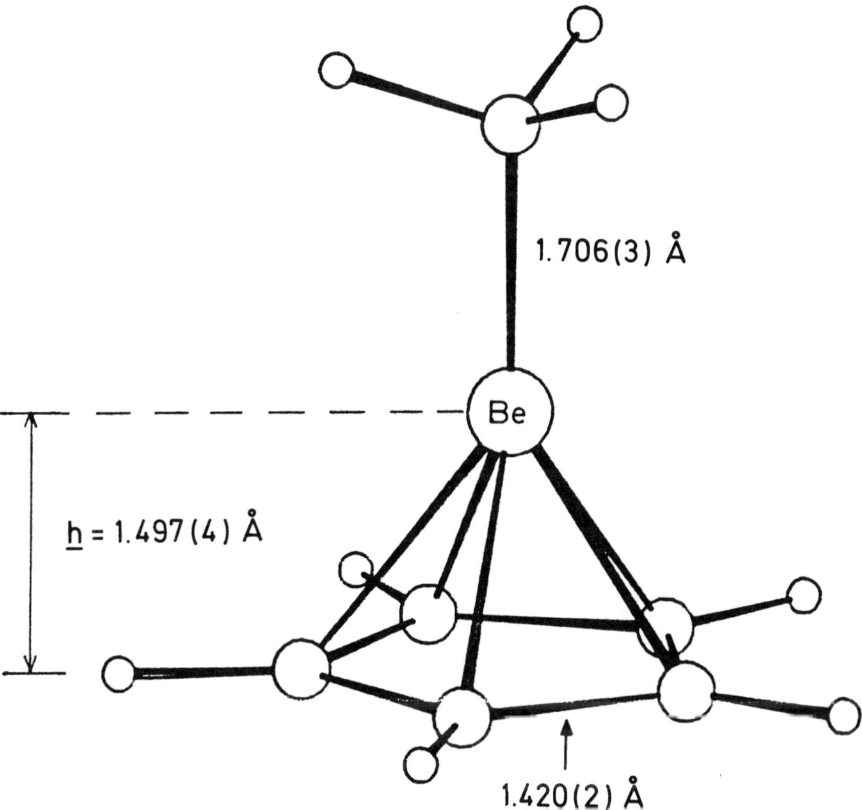

Figure 8-20. The molecular structure of $CH_3Be(\eta^5\text{-}C_5H_5)$.[182]

center of the Cp ring and interacts with the a_1 ring π orbital. Two unhybridized p orbitals on Be interact with the two e_1 ring π orbitals.

A microwave spectroscopic study of HBeCp[183] confirmed that the molecule is a symmetric top. The C—C bond distance is 1.423(1) Å, and Be—H = 1.32(1) Å. The dipole moment is 2.08(1) debye.

Ab initio MO calculations on HBeCp give results in agreement with the qualitative molecular orbital picture just described, and indicate that the direction of the dipole moment corresponds to $(CpBe)^{\delta+}H^{\delta-}$.[184,185] Both gas-phase electron diffraction[186] and microwave spectroscopy[187] have been used in the investigation of CpBeCl. The structure of the CpBe fragment is indistinguishable from that of the methyl analog: the Be—Cl distance (1.837(7) Å by GED and 1.839 Å by MW) is about 0.09 Å longer than in gaseous, monomeric $BeCl_2$.[188] The exceptionally small quadrupole coupling constant on Cl^{187} suggests that the Be—Cl bond is very polar. The dipole moment is 4.3 debye, the negative pole is presumably at Cl. The long Be—Cl bond and the apparent high negative charge on the halogen atom was rationalized by assuming that π backbonding from Cl to Be is prevented by Cp—Be π bonding.

Other cyclopentadienylberyllium compounds investigated by GED are CpBeBr,[189] CpBeC≡CH,[189] and CpBeBH$_4$.[190] The gas-phase structure of CpZnMe is analogous to that of CpBeMe.[191]

A magnesium compound, CpMgMe, has been synthesized only as a donor-acceptor complex with diethylether. Removal of the ether at room temperature leads to disproportionation to MgCp$_2$ and polymeric MgMe$_2$.[192] Recently CpMgCH$_2$CMe$_3$ has been synthesized by fusion of Cp$_2$Mg and Mg(CH$_2$CMe$_3$)$_2$, and the structure, which is analogous to that of CpBeMe, was determined by GED,[193] (see Figure 8-21). Table 8-6 compares M—C bond distances and vibrational amplitudes in the complexes CpBeMe, CpMgCH$_2$CMe$_3$, and CpZnMe. If the $3d$ electrons of Zn are disregarded, the metal atom in each molecule is surrounded by an electron octet. For each of the three complexes, the M—C alkyl bond distance and vibrational amplitude are indistinguishable from those of the corresponding dialkyl. The M—C(Cp) bond distance is about 0.22 Å greater than the M—C(R) single-bond distance for Be,

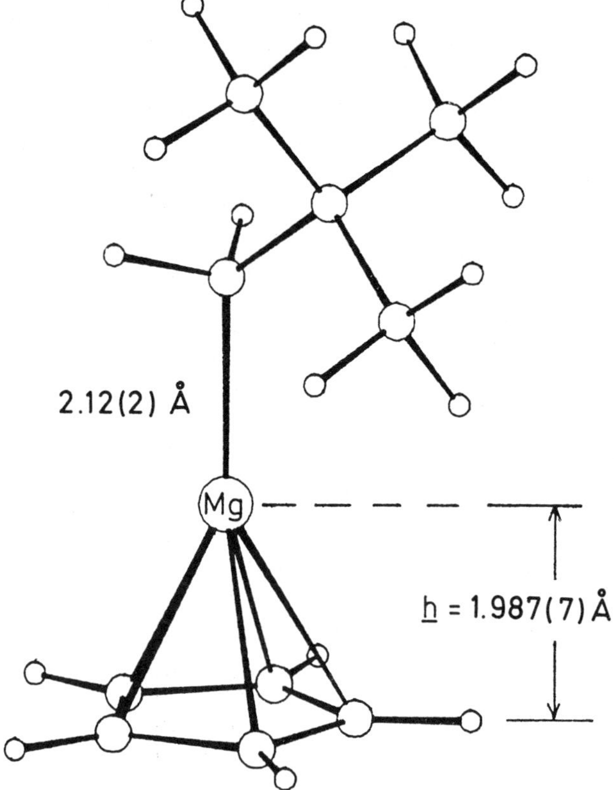

Figure 8-21. The molecular structure of $(\eta^5\text{-}C_5H_5)MgCH_2CMe_3$.[193]

Table 8-6. BOND DISTANCES AND ROOT-MEAN-SQUARE
VIBRATIONAL AMPLITUDES, l OF MeBe(η^5-Cp),
Me$_3$CH$_2$CMg(η^5-Cp), AND MeZn(η^5-Cp) DETERMINED BY
GED

Parameter	Be[a]	Mg[b]	Zn[c]
M—C(R), Å	1.706(3)	2.12(2)	1.903(12)
M—C(Cp), Å	1.923(3)	2.328(7)	2.280(9)
l(M—C(R)), Å	0.061(10)	0.08(2)	0.073(13)
l(M—C(Cp)), Å	0.075(3)	0.083(7)	0.147(11)
C—C (in ring), Å	1.420(1)	1.426(4)	1.422(3)
l(C—C), Å	0.051(1)	0.046(7)	0.052(5)

[a] Data from reference 182.
[b] Data from reference 193.
[c] Data from reference 191.

about 0.20 Å for Mg, and about 0.38 Å for Zn. In *percent* these elongations are
13% for Be, 10% for Mg, and 20% for Zn.

The M—C(Cp) vibrational amplitudes may be compared to the vibrational
amplitudes of the M—C single bonds in the dialkyls Be(CMe$_3$)$_2$,
Mg(CH$_2$CMe$_3$)$_2$, and ZnMe$_2$: the Be—C(Cp) vibrational amplitude in
MeBeCp is about 50% higher than the Be—C amplitude in Be(CMe$_3$)$_2$, the
Mg—C(Cp) vibrational amplitude in Me$_3$CCH$_2$MgCp is indistinguishable
from the Mg—C amplitude in Mg(CH$_2$CMe$_3$)$_2$, while the Zn—C(Cp) ampli-
tude in MeZnCp is more than twice as large as the Zn—C amplitude in ZnMe$_2$.
Taken together, the data in Table 8-6 strongly suggest that the strength of
M—Cp π bonding increases in the order: Zn $\ll$ Be $<$ Mg.

In this context it may be pertinent to recall that the infrared spectra of
CpHgMe strongly suggest that the Cp ring is σ-bonded[194] and that the presence
of a σ-bonded pentachlorocyclopentadienyl ring was demonstrated in a recent
X-ray study of PhHg(C$_5$Cl$_5$).[195]

The cyclopentadienyl ligand forms stable complexes with monovalent In and
Tl. Both CpIn and CpTl are monomeric in the gas phase; their molecular
structures have been determined, respectively, by GED[196] and by MW
spectroscopy.[197] CpIn and CpTl are open-faced sandwiches with C_{5v} symmetry.
The M—C bond distances for In—C and Tl—C, are 2.619(5) and 2.705(10) Å,
respectively. If the Cp rings are assumed to act as five-electron ligands, In and Tl
are both surrounded by an electron octet. The metal atoms may again be
regarded as *sp*-hybridized; a lone electron pair occupies the *sp* hybrid pointing
away from the ring. The remaining metal atomic orbitals interact with ligand π
orbitals in the manner already suggested for CpBeMe. The results of *ab initio*
calculations on CpIn are described in reference 198.

We now turn our attention to two cyclopentadienyl compounds in which the
electron count appears to exceed eight.

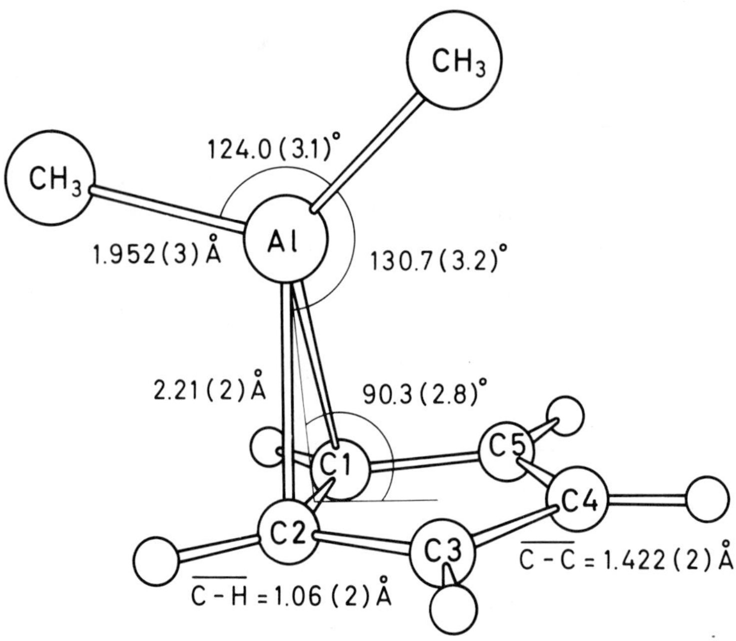

Figure 8-22. The molecular structure of $(CH_3)_2Al(\eta^2\text{-}C_5H_5)$.[199,200]

The gas-phase structure of dimethyl(cyclopentadienyl)aluminum is shown in Figure 8-22.[199,200] The C_5 ring has D_{5h} symmetry or very nearly so. The Al atom occupies a position above the periphery of the ring at a perpendicular distance of 2.10(2) Å above the ring plane. The orientation of the Me_2Al fragment is such that the $AlC(Me)_2$ plane bisects the C_5 ring. Both CNDO[199] and *ab initio* [200]MO calculations indicate an equilibrium conformation in which the Al atom resides above the midpoint of a CC bond as shown in Figure 8-22. This conformer may be described as containing a *dihapto* η^2-Cp ring. The calculations indicate, however, that the energy of an η^3- conformer obtained by rotating the Cp ring 36° about the ring center is less than 10 kJ mol^{-1} higher. The M—Cp bonding in this complex has been described as "peripheral," as "asymmetric π," or as "η^2/η^3".

We have already noted that the structure of the MCp fragment in Figure 8-22 represents a possible model for the transition state in the migration of Me_3M groups in Me_3MCp (M = Si, Ge, Sn).

The ^{1}H NMR spectrum of Me_2AlCp in benzene consists of two singlets corresponding to six Me protons and five Cp protons, respectively.[201] The spectrum is thus consistent with a low barrier to ring rotation. Interpretation of the spectrum is, however, not straightforward, since cryoscopic molecular weights indicate partial association.[202]

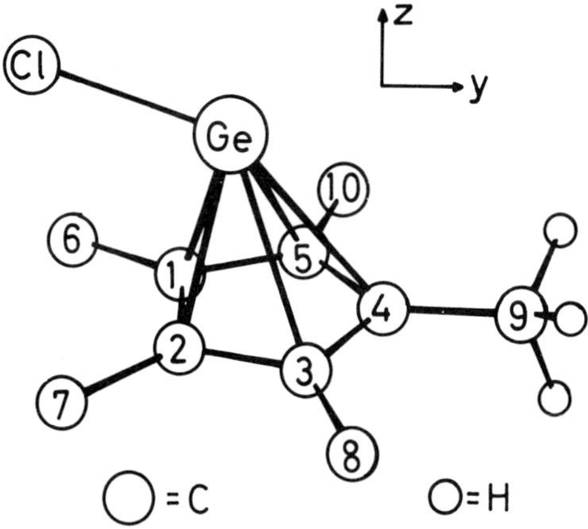

Figure 8-23. The molecular structure of $(C_5Me_5)GeCl$.[203]

The gas-phase structure of Cp*GeCl (Cp* = C_5Me_5) is shown in Figure 8-23.[203] The GED data are consistent with a model where the ring has C_{5v} symmetry. The Ge atom resides $h = 2.11(2)$ Å above the C_5 ring plane at a distance $\delta = 0.43(6)$ Å from the fivefold ring axis, ie, about one third of the distance from the center to the periphery; Ge—C distances range from 2.30 to 2.70 Å. We would describe the structure as corresponding to asymmetric π bonding.

The Ge—Cl bond vector is contained in a plane bisecting the C_5 ring. The angle between the Ge—Cl bond and the ring normal through Ge is $\angle h$, GeCl = 110(2)°. The Ge—Cl bond distance is 2.258(12) Å, about 0.07 Å greater than in $GeCl_2$.[204]

If the cyclopentadienyl ligands contribute five electrons and the Me groups and Cl atom one electron each, Me_2AlCp and Cp*GeCl must be regarded as 10-electron complexes. The structure of Me_2AlCp may be rationalized by assuming the Al atom to be sp^2-hybridized. Two (sp^2) hybrids interact with sp^3 hybrids of the Me group C atoms, the third interacts with the a_1 and one of the e_1 ring π orbitals. The unhybridized p orbital interacts with the other e_1 π orbital.

The dominant stereochemical feature of CpGeCl is the presence of a lone electron pair on Ge. *Ab initio* MO calculations[205] on a C_{5v} structure similar to that of CpBeCl show that the electron pair occupies a $4s$ orbital strongly destabilized through antibonding interactions with both ligands. When the molecule is bent, ie, when $\angle h$, Ge—Cl is reduced from 180°, the lone-pair orbital is gradually transformed to an sp hybrid pointing away from both ligands, and the antibonding interactions are reduced (but not eliminated). The displacement of the Ge atom from the position above the ring center may also be regarded as a result of ring tilting. However described, this deformation leads to a further reduction of antibonding interactions between the lone-pair and the ring π electrons:

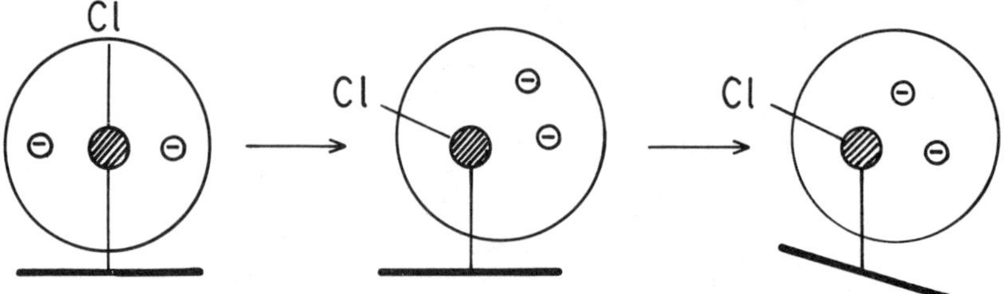

Similarly the position of the Al atom above the ring periphery in Me_2AlCp may be that which minimizes antibonding interactions between the Al—C(Me) bond pairs and the ring π electrons.

If the Cp rings in H_3GeCp,[177] Me_3GeCp,[206] or Me_3SnCp[207] were symmetrically π-bonded, the metal atoms would presumably have to accommodate 12 electrons in the valence shell; the Cp rings in these compounds are, as we have seen, σ-bonded.

Finally we wish to point out that $MeZnCp$,[208] Me_2AlCp,[209] $CpIn$,[210] and $CpTl$[210] are polymeric in the solid phase, and that association occur via bridging cyclopentadienyl rather than bridging alkyl groups.

Dicyclopentadienyl Compounds

The molecular structure of dicyclopentadienylmagnesium, Cp_2Mg, as determined by gas-phase electron diffraction is shown in Figure 8-24.[211] The best fit between observed and calculated intensities was obtained with a model of D_{5h} symmetry, ie, with eclipsed ligand rings as shown in Figure 8-24. A model with D_{5d} symmetry, ie, with staggered ligand rings as found in the solid state,[212] could not be ruled out, however, the barrier to internal rotation of the Cp ring is probably less than the thermal energy available under the experimental conditions, $RT = 3 \text{ kJ mol}^{-1}$.

Ab initio MO calculations with a very large basis set indicate that the energy of the staggered conformation is only 0.12 kJ mol^{-1} higher than the energy of the eclipsed conformation.[213]

If the Cp rings are considered to be $5e$ (e = electron) ligands, the valence shell of Mg must accommodate 12 electrons. It is perhaps surprising that the Mg—C(Cp) bond distance is indistinguishable from that in the $8e$ complex $CpMgCH_2CMe_3$. The vibrational amplitude of the M—C(Cp) distance in Cp_2Mg may, however, be 0.01 or 0.02 Å greater, $l = 0.103(3)$ vs $0.083(7)$ Å.

The *ab initio* calculations suggest that the Mg atom carries a positive charge close to $+2$. The charge distribution may therefore be described as ionic.[213] It should not be inferred, however, that metal–ligand bonding is weak: the M—Cp stretching force constant and the mean M—Cp dissociation energy are similar to those of Cp_2Ni.[211]

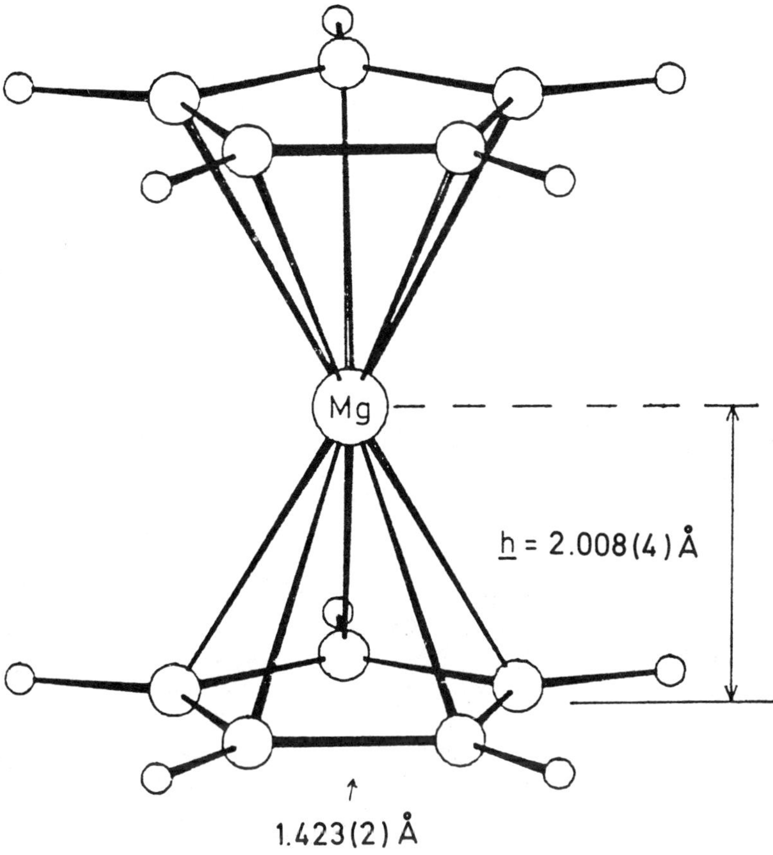

Figure 8-24. The gas-phase molecular structure of $Mg(\eta^5\text{-}C_5H_5)_2$.[211]

Beryllocene, Cp_2Be, was first investigated by GED in 1963, and a report was published[214] in the following year. Two molecular models had been proposed on the basis of a determination of the crystal space group (isomorphous with ferrocene), a nonzero dipole moment in hydrocarbon solution, and the infrared spectra: a D_{5d} symmetric sandwich model (I in Figure 8-25) or a model with one π-bonded and one σ-bonded ring (IV). Neither model was consistent with the available information. Both models proved inconsistent with the GED data. These showed that both rings had D_{5h}- or near D_{5h}-symmetry, and that the Be atom was symmetrically π-bonded to one of them at a perpendicular distance of 1.49 Å, similar to the distance later found in CpBeMe. The position of the second ring relative to the tightly bonded Be(π-Cp) fragment was difficult to determine, but a C_{5v} model (II), in which the second ring was π-bonded to Be but at a greater distance, was consistent with the data. The possibility that the second ring had slipped sideways as in model III was without precedent and was not considered.

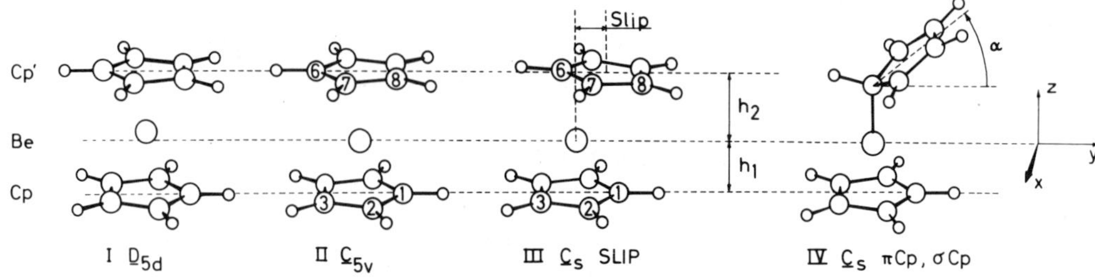

Figure 8-25. Four models of beryllocene, $Be(C_5H_5)_2$.

Several years later Wong and co-workers published the results of X-ray crystallographic investigations of $BeCp_2$ at room temperature[215] and at $-120°C$.[216] The crystals were found to be disordered, but the electron density in the asymmetric unit corresponded to the supposition of two slip sandwich molecules: see model III in Figure 8-25.

We carried out refinement of a slip sandwich structure based on the 1963 GED data,[214] and concluded that they were incompatible.[217] This erroneous conclusion was due to incorrect treatment of the "background" of atomic intensities. With the availability of better programs for background elimination, we did not repeat the error 6 years later, and it would not be repeated by any competent worker in the field today.

In 1977–1978 no fewer than four groups published the results of extensive MO calculations (PRDDO,[218] MNDO,[219] and *ab initio*[220,221]) on $BeCp_2$. These calculations all indicated that the two models suggested prior to our 1963 GED study, namely I and IV in Figure 8-25, were energetically more stable than the C_{5v} model, and it was suggested that the gas might consist of a mixture of the two.

In 1979 we published the result of a new GED investigation.[222] Models I and IV were again found to be inconsistent with the GED data, as was any mixture of the two. The C_{5v} model was consistent with the data, but a model in which the second ring had slipped sideways (III) gave even better fit.

The best model is shown in Figure 8-26, along with the disordered structure obtained in a recent X-ray diffraction investigation at 130 K.[223] The structures are in good qualitative agreement, but there are two differences. In the gas phase the magnitude of the parameter slip places the Be atom two-thirds of the way from a position below the center of the second ring to a position directly below the periphery; in the solid phase the Be atom occupies the latter position. In the gas phase both rings appear to have D_{5h} symmetry or nearly so; the X-ray structure offers evidence for partial localization of double bonds in one ring, presumably the one that is asymmetrically bonded to Be.

^{1}H NMR spectra of $BeCp_2$ in noncoordinating solvents consists of one single sharp peak over the entire temperature range from $+50$ to $-135°C$.[224] Clearly exchange of the symmetrically and unsymmetrically π-bonded rings (alterna-

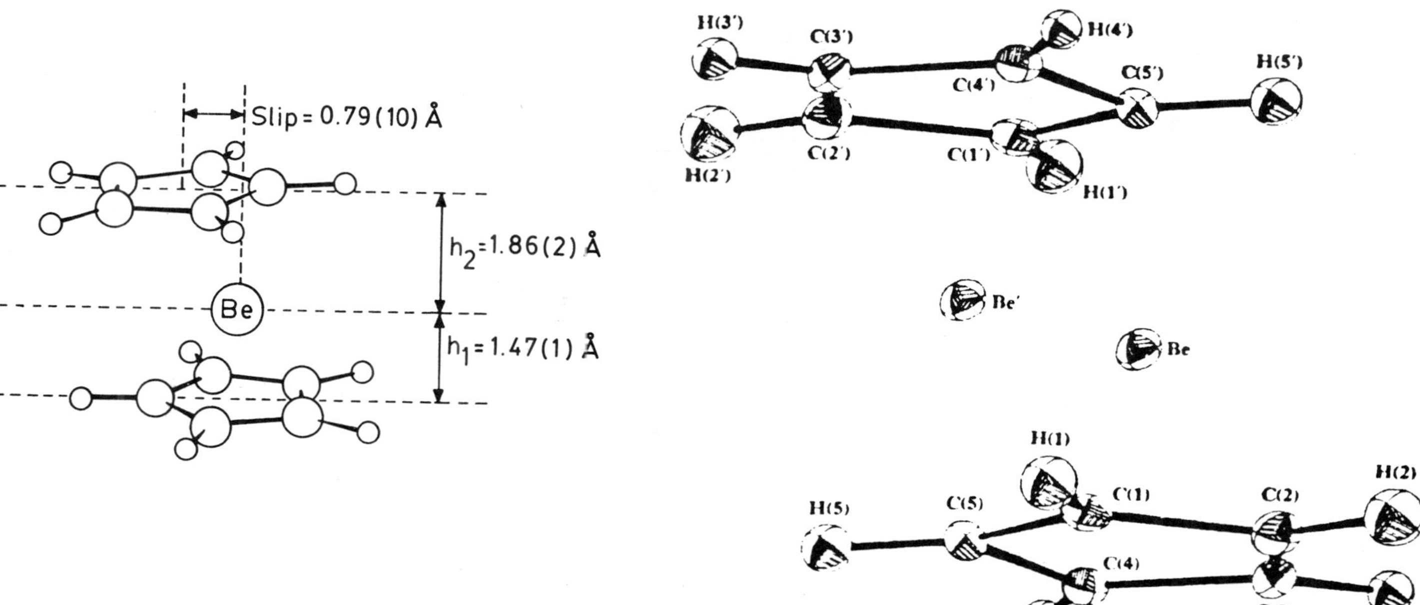

Figure 8-26. Gas-phase structure of $Be(C_5H_5)_2$ as determined by GED (left)[222] and solid-state structure as determined by XR (right).[223] The crystal structure is disordered, and two alternative positions of Be are indicated. (Reproduced with permission from reference 223.)

tively motion of the Be atom between the two disordered positions in model II of Figure 8-25) is rapid on the NMR time scale.

Dicyclopentadienylzinc is polymeric in the solid phase,[225] and the vapor pressure/thermal stability is too low to allow determination of the gas-phase structures by GED. Bis(pentamethylcyclopentadienyl)zinc, Cp_2^*Zn, is more volatile and more stable. The gas-phase structure determined by GED is shown in Figure 8-27A.[226] The Zn atom is symmetrically π-bonded to one ring with a $Zn—C(\eta^5)$ bond distance of 2.28(2) Å, similar to the $Zn—C(Cp)$ bond distance in CpZnMe. The second ring is approximately parallel to the first. The Zn atom is located at or just outside the periphery of this ring, the angle between the $Zn—C(5)$ bond and the ring plane being 84(4)°. The $Zn—C(5)$ bond distance is very inaccurately determined at 2.04(6) Å, but it is probably greater than the single $Zn—C(Me)$ bond distance in CpZnMe. This molecule is much too large to allow determination of individual CC distances, but a model with CC bond distances in the η^1-bonded ring corresponding to localized single and double bonds is in significantly better agreement with the GED data than a model in which the five CC bonds are equal.

The molecular structure of $Zn(C_5H_4SiMe_3)_2$, $ZnCp_2^\dagger$, is shown in Figure 8-27B.[226] The predominant conformer is that in which the $(\eta^5\text{-}Cp)$ Zn fragment and the $SiMe_3$ group are bonded to the same C atom, C(5). The presence of other conformers has, however, not been ruled out. The $SiMe_3$ group appears to stabilize a more pure η^1 attachment; the angle between the $Zn—C(5)$ bond and the C_5 ring plane is now 53(7)°.

NMR spectra show that exchange of η^1- and η^5-bonded rings is rapid on the NMR time scale at room temperature, but ^{13}C spectra of $ZnCp_2^\dagger$ show that in this compound the exchange is frozen out at $-37°C$.[226]

The structures of the three compounds $MgCp_2$, $BeCp_2$, and $ZnCp_2^*$ reflect the decreasing strength of $M—Cp$ π bonding noted in the monocyclopentadienyl compounds. Taken together, the series of gas-phase structures $MgCp_2$, $BeCp_2$, $ZnCp_2^*$, $ZnCp_2^\dagger$ presents us with a gradual transition from an $(\eta^5\text{-}Cp)_2M$ to an $(\eta^5\text{-}Cp)M(\eta^1\text{-}Cp)$ structure.

We now move on to dicyclopentadienyl compounds of divalent Group 14 metals, ie, to Cp_2Ge, Cp_2Sn, Cp_2Pb, and their derivatives. The "bent metallocene" structures of $SnCp_2$ and $PbCp_2$, similar to the structures of $(C_5H_4Me)_2M$, M = Ge and Sn, as shown in Figure 8-28, were first demonstrated by GED studies published in 1967.[227] The GED study of Cp_2Sn has been superseded by a more accurate X-ray study.[228] The molecule Cp_2Pb is polymeric in the solid phase,[229] but an X-ray structure of Cp_2^*Pb is available.[228]

The first published germanocene structure was that of Cp_2^*Ge.[230] This proved, however, to be a case of mistaken identity: elemental analysis showed that the sample consisted of the then-unknown compound Cp^*GeCl, and the structure was consequently withdrawn.[231] Later we published the structures of $(C_5H_4Me)_2Ge$[232] and the structure of Cp_2^*Ge determined with an authentic sample.[203] The crystal structure of unsubstituted Cp_2Ge has been determined by du Mont and co-workers.[233] For a discussion of why it was possible to bring

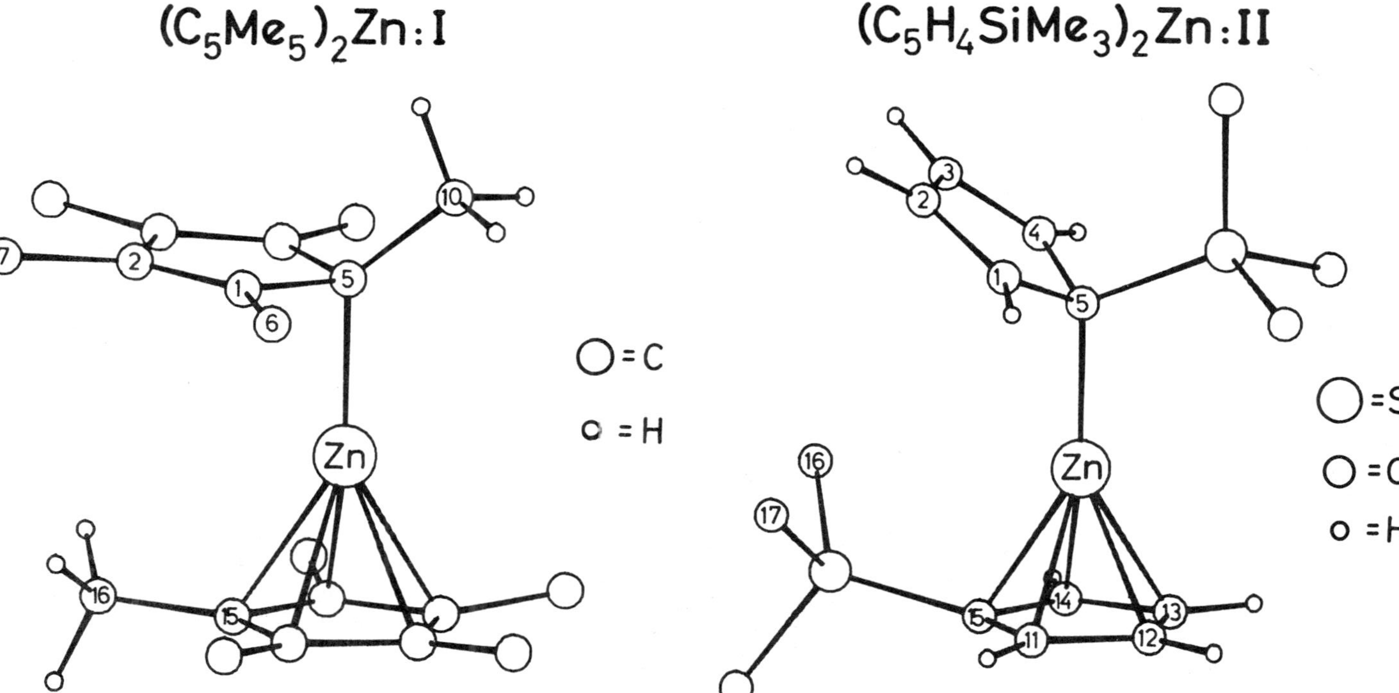

Figure 8-27. The molecular structures of $ZnCp_2^*$ (I) and $ZnCp_2^\dagger$ (II).[226] Some hydrogen atoms have been omitted for clarity.

Table 8-7. Dicyclopentadienyl Compounds of Ge(II), Sn(II), and Pb(II) Studied by GED and X-Ray Crystallography (XR)

Compound	h (Å)[a]	R(M—C) (Å)[b]	$\angle C_5,C_5$ (deg)[c]	Method	Ref.
$Ge(C_5H_5)_2$	—	2.52	50	XR	233
$Ge(C_5H_4Me)_2$	2.22	2.53	34(7)	GED	232
$Ge(C_5Me_5)_2$	2.21	2.52	22(2)	GED	203
$Sn(C_5H_5)_2$	2.42	2.71	($\sim$55)	GED	227
$Sn(C_5H_5)_2$[d]	2.41	2.68	47	XR	228
$Sn(C_5H_4Me)_2$	2.40	2.69	50(6)	GED	232
$Sn(C_5Me_5)_2$[d]	2.39	2.68	36	XR	234
$Sn(C_5Ph_5)_2$	2.40	2.69	0	XR	235
$Pb(C_5H_5)_2$	2.50	2.78	45(5)	GED	227
$Pb(C_5Me_5)_2$	2.48	2.79	39	XR	228

[a] Perpendicular metal-to-ring distance.
[b] Mean bond distances.
[c] Angle between ring planes.
[d] Average of two crystallographically independent molecules.

a model of Cp_2^*Ge into agreement with data obtained with Cp*GeCl, we refer the reader to reference 203.

In Table 8-7 we have collected structure parameters of some Group 14 metallocenes. As shown in Figure 8-28, the metal atom is pentahapto-bonded to both rings, but with the notable exception of $(C_5Ph_5)_2Sn^{235}$, the ligand rings are not parallel as in Cp_2Mg.

If the Cp rings are regarded as 5e ligands, each metal atom is surrounded by 14 electrons. Since the Zn atom in Cp_2^*Zn appears unable to accommodate 12 electrons in the valence shell (and to form symmetric π bonds to both rings), it is perhaps surprising that the metal atom in Cp_2^*Ge can accommodate 14. Comparison of the perpendicular metal-to-ring distances in the germanocenes in Table 8-7 (2.21 or 2.22 Å) with the perpendicular M—Cp (η^5) distance in Cp_2^*Zn (1.93 Å) or with the perpendicular M—Cp* distance in the less electron-rich Cp*GeCl ($h = 2.11$ Å) indicates that M—Cp bonding in germanocenes is exceptionally weak. Comparison of mean M—C bond distances in Cp_2Sn (2.68–2.71 Å) with the M—C bond distance in the 8e compound InCp (2.62 Å) and of the mean M—C bond distances in Cp_2Pb and TlCp (2.78 vs 2.71 Å) suggests a similar conclusion.

Ab initio MO calculations on $(C_5H_5)_2Ge$ in a D_{5h} conformation, ie, with parallel ligand rings,[232] show that the lone electron pair on Ge occupies a 4s orbital destabilized by significant antibonding interaction with the ring a_1 π orbitals: if the perpendicular M—Cp distance is reduced, the energy of this orbital increases rapidly. Bonding between the metal and the rings seems to be due to interaction of the ring e_1 π orbitals with the appropriate 4p orbitals on Ge. When the molecule is bent (ie, when the ring–metal–ring angle is reduced

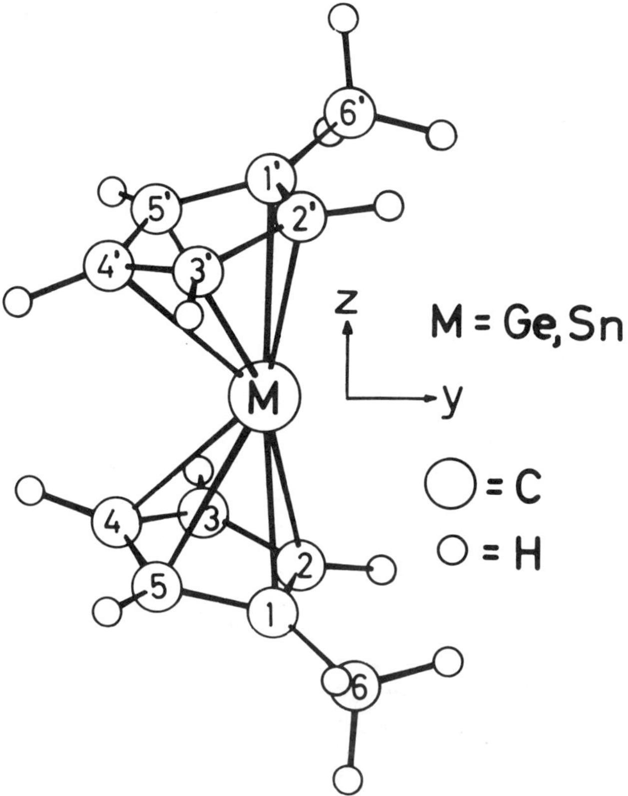

Figure 8-28. The molecular structures of $M(\eta^5\text{-}C_5H_4Me)_2$ for Ge and Sn.[232]

while the two CpGe fragments retain C_{5v} symmetry), the energy of the lone-pair orbital falls. At the same time, the $4s$ orbital becomes polarized in the direction of the opening between the ligand rings through admixture of a small amount of $4p_y$: like the angular structure of the dialkyl compounds R_2M, the angular structure of the metallocenes is an effect of the presence of a lone electron pair on M. In both R_2M and Cp_2M the lone pairs may be described as polarized $4s$ electrons.

Deformation of the metallocenes is clearly limited by repulsion between the ligands: on going from $(C_5H_5)_2Ge$ to $(C_5Me_5)_2Ge$, the angle between the ligand ring planes is reduced from 50° to about 22°; on going from $(C_5H_5)_2Sn$ to $(C_5Me_5)_2Sn$ to $(C_5Ph_5)_2Sn$, $\angle C_5,C_5$ is reduced from 46° to 36° to 0°.

Though the bonding is best described as pentahapto, the X-ray studies show that the M atom is displaced from the intersection of the ring symmetry axes by 0.20 or 0.30 Å in the direction away from the opening between the ligand rings. The observed geometries may be described as resulting from bending followed by tilting. Thus the geometry of Cp_2Ge may be described as due to bending by about 28° followed by 11° tilt of each ring, and the geometry of Cp_2Sn as due to bending by about 33° followed by about 7° tilt of each ring:

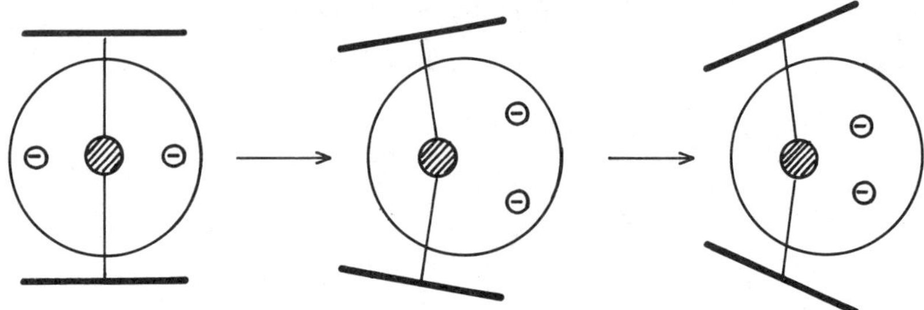

Let us now return to group 2. Following the recent report that Cp_2^*Sm[236] is bent ($\angle C_5,C_5 = 40°$) in the solid phase, we decided to investigate Cp_2^*Yb (Yb has the electron configuration of Ba plus a filled $4f$ shell), and Cp_2^*Ca.[237] The values obtained for the metal–ring-centroid distance d, the ring-centroid-M-ring-centroid angle α, and the ring tilt angle θ are listed in Table 8-8.

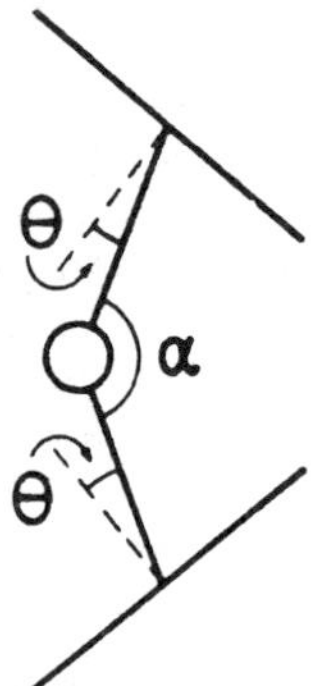

Like Cp_2^*Sm in the solid phase, the thermal average structures of Cp_2^*Yb and Cp_2^*Ca in the gas phase are *bent*, the ring-centroid-M-ring-centroid angles being 158(4) and 154(3)°, respectively. As a consequence of this bending, the shortest distance between Me groups in different rings is about 4.0 Å, equal to the accepted van der Waals diameter of an Me group. In a simultaneous study of

Table 8-8. Selected Structure Parameters of
$(C_5Me_5)_2M$

	M		
Parameter	Mg	Ca	Yb
d Å	2.019(9)	2.313(6)	2.326(5)
α, deg	180	154(3)	158(4)
θ, deg	0	$3(1_5)$	$1(2_5)$

Source: Reference 237.

Cp_2^*Mg, the ligand rings were found to be parallel or very nearly so. The perpendicular metal-to-ring distance in this compound is such that interring Me...Me contacts are about 4.0 Å when the ligands are parallel.

Even though the thermal average structures of Cp_2^*Yb and Cp_2^*Ca are bent, it does not follow that the *equilibrium* structures are so. *If*, however, the equilibrium conformation is one in which the rings are parallel, the energy required to bend the molecule 20° must be less than the thermal energy available, $4\,kJ\,mol^{-1}$.

In addition to van der Waals attraction between Me groups on different rings, a bent form may also be stabilized through polarization of the Ca^{2+} ion, just as the bent form of Cp_2^*Ge is stabilized through polarization of the Ge^{2+} ion. The polarization of Ca^{2+} is however not large enough to introduce detectable tilting of the rings.

Concluding Remarks

In addition to the organometallic compounds of main group elements covered by this chapter, about 50 organometallic compounds of transition elements have been studied by GED. The author regrets that lack of time prevents their inclusion here. A bibliography that covers the structures published before the end of 1980 can be found in reference 2.

Most of the organometallic compounds of transition elements are mononuclear carbonyl and cyclopentadienyl derivatives. The structures of ferrocene and the other $3d$ metallocenes were reviewed in 1979.[238] The structure of the first homoleptic transition metal alkyl, *high-spin* bis(neopentyl)manganese, was published in 1985.[239]

Gas-phase electron diffraction is the best method for providing accurate structure parameters for small molecules like $Ga(CH_3)_3$ or $Ge(CH_3)_4$. In the realm of organometallic chemistry its main importance has probably lain in the studies of species that are associated in the solid phase as dimethylberyllium, trimethylaluminum, dialkyls of Mg, Ge(II), and Sn(II), and as the cyclopentadienyl compounds of CpZnMe, CpIn, $CpAlMe_2$, or Cp_2Pb. GED may also be the method of choice for the study of compounds that have proven difficult to crystallize. In the realm of organic chemistry, the main importance of GED probably lies in its ability to give information on mixtures of conformers. This ability has yet to be exploited in organometallic chemistry.

References

1. See Chapter 1, "Boron and Silicon Compounds", by V. S. Mastryukov in the present volume.
2. Bruce. M. I. In "Comprehensive Organometallic Chemistry", Vol. 8, Wilkinson, G.; Stone, F.G.A.; Abel, E. W., Eds.; Pergamon Press: Elmsford, NY, 1982, p. 1209.
3. Haaland, A. Organometallic compounds studied by gas-phase electron diffraction, *Top. Curr. Chem.* **1975**, *53*, 1.

4. Hargittai, M.; Hargittai, I. "The Molecular Geometries of Coordination Compounds in the Vapour Phase". Elsevier: Amsterdam, 1977.

5. Vilkov, L. V.; Mastryukov, V. S.; Sadova, N. I. "Determination of the Geometrical Structure of Free Molecules". Mir: Moscow, 1983.

6. Hencher, J. L.; Mustoe, F. J. *Can. J. Chem.* **1975**, *53*, 3542.

7. Durig, J. R.; Chen, M. M.; Lee, Y. S.; Turner, J. B. *J. Phys. Chem.* **1973**, *77*, 227.

8. Thomas, E. C.; Laurie, V. W. *J. Chem. Phys.* **1969**, *50*, 3512.

9. Laurie, V. W. *J. Chem. Phys.* **1959**, *30*, 1210.

10. Kattenberg, H. W.; Gabes, W.; Oskam, A. *J. Mol. Spectrosc.* **1972**, *44*, 425.

11. Nagashima, M.; Fujii, H.; Kimura, M. *Bull. Chem. Soc. Japan* **1973**, *46*, 3708.

12. Beagley, B.; McAloon, K.; Freeman, J. M. *Acta Crystallogr.* **1974**, *B30*, 444.

13. Lide, D. R. *J. Chem. Phys.* **1951**, *19*, 1605.

14. Kattenberg, H. W.; Oskam, A. *J. Mol. Spectrosc.* **1974**, *51*, 377.

15. Oyamada, T.; Iijima, T.; Kimura, M. *Bull. Chem. Soc. Japan* **1971**, *44*, 2638.

16. Drake, J. E.; Hemmings, R. T.; Hencher, J. L.; Mustoe, F. M.; Shen, Q. *J. Chem. Soc. Dalton Trans.* **1976**, 394.

17. Durig, J. R.; Hellams, K. L. *J. Mol. Struct.* **1975**, *29*, 349.

18. Drake, J. E.; Hencher, J. L.; Shen, Q. *Can. J. Chem.* **1977**, *55*, 1104.

19. Vajda, E.; Hargittai, I. *Acta Chim. Acad. Hung.* **1976**, *91*, 185.

20. Morino, Y.; Nakamura, Y.; Iijima, T. *J. Chem. Phys.* **1960**, *32*, 643.

21. Li, Y. S.; Durig, J. R. *Inorg. Chem.* **1973**, *12*, 306.

22. Drake, J. E.; Hemmings, R. T.; Hencher, J. L.; Mustoe, F. M.; Shen, Q. *J. Chem. Soc. Dalton Trans.* **1976**, 811.

23. Souza, G.G.B.; Wieser, J. D. *J. Mol. Struct.* **1975**, *25*, 442.

24. Fujii, H.; Kimura, M. *Bull. Chem. Soc. Japan* **1971**, *44*, 2443.

25. Fujii, H.; Kimura, M. *Bull. Chem. Soc. Japan* **1970**, *43*, 1933.

26. Durig, J. R.; Kizer, K. L.; Lee, Y. S. *J. Am. Chem. Soc.* **1974**, *96*, 7400.

27. Thomas, E. C.; Laurie, V. W. *J. Chem. Phys.* **1966**, *44*, 2602.

28. Belyakov, A. V.; Khaikin, L. S.; Vilkov, L. V.; Apalkova, G. M.; Nikitin, V. S.; Bogoradovskii, E. T.; Zavgorodnii, V. S. *J. Mol. Struct.* **1986**, *129*, 17.

29. Khaikin, L. S.; Novikov, V. P.; Vilkov, L. V. *J. Mol. Struct.* **1977**, *42*, 129.

30. Khaikin, L. S.; Belyakov, A. V.; Vilkov, L. V.; Bogoradovskii, E. T.; Zavgorodnii, V. S. *J. Mol. Struct.* **1980**, *66*, 149.

31. Belyakov, A. V.; Khaikin, L. S.; Vilkov, L. V.; Bogoradovskii, E. T.; Zavgorodnii, V. S. *J. Mol. Struct.* **1980**, *61*, 149.

32. Novikov, V. P.; Khaikin, L. S.; Vilkov, L. V. *J. Mol. Struct.* **1977**, *42*, 139.

33. Khaikin, L. S.; Novikov, V. P.; Vilkov, L. V. *J. Mol. Struct.* **1978**, *44*, 43.

34. Belyakov, A. V.; Khaikin, L. S.; Vilkov, L. V.; Bogoradovskii, E. T.; Zavgorodnii, V. S. *J. Mol. Struct.* **1981**, *72*, 233.

35. Oberhammer, H.; Eujen, R. *J. Mol. Struct.* **1972**, *51*, 211.

36. Eujen R.; Bürger, H.; Oberhammer, H. *J. Mol. Struct.* **1981**, *71*, 109.

37. Pilcher, G.; Skinner, H. A. In "The Chemistry of the Metal–Carbon Bond", Hartley, F. R.; Patai, S.; Eds.; Wiley: New York, 1982, p. 43.

38. Yokozeki, A.; Bauer, S. H. *Top. Curr. Chem.* **1975**, *53*, 71.

39. Bent, H. A. *J. Chem. Educ.* **1960**, *37*, 616.

40. Gillespie, R. J. *J. Chem. Educ.* **1970**, *47*, 18.

41. Shen, Q.; Kapfer, C. A.; Boudjouk, P.; Hilderbrandt, R. L. *J. Mol. Struct.* **1979**, *54*, 295.

42. Dakkouri, M. *J. Mol. Struct.* **1985**, *130*, 289.

43. Volpin, M. E.; Koreshkov, Y. D.; Gulova, V. G.; Kursanov, D. N. *Tetrahedron* **1962**, *18*, 107.

44. Vilkov, L. V.; Mastryukov, V. S.; Shcherbik, L. K.; Dulova, V. G. *J. Struct. Chem.* **1970**, *11*, 3.

45. Johnsen, F.; Golka, R. S. *Tetrahedron Lett.* **1962**, 1291.

46. Volpin, M. E.; Dulova, V. G.; Struchkov, Y. T.; Bokiy, N. K.; Kursanov, D. N. *J. Organomet. Chem.* **1967**, *8*, 87.

47. Krebs, A.; Bernd, J. *Tetrahedron Lett.* **1983**, *24*, 4083.

48. Egorov, M. P.; Kolesnikov, S. P.; Struchkov, Y. T.; Antipin, M. Y.; Sereda, S. V.; Nefedov, O. M. *J. Organomet. Chem.* **1985**, 290, C27.

49. Glidewell, C.; Rankin, D.W.H.; Robiette, A. G.; Sheldrick, G. M.; Beagley, B.; Cradock, S. *J. Chem. Soc. A* **1970**, 315.

50. Vilkov, L. V. *Kem. Kozlem.* **1971**, *35*, 375.

51. Murdoch, J. D.; Rankin, D.W.H.; Glidewell, C. *J. Mol. Struct.* **1971**, *9*, 17; erratum *10*, 496.

52. Glidewell, C.; Rankin, D.W.H.; Robiette, A. G. *J. Chem. Soc. A* **1970**, 2935.

53. Rankin, D.W.H.; Robiette, A. G.; Sheldrick, G. M.; Beagley, B.; Hewitt, T. G. *Inorg. Nucl. Chem.* **1969**, *31*, 2351.
54. Rozsondai, B.; Hargittai, I. *J. Mol. Struct.* **1973**, *17*, 53.
55. Khaikin, L. S.; Belyakov, A. V.; Kaptev, G. S.; Golubinskii, A. V.; Vilkov, L. V.; Girbasova, N. V.; Bogoradovskii, E. T.; Zavgorodnii, V. S. *J. Mol. Struct.* **1980**, *66*, 191.
56. Pauling, L. "The Nature of the Chemical Bond", 3rd. ed. Cornell University Press: Ithaca, NY, 1960.
57. Blom, R.; Haaland, A. *J. Mol. Struct.* **1985**, *128*, 21.
58. Bartell, L. S. *J. Chem. Educ.* **1968**, *45*, 754.
59. Glidewell, C. *Inorg. Chim. Acta* **1975**, *12*, 219.
60. Glidewell, C. *Inorg. Chim. Acta* **1979**, *36*, 135.
61. (a) Davidson, P. J.; Lappert, M. F. *J. Chem. Soc. Chem. Commun.* **1973**, 317. (b) Goldberg, D. E.; Harris, D. H.; Lappert, M. F.; Thomas, K. M. *J. Chem. Soc. Chem. Commun.* **1976**, 261. (c) Davidson, J.; Harris, D. H.; Lappert, M. F. *J. Chem. Soc. Dalton Trans.* **1976**, 2268.
62. (a) Hitchcock, P. B.; Lappert, M. F.; Miles, S. J.; Thorne, A. J. *J. Chem. Soc. Chem. Commun.* **1984**, 480. (b) Goldberg, D. E.; Hitchcock, P. B.; Lappert, M. F.; Thomas, K. M.; Thorne, A. J.; Fjeldberg, T.; Haaland, A.; Schilling, B.E.R. *J. Chem. Soc. Dalton Trans.* **1986**, 2387.
63. (a) Fjeldberg, T.; Haaland, A.; Lappert, M. F.; Schilling, B.E.R.; Seip, R.; Thorne, A. J. *J. Chem. Soc. Chem. Commun.* **1982**, 1407. (b) Fjeldberg, T.; Haaland, A.; Lappert, M. F.; Schilling, B.E.R.; Thomas, A. J.; Volden, H. V. *J. Organomet. Chem.* **1985**, *280*, C43. (c) Fjeldberg, T.; Haaland, A.; Schilling, B.E.R.; Lappert, M. F.; Thorne, A. J. *J. Chem. Soc. Dalton Trans.* **1986**, 1551.
64. Bartelat, J.-C.; Roch, B. S.; Trinquier, G.; Satge, J. *J. Am. Chem. Soc.* **1980**, *102*, 4080.
65. Bleckmann, P.; Maly, H.; Minkwitz, R.; Neumann, W. P.; Watta, B.; Olbrich, G. *Tetrahedron Lett.* **1982**, *23*, 4655.
66. Huffman, J. C.; Streib, W. E. *J. Chem. Soc. Chem. Commun.* **1971**, 911.
67. Amma, E. M.; Rundle, R. E. *J. Am. Chem. Soc.* **1958**, *80*, 4141.
68. Sheldrick, G. M.; Sheldrick, W. S. *J. Chem. Soc. A* **1970**, 28.
69. Henrickson, C. H.; Eyman, D. P. *Inorg. Chem.* **1967**, *6*, 1461.
70. Almenningen, A.; Halvorsen, S.; Haaland, A. *Acta Chem. Scand.* **1971**, *25*, 1937.
71. Bartell, L. S.; Carroll, B. L. *J. Chem. Phys.* **1965**, *42*, 3076.
72. Beagley, B.; Schmidling, D. G.; Steer, I. A. *J. Mol. Struct.* **1974**, *21*, 437.
73. Fjeldberg, T.; Haaland, A.; Seip, R.; Shen, Q.; Weidlein, J. *Acta Chem. Scand.* **1982**, *A36*, 495.
74. Herzberg, G.; Johns, J.W.C. *Proc. R. Soc. (London) Ser A* **1967**, *298*, 142.
75. Herzberg, G.; Johns, J.W.C. In Molecular Spectra and Molecular Structure, Vol. III, "Polyatomic Molecules", Herzberg, G., Ed.; Van Nostrand: New York, 1966, pp. 490, 583.
76. Yamamoto, O.; Yanagisawa, M. *Bull. Chem. Soc. Japan* **1978**, *51*, 1231.
77. Smith, M. B. *J. Organomet. Chem.* **1972**, *46*, 31.
78. Smith, M. B. *J. Chem. Phys.* **1967**, *71*, 364.
79. Smith, M. B. *J. Organomet. Chem.* **1970**, *22*, 273.
80. Wartic, T.; Schlessinger, H. I. *J. Am. Chem. Soc.* **1953**, *75*, 835.
81. Almenningen, A.; Anderson, G. A.; Forgaard, F. R.; Haaland, A. *Acta Chem. Scand.* **1972**, *26*, 2315.
82. Almenningen, A.; Gundersen, G.; Haugen, T.; Haaland, A. *Acta Chem. Scand.* **1972**, *26*, 3928.
83. Malone, J. F.; McDonald, W. S. *J. Chem. Soc.* **1972**, 2649.
84. Albright, M. J.; Butler, W.; Anderson, T. J.; Glick, M. D.; Oliver, J. P. *J. Am. Chem. Soc.* **1976**, *98*, 3995.
85. Oliver, J. P.; Stevens, L. G. *J. Inorg. Nucl. Chem.* **1962**, *24*, 953.
86. Fries, W.; Sille, K.; Weidlein, J.; Haaland, A. *Spectrochim. Acta* **1980**, *A36*, 611.
87. Fjeldberg, T.; Haaland, A.; Seip, R.; Weidlein, J. *Acta Chem. Scand.* **1981**, *A35*, 637.
88. Mole, T.; Surtees, J. R. *Aust. J. Chem.* **1964**, *17*, 1229.
89. Jeffery, E. A.; Mole, T.; Saunders, J. K. *Aust. J. Chem.* **1968**, *21*, 137.
90. Ham, N. S.; Jeffery, E. A.; Mole, T. *Aust. J. Chem.* **1968**, *21*, 2687.
91. Stucky, G. D. McPherson, A. M.; Rhine, W. E.; Eisch, J. J.; Considine, J. L. *J. Am. Chem. Soc.* **1974**, *96*, 1941.
92. Almenningen, A.; Fernholt, L.; Haaland, A. *J. Organomet. Chem.* **1978**, *155*, 245.
93. Fjeldberg, T.; Haaland, A.; Seip, R.; Weidlein, J. *Acta Chem. Scand.* **1981**, *A35*, 437.
94. Tecle, B.; Ilsley, W. H.; Oliver, J. P. *Inorg. Chem.* **1981**, *20*, 2335.
95. Fries, W.; Schwarz, W.; Hausen, H.-D.; Weidlein, J. *J. Organomet. Chem.* **1978**, *159*, 373.
96. Barlow, M. T.; Downs, A. J.; Thomas, P.D.P.; Rankin, D.W.H. *J. Chem. Soc. Dalton Trans.* **1979**, 1793.

97. Barlow, M. T.; Dain, C. J.; Downs, A. J.; Thomas, P.D.P.; Rankin, D.W.H. *J. Chem. Soc. Dalton Trans.* **1980**, 1374.

98. Almenningen, A.; Gundersen, G.; Haaland, A. *Acta Chem. Scand.* **1968**, *22*, 328.

99. Dain, C. J.; Downs, A. J.; Rankin, D.W.H. *J. Chem. Soc. Dalton Trans.* **1981**, 2465.

100. Henrickson, C. H.; Eyman, D. P. *Inorg. Chem.* **1967**, *6*, 1461.

101. Henrickson, C. H.; Duffy, D.; Eyman, D. P. *Inorg. Chem.* **1968**, *7*, 1047.

102. Anderson, G. A.; Forgaard, F. R.; Haaland, A. *Acta Chem. Scand.* **1972**, *26*, 1947.

103. Almenningen, A.; Haaland, A.; Haugen, T.; Novak, D. P. *Acta Chem. Scand.* **1973**, *27*, 1821.

104. Almenningen, A.; Fernholt, L.; Haaland, A.; Weidlein, J. *J. Organomet. Chem.* **1978**, *145*, 109.

105. Haaland, A.; Samdal, S.; Stokkeland, O.; Weidlein, J. *J. Organomet. Chem.* **1977**, *134*, 165.

106. Fernholt, L.; Haaland, A.; Hargittai, M.; Seip, R.; Weidlein, J. *Acta Chem. Scand.* **1981**, *A35*, 529.

107. Golubinskaya, L. M.; Golubinskii, A. V.; Mastryukov, V. S.; Vilkov, L. V.; Bregadze, V. I. *J. Organomet. Chem.* **1976**, *117*, C4. Golubinskaya, L. M.; Bregadze, V. I.; Golubinskii, A. V.; Mastryukov, V. S.; Novikov, V. P.; Vilkov, L. V. *Abstracts*, Twenty-Second International Conference on Coordination Chemistry, Budapest, 1982, p. 788.

108. Baxter, P. L.; Downs, A. J.; Rankin, D.W.H. *J. Chem. Soc. Dalton Trans.* **1984**, 1755.

109. Durig, J. R.; Chatterjee, K. K.; Li, Y. S.; Jalilian, M.; Zozulin, A. J.; Odom, J. D. *J. Chem. Phys.* 1980, *73*, 21.

110. Carter, J. C.; Jugie, G.; Enjalbert, R.; Galy, J. *Inorg. Chem.* **1978**, *17*, 1248.

111. Novikov, V. P.; Golubinskii, A. V.; Mastryukov, V. S.; Vilkov, L. V.; Golubinskaya, L. M.; Bregadze, V. I. *J. Struct. Chem. (Russ.)* **1985**, *26*, 34.

112. Grant, D. F.; Killean, R.C.G.; Lawrence, J. L. *Acta Crystallogr.* **1969**, *B25*, 377.

113. Beagley, B.; Hewitt, T. G. *Trans. Faraday Soc.* **1968**, *64*, 2561.

114. Bartell, L. S.; Brockway, L. O. *J. Chem. Phys.* **1960**, *32*, 512.

115. Medley, J. H.; Franczek, F. R.; Ahmad, N.; Day, M. C.; Rogers, R. D.; Kerr, C. R.; Atwood, J. L. *J. Cryst. Spectrosc. Res.* **1985**, *15*, 99.

116. McLean, W. C.; Jeffrey, G. A. *J. Chem. Phys.* **1967**, *47*, 414.

117. Tamagawa, K.; Takemura, M.; Konaka, S.; Kimura, M. *J. Mol. Struct.* **1984**, *125*, 131.

118. Blake, A. J.; Ebsworth, E.A.V.; Rankin, D.W.H.; Robertson, H. E.; Smith, D. E.; Welch, A. J. *J. Chem. Soc. Dalton Trans.* **1986**, 91.

119. Gropen, O.; Johansen, R.; Haaland, A.; Stokkeland, O. *J. Organomet. Chem.* **1975**, *92*, 147.

120. Heitsch, C. W.; Nordman, C. E.; Parry, R. W. *Inorg. Chem.* **1963**, *2*, 508.

121. Mastryukov, V. S.; Golubinskii, A. V.; Vilkov, L. V. *J. Chem. Soc.* **1979**, *20*, 788.

122. Gundersen, G.; Haugen, T., Haaland, A. *J. Organomet. Chem.* **1973**, *54*, 77.

123. Oberhammer, H.; Zeil, W.; Fogarasi, G. *J. Mol. Struct.* **1973**, *18*, 309.

124. Drew, D. A.; Haaland, A.; Weidlein, J. *Z. Anorg. Allg. Chem.* **1973**, *398*, 241.

125. Jeffery, E. A.; Mole, T. *Aust. J. Chem.* **1968**, *21*, 2683.

126. Schmidbaur, H.; Schindler, G. *Chem. Ber.* **1966**, *99*, 2178.

127. Haaland, A.; Stokkeland, O. *J. Organomet. Chem.* **1975**, *94*, 345.

128. See Vilkov, L.; Golubinskii, A. V.; Mastryukov, V. S.; Sadova, N. I.; Khaikin, L. S.; Hargittai, I. In "Sovremenniye Problemi Fizicheskoi Khimii", Vol. 11, Gerasimov, Y. I.; Akishin, P. A., Eds.; Izd. Moskov Univ.: Moscow, 1979, pp. 59–112.

129. Ouzounis, K.; Riffel, H.; Hess, H.; Kohler, U.; Weidlein, J. *Z. Anorg. Allg. Chem.* **1983**, *504*, 67.

130. Hess, H.; Hinderer, A.; Steinhauser, S. *Z. Anorg. Allg. Chem.* **1970**, *377*, 1.

131. Turova, N. Y.; Kozunov, V. A. *J. Inorg. Nucl. Chem.* **1979**, *41*, 5.

132. Jeffrey, G. A.; Parry, G. S.; Mozzi, R. L. *J. Chem. Phys.* **1956**, *25*, 1024.

133. McDonald, T.R.R.; McDonald, W. S. *Acta Crystallogr.* **1972**, *B28*, 1619.

134. Piero, G. D.; Cesari, M.; Dozzi, G.; Mazzei, A. *J. Organomet. Chem.* **1977**, *129*, 281.

135. Sheldrick, G. M.; Sheldrick, W. S. *J. Chem. Soc. A* **1969**, 2279.

136. Cesari, M.; Perego, G.; del Piero, G.; Cucinella, S.; Cernia, E. *J. Organomet. Chem.* **1974**, *78*, 203.

137. Baxter, P. L.; Downs, A. J.; Rankin, D.W.H.; Robertson, H. E. *J. Chem. Soc. Dalton Trans.* **1985**, 807.

138. Brendhaugen, K.; Haaland, A.; Novak, D. P. *Acta Chem. Scand.* **1974**, *A28*, 45.

139. Haaland, A.; Stokkeland, O.; Weidlein, J. *J. Organomet. Chem.* **1975**, *94*, 353.

140. Brauer, D. J.; Stucky, G. D. *J. Am. Chem. Soc.* **1969**, *91*, 5462.

141. Snow, A. I.; Rundle, R. E. *Acta Crystallogr.* **1951**, *4*, 348.

142. Almenningen, A.; Haaland, A.; Morgan, G. L. *Acta Chem. Scand.* **1969**, *23*, 2921.

143. Almenningen, A.; Haaland, A.; Nilsson, J. E. *Acta Chem. Scand.* **1968**, *22*, 972.

144. Weiss, E. *J. Organomet. Chem.* **1964**, *2*, 314.

145. Ashby, E. C.; Fernholt, L.; Haaland, A.; Seip, R.; Smith, R. S. *Acta Chem. Scand.* **1980**, *A34*, 213.

146. Andersen, R. A.; Wilkinson, G. *J. Chem. Soc. Dalton Trans.* **1977**, 809.

147. Rao, R. S.; Stoicheff, B. P.; Turner, R. *Can. J. Phys.* **1960**, *38*, 1516.

148. Almenningen, A.; Helgaker, T. U.; Haaland, A.; Samdal, S. *Acta Chem. Scand.* **1982**, *A36*, 159.

149. Kashiwabara, K.; Konaka, S.; Iijima, T.; Kimura, M. *Bull. Chem. Soc. Japan* **1973**, *46*, 407.

150. Wells, C.; Lister, D. G.; Sheridan, J. *J. Chem. Soc. Faraday Trans. 2* **1975**, *71*, 1091.

151. Kashivabara, K.; Konaka, S.; Kimura, M. *Bull. Chem. Soc. Japan* **1973**, *46*, 410.

152. Gershikov, A. G.; Spiridonov, V. P. *J. Mol. Struct.* **1981**, *75*, 291.

153. Oberhammer, H. *J. Mol. Struct.* **1978**, *48*, 389.

154. Günther, H.; Oberhammer, H.; Eujen, R. *J. Mol. Struct.* **1980**, *64*, 249.

155. Vilkov, L. V.; Anashkin, M. G.; Mamaeva, G. I. *J. Struct. Chem.* **1968**, *9*, 372.

156. Vilkov, L. V.; Anashkin, M. G. *J. Struct. Chem.* **1968**, *9*, 690.

157. Pyyköö, P.; Desclaux, J.-P. *Acc. Chem. Res.* **1979**, *12*, 276.

158. Pitzer, K. S. *Acc. Chem. Res.* **1979**, *12*, 271.

159. Atwood, J. L.; Whitt, C. D. *J. Organomet. Chem.* **1971**, *32*, 17.

160. Toney, J.; Stucky, G. D. *J. Organomet. Chem.* **1970**, *22*, 241.

161. Thiele, K.-H. *Z. Anorg. Allg. Chem.* **1963**, *325*, 156.

162. Boersma, J.; Haaland, A.; Hougen, J. Unpublished result.

163. Thiele, K.-H. *Z. Anorg. Allg. Chem.* **1962**, *319*, 183.

164. Boersma, J.; Fernholt, L.; Haaland, A. *Acta Chem. Scand.* **1984**, *A38*, 523.

165. Dekker, J.; Boersma, J.; Fernholt, L.; Haaland, A.; Spek, A. L. *Organometallics* **187**, *6*, 1202.

166. Haaland, A.; Hedberg, K.; Power, P. P. *Inorg. Chem.* **1984**, *23*, 1972.

167. Davis, M. I.; Speed, C. S. *J. Organomet. Chem.* **1970**, *21*, 401.

168. St. Denis, J.; Oliver, J. P.; Dolzine, T. W.; Smart, J. B. *J. Organomet. Chem.* **1974**, *71*, 315.

169. Haaland, A.; Lehmkuhl, H.; Nehl, H. *Acta Chem. Scand.* **1984**, *A38*, 547.

170. Gropen, O.; Haaland, A.; deFrees, D. *Acta Chem. Scand.* **1985**, *A39*, 367.

171. Fjeldberg, T.; Lappert, M. F.; Thorne, A. J. *J. Mol. Struct.* **1985**, *127*, 95.

172. Schleyer, P.v.R.; *Pure Appl. Chem.* **1983**, *55*, 355; **1984**, *56*, 151.

173. Atwood, J. L.; Fjeldberg, T.; Lappert, M. F.; Luong-Thi, N. T.; Shakir, R.; Thorne, A. J. *J. Chem. Soc. Chem. Commun.* **1984**, 1163.

174. Haaland, A.; Hougen, J.; Puddephatt, R. J.; Volden, H. V. Unpublished result.

175. Huber, K. P.; Herzberg, G. Molecular Spectra and Molecular Structure, Vol. IV, "Constants of Diatomic Molecules", Van Nostrand: New York, 1979.

176. Haaland, A. *J. Mol. Struct.* **1983**, *97*, 115.

177. Barrow, M. J.; Ebsworth, E.A.V.; Harding, M. M.; and Rankin, D.W.H. *J. Chem. Soc. Dalton Trans.* **1980**, 603.

178. Damiani, D.; Ferretti, L.; Gallinella, E. *Chem. Phys. Lett.* **1976**, *37*, 265.

179. Angus, P. C.; Stobart, S. R. *J. Chem. Soc. Dalton Trans.* **1973**, 2374.

180. Sergeyev, N. M. Avramenko, G. I.; Kisin, A. V.; Korenevsky, V. A.; Ustynyuk, Y. A. *J. Organomet. Chem.* **1971**, *32*, 55.

181. Kisin, A. V.; Korenevsky, V. A.; Sergeyev, N. M.; Ustynyuk, Y. A. *J. Organomet. Chem.* **1972**, *34*, 93.

182. Drew, D. A.; Haaland, A. *Acta Chem. Scand.* **1972**, *26*, 3079.

183. Bartke, T. C.; Björseth, A.; Haaland, A.; Marstokk, K.-M.; Möllendal, H. *J. Organomet. Chem.* **1975**, *85*, 271.

184. Jemmis, E. D.; Alexandratos, S.; Schleyer, P.v.R.; Streitwieser, A., Jr.; Schaefer, H. F. III *J. Am. Chem. Soc.* **1978**, *100*, 5695.

185. Haaland, A.; Nielsen, E. W.; Schilling, B.E.R. Unpublished results.

186. Drew, D. A.; Haaland, A. *Acta Chem. Scand.* **1972**, *26*, 3351.

187. Björseth, A.; Drew, D. A.; Marstokk, K. M.; Möllendal, H. *J. Mol. Struct.* **1972**, *13*, 233.

188. Akishin, P. A.; Spiridonov, V. P. *Kristallografiya* **1957**, *2*, 475.

189. Haaland, A.; Novak, D. P. *Acta Chem. Scand.* **1974**, *A28*, 153.

190. Drew, D. A.; Gundersen, G.; Haaland, A. *Acta Chem. Scand.* **1972**, *26*, 26.

191. Haaland, A.; Samdal, S.; Seip, R. *J. Organomet. Chem.* **1978**, *153*, 187.

192. Parris, G. E.; Ashby, E. C. *J. Organomet. Chem.* **1974**, *72*, 1.

193. Andersen, R. A.; Blom, R.; Haaland, A.; Schilling, B.E.R.; Volden, H. V. *Acta Chem. Scand.* **1985**, *A39*, 563.

194. Lorberth, J.; Weller, F. *J. Organomet. Chem.* **1971**, *32*, 145.

195. Davies, A. G.; Goddard, J. P.; Hursthouse, M. B.; Walker, N.P.C. *J. Chem. Soc. Dalton Trans.* **1985**, 471.

196. Shibata, S.; Bartell, L. S.; Gavin, R. M. *J. Chem. Phys.* **1964**, *41*, 717.
197. Tyler, J. K.; Cox, A. P.; Sheridan, J. *Nature* **1959**, *183*, 1182.
198. Canadell, E.; Eisenstein, O.; Rubio, J. *Organometallics* **1984**, *3*, 759.
199. Drew, D. A.; Haaland, A. *Acta Chem. Scand.* **1973**, *27*, 3735.
200. Gropen, O.; Haaland, A. *J. Organomet. Chem.* **1975**, *92*, 157.
201. Kroll, W. R.; Naegele, *J. Chem. Soc. Chem. Commun.* **1969**, 246.
202. Haaland, A.; Weidlein, J. *J. Organomet. Chem.* **1972**, *40*, 29.
203. Fernholt, L.; Haaland, A.; Jutzi, P.; Kohl, F. X.; Seip, R. *Acta Chem. Scand.* **1984**, *A38*, 211.
204. Schultz, G.; Tremmel, J.; Hargittai, I.; Berecz, I.; Bohatka, S.; Kagramanov, N. D. Maltsev, A. K.; Nefedov, O. M. *J. Mol. Struct.* **1979**, *55*, 207.
205. Haaland, A.; Schilling, B.E.R. *Acta Chem. Scand.* **1984**, *A38*, 217.
206. Veniaminov, N. N.; Ustynyuk, Y. A. Struchkov, Y. T.; Alekseev, N. V.; Ronova, I. A. *J. Struct. Chem.* **1970**, *11*, 111.
207. Veniaminov, N. N.; Ustynyuk, Y. A.; Alekseev, N. V.; Ronova, I. A. *Dokl. Chem.* **1971**, *199*, 577.
208. Aoya, T.; Shearer, H.M.M.; Wade, K.; Whitehead, G. *J. Organomet. Chem.* **1978**, *146*, C29.
209. Tecle, B.; Corfield, P.W.R.; Oliver, J. P. *Inorg. Chem.* **1982**, *21*, 458.
210. Frasson, E.; Menegus, F.; Panattoni, C. *Nature* **1963**, *199*, 1087.
211. Haaland, A.; Lusztyk, J.; Brunvoll, J.; Starowieyski, K. B. *J. Organomet. Chem.* **1975**, *85*, 279.
212. Bünder, W.; Weiss, E. *J. Organomet. Chem.* **1975**, *92*, 1.
213. Faegri, K., Jr.; Almlöf, J.; Lüthi, H. F. *J. Organomet. Chem.* **1983**, *249*, 303.
214. Almenningen, A.; Bastiansen, O.; Haaland, A. *J. Chem. Phys.* **1964**, *40*, 3434.
215. Wong, C.-H.; Lee, Y. T.; Chang, T. W.; Liu, C. S. *Inorg. Nucl. Chem. Lett.* **1973**, *9*, 667.
216. Wong, C.-H.; Lee, Y. T.; Chao, K. T.; Lee, S. *Acta Crystallogr.* **1972**, *B28*, 1662.
217. Drew, D. A.; Haaland, A. *Acta Crystallogr.* **1972**, *B28*, 3671.
218. Marynick, D. S. *J. Am. Chem. Soc.* **1977**, *99*, 1436.
219. Dewar, M.J.S.; Rzepa, H. S. *J. Am. Chem. Soc.* **1978**, *100*, 777.
220. Chiu, N. S.; Schäfer, L. *J. Am. Chem. Soc.* **1978**, *100*, 2604.
221. Jemmis, E. D.; Alexandratos, S.; Schleyer, P.v.R.; Streitwieser, A.; Schaefer, H. F. III *J. Am. Chem. Soc.* **1978**, *100*, 5695.
222. Almenningen, A.; Haaland, A.; Lusztyk, J. *J. Organomet. Chem.* **1979**, *170*, 271.
223. Nugent, K. W.; Beattie, J. K. Hambley, T. W.; Snow, M. R. *Aust. J. Chem.* **1984**, *37*, 1601.
224. Wong, C.-H.; Wang, S.-M. *Inorg. Nucl. Chem. Lett.* **1975**, *11*, 677.
225. Budzelaar, P.H.M.; Boersma, J.; Kerk, G.J.M.v.d.; Spek, A. L.; Duisenberg, A.J.M. *J. Organomet. Chem.* **1985**, *281*, 123.
226. Blom, R.; Boersma, J.; Budzelaar, P.H.M.; Fischer, B.; Haaland, A.; Volden, H. V.; Weidlein, J. *Acta Chem. Scand.* **1986**, *A40*, 113.
227. Almenningen, A.; Haaland, A.; Motzfeldt, T. *J. Organomet. Chem.* **1967**, *7*, 97.
228. Atwood, J. L.; Hunter, W. E.; Cowley, A. H.; Jones, R. A.; Stewart, C. A. *J. Chem. Soc. Chem. Commun.* **1981**, 925.
229. Panattoni, C.; Bambieri, G.; Croatto, U. *Acta Crystallogr.* **1966**, *21*, 823.
230. Fernholt, L.; Haaland, A.; Jutzi, P.; Seip, R. *Acta Chem. Scand.* **1980**, *A34*, 585.
231. Fernholt, L.; Haaland, A.; Jutzi, P.; Seip, R.; Almlöf, J.; Faegri, K., Jr.; Kvåle, E.; Lüthi, H. P.; Schilling, B.E.R.; Taugbøl, K. *Acta Chem. Scand.* **1982**, *A36*, 93.
232. Almlöf, J.; Fernholt, L.; Faegri, K., Jr.; Haaland, A.; Schilling, B.E.R.; Seip, R.; Taugbøl, K. *Acta Chem. Scand.* **1983**, *A37*, 131.
233. Grenz, M.; Hahn, E.; du Mont, W.-W.; Pickardt, J. *Angew. Chem.* **1984**, *96*, 69.
234. Jutzi, P., Kohl, F.; Hofmann, P.; Krüger, C.; Tsay, Y.-H. *Chem. Ber.* **1980**, *113*, 757.
235. Heeg, M. J.; Janiak, C.; Zuckerman, J. J. *J. Am. Chem. Soc.* **1984**, *106*, 4259.
236. Evans, W. J.; Hughes, L. A.; Hanusa, T. P. *J. Am. Chem. Soc.* **1984**, *106*, 4270.
237. Andersen, R. A.; Boncella, J. M.; Burns, C. J.; Blom, R.; Haaland, A.; Volden, H. V. *J. Organomet. Chem.* **1986**, *312*, C49.
238. Haaland, A. *Acc. Chem. Res.* **1979**, *11*, 415.
239. Andersen, R. A.; Haaland, A.; Rypdal, K.; Volden, H. V. *J. Chem. Soc. Chem. Commun.* **1985**, 1807.

METAL HALIDES

Magdolna Hargittai

STRUCTURAL CHEMISTRY RESEARCH GROUP
HUNGARIAN ACADEMY OF SCIENCES
EOTVÖS UNIVERSITY
BUDAPEST, HUNGARY

CONTENTS

INTRODUCTION 384
    Peculiarities of Structural Studies on Metal Halides 384
    Determination of the Shape of Simple Halides 389
HALIDES OF MAIN GROUP METALS 393
    Group I Halides 393
    Group IIA Dihalides 398
    Group IIIA Halides 401
    Group IVA Halides 408
    Group VA Halides 410
TRANSITION METAL HALIDES 412
    Monohalides 412
    Dihalides 413
    Trihalides 422
    Tetrahalides 425
    Pentahalides 431
    Hexahalides and Heptahalides 438
COMPLEX METAL HALIDES 440
DONOR–ACCEPTOR COMPLEXES 443
    Acceptor Geometry 447
    Donor Geometry 448
REFERENCES 448

Introduction

The determination of metal halide molecular structures by electron diffraction differs from the study of other substances in several ways. Therefore, before beginning our systematic discussion, some peculiarities of such work will be mentioned.

Our discussion aims to be comprehensive (through 1985) and critical as regards metal halide structures. However, not all papers will be mentioned in cases of repeated studies. Metal halide molecules were selected for detailed presentation, involving the elements listed in Figure 9-1, while other important classes of compounds, like oxyhalides, were omitted. The present work was compiled following two goals: first, to provide a source for metal halide structural information; and second, to demonstrate the possibilities and limitations of structure analysis of an important class of compounds also as a model of current structural work.

Peculiarities of Structural Studies on Metal Halides

Experimental Difficulties. Most metal halides have low volatility, and their gas-phase investigation requires high-temperature experimental conditions (electron diffraction[1,2] and spectroscopy[3]).

The highly corrosive nature of metal halides adds to the experimental difficulties. The sample may react with the container, producing further vapor-phase halides. The good complexing ability of many metal chlorides is a disadvantage in this respect. Aluminum trichloride and iron trichloride are good complexing agents with most other metal halides and thus enhance the volatility of those halides by many orders of magnitude. For this reason stainless steel, otherwise a common and accessible nozzle material, is often inadequate for high-temperature experiments.

Small traces of other halides in a given sample may also evaporate either by themselves or in the form of mixed complex halides, and may be present in higher concentration in the vapor than in the original solid sample. Such traces may falsify the experimental results, if they remain undetected.

These problems already indicate the importance of the control of vapor composition. For this reason the electron diffraction apparatus has been coupled with a quadrupole mass spectrometer in some laboratories, including our Budapest laboratory.[4] This permits the purity of the sample and the suitability of the container to be checked and the experimental conditions to be optimized before the diffraction experiments are run.

The complicated and easily changing vapor composition (cf, eg, references 5–7) also poses experimental difficulties. The vapors of metal halides are often rich in a variety of different molecular species. Besides, the vapor composition is sensitive to temperature. Dimers and higher associates are often the predominant species of the vapor at lower temperatures, and their relative abundances

IA	IIA	IIIB	IVB	VB	VIB	VIIB	VIIIB			IB	IIB	IIIA	IVA	VA
Li														
Na	Mg											Al	Si	
K	Ca	Sc	Ti	V	Cr	Mn	Fe	Co	Ni	Cu	Zn	Ga	Ge	As
Rb	Sr		Zr	Nb	Mo						Cd		Sn	Sb
Cs	Ba	La	Hf	Ta	W	Re	Os	Ir		Au	Hg	Tl	Pb	

Ce	Pr	Nd			Gd	Tb		Ho				Lu
		U	Np	Pu								

Figure 9-1. Metals and semimetals whose halides are discussed in this chapter.

decrease with increasing temperature. Aluminum trichloride, iron trichloride, and cuprous chloride are examples. Other substances, such as the first-row transition metal dihalides, begin to evaporate predominantly as monomers and the relative dimer content may increase with increasing temperature. This phenomenon, described by Brewer's rule,[8] is rigorously valid for closed systems. Although the electron diffraction evaporation procedure does not take place in quite such a system, the possibility of the presence of dimers in the vapor cannot be excluded.

The presence of different species in the vapor was not routinely considered in the early days of electron diffraction, as noted in the section on Group I halides. This problem is often ignored even today if independent experiments (mass spectrometry, vapor pressure measurements) indicate that monomers are the predominant species of the vapor. As will be shown in connection with alkaline earth and transition metal dihalides, however, even a small amount of dimeric species can influence the determination of the monomer geometrical parameters. This fact should be taken into consideration, at least, in error estimation. Mass spectra from a coupled electron diffraction/mass spectrometric experiment[4] greatly facilitate the choice of optimal nozzle temperature and provide information on the vapor composition, which is further refined in the electron diffraction analysis.

Interpretational Difficulties. The possibility of complicated vapor composition is the first thing to consider, even if there is no direct control of vapor composition. If a relatively large amount of an additional molecular species is present, it has a large relative contribution to the electron scattering pattern and will probably be detected during the analysis. Small amounts of additional species (eg, 10% of dimeric molecules with MX_2 dihalides) are more difficult to detect. Table 9-1 presents some examples to show the severity of this problem in the determination of the monomer bond length for alkaline earth and transition

Table 9-1. CONSEQUENCES OF THE DIMER (D) PRESENCE ON THE MONOMER BOND LENGTHS AS DETERMINED FROM ELECTRON DIFFRACTION

Molecule	Refined for monomer + dimer			Refined for monomer only		$\dfrac{\Delta r}{r}\,(\%)^c$
	$D\,(\%)^a$	$r_g(M\!-\!X)$ (Å)	$R\,(\%)^b$	$r_g(M\!-\!X)$ (Å)	$R\,(\%)^b$	
$CaCl_2$	1.9(8)	2.483(6)	5.9	2.487(5)	6.2	0.2
$CaBr_2$	5.3(15)	2.616(7)	7.6	2.630(6)	8.3	0.5
CaI_2	2.4(7)	2.840(7)	9.9	2.847(7)	10.7	0.25
$FeCl_2$	5.7(8)	2.151(5)	5.4	2.161(4)	6.9	0.5
$MnBr_2$	5.3(8)	2.344(6)	5.6	2.354(5)	7.6	0.4
VCl_2	9.2(2)	2.172(6)	3.4	2.190(4)	5.8	0.8

[a] Dimer contents.
[b] Agreement factor between experimental and theoretical intensities (cf Chapter 1 of Part A).
[c] Relative error in the determination of the monomer bond length caused by ignoring the presence of dimers.

metal dihalides. The difference in the monomer bond length caused by ignoring the presence of dimers may easily exceed the usual experimental errors. Experience shows that for each percent of ignored dimer presence, the bond length of the monomer refines to a value 0.1 % higher than the bond length determined when the vapor composition is properly considered. The shift is always in the same direction; ie, ignoring the dimers *increases* the apparent monomer bond distance. The origin of this change is seen in Figure 9-2, which shows the principal peak of the radial distribution curve of VCl_2. The dimers have terminal bonds of about the same length as the monomer bonds, but they also have considerably longer bridging bonds whose contribution appears

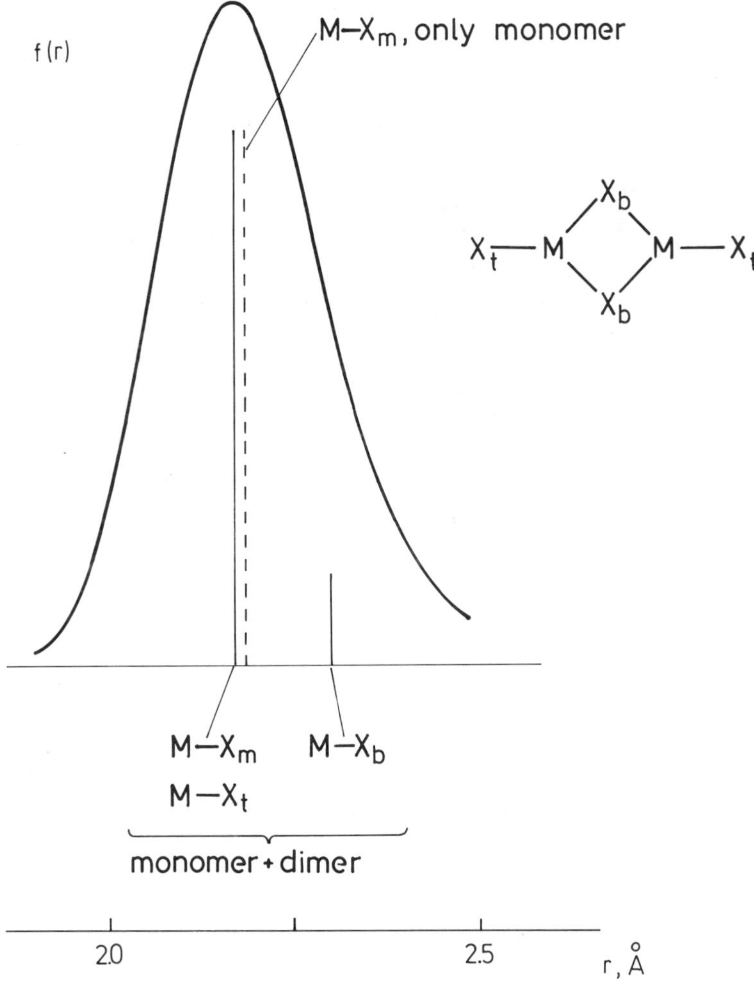

Figure 9-2. Part of the radial distribution curve of VCl_2 to show the influence of presence (or ignorance) of dimers in the vapor of metal dihalides.

under the same maximum. If the dimer presence is ignored, only one bond length is refined, which then represents the weighted average of all bond lengths. Naturally the other parameters are also influenced.

The complexity of vapor composition may make electron diffraction analysis difficult and sometimes even prevents the determination of any reliable information.

Examples. *The Case of Cuprous Chloride.* This compound was stuaied in 1957 by Wong and Schomaker,[9] who found that the vapor at 723 K consisted of trimeric species with a Cu—Cl bond length of 2.160(15) Å, and bond angles Cu—Cl—Cu around 90° and Cl—Cu—Cl around 150°. The presence of tetrameric species was considered, but the visual technique (see Chapter 1, Part A) of the 1950s did not allow a quantitative evaluation.

The experiment was recently repeated[10] in the hope of a more reliable geometry determination, at Professor Schomaker's suggestion. The lowest possible temperature at which diffraction photographs could be recorded was 693 K, and even at this temperature the mass spectra indicated the presence of both trimeric and tetrameric species. Their relative abundance was about 80% trimers and 20% tetramers according to the electron diffraction analysis. The tetrameric species hindered the determination of the trimer structure without permitting the determination of the tetramer structure. The bond distances could not be distinguished, and only a mean Cu—Cl distance was determined. The large number of a priori possible shapes for both associates presented additional problems. The trimer has a six-membered ring structure, in agreement with the early electron diffraction results.[9] As averaged over molecular vibrations, its conformation could be either chair or boat, less probably planar; however, the equilibrium conformation may in turn be planar. Eight-membered ring structures and a cubelike cage structure were considered for the tetramer. It was not possible to choose among the many possible eight-membered ring forms, but the cubelike cage could be rejected.

The experiment has been repeated at a considerably higher temperature (1333 K) in the hope of eliminating the tetrameric species. This did not happen, however, the vapor composition of this experiment was even more complicated than the one at 693 K, and the only parameter that could be determined again was the mean Cu-Cl distance.[10] These mean distances will be given in the discussion of transition metal monohalides.

The Case of Iron Trichloride. Iron trichloride is a relatively volatile compound; it gives enough vapor pressure for electron diffraction at around 463 K. The vapor consists entirely of dimeric species at this temperature.[11] The geometry of the dimer has been repeatedly investigated and will be discussed later.

The amount of monomeric species increases with increasing vapor temperature. To determine the monomer geometry, an additional experiment was carried out at 648 K, and the ratio of monomer to dimer was found to be about 45:55.[12] It was not possible, however, to determine the monomer geometry because of the strong correlation among the parameters; the monomer Fe—Cl distance and the dimer terminal Fe—Cl distance could not be resolved. The geometry of the dimer could not be assumed from the lower temperature experiment, since the effect of the temperature was supposed to be considerable.

The experiment was repeated at an even higher temperature, 723 K, where the monomer–dimer ratio was about 65:35. This proved, however, still insufficient for the determination of the two geometries. Further temperature increase was not feasible because iron trichloride begins to reduce into $FeCl_2$, whose presence further complicates the analysis. The lack of electron diffraction information on the geometry of monomeric iron trichloride is unfortunate, since the available spectroscopic results are controversial. The molecule is planar with D_{3h} symmetry according to an infrared spectroscopic study,[13] while it is pyramidal with C_{3v} symmetry according to a work using Raman spectroscopy.[14]

Determination of the Shape of Simple Halides

The example above brings us to the curious problem that available information on the shape and symmetry of even the simplest metal halide molecules is sometimes ambiguous. This is intriguing in an age when the structures of extremely complicated macromolecules are known, often with great accuracy. Spectroscopic results, though themselves not without controversy, point to a linear arrangement for all first-row transition metal dichlorides, while according to a recent electron diffraction study, VCl_2 and $CrCl_2$ are bent.[15]

"The question as to whether a given triatomic molecule is linear or bent is certainly among the simplest problems in polyatomic geometry." This statement is cited from a paper on the geometry of high temperature molecular species published more than two decades ago.[16] Ensuing progress has proved this question not so simple after all. Each technique has its difficulties in this respect.

Spectroscopic Methods. These are generally ideal tools for the determination of molecular symmetry. For MX_2 and MX_3 metal halides, the selection rules are different for linear versus bent and planar versus pyramidal configurations both in infrared and Raman spectroscopy. Nevertheless, several problems may make the determination of the symmetry of such molecules ambiguous. The relatively broad lines make the assignment of high-temperature gas-phase spectra difficult and less reliable. These problems do not arise in matrix isolation spectra. There are there two other problems, though. One is the possibility of complex formation between the sample metal halide and either the matrix or some other species present. An example is an early matrix isolation infrared spectroscopic investigation of $AlCl_3$, where an addition compound, $AlCl_3 \cdot N_2$, was formed between aluminum trichloride and nitrogen. Nitrogen was the carrier gas in the purification of the sample.[17,18] This unfortunate compound formation gave rise to the appearance of what was supposed to be the v_1 stretching frequency of $AlCl_3$, and consequently, to a misinterpretation of the spectra.

Another problem is the shift of fundamentals due to matrix effects. With higher frequencies this is not so crucial as with very low frequencies, which are often characteristic for small metal halide molecules. The relative magnitude of these shifts may be considerable. The bending frequency of a first-row transition metal dihalide, eg, falls into the region of about 50–100 cm^{-1} and the magnitude of the matrix shift may easily be about 10–20 cm^{-1} (see also later). Thermodynamic calculations, eg, are sensitive to the value of the v_2 bending frequency. Moreover, these bending frequencies are often unavailable, since they fall into the far-infrared region. The presence of different molecular species in the vapor may also hinder the interpretation of the spectra. The bands of monomers and dimers often overlap. Taking spectra at different temperatures and examining the changes in intensities may facilitate their assignment.

The bond angle of a bent molecule can be determined by isotopic substitution from isotopic shifts of the asymmetric stretching frequency v_3 in the infrared spectrum.[19] This method has been shown, however (cf, eg, references 3a and 19), to be insufficiently sensitive for *nearly linear* configurations. Also, the absence of the v_1 symmetric stretching frequency in the infrared spectrum does not necessarily mean that the molecule is linear; sometimes the respective band is simply too weak to be identified.

Electron Diffraction. Many of the early electron diffraction studies on metal halides determined molecular symmetry erroneously (see, eg, alkaline earth dihalides) due to inadequacies of the contemporary technique. Some X...X contributions could not be measured because of the large difference in the atomic numbers of the metal and the halogen. The determination of the symmetry for such molecules as MX_2 or MX_3, however, still presents a challenge for modern electron diffraction.

This technique determines the atom–atom contributions to electron scattering, and these contributions are thermal averages. The metal halide molecules generally have very low bending or puckering mode frequencies and undergo large-amplitude motion. They are often referred to as "floppy" molecules.[3b] Electron diffraction provides an average picture of the molecules being distributed on the bending or puckering potential surfaces; thus even a linear or a planar molecule will appear to be bent or pyramidal. As Beattie and Greenhalgh pointed out,[3b] this is also a philosophical question. Chemists wish to conveniently categorize as linear or planar molecules that actually spend most of their time in a bent or pyramidal configuration. Figure 9-3 illustrates this for a linear molecule.

As a consequence of bending vibrations, the X...X distance of a linear MX_2 molecule appears to be shorter than twice the M—X bond length. This is called the *shrinkage effect*,[20] and it makes the *apparent* bond angle of a linear MX_2 molecule smaller than 180°. Similarly, the puckering vibrations make the apparent bond angle of a planar MX_3 molecule smaller than 120°. Thus it is impossible to distinguish by electron diffraction alone between truly linear and planar molecules on one hand and bent and pyramidal molecules on the other.

This disadvantage may be turned into an advantage if even a limited amount of additional data is available. Information on bending and puckering frequencies can facilitate an unambiguous choice between linear and bent as well as

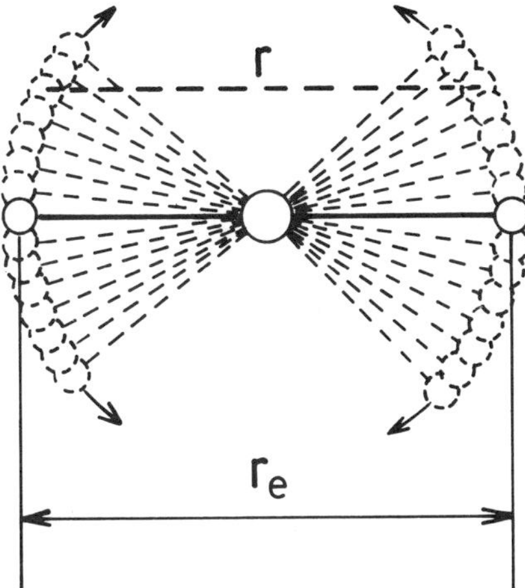

Figure 9-3. Illustration of the origin of shrinkage effect in linear metal dihalides.

between planar and pyramidal geometries. Conversely, unambiguous knowledge of a linear equilibrium configuration may lead to the estimation of vibrational quantities from the electron diffraction data. Several examples will illustrate this approach.

Electric Deflection of Molecular Beam. Another, qualitative method for the determination of the symmetry of simple high-temperature molecules is the molecular beam electric deflection experiment.[16] It has been mostly successful, with some notable exceptions (see transition metal dihalides). The possible origin of a dipole moment appearing for a highly symmetric molecule has been discussed.[21]

Quantum Chemical Calculations. This method is rapidly gaining recognition in the determination of molecular structure. Unfortunately, the size of the atoms often poses a problem in the study of metal halides. The larger the size and number of atoms in a molecule, the more complicated and time-consuming the calculation of molecular energy and geometry will be. The problem of electron correlation becomes also increasingly important. There are many ways to construct basis function sets for these molecules, and the results may strongly depend on the type of set selected.[22] Systematic studies have already been carried out for lighter atoms, and a number of generally applicable methods and basis sets have been developed. Another difficulty arises for transition metal compounds due to the open-shell electronic structure of the metal atom. No systematic investigations have been carried out on these systems, and generally acceptable calculation procedures and optimal basis sets have not yet been developed. New development is anticipated in the near future in this area.

Of course, the physical meaning of the geometrical parameters determined by calculation is different from the electron diffraction geometry. Quantum chemical calculations yield the symmetry and geometry of the hypothetical,

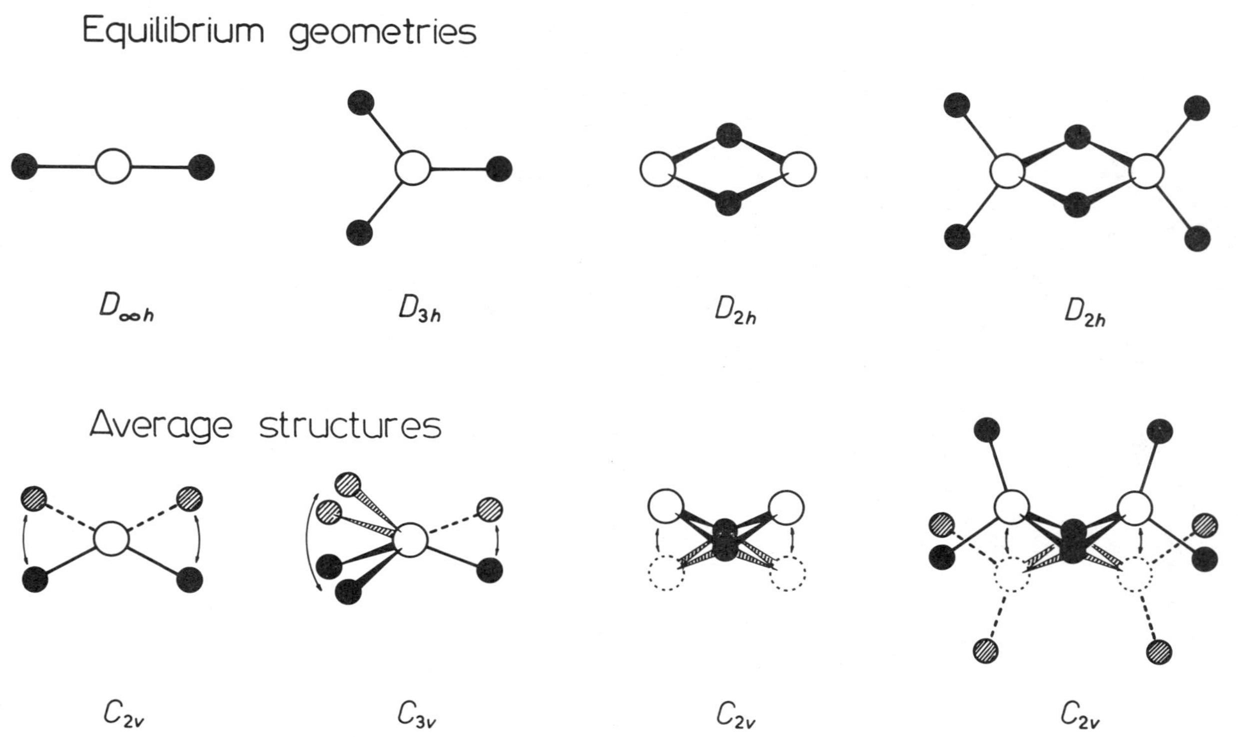

Figure 9-4. Equilibrium and average structures for some high-temperature metal halide molecular species.

motionless molecule, corresponding to the minimum position of the potential energy function. This is called the *equilibrium geometry.*

The electron diffraction geometry is a thermal average, effective structure. The symmetry of high-temperature molecules often appears to be lower from electron diffraction than the equilibrium symmetry, due to shrinkage effects. In addition to previous examples, the consequences of puckering in different planar halogen-bridged ring structures are mentioned here for Al_2Cl_6, Fe_2Cl_4, and Cu_3Cl_3. Figure 9-4 illustrates the possible lowering of symmetry as it appears in electron diffraction determinations of various molecule types.

Combination of techniques seems to be the most promising possibility for investigating metal halide molecules. This is especially so because the information they provide for the same molecule is truly complementary.

A possible scenario for a complete determination of the geometry of, say, MX_2 dihalides with some amount of dimers in their vapor, could be envisaged as follows:

1. Determine the difference of the dimer terminal and bridging bonds by calculation. Although the calculation may not be reliable enough for the absolute value of the distances, their *difference* can be reliably determined.

2. Use this difference as a constraint in the electron diffraction analysis.

3. Take the available frequencies from spectroscopy and calculate the unknown frequencies from the electron diffraction vibrational parameters. Spiridonov and co-workers,[23] for instance, use a program for linear triatomic molecules in which a simultaneous refinement of electron diffraction and spectroscopic data produces geometrical parameters as well as force constants and vibrational parameters.

A final comment concerns the quoted errors in this chapter. Uncertainties, which are given in parentheses, are generally estimated total errors taken from the original work and are expressed in units of the last digit of the parameter value.

Halides of Main Group Metals

Group I Halides

Most alkali metal halides were first investigated by electron diffraction in the pioneering days of the technique.[24] As simple diatomic molecules, they were thought to be among the simplest possible objects. It was only during the 1950s that the complexity of the vapors of alkali halides was discovered by different techniques (for literature, see reference 25). According to these studies, the tendency for association increases from cesium halides to lithium halides. There are mainly dimers in the vapor, but higher associates have also been identified.

The M—X distances from the early works[24] do not correspond to the monomers but to a weighted mean M—X distance of the species present. The same is true for the results of the reanalysis of all alkali halides at the end of the 1950s.[26,27] Comparing these data with the M—X bond distances determined for the monomers by microwave spectroscopy,[28] the differences decrease from Li to Cs, with increasing relative monomer content of the vapor. The available data have been reviewed.[21,25] Only three more recent electron diffraction investigations appeared on alkali halides, and they are considered here in some detail.

Lithium Fluoride. The electron diffraction reinvestigation of Li-F by Solomonik and co-workers[29] was carried out at a nozzle temperature of 1360 ± 50 K. The diffraction patterns were recorded at one camera range only, yielding intensity data in the interval of $2.0 < s < 15.00$ Å^{-1}.

According to other experimental evidence,[30] monomeric, dimeric, and trimeric molecules had to be considered as components of the vapor, with the dimers prevailing. The low scattering power of lithium presented an additional difficulty for the structure analysis. Several constraints were introduced in order to obtain the geometry of the dimer. They concerned the vapor composition and the geometry of the monomeric and trimeric species. The Li...Li nonbonded distance of the dimer was calculated for a diamond-shaped structure (**1**). The best agreement was found for a composition of 30% monomer, 60% dimer, and 10% trimer.

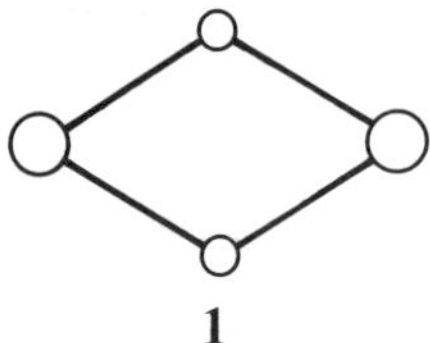

1

The rings of the dimer and trimer should be expected to be nonplanar at high-temperature experimental conditions. However, since the Li atoms do not contribute much to the electron diffraction picture, very little can be learned about the ring conformation. The following parameters have been determined[29]:

$$r_g(\text{Li}-\text{F}) = 1.746 \pm 0.015 \text{ Å}$$

$$l(\text{Li}-\text{F}) = 0.102 \pm 0.011 \text{ Å}$$

$$r_g(\text{F}...\text{F}) = 2.76_5 \pm 0.02 \text{ Å}$$

$$l(\text{F}...\text{F}) = 0.21 \pm 0.02 \text{ Å}$$

The calculated F—Li—F angle is $104.7 \pm 2.3°$.

The structure and geometry of alkali halides have been studied by theoretical methods.[32] Lithium fluoride, the lightest among them, has been subject to most theoretical work. Table 9-2 presents some of the results for Li_2F_2. While the Li—F distance from electron diffraction compares well with the theoretical

Table 9-2. Geometrical Parameters of Li_2F_2 as Determined by Different Techniques

Technique	r_e(Li—F) (Å)	r_e(F...F) (Å)	$\angle$ F—Li—F (deg)	Ref.
I[a]	1.72			29
II[b]	1.728	2.65	100.1	33
	1.714	2.550	96.1	34
	1.70	2.56	98.0	35
	1.75	2.60	96.2	
III[c]	1.69	2.50	95	36
	1.68	2.52	97	37

[a] Calculated from electron diffraction r_g parameters using spectroscopic constants.
[b] *Ab initio* results; different level basis sets.
[c] Calculated: electrostatic models.

results, the experimental bond angle appears to be larger than the calculated one. It is not known whether this difference is due to vibrational effects in the experimental geometry or to inadequacy in the calculations.

Sodium Chloride. The electron diffraction investigation of NaCl has been carried out by Miki and colleagues[38] at a nozzle temperature of 1130 K. Intensity data were recorded at two different camera ranges in the interval of $2 < s < 17 \text{ Å}^{-1}$. Since the atoms in sodium chloride vapor are considered to be highly ionized, scattering factors for the ions Na^+ and Cl^- were calculated. Two independent analyses were carried out, using neutral atomic and ionic scattering factors, respectively. Two peaks appear on the radial distribution curve, indicating the presence of more than just the monomers, and both monomers and dimers were considered in the analysis. The dimers were assumed to have a planar diamond shape with alternating Na and Cl atoms. Several constraints were applied; the parameters were varied in subsequent steps rather than simultaneously. Many possible sources were considered in the error estimation. The results are cited in Table 9-3; sets 1 and 2 correspond to the refinements with neutral atomic and ionic scattering factors, respectively.

Two features of this analysis deserve further comment. The NaCl bond distance of the dimer was a dependent parameter determined from the Na...Na and Cl...Cl nonbonded distances by their relation in a planar diamond shape rhomboid. Spectroscopic evidence[39] indicated a planar D_{2h} symmetry rhombic structure for the Na_2Cl_2 dimer. This ring is expected to appear quite puckered with C_{2v} symmetry, however, at high-temperature experimental conditions. The small out-of-plane bending vibrational frequency, 108 cm^{-1},[39] supports this possibility (see also thallous fluoride and the dimers of third-group halides). This experimental out-of-plane bending frequency corresponds to a 20° out-of-plane bending amplitude for Na_2Cl_2. Accordingly, the molecule would, indeed, appear with a puckered ring from electron diffraction, while the Na—Cl bond distance of the dimer calculated from nonbonded distances[38] suffers from the assumption of a planar ring. Figure 9-5 shows the relationship of Na...Na and

Table 9-3. PARAMETERS OF SODIUM CHLORIDE

	Electron diffraction[a]				
	At 1130 K (Ref. 38)		At 943 K (Ref. 266)		Calculation
Parameter	Set 1[c]	Set 2[d]	Set 3[e]	Set 4[e]	(Ref. 37)[b]
Monomer					
Na—Cl, Å	2.396(18)	2.392(28)	2.388(8)	2.391(12)	2.36
Dimer					
Na—Cl, Å	2.538(18)[f]	2.515(17)[f]	2.584(34)	2.569(34)	2.55
	2.553[g]	2.530[g]			
Na...Na, Å	3.227(46)	3.184(45)	3.272(110)	3.233(120)	
Cl...Cl, Å	3.917(24)	3.893(21)	4.000(66)	3.993(54)	
∠Cl—Na—Cl, deg	101.0(1.2)	101.4(1.1)	101.4(2.4)	102.0(2.5)	101
X[h]	0.50(4)	0.88(8)	0.20[i]	0.27[i]	
Dimer content, % of total	33.3[i]	46.8[i]	16.6(6.6)	21.0(9.0)	

[a] r_a.
[b] r_e, simple shell model.
[c] Analysis with neutral atomic scattering factors.
[d] Analysis with ionic scattering factors.
[e] Different constraints applied in the analyses.[266]
[f] Calculated from the other parameters with the assumption of a planar ring.[38]
[g] Calculated by us with an out-of-plane bending amplitude from spectroscopy (see text).
[h] X = number of dimers/number of monomers.
[i] Calculated from communicated data.

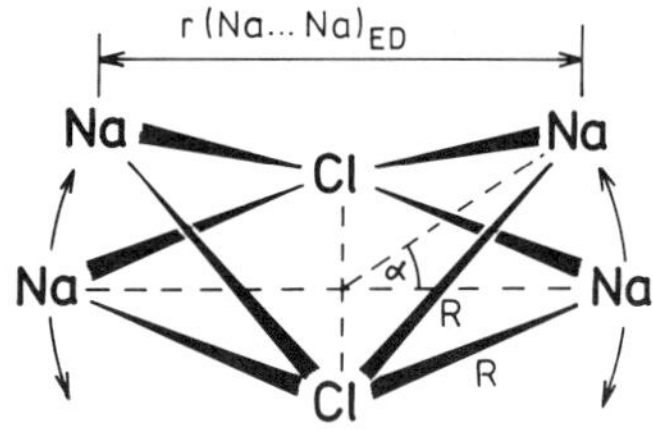

$$R = \frac{1}{2}\left[r(Cl...Cl)^2 + \frac{r(Na...Na)^2_{ED}}{\cos^2 \alpha} \right]^{1/2}$$

Figure 9-5. Relationship of the nonbonded distances Na...Na and Cl...Cl to the Na—Cl bond distance in the puckered ring of Na_2Cl_2.

Cl...Cl distances to the bond distance in a puckered structure. The 20° out-of-plane bending amplitude corresponds to a puckering angle, α, of 10° in Figure 9-5. Assuming the Na...Na and Cl...Cl distances to remain unchanged, the dimer bond length, r_a(Na—Cl), is 2.553 Å and 2.530 Å for sets 1 and 2 of Table 9-3, respectively. Comparing the electron diffraction results[38] with the calculated bond lengths of NaCl and Na_2Cl_2[37] (Table 9-3), a longer Na—Cl bond in the dimer seems to be reasonable. The results of a recent electron diffraction study are also given in Table 9-3.

The ratio of dimers to monomers differs much in the two 1130 K parameter sets of Table 9-3. According to the authors, set 1, obtained with neutral atomic scattering factors, is more consistent with the experimental data on the vapor composition[40] at this temperature. Of course, the experimental conditions may have been quite different in the two experiments, and this could account for some of the difference. The authors prefer set 2, however, since bonding in the dimer can be highly ionic.

Sodium, Potassium, Rubidium, and Cesium Chlorides. The latest electron diffraction study of alkali halides was carried out by Fink and co-workers[266] at 943 (NaCl), 964 (KCl), 898 (RbCl), and 837 (CsCl) K. The geometrical parameters have been compared to calculated geometries and were found to be closest to the results of Welch and colleagues[37] obtained by the so-called simple shell model. The data are summarized in Tables 9-3 and 9-4. The monomer Na—Cl distances determined in the two electron diffraction studies[38,266] agree within experimental errors. The dimer Na—Cl bond is much longer according to the new study, supporting the doubts about its earlier determination.[38] The Cl—M—Cl bond angle of the dimer decreases as the size of the alkali metal M increases.

Table 9-4. Parameters of Some Alkali Halides[a]

Parameter	KCl ED	KCl Calc.	RbCl ED[b]	RbCl Calc.	CsCl ED[b]	CsCl Calc.
Monomer						
M—Cl, Å	2.703(8)	2.64	2.817(4)	2.79	2.940(12)	2.94
			2.825(6)		2.946(16)	
Dimer						
M—Cl, Å	2.950(54)	2.84	3.008(22)	2.99	3.017(32)	3.17
			2.984(20)		3.024(32)	
∠Cl—M—Cl, deg	96.0(2.4)	93	88.2(2.0)	90	83.6(1.3)	85
			86.8(2.3)		83.8(1.5)	
Dimer content, % of total	9.5(5.6)		12.4(3.4)		17.6(9.2)	
			15.8(3.6)		16.0(9.8)	

[a] ED: electron diffraction, r_a, data from reference 266; calculation: simple shell model, r_e, data from reference 37.
[b] The two different values are results of refinements with different constraints.[266]

Group IIA Dihalides

The first electron diffraction studies of all Group IIA dihalides were reported in the 1950s by the Moscow Laboratory.[41] Several have since been reinvestigated. The agreement between old and new results is good in some cases, while the differences in bond lengths are well outside the error limits in others. Already the early works gave the expected trends in bond length variations and have been used for various empirical relationships.[42] Our discussion will focus on the more recent results.

Dihalides of the Group IIA metals have very low volatility. Table 9-5 gives the temperatures of the electron diffraction experiments.

According to mass spectrometry,[49,50] all Group IIA dihalides have a small amount of dimeric species in their vapors. The presence of dimers has been considered only in the investigation of calcium dihalides.[45] The results of mass spectrometry[50] and of the coupled electron diffraction/quadrupole mass spectrometric experiment[4,45] are consistent. The mass spectra of strontium and barium halides[50] recorded such small amounts of dimers that they could be justifiably ignored in the electron diffraction analysis.[46–48] The vapors of magnesium dihalides, however, may have contained several percent of dimers in the diffraction experiment according to independent mass spectrometric data.[49] Their presence, however, was ignored in the structure analyses.

Table 9-6 gives the M—X and X...X distances from the reinvestigations.[43–48] The variation of the M—X bond distances and respective ionic radii have been used to predict the bond distances of radium dihalides.[42]

The investigation of BaI_2 was among the first by Spiridonov and associates[48] using an extended expression for electron scattering intensities incorporating more detailed dependence on vibrational parameters than is customary. The

Table 9-5. NOZZLE TEMPERATURES
OF ELECTRON DIFFRACTION
EXPERIMENTS ON GROUP IIA
DIHALIDES

Molecule	Temperature (K)	Ref.
MgF_2	1750	43
$MgCl_2$	1150	44
$CaCl_2$	1433	45
	1243	46
$CaBr_2$	1383	45
CaI_2	1183	45
	1300	47
$SrCl_2$	1237	46
SrI_2	1250	47
$BaCl_2$	1363	46
BaI_2	1100	48

Table 9-6. Bond Lengths and Nonbonded
Distances (Å) in Group IIA Dihalides

Molecule	r_g(M—X)	r_g(X...X)	Ref.
MgF_2	1.771(10)	3.436(36)	43
$MgCl_2$	2.186(11)	4.259(24)	44
$CaCl_2$	2.483(6)	4.746(18)	45
$CaBr_2$	2.616(7)	5.082(27)	45
CaI_2	2.830(7)	5.458(17)	45
$SrCl_2$	2.616(6)[a]	[b]	46
SrI_2	3.009(15)	5.727(33)	47
$BaCl_2$	2.768(9)[a]	[b]	46
BaI_2	3.150(4)	5.894(6)	48

[a] Here the errors were increased for consistency, as we
supposed that the original work quoted least-
squares standard deviations only.
[b] Not given in reference 46.

structure analysis was first carried out in the usual way, yielding the geometrical parameters indicated in Table 9-6 and a bond angle I—Ba—I 138.9°. The new scheme of analysis yielded a so-called equilibrium geometry in harmonic approximation r_e^h(Ba—I) = 3.150(7) Å and $\angle_e^h$ I—Ba—I = 148.0(9)°. No spectroscopic results are available for comparison. The unusually large amplitude of vibration of the I...I distance, 0.438(11) Å, indicates a low-frequency bending for this molecule, making the effective I—Ba—I bond angle smaller than the equilibrium angle. The r_g and r_e^h bond distances appear similar, as several terms cancel in the expression of r_g for symmetrical triatomic MX_2 molecules due to the curvilinear motion of the terminal atoms. This results in an approximate equality of these two kinds of distance parameter.[48]

The Shape of Alkaline Earth Dihalides. The most intriguing question about the geometry of alkaline earth dihalides is whether these molecules are linear or bent. There is general agreement that the beryllium and magnesium dihalides are linear,[16,43,44,53,54] and all barium dihalides are bent.[16,46,48,52,54–57] Calcium difluoride is bent, and all the other calcium dihalides are linear.[16,45,52,54,56,57] The difluoride and the dichloride of strontium are bent, and SrI_2 is most probably linear.[46,47,52,54,56–58] The $SrBr_2$ molecule has not been much studied; molecular beam deflection measurements predict it to be linear,[56] while according to theoretical calculations[54,55] it is bent. The bond angles of alkaline earth dihalides are collected in Table 9-7.

Structural variations of Group IIA dihalides can be understood by supposing ionic nature of bonding. Several empirical relationships were communicated,[42,51] correlating geometrical and spectroscopic features involving also polarizabilities and ionic radii.

The calculations in References 54 and 55 represent two different approaches accounting for the bond angle variations of alkaline earth dihalides.[54,55] One

Table 9-7. Bond Angles of Alkaline Earth Dihalides (deg)

	F		Cl		Br		I	
	Exp.	Calc.	Exp.	Calc.	Exp.	Calc.	Exp.	Calc.
Be	180[52]	180[54]		180[54]		180[54]		180[54]
Mg	180[43,53]	180[54]	180[44,53,58]	180[54]	180[53]	180[54]	180[53]	180[54]
	155[52]							
Ca	140[57]	133[54]	180[45,56,58]	180[54]	180[45,56]	180[54,55]	180[45]	180[54,55]
	155[52]	145[55]		173[55]				
Sr	108[57]	115[54]	120[58]	143[54]	180[56]	164[54]	180[47,56]	180[54]
	135[52]	127[55]	142(8)[46]	133[55]		133[55]		161[55]
Ba	100[57]	103[54]	100[58]	125[54]		138[54]	148.0(9)[48]	154[54]
	115[52]	109[55]	127(8)[46]	103[55]		95[55]		102[55]

assumes ionic character of bonding and uses the Rittner polarizable ion model to calculate equilibrium bond angles.[54] The Rittner model was shown[51] to be sensitive to the values of metal ion polarizabilities and internuclear distances, and it has limitations for quantitative predictions. Nevertheless, the calculated values agree surprisingly well with the experimental results, supporting the assumption of ionic bonding in these molecules. On the other hand, the valence-shell electron-pair repulsion (VSEPR) model,[59] predicts linear shapes for all these molecules. Simple Walsh diagrams[60] and semiempirical calculations with only s-p basis sets[55] give the same results. A modified Walsh diagram,[61] including d orbitals on the metal atom, accounts for the bent geometries of some alkaline earth dihalides. Semiempirical,[55] as well as *ab initio* calculations[62] show the low-lying metal orbitals essential in the bonding of these molecules. Their shape is then determined by the extent of metal d-orbital participation. The results of semiempirical calculations[55] follow the general trend in the variation of bond angles; however, the actual values were sensitive to the d-orbital exponents used in the calculation.

Returning to the ionic model of MX_2 dihalides, a bent geometry is favored by large, more polarizable metal ions and small, more electronegative halide ions.[51] Accordingly, the bond angle decreases in the following order, $MgF_2 > CaF_2 > SrF_2 > BaF_2$. For the completely bent barium halides, the bond angles are predicted to decrease with increasing electronegativity and decreasing ligand size: $BaI_2 > BaBr_2 > BaCl_2 > BaF_2$. The polarizable ion model calculations follow this trend, and the predictions agree with available experimental data within experimental error. Figure 9-6 summarizes the variation of shapes of alkaline earth dihalides.

Determination of the Shape of MX_2 Molecules from Electron Diffraction. Even linear MX_2 molecules appear to be bent in an electron diffraction structure analysis, as pointed out in the introduction. Thus the bond angles of magnesium and calcium dihalides appeared to be around 150°. Other experimental evidences from spectroscopy and molecular beam deflection measurements, how-

	F	Cl	Br	I
Be	l	l	l	l
Mg	l	l	l	l
Ca	b	l	l	l
Sr	b	b	(l)	l
Ba	b	b	b	b

l linear

b bent

( ) uncertain

Figure 9-6. Shapes of alkaline earth dihalides.

ever, indicated linearity for these molecules.[16,53–56] Fortunately, some bending vibrational frequencies available from spectroscopy provide some idea about their magnitude even for molecules whose spectra have not yet been investigated.

Supposing linear equilibrium configuration, the shrinkage

$$\delta_g = 2r_g(\mathrm{M{-}X}) - r_g(\mathrm{X...X}) \tag{9-1}$$

can be calculated from electron diffraction data. This shrinkage is directly related to the amplitude of perpendicular vibrations of the M—X bond distance and to the v_2 bending frequency.[63] Figure 9-7 shows this relationship for the CaX_2 dihalides, with each curve calculated at the experimental temperature. This relationship allows an electron diffraction estimation of the v_2 frequency. Such estimates may be less accurate than spectroscopic measurements. However, they are valuable because these frequences belong to the far-infrared region and their spectroscopic determination is rather difficult. The results from matrix isolation spectroscopy may suffer from matrix shifts of comparable amounts to the uncertainty of the electron diffraction estimation (see Group IIIA halides). Table 9-8 compares available spectroscopic and electron diffraction v_2 bending frequencies for alkaline earth dihalides.

Group IIIA Halides

Trihalides. Trihalides of Group IIIA elemensts exist as monomers and dimers in the vapor phase.[64,65] They have been studied extensively by spectroscopic methods, both in the gas phase and in inert gas matrices.[66–68] Early electron diffraction studies were reported by Palmer and Elliott[69] on aluminum halides, by Brode[70] on some aluminum, gallium, and indium halides, by

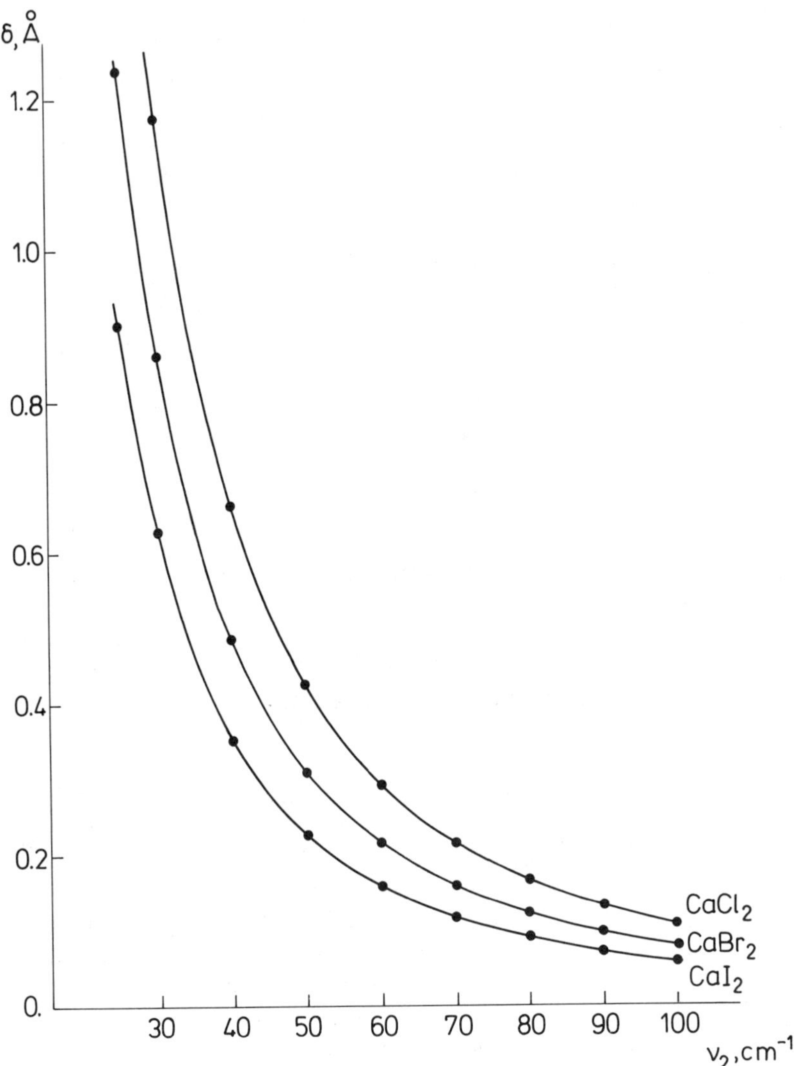

Figure 9-7. Shrinkage versus ν_2 bending vibrational frequency relationship for calcium dihalides at 1433, 1383, and 1182 K for $CaCl_2$, $CaBr_2$, and CaI_2, respectively.

Stevenson and Schomaker[71] on gallium and indium halides, and by Akishin and co-workers[72,73] on all aluminum and gallium halides.

More recent information concerns the structure of dimeric and monomeric species of aluminum and gallium halides.[74–78] Numerous techniques,[64–66,68–70,72–74] showed the presence of dimeric species in the vapors of aluminum and gallium chlorides and bromides at lower temperatures. With increasing temperature, the dimers dissociate into monomers.

Table 9-8. BENDING FREQUENCIES OF LINEAR
ALKALINE EARTH DIHALIDES DETERMINED
BY INFRARED SPECTROSCOPY (SP) AND BY
ELECTRON DIFFRACTION (ED)

Molecule	SP		ED	
	$v^2(cm^{-1})$	Ref.	$v_2(cm^{-1})$	Ref.
MgF_2	165^a	52	160(30)	43
	249^b	53		
$MgCl_2$	87.7^b	58	110(20)	44
	93^b	53		
$CaCl_2$	63.6^b	58	69(10)	45
$CaBr_2$			70(20)	45
CaI_2			50(10)	45
SrI_2			30(10)	47

^{a} Gas phase.
^{b} Matrix isolation.

Dimers. Shen[74] reinvestigated some Group IIIA trihalides in the Corvallis laboratory using nozzle temperatures about 423, 433, 503, and 393 K for aluminum chloride, aluminum bromide, aluminum iodide, and gallium chloride, respectively. For gallium bromide, the electron diffraction experiments were carried out at three different temperatures: 436, 465, and 499 K.

The possible presence of monomeric species was checked in each case. Only dimeric species were found in the vapor of aluminum chloride and bromide, and gallium chloride.

The dissociation energy of dimers decreases from the chlorides toward the iodides, and both monomeric and dimeric species were identified in the vapors of the electron diffraction experiment for aluminum iodide and gallium bromide. The structure analyses faced difficulties similar to those experienced in other cases with complicated vapor composition, and several constraints had to be applied.

The enthalpy and entropy of dissociation of Ga_2Br_6 could be calculated from the varying relative dimer concentration at different temperatures: $\Delta H° = 81.6(14.6)$ kJ mol^{-1} and $\Delta S° = 147.7(27.2)$ J molK^{-1}.

A geometrical model of M_2X_6 dimeric molecules is shown in Figure 9-8. The two monomeric units share two halogens (one from each) in a four-membered ring. The molecules can also be looked at as two tetrahedra sharing a common edge. The metal atoms are located at the centers of the tetrahedra and the halogen atoms at the apices. The possibility of other arrangements with one or three halogen bridges was investigated and ruled out by *ab initio* calculations for Al_2F_6.[79]

The symmetry of the M_2X_6 molecules is D_{2h} according to all spectroscopic studies.[66,68] This implies a planar four-membered ring. The ring puckering mode of all these molecules, however, is calculated[66] to be smaller than 40 cm^{-1},

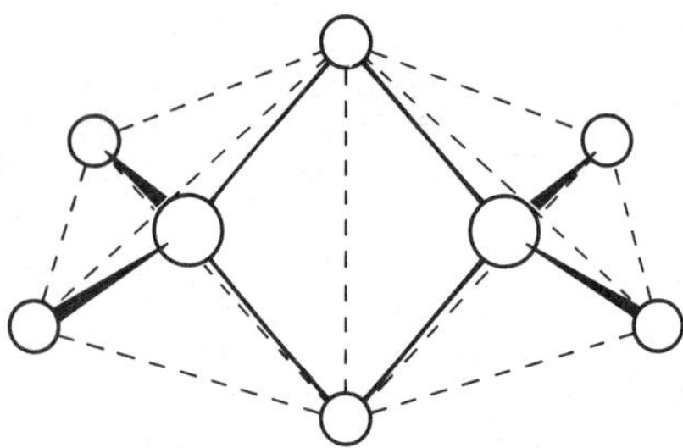

Figure 9-8. Geometrical arrangement of dimeric metal trihalides, M_2X_6.

predicting very flexible rings. Such low frequencies have not yet been observed experimentally for these molecules. The M_2X_6 dimers appear to be puckered indeed at the electron diffraction experimental conditions.[74] This symmetry lowering, from D_{2h} to C_{2v}, is the consequence of low-frequency, large-amplitude puckering motion of the molecule around the axis connecting the two bridging halogen atoms (see Figure 9-9 and also Figure 9-4). Consequently, the effective internuclear distances measured by electron diffraction between the two terminal $AlCl_2$ units will be shorter than what would correspond to a D_{2h} symmetry geometry. This is also a shrinkage effect. The geometrical parameters of the dimers are collected in Table 9-9 with the results of minimum basis *ab initio* calculations[79] included for comparison.

The bridging M—X bonds are considerably longer than the terminal ones. This is consistent with the spectroscopic results[66] according to which the stretching frequencies of the terminal bonds are about 2.4 times larger than those of the bridging bonds. There is a remarkable constancy in the differences and ratios of the lengths of these two bond types: the bridging bonds are about 0.19 Å longer than the terminal bonds and the ratio of terminal bond to bridge bond is about 0.92.

The variation of the X_b—M—X_b bond angle in the rings of molecules with the same metal reflects the change in the size of the halogen atom: the X_b—M—X_b angle increases with increasing halogen bulkiness.

Monomers. Aluminum trifluoride has much lower volatility[80] than the other aluminum trihalides. Its electron diffraction investigation[75] was carried out at about 1320 K. The geometrical parameters are given in Table 9-10.

Zasorin and Rambidi[76] used an overheating technique to obtain a high enough concentration of monomeric species of aluminum trichloride. The

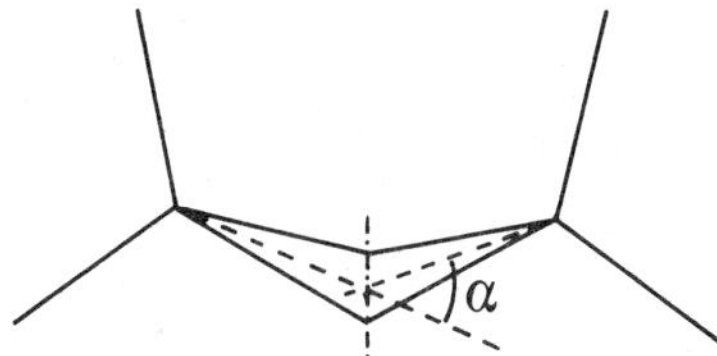

Figure 9-9. Large-amplitude ring puckering vibration of M_2X_6 bridged molecules.

Table 9-9. GEOMETRICAL PARAMETERS OF DIMERIC GROUP IIIA TRIHALIDES

| Molecule | Bond lengths (Å) | | Bond angles (deg) | | | |
	$r_g(M—X)_t$	$r_g(M—X)_b$	$X_t—M—X_t$	$X_b—M—X_b$	φ^a	Ref.
Al_2F_6	1.60^b	1.72^b	124.6	80.0		79
Al_2Cl_6	2.05^b	2.24^b	122.8	94.7		79
	2.066(2)	2.254(4)	123.5(1.6)	91.0(0.5)	23.4(6.0)	74
Al_2Br_6	2.223(5)	2.417(8)	122.8(3.3)	92.2(0.9)	23.3(14.0)	74
Al_2I_6	2.451(13)	2.637(29)	115.1(7.4)	99.5(4.5)	25.0^c	74
Ga_2Cl_6	2.100(2)	2.303(3)	124.7(1.8)	88.3(0.8)	20.5(3.3)	74
Ga_2Br_6	2.246(3)	2.449(9)	128.2(3.0)	91.1(2.2)	25.0^c	74

[a] Root-mean-square amplitude of ring puckering.
[b] r_e, *ab initio* calculation.
[c] Assumed value.

geometrical parameters determined in this study are given in Table 9-10. The Al—Cl and Cl...Cl distances correspond to a bond angle of 118.5°, which characterizes a rather flat pyramidal structure. The deviation from planarity, however, was considered[76] to be a consequence of shrinkage[20] due to large-amplitude, out-of-plane deformation vibrations. Spectra were also interpreted[66–68] in terms of planar geometry. As mentioned before, a matrix isolation infrared spectroscopic study,[17] however, recorded a frequency that could be attributed to the ν_1 symmetric stretching mode. Since this is inactive in the infrared region for D_{3h} symmetry, a pyramidal structure was suggested with a 110–112° bond angle. Subsequent spectroscopic results,[67,68] quantum chemical calculations,[79,81,82] and structural considerations,[83] however, all provided further evidence for the planarity of $AlCl_3$. As was determined later,[18] complex formation during the experiment resulted in the appearance of the ν_1 symmetric stretching mode in the above-cited spectroscopic study.[17]

A recent electron diffraction reinvestigation of $AlCl_3$[77] yielded a revised set of structural parameters, also given in Table 9-10. Joint analysis of two different

Table 9-10. GEOMETRICAL PARAMETERS OF GROUP IIIA TRIHALIDE MONOMERS

| Parameter | AlF_3 Ref. 75 | $AlCl_3$ | | | GaI_3 Ref. 78 |
		Ref. 76	Ref. 77	Ref. 77	
T, K		800	1150	1410	528
$r_g(M—X)$, Å	1.631(3)	2.06(1)	2.068(4)	2.074(4)	2.458(5)
$l(M—X)$, Å	0.070(2)	0.060(5)	0.074(3)	0.083(2)	0.067(4)
$r_g(X...X)$, Å	2.804(10)	3.53(1)	3.571(9)	3.567(10)	4.256(18)
$l(X...X)$, Å	0.150(8)	0.180(10)	0.187(6)	0.198(7)	0.163(10)
$r_e(Al—X)$, Å^a	1.60	2.05			
$\angle_e X—Al—X$, deg.^{a}	120	120			

[a] *Ab initio*, minimal basis set.[79]

Table 9-11. FUNDAMENTAL FREQUENCIES AND HARMONIC FORCE
FIELD FOR ALUMINUM CHLORIDE MONOMER

Parameter	Gas-phase spectroscopy[a]		Electron diffraction[b]
	IR	Raman	
$\nu_3(E')$, cm^{-1}	616	610	643(27)
$\nu_1(A_1')$, cm^{-1}		375	391(6)
$\nu_2(A_2'')$, cm^{-1}	214		164(10)
$\nu_4(E')$, cm^{-1}	151	148	159(9)
F_{33}, mdyn Å^{-1}	2.692		2.88(0.12)
F_{34}, mdyn rad^{-1}	−0.141(0.081)		−0.097(0.012)
F_{44}, mdyn rad^{-2}	0.390(0.009)		0.436(0.038)
F_{22}, mdyn rad^{-2}	0.275(0.005)		0.163(0.017)

[a] Data from reference 68.
[b] Data from reference 77.

temperature data sets resulted in a so-called harmonic equilibrium bond
distance, r_e^h(Al—Cl), and predicted harmonic force constants. Table 9-11
compares these data with the latest gas-phase spectroscopic results[68] on
monomeric $AlCl_3$. There is considerable discrepancy in the frequency of
out-of-plane bending.

Gallium iodide was studied by Morino and co-workers[78] at a nozzle
temperature of 528 K. Independent vapor pressure and mass spectrometric
evidence showed the vapor to contain about 97% monomers at this temperature. The structural parameters are given in Table 9-10. The molecule, like the
other Group IIIA trihalides, is planar.

Lower Valence Halides. The stability of lower oxidation states increases in
the sequence Al < Ga < In < Tl.[84] Different valence state indium chlorides
have been investigated by gas-phase spectroscopy,[85] whereas no electron
diffraction studies have been reported yet.

Thallium(I) is the most stable oxidation state for the halides of this element.
Thallous fluoride has been extensively studied (cf Reference 2). According to
mass spectrometric and vapor pressure measurements,[86,87] dimeric species
prevail in its vapor. Thallous halides have also been subject of vibrational[88] and
photoelectron[89] spectroscopic studies. Three possible models displayed in
Figure 9-10 have been considered, with the linear geometry favored eventually.

A low-energy electron scattering experiment[90] indicated considerable instantaneous dipole moment for Tl_2F_2 that cannot be explained by a symmetrical
rigid linear or rhombic structure. Molecular beam deflection,[91] on the other
hand, was consistent with a nonpolar dimeric structure of high symmetry. Both
the linear and the planar rhombic models confirm this observation.

The electron diffraction data on thallous fluoride dimer at 700 K are
consistent with a rhombic structure, and the linear model could be ruled out.[92]
Diffraction photographs were recorded at one camera distance only, in the

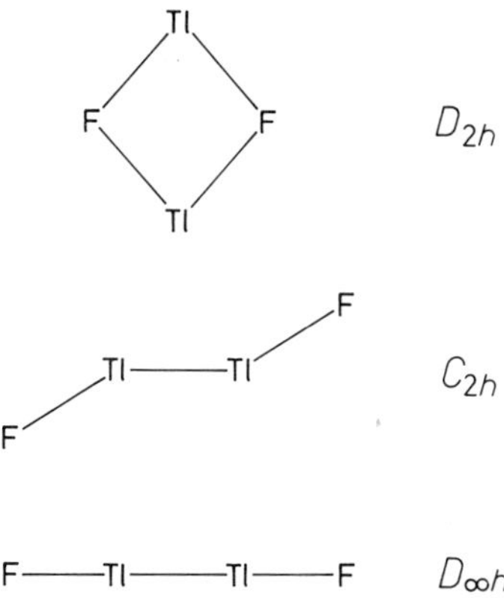

Figure 9-10. Models considered for the Tl_2F_2 molecule.

range of $2 < s < 15\ \text{Å}^{-1}$. A number of constraints were applied in the analysis, similarly to the study of other M_2X_2 species. The results of the analysis were as follows:

$r_g(\text{Tl}—\text{F}) = 2.29(2)\ \text{Å}$
$r_g(\text{Tl}...\text{Tl}) = 3.68(1)\ \text{Å}$
$r_g(\text{F}...\text{F}) = 2.61(18)\ \text{Å}$
$l(\text{Tl}—\text{F}) = 0.095(15)\ \text{Å}$
$l(\text{Tl}...\text{Tl}) = 0.14(15)\ \text{Å}$
$l(\text{F}...\text{F}) = 0.19(15)\ \text{Å}$

The dimer content was around 80%

The molecule was assumed to have a planar four-membered ring of D_{2h} symmetry. An F—Tl—F angle of 73(2)° is calculated for the planar model from the measured Tl—F and Tl...Tl distances. This is consistent with the results from molecular beam deflection[91] and also with a detailed analysis of the infrared and Raman spectra of matrix-isolated thallous fluoride.[93] The electron diffraction data, however, indicate a puckered ring and C_{2v} symmetry for the thermal average structure of the dimer. A puckering angle of about 17° can be calculated from the measured distances (cf Figure 9-5). A 34° out-of-plane bending vibrational amplitude for the Tl—F—Tl unit around the Tl...Tl axis corresponds to such puckering. This agrees well with the large amplitude of 31° calculated for the out-of-plane bending coordinate[93] for the temperature of the experiment. The measured[93] out-of-plane bending mode is 81 cm^{-1}, indicating great flexibility of the ring. The large instantaneous dipole moment[90] can be understood in view of this low-frequency puckering. Since the interaction time of the electron beam and the molecules is very short, "frozen" nuclear configurations are detected by the low-energy electron scattering experiment. Lesiecki and Nibler[93] commented on the possible bending of the linear $D_{\infty h}$ symmetry structure to account for the large dipole moments. The π_u bending mode in the spectral assignment[88] for a linear structure yields a bending amplitude of about 2°, which gives a dipole of 0.8 debye only.[93] A planar rhombic model with a 31° bending amplitude gives a dipole moment of 3.7 debye, in agreement with the experimental data.[90]

The geometry of Tl_2F_2 dimers thus seems to be well understood; it is planar rhombic with alternating Tl and F atoms. This four-membered ring, however, undergoes large-amplitude, out-of-plane deformation motion at high temperatures.

Group IVA Halides

There is a reverse tendency in the variation of the stability of dihalides and tetrahalides of Group IVA elements.[84] While the stability of tetrahalides decreases down the group, the sequence for the dihalides' stability is $CX_2 \ll SiX_2 < GeX_2 < SnX_2 < PbX_2$.

Tetrahalides. All tetrahalides have regular tetrahedral arrangement. The geometrical parameters of $GeCl_4$[94] and $SnCl_4$[95] from electron diffraction are as follows:

$r_g(\text{Ge}—\text{Cl}) = 2.113(3)$ Å
$r_g(\text{Cl}...\text{Cl}) = 3.444(6)$ Å
$r_g(\text{Sn}—\text{Cl}) = 2.281(4)$ Å
$r_g(\text{Cl}...\text{Cl}) = 3.717(8)$ Å.

Dihalides. Several of the Group IVA dihalides have been investigated by electron diffraction. Although silicon compounds are covered in detail in Chapter 1, their relevant halides are mentioned here for comparison.

The electron diffraction investigation of silicon and germanium dihalides involved gas/solid reaction carried out in the diffraction chamber to produce these highly reactive carbene analogs. Good use was made of the combined electron diffraction/mass spectrometric technique.[4] The experiments required relatively high temperature, as indicated in Table 9-12.

All molecules have a bent geometry, in agreement with the predictions of the VSEPR model. The geometrical parameters are presented in Table 9-13. The bond angles increase in the order F < Cl < Br < I with a given central atom, again in accordance with the VSEPR model. Another trend is observed down the group; the angles decrease in order C > Si > Ge > Sn > Pb considering the same halides. Both variations can be interpreted with nonbonded interactions as well. Increasing the ratio of ligand size to central atom size results in increased nonbonded interactions and larger bond angles.

Let us compare now the geometries of tetrahalides and dihalides of the same central atom. The bond lengths of the chlorides of Si [$SiCl_4$[106]: $r_g(\text{Si}—\text{Cl})$ 2.019(3) Å], Ge, and Sn are about 0.07 Å longer in the dihalides than in the tetrahalides. The bond angles in the dihalides are smaller than in the tetrahalides. It seems as if the nonbonding electrons in the dihalides exert greater

Table 9-12. NOZZLE TEMPERATURE OF
ELECTRON DIFFRACTION EXPERIMENTS ON
GROUP IVA DIHALIDES

Molecule	Temperature (K)	Ref.
$SiCl_2$	1473	96
$SiBr_2$	1473	96
$GeCl_2$	933	97
$GeBr_2$	647	98
$SnCl_2$	683	99
$SnBr_2$	550	100
SnI_2	600	100
PbF_2	1000	100
$PbCl_2$	853	101
	963	101
$PbBr_2$	720	100
PbI_2	750	100

Table 9-13. GEOMETRICAL PARAMETERS OF GROUP IVA DIHALIDES

Parameter	F	Cl	Br	I
Carbon				
r, Å	1.304^a 1.291^b	1.756^b		
$\angle$, deg	104.8 104.7	109.4		
Silicon				
r, Å	1.590^c	$2.089(4)^d$	$2.249(5)^d$	
$\angle$, deg	100.8	103.1(6)	102.9(3)	
Germanium				
r, Å	1.732^e	$2.186(4)^f$	$2.337(13)^g$	
$\angle$, deg	97.2	100.4(4)	101.4(9)	
Tin				
r, Å		$2.346(7)^h$	$2.512(3)^i$	$2.706(4)^i$
$\angle$, deg		99(1)	100.0(7)	103.8(7)
Lead				
r, Å	$2.033(3)^i$	$2.445(5)^j$	$2.597(3)^i$	$2.804(4)^i$
$\angle$, deg	97.8(2.0)	97.6(1.5)	98.8(1.5)	99.7(8)

[a] MW, r_o, reference 102.
[b] *Ab initio*, r_e, reference 103.
[c] MW, r_e, reference 104.
[d] ED, r_g, reference 96.
[e] MW, r_e, reference 105.
[f] ED, r_g, reference 97.
[g] ED, r_g, reference 98.
[h] ED, r_a or r_g, reference 99.
[i] ED, r_g, reference 100.
[j] ED, r_g, reference 101.

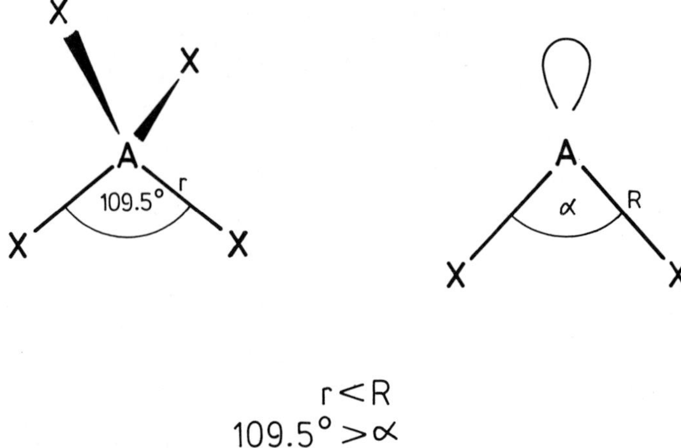

Figure 9-11. Comparison of geometries of Group IVA tetrahalides and dihalides.

repulsions than do the bonding pairs in the tetrahalides (Figure 9-11); the consequences of greater repulsions are the smaller angles and longer bonds in AX_2 as compared to AX_4. The stretching force constants of the tetrahalides are larger than those of the dihalides.[97]

Group VA Halides

Arsenic and antimony can be considered to be semimetals, and their halides will be discussed. Bismuth is a metal, but its halides have not yet been studied by electron diffraction.

Trihalides. The geometrical parameters of the Group VA trihalides studied so far are collected in Table 9-14. All molecules are pyramidal, as expected due to the presence of the lone electron pair on their central atom. The bond angles are all smaller than the ideal tetrahedral in accordance with the VSEPR theory.[59] Their variation can also be accounted for by this model. The XMX angle of the same central atom increases as the electronegativity of the ligands decreases. Nonbonded interactions also explain the bond angle variation. The nonbonded interactions strengthen with increasing ratio of ligand size to central atom size, resulting in larger bond angles. These interactions can also account for the bond angle variation down the group. Take the chloride as an example, including the bond angles of NCl_3[113] and PCl_3,[114] 107.8(4) and 100.3(1)°, respectively. The angles decrease from N to Sb. All bond angle variations follow the same trends as those observed for the Group IVA dihalides.

Electron diffraction experiments of $AsBr_3$[109] were run at two temperatures. The vibrational amplitudes from the two experiments were combined with experimental vibrational frequencies yielding a reliable force field.[109]

Table 9-14. GEOMETRICAL PARAMETERS OF GROUP
VA TRIHALIDES

Molecule	r_g(M—X), (Å)	$\angle$ X—M—X (deg.)	Ref.
AsF_3	1.710(3)	96.0(3)[a]	107
	1.706(2)	96.2(2)	108
$AsCl_3$	2.166(3)	98.6(4)[a]	107
$AsBr_3$[b]	2.326(2)	99.6(1)	109
$AsBr_3$[c]	2.329(2)	99.6(2)	109
AsI_3	2.557(5)	100.2(4)	78
$SbCl_3$	2.333(3)	97.2(10)	110a
	2.334(4)	97.1(10)	110b
$SbBr_3$	2.490(3)	98.2(6)[a]	111
SbI_3[d]	2.719(5)	99.1(10)	112

[a] $\angle_\alpha$ value.
[b] At 373 K.
[c] At 466 K.
[d] Unidentified, possibly r_a.

Pentahalides. Only three molecules have been investigated so far: AsF_5,[108] SbF_5,[115] and $SbCl_5$.[116] A trigonal bypiramidal configuration (**2**) is expected for these molecules by analogy with phosphorus pentahalides.

2

The vapors of AsF_5 and $SbCl_5$ consist of monomeric molecules, which indeed have a trigonal bypiramidal configuration. The geometrical parameters are as follows:

r_g	r_e^h
$(As-F)_{mean} = 1.678(2)$ Å	
$As-F_{ax} = 1.711(5)$ Å	$Sb-Cl_{ax} = 2.338(7)$ Å
$As-F_{eq} = 1.656(4)$ Å	$Sb-Cl_{eq} = 2.277(5)$ Å

The axial bonds are longer than the equatorial ones, in accordance with the VSEPR model.

The vapors of SbF_5, on the other hand, consist mainly of trimeric molecules in the gas phase. The molecular shape and the geometrical parameters are shown in Figure 9-12. The Sb atoms are six-coordinated, with three different types of Sb—F bond. The axial and terminal bond lengths could not be resolved

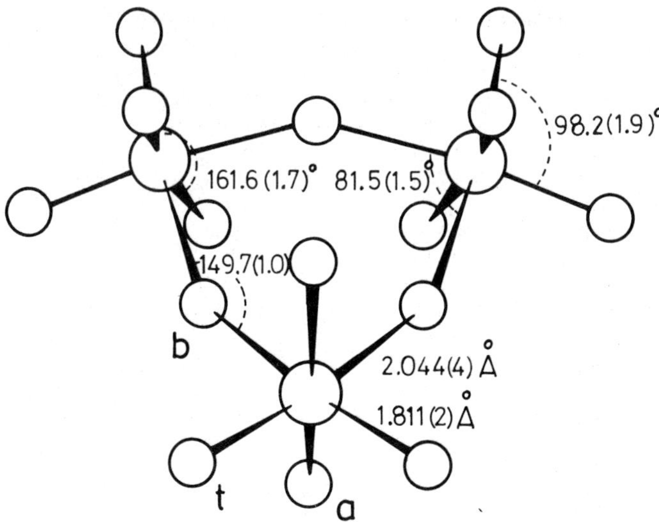

Figure 9-12. Atomic arrangement and geometrical parameters of trimeric antimony pentafluoride $(SbF_5)_3$.[115]

and were assumed to be equal; both are much shorter than the bridging bond. The six-membered ring deviates considerably from planarity. Two models, one with C_{3v} and the other with C_s symmetry, gave about equally good agreement with the experiment. The important geometrical parameters are practically the same in the two models, except the dihedral angles.

Trimeric $(TaF_5)_3$ and $(NbF_5)_3$ have similar geometries, except that their six-membered rings are planar.[117,115] It has been suggested[118] that d-orbital participation in the bonding of tantalum and niobium fluorides may provide a π-type delocalized molecular orbital over the ring, favoring planar conformation. On the other hand, such molecular orbitals would not be favorable in $(SbF_5)_3$ with the full $4d$ orbitals of the Sb atom.

Transition Metal Halides

Monohalides

The principal components of the vapor in equilibrium with solid cuprous chloride are trimers and tetramers according to mass spectrometric studies[119] and vapor pressure measurements.[120] Trimers were found to be the main component for cuprous iodide.[120,121] Cuprous chloride was first investigated by electron diffraction in the 1950s,[9] and a reinvestigation was carried out recently[10] (see Introduction) yielding a limited amount of information shown in Table 9-15. The mean Cu—Cl distances in the two new experiments differ by

Table 9-15. GEOMETRICAL PARAMETERS OF CUPROUS HALIDES

| Parameter | Cuprous chloride | | | Cuprous iodide | |
| | At 723 K (Ref. 9) | Ref. 10 | | At 693 K (Ref. 122) | |
		At 693 K	At 1333 K		
r_g(Cu—X)$_{mean}$, Å	2.160(15)[a]	2.155(5)	2.179(5)	2.447(5)	2.447(5)
Angles of the trimer					
∠Cu—X—Cu, deg	~90	75–85	75–85	106(1)	108(1)
∠X—Cu—X, deg	~150	155–165	155–165	125(1)	127(2)
Polymeric species considered	Trimer	70–90% trimer 10–30% tetramer	60–70% trimer Small amounts of monomers, dimers, and tetramers	Trimer	Trimer
Configuration	Planar (supposed)	Different nonplanar		Chair	Boat

[a] r_a.

about 0.02 Å. This difference cannot be ascribed to the effect of temperature alone because of the different vapor compositions in the two experiments.

Cuprous iodide has also been studied by electron diffraction.[122] The presence in the vapor of species other than trimeric was not considered. A "zigzag" open chain geometry could be ruled out, and the D_{3h} symmetry planar six-membered ring structure gave somewhat worse agreement with experiment than the chair and boat conformers, while the latter two were about equally satisfactory. The geometrical parameters for these two models are also given in Table 9-15.

Dihalides

Group IIB Dihalides. The Zn group dihalides are more volatile than the Group IIA dihalides. The electron diffraction nozzle temperatures are given in Table 9-16. According to the combined electron diffraction/mass spectrometric analysis, no dimeric species were present in the vapors of $ZnCl_2$, $ZnBr_2$, and ZnI_2.[123] The dimer content has not been checked in the other studies.

The geometrical parameters are collected in Table 9-17. All molecules were assigned linear geometries from electron diffraction. The spectroscopic studies[128–131] are inconclusive as to the shape of zinc and cadmium dihalides, and they point to bent geometry for several mercury dihalides. Molecular beam deflection studies, on the other hand, indicate linearity for both zinc and mercury dihalides.[132] Let us now consider the geometry of zinc dihalides in some detail. The apparent bond angles of about 165° from electron diffraction would seem to be in good agreement with the 148–180° angle intervals given by

Table 9-16. TEMPERATURE OF ELECTRON DIFFRACTION EXPERIMENTS ON GROUP IIB DIHALIDES

Molecule	Temperature (K)	Ref.
$ZnCl_2$	656	123
$ZnBr_2$	614	123
ZnI_2	580	123
$CdBr_2$	743	124
	663	125
$HgCl_2$	710	126
HgI_2	683	127

Givan and Loewenschuss[131] on the basis of the Raman spectra. They emphasized that the linear geometry was never outside the acceptable range of parameters for these molecules. At the same time, the electron diffraction results must be considered in conjunction with the spectroscopic data, especially with the bending vibrational frequencies. Supposing linearity for the molecules, their v_2 bending frequencies were estimated from electron diffraction and are shown together with the spectroscopic ones in Table 9-17. The quoted errors reflect the uncertainties of the observed shrinkages, and not the error of the harmonic approximation used. The δ_g/v_2 relationship is most fortunate to use for estimating bending frequencies from electron diffraction shrinkages for high-temperature data and for low v_2 values, where it becomes especially steep. Among the estimates for zinc halides, the worst agreement with experiment is seen for $ZnCl_2$, and there is gradual improvement towards ZnI_2. The spectroscopic values, of course, also have uncertainties (eg, from possible matrix effects),

Table 9-17. SOME GEOMETRICAL AND VIBRATIONAL PARAMETERS OF GROUP IIB DIHALIDES DETERMINED BY ELECTRON DIFFRACTION AND BY SPECTROSCOPY (SP)

Molecule	r_g(M—X) (Å)	r_g(X...X) (Å)	Ref.	v_2(cm^{-1}) ED	v_2(cm^{-1}) SP	Ref. (SP)
$ZnCl_2$	2.062(4)	4.115(10)	123	123(15)	103.0	128
$ZnBr_2$	2.204(5)	4.370(10)	123	86(8)	71.0	128
ZnI_2	2.401(5)	4.753(10)	123	67(4)	61.0	128
$CdBr_2$	2.394(2)	4.743(4)	125	66(7)	62.0	129
$HgCl_2$	2.252(5)	4.480(44)	126		107.0	130
HgI_2	2.568(4)	a	127		63.0	130

[a] Not given in reference 127.

Table 9-18. RESULTS OF THE COMBINED ELECTRON
DIFFRACTION VIBRATIONAL SPECTROSCOPIC
ANALYSIS OF HgI_2

$r_e^h(Hg-I)$, Å	2.554(3)
f_r, mdyn Å^{-1}	1.838(5)
f_{rr}, mdyn Å^{-1}	$-0.017(1)$
f_α, mdyn Å^{-1}	0.0429(1)
v_1, cm^{-1}	156
v_2, cm^{-1}	51
v_3, cm^{-1}	237

and typically an agreement between our estimates and the experimental v_2 values of about 10–20 cm^{-1} is considered satisfactory. It is noteworthy that all electron diffraction v_2 values for zinc dihalides are higher than the spectroscopically measured ones, making the assumption of linearity secure: according to the estimation, even somewhat higher frequency bending vibrations would lead to the observed deviation from linearity in the effective structures directly determined from electron diffraction data at the experimental nozzle temperature. Conversely, had the electron diffraction estimates approached the spectroscopic v_2 values from the lower side, the difference might have been taken to indicate a geometry between the effective bent and the linear structures.

Two papers[124,125] were published on the geometry of $CdBr_2$, with markedly different results. Petrov and co-workers[125] concluded that the previous study[124] probably suffered from considerable experimental error.

Table 9-18 gives results of a combined electron diffraction/vibrational spectroscopic analysis[127] of HgI_2. All three frequencies were available from gas-phase spectra. The experimental conditions and the r_g parameters, except for the Hg—I bond, were not communicated.

First-Row Transition Metal Dihalides. Most of the molecules belonging to this group were investigated in the Budapest laboratory; with MnF_2,[133] NiF_2,[134] and $NiCl_2$[135] as exceptions. This series of molecules illustrates the importance of direct vapor control for the electron diffraction experiments. In the earlier studies ($MnCl_2$,[136] $CoCl_2$,[137] and $NiBr_2$[138]), there was no simultaneous vapor control, and monomers were assumed to be present alone on the basis of independent mass spectrometric evidence.[139] These early works are being repeated now, employing our improved experimental techniques.

Preliminary results[140] show that the vapors of $MnCl_2$ do not contain any detectable amount of dimers, while there is about 4% of dimers present for $CoCl_2$. The consequences of ignoring a dimer content were discussed in the Introduction. The experimental temperatures, monomer amounts of the vapor, and references to the electron diffraction studies are collected in Table 9-19.

Monomers. The most interesting question about the structure of transition metal dihalides is again their shape. Although most of the literature data agree on the linearity of these molecules, there are also contradictory reports.

Table 9-19. SOME CHARACTERISTICS OF THE ELECTRON DIFFRACTION EXPERIMENTS OF FIRST-ROW TRANSITION METAL DIHALIDES[a]

Molecule	Temperature (K)	Monomer amount (%)	Ref.
MnF_2	1400	[b]	133
NiF_2	1473	[b]	134
VCl_2	1330	[c]	15
$CrCl_2$	1169	[c]	15
$MnCl_2$	1073	100	140
$FeCl_2$	898	94.3(8)	140
$CoCl_2$	1010	96.1(8)	140
$NiCl_2$	1073	[b]	135
$MnBr_2$	881	94.7(8)	142
$FeBr_2$	898	~90	143
$CoBr_2$	908	94.2(7)	144
$NiBr_2$	943	[b]	138

[a] The presence of monomers and dimers was considered in the analysis unless otherwise indicated.
[b] Only monomers considered.
[c] Recent reinvestigation indicates, that associates higher than dimeric may also be present in substancial amounts.

Molecular beam deflection measurements and vibrational spectroscopy were the two most important tools for the determination of shape and symmetry. Molecular beam deflection studies[132] indicated linear geometry for the first-row transition metal fluorides from manganese on. According to spectroscopic results, TiF_2 is bent while the other difluorides are linear or nearly linear.[145] As has been mentioned before, the isotope shift measurements, which can in principle be used for angle determination, are not sensitive enough for *nearly linear* geometries[3a, 19] within about 20°. All dichlorides were found to be linear in a series of spectroscopic studies,[145] while one investigation[146] reported bent geometries for the iron, cobalt, and nickel dichlorides.

The electron diffraction study[15] of VCl_2 and $CrCl_2$ involved parallel quadrupole mass spectrometric control of the vapor composition. The presence of a considerable amount of other species than monomers was indicated by the mass spectra. The structure analysis was carried out under the assumption of having monomers and dimers present. In addition, for vanadium dichloride, the possible presence of VCl_4 was also indicated and checked by the analysis. In retrospect, however, it was a mistake to ignore the possible presence of higher than dimeric associates in our electron diffraction analysis of VCl_2 and $CrCl_2$.[15] The assumption of monomers and dimers only, resulted in the conclusion of bent structures for both vanadium dichloride and chromium dichloride. This conclusion can very well be erroneous. As the analyses are currently being repeated, their linear configuration cannot be excluded at all, provided a more complex vapor composition is supposed than was presumed originally. Figure

Table 9-20. CALCULATED MEAN
AMPLITUDES OF VIBRATION
$l(Cl \ldots Cl)$ FOR LINEAR AND BENT
MODELS OF VCl_2 AND $CrCl_2$

	$l(Cl \ldots Cl)$, Å	
Shape	I[a]	II[b]
VCl_2		
Linear	0.120	0.146
Bent	0.241	0.273
$CrCl_2$		
Linear	0.121	0.140
Bent	0.280	0.307

[a] Calculated by the conventional method.[63]
[b] Calculated by a curvilinear approximation.

9-13 compares radial distributions for a linear model of $MnCl_2$ and highly bent models of VCl_2 and $CrCl_2$ with considerable dimer amounts included for the latter two. Curve b for $CrCl_2$ indicates that assuming linearity for the monomer, higher associates must also be present. In view of the above mentioned uncertainties, it is not recommended to use the parameters reported by us for VCl_2 and $CrCl_2$,[15] although the V—Cl and Cr—Cl distances are perhaps safe to consider of being 2.17 ± 0.01 Å and 2.20 ± 0.02 Å long. Calculated vibrational amplitudes are given for both linear and bent models of the VCl_2 and $CrCl_2$ molecules in Table 9-20.

All the other transition metal dihalides were found to be linear by electron diffraction. The shape of the iron, cobalt, and nickel dichlorides deserves comment, since a recent matrix isolation infrared spectroscopic study[146] concluded that these molecules are bent with bond angles of 161, 157, and 161°, respectively. It was discussed before that electron diffraction alone is not capable of solving the question of linearity versus nonlinearity for molecules of these types. However, this can be done in conjunction with spectroscopic information. Figures 9-14 and 9-15 show the variation of two amplitude parameters with the ν_2 bending frequency. The two amplitude parameters are the mean parallel amplitude for the Cl...Cl distance and another parameter characteristic for the perpendicular vibrations of the metal-chlorine bond, $K(M—Cl)$. The general expression for this term is:

$$K_{ij} = \frac{\langle (x_{ij})^2 \rangle + \langle (y_{ij})^2 \rangle}{2r_{ij}} \tag{9-2}$$

where x_{ij} and y_{ij} are the perpendicular displacements and r_{ij} is the internuclear distance. Both $l(Cl...Cl)$ and $K(M—Cl)$ strongly depend on the bond angle and

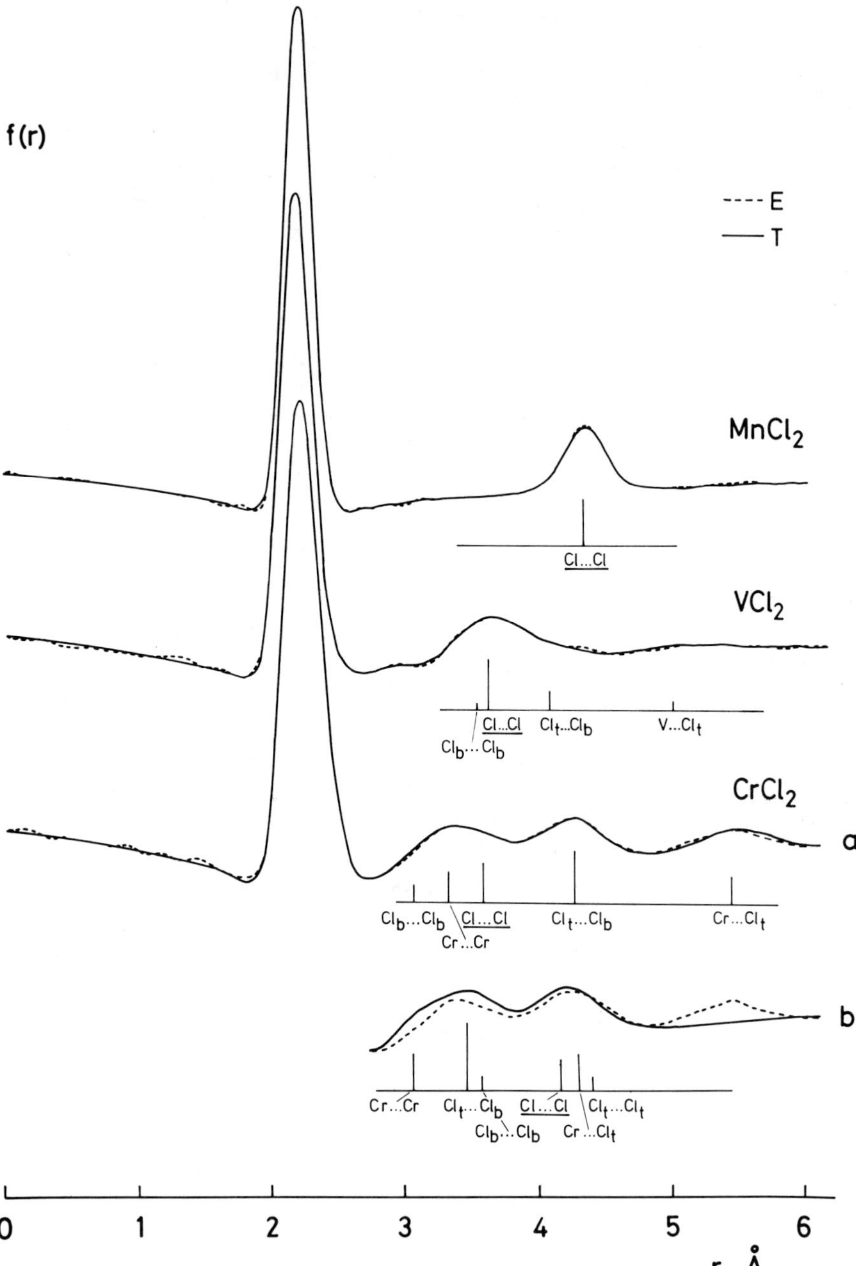

Figure 9-13. Comparison of radial distributions for linear model of $MnCl_2$, bent model of VCl_2, and linear and bent models of $CrCl_2$. The latter three include considerable amounts of dimers. (a) $CrCl_2$ bent model plus dimers; (b) $CrCl_2$ linear model plus dimers. On this curve the distinct peak at about 5.5 Å can be taken as indication of higher associates, provided that the monomeric molecule is linear.

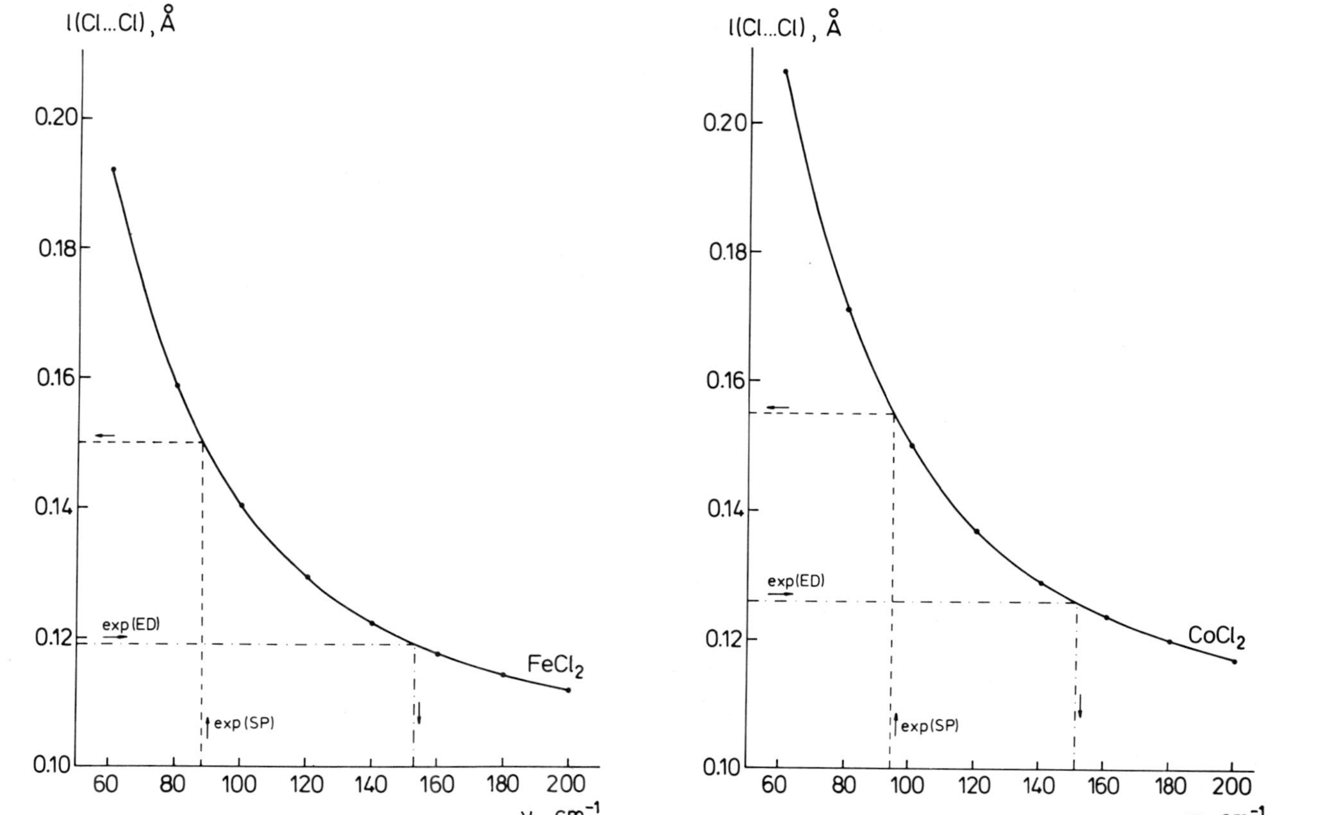

Figure 9-14. Variation of $l(\mathrm{Cl}\ldots\mathrm{Cl})$ with ν_2 bending vibrational frequency for bent $FeCl_2$ (temperature 898 K, bond angle 157.5°, ν_1 350 cm⁻¹, ν_3 494 cm⁻¹) and $CoCl_2$ (temperature 1010 K, bond angle 156.3°, ν_1 359 cm⁻¹, ν_3 493.4 cm⁻¹).

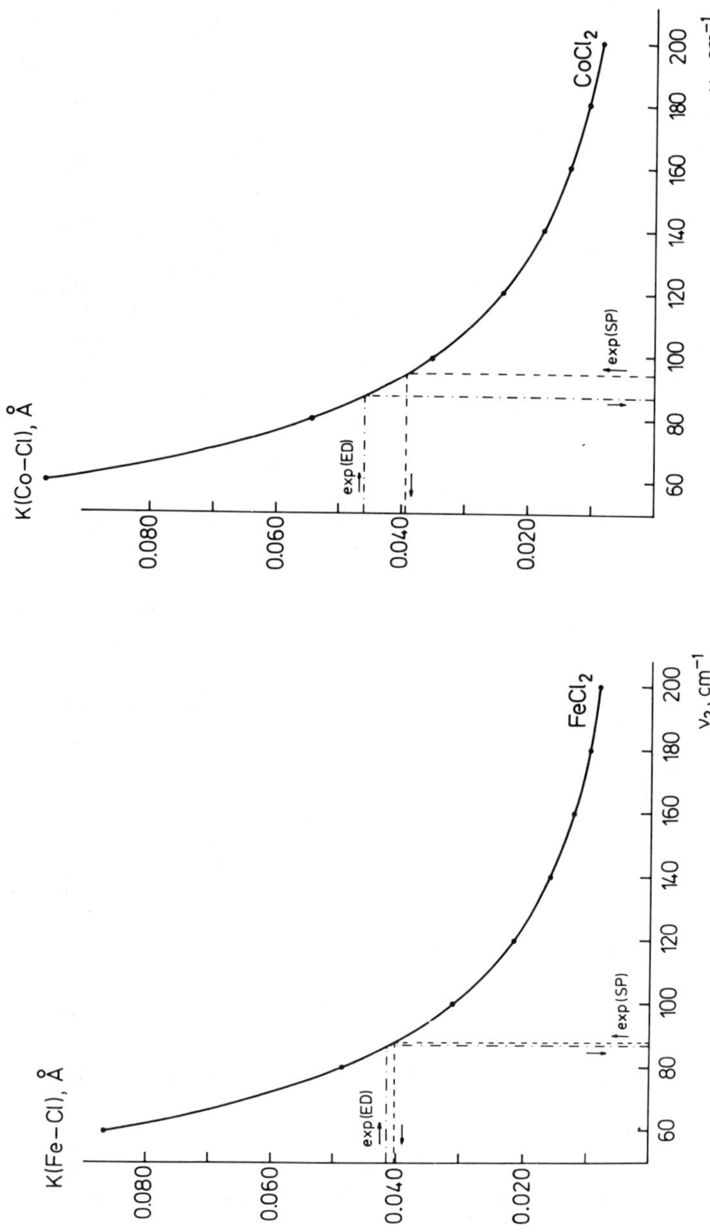

Figure 9-15. Variation of $K(M—Cl)$ with v_2 bending vibrational frequency for linear $FeCl_2$ at 898 K and $CoCl_2$ at 1010 K.

Table 9-21. SOME GEOMETRICAL AND VIBRATIONAL PARAMETERS OF
LINEAR FIRST-ROW TRANSITION METAL DIHALIDES[a]

Molecule	r_g(M—X) (Å)	r_g(X...X) (Å)	ν_2(ED) (cm^{-1})	ν_2(SP) (cm^{-1})	Ref. (SP)
MnF_2	1.813(5)	3.615(50)		124.8	145
NiF_2	1.729(4)	3.400(34)	173[b]	139.7	145
$MnCl_2$	2.202(4)	4.324(10)	91	83	148
$FeCl_2$	2.151(7)	4.219(11)	87(8)	88	148
$CoCl_2$	2.113(5)	4.133(10)	87(15)	94.5	148
$NiCl_2$	2.056(4)	4.018(76)	90[b]	85	148
$MnBr_2$	2.344(6)	4.602(11)	71(5)		
$FeBr_2$	2.310(5)	4.464(9)	53(4)		
$CoBr_2$	2.241(5)	4.387(9)	68(8)		
$NiBr_2$	2.212(5)	4.326(13)	69(10)	69	148

[a] For references of ED data see Table 9-19.
[b] Estimated by us.

the ν_2 frequency. The agreement for the linear cases and the marked disagreement for the bent cases shows that for $FeCl_2$[141] as well as for $CoCl_2$[140] the electron diffraction results, in conjunction with the spectroscopic results, are not compatible with bent geometry.

The ν_2 bending frequencies were estimated for all linear molecules from electron diffraction. They are given together with other structural parameters and available spectroscopic bending frequencies in Table 9-21.

The bond length variations of firt-row transition metal dihalides have been correlated with electronic structure.[151]

Dimers. The presence of M_2X_4 dimers has been indicated in many experiments. Their geometry, however, could not be determined accurately. They have a four-membered, halogen-bridged structure (Figure 9-16) similar to the M_2X_6 dimers. Their apparent puckered ring geometry under electron diffraction experimental conditions probably corresponds to D_{2h} equilibrium symmetry with a planar ring. The D_{2h}/C_{2v} symmetry lowering is the result of large-amplitude deformation vibrations.

The dimers have two different M—X distances, bridging and terminal. It was not possible to distinguish between the monomer and the dimer terminal bond distances; they are probably very similar. The dimer bridging bond distance is, however, considerably longer than the terminal and the monomer bond distances; the difference is about 0.15–0.20 Å.

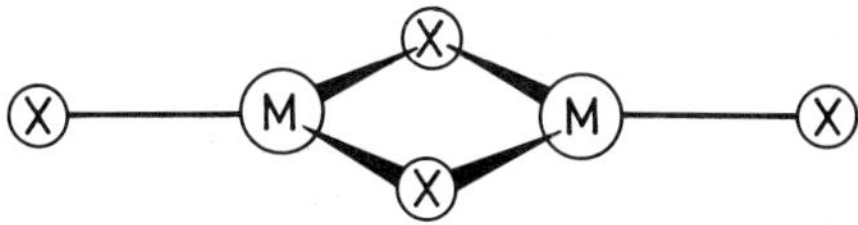

Figure 9-16. Geometrical arrangement of dimeric metal dihalides M_2X_4.

Trihalides

Lanthanide (Rare Earth) Trihalides. The two main questions about the structure of these rather extensively studied molecules involve molecular shape and bond length variation.

While the monomeric Group IIIA trihalides are all planar, experimental and theoretical evidence points to the nonplanarity of the rare earth trihalides. Molecular beam deflection measurements[152] as well as infrared spectroscopic studies [153,154] have indicated pyramidal geometries for these molecules. Calculations also predict C_{3v} symmetry for the trihalides of lanthanides.[155] The molecules studied so far by electron diffraction alone with experimental temperatures, bond angles, and references are listed in Table 9-22. Based on other experimental evidence, the monomers were assumed to be the predominant species in these studies. Possible dimer content was checked but also ruled out for $LuCl_3$.

As already discussed, the apparent bond angle may be smaller than 120° even for a planar MX_3 molecule from electron diffraction. If the molecule appears to be pyramidal even after applying the shrinkage corrections, it is an indication that the equilibrium symmetry is C_{3v}. This conclusion may still not be entirely unambiguous,[163] since the anharmonicity of the bending vibration seldom is taken into account in such corrections. The barrier separating two equivalent minima of the potential energy curve at the pyramidal configuration may be very low for very low bending frequencies, resulting in extremely large-amplitude vibrations. Vibronic interactions with low-lying electronic states may cause further distortion.[163]

A simple polarization model was used by Drake and Rosenblatt[51] to account for the shapes of these molecules as well as for the shapes of alkaline earth

Table 9-22. EXPERIMENTAL TEMPERATURES AND
BOND ANGLES IN RARE EARTH TRIHALIDES

Molecule	T(K)	X—M—X (deg)	Ref.
ScF_3	1600(100)	110.0(2.5)	156
$LaCl_3$	1250(50)	112.5(2.5)	157
$PrCl_3$	1250(50)	110.8(3.0)	158
$GdCl_3$	1325(50)	113.0(1.5)	159
$TbCl_3$	1230(50)	110.5(1.7)	160
$HoCl_3$	1200(50)	111.2(2.0)	158
$LuCl_3$	1250(50)	111.5(2.0)	156
$LaBr_3$	1300(100)	115.5(2.0)	161
$GdBr_3$	1150(100)	113.7(1.8)	162
$LuBr_3$	1100(100)	114.5(1.8)	162
PrI_3	1050(30)	112.7(2.2)	163
NdI_3	1070(30)	111.8(1.8)	163
GdI_3	1060(30)	108.0(2.0)	163
LuI_3	1015(30)	114.5(2.1)	163

dihalides. According to this model, the large, more polarizable central atoms are the prime origin for pyramidal geometry in contrast to Group IIIA trihalides, which are planar with their smaller central atoms.

The bottom curve in Figure 9-17 shows the variation of the M^{3+} ionic radii[84] in going from La through the whole lanthanide series. This is the well-known "lanthanide contraction," which is the consequence of the imperfect shielding of $4f$ electrons from the nucleus. This contraction is similar to the one observed for the first-row transition metals or for each row of the periodic table, for that

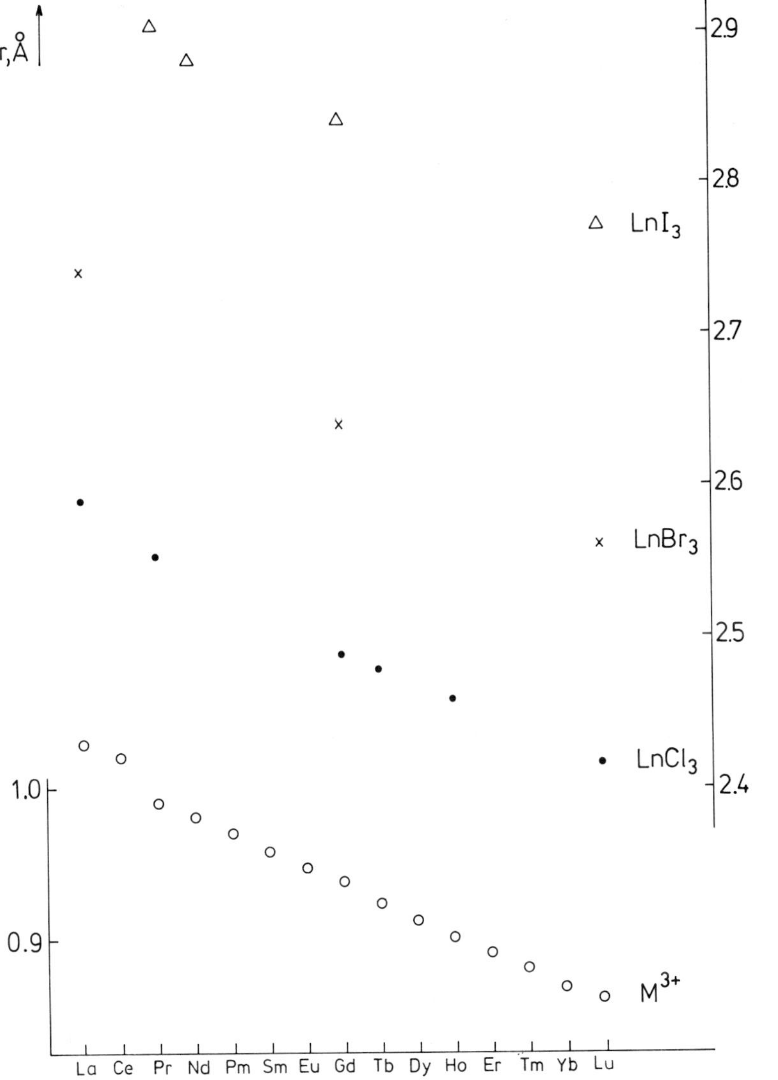

Figure 9-17. Variation of M^{3+} ionic radii and bond lengths of lanthanide trihalides.

matter. One interesting consequence of the lanthanide contraction has bearing on the structure of molecules of some other metals as well.[84] There is a certain expansion of atomic and ionic radii when going from the first to the second transition metal series. The same would be expected in going from the second to the third series. However, the lanthanide elements stand at the beginning of the third series, and the amount of total lanthanide contraction cancels the otherwise expected increase. This is why the second and third members in each group of the transition elements in the periodic table have very similar sizes, compounds, and chemical properties.[84]

Figure 9-17 displays also the bond lengths of lanthanide trihalides determined so far. Although their number is still small, it is seen that their variation closely follows that of the M^{3+} ionic radii. The difference of the metal–halogen distances and the ionic radii with their standard deviations are 1.557(4), 1.703(4), and 1.906(6) Å for the chlorides, bromides, and iodides, respectively. Using these differences, the yet unknown bond lengths of other lanthanide trihalides can be estimated. These are given in Table 9-23 together with the experimental data. Similar predictions based on fewer data appeared in reference 164.

Other Transition Metal Trihalides.

Iron Trichloride. The investigation of iron trichloride was referred to in the Introduction.

The geometry of the dimeric iron trichloride was first investigated by Zasorin and co-workers.[165] They proposed a halogen-bridged configuration (cf. Figure

Table 9-23. BOND LENGTHS OF LANTHANIDE TRIHALIDES (Å): EXPERIMENTAL (r_g, in italics) AND ESTIMATED

| | Halides | | |
Lanthanides	Chlorides	Bromides	Iodides
La	*2.590(6)*	*2.741(5)*	2.938(10)
Ce	2.577(10)	2.723(10)	2.926(10)
Pr	*2.553(6)*	2.693(10)	*2.904(6)*
Nd	2.540(10)	2.686(10)	*2.881(5)*
Pm	2.527(10)	2.673(10)	2.876(10)
Sm	2.515(10)	2.661(10)	2.864(10)
Eu	2.504(10)	2.650(10)	2.853(10)
Gd	*2.489(6)*	*2.640(10)*	*2.841(5)*
Tb	*2.478(5)*	2.626(10)	2.829(10)
Dy	2.469(10)	2.615(10)	2.818(10)
Ho	*2.459(6)*	2.604(10)	2.807(10)
Er	2.447(10)	2.593(10)	2.796(10)
Tm	2.437(10)	2.583(10)	2.786(10)
Yb	2.425(10)	2.571(10)	2.774(10)
Lu	*2.417(6)*	*2.506(1)*	*2.771(6)*

9-8). The geometrical parameters from an electron diffraction reinvestigation[11] are as follows:

$$r_g(\text{Fe—Cl})_t = 2.129(4) \text{ Å}$$
$$r_g(\text{Fe—Cl})_b \ 2.329(5) = \text{Å}$$
$$\angle_\alpha \text{Cl}_t\text{—Fe—Cl}_t = 124.3(7)°$$
$$\angle_\alpha \text{Cl}_b\text{—Fe—Cl}_b = 90.7(4)°$$

Similarly to the Group IIIA M_2X_6 molecules, Fe_2Cl_6 appears to be of C_{2v} rather than D_{2h} symmetry, due to the puckering of the four-membered ring. The puckering angle (see Figure 9-9) is 16.7(1.0)°. The bridging bond is 0.2 Å longer than the terminal one, again in accordance with the data of Table 9-9. The bond angles are also similar.

Rhenium Tribromide. According to an electron diffraction study,[166] the $ReBr_3$ molecule exists in trimeric form in the vapor. This is consistent with mass spectrometric studies of rhenium halide vapors and X-ray diffraction studies of crystalline rhenium halides.[25] The molecular configuration and the main geometrical parameters are shown in Figure 9-18. The short Re—Re distance, compared to the covalent radius of 1.283 Å,[167] indicates direct chemical bond between the rhenium atoms.

Tetrahalides

Tetrahedral Tetrahalides

Titanium Subgroup. All tetrahalides of the titanium subgroup are relatively rigid, tetrahedral molecules. This is one of the best studied groups; 10 of the 12 tetrahalides have been investigated by the Ivanovo electron diffraction group. A short survey and references to earlier works can be found in reference 168. Some of the structural parameters and experimental temperatures are given in Table 9-24. Two VX_4 molecules are also included for comparison.

The ionic radii of Zr and Hf are practically identical,[84] and their chemical behavior is similar as a result of the lanthanide contraction. The bond lengths of

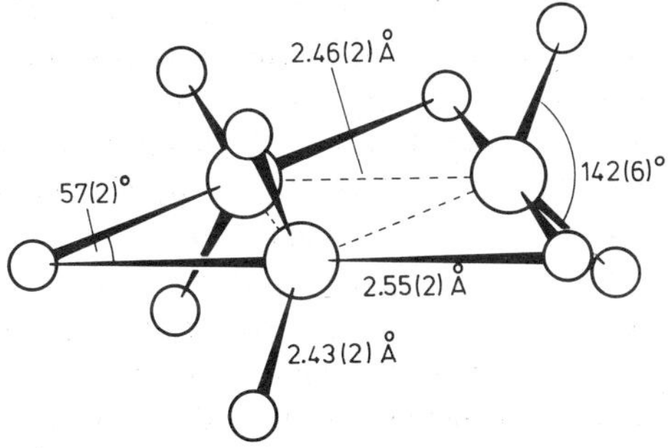

Figure 9-18. Molecular configuration and geometrical parameters of trimeric rhenium tribromide Re_3Br_9.

Table 9-24. GEOMETRICAL PARAMETERS OF TETRAHEDRAL TRANSITION METAL TETRAHALIDES

Molecule	T (K)	r_g(M—X) (Å)	r_g(X...X) (Å)	l(M—X) (Å)	l(X...X) (Å)	Ref.
Fluorides						
TiF_4	475(15)	1.754(3)	2.859(11)	0.044(2)	0.129(7)	169
	689(20)	1.756(3)	2.856(11)	0.049(2)	0.141(7)	170
ZrF_4	975(50)	1.902(4)	3.100(22)	0.059[a]	0.178(13)	171
HfF_4	1025(50)	1.909(5)	3.140(34)	0.059[a]	0.187(23)	171
Chlorides						
$TiCl_4$	293	2.170(2)	3.539(4)	0.049(3)	0.116(4)	172
VCl_4	293	2.138(2)	3.485(5)	0.051(2)	0.135(4)	172
$ZrCl_4$		2.32(2)[b]				173
$HfCl_4$	470(15)	2.316(5)	3.759(13)	0.058(5)	0.164(11)	174
Bromides						
$TiBr_4$	300(5)	2.339(5)	3.812(10)	0.043(5)	0.134(6)	175
VBr_4	390(5)	2.276(4)[c]		0.079(4)	0.214(11)	176
$ZrBr_4$	473(15)	2.465(4)	4.012(13)	0.062(3)	0.192(12)	175
$HfBr_4$	473(15)	2.450(4)	3.981(10)	0.054(3)	0.177(10)	175
Iodides						
TiI_4	403(15)	2.546(4)	4.138(8)	0.052(6)	0.169(8)	175
ZrI_4	493(20)	2.660(10)	4.351(24)	0.060(6)	0.215(20)	175
HfI_4	543(30)	2.662(8)	4.346(13)	0.072(3)	0.212(14)	175

[a] Calculated from the frequencies, see reference 171.
[b] Unidentified, possibly r_a.
[c] r_α.

the corresponding Zr and Hf tetrahalides are the same within experimental error, except for the bromides, where further contraction is observed.

Now let us compare the titanium and vanadium tetrahalides. According to the generally observed contraction in a period, the V—X bonds are shorter than the corresponding Ti—X bonds. Vanadium tetrahalides will be discussed later.

The electron diffraction experiment with TiF_4 was carried out at two different temperatures to get more reliable force constants and vibrational frequencies.[169,170] The vibrational frequencies compare well with the gas-phase Raman spectroscopic data. The vibrational frequencies for ZrF_4 and HfF_4 have been estimated from electron diffraction.[177] The so-called harmonic equilibrium internuclear distances have also been calculated.

The combination of electron diffraction and spectroscopic data made possible the determination of Coriolis coupling constants for TiF_4[169] and for all tetrabromides[175] and tetraiodides.[175] The determination of Coriolis coupling constants for molecules with heavy atoms is difficult and uncertain with the usual spectroscopic techniques. In such cases, according to Girichev and co-workers,[175] the accuracy of the electron diffraction results is comparable to that of the spectroscopic data, if the latter are obtainable at all. Thus the application of electron diffraction extends in this direction.

Vanadium Tetrahalides. From among the Group VB elements only the tetrahalides of vanadium have been studied so far: VCl_4 by Morino and Uehara[172] and by Spiridonov and Romanov,[178] and VBr_4 by Spiridonov and Romanov[178] and by Ivashkevich and colleagues.[176] Pure VCl_4 was evaporated in one experiment,[172] while the other three experiments utilized the following disproportionation reaction for the preparation of vanadium tetrahalides:

$$2\ VX_3(\text{solid}) \rightarrow VX_2(\text{solid}) + VX_4(\text{gas})$$

Some structural parameters are presented in Table 9-24. The metal–halogen bond lengths from the two independent studies agree within the large experimental errors of reference 178 [V—Cl, 2.14(2) Å; V—Br, 2.30(2) Å]

The vanadium halides are interesting objects for structure determination, since they should be subject to Jahn–Teller distortions.[179] Both VCl_4 and VBr_4 are believed to be in a doubly degenerate ground electronic state in a tetrahedral configuration. The electronic and nuclear motions in the molecule are essentially mixed in such cases. The doubly degenerate deformation vibration, $v_2(E)$, should be strongly involved in vibronic interactions with the degenerate ground electronic state, and they would cause a distortion of the molecules from the regular tetrahedral configuration. The electron diffraction results, however, do not show any deviation from the regular tetrahedral geometry within experimental error. Different distorted models (D_{2d}, D_2, C_{2v}) have been tested in the structure analysis of VBr_4, but they all gave poorer agreement with the experiment than the regular tetrahedral model. Thus, there is no indication of a Jahn–Teller distortion. This absence of evidence, however, refers only to the so called *static Jahn–Teller effect.* As was shown by Bersuker,[180] the static Jahn–Teller effect can be observed only in the presence of external influence. Jahn–Teller distortions displayed by systems without any external influence are *dynamic.* There may be many minimum-energy distorted structures in such systems. Whether an experiment will or will not detect such a dynamic Jahn–Teller effect depends on the relationship between the time scale of the physical measurement used for the investigation and the mean lifetime of the distorted configurations.

Among the electron diffraction results, a marked increase in the mean amplitudes of vibration compared to the calculated amplitudes is an indication of the possible presence of several distorted configurations. If this is the case, it is still only a thermal average that appears from the analysis.

According to Bersuker,[180] the potential energy surface in such cases has the form of a "Mexican hat" illustrated in Figure 9-19. Here V is the potential energy (adiabatic potential), Q_{2a} and Q_{2b} are the two degenerate normal coordinates corresponding to the v_2 deformation vibrations, E_{JT} is the Jahn–Teller stabilization energy, ρ_0 is the radius of circumference comprising the set of minimum points, and $|a_E|$ is the first-order vibronic interaction constant. The physical meaning of the Jahn–Teller stabilization energy E_{JT}, is the energy

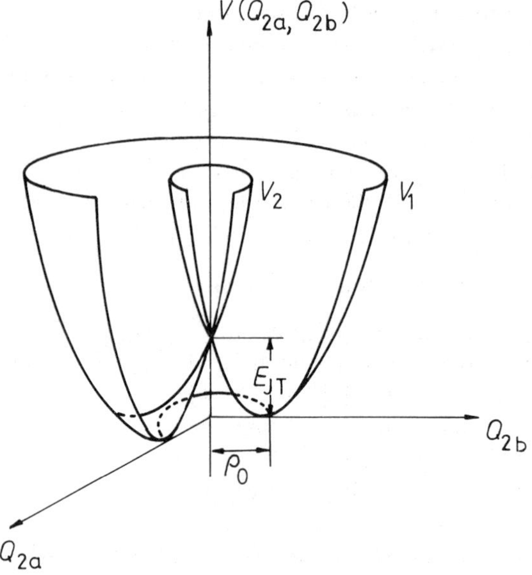

Figure 9-19. "Mexican hat" potential energy surface for the two degenerate normal coordinates of the v_2 deformation vibration of an MX_4 tetrahedral molecule. Reproduced with permission from reference 176.

decrease due to vibronic interactions, or in other words, the energy decrease due to the dynamic Jahn–Teller effect. These parameters, determined from the electron diffraction data of VBr_4 by Ivashkevich and co-workers[176] and of Morino and Uehara,[172] are as follows:

	VBr_4	VCl_4
$|a_E|$	1.36 ± 0.07	2.1 ± 0.7
E_{JT}	62 ± 5 cm^{-1}	73 ± 43 cm^{-1}
ρ_0	0.183 ± 0.007 Å	0.16 ± 0.06 Å

A final note concerns the reliability of these results. The value of the crucial v_2 frequency had to be assumed in both studies.[172,176] At the time of the VCl_4 work,[172] the gas-phase spectra of VCl_4 and $TiCl_4$ were already available, except that the v_2 frequency of the former was missing. From the similarity of the other three frequencies, however, Morino and Uehara concluded that the v_2 frequency of the two compounds should also be similar. Later Raman spectra of VCl_4 solutions[181] confirmed this; all four frequencies of VCl_4 are very close to those of $TiCl_4$ (the v_2 was found to be 119 cm^{-1} compared to the value of 118 cm^{-1} used by Morino and Uehara). Since there were no experimental frequencies available for VBr_4, Ivashkevich and co-workers[176] used the $TiBr_4$ gas-phase frequencies[182] in their calculation.

Nontetrahedral Tetrahalides

Lanthanide and Actinide Tetrahalides. While the configuration of simple main group molecules can usually be predicted by simple theories, such as the VSEPR model, the structures of transition metal halides often pose puzzles. Several lanthanide and actinide tetrahalides have been investigated so far, and there is indication of some symmetry lowering from a regular tetrahedral geometry. Ezhov and co-workers[183] concluded that UX_4 and ThX_4 tetrahalides (X = F, Cl, Br) have lower, possibly C_{2v} symmetry. The $ThCl_4$[184] and $ThBr_4$[185] molecules were also found to have a C_{2v} symmetry structure in the crystal according to early X-ray diffraction studies. In an earlier electron diffraction work[186] a distorted tetrahedral, C_{2v} symmetry geometry was suggested for UBr_4 with U—Br = 2.64(1) Å.

Later UF_4 was reinvestigated by Girichev and colleagues[187] to establish the shape of the molecule. Mass spectrometric[188] measurements suggested that UF_4 probably has a lower than T_d symmetry. Photoelectronic spectra[189] as well as matrix isolation infrared spectra[190] were interpreted in terms of regular tetrahedral symmetry, although small deviations could not be ruled out. Several lower symmetry models (Figure 9-20) have also been tested[187] and the C_{4v} and D_{4h} symmetry arrangements could be ruled out. The C_{3v} and D_{2d} geometries gave about equally good agreement with the experiment and also somewhat better than the tetrahedral configuration. The latter was ruled out on the ground

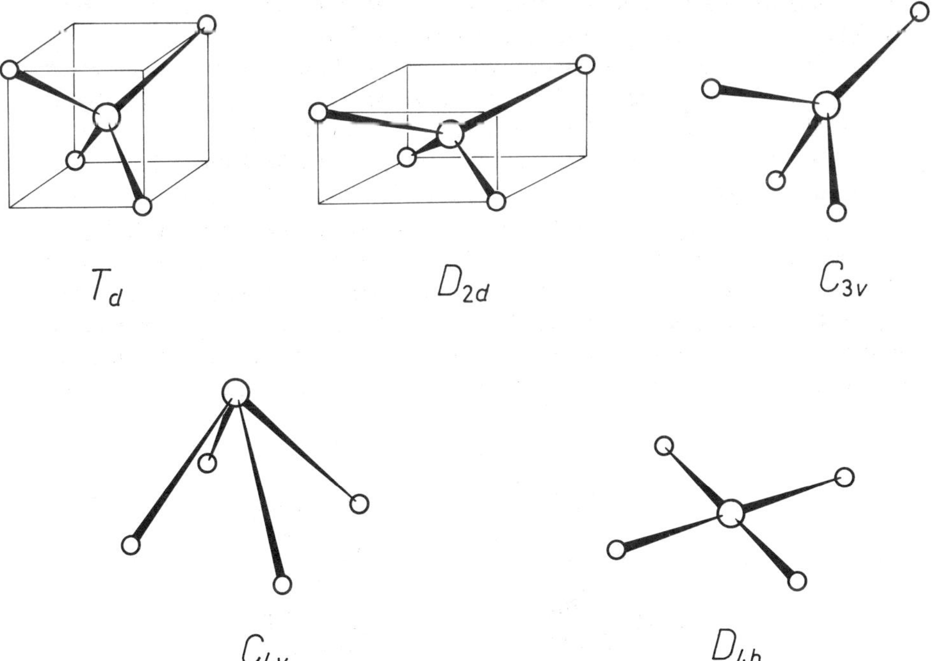

Figure 9-20. Models with different symmetries tested in the electron diffraction analysis of UF_4.[187]

Table 9-25. Parameters of Some MX_4 Halides

Molecule	T(K)	r(M—X) (Å)	Ref.
ThF_4	1530(50)	2.14(2)[a]	183
$ThCl_4$	930(50)	2.58(2)[a]	183
$ThBr_4$	910(50)	2.72(2)[a]	183
UF_4	1340(50)	2.06(2)[a]	183
	1300(50)	2.069(4)[b]	187
UCl_4	840(50)	2.53(2)[a]	183
UBr_4	800(50)	2.66(2)[a]	183
CeF_4	1180(50)	2.035(5)[b]	191
$MoCl_4$		2.23(2)[c]	192
$MoBr_4$		2.39(2)[c]	192
WCl_4	418	2.248(6)[d]	193
		2.202(6)[e]	193
		2.295(6)[e]	193

[a] r_a; uncertainties are twice the original data.
[b] r_g
[c] Unidentified, possibly r_a.
[d] Mean W—Cl distance.
[e] See Figure 9-21 for definition.

of an unrealistically large amplitude of vibration for the F...F nonbonded distances needed to approximate the experimental data.

The CeF_4 molecule has also been investigated,[191] and the conclusions about its geometry were essentially the same as those for UF_4. The molecule has probably C_{3v} or D_{2d} (or C_{2v}?) symmetry rather than T_d. The geometrical parameters for all the foregoing molecules are collected in Table 9-25.

Incidentally, the atomic scattering factors for U in UF_4 were extrapolated from those of other atoms and the factors of La were used for Ce in CeF_4.

Molybdenum and Tungsten Tetrahalides.   A recent electron diffraction study of WCl_4 reported[193] a C_{2v} configuration; the agreement with experimental data was only slightly poorer for the T_d model. The geometrical parameters are shown in Figure 9-21. The structure is compared with that of SF_4.[194] This may be of interest because both central atoms belong to Group VI. The SF_4 molecule

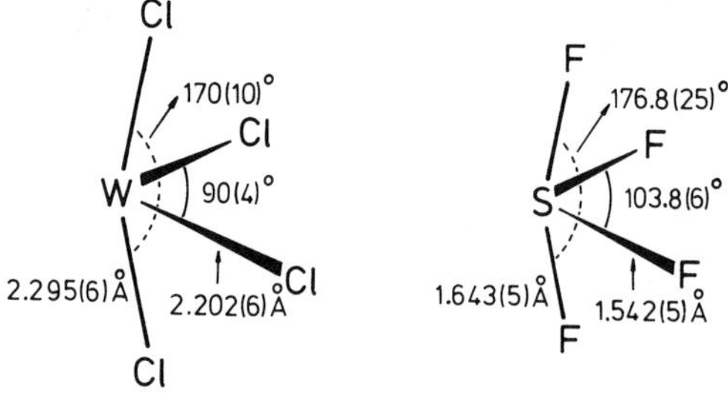

Figure 9-21. Supposed[193] geometry of WCl_4; comparison with the structure of SF_4.[194]

is a classic example of an EAX_4 type structure, where E is a lone electron pair and X is a ligand. The VSEPR model[59] correctly predicts a trigonal bipyramidal electron pair configuration for SF_4 with the lone pair in equatorial position, and, accordingly a C_{2v} bond configuration. Whether the lone electron pair of a transition metal is as active stereochemically as the lone pairs of main group elements remains a question. In the light of such a VSEPR-type WCl_4 geometry, the shapes of the $O{=}WCl_4$, $S{=}WCl_4$, and $Se{=}WCl_4$ molecules would also be expected to be trigonal bipyramidal. However this is not the case; rather, they have a tetragonal pyramidal configuration with C_{4v} symmetry,[195,196] as shown in **3**. It is true, however, that the trigonal bipyramidal configuration is only slightly favored over the tetragonal pyramidal structure for AX_5 molecules by

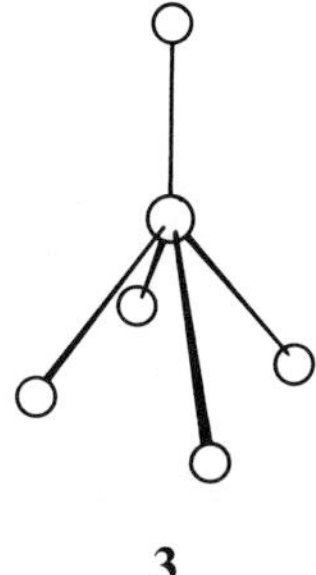

3

the VSEPR model.[59] Finally, structural studies on $MoCl_4$ and $MoBr_4$ are mentioned here for their analogy with WCl_4, although the conclusions of the original work[192] were for T_d symmetry structures.

Pentahalides

The $+5$ oxidation state is most natural for Group V of the periodic table, and its stability increases down the group.[84] Vanadium has only one known pentahalide, VF_5, while all pentahalides have been obtained for niobium and tantalum.

Group VI transition metals also have stable pentahalides, some of which have been investigated.

Monomers

Vanadium pentafluoride. The VF_5 molecule has been repeatedly investigated. A trigonal bipyramidal configuration was established in the first investigation[197] with an average V—F bond distance of 1.71(1) Å. The axial and equatorial bonds were found not to differ more than 0.1 Å.

A new study[198] of VF_5 aimed at the accurate determination of the bond lengths and of any deviation from D_{3h} symmetry. Molecular beam deflection experiments[199] suggested such distortion. However, the electron diffraction results were consistent with D_{3h} symmetry; deviations should not exceed a few degrees in any bond angle. The geometrical parameters are given in Table 9-26.

Table 9-26. PARAMETERS OF SOME PENTAHALIDES

	Molecules			
Parameter	VF_5 (Ref. 198)	$NbCl_5$ (Ref. 204)	$TaCl_5$ (Ref. 204)	$TaBr_5$ (Ref. 205)
$r_\alpha(M{-}X)_{eq}$, Å	1.704(5)	2.241(4)	2.227(3)	2.412(4)
$r_\alpha(M{-}X)_{ax}$, Å	1.732(7)	2.338(6)	2.369(4)	2.473(8)
$r_\alpha(X_{ax}{\ldots}X_{eq})$, Å	2.429(3)	3.238(4)	3.251(4)	3.454(6)
$r_\alpha(X_{eq}{\ldots}X_{eq})$, Å	2.951(8)	3.882(7)	3.857(5)	4.178(8)
$r_\alpha(X_{ax}{\ldots}X_{ax})$, Å	3.463(14)	4.676(12)	4.737(8)	4.947(9)
Δ_{ax-eq}, Å^a	0.028(11)	0.097(9)	0.142(5)	0.061(10)
V_0, kJmol^{-1}		6.3(2.9)	5.0(2.5)	5.4(2.5)

$^a \Delta_{ax-eq} = r(M{-}X)_{ax} - r(M{-}X)_{eq}$.

The possibility of Berry-type[200] permutational isomerism ("pseuodorotation") was also investigated for VF_5. However, the diffraction picture did not indicate the presence of intermediate geometries that would occur along a supposed inversion coordinate.

Niobium and Tantalum Pentahalides. The fluorides of Nb and Ta are trimers in the gas phase at lower temperatures, while their chlorides and bromides are monomers.[201–203] Recent electron diffraction studies[204,205] determined a D_{3h} symmetry trigonal bipyramidal geometry for these monomers with the axial bonds longer than the equatorial ones. The geometrical parameters are collected in Table 9-26. The barriers to "pseudorotation," which were also estimated from the electron diffraction data, are also given in Table 9-26. They compare well with the barriers calculated for similar molecules.

The mean M—X distances agree with earlier electron diffraction results on the same molecules (Table 9-27). Only an early result is available for the mean Nb—Br bond distance in $NbBr_5$.[207] The mean bond distances in $NbCl_5$ and $TaCl_5$ are nearly identical, which is again a consequence of lanthanide contraction.

That the differences of the axial and equatorial bonds in trigonal bipyramidal molecules are larger in the pentachlorides than in the pentafluorides (Table

Table 9-27. MEAN BOND DISTANCES FOR NIOBIUM AND TANTALUM PENTAHALIDES

Molecule	Mean M—X distance (Å)	Ref.
$NbCl_5$	2.280(3)	204
	2.28(2)	206
$TaCl_5$	2.284(2)	204
	2.27(2)	206
$TaBr_5$	2.436(3)	205
	2.44(2)	207
$NbBr_5$	2.45(2)	207

Table 9-28. DIFFERENCES IN AXIAL AND
EQUATORIAL BOND LENGTHS IN TRIGONAL
BIPYRAMIDAL PENTACHLORIDES VERSUS
PENTAFLUORIDES

Molecule	$r(M - X)_{ax} - r(M - X)_{eq}$ (Å)	Ref.
$NbCl_5$	0.097	204
$TaCl_5$	0.142	204
PCl_5	0.104	208
PF_5	0.043	209
AsF_5	0.055	108
VF_5	0.027	198

9-28) could be interpreted by weaker repulsions of the bonding pairs to
fluorine due to the greater electronegativity of this ligand. The difference for
$TaBr_5$ is again rather small, 0.061 Å.

Pentahalides of the Sixth Group. The $+5$ oxidation state is an interesting
phenomenon for the Group VI transition metals. Since they have a d^1 electronic
configuration, it is degenerate with the one electron occupying an e'' orbital
in a supposed D_{3h} symmetry structure. Therefore, these molecules should be
subject to Jahn—Teller distortion just as VCl_4 is in a tetrahedral configuration.

Indeed, CrF_5 was found[210] to have a lower, C_{2v} symmetry geometry rather
than the more symmetric D_{3h} structure. The parameters corresponding to the
two models are given in Table 9-29. The C_{2v} geometry is derived from the D_{3h}
structure by opening one of the equatorial angles and then by bending the two
axial bonds toward this widening angle (see Figure 9-22).

According to Jacob and co-workers,[210] although the C_{2v} symmetry for the
equilibrium structure of CrF_5 is not unlikely, the static model emerging from the

Table 9-29. PARAMETERS OF CrF_5 FOR TWO
MODELS

Parameter	C_{2v} model	D_{3h} model
$r_g(Cr—F)_{av}$, Å[a]	1.708(2)	1.708(2)
$\Delta(Cr—F)$, Å[b]	0.048(16)	0.036(22)
$r_g(Cr—F)_{eq}$, Å	1.695(6)	1.700(9)
$r_g(Cr—F)_{ax}$, Å	1.742(10)	1.734(14)
$\angle F_1—Cr—F_2$, deg[c]	115.9(9)	120
$\angle F_2—Cr—F_3$, deg	128.1(18)	120
$\angle F_1—Cr—F_4$, deg	95.8(4)	90
$\angle F_2—Cr—F_4$, deg	87.5(2)	90

[a] Weighted average bond length.
[b] $r(Cr—F)_{ax} - r(Cr—F)_{eq}$.
[c] For the numbering of atoms see Figure 9-22.

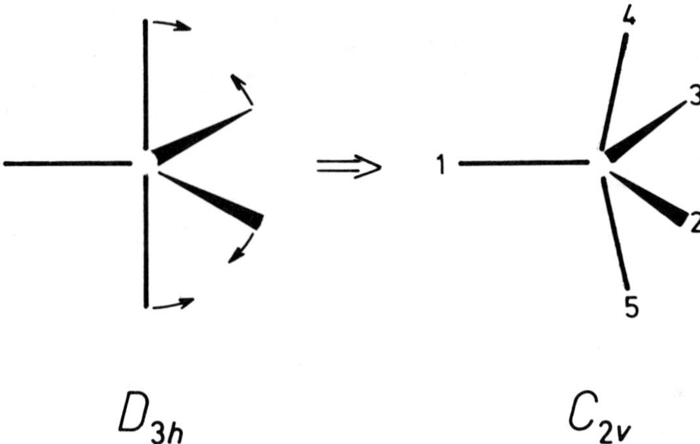

Figure 9-22. Symmetry lowering due to Jahn–Teller distortion of the D_{3h} symmetry structure of CrF_5.[210]

electron diffraction analysis is not realistic. It is more probable, that CrF_5 undergoes a dynamic Jahn–Teller distortion making the molecule appear as a C_{2v} symmetry structure on the average.

The $MoCl_5$ and WCl_5 molecules are also monomeric in the gas phase and have been repeatedly studied by electron diffraction. The early studies[211,212] could determine only a mean metal–halogen bond length under assumed D_{3h} symmetry. Ezhov and Sarvin concluded that both $MoCl_5$[213] and WCl_5[214] have D_{3h} symmetry at the electron diffraction experimental temperature. Nevertheless, there is a possibility that the equilibrium configuration is actually C_{4v} with a low barrier corresponding to D_{3h} symmetry. The mean M—Cl distances determined in these studies agree with earlier results within experimental error: $r_a(\text{Mo—Cl})_{\text{mean}} = 2.27(2)$ and $2.271(4)$ Å from references 211 and 213, respectively, and $r_a(\text{W—Cl})_{\text{mean}} = 2.26(2)$ and $2.260(3)$ Å (references 212 and 214). The upper limit for the difference between the axial and equatorial bond lengths was put at 0.1 Å by Spiridonov and Romanov[212] and at 0.04 Å by Ezhov and Sarvin.[213,214]

The $MoCl_5$ molecule was also studied by Brunvoll and co-workers,[215] who concluded that the vapor either is a mixture of molecules with D_{3h} and C_{4v} symmetry (see the models in Figure 9-23) or is such a mixture with some dimers present in addition. The first model could be accepted supposing large-amplitude motion. On the other hand, employing a relatively rigid model, a certain amount of dimers had to be introduced. The geometrical parameters of both models are given in Table 9-30. The dimer consists of two C_{4v} symmetry units connected by a Mo–Mo bond as shown in Figure 9-23; D_{4d} symmetry was assumed for the molecule.

Jacob and colleagues[210] pointed out that while the foregoing geometries were not interpreted in terms of Jahn–Teller effect, there is evidence for some

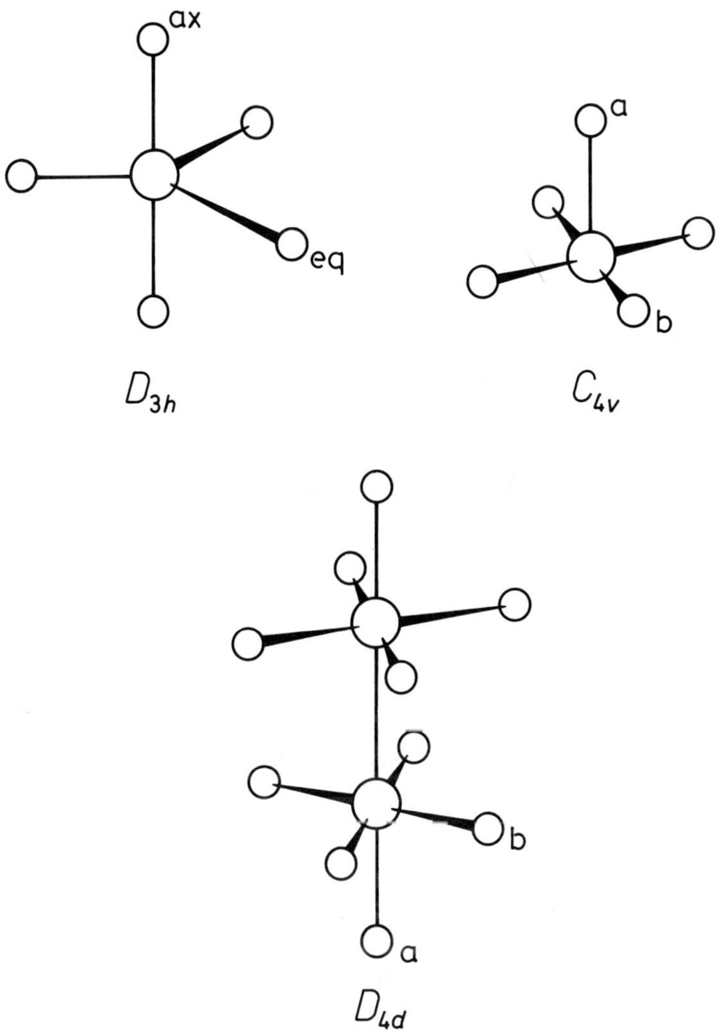

Figure 9-23. Molecular models considered in the structure analysis of $MoCl_5$.[215]

distortion in these molecules as well as in CrF_5. In the light of this, model 1 of Table 9-30 may be more probable for $MoCl_5$ than model 2.

Polymeric Species. The vapors of NbF_5 and TaF_5 at 300–320 K were found to contain tetrameric species in their first electron diffraction study, while the experiments at 473 K were interpreted with monomeric molecules.[216] The tetramers at lower temperature were related to the corresponding crystal structures.[217] However, mass spectrometric studies[218] of TaF_5 indicated the presence of primarily trimers in the vapor. A subsequent electron diffraction investigation[115,219] of both compounds provided conclusive evidence for trimeric species in the vapors under the experimental conditions.

Table 9-30. GEOMETRICAL PARAMETERS OF $MoCl_5$

Molecule, configuration	Parameter	Model 1	Model 2
Monomer, C_{4v}	%	55.7(3.1)	50.7(4.3)
	$r_a(Mo—Cl)_{av}$, Å[a]	2.230(7)	2.227(5)
	$\angle Cl_a—Mo—Cl_b$, deg[b]	88.1(9)	83.6(9)
Monomer, D_{3h}	%	44.3(3.1)	25.6(5.3)
	$r_a(Mo—Cl)_{av}$, Å[c]	2.297(16)	2.300(20)
	Δ_{ax-eq}, Å[d]	0.064(31)	
	$r_a(Mo—Cl)_{eq}$, Å	2.271(16)	
	$r_a(Mo—Cl)_{ax}$, Å	2.335	
Dimer,[e] D_{4d}	%		23.6(2.6)
	$r_a(Mo—Cl)_{av}$, Å		2.276(24)
	$r_a(Mo—Mo)$, Å		2.600(38)
	$\angle Cl_a—Mo—Cl_b$, deg		84.4(8)

[a] Weighted average of the Mo—Cl distances in a tetragonal pyramid (see Figure 9-23).
[b] For definition see Figure 9-23.
[c] Weighted average of the Mo—Cl distances in a trigonal bipyramid.
[d] $\Delta_{ax-eq} = r(Mo—Cl)_{ax} - r(Mo—Cl)_{eq}$.
[e] See Figure 9-23.
Source: Reference 215.

The $(NbF_5)_3$ and $(TaF_5)_3$ trimers have a six-membered ring structure with alternating metal and fluorine atoms in the ring and metal–halogen bonds of three different types—axial, terminal, and bridging—as indicated in Figure 9-24. These rings were found to be planar, in contrast to $(SbF_5)_3$, whose ring is nonplanar. The geometrical parameters are given in Table 9-31.

Molybdenum Pentafluoride. This compound was also found to be trimeric in the vapor at lower temperatures by mass spectrometric studies.[220] The electron diffraction analysis,[221] however, could not reach acceptable agreement between the experimental and theoretical distributions when allowing only for the

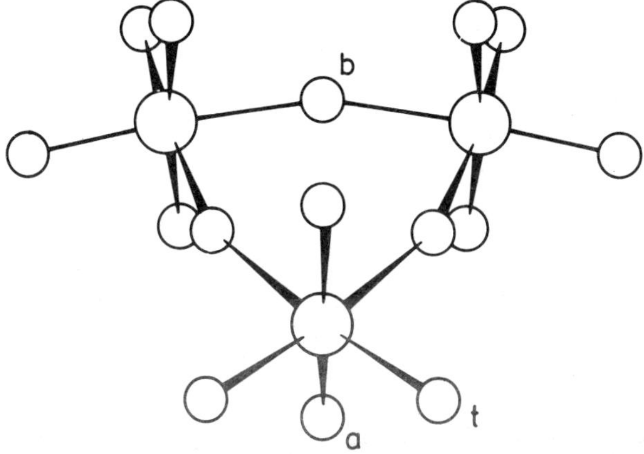

Figure 9-24. Molecular configuration of $(NbF_5)_3$ and $(TaF_5)_3$.

Table 9-31. GEOMETRICAL PARAMETERS OF TRIMERIC PENTAHALIDES

	Molecules			
Parameter	$(NbF_5)_3$ (Ref. 115)	$(TaF_5)_3$ (Ref. 219)	$(MoF_5)_3$ (Ref. 221)[a]	$(AuF_5)_3$ (Ref. 223)
$r_\alpha(M{-}F)_t$, Å[b]	1.810(2)	1.823(5)	1.804(35)	1.822(8)
$r_\alpha(M{-}F)_a$, Å		1.846(5)	1.821(30)	1.889(9)
$r_\alpha(M{-}F)_b$, Å	2.046(4)	2.062(2)	2.012(10)	2.030(7)
$\angle_\alpha F_a{-}M{-}F_a$, deg	162.5(14)	173.1(21)	160.1(10)	193.1(32)
$\angle_\alpha F_t{-}M{-}F_t$, deg	102.9(12)	96.4(15)	100.5(21)	75.3(65)
$\angle_\alpha F_b{-}M{-}F_b$, deg	82.0(10)	83.5(6)	79.4(11)	115.7(11)
$r_\alpha(M{...}M)$, Å[c]	4.017	4.038	3.966	3.590

[a] r_a.
[b] For definition of the different bond distances see Figure 9-24.
[c] Calculated from the other parameters.

presence of $(MoF_5)_3$ molecules. Girichev and co-workers[221] supposed that hydrolysis may have taken place during the diffraction experiment and the discrepancy was caused by the presence of hydrolysis product $MoOF_4$. They assumed a $MoOF_4$ geometry[222] and refined the vapor composition. Inclusion of about 28% of $MoOF_4$ considerably improved the agreement between the models and the experiment. The $(MoF_5)_3$ molecule has the same kind of six-membered ring structure as the other trimeric pentahalides. Deviation from planarity of the ring was apparently not investigated. The geometrical parameters are cited in Table 9-31.

Gold Pentafluoride. Among the Group IB elements, only gold forms a pentahalide. The electron diffraction patterns of AuF_5 were recorded at 493 K.[223] Mass spectra at 360 K[224] indicated dimers as the main component of the vapor. Electron diffraction detected small amounts of trimer in addition to the predominant dimeric species. Both polymeric forms have a ring structure. The dimer is shown in Figure 9-25, together with the most important geometrical parameters. The trimer has the same form as the other trimeric pentahalides (Fig. 9-24); its geometrical parameters are given in Table 9-31. Both the dimer and the trimer rings were found to be planar, with D_{2h} and D_{3h} symmetry,

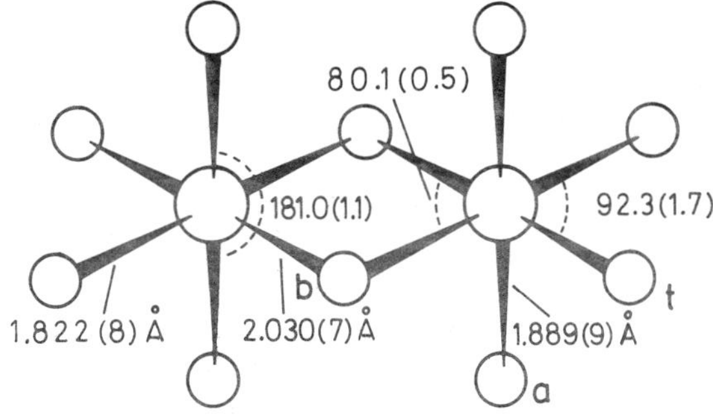

Figure 9-25. Molecular configuration and geometrical parameters of dimeric gold pentafluoride, $(AuF_5)_2$.[223]

respectively. The mean vibrational amplitudes were calculated from experimental frequencies[224] and were mostly kept unchanged in the analysis. The corresponding Au—F bonds in the dimer and trimer could not be distinguished and were assumed to be equal. The terminal bonds are shorter than the axial ones in accordance with the VSEPR theory,[59] and the bridging bonds are the longest.

Comparing the geometry of $(AuF_5)_3$ to those of $(SbF_5)_3$, $(NbF_5)_3$, $(TaF_5)_3$, and $(MoF_5)_3$, the most important difference is in the very short metal...metal distance of $(AuF_5)_3$. The relatively large F_b—Au—F_b angle in the ring—115.7(1.1)°, compared to the same angles of around 80° in the other trimers—is consistent with this short Au...Au separation.

Hexahalides and Heptahalides

All hexahalides studied so far have regular octahedral arrangement without any detectable deviation. Some of the hexahalides, such as ReF_6 and OsF_6, have a degenerate electronic ground state, which could allow Jahn–Teller distortions, although such distortions have not been detected. As was discussed before, the dynamic nature of these distortions may keep them concealed. The bond distances are collected in Table 9-32 for all MX_6 molecules studied to date. Their variations correspond to expectation.

The only heptahalide investigated so far is ReF_7.[229] The mean Re—F bond length is 1.835(5) Å. The most symmetrical model for this molecule would be a pentagonal bipyramid with D_{5h} symmetry. Vibrational spectra[230] were interpreted with this symmetry. Molecular beam deflection,[231] however, was not compatible with such a high symmetry.

The apparent molecular symmetry from electron diffraction is lower, C_2 or C_s, due to "pseudorotation." Two vibrations, the e'_1 axial bend and the e''_2 ring puckering modes couple in phase in this dynamic model as indicated schematically in Figure 9-26. Vibrational displacements carry the molecule from C_2 to C_s and again to C_2 symmetry. This motion closely resembles the pseudorotation of cyclopentane.

Table 9-32. BOND LENGTHS (Å) IN METAL
HEXAHALIDES

Molecule	r_g(M—X)	r_a(M—X)	Ref.
MoF_6	1.820(3)		225
WF_6	1.832(3)		225
		1.833(8)	226
ReF_6	1.832(4)		227
OsF_6		1.831(8)	226
IrF_6		1.830(8)	226
UF_6	1.999(3)		228
		1.996(8)	226
NpF_6		1.981(8)	226
PuF_6		1.971(10)	226

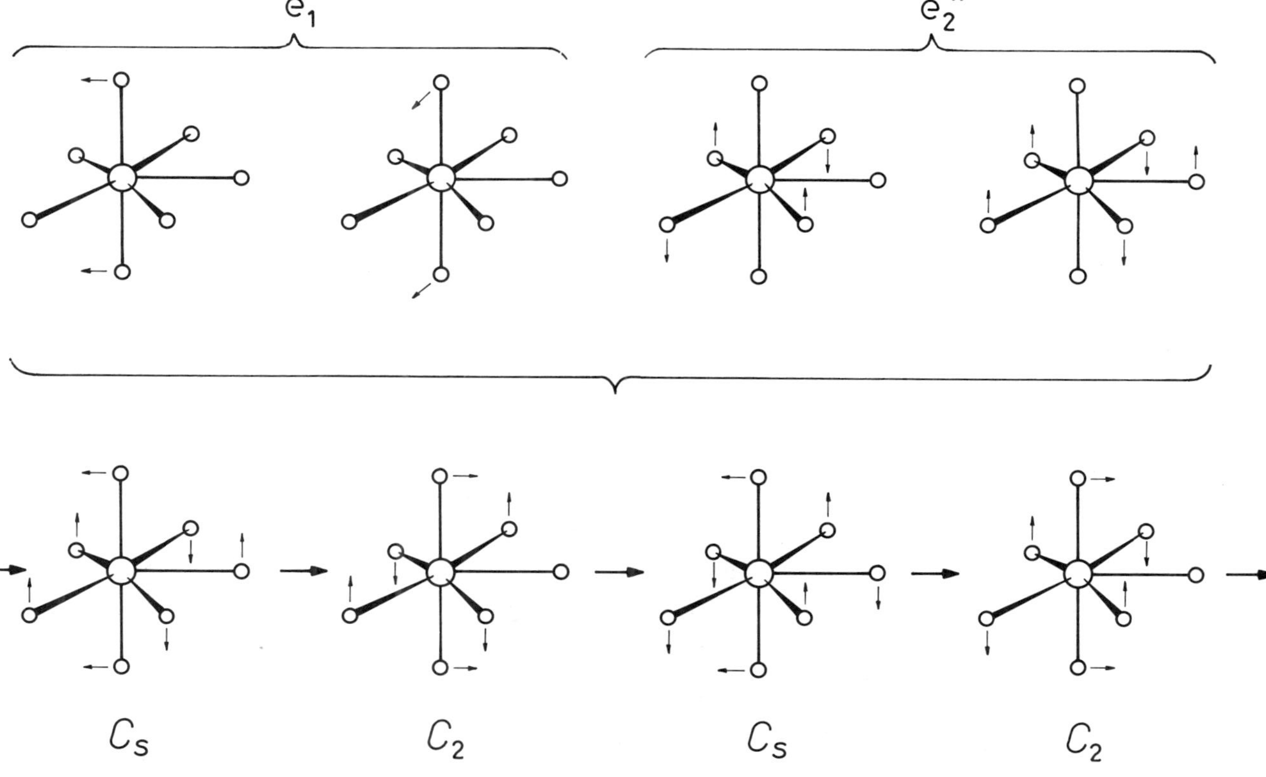

Figure 9-26. Schematic coupling in the dynamic description of the ReF_7 structure.[229]

Complex Metal Halides

Aluminum chloride and other Lewis acid halides form volatile metal halide complexes with almost all metal halides. These vapor complexes have been extensively studied by different techniques due to their general practical importance in halogen metallurgy.[232–237]

In spite of the great variety of metal halide complexes, few of them are abundant enough in the vapor phase for electron diffraction analysis. Some trifluoroberyllates, $AlkBeF_3$ (Alk = Li,[238] Na,[239] K[239]), and several tetrahalo-aluminates (see later) and related compounds ($KYCl_4$,[240] $TlInCl_4$[241]) have been investigated. The results published before 1974 have been reviewed.[25] From among the models tested for $AlkBeF_3$ and shown in Figure 9-27, model I gave the best agreement with experiment.

Possible models for the $AlkAlX_4$ complexes are shown in Figure 9-28. All three agree in having an X_4 tetrahedron around the aluminum atom; only the linkages to the alkali atom are different. Fluoride and chloride complexes have been studied recently, and their geometrical parameters are given in Tables 9-33 and 9-34, respectively.

The general conclusion for the configuration of all $AlkAlX_4$ molecules studied so far is a bidentate structure with C_{2v} symmetry (model I in Figure 9-28). A C_s configuration was found as best static model for $KAlF_4$ and $NaAlF_4$, with the F_b—Alk—F_b triangle twisted with respect to the F_b—Al—F_b plane by 26°, as shown in **4**. This twist is supposed to be again a consequence of large-amplitude vibrations of the four-membered ring and is not at variance with a C_{2v} symmetry equilibrium structure for these molecules.

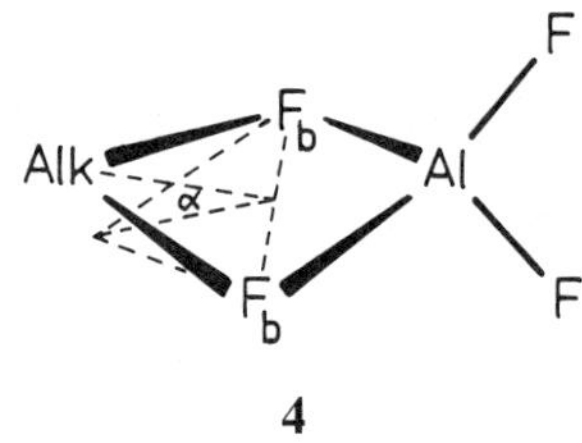

4

The bonds between the alkali metal and the bridging halogen are longer than the bonds in the corresponding free alkali halides. The difference is considerable; 0.2–0.5 Å. They are also longer than the Alk—X bonds of the Alk_2X_2 dimers. The Alk—X_b distances have unusually large amplitudes of vibration for a chemical bond: 0.15—0.25 Å. All this is in agreement with the spectroscopic results on the alkali tetrafluoroaluminates,[236] indicating very small stretching force constants for these bonds, viz, 0.66, 0.36, 0.25, 0.18, and 0.15 mdyn $Å^{-1}$ for the Li, Na, K, Rb, and Cs derivatives, respectively. These features point to a rather weak linkage between the alkali and halogen atoms. This "weakness," however, does not necessarily mean thermodynamic instability.[2]

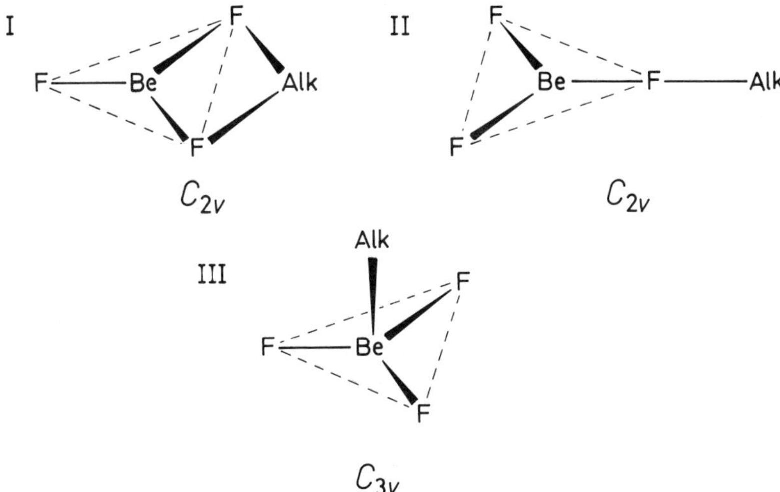

Figure 9-27. Molecular models for alkali trifluoroberyllates.

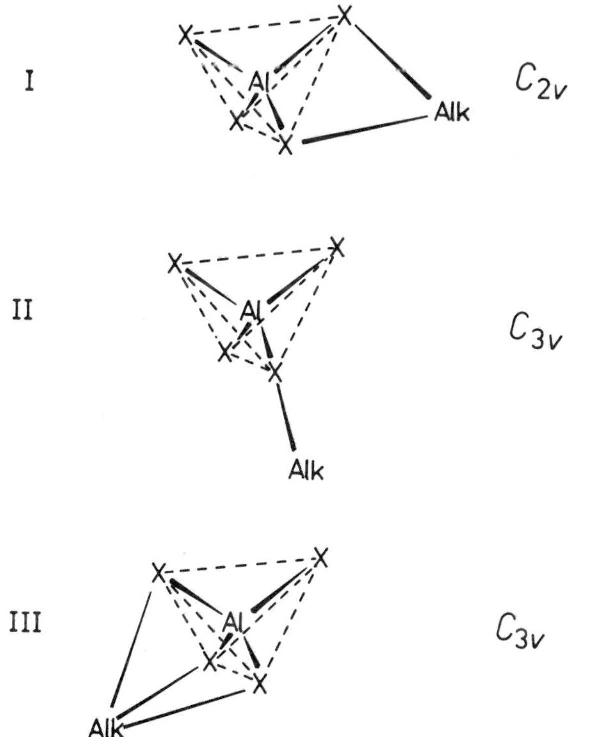

Figure 9-28. Possible models of alkali tetrahaloaluminates.

Table 9-33. Geometrical Parameters of Alkali Tetrafluoroaluminates, AlkAlF$_4$

	Alkali atom				
Parameter	Li (Ref. 242)[a]	Na (Ref. 243)[b]	K (Ref. 244)[c]	Rb (Ref. 245)[c]	Cs (Ref. 245)[c]
T(K)		1150	1000	983	963
r(Al—F)$_t$, Å	1.673	1.69(2)	1.695(8)	1.696(5)	1.695(6)
r(Al—F)$_b$, Å	1.764		1.695(10)		
r(Alk—F), Å	1.790	2.11(2)	2.525(14)	2.64(3)	2.84(9)
r_e(Alk—F), Å[d]	1.573	1.926	2.171	2.266	2.345
	1.564				
r_e(Alk—F), Å[e]	1.68	1.90	2.25	2.35	2.55
r(Alk...Al), Å	2.647	2.58(3)	3.137(14)	3.32(7)	3.51(8)
∠F$_b$—Al—F$_b$, deg	84.5	f	103.1(11)	f	f
∠F$_t$—Al—F$_t$, deg	118.8	f	117.9(8)	f	f
∠F$_b$—Alk—F$_b$, deg	83.0	81.7(2.0)	63.4(3)	62.8(1.3)	57.5(3.5)

[a] r_e, *ab initio* calculation.
[b] r_a.
[c] r_g.
[d] Distance in the gaseous alkali halide, LiF upper value, *ab initio*,[242] lower value, microwave spectroscopy,[246] all others microwave spectroscopy.[27]
[e] Distance in the Alk$_2$X$_2$ dimer, from calculation.[37]
[f] AlF$_4$ unit supposed to be tetrahedral.

The structural findings for these molecules have been rationalized in terms of a dynamic rather than static model involving polytopic bonds.[2,250] The alkali atom has great mobility and belongs to all four halogen atoms in this description, while the AlX$_4$ unit is rather rigid and has T_d symmetry. Unperturbed T_d symmetry of the AlX$_4$ fragment, however, is not supported by other experimental and theoretical evidences. There are indications for a distorted AlX$_4$ fragment in these molecules. Unfortunately the terminal and bridging bonds could not be distinguished in the electron diffraction analyses; the mean lengths are about the same in all groups and are larger than the corresponding

Table 9-34. Bond Distances (Å) in AlkAlCl$_4$ Molecules

	Alkali atom		
Bond distance	K (Ref. 247)[a]	Rb 893 K (Ref. 248)	Cs 843 K (Ref. 248)
r_g(Al—Cl)	2.153(6)	2.151(6)	2.149(9)
r_g(Alk—Cl)	2.98(5)	3.16(3)	3.31(7)
r_e(Alk—Cl)[b]	2.667	2.787	2.906
r_e(Alk—Cl)[c]	2.84	2.99	3.17
r_g(Alk...Al)	3.71(7)	3.80(5)	3.85(10)

[a] For an earlier study of this molecule see reference 249.
[b] Distance in the gaseous alkali halide, microwave spectroscopy.[27]
[c] Distance in the Alk$_2$X$_2$ dimer, from calculation.[37]

bond lengths in the monomers (cf. Table 9-10). The mean Al—Cl distances in the tetrachloroaluminates are the same as the mean of the terminal and bridging bond lengths of the Al_2Cl_6 dimer (Table 9-9). The force constants are quite different,[236] indicating also a weaker bridging bond similar to that in the Al_2X_6 type dimers. *Ab initio* calculations on $LiAlF_4$ also predicted a longer Al—F bridging bond. The bond angles in $KAlF_4$ from electron diffraction and in $LiAlF_4$ from calculation also indicate a distortion similar to the one described above.

The interaction between the alkali cation and the tetrahaloaluminate anion was found to be very weak in melts.[251] The anions have a regular tetrahedral shape. According to Huglen and co-workers,[236] this regular tetrahedral arrangement undergoes C_{2v} distortion in the vapor phase. This is indicated by the splitting of the v_3 and the occurrence of the v_1 stretching frequencies in the infrared spectra. Moreover, this distortion depends on the cation; the more polarizable it is, the larger is the distortion. This is expressed by increased splitting of the v_3 stretching frequencies as going from Cs to Li.[236]

A splitting of frequencies due to the large polarizing power of the small cations can also be observed, although to a lesser extent, for alkali haloaluminate melts.[251] The simple model[250] of polytopic bonds of these structures, however, does not make use of cation polarizability.

Structure analysis combining different techniques will facilitate a better understanding of these interesting geometries. They are expected to become subject of intensive research in the future.

Aluminum halide complexes with other than alkali metals have been extensively studied by different spectroscopic[237] techniques as well as by *ab initio* calculations.[252] The most common composition of these complexes is $MX_2 \cdot Al_2Cl_6$. Electron diffraction has not yet been applied to them.

Donor–Acceptor Complexes

There are two partner molecules in the complexes often called addition compounds or charge–transfer complexes: an electron donor such as an amine or phosphine and an electron acceptor (eg, boron or aluminum compounds).

Only very few such molecules will be mentioned here, ie, those with a metal halide as acceptor: $AlCl_3 \cdot NH_3$,[253] $AlCl_3 \cdot N(CH_3)_3$,[254] $AlBr_3 \cdot NH_3$,[255] $GaCl_3 \cdot NH_3$,[256] and $GaBr_3 \cdot NH_3$.[255]

All these molecules have an ethanelike shape with C_{3v} symmetry as shown in Figure 9-29. The vibrational spectra of all four NH_3 complexes ($AlCl_3 \cdot NH_3$,[257,258] $AlBr_3 \cdot NH_3$,[257] $GaCl_3 \cdot NH_3$,[259] and $GaBr_3 \cdot NH_3$[259]) are consistent with a staggered conformation. *Ab initio* calculations, which have been carried out for $AlCl_3 \cdot NH_3$[257,258] and $AlF_3 \cdot NH_3$,[257] predicted the staggered conformation to be the most stable for the former. For $AlF_3 \cdot NH_3$ the eclipsed form was found to be slightly more favorable than the staggered one,

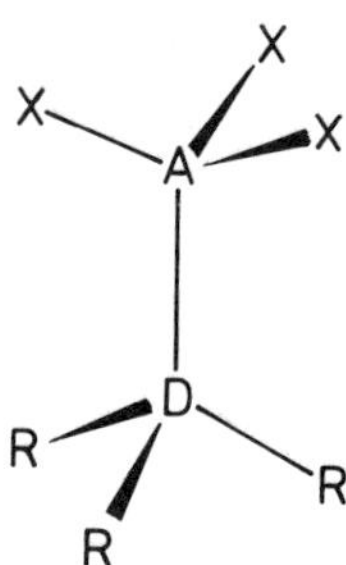

Figure 9-29. Ethanelike shape of $AX_3 \cdot NR_3$ donor–acceptor complexes (symmetry: C_{3v}).

but the calculations were not considered accurate enough for deciding conformational preferences.[257]

The most important geometrical parameters of the complexes are given in Tables 9-35 and 9-36, where the geometries of the corresponding free donor and acceptor molecules are also indicated.

Several matters are of interest: the change of the donor and acceptor geometries upon complexation, the length of the coordination linkage, and the geometrical consequences of substituent changes.

The acceptor geometries undergo the most drastic changes. As is seen from Tables 9-35 and 9-36, the planar acceptor molecules become pyramidal and their bonds lengthen. The changes of the donor parts are less pronounced. Unfortunately, the relatively small scattering power of the ammonia hydrogen atoms made the determination of the donor geometries less accurate.

As was suggested before,[25] all geometrical changes can be interpreted by VSEPR theory and nonbonded interactions. Figure 9-30 helps to visualize these effects.

Consider first the acceptor geometry. Due to the formation of the coordination linkage, a new bonding electron pair will be repelling the other bonding pairs of the acceptor, and thus the originally planar molecule will be replaced by a tetrahedral configuration with longer bonds than those in the uncomplexed acceptor. Complex formation brings about new nonbonded interactions for the acceptor ligands, viz, those between the acceptor ligands on one hand and all atoms of the donor part, on the other hand. These additional repulsions also tend to decrease the angles and lengthen the bonds of the acceptor. Thus the geometry of the acceptor part of the complex will be influenced by two effects, both acting in the same direction. This is why the acceptor geometry changes considerably upon complexation, and this change always occurs as a decrease in its angles and a lengthening of its bonds.

Consider now the donor part. The new nonbonded interactions occurring upon complexation work in the same way as for the acceptor part: they tend to decrease the bond angles and lengthen the bonds of the donor. The consequences of the changes in the electron-pair repulsions, however, will influence the donor geometry in the opposite direction. There is a lone pair of electrons in

Table 9-35. Geometrical Parameters of Donor-Acceptor Complexes of Aluminum[a] and Related Free Molecules

Parameter	$AlF_3 \cdot NH_3$ Calculated, eclipsed (Ref. 257)	$AlCl_3 \cdot NH_3$, staggered Calculated (Ref. 257)	$AlCl_3 \cdot NH_3$, staggered Calculated (Ref. 253)	$AlCl_3 \cdot NH_3$, staggered Experimental (Ref. 253)	$AlCl_3 \cdot N(CH_3)_3$ Experimental, staggered (Ref. 254)	$AlBr_3 \cdot NH_3$ Experimental, staggered (Ref. 255)
$AlX_3 \cdot NR_3$						
$r(Al{-}X)$, Å	1.62	2.08	2.024	2.102(5)	2.122(4)	2.266(5)
$r(Al{-}N)$, Å	1.99	1.94	1.943	1.998(19)	1.949(35)	1.999(19)
$r(N{-}R)$, Å	1.03	1.03	1.033	[1.030][b]	1.517(12)	1.063(33)
$\angle X{-}Al{-}X$, deg	117.0	116.4	116.2	116.9(4)[c]	113.6[d]	116.1(3)[c]
$\angle R{-}N{-}R$, deg	106.6	106.3	107.0	112.8(35)	106.2[d]	114.5(40)
Free acceptor						
$r(Al{-}X)$, Å	1.60[e]	2.05[e]		2.068(4)[f]		2.223(5)[g]
$\angle X{-}Al{-}X$, deg	120[e]	120[e]		120[f]		120[h]
Free donor						
$r(N{-}R)$, Å	1.03[i]			1.030(2)[j]	1.454(4)[k]	
$\angle R{-}N{-}R$, deg	104.2[i]			107.5[j]	110.6(6)[k]	

[a] Calculated distances r_e, experimental distances r_g.
[b] Assumed value.
[c] $\angle_\alpha$.
[d] Calculated from the other angles by virtue of C_{3v} symmetry.
[e] Calculated.[79]
[f] Reference 77.
[g] Terminal distance of the dimer Al_2Br_6.[74]
[h] From spectroscopy.
[i] Calculated.[260]
[j] Reference 261.
[k] Reference 262.

Table 9-36. GEOMETRICAL PARAMETERS OF
DONOR-ACCEPTOR COMPLEXES OF GALLIUM
AND RELATED FREE MOLECULES

Parameter	$GaCl_3 \cdot NH_3$ (Ref. 256)	$GaBr_3 \cdot NH_3$ (Ref. 255)
$GaX_3 \cdot NH_3$		
r_g(Ga—X), Å	2.144(5)	2.290(5)
r_g(Ga—N), Å	2.058(11)	2.082(23)
r_g(N—H), Å	1.030(12)	1.063(33)
∠X—Ga—X, deg	117.1(3)[a]	116.6(3)
∠H—N—H, deg	114.3(1.2)	115.6(4.1)
Free acceptor		
r_g(Ga—X), Å	2.100(2)[a]	2.246(3)[a]
∠X—Ga—X, deg	120[b]	120[b]
Free donor		
r_g(N—H), Å	1.030(2)[c]	
∠H—N—H, deg	107.5[c]	

[a] Terminal distance of the dimer.[74]
[b] From spectroscopy.
[c] Reference 261.

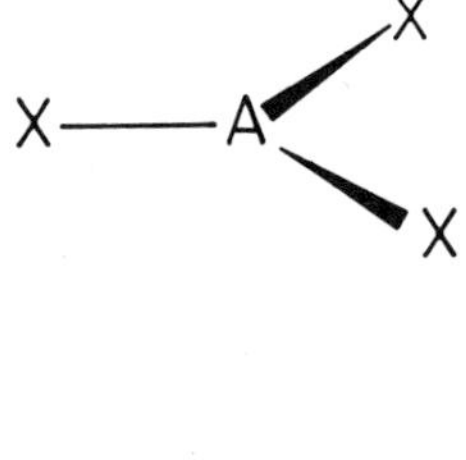

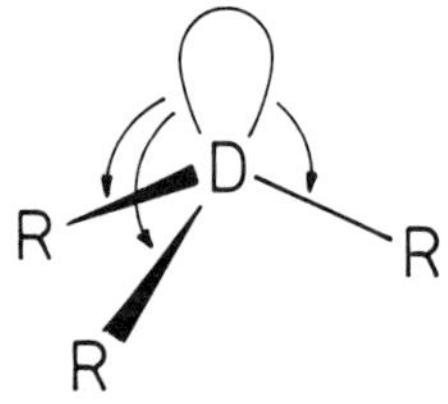

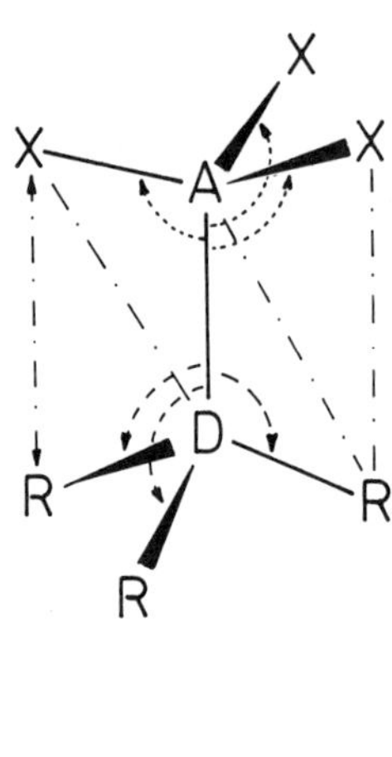

Figure 9-30. Illustration of the geometrical changes upon complexation: solid arrows, strong repulsions from the lone electron pair; dashed arrows, weaker repulsions from the new bonding pair; dotted arrows, new repulsions from the new bonding pair towards the acceptor bonds; mixed (dot–dash) arrows, some of the new nonbonded repulsions.

the donor originally, which is replaced by a bonding pair upon complexation. Thus the electron-pair repulsions decrease, and this would tend to allow an increase in the donor bond angles and a shortening of its bonds. The net change for the donor will thus be the result of two competing effects. For this reason the direction of the geometrical changes in the donor part cannot be predicted; their magnitude is usually rather small due to the canceling effects of the changes. The geometrical data of Tables 9-35 and 9-36 are in complete agreement with the foregoing reasoning.

Acceptor Geometry

The acceptor geometries changed considerably upon complexation, and the geometry of $AlBr_3$ changed even more than that of $AlCl_3$ in the two ammonia complexes. This demonstrates that with identical donors, the strongest acceptor undergoes the largest structural changes. In other words, the structure of the stronger acceptor is more easily distorted by the same donor upon complex formation than that of the weaker acceptor. The acceptor strength of aluminum halides increases in the order: $AlCl_3 < AlBr_3 < AlI_3$.

The donor–acceptor bonds have the same lengths in the two complexes. This is also an indication of the stronger acceptor character of $AlBr_3$. Both electron affinity and substituent size increase when going from $AlCl_3$ to $AlBr_3$. These represent two opposite effects for the donor–acceptor bond. The force constants of the Al—N bonds in AlX_3 molecules were calculated to be 1.61, 1.50, and 1.45 mdyn $Å^{-1}$, for the X = F, Cl, and Br derivatives, respectively.[257] This was consistent with a weaker Al—N bond in $AlBr_3 \cdot NH_3$ than in $AlCl_3 \cdot NH_3$, with *no* apparent bond length change. Similar phenomena have been observed and rationalized[263] for several boron halide addition compounds. Bond strength and stability as well as bond strength and bond length do not necessarily display simple correlations. While stability is determined by the depth of the minimum of the potential energy function, the bond-stretching force constant is the second derivative of this function and is related to the rigidity or compliance of the bond rather than to its strength. The above-mentioned difference in the lengths of the coordination linkages is determined by joint influence of acceptor strengths and nonbonded interactions.

The results of *ab initio* calculations on $AlF_3 \cdot NH_3$[257] and $AlCl_3 \cdot NH_3$[257,258] cited in Table 9-35 are in good agreement with the experimental results on $AlCl_3 \cdot NH_3$. The H—N—H bond angles differ considerably, but the opening of this angle, as compared to the free NH_3 molecule, is correctly predicted, notwithstanding the large numerical discrepancies.

The influence of different donors on the same acceptor can be investigated on the pair of $AlCl_3 \cdot NH_3$ and $AlCl_3 \cdot N(CH_3)_3$. The stronger donor causes larger structural changes on the same acceptor, as expected. The $AlCl_3$ geometry changes more in the trimethylamine complex than in the other; and the Al—N bond is shorter in $AlCl_3 \cdot NH_3$ than in $AlCl_3 \cdot N(CH_3)_3$.

Table 9-37. GEOMETRICAL PARAMETERS OF DIFFERENT
$AlX_3 \cdot N(CH_3)_3$ COMPLEXES

Parameter	$AlCl_3 \cdot N(CH_3)_3$ (Ref. 254)	$AlH_3 \cdot N(CH_3)_3$ (Ref. 264)	$Al(CH_3)_3 \cdot N(CH_3)_3$ (Ref. 265)
r_g(Al—X), Å	2.122(4)	1.565(11)	1.989(5)
r_g(Al—N), Å	1.949(35)	2.066(8)	2.102(10)
r_g(N—C), Å	1.517(12)	1.477(3)	1.475(3)
∠X—Al—X, deg[a]	113.6	114.1	115.6
∠X—Al—N, deg	104.9(7)	104.3(1.1)	102.3(3)
∠C—N—C, deg[a]	106.2	109.9	109.6
∠C—N—Al, deg	112.6(1.5)	109.0(3)	109.3(4)

[a] Calculated from the other angle by virtue of C_{3v} symmetry.

Donor Geometry

The geometry of the donor part in the ammonia complexes could be determined only with large uncertainty. The only geometrical change that can be discussed is the increase of the H—N—H angle compared to free ammonia. This suggests the electron-pair repulsions to prevail over nonbonded interactions here. Opposite geometrical changes are observed in $N(CH_3)_3$ upon complexation. The C—N—C angles decrease and the N—C bonds lengthen as compared to free trimethylamine. Thus nonbonded interactions prevail as a consequence of bulkier ligands on nitrogen.

Table 9-37 compares the geometries of $AlCl_3 \cdot N(CH_3)_3$ and two related molecules, $AlH_3 \cdot N(CH_3)_3$ and $Al(CH_3)_3 \cdot N(CH_3)_3$. These are complexes of the same donor with acceptors of decreasing strength in the order: $AlCl_3 > AlH_3 > Al(CH_3)_3$. The gradually longer Al—N coordination linkage is in accordance with this change in acceptor strength as well as with its increasing X—Al—X angle. The angle variation indicates that a weaker acceptor changes less upon complexation with the same donor than does a stronger acceptor.

The geometrical changes of the trimethylamine parts also reflect gradually decreasing acceptor strength; the largest change occurs in the $AlCl_3$ complex.

Structural variations of further donor-acceptor complexes are discussed in Chapters 1 and 8 of this volume.

References

1. Tremmel, J.; Hargittai, I. *Hung. Sci. Instrum.* **1980**, *50*, 43; see also Chapter 6 in volume A.
2. Spiridonov, V. P.; Zasorin, E. Z. Modern high-temperature electron diffraction. In "Proceedings of the Tenth Materials Research Symposium on Characterization of High Temperature Vapors and Gases", National Bureau of Standards Special Publication No. 561, Hastie, J. W., Ed.; National Bureau of Standards: Washington, DC, 1979, p. 711.
3. Several of the papers in "Proceedings of the Tenth Materials Research Symposium on Characterization of High Temperature Vapors and Gases", National Bureau of Standards Special Publication No. 561, Hastie, J. W, Ed.; National Bureau of Standards: Washington, DC, 1979, dealt with this question. See, eg, (a) Hauge, R. H.; Margrave, J. L. Matrix isolation

studies of high temperature species, p. 495. (b) Beattie, I.; Greenhalgh, D. A. The determination of the molecular shapes of high temperature species, p. 597. (c) Drake, M. C.; Rosenblatt, G. M. "Raman Spectroscopy in High Temperature Chemistry, p. 609. (d) Torring, T.; Tiemann, E. Rotational spectroscopy of high temperature molecules, p. 695.

4. Hargittai, I.; Bohátka, S.; Tremmel, J.; Berecz, I. *Hung. Sci. Instrum.* **1980**, *50*, 51.

5. Margrave, J. L., Ed. "The Characterization of High Temperature Vapors". Wiley: New York, 1967.

6. Hastie, J. W. "High Temperature Vapors: Science and Technology. Academic Press: New York, 1975.

7. Hildenbrand, D. L.; Cubicciotti, D. D., Eds. "Proceedings of the Symposium on High Temperature Metal Halide Chemistry". The Electrochemical Society: Princeton, NJ, 1978.

8. See, eg, Brewer, L. In reference 7: Behavior of halides at high temperatures, p. 177.

9. Wong, C.-H.; Schomaker, V. *J. Phys. Chem.* **1957**, 61 358.

10. Vajda, E.; Hargittai, M.; Tremmel, J.; Hargittai, I. Unpublished results.

11. Hargittai, M.; Tremmel, J.; Hargittai, I. *J. Chem. Soc. Dalton Trans.* **1980**, 87.

12. Hargittai, M.; Tremmel, J. Unpublished results.

13. Frey, R. A.; Werder, R. D.; Günthard, H. H. *J. Mol. Spectrosc.* **1970**, *35*, 260.

14. Givan, A.; Loewenschuss, A. *J. Raman Spectrosc.* **1977**, 6, 84, Loewenschuss, A.; Givan, A. *Ber. Bunsenges. Phys. Chem.* **1978**, *82*, 74.

15. Hargittai, M.; Dorofeeva, O. V.; Tremmel, J. *Inorg. Chem.* **1985**, *24*, 3963.

16. Büchler, A.; Stauffer, J. L.; Klemperer, W. *J. Am. Chem. Soc.* **1964**, *86*, 4544.

17. Lesiecki, M. L.; Shirk, J. S. *J. Chem. Phys.* **1972**, *56*, 4171.

18. Shirk, J. S.; Shirk, A. E. *J. Chem. Phys.* **1976**, *64*, 910.

19. Hauge, R. H.; Margrave, J. L.; Kanáan, A. S. *J. Chem. Soc. Faraday* **1975**, *71*, 1082.

20. See, eg, Kuchitsu, K. In "Diffraction Studies on Non-Crystalline Substances", Hargittai, I.; Orville-Thomas, W. J., Eds.; Elsevier: Amsterdam, 1981.

21. Szpiridonov, V. P. *Kém. Közl.* **1972**, *37*, 399, and references therein.

22. Schaefer, H. F. *J. Mol. Struct.* **1981**, *76*, 117.

23. Spiridonov, V. P; Gershikov, A. G.; Zasorin, E. Z.; Butayev, B. S. In "Diffraction Studies on Non-Crystalline Substances", Hargittai, I.; Orville-Thomas, W. J., Eds.; Elsevier. Amsterdam, 1981.

24. Maxwell, L. R.; Hendricks, S. B.; Mosley, V. M. *Phys. Rev.* **1937**, *52*, 968.

25. Hargittai, M.; Hargittai, I. "The Molecular Geometries of Coordination Compounds in the Vapour Phase". Akadémiai Kiadó: Budapest; Elsevier: Amsterdam, 1977.

26. Akishin, P. A.; Rambidi, N. G.; Kuznetsov, G. N.; Matrosov, E. I. *Zh. Neorg. Khim.* **1957**, *2*, 1699, Akishin, P. A.; Rambidi, N. G. *Vestn. Mosk. Univ. Ser. Khim.* **1958** (6), 223; *Zh. Neorg. Khim.* **1958**, *3*, 2599; **1959**, *4*, 718; **1960**, *5*, 23.

27. Akishin, P. A.; Rambidi, N. G. *Z. Phys. Chem.* **1960**, *213*, 111.

28. Lide, D. R., Jr. Cahill, P.; Gold, L. P. *J. Chem. Phys.* **1964**, *40*, 156, and references therein.

29. Solomonik, V. G.; Krasnov, K. S.; Girichev, G. V.; Zasorin, E. Z. *Zh. Strukt. Khim.* **1979**, *20*, 427.

30. Chao, J. *Termochim. Acta* **1970**, *1*, 71.

31. Snelson, A. *J. Chem. Phys.* **1967**, *46*, 3652, Solomonik, V. G.; Danilova, T. G. *Zh. Fiz. Khim.* **1973**, *47*, 1063.

32. Jordan, K. D. "Structure of alkali halides: Theoretical methods. In "Alkali Halide Vapors: Structure, Spectra, and Reaction Dynamics". Davidovits, P.; McFadden, W. H., Eds.; Academic Press: New York, 1979, Chap. 15.

33. Baskin, C. P.; Bender, C. F.; Kollmann, P. A. *J. Am. Chem. Soc.* **1973**, 95, 5868.

34. Swepston, P. N.; Sellers, H. L.; Schäfer, L. *J. Chem. Phys.* **1981**, *74*, 2372.

35. Boldyrev, A. I.; Solomonik, V. G.; Zakzhevskii, V. G.; Charkin, O. P. *Chem. Phys. Lett.* **1980**, *73*, 58.

36. Milne, A.; Cubicciotti, D. *J. Chem. Phys.* **1958**, *29*, 846.

37. Welch, D. O.; Lazareth, O. W.; Dienes, G. J. *J. Chem. Phys.* **1976**, *64*, 835.

38. Miki, H.; Kakumoto, K.; Ino, T.; Kodera, S.; Kakinoki, J. *Acta Crystallogr.* **1980**, *A36*, 96.

39. Martin, T. P.; Schaber, H. *J. Chem. Phys.* **1978**, *68*, 4299.

40. Miller, R. C.; Kusch, P. *J. Chem. Phys.* **1956**, *25*, 860.

41. Akishin, P. A.; Spiridonov, V. P.; Sobolev, G. A. *Dokl. Akad. Nauk SSSR* **1958**, *118*, 1134, Akishin, P. A.; Spiridonov, V. P. *Kristallografiya* **1957**, *2*, 475, Akishin, P. A.; Spiridonov, V. P.; Sobolev, G. A. *Zh. Fiz. Khim.* **1957**, *31*, 648, and references therein.

42. Spoliti, M.; De Maria, G.; D'Alessio, L.; Maltese, M. *J. Mol. Struct.* **1980**, *67*, 159.

43. Kasparov, V. V.; Ezhov, Y. S.; Rambidi, N. G. *Zh. Strukt. Khim.* **1980**, *21*, 41.
44. Kasparov, V. V.; Ezhov, Y. S.; Rambidi, N. G. *Zh. Strukt. Khim.* **1979**, *20*, 260.
45. Vajda, E.; Hargittai, M.; Brunvoll, J.; Tremmel, J.; Hargittai, I. *Inorg. Chem.* **1987**, *26*, 1171.
46. Spiridonov, V. P..; Altman, A. B.; Gershikov, A. G.; Romanov, G. V. *Abstracts*, Second Conference on the Determination of Molecular Structure by Microwave Spectroscopy and Electron Diffraction, Tübingen, Sept. 16–19, 1980.
47. Kasparov, V. V.; Ezhov, Y. S.; Rambidi, N. G. *Zh. Strukt. Khim.* **1979**, *20*, 341.
48. Spiridonov, V. P.; Gershikov, A. G.; Altman, A. B.; Romanov, G. V.; Ivanov, A. A. *Chem. Phys. Lett.* **1981**, *77*, 41.
49. Berkowitz, J.; Marquart, J. R. *J. Chem. Phys.* **1962**, *37*, 1853.
50. Emons, von H. H.; Kiessling, D.; Horlbeck, W. *Z. anorg. allg. Chem.* **1982**, *488*, 219.
51. Drake, M. C.; Rosenblatt, G. M. *J. Electrochem. Soc.* **1979**, *126*, 1387.
52. Baikov, V. I. *Opt. Spektrosk.* **1968**, *25*, 356.
53. Lesiecki, M. L.; Nibler, J. W. *J. Chem. Phys.* **1976**, *64*, 871.
54. Guido, M.; Gigli, G. *J. Chem. Phys.* **1976**, *65*, 1397.
55. Hase, Y.; Takahata, Y. *An. Acad. brasil. Cienc.* **1982**, *54*, 502; **1983**, *55*, 23.
56. Wharton, L.; Berg, R. A.; Klemperer, W. *J. Chem. Phys.* **1963**, *39*, 2023.
57. Calder V.; Mann, D. E.; Seshadri, K. S.; Allavena, M.; White, D. *J. Chem. Phys.* **1969**, *51*, 2093.
58. White, D.; Calder, G. V.; Hemple, S.; Mann, D. E. *J. Chem. Phys.* **1973**, *59*, 6645.
59. Gillespie, R. "Molecular Geometry". Van Nostrand Reinhold: London, 1972.
60. Walsh, A. D. *J. Chem. Soc.* **1953**, 2266.
61. Hayes, E. F. *J. Phys. Chem.* **1966**, *70*, 3740.
62. Gole, J. L.; Siu, A.K.Q.; Hayes, E. F. *J. Chem. Phys.* **1973**, *58*, 857.
63. Cyvin, S. J. "Molecular Vibrations and Mean Square Amplitudes". Universitetsforlaget: Oslo; Elsevier: Amsterdam, 1968.
64. Vrieland, G. E.; Stull, D. R. *J. Chem. Eng. Data* **1967**, *12*, 532.
65. Porter, R. F.; Zeller, E. E. *J. Chem. Phys.* **1960**, *33*, 858.
66. Sjøgren, C. E.; Klaeboe, P.; Rytter, E. *Spectrochim. Acta* **1984**, *40A*, 457, and references therein.
67. Perov, P. A.; Nedjak, S. V.; Malcev, A. A. *Vestn. Mosk. Univ. Ser. Khim.* **1974**, 201, Beattie, I. R.; Blayden, H. E.; Ogden, J. S. *J. Chem. Phys.* **1976**, *64*, 909, Beattie, I. R.; Blayden, H. E.; Hall, S. M.; Jenny, S. N.; Ogden, J. S. *J. Chem. Soc. Dalton Trans.*, **1976**, *666*, Schnökel, v. H. *Z. anorg. allg. Chem.* **1976**, *424*, 203.
68. Tomita, T.; Sjøgren, C. E.; Klaeboe, P.; Papatheodorou, G. N.; Rytter, E. *J. Raman Spectrosc.* **1983**, *14*, 415.
69. Palmer, K. J.; Elliott, N. *J. Am. Chem. Soc.* **1938**, *60*, 1852.
70. Brode, H. *Ann. Phys.* **1940**, *37*, 344.
71. Stevenson, D. P.; Schomaker, V. *J. Am. Chem. Soc.* **1942**, *64*, 2514.
72. Akishin, P. A.; Naumov, V. A.; Tatevskii, V. M. *Kristallografiya* **1959**, *4*, 194.
73. Akishin, P. A.; Rambidi, N. G.; Zasorin, E. Z. *Kristallografiya* **1959**, *4*, 186.
74. Shen, Q. Ph. D. thesis. Oregon State University, Corvallis, 1974.
75. Girichev, G. V.; Utkin, A. N.; Giricheva, N. I. *Izv. Vys. Uch. Zav. Khim. Khim. Teckhnol.* **1983**, *26*, 634.
76. Zasorin, E. Z.; Rambidi, N. G. *Zh. Strukt. Khim.* **1967**, *8*, 391.
77. Spiridonov, V. P.; Gershikov, A. G.; Zasorin, E. Z.; Popenko, N. I.; Ivanov, A. A.; Ermolayeva, L. I. *High Temp. Sci.* **1981**, *14*, 285.
78. Morino, Y.; Ukayi, T.; Ito, T. *Bull. Chem. Soc. Japan* **1966**, *39*, 71.
79. Curtiss, L. A. *Int. J. Quantum Chem.* **1978**, *14*, 709.
80. Büchler, A.; Marram, E. P.; Stauffer, J. L. *J. Phys. Chem.* **1967**, *71*, 4139.
81. Companion, A. L. *J. Chem. Phys.* **1972**, *57*, 1807.
82. So, S. P.; Richards, W. G. *Chem. Phys. Lett.* **1975**, *32*, 231.
83. Hargittai, I.; Hargittai, M. *J. Chem. Phys.* **1972**, *57*, 1807.
84. Greenwood, N. N.; Earnshaw, A. "Chemistry of the Elements". Pergamon Press: Oxford, 1984.
85. Radloff, P. L.; Papatheodorou, G. N. *J. Chem. Phys.* **1980**, *72*, 992.
86. Cubicciotti, D. *High Temp. Sci.* **1970**, *2*, 65.
87. Keneshea, F. J.; Cubicciotti, D. *J. Phys. Chem.* **1965**, *69*, 3910; **1967**, *71*, 1958.
88. Brom, J. M.; Fransen, H. F. *J. Chem. Phys.* **1971**, *54*, 2874, and references therein.
89. Dehmer, J. L.; Berkowitz, J.; Cusachs, L. C. *J. Chem. Phys.* **1973**, *58*, 5681.
90. Fickes, M. G.; Slater, R. C.; Becker, W. G.; Stern, R. C. *Chem. Phys. Lett.* **1974**, *24*, 105.

91. Muenter, J. S. *Chem. Phys. Lett.* **1974**, 26, 97.
92. Solomonik, V. G.; Zasorin, E. Z.; Girichev, G. V.; Krasnov, K. S. *Izv. Vys. Uch. Zav. Khim. Khim. Tekhnol.* **1974**, *17*, 136.
93. Lesiecki, M. L.; Nibler, J. W. *J. Chem. Phys.* **1975**, *63*, 3452.
94. Morino, Y.; Nakamura, Y.; Iijima, T. *J. Chem. Phys.* **1960**, *32*, 643.
95. Fujii, H.; Kimura, M. *Bull. Chem. Soc. Japan* **1970**, *43*, 1933.
96. Hargittai, I.; Schultz, G.; Tremmel, J.; Kagramanov, N. D.; Maltsev, A. K.; Nefedov, O. M. *J. Am. Chem. Soc.* **1983**, *105*, 2895.
97. Schultz, G.; Tremmel, J.; Hargittai, I.; Berecz, I.; Bohátka, S.; Kagramanov, N. D.; Maltsev, A. K.; Nefedov, O. M. *J. Mol. Struct.* **1979**, *55*, 207.
98. Schultz, G.; Tremmel, J.; Hargittai, I.; Kagramanov, N. D.; Maltsev, A. K.; Nefedov, O. M. *J. Mol. Struct.* **1982**, *82*, 107.
99. Ischenko, A. A.; Ivashkevich, L. S.; Zasorin, E. Z.; Spiridonov, V. P.; Ivanov, A. A. Sixth Austin Symposium on Gas-Phase Molecular Structure, Austin, TX, 1976, Paper No. A15. Nasarenko, A. Y.; Spiridonov, V. P.; Butayev, B. S.; Zasorin E. Z. *J. Mol. Struct.* **1985**, *119*, 263.
100. Demidov, A. V.; Gershikov, A. G.; Zasorin, E. Z.; Spiridonov, V. P.; Ivanov, A. A. *Zh. Strukt. Khim.* **1983**, *24*, 9.
101. Hargittai, I.; Tremmel, J.; Vajda, E.; Ishchenko, A. A.; Ivanov, A. A.; Ivashkevich, L. S.; Spiridonov, V. P. *J. Mol. Struct.* **1977**, *42*, 147.
102. Kirchhoff, W. H.; Lide, D. R.; Powell, F. X. *J. Mol. Spectrosc.* **1973**, *47*, 491.
103. Bauschlicher, C. W.; Schaefer, H. F.; Bagus, P. S. *J. Am. Chem. Soc.* **1977**, *99*, 7106.
104. Shoji, H.; Tanaka, T.; Hirota, E. *J. Mol. Spectrosc.* **1973**, *47*, 268.
105. Takeo, H.; Curl, R. F.; Wilson, P. W. *J. Mol. Spectrosc.* **1971**, *38*, 464.
106. Morino, Y.; Murata, Y. *Bull. Chem. Soc. Japan* **1965**, *38*, 104.
107. Konaka, S. *Bull. Chem. Soc. Japan* **1970**, *43*, 3107.
108. Clippard, F. B.; Bartell, L. S. *Inorg. Chem.* **1970**, *9*, 805.
109. Samdal, S.; Barnhart, D. M.; Hedberg, K. *J. Mol. Struct.* **1976**, *35*, 67.
110. (a) Konaka, S.; Kimura, M. *Bull. Chem. Soc. Japan* **1973**, *46*, 404, (b) Ugarov, V. V.; Kalaichev, Y. S.; Petrov, K. P. *Zh. Strukt. Khim.* **1985**, *26*(2), 170.
111. Konaka, S.; Kimura, M. *Bull. Chem. Soc. Japan* **1973**, *46*, 413.
112. Almenningen, A.; Bjorvatten, T. *Acta Chem. Scand.* **1963**, *17*, 2573.
113. Bürgi, H. B.; Stedman, D., Bartell, L. S. *J. Mol. Struct.* **1971**, *10*, 31.
114. Hedberg, K.; Iwasaki, M. *J. Chem. Phys.* **1962**, *36*, 589.
115. Brunvoll, J.; Ischenko, A. A.; Miakshin, I. N.; Romanov, G. V.; Spiridonov, V. P.; Strand, T. G.; Sukhoverkhov, V. F. *Acta Chem. Scand.* **1980**, *A34*, 733.
116. Ivashkevich, L. S.; Ischenko, A. A.; Spiridonov, V. P.; Strand, T. G.; Ivanov, A. A.; Nikolaev, A. N. *Zh. Strukt. Khim.* **1982**, *23* (2), 144.
117. Brunvoll, J.; Ischenko, A. A.; Miakshin, I. N.; Romanov, G. V.; Sokolov, V. B.; Spiridonov, V. P.; Strand, T. G. *Acta Chem. Scand.* **1979**, *A33*, 775.
118. Mitchell, K.A.R. *Chem. Rev.* **1969**, *69*, 157.
119. Guido, M.; Balducci, G.; Gigli, G.; Spoliti, M. *J. Chem. Phys.* **1971**, *55*, 4566.
120. Krabbes, v. G.; Oppermann, H. *Z. anorg. allg. Chem.* **1977**, *435*, 33.
121. Joyce, T. E.; Rolinski, E. J. *J. Phys. Chem.* **1972**, *76*, 2310.
122. Butayev, B. S.; Gershikov, A G.; Spiridonov, V. P. *Vestn. Mosk Univ. Ser. Khim.* **1978**, *19*, 734.
123. Hargittai, M.; Tremmel, J.; Hargittai, I. *Inorg. Chem.* **1986**, *25*, 3163.
124. Kulikov, V. A.; Ugarov, V. V.; Rambidi, N. G. *Zh. Strukt. Khim.* 1980, *21* (4), 201.
125. Petrov, V. M.; Utkin, A. N.; Girichev, G. V.; Ivanov, A. A. *Zh. Strukt. Khim.* **1985**, *26*(2), 52.
126. Kashiwabara, K.; Konaka, S.; Kimura, M. *Bull. Chem. Soc. Japan* **1973**, *46*, 410.
127. Spiridonov, V. P..; Gershikov, A. G.; Butayev, B. S. *J. Mol. Struct.* **1979** *52*, 53; Gershikov, A. G. *Zh. Strukt. Khim.* **1984**, *25*, 30.
128. Loewenschuss, A.; Givan, A. *Ber. Bunsenges. Chem.* **1978**, *82*, 69.
129. Loewenschuss, A.; Ron, A.; Schnepp, O. *J. Chem. Phys.* **1968**, *49*, 272; **1969**, *50*, 2502.
130. Givan, A..; Loewenschuss, A. *J. Chem. Phys.* **1976**, *65*, 1851.
131. Givan, A.; Loewenschuss, A. *J. Mol. Struct.* **1978**, *48*, 325.
132. Büchler, A.; Stauffer, J. L.; Klemperer, W. *J. Chem. Phys.* **1964**, *40*, 3471.
133. Girichev, G. V.; Subbotina, N. Y.; Giricheva, N. I. *Zh. Strukt. Khim.* **1983**, *24*, 158.
134. Girichev, G. V.; Subbotina, N. Y.; Krasnov, K. S.; Ostropikov, V. V. *Zh. Strukt. Kim.* **1984**, *25*, 170.

135. Eddy, L. *Dissertation*, Oregon State University, Corvallis, 1973.
136. Hargittai, I.; Tremmel, J.; Schultz, G. *J. Mol. Struct.* **1975**, *26*, 116.
137. Tremmel, J.; Ivanov, A. A.; Schultz, G.; Hargittai, I.; Cyvin, S. J.; Eriksson, A. *Chem. Phys. Lett.* **1973**, *23*, 533.
138. Molnár, Z.; Schultz, G.; Tremmel, J.; Hargittai, I. *Acta Chim. Acad. Sci. Hung.* **1975**, *86*, 223.
139. Schoonmaker, R. C.; Friedman, A. H.; Porter, R. F. *J. Chem. Phys.* **1959**, *31*, 1586.
140. Hargittai, M.; Kolonits, M.; Tremmel, J.; Hargittai, I. To be published.
141. Hargittai, M.; Hargittai, I. *J. Mol. Spectrosc.* **1984**, *108*, 155.
142. Hargittai, M.; Hargittai, I.; Tremmel, J. *Chem. Phys. Lett.* **1981**, *83*, 207.
143. Vajda, E.; Tremmel, J.; Hargittai, I. *J. Mol. Struct.* **1978**, *44*, 101.
144. Hargittai, M.; Dorofeeva, O. V.; Tremmel, J. *Inorg. Chem.* **1985**, *24*, 245.
145. Hastie, J. W.; Hauge, R. H.; Margrave, J. L. *Ann. Rev. Phys. Chem.* **1970**, *21*, 475; *J. Chem. Soc. Chem. Commun.* **1969**, 1452; *High Temp. Sci.* **1969**, *1*, 76; **1971**, *3*, 257.
146. Green, D. W.; McDermott, D. P.; Bergman, A. *J. Mol. Spectrosc.* **1983**, *98*, 11.
147. Papatheodorou, G. N. In "Proceedings of the Tenth Materials Research Symposium on Characterization of High Temperature Vapors and Gases", National Bureau of Standards Special Publication No. 561, Hastie, J. W., Ed.; National Bureau of Standards: Washington, DC, 1979, pp. 647–678.
148. Thompson, K. R.; Carlson, K. D. *J. Chem. Phys.* **1968**, *49*, 4379.
149. Jacox, M. E.; Milligan, D. E. *J. Chem. Phys.* **1969**, *51*, 4143.
150. Hargittai, M. *Inorg. Chim. Acta* **1981**, *53*, L111.
151. Hargittai, I.; Hargittai, M. "Symmetry Through the Eyes of a Chemist". VCH: Weinheim, 1986.
152. Kaiser, E. W.; Falconer, W. E.; Klemperer, W. *J. Chem. Phys.* **1972**, *56*, 5392.
153. Hastie, J. W.; Hauge, R.H.; Margrave, J. L. *J. Less Common Metals* **1975**, *39*, 309, Hauge, R. H.; Hastie, J. W.; Margrave, J. L. *ibid* **1971**, *23*, 359.
154. Wells, J. W., Jr., Gruber, J. B.; Lewis, M. *Chem. Phys.* **1977**, *24*, 391.
155. Charkin, O. P.; Dyatkina, M. E. *Zh. Strukt. Khim.* **1964**, *5*, 921.
156. Giricheva, N. I.; Zasorin, E. Z.; Girichev, G. V.; Krasnov, K. S.; Spiridonov, V. P. *Zh. Strukt. Khim.* **1976**, *17*, 797.
157. Krasnov, K. S.; Girichev, G. V.; Giricheva, N. I.; Petrov, V. M.; Danilova, T. G.; Zasorin, E. Z.; Popenko, N. I. *Abstracts*, Seventh Austin Symposium on Gas-Phase Molecular Structure, Austin, TX, 1978, p. 89.
158. Girichev, G. V.; Danilova, T. G.; Giricheva, N. I.; Krasnov, K. S.; Petrov, V. M.; Utkin, A. N.; Zasorin, E. Z. *Izv. Vys. Uch. Zav. Khim. Khim. Technol.* **1978**, , *21*, 627.
159. Danilova, T. G.; Girichev, G. V.; Giricheva, N. I.; Krasnov, K. S.; Zasorin, E. Z. *Izv. Vys. Uch. Zav. Khim. Khim. Technol.* **1977**, *20*, 1069.
160. Girichev, G. V.; Danilova, T. G.; Giricheva, N. I.; Krasnov, K. S.; Zasorin, E. Z. *Izv. Vys. Uch. Zav. Khim. Khim. Technol.* **1977**, *20*, 1233.
161. Giricheva, N. I.; Zasorin, E. Z.; Girichev, G. V.; Krasnov, K. S.; Spiridonov, V. P. *Izv. Vys. Uch. Zav. Khim. Khim. Technol.* **1977**, *20*, 284.
162. Giricheva, N. I.; Zasorin, E. Z.; Girichev, G. V.; Krasnov, K. S.; Spiridonov, V. P. *Izv. Vys. Uch. Zav. Khim. Khim. Technol.* **1974**, *17*, 762.
163. Popenko, N. I.; Zasorin, E. Z.; Spiridonov, V. P.; Ivanov, A. A. *Inorg. Chim. Acta* **1978**, *31*, L371.
164. Krasnov, K. S.; Giricheva, N. I.; Girichev, G. V. *Zh. Strukt. Khim.* **1976**, *17*, 667.
165. Zasorin, E. Z.; Rambidi, N. G.; Akishin, P. A. *Zh. Strukt. Khim.* **1963**, *4*, 910.
166. Ugarov, V. V.; Vinogradov, V. S.; Zasorin, E. Z.; Rambidi, N. G. *Zh. Strukt. Khim.* **1971**, *12*, 315.
167. Pauling, L. "The Nature of the Chemical Bond", 3rd ed. Cornell University Press: Ithaca, NY, 1960.
168. Krasnov, K. S. *Zh. Strukt. Khim.* **1983**, *24*, 3.
169. Petrov, V. M.; Girichev, G. V.; Giricheva, N. I.; Krasnov, K. S.; Zasorin, E. Z. *Zh. Strukt. Khim.* **1979**, *20*, 55.
170. Girichev, G. V.; Petrov, V. M.; Giricheva, N. I.; Krasnov, K. S. *Zh. Strukt. Khim.* **1982**, *23*, 56.
171. Petrov, V. M.; Girichev, G. V.; Giricheva, N. I.; Shaposhnikova, O. K.; Zasorin, E. Z.; *Zh. Strukt. Khim.* **1979**, *20*, 136.
172. Morino, Y.; Uehara, H. *J. Chem. Phys.* **1966**, *45*, 4543.
173. Spiridonov, V. P.; Akishin, P. A.; Tsirel'nikov, V. I. *Zh. Strukt. Khim.* **1962**, *3*, 329.
174. Girichev, G. V.; Petrov, V. M.; Giricheva, N. I.; Utkin, A. N.; Petrova, V. N. *Zh. Strukt. Khim.* **1981**, *22*, 65.

175. Girichev, G. V.; Zasorin, E. Z.; Giricheva, N. I.; Krasnov, K. S.; Spiridonov, V. P. *Zh. Strukt. Khim.* **1977**, *18*, 42.

176. Ivashkevich, L. S.; Ischenko, A. A.; Spiridonov, V. P.; Romanov, G. V. *J. Mol. Struct.* **1979**, *51*, 217.

177. Girichev, G. V.; Giricheva, N. I.; Malysheva, T. N. *Zh. Strukt. Khim.* **1982**, *56*, 1833.

178. Spiridonov, V. P.; Romanov, G. V. *Zh. Strukt. Khim.* **1967**, *8*, 160.

179. Jahn, H. A.; Teller, E. *Proc. R. Soc. London* **1937**, *161*, 220.

180. Bersuker, I. B. *Coord. Chem. Rev.* **1975**, *14*, 357. Bersuker, I. B. "The Jahn-Teller Effect and Vibronic Interactions in Modern Chemistry". Plenum Press: New York, 1984.

181. Parameswaran, T.; Koningstein, J. A.; Haley, L. V. *J. Mol. Spectrosc.* **1977**, *66*, 350.

182. Clark, J. H.; Hunter, B. K.; Rippon, D. M. *Inorg. Chem.* **1972**, *11*, 56.

183. Ezhov, Y. S.; Akishin, P. A.; Rambidi, N. G. *Zh. Strukt. Khim.* **1969**, *10*, 763.

184. Mooney, R.C.L. *Acta Crystallogr.* **1949**, *2*, 189.

185. D'Eye, R. W. M. *J. Chem. Soc.* **1950**, 2764.

186. Rambidi, N. G.; Akishin, P. A.; Zasorin, E. Z. *Zh. Strukt. Khim.* **1961**, *35*, 1171.

187. Girichev, G. V.; Petrov, V. M.; Giricheva, N. I.; Zasorin, E. Z.; Krasnov, K. S.; Kiseljev, Y. M. *Zh. Strukt. Khim.* **1983**, *24*, 70.

188. Hildenbrand, D. L. *J. Chem. Phys.* **1977**, *66*, 4788.

189. Dyke, J. M.; Fayad, N. K.; Morris, A.; Trickle, I. R.; Allen, G. C. *J. Chem. Phys.* **1980**, *72*, 3822.

190. Kunze, K. R.; Hauge, R. H.; Hamill, D.; Margrave, J. L. *J. Phys. Chem.* **1977**, *81*, 1664.

191. Petrov, V. M.; Girichev, G. V.; Giricheva, N. I.; Petrova, V. N.; Krasnov, K. S.; Zasorin, E. Z.; Kiseljev, Y. M. *Dokl. A.N. SSSR* **1981**, *259*, 1399.

192. Spiridonov, V. P.; Romanov, G. V. *Vestn. Mosk. Univ. Ser. Khim.* **1967**, *22*, 118.

193. Ezhov, Y. S.; Komarov, S. A. *Zh. Strukt. Khim.* **1983**, *24*, 156; **1984**, *25*, 82.

194. Kimura, K.; Bauer, S. H. *J. Chem. Phys.* **1963**, *39*, 3172.

195. Iijima, K.; Shibata, S. *Bull. Chem. Soc. Japan* **1974**, *47*, 1393.

196. Page, E. M; Rice, D. A.; Hagen, K.; Hedberg, L.; Hedberg, K. *Inorg. Chem.* **1982**, *21*, 3280.

197. Romanov, G. V.; Spiridonov, V. P. *Zh. Strukt. Khim.* **1966**, *7*, 882.

198. Hagen, K.; Gilbert, M. M.; Hedberg, L.; Hedberg, K. *Inorg. Chem.* **1982**, *21*, 2690.

199. Dyke, T. R.; Muenter, A. A.; Klemperer, W.; Falconer, W. E. *J. Chem. Phys.* **1970**, *53*, 3382.

200. Berry, R. S. *J. Chem. Phys.* **1960**, *32*, 933.

201. Preiss, H. *Z. anorg. allg. Chem.* **1972**, *389*, 280.

202. Bennett, L. S.; Margrave, J. L.; Franklin, J. L. *J. Inorg. Nucl. Chem.* **1975**, *37*, 937.

203. Berdonosov, S. S.; Lapitskij, P. V.; Bakov, E. K. *Zh. Neorg. Khim.* **1965**, *10*, 322.

204. Ischenko, A. A.; Strand, T. G.; Demidov, A. V.; Spiridonov, V. P. *J. Mol. Struct.* **1978**, *43*, 227.

205. Demidov, A. V.; Ivanov, A. A.; Ivashkevich, L. S.; Ischenko, A. A.; Spiridonov, V. P. *Chem. Phys. Lett.* **1979**, *64*, 528.

206. Spiridonov, V. P.; Romanov, G. V. *Vestn. Mosk. Univ. Ser. Khim.* **1968**, *23*, 10.

207. Spiridonov, V. P.; Romanov, G. V. *Vestn. Mosk. Univ. Ser. Khim.* **1966**, *21*, 109.

208. McClelland, B. W.; Hedberg, L.; Hedberg, K. *J. Mol. Struct.* **1983**, *99*, 309.

209. Hansen, K. W.; Bartell, L. S. *Inorg. Chem.* **1965**, *4*, 1775.

210. Jacob, E. J.; Hedberg, L.; Hedberg, K.; Davis, H.; Gard, G. L. *J. Phys. Chem.* **1984**, *88*, 1935.

211. Spiridonov, V. P.; Romanov, G. V. *Vestn. Mosk. Univ. Ser. Khim.* **1967**, *22*, 98.

212. Spiridonov, V. P.; Romanov, G. V. *Vestn. Mosk. Univ. Ser. Khim.* **1968**, *23*, 10.

213. Ezhov, Y. S.; Sarvin, A. P. *Zh. Strukt. Khim.* **1983**, *24*, 57.

214. Ezhov, Y. S.; Sarvin, A. P. *Zh. Strukt. Khim.* **1983**, *24*, 149.

215. Brunvoll, J.; Ischenko, A. A.; Spiridonov, V. P.; Strand, T. G. *Acta Chem. Scand.* **1984**, *A38*, 115.

216. Romanov, G. V.; Spiridonov, V. P. *Vestn. Mosk. Univ. Ser. Khim.* **1968**, *23*, 7.

217. Edwards, A. J. *J. Chem. Soc.* **1964**, 3714.

218. Gusarov, A. V.; Gorokhov, L. N.; Gotkis, I. S. *Adv. Mass-Spectrom.* **1978**, *7*, 666.

219. Brunvoll, J.; Ischenko, A. A.; Miakshin, I. N.; Romanov, G. V.; Sokolov, V. B.; Spiridonov, V. P.; Strand, T. G. *Acta Chem. Scand.* **1979**, *A33*, 775.

220. Gotkis, I. S.; Gusarov, A. V.; Pervov, V. S.; Butskii, V. D. *Koord. Khim.* **1978**, *4*, 720.

221. Girichev, G. V.; Petrova, V. N.; Petrov, V. M.; Krasnov, K. S.; Goncharuk, V. P. *Zh. Strukt. Khim.* **1983**, *24*, 54.

222. Iijima, K. *Bull. Chem. Soc. Japan* **1977**, *50*, 373.

223. Brunvoll, J.; Ischenko, A. A.; Ivanov, A. A.; Romanov, G. V.; Sokolov, V. B.; Spiridonov, V. P.; Strand, T. G. *Acta Chem. Scand.* **1982**, *A36*, 705.

224. Vasile, M. J.; Richardson, T. J.; Stevie, F. A.; Falconer, W. E. *J. Chem. Soc. Dalton Trans.* **1976**, *351*.

225. Seip, H. M.; Seip, R. *Acta Chem. Scand.* **1966**, *20*, 2698.

226. Kimura, M.; Schomaker, V.; Smith, D. W.; Weinstock, B. *J. Chem. Phys.* **1968**, *48*, 4001.

227. Jacob, E. J.; Bartell, L. S. *J. Chem. Phys.* **1970**, *53*, 2231.

228. Seip, H. M. *Acta Chem. Scand.* **1965**, *19*, 1955.

229. Jacob, E. J.; Bartell, L. S. *J. Chem. Phys.* **1970**, *53*, 2235.

230. Claassen, H. H.; Gasner, E. L.; Selig, H. *J. Chem. Phys.* **1968**, *49*, 1803, Claassen, H. H.; Selig, H. *J. Chem. Phys.* **1965**, *43*, 103.

231. Kaiser, E. W.; Muenter, J. S.; Klemperer, W.; Falconer, W. E. *J. Chem. Phys.* **1970**, *53*, 53.

232. Schäfer, H. *Angew. Chem. Int. Ed. Engl.* **1976**, *15*, 713.

233. Hargittai, M. *Kém. Közl.* **1978**, *50*, 489.

234. Novikov, G. I..; Gavryuchenkov, F. G. *Usp. Khim.* **1967**, *36*, 399.

235. Solc, V. B.; Sidorov, L. N. *Vestn. Mosk. Univ. Ser. Khim.* **1972**, *13*, 371.

236. Huglen, R.; Cyvin, S. J.; Øye, H. A. Spectroscopic studies of alkali tetrafluoroaluminates. In "High Temperature Metal Halide Chemistry", Hildenbrand, D. L.; Cubicciotti, D. D.; Eds.; The Electrochemical Society; Princeton, NJ, 1978.

237. Brooker, M. H.; Papatheodorou, G. N. Vibrational spectroscopy of molten salts and related glasses and vapors. In "Advances in Molten Salt Chemistry", Vol. 5, Mamantov, G., Ed.; Elsevier: Amsterdam, 1983.

238. Brezgin, Y. A. Thesis, Moscow State University, 1972.

239. Spiridonov, V. P.; Erokhin, E. V.; Brezgin, Y. A. *Zh. Strukt. Khim.* **1972**, *13*, 321.

240. Spiridonov, V. P.; Brezgin, Y. A.; Shakhparonov, M. I. *Zh. Strukt. Khim.* **1971**, *12*, 1080.

241. Spiridonov, V. P.; Brezgin, Y. A.; Shakhparonov, M. I. *Zh. Strukt. Khim.* **1972**, *13*, 320.

242. Curtiss, L. A. *Chem. Phys. Lett.* **1979**, *68*, 225.

243. Spiridonov, V. P.; Erokhin, E. V. *Zh. Neorg. Khim.* **1969**, *14*, 636.

244. Vajda, E.; Hargittai, I.; Tremmel, J. *Inorg. Chim. Acta* **1977**, *25*, L143.

245. Kalaichev, Y. S.; Petrov, K. P.; Ugarov, V. V. *Zh. Strukt. Khim.* **1983**, *24*, 176.

246. Veazy, S. E.; Gordy, W. *Phys. Rev.* **1965**, *138*, 1303.

247. Kalaichev, Y. S.; Petrov, K. P.; Ugarov, V. V. *Zh. Strukt. Khim.* **1983**, *24*, 173.

248. Kalaichev, Y. S. ; Petrov, K. P.; Ugarov, V. V. *Zh. Strukt. Khim.* **1983**, *24*, 179.

249. Spiridonov, V. P.; Erokhin, E. V.; Lutoshkin, B. I. *Vestn. Mosk. Univ. Ser. Khim.* **1971**, *12*, 296.

250. Rambidi, N. G. *J. Mol. Struct.* **1975**, *28*, 77, 88.

251. Rytter, E.; Øye, H. A.; Cyvin, S. J.; Cyvin, B. N.; Klaeboe, P. *J. Inorg. Nucl. Chem.* **1973**, *35*, 1185.

252. Curtiss, L. A.; Heinricher, A. *Int. J. Quantum Chem.* **1982**, *16*, 285; *Chem. Phys. Lett.* **1982**, *86*, 467.

253. Hargittai, M.; Hargittai, I.; Spiridonov, V. P.; Pelissier, M.; Labarre, J.-F. *J. Mol. Struct.* **1975**, *24*, 27.

254. Almenningen, A.; Haaland, A.; Haugen, T.; Novak, D. P. *Acta Chem. Scand.* **1973**, *27*, 1821.

255. Hargittai, M.; Hargittai, I.; Spiridonov, V. P.; Ivanov, A. A. *J. Mol. Struct.* **1977**, *39*, 225.

256. Hargittai, M.; Hargittai, I.; Spiridonov, V. P. *J. Mol. Struct.* **1976**, *30*, 31.

257. Papatheodorou, G. N.; Curtiss, L. A.; Maroni, V. A. *J. Chem. Phys.* **1983**, *78*, 3303.

258. Sjøgren, C. E.; Rytter, E. *Spectrochim. Acta* **1985**, *41A*, 1277.

259. Taylor, M. J.; Riethmiller, S. *J. Raman Spectrosc.* **1984**, *15*, 370.

260. Lathan, W. A.; Hehre, W. J.; Curtiss, L. A.; Pople, J. A. *J. Am. Chem. Soc.* **1971**, *93*, 6377.

261. Kuchitsu, K.; Guillory, J. P.; Bartell, L. S. *J. Chem. Phys.* **1968**, *49*, 2488.

262. Beagley, B.; Hewitt, T. G. *Trans. Faraday Soc.* **1968**, *64*, 2561.

263. Hargittai, M. Thesis, Hungarian Academy of Sciences, Budapest, 1977.

264. Almenningen, A.; Gundersen, G.; Haugen, T.; Haaland, A. *Acta Chem. Scand.* **1972**, *26*, 3928.

265. Anderson, G. A.; Forgaard, F. R.; Haaland, A. *Acta Chem. Scand.* **1972**, *26*, 1947.

266. Mawhorter, R. J.; Fink, M.; Hartley, J. G. *J. Chem. Phys.* **1985**, *83*, 4418.

10

INTERACTION OF THEORETICAL CHEMISTRY WITH GAS-PHASE ELECTRON DIFFRACTION

James E. Boggs

Department of Chemistry
University of Texas
Austin, Texas

CONTENTS

INTRODUCTION 455
HOW IS A MOLECULAR STRUCTURE DETERMINED FROM QUANTUM
THEORY? 457
HOW CAN SUFFICIENT ACCURACY BE ATTAINED? 459
HOW IS A MOLECULAR FORCE FIELD OBTAINED FROM QUANTUM
THEORY? 462
THE INTERACTION OF QUANTUM CHEMISTRY WITH EXPERIMENTAL
STRUCTURE DETERMINATION 466
PROJECTIONS FOR THE NEAR FUTURE 473
REFERENCES 474

Introduction

In recent years there has been an explosive growth in the ability to obtain even very fine details of molecular structure by purely theoretical computation. The proliferation of papers reporting structural results obtained by a wide variety of computational techniques, often with little or no comment on the expected accuracy and reliability of the results, can lead to great uncertainty in attempting an evaluation of the field and its interactions with classic experimental methods. Will computation soon make the practice of electron diffraction obsolete so that any structure can be obtained at will from theory? Can theory only serve as an adjunct and guide to experiment, making predictions that must be verified experimentally before they can be relied on? In all the cornucopia of products sold under the name "theoretical chemistry," which methods and levels of calculation are reliable, and to what degree can they be believed in the absence of experimental confirmation?

Before examining the areas of interaction between electron diffraction and the quantum chemical evaluation of molecular structure, it is necessary to gain some insight into the procedures used in the theoretical approach so that their applicability and limitations can be appreciated. This chapter therefore begins with a nonmathematical summary of the basis of the theoretical method for determining molecular structures and the factors that determine the accuracy of the results. Next, the theoretical methods available for determination of molecular vibrational force fields are discussed. These are important in the present context, since any accurate comparison of the structural results coming from different experimental or theoretical methods must consider the vibrational behavior of the molecule and the manner in which this is averaged by the particular experiment used. The ways in which quantum chemical calculations rely on accurate experimental data from gas-phase electron diffraction and from other experimental methods are discussed, as well as the types of assistance that theoretical calculations can provide to the experimentalist. These interactions are illustrated by numerous examples, perhaps excessively taken from the work of the author and his colleagues, but the results given are intended only to illustrate the main points discussed rather than to serve as an exhaustive compilation of all the work that has been reported in the literature. Finally, a few projections are offered for future work in this field. The entire discussion is directed toward molecules in the range of sizes amenable to study by electron diffraction, and emphasis is placed on the desirability of accuracy and reliability at the levels which electron diffraction achieves. The chapter is therefore organized into the following sequence of topics:

How is a molecular structure determined from quantum theory?
How can sufficient accuracy be attained?
How is a molecular force field obtained from quantum theory?
The interaction of quantum chemistry with experimental structure determinations
Projections for the near future

The entire discussion is maintained at a nonmathematical level, since the purpose of this volume is not to train theoretical chemists but to provide an understanding of the state of the art and its implications for those in other fields of chemistry. Hopefully, practitioners in other areas can gain heightened awareness of whatever utility present-day theory may have in their own fields and become able to evaluate the results presented in theoretical papers with greater confidence.

How Is a Molecular Structure Determined from Quantum Theory?

In outline, the quantum mechanical evaluation of the structure of a molecule is straightforward:

1. An initial trial set of nuclear coordinates is assumed.
2. With that set of nuclear positions and assumption of the Born–Oppenheimer separation of electronic and nuclear motions, the time-independent electronic Schrödinger equation is solved at some chosen level of approximation.
3. From the resulting approximate wave function, the first derivative of energy with respect to all nuclear coordinates (the force acting on all nuclei) is calculated analytically.
4. A set of molecular force constants is assumed and all nuclei are moved as suggested by the forces and force constants.
5. Steps 2–4 are repeated until all forces become as near zero as desired

A number of comments must be made on this procedure. First, it is important to realize that the assumptions made in steps 1 and 4 have no effect on the accuracy of the final result; they merely affect how many cycles may be necessary to reach it. In the end, all forces acting on the nuclei are zero, and this by definition is the equilibrium molecular geometry. The important consequence therefore develops that, unlike all experimental methods, computation yields directly an equilibrium structure. The equilibrium structure does not appear as such in nature, but neither do any of the various types of vibrationally averaged structure that result from different experiments. The equilibrium structure has a precise physical definition (a minimum on the potential energy surface) unlike, for example, the r_s structure that can be obtained from microwave spectroscopy. Given sufficient information about the molecular vibrational potential, the r_e (equilibrium) structure can be converted to any of the other types of vibrationally averaged structure that do have a physical meaning.

Since computer time is generally important, careful thought given to the assumptions in steps 1 and 4 above can be well rewarded. A number of helpful suggestions have been made in a paper dealing with the general procedure.[1] It is especially useful to adopt a complete and nonredundant set of internal coordinates rather than to perform the geometry optimization in, for example, Cartesian coordinates. Certain computer programs carry out the cyclic procedure automatically, but it is almost always possible to save appreciable computer time if thoughtful consideration is given to the results at the end of each cycle.

It must be remembered that a force is zero at a maximum or saddle point in the potential as well as at a minimum, so care must be taken to ensure that the zero-force geometry found is not that of a transition state. If the computer program used is one that calculates second derivatives of the energy analytically, one can check to see that the force constant matrix is positive definite. Otherwise, it is a simple matter to be sure that the molecular energy increases with a slight displacement of any relevant nuclear coordinate.

If the molecular potential energy surface has multiple minima, the procedure will normally converge to the one nearest the initial assumed geometry. The convergence is not like the least-squares methods used in electron diffraction analyses in that it does not lead to false minima. All minima found correspond to real conformations of the molecule, but they need not be the ones desired. Consequently, if information about all possible molecular conformations is desired, the computation must be started with a geometry near each prospective minimum, in turn.

Step 3 in the outline, the computation of the energy gradient, is the key that opened the application of quantum chemical methods for determining structures and force constants to molecules containing more than a very few atoms. It is essential to compute the total force, not just the Hellmann–Feynman force, as was emphasized by Péter Pulay in the 1969 article[2] in which he gave the first complete development of the gradient method. Building on the very important early systematic studies of Pople and his co-workers (see reference 3 for a recent review), the gradient procedure has become standard and is now used in a great variety of formulations throughout the world.[4,5]

All the error involved in the theoretical determination of molecular structure arises from the approximations made in step 2, the solution of the electronic Schrödinger equation. We turn now to a consideration of the approximations that are commonly adopted and to their consequences:

1. The Born–Oppenheimer separation of electronic and nuclear motion is assumed. This causes no difficulty at the level of accuracy that can be achieved with the other approximations used.

2. Nonrelativistic methods are commonly used, and no spin terms are included in the Hamiltonian. This approximation is negligible when working with elements lighter than the transition metals but becomes serious for heavier elements. Because of the great practical importance of such systems, extensive work is currently in progress to find techniques that give accurate information for the heavier elements but still have practical requirements for computational facilities. Organometallic compounds containing both heavy atoms and an appreciable number of lighter ones are of great interest, as are metal clusters, but both kinds of system are still at the frontier of theory. A detailed consideration of current progress in this field is beyond the scope of the present survey. The limitation, hopefully a temporary one, eliminates a large number of substances of interest to electron diffractionists, but it still leaves access to most organic and many inorganic substances.

3.   Next, it is very common to use the Hartreee-Fock self-consistent field (SCF) method with complete neglect of the effect of dynamic electron correlation. Unless correction is made for this neglect, serious errors result in computed energies, molecular geometries, vibrational force constants, and other molecular properties. The errors, especially in computed bond lengths, are larger than is generally considered acceptable in modern electron diffraction work. Methods to deal with this problem are discussed below.

4. In the Hartree-Fock method, the individual molecular orbitals are expanded in some chosen orthonormal basis set and the expansion coefficients are determined in accordance with the variation principle. If the basis set could be infinite, the method would be exact, but in practice the basis is not only finite but often rather small. This produces the second major cause of error in the determination of the structural parameters for molecules of the lighter elements. A delicate balance exists between the requirement of high accuracy, the desirability of applicability of the method to ever larger molecules, and the limits imposed by existing computers. The balance and the necessary compromises are discussed and illustrated below.

5. As a further concession to computer limitations, it is possible to avoid computation of some of the more difficult integrals by fitting appropriate parameters to various measured molecular properties. This approach leads to the variety of semiempirical methods (CNDO, MINDO, MNDO, etc) in widespread use. Such methods are fully adequate for many purposes, and they have made important contributions both in the understanding of organic reaction mechanisms and in more applied areas such as the design of pharmaceuticals, where purely empirical methods such as molecular mechanics are also valuable. However, for the evaluation of molecular structure and vibrational dynamics, semiempirical methods cannot reliably achieve the level of accuracy reached by high-level *ab initio* calculations or by the better experimental methods. Use of these approximations, valuable as they are for other purposes, will therefore not be considered further in this chapter.

In summary, then, this chapter deals with *ab initio* gradient methods for calculating geometries and force fields of molecules containing up to about two dozen atoms, with attention limited to elements lighter than the transition metals. The effects of finite basis set size and of electron correlation are of major concern.

How Can Sufficient Accuracy Be Attained?

The most obvious way to avoid the errors introduced by neglect of electron correlation and use of an excessively small basis set is not to make these approximations. Useful tests of accuracy can be made on very small molecules for which both experiments and calculations of the highest quality can be

Table 10-1. Experimental and Theoretical Equilibrium
Bond Lengths (Å) in HCN

Parameter	Experimental[a]	Large calculation[b]	Small calculation[c]
r_e(C—H)	1.0655	1.0656	1.0504
r_e(C—N)	1.1532	1.1538	1.1368

[a] Reference 6.
[b] 6-311G^4* basis plus single and double configuration interaction (except only single excitation from the core orbitals) plus Davidson coupled cluster correction (reference 7).
[c] 4-21 basis, no electron correlation treatment (reference 7).

performed. As one illustration, Table 10-1 shows such a comparison for HCN. The "large calculation" is not at the limit of modern computer capabilities, but it is near enough to that limit to make such a computation impossible for any other molecule appreciably larger than HCN. Here, and in the few other cases where such comparisons can be made, it appears that the deviation between the best theory and the best experiments is in the 10^{-4} Å range.

As an illustration of what can be done with larger molecules, Table 10-1 also shows a "small calculation" on HCN at a level that can practically be applied to a molecule as large, for example, as quinoline. The error is now in the 10^{-2} Å range, and although this still represents 1% accuracy, it is large enough to mask many chemical effects of interest. Calculations at quantum mechanical levels intermediate between the two shown in Table 10-1 can produce slight variations in the accuracy, but if electron correlation is ignored, close agreement between theory and experiment normally reflects a fortuitous cancellation of errors. One may correctly conclude that highly accurate structures can be calculated *ab initio* with high reliability for very small molecules, but if similar accuracy is desired from larger molecules, there is no direct method of calculation that is reliable.

One way around the problem can be understood by considering how the two important computational sources of error—use of too small a basis set and neglect of electron correlation—separately affect the resulting computational error. A great deal of experience in computing geometrical parameters at various levels of sophistication is summarized schematically in Figure 10-1.[8] This figure sketches the absolute error in the calculation of some one type of geometrical parameter, for example, C—H bond lengths, in studies on a family of related molecules. If the various calculations are done with different sized basis sets and without allowance for electron correlation, the errors are found to fall within the shaded area of the diagram. The convergence limit for a Hartree-Fock calculation with a very large basis set still shows an error due to neglect of correlation. The diagram presents the empirical observation that this electron correlation error in geometrical parameters is independent of the

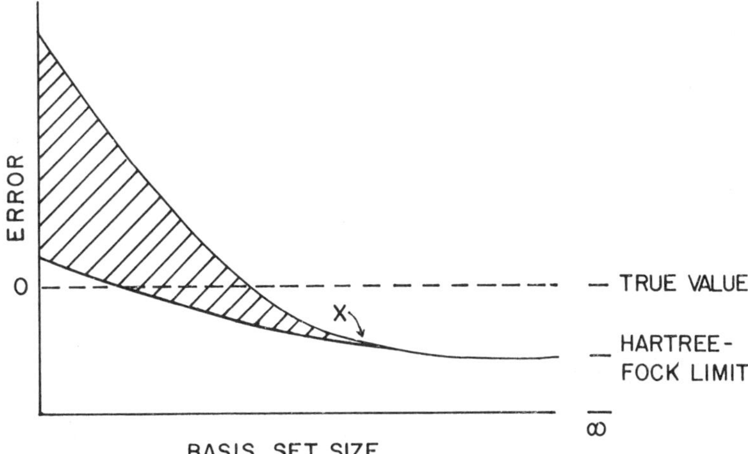

Figure 10-1. Schematic representation of the error in calculation of some single structural parameter (such as C—H distance) as a function of basis set size in a variety of molecular environments. For large families of systems, the error is expected to fall in the shaded area; X marks the level of calculation that combines economy with consistency.

molecular environment within the accuracy allowed by comparison with existing experimental data. Furthermore, there is a convergence in the error due to finite basis set size. These observations of consistency in the computational error support the idea that *relative* values of a given bond length or angle in different environments can be calculated with very high accuracy even though there is a constant small error in the *absolute* value.

Fortunately, nearly all the chemical questions that can be resolved by knowing structural information require relative rather than absolute data. There is very little information content in knowing the absolute length of some one internuclear distance in one molecule. Chemical problems normally involve the subtle shifts in molecular electron density distribution that follow from substitution or addition of groups to a molecule. Structural clues to such effects come from observation that one bond length or angle is altered relative to another similar one in the same molecule or in a related molecule. Thus, it is only a small handicap that absolute structural parameters cannot be computed directly with high accuracy, and it is one of the greatest advantages of the computational method that errors are correlated in such a way that relative values can be computed fairly readily with high precision. This is in contrast to all experimental methods in which the errors in measuring a C—C bond length in one molecule, for example, are largely uncorrelated with those in a similar measurement on different, though related molecules.

Efforts to determine how accurately the generalizations summarized in Figure 10-1 are reflected in real data are limited by the accuracy of experimental information available for comparison. For C—H bond lengths, there is evidence that the uncertainty is in the 10^{-4} Å range.[9] For some other bond lengths it may well be in the 10^{-3} Å place.

The absolute value of a geometrical parameter is more often of interest for bond angles than for bond lengths. Fortunately, the offset value (ie, the difference between the calculated value and the true value) for valence angles around elements such as C, B, or Si is essentially zero when the calculations are made with a double-zeta or 4-21 basis set. That is, the computed angles are within experimental error of present experimental values when comparison is made for a number of compounds.[1] For valence angles around elements with a lone pair of electrons in the outer shell, such as N, O, S, or Cl, d functions must be added to the basis set for the central atom or else a small offset value must be used.[8] For torsional angles, the calculation is often more delicate and a simple generalization is not possible.[10]

The situation with bond lengths is less fortunate, since in this case the correlation error is more serious so that even increasing the basis set to achieve the Hartree–Fock limit leaves an error, as illustrated in Figure 10-1. An effort has been made to establish standard offset values to correct computed bond lengths to true equilibrium, or r_e, values.[1] The usual reason for interest in an absolute bond length is for comparison with some experiment, and there the problem becomes more complicated, since various experiments measure various sorts of vibrationally averaged distances. If the experimental measurement can be converted to something approximating an r_e structure, the comparison is straightforward. Alternatively, Schäfer[11] and others[12–14] have used the approximation that $r_0 - r_e$ is nearly constant for a given type of bond to compare computed equilibrium structures with r_0 structures that arise from certain types of experiment. This approximation is, of course, not true in general, but it seems to be quite reliable for terminal groups, such as a C—H bond.[9]

The use of offset corrections appears to be extremely reliable for bond types, such as C—H, where the offset error is small. It appears that it is less accurate for bonds such as C=O where the correction is larger. This is reasonable, of course, since transferability of a correction must improve as the magnitude of the correction becomes smaller. Some examples of comparisons between calculation and experiment are given in subsequent sections.

How Is a Molecular Force Field Obtained from Quantum Theory?

The molecular energy can be expanded as a function of the nuclear coordinates q_i in the series:

$$E = E_{\text{ref}} + \Sigma_i g_i q_i + \tfrac{1}{2}\Sigma_{ij} F_{ij} q_i q_j + \tfrac{1}{6}\Sigma_{ijk} q_i q_j q_k + \cdots \tag{10-1}$$

At the equilibrium geometry all the gradients g_i are zero. The second derivatives $F_{ij} = \partial^2 E/\partial q_i \partial q_j$ are the ordinary harmonic force constants if $i = j$ and harmonic coupling constants if $i \neq j$. Methods have been developed[2,5] for the analytical evaluation of these constants from approximate wave functions of a variety of types. Alternatively, the first derivatives g_i may be evaluated analytically and the second derivatives by numerical differentiation.[15,16] Similarly, the anharmonicity constants, F_{ijk}, F_{ijkl}, etc, can be evaluated by known methods. Relatively little use has been made of the cubic, quartic, and higher constants, and most existing work has been within the harmonic oscillator approximation, which terminates the series after the quadratic term. This situation may well change in the future.

The harmonic constants F_{ij} are conveniently ordered as a matrix, $\mathbf{F}$, with the force constants F_{ii} as the diagonal terms and the coupling constants F_{ij} as the off-diagonal terms. Once the $\mathbf{F}$ matrix has been determined in any complete, nonredundant set of coordinates q_i, it can be transformed to any other desired coordinate system by standard methods.[17] The spectrum of fundamental vibrational transitions can be predicted in the harmonic oscillator approximation.

As in the case of structural calculations, the interesting question for the chemist is how accurately and reliably these predictions can be made and to what extent the computations can supplement or perhaps replace experimental measurements. Before proceeding to a comparison of computation with experiment, it is worthwhile to consider briefly the limitations of purely empirical methods. In a conventional spectrometer, band frequencies can be measured to a fraction of a wavenumber, and far greater precision is now possible with some of the newer forms of spectroscopy. Great difficulties arise, however, in attempts to transfer this precision to properties of the molecule under investigation. First, it is commonly impossible to assign all the observed bands in the spectrum except for extremely small molecules. In a molecule with, say, a dozen atoms, it is typically possible to give a definite assignment for a few of the 30 fundamental modes and make only plausible guesses for many of the remainder. The situation can be helped if data from isotopic species are also available, and if the molecule has some form of symmetry, classification of the vibrations into symmetry classes is very useful. Even if all the fundamental modes can be assigned, use of the isolated harmonic oscillator approximation makes impossible the quantitative treatment of Fermi resonances or any other type of intermode interaction.

Information about the mechanical properties of a molecule is contained in the energy expansion given above and, within the harmonic oscillator approximation, in the quadratic force matrix $\mathbf{F}$. It is an unfortunate fact that even with very exhaustive data, there is never enough information to determine this complete matrix except for very small molecules. For example, the molecule referred to above with a dozen atoms and 30 fundamental modes with no symmetry restrictions has 465 elements in the quadratic force field. The number of elements of types F_{ijk}, F_{ijkl}, etc, quickly becomes astronomical. The usual empirical procedure is to construct a harmonic force field by ignoring most or all of the off-diagonal constants, transferring some of the constants unchanged from related molecules, and fitting the remainder to the available experimental data. Such a force field is obviously not unique, and it is a common experience to find widely divergent force fields that give equally good fits to all the measured vibrational frequencies. In any case, use of the harmonic oscillator approximation limits accuracy in the interconversion between vibrational frequencies and force constants to about 2 % in the force constants, corresponding to 1 % in the frequencies.

Turning now to computed results, we see in Table 10-2 the harmonic and selected cubic and quartic force constants for HCN, calculated at the same two levels of quantum chemical approximation as used for the geometries shown in Table 10-1. It can be seen from Table 10-2 that the high-level calculation that gave accuracy of parts in 10^5 in structure parameters does not quite reach the 2 % accuracy level in diagonal force constants, which we assume to be a practical target for accuracy when the vibrations are to be treated as independent harmonic oscillations. Considering that this level of computation is impossible for any molecule much larger than HCN, it is clear that direct computation cannot alone give accuracy in force constants and vibrational frequencies comparable to that obtained from the best experiments. The calculations with a 4-21 basis set and neglect of electron correlation are much farther off, by about 14 % for the C≡N diagonal stretching constant.

Surprisingly enough, the situation is much better for the off-diagonal harmonic and the cubic and quartic constants. The higher level computation gives

Table 10-2. Experimental and Theoretical Harmonic, Cubic, and Diagonal Quartic Force Constants of HCN

Parameter[a]	Experimental[b]	Large calculation[c]	Small calculation[d]
F_{rr}	6.251(4)	6.301	6.507
F_{RR}	18.703(12)	19.422	21.353
$F_{\alpha\alpha}$	0.2596(4)	0.2720	0.4013
F_{rR}	$-0.200(4)$	-0.197	-0.323
F_{rrr}	$-35.4(24)$	-35.8	-37.8
F_{RRR}	$-126.0(52)$	-127.5	-138.2
F_{rRR}	0.41(240)	-0.05	-0.10
F_{rrR}	0.04(80)	0.38	0.48
$F_{r\alpha\alpha}$	$-0.19(50)$	-0.175	-0.113
$F_{R\alpha\alpha}$	$-0.65(40)$	-0.682	-0.770
F_{rrrr}	181	185	202
F_{RRRR}	580	676	757

[a] r is the H—C stretch, R is the C≡N stretch, α is the angle bend. (Units are such that $F_{ij}...\Delta q_i \Delta q_j...$ is in millidynes per angstrom (mdyn Å^{-1}), distances in angstroms, and angles in radians.)
[b] Reference 18.
[c] 6-311G^{4*} basis plus single and double configuration interaction (except only single excitation from the core orbitals) plus Davidson coupled cluster correction (Reference 7).
[d] 4-21 basis, no electron correlation treatment (reference 7).

results within experimental error for all these and even the lower level calculation comes within experimental error of all of the anharmonic constants except for the cubic $C\equiv N$ stretch. This situation results from the greatly increased error in the experimental determination of these constants coupled with a roughly unchanged percentage error in the computed results. These conclusions are borne out by similar comparisons in the few other cases in which experimental results of the quality shown in Table 10-2 exist.

In the calculation of vibrational frequencies, the off-diagonal harmonic and the anharmonic constants make much smaller contributions than do the diagonal harmonic force constants. It would therefore seem to be a practical procedure to combine experimental and computational results by using computed coupling constants and anharmonic constants and obtaining the more important diagonal constants by a combined fit to the available experimental data. This procedure has not yet been used as much as it would seem to merit.

A more common approach has been based on the observation that there is a consistency in the error in the calculation of force constants, just as there is in the calculation of geometrical parameters. If the energy expansion is expressed in a set of internal coordinates, it is found that a given type of vibration is affected in a similar way in different molecules by neglect of electron correlation and by the necessary use of too small a basis set. There are still differences of opinion regarding the best manner for compensating for this consistent error. Some workers simply multiply all computed frequencies by a single factor, usually 0.9 if the calculation is done at the double-zeta or 3-21 Hartree–Fock level. Others make the correction on the force field, which is more fundamental in the sense that it can be used to compute other molecular properties in addition to vibrational frequencies. It has been argued that use of only one scale factor for all types of vibration is no more justified than is the use of only one offset correction for correcting all types of geometrical parameters. A systematic procedure has been proposed[19] that makes use of a small group of scale factors to correct related types of diagonal force constants, with the off-diagonal elements being corrected by:

$$F_{ij}^{\text{scaled}} = (c_i c_j)^{1/2} F_{ij}^{\text{calc}}$$

where the c_i are diagonal scale factors. This approach has been shown to be capable of remarkably high precision in numerous cases, among which may be mentioned the distinction between the very similar experimentally observed spectra of *cis*- and *trans*-acrylic acids.[20] The importance of using more than a single scale factor is emphasized in studies such as one on pyrrole[21] in which the optimized scale factors were found to range between 0.451 for the $N{-}H$ wag and 0.930 for the $C{-}C$ stretch.

Just as bond length offset values are transferable between similar bonds in different molecules, so the force constant scale factors are transferable. Transfer of such scale factors is quite a different matter from transfer of the force constants themselves, as is sometimes done in the development of empirical force fields. It is not assumed that the force constants are identical in different molecules, only that the *error* in computing the force constants is the same. The assumption obviously becomes progressively better as the computation is made at higher quantum chemical levels of approximation so that the scale factors approach nearer to unity.

The Interaction of Quantum Chemistry with Experimental Structure Determination

In modern chemistry, the intermixture of theory and experiment is so intimate that it is almost artificial to separate them. Theory is fundamentally dependent on observation for inspiration, guidance, and confirmation. On the other hand, there is a great deal of theory involved in the interpretation of any experiment. Theory intervenes in many ways between a direct observation that a certain meter needle swings by $5°$ and some conclusion about the nature of molecules. Here, we shall assume the existence of all these more fundamental interactions and concentrate on the very limited area to which the path has been paved in the preceding sections. Specifically, we shall concentrate on the ways in which the theoretical chemist draws on the results of electron diffraction and other experimental procedures in structural chemistry and the ways in which the practicing structural experimentalist can profit from recent developments in computational quantum chemistry. We shall find a symbiotic relationship in which both kinds of chemists can profit immensely from a deeper understanding of the accomplishments and limitations of the other methodology.

From the beginning, computational quantum chemistry has relied on experiment as a standard against which various new methods can be assessed. As indicated in the preceding sections of this chapter, many approximations must be used in obtaining a molecular structure by the solution of the Schrödinger equation, and the validity of these approximations can be checked only by comparison with known numbers. The long period of development and refinement of the computational techniques has even led to the rather unfair charge that computational chemistry can only "predict" numbers that are already known, a charge that was encouraged by the very limited reliability of early methods. The situation can be understood more clearly if we consider briefly the basis for "belief" in a scientific result.

We commonly say that chemistry is an experimental science and that we fundamentally believe only results that come from observation. A moment's thought reminds us, though, that observations and experiments are subject to many errors and to even more misinterpretations. We have learned that we should not profess absolute and permanent belief in any conclusion that we might make based on experimental evidence—a single sustainable contrary experiment could cause us to alter our belief in any existing scientific conclusion.

Perhaps when we say that we believe an experimental result, we mean that we are convinced it has been made by a reliable investigator using a method that long experience has shown leads to results that have a high and more-or-less known probability of lying within some approximately known range around the "true" result. We also require that the new result fit in a logical manner with the preceding body of science, which we have accepted. Discordant results may be more interesting ultimately, but they certainly require greater corroboration before we believe them.

The highly tentative and cautious belief which we properly accord to an experimental result can also apply to a result coming from theory. How much faith we place in the result must follow from a skeptical evaluation in the same manner as described above for an experimental number. We feel that we understand some theory very well, in terms of both its use and its limitations. If we know the diameter of a circle, few of us would quibble about using theory to compute the area—we understand the limitations of the common formula to a flat surface in Euclidean space. Our concerns normally arise from more recently developed theory, where the approximations and limitations are less well understood, and in these cases a very cautious skepticism is certainly valid, as it is for a new experimental method for which all the ways of making mistakes have not yet become apparent.

The foregoing comments should make clear the absolute dependence of computational quantum chemistry on the existence of accurate and reliable numbers from experiment. It is essential not only that the numbers be reliable (ie, capable of eventual reproduction by other, independent methods) but also that the range of uncertainty in the numbers be clearly and reliably given. In general, electron diffractionists have been better in the latter respect than have practitioners in some other experimental structural techniques. In turn, the theoreticians have an obligation to make note of the uncertainties and consider them in whatever comparisons they choose to make. It is worse than meaningless to exult that a computed bond length agrees within 0.001 Å with experiment when a perusal of the experiment would show that the authors believe the latter to be accurate to only 0.01 Å.

Even higher accuracy and reliability than now exists in experimental structural parameters would be valuable for evaluation of theoretical approximations. A great deal of chemistry is reflected in bond length and angle differences on the order of 0.001 Å and 0.1°. Using these variations to learn about the underlying chemical effects requires accuracy an order of magnitude lower. There is plausible basis for hope that computational methods can be developed to provide such reliability for molecules of the size being considered here (half a dozen to two dozen atoms), but how can we know? Tests of self-consistency can be applied, but what is really needed is a body of experimental data of comparable accuracy. The challenge to experimentalists is severe, but the reward in increased chemical understanding of bonding and reactivity should be great.

So far, emphasis has been placed on the need for reliable experimental data to develop experience and confidence in theoretical methods. The time has already arrived, however, when the interaction also works in the other direction. There are numerous examples in the literature in which computed molecular geometries have led to the correction of reported experimental structures that were based on incorrect measurements or, more commonly, incorrect interpretations. These are instances in which further experimental work verified the error in the original experimental result. It is so customary to think that a discrepancy between a computed result and experiment means that the computation is wrong that it may be worthwhile to examine a few cases in which the computation was of help in discovering the experimental mistake. It is hoped that the original experimentalists will not take the following discussion as criticism, but rather as illustrations of the clarification that computational techniques have occasionally been able to bring to bear on difficult experimental problems.

Occasionally, the experimental difficulty has been a simple misinterpretation of the data. An example is the microwave study[22] of the structure of CH_3POF_2 for which a later quantum chemical study[23] gave quite different results. After the problem with the microwave analysis was pointed out, a further and more extensive microwave investigation gave fully reliable results.[24] Similarly, a quantum theoretical calculation[25] of the structure of cyclopropyl amine gave a C_1-C_2 bond distance of 1.508 Å, in sharp disagreement with the value of 1.535(6) Å reported as an r_s structure in a previous microwave spectroscopic investigation.[26] The theoretical paper suggested that the discrepancy might be due to a misassignment of the ^{13}C isotopic species in the microwave study. A subsequent reinvestigation[27] showed that this was indeed the case and reported a much shorter C_1-C_2 bond distance of 1.486(8) Å.

Of more general interest are cases in which a computational study can bring to light an interpretational hazard that had not been fully appreciated by experimentalists. The successful resolution of such conflicts can bring a deeper level of understanding to the experimental analysis, leading to greater reliability of future results. A good example arose in a series of studies of 1,1-dichlorocyclopropane. An exceptionally thorough microwave spectroscopic study of this compound[28] using many isotopic species was carried out in an effort to detect a difference between the principal axis of the chlorine nuclear quadrupole coupling tensor and the C—Cl direction, which would give experimental support to the concept of "bent bonds." As a by-product of this main purpose, a complete r_s structure of the molecule was obtained and it was found that the C—C bond opposite the point of substitution was only about 0.002 Å longer than the C—C bond adjacent to the substitution. A strikingly different result was obtained by an early structural computation,[29] which claimed that the difference between the two bond lengths should be 0.067 Å. The conflict became especially disturbing as more evidence from electron diffraction, microwave spectroscopy, and computational quantum mechanics converged to agreement on a common pattern to interpret the relative lengths of these ring bonds as a function of electronic interactions with the substituent. The correlation predicted that the bond length difference should be considerably larger than was found in the microwave study.

To seek a resolution of the problem, an electron diffraction study of 1,1-dichlorocyclopropane was carried out[30] using modern techniques. Corrections were made for the difference in vibrational averaging seen by an electron diffraction experiment and in the ground state r_0, microwave rotational constants so that the scattering data and the accurately known rotational constants could be used in a simultaneous structure refinement. Various weights given to the two sets of data produced nearly identical results, with the important C—C bond length difference being 0.041(11) Å. The same paper[30] reported a more thorough computational study of the molecule, using complete geometry optimization by the gradient method, which was not available at the time of the earlier study.[29] The theoretical bond length difference was 0.038 Å, identical with the electron diffraction value within the experimental error of the latter.

Since the inclusion of the microwave rotational constants in the analysis of the diffraction data of 1,1-dichlorocyclopropane made little difference in the results, it was concluded that the rotational constants were accurate and the error of 0.04 Å in the microwave r_s structure must have arisen from a fundamental flaw in the substitution method of analysis. The problem has been thoroughly investigated by D. Reuter and W. D. Gwinn,[31] and the reason for the error is now understood, although, regrettably, the conclusions have not been published. The new analysis of the same microwave data resulted in an r_0 bond length difference of 0.037 Å, in full agreement with the electron diffraction and the computational result. It has not always been recognized that the full microwave substitution method can result in errors this large for heavy-atom distances in especially unfavorable circumstances.

Similar effects involving large errors in experimental r_s structures involving bonds to hydrogen atoms are more common. One example was pointed out in a study[32] of the extent of asymmetry of the methyl group in compounds of the type CH_3YH_2X, where Y is C or Si and X is a halogen atom. An even more extreme example was found in a study[33] of the structures of a family of substituted dimethyl ethers. A lengthy series of microwave studies[34] of the gauche conformers of compounds of the type CH_3OCH_2X exhibited one remarkable feature: an extreme apparent lengthening in the r_s structure of the methyl group C—H bond that is approximately parallel to the substituent C—X bond. In the most extreme case, one of the C—H bonds in the $-OCH_2-$ group of chloromethyl ethyl ether was found to be an impossible 0.133 Å longer than the other. The longer bond had an r_s length of 1.225(15) Å, phenomenal for a C—H linkage. The microwave spectroscopists recognized that their results were impossible but pointed out that they had followed rigorously from the standard application of the substitution methods. It should be mentioned that none of these cases involved the well-understood problem that arises when one of the atoms is near an inertial axis. The theoretical study[33] gave plausible bond distances and suggested relationships between shifts in molecular electron density and the computed, rather small, bond length variations, but of course could shed no direct light on the problems in the substitution technique that led to the erroneous experimental results.

The immediately preceding examples illustrate the value of having a completely independent source of structural information to complement experimental methods. Even the best experimental data can sometimes lead to erroneous conclusions because of unsuspected flaws in the methods used for interpretation.

In addition to criticism of computed results by comparison with experiment or criticism in the reverse direction, it is also possible to have more directly constructive interaction between the various techniques. An illustration of direct assistance by computation in an experimental structure determination is provided by a study[35] of S_2F_4. Although this molecule is formed by dimerization of SF_2, it was believed, and later proved, to have a structure F_3S—SF. In spite of the necessary migration of a fluorine atom, the dimerization is reversible, and dissociation back to SF_2 occurs readily.

The original analysis of the electron scattering pattern[35] from S_2F_4 did not yield good agreement with any model tested, all of which were based on a central bipyramidal skeleton with local C_s symmetry of the SF_3 group. For conclusive evidence, the microwave spectrum was observed and an assignment was sought based on rotational constants predicted by the best electron diffraction model. The effort was in vain and no assignment of the microwave spectrum was forthcoming. Recourse was then had to an *ab initio* computation of the geometry.[35] Only one stable conformation was found, a grossly distorted version of one of the models tested in the electron diffraction experiment. The structure around the SF_3 group may be described as a trigonal bipyramid, with two fluorine atoms in the axial positions and the sulfur atom, the remaining fluorine atom, and a lone pair in the equatorial positions. One of the axial fluorines, however, is bent very strongly toward the other sulfur atom; indeed, the SSF angle is only 76°. This fluorine has not moved very far in the dimerization process to attach itself to the other sulfur atom. Consideration of the very large distortions revealed by the computation permitted a good minimum to be found in the least-squares analysis of the electron diffraction data and also guided the way to an assignment of the microwave spectrum. Ultimately, excellent agreement was found between the results of the three quite independent methods.[35]

The multiple minima often produced in the analysis of a structure from electron diffraction data do not, of course, often or necessarily indicate alternative true minima in the potential surface and therefore different real conformations of a molecule. They are merely different minima in the least-squares analysis of the data and, generally, only one of them has physical meaning. The correct minimum is usually chosen as the one having the least residual error, but the differences in error are sometimes small and the decision is potentially questionable. At this point, information from an alternative, independent source can be of great help. One illustration of this is found in the case of CF_3OOCF_3. An electron diffraction study by Marsden and co-workers[36] led to three least-squares minima, of which the authors chose one on the basis of what they believed to be clear evidence. A later investigation, which has not been published, analyzed new electron diffraction data and found a structure very near one of the other minima in the earlier study. To resolve the question, an *ab initio* geometry determination was carried out[37] which gave a solid decision in favor of the earlier structure. The structural differences are often so great between electron diffraction minima that are nearly equivalent in the least-squares sense that distinction between them by computation requires no great accuracy.

In the S_2F_4 example cited above, the initial problem in analysis of the electron diffraction data came from the necessity to make arbitrary constraints in the structure because of limitations in the data. Such problems arise very commonly in studies of alternative real molecular conformations and in studies of molecular motions having several minima. For example, analysis of the electron diffraction data from a compound that exists as a mixture of two conformers almost invariably is based on the assumption that all the bond distances and the angles that are not involved in the conformational change are identical in the two forms. This is known not to be true, and in some cases the structural differences between the two conformers are considerable. Similarly, microwave spectroscopic studies of internal rotation are nearly always reduced to a one-dimensional problem of rotation around a particular bond with all other geometrical parameters held constant.

One example of the use of computational information to make possible a more thorough analysis of a conformation change came in a recent study of ethene thiols of the type XCH=CHSH. Such molecules exist in two forms with respect to rotation of the thiol group around the C—S bond. One is a planar syn conformer, which is generally the more stable, and a "quasi-planar" anti conformer. The latter really consists of a pair of equivalent gauche conformers separated by a very small barrier at the anti position. Early studies of the internal rotation showed that good agreement between the experimental data and a description of the molecular behavior could not be obtained without inclusion of extensive relaxation of the heavy-atom framework of the molecule during the rotation motion. A computational study of the family of molecules XCH=CHSH (X = H, CH_3, CN, F) gave the needed information.[38] It was found that during rotation around the C—S bond, the CCS angle varied by as much as 7° and the C—S bond length by as much as 0.03 Å. Other structural variations were also found, but of a smaller size. This information was incorporated into a more fundamental analysis[39] of the microwave data, including a consideration of the variation of the CCS angle and the C—S bond distance as a function of the torsional angle around the C—S bond. The result was an excellent fit of the variation of the rotational constants with torsional excitation and an overall satisfactory description of the molecular motion.

The world's record for change in other geometrical parameters during a transformation between two stable conformations may well be held by the cis and trans forms of azomethane, $CH_3N=NCH_3$. The more stable conformation has the two methyl groups trans to each other across the double bond. Although the cis form has extreme crowding between the methyl groups, it exists as a stable conformer about 54 kJ mol^{-1} higher in energy, as shown both by computation[40] and by microwave spectroscopy.[41] Depending on the relative rotational orientation of the methyl groups, the cis form has $N=N—C$ angles larger than those in the trans form by 7-10°. If the methyl groups have a hydrogen atom in the eclipsed conformation (with respect to the double bond), the H—C—N angle bends back an additional 5-7°. Considering that this happens on both sides of the molecule, there is a total angle bending of about 30° to relieve the methyl group crowding.

Computational quantum chemistry can influence experiment additionally through the ability of theory to sweep relatively rapidly through an entire family of related molecules and determine whether there are important trends or puzzling anomalies that merit detailed experimental study, or perhaps that the entire series is rather dull. An example of such an approach, with a favorable outcome, is found in *ab initio* geometry studies of a family of monosubstituted cyclobutanes.[42,43] For these compounds, it was possible to deduce simple relationships between the electronegativity of the substituent and such properties as the relative stability of axial or equatorial positions of the substituent or a number of geometric parameters of the ring. The results stimulated one of the authors, Torgeir Jonvik, to undertake an extensive series of electron diffraction studies to confirm and extend the results. Similar interrelationships, but with more complex conclusions, are shown in the history of studies of substituted cyclopropanes. In this case, however, the experimental studies preceded and later were commingled with the theoretical studies.

The most formal incorporation of quantum theoretical information into the analysis of electron diffraction results has been in the MOCED (molecular orbital constrained electron diffraction) procedure of Lothar Schäfer, which is described in Chapter 9 of Part A. The method appears to be very successful in extending the range of molecular complexity that can be handled by electron diffraction. A similar concept is behind the use of theoretical constraints in the analysis of microwave spectroscopic data or the use of microwave rotational constants in the optimization of parameters from electron diffraction studies. While some critics feel that the use of the different methods should be kept independent so that their conclusions can be contrasted a posteriori, there would seem to be good arguments for combining them from the beginning if results can be obtained in this fashion that would be impossible to get otherwise.

This discussion of comparison of theoretical and experimental results should not be closed without a final example showing a case in which experiment was right and the computational result was wrong. The experimental preparation of a compound, $CF_3—C≡SF_3$, having an apparent $C≡S$ triple bond inspired a theoretical study[44] to determine from the bond length if it could properly be described as a triple bond and also to examine the geometrical structure around the sulfur atom. The calculations, using a split-valence basis set with added polarization functions on S, gave apparently valid information on these factors. The SF_3 group has nearly exact octahedral symmetry with the triple bond symmetrically formed from the three remaining S orbitals. The $S\equiv C-C$ linkage was found to be linear, as might be expected. A simultaneous electron diffraction experiment, however, persisted in indicating that this linkage was bent. Only after it was concluded that the diffraction results[45] rested on quite a solid foundation did more detailed computational studies provide confirmation. It has very recently been shown[46] that this is one of the very rare cases where inclusion of a treatment of electron correlation, at least at the MP2 level, is necessary to describe a major qualitative feature of a molecule. The revised theoretical study[46] agreed with the earlier one in saying the $S\equiv C-C$ group is linear if the calculation is made at the SCF level, but the bending is reproduced when correlation is introduced.

Projections for the Near Future

Prediction is, at best, a hazardous sport. It is particularly hazardous in a field like computational chemistry, where both the hardware and the scientific methodology are changing so rapidly. Provided the projection is not made for too long in the future, however, it may be possible to distinguish a few trends.

First, it appears clear that computational methods are rapidly gaining a respectable place as a source of structural information. The few theorists who still decry computation as mere "number crunching" and especially object to any contamination of the theoretical results by the incorporation of experimental information are decreasing in number. Similarly, the experimentalists who accept theory in the interpretation of an experiment but reject it as an independent source of information are becoming fewer. As practitioners of various experimental methods and of computational chemistry become more closely familiar with one another's accomplishments and limitations, the interweaving of techniques will surely become more commonplace and more emphasis will be placed on the chemical reason for seeking the information rather than on the particular combination of techniques chosen to obtain it.

Some of the present limitations of theory are likely to be overcome in the near future, but others appear more intractable. It would seem likely that the current research into relativistic and spin corrections will greatly increase the accuracy of treatments of molecules containing heavy atoms. Considering the importance of studies of organometallic compounds and adsorption on metal surfaces, the importance of such an advance would be hard to overemphasize. Progress is currently rapid, and convenient methods are likely to be available soon.

The problem of extending highly accurate geometry and force field calculations to larger molecules is more difficult. The problem comes from the roughly n^6 dependence of computer time for a configuration interaction treatment of electron correlation on the number of electrons in the molecule. The more accurate perturbation methods, such as MP4, have a similar steep dependence. The result is that the very accurate structural results that can now be obtained for small molecules cannot be achieved with present techniques for large molecules, even with any imaginable increase in computer speed. A thousand-fold increase in computer speed would increase the size of a molecule on which such a calculation could be done by only a factor of 3.2. While faster computers will obviously be of help, mere speed is not the answer.

Various methods are currently being explored to adapt the computational procedures to the newly developed parallelism in computer architecture. Perhaps the reason SCF procedures have developed as they have, adding one electron at a time to build up a molecule in a serial fashion, is that humans think in a serial manner and our logic says to do arithmetic that way. Nature treats the interparticle interactions in parallel, however, and the highly parallel architecture of some new computer designs may inspire some very bright person to revise completely the way in which the physical problem is approached.

Less profound changes in the computational approach are also being investigated. One of these is the treatment of correlation in a basis set of localized orbitals.[47] Without going into the technical details, this approach has the promise of greatly reducing the n^6 dependence of high-level correlation treatments on molecular size. The importance of such efforts to other chemists does not lie in the treatment of ground states of stable molecules, for which the present treatments with empirically derived offset values and scale factors, as described above, appear to be quite adequate with rare exceptions. Rather, the value will come in the ability to treat structures and vibrational properties of reaction transition states, excited states, and unusual species for which comparison with known structures and spectra is not possible. A combination of faster computers, including faster access to larger memories, and techniques to reduce the n^6 dependence would expand the applicability of computational methods to chemistry enormously. It should be emphasized, however, that constant comparison with solid experimental data is essential to the development of this field. A computational result requires confirmation by experiment, just as an experimental result needs confirmation by an independent experiment or by theory before it can be firmly believed.

References

1. Pulay, P.; Fogarasi, G.; Pang, F.; Boggs, J. E. *J. Am. Chem. Soc.* **1979**, *101*, 2550.
2. Pulay, P. *Mol. Phys.* **1969**, *17*, 197.
3. Pople, J. A. In "Modern Theoretical Chemistry", Vol. 4, Schafer, H. F., III, Ed.; Plenum Press: New York, 1977, pp. 1–27.
4. Pulay, P. In "Modern Theoretical Chemistry", Vol. 4, Schaefer, H. F., III, Ed.; Plenum Press: New York, 1977, pp. 153–185.
5. Pulay, P. In "The Force Concept in Chemistry", Deb, B. M., Ed.; Van Nostrand Reinhold: New York, 1981.
6. Douglas, A. E.; Sharma, E. *J. Chem. Phys.* **1953**, *21*, 448.

7. Pulay, P.; Lee, J. G.; Boggs, J. E. *J. Chem. Phys.* **1983**, *79*, 3382.
8. Boggs, J. E.; Cordell, F. R. *J. Mol. Struct.* **1981**, *76*, 329.
9. McKean, D. C.; Boggs, J. E.; Schäfer, L. *J. Mol. Struct.* **1984**, *116*, 313.
10. Boggs, J. E.; Niu, Z. *J. Comput. Chem.* **1985**, *6*, 46.
11. Schäfer, L. *J. Mol. Struct.* **1983**, *100*, 51.
12. Boggs, J. E.; Pang, F.; Pulay, P. *J. Comput. Chem.* **1982**, *3*, 344.
13. Pang. F.; Pulay, P.; Boggs, J. E. *J. Mol. Struct. Theochem* **1982**, *88*, 79.
14. Boggs, J. E.; Pang, F. *J. Heterocycl Chem.* **1984**, *21*, 1801.
15. Fogarasi, G.; Pulay, P. In "Annual Reviews of Physical Chemistry", Vol. 34, Rabinovitch, B. S., Ed.; Annual Reviews, Inc.: Palo Alto, CA, 1984, p. 191.
16. Fogarasi, G.; Pulay, P. In "Vibrational Spectra and Structure", Vol. 14, Durig, J. R., Ed.; Elsevier: Amsterdam, 1985.
17. Wilson, E. B., Jr.; Decius, J. C.; Cross, P. C. "Molecular Vibrations". McGraw-Hill: New York, 1955.
18. Strey, G.; Mills, I. M. *Mol. Phys.* **1973**, *26*, 129.
19. Pulay, P.; Fogarasi, G.; Pongor, G.; Boggs, J. E.; Vargha, A. *J. Am. Chem. Soc.* **1983**, *105*, 7037.
20. Fan, K.; Boggs, J. E. *J. Mol. Struct.* **1987**, *157*, 31.
21. Xie, Y.; Fan, K.; Boggs, J. E., *Mol. Phys.* **1986**, *58*, 401.
22. Durig, J. R.; Kalasinsky, K. S.; Kalasinsky, V. F. *J. Mol. Struct.* **1976**, *34*, 9.
23. von Carlowitz, S.; Zeil, W.; Pulay, P.; Boggs, J. E., *J. Mol. Struct. Theochem* **1982**, *87*, 113.
24. Durig, J. R.; Stanley, A. E.; Li, Y. S. *J. Mol. Struct.* **1982**, *78*, 247.
25. Skancke, A.; Boggs, J. E. *J. Mol. Struct.* **1978**, *50*, 173.
26. Harmony, M. D.; Bostrom, R. E.; Hendricksen, D. K. *J. Chem. Phys.* **1975**, *62*, 1599.
27. Mathew, S. N.; Harmony, M. D. *J. Chem. Phys.* **1978**, *69*, 4316.
28. Flygare, W. H.; Narath, A.; Gwinn, W. D. *J. Chem. Phys.* **1962**, *36*, 200.
29. Skancke, A. *J. Mol. Struct.* **1977**, *42*, 235.
30. Hedberg, L.; Hedberg, K.; Boggs, J. E. *J. Chem. Phys.* **1982**, *77*, 2996.
31. Reuter, D.; Gwinn, W. D. Private communication. Preliminary results were reported by Gwinn in Paper No. WM4, presented at the Eighth Austin Symposium on Molecular Structure, Austin, TX, March 3–5, 1980.
32. Boggs, J. E.; von Carlowitz, M.; von Carlowitz, S. *J. Phys. Chem.* **1982**, *86*, 157.
33. Boggs, J. E.; Altman, M.; Cordell, F. R.; Dai, Y. *J. Mol. Struct. Theochem* **1983**, *94*, 373.
34. Hayashi, M.; Kato, II. *Bull. Chem. Soc. Japan.* **1980**, *53*, 2701, and references therein.
35. von Carlowitz, M.; Oberhammer, H.; Willner, H.; Boggs, J. E. *J. Mol. Struct.* **1983**, *100*, 161.
36. Marsden, C. J.; Bartell, L. S.; Diodati, F. P. *J. Mol. Struct.* **1977**, *39*, 253.
37. Gase, W.; Boggs, J. E. *J. Mol. Struct.* **1984**, *116*, 207.
38. Plant, C.; Macdonald, J. N.; Boggs, J. E. *J. Mol. Struct.* **1985**, *128*, 353.
39. Almond, V.; Permanand, R. R.; Macdonald, J. N. *J. Mol. Struct.* **1985**, *128*, 337.
40. Flood, E.; Pulay, P.; Boggs, J. E. *J. Mol. Struct.* **1978**, *50*, 355.
41. Stevens, J. F.; Engle, P. S.; Curl, R. F. *J. Phys. Chem.* **1979**, *83*, 1432.
42. Jonvik, T.; Boggs, J. E. *J. Mol. Struct. Theochem* **1981**, *85*, 293.
43. Jonvik, T.; Boggs, J. E. *J. Mol. Struct. Theochem* **1983**, *105*, 201.
44. Boggs, J. E. *Inorg. Chem.* **1984**, *23*, 3577.
45. Oberhammer, H. Personal communication.
46. Marsden, C. J. Eleventh Austin Symposium on Molecular Structure, Austin, TX, March 3–5, 1986, Paper No. TM6.
47. Saebø, S.; Pulay, P. *Chem. Phys. Lett.* **1985**, *113*, 13.

AUTHOR INDEX

A

Aanonsen, J.E.   270
Abe, M.   116, 117
Abe, T.   195
Abe, Y.   254
Abdesaken, F.   18
Abel, E.W.   326, 377
Abrabenkov, A.V.   72
Abraham, R.J.   40, 273
Acha, L.   74, 173
Adachi, N.   8
Adams, W.J.   136, 138, 197, 232
Agopovich, J.W.   158
Ahlrichs, R.   183
Ahmad, N.   342, 344
Aihara, J.   12
Airey, W.   106, 163
Akishin, P.A.   3, 20, 349, 363, 394, 398, 402,
   424, 426, 429, 430, 442
Akulinin, O.B.   136
Al-Aidah, G.N.D.   152
Albright, M.J.   339
Alderliesten, P.   40, 99
Aleksandrova, I.A.   253
Alekseev, N.V.   19, 22, 110, 163, 233, 234, 253,
   368
Alexander, J.   158
Alexandratos, S.   370
Allavena, M.   399, 400
Allen, G.C.   429
Allinger, N.L.   27, 215, 216, 225, 227, 231,
   233–237, 258
Almenningen, A.   5, 15, 16, 25, 37, 39, 40, 43,
   49, 54, 55, 95, 96, 99, 106, 108, 109, 113,
   118, 119, 136, 137, 150, 157, 158, 163, 164,
   167, 181, 231, 232, 235, 236, 243–245,
   254–256, 258, 259, 269–271, 293, 296, 297,
   311–313, 315–317, 336, 337, 339, 340,
   342–344, 353–356, 369–372, 374, 411, 443,
   445, 448
Almlöf, J.   3, 4, 316, 317, 368, 372, 374, 375

Almond, V.   471
Altman, A.B.   196, 398–400
Altman, M.   469
Alyea, E.C.   21
Amma, E.M.   335
Amir-Ebrahimi, V.   305–307
Amrein, J.   299, 301
Anashkin, M.G.   75, 356
Andersen B.   181
Andersen, P.   117
Andersen, R.A.   21, 353, 364, 365, 376, 377
Anderson, D.   198
Anderson, D.W.W.   156, 167
Anderson, G.A.   337, 341, 342, 343, 351, 448
Anderson, T.J.   339
Andreassen, A.L.   113, 116, 117, 156, 168, 250
Andrianov, V.G.   61, 62
Andrutskaya, L.G.   37, 48, 52, 65, 66, 80, 180
Anfinsen, I.M.   43, 167, 243–245, 259
Angus, P.C.   362
Angyal, S.J.   215, 216, 227, 233, 258
Antipin, M.Y.   49, 332
Antonio, P.D.   250
Aomar, A.M.   101, 103
Aoya, T.   368
Apalkova, G.M.   271, 328
Apeloig, Y.   162
Appelman, E.H.   197
Arbuzov, B.A.   72, 125
Arnesen, S.P.   270
Arnold, D.E.J.   48, 68, 70, 80, 108, 109, 170,
   171, 180
Aror, H.S.   273
Ashby, E.C.   353, 355, 356, 364
Ashe, A.J., III.   62
Askari, M.   114, 116
Astoja-Starzewski, K.A.   76
Astrup, E.E.   43, 76, 101–103, 117, 139, 140,
   181
Atavin, E.G.   12, 15, 16, 22
Aten, C.F.   262
Atwood, J.L.   342, 344, 357, 359, 372, 374

Auberg, E.   113
Auner, N.   161, 162
Avramenko, G.I.   362

B

Bagus, P.S.   409
Baikov, V.I.   399, 400, 403
Bak, B.   37, 39, 43, 136
Bakhe, J.M.   273
Bakken, P.   37, 135, 254, 255, 258–260,
    269–271, 273, 274
Bakov, E.K.   432
Balducci, G.   412
Bally, T.   24
Bambieri, G.   372
Banki, J.   100, 114, 266
Banks, R.E.   134, 158
Barlow, M.T.   11, 337, 341
Barnhart, D.M.   410, 411
Barrow, M.J.   112, 361, 362, 368
Bartelat, J.-C.   334
Bartell, 9, 11, 19, 20, 24, 25, 50, 62, 63, 68, 71,
    79, 110, 114, 115, 121, 122, 132, 133, 136,
    138, 149, 166, 169, 171, 172, 174, 175, 179,
    182, 183, 185, 193, 197, 198, 213–229, 232,
    274, 333, 336, 344, 365, 410, 411, 433, 438,
    439, 445, 446, 471
Bartke, T.C.   363
Bartók, M.   135
Barton, T.J.   23
Bashkirova, S.A.   22
Baskin, C.P.   395
Baskir, E.G.   19
Bastiansen, O.   16, 23, 24, 85, 95, 96, 106, 118,
    119, 157, 158, 163, 230–236, 256, 270, 297,
    312, 315, 316, 369, 370,
Bats, J.W.   159
Batyukhnova, O.G.   54, 55, 312
Bauder, A.   99, 183, 273
Bauer, G.   17, 25, 27
Bauer, S.H.   2, 8, 15, 42–44, 50, 51, 95, 98,
    113, 116, 117, 122–124, 136, 139, 148–152,
    154, 156–159, 164–168, 186, 190–192, 198,
    246, 250, 255, 256, 259, 263, 269,282, 297,
    330, 430
Baumik, A.   95, 98
Bauschlicher, C.W.   409
Baxter, P.L.   187, 343, 350
Beach, A.L.   177
Beach, J.Y.   16
Beagley, B.   4, 8, 18–21, 29, 39, 40, 44, 45, 64,
    70, 71, 105, 106, 108, 122, 134, 150, 152,

153, 158, 160, 161, 163–165, 169, 170, 172,
    185, 196, 197, 222, 223, 253, 271, 328, 332,
    336, 342, 343, 445
Beattie, I.   384, 390, 401, 405
Beattie, J.K.   370, 371
Beaudet, R.A.   12, 16
Becker, G.   60
Becker, W.G.   406–408
Beckers, H.   162
Beecher,J.F.   165, 184
Belyakov, A.V.   19, 47, 271, 328, 330, 332
Bender, C.F.   395
Bennett, L.S.   432
Bent, H.A.   20, 287, 331
Bentham, J.E.   263
Berdnikov, E.A.   126, 253
Berdonosov, S.S.   432
Berecz, I.   17, 367, 384, 386, 398, 408–410
Berg, R.A.   399–401
Bergman, A.   416, 417
Berkowitz, J.   398, 406
Bernd, J.   332
Bernstein, L.S.   165, 184
Berry, R.S.   432
Bersuker, I.B.   427
Bertram, M.   160
Bett, W.   112
Bevan, J.W.   123
Bezzubov, V.M.   72, 77
Bhaumik, A.   155
Bicerano, J.   12
Bijen, J.M.J.M.   95, 96, 101, 103, 108, 109,
    119, 120, 157, 251, 270
Biskenmeier, J.A.   95
Bjorvatten, T.   411
Björseth, A.   363
Blair, P.D.   39, 171
Blake, A.J.   345
Blankenship, C.S.   23
Blaschette, A.   122
Blaschke, G.   75, 76
Blayden, H.E.   401, 405
Bleckmann, P.   334
Bliefert, C.   126, 128, 130, 190, 191
Blom, R.   20, 109, 332, 333, 364, 365, 372, 373,
    376
Blount, J.F.   24
Blow, R.   184
Boates, T.L.   24, 25, 224–226
Bobkova, R.   61, 62
Bock, C.W.   283, 295, 298, 299, 301–306, 309,
    310
Bock, H.   132, 192
Boersma, J.   357, 358, 372, 373

Boggs, J.E.   3, 4, 27, 58, 61, 62, 67, 85, 122,
    126, 130–132, 157, 162, 168, 173, 176, 177,
    183, 184, 187, 189–192, 295, 298, 305, 306,
    455, 457, 460, 462, 464, 465, 468–472
Bogoradovskii, E.T.   47, 271, 328, 330, 332
Bohátka, S.   17, 367, 384, 386, 398, 408, 409,
    410
Bohn, M.D.   15, 16
Bohn, R.K.   15, 16, 43, 101, 103, 165, 166,
    198, 311, 312
Boiko, Y.A.   136
Bojensen, I.   282, 283, 287, 295, 297, 298, 307
Bokiy, N.G.   17
Bokiy, N.K.   331
Boldyrev, A.I.   395
Boncella, J.M.   376
Bondybey, V.V.   42
Bonham, R.A.   117, 215–219
Boogaard, A.   113
Boonstra, L.H.   18, 106
Borgan, A.   117
Bostrom, R.E.   468
Boudjouk, P.   23, 331
Boulet, G.A.   121
Bouzga, A.M.   43, 76,
Bowater, I.C.   186
Bowen, H.J.M.   170
Boyd, A.S.F.   74, 173
Braathen, G.O.   266
Bradford, W.F.   215, 216, 222, 223, 229, 274
Bragin, J.   16
Brake, J.H.M.   117, 121
Brandes, D.   122
Brauer, D.J.   353
Bredikhin, A.A.   49
Breed, H.E.   138
Bregadze, V.I.   79, 343, 346
Brendhaugen, K.   11, 137, 158, 350, 352
Brewer, L.   386
Brezgin, Y.A.   440
Brinker, W.D.   45
Britt, C.O.   126, 130, 131, 168, 190
Brittain, A.H.   169, 170
Brockway, L.O.   2, 16, 169, 196, 233, 234, 344
Brode, H.   401, 402
Brom, J.M.   406, 408
Brook, A.   18
Brooker, M.H.   440, 443
Brooks, W.V.F.   95, 98, 134, 155
Brown, D.E.   152, 153
Brown, D.P.   160, 161, 222, 223
Brown, H.C.   86
Brown, L.D.   12
Brown, R.D.   165, 186, 170

Brownlee, R.T.C.   289
Bruce, M.I.   326, 277
Brunvoll, J.   39, 54, 55, 81, 101, 103, 126, 128,
    130, 131, 152, 190, 191, 253, 293, 296,
    311–313, 315–317, 368, 369, 398–400, 403,
    411, 412, 434–437
Buck, I.   2
Buckton, K.S.   165
Budenz, R.   187
Budzelaar, P.H.M.   372, 373
Bulcke, P.   126, 253
Bunker, P.R.   45
Burden, F.R.   165, 186, 190
Burg, A.B.   12, 172
Burkert, V.   27, 231, 235–237
Burns, C.J.   376
Burns, G.T.   23
Buslaev, Y.M.   193
Butayev, B.S.   393, 409, 413–415
Butcher, S.S.   182
Butenko, G.G.   126, 128, 130
Butler, W.   339
Butskii, V.D.   436
Buys, H.R.   233, 234, 255
Büchler, A.   389, 391, 399, 401, 404, 413, 416
Bünder, W.   368
Bürger, H.   43, 45, 46, 149, 156, 162–165, 167,
    183, 250, 328, 330
Bürgi, H.B.   227, 228, 410

C

Cabanna, A.   60
Cahill, P.   394
Calder, V.   399, 400, 403
Caminiti, W.J.   58, 85
Canadell, E.   365
Cardillo, M.J.   50, 166
Carleer, R.   22, 105
Carlos, J.L.   154, 246, 259
Carlotti, M.   166
Carlson, K.D.   421
Carreira, L.A.   71, 172
Carroll, B.L.   9, 11, 336
Carter, J.C.   343, 346
Carter, O.L.   283, 287
Carter, R.P., Jr.   177
Cartwight, G.J.   160
Casado, J.   282, 299, 301, 307
Catalano, D.   302, 303
Cavell, R.G.   175–177
Centofani, L.F.   173
Cernia, E.   351

Cesari, M.   350, 351
Chadaeva, N.A.   86
Chang, C.H.   43, 256
Chang, S.C.   166
Chang, T.W.   370
Chang, Y.-M.   18
Chantrell, S.J.   44
Chanussot, J.   196
Chao, J.   394
Chao, K.T.   370
Charkin, O.P.   395, 422
Charpentier, L.   110, 185, 188
Chatterjee, K.K.   343
Chen, M.M.   328
Chen, M.M.L.   188
Cherkasov, R.A.   78, 85
Chernick, C.L.   198
Chernyshev, E.A.   22, 263
Cheung, C.C.S.   16
Chevalier, R.   299
Chiang, C.H.   149, 156, 164, 167
Chiang, J.F.   42, 158, 159, 255–257, 263, 264, 269
Chiang, R.L.   42, 256
Chikaraishi, T.   169
Chin, N.S.   114, 115, 370
Chmutova, G.A.   103
Choplin, A.   305–307
Christe, K.O.   39, 167, 192, 196
Christen, D.   39, 40, 157, 165–168, 183, 189, 193
Christensen, D.   136
Claassen, H.H.   438
Clark, A.H.   17, 139, 196
Clark, J.H.   428
Clark, W.W., III.   196
Clippard, F.B., Jr.   24, 25, 169, 174, 197, 411, 433
Codding, E.G.   171
Coffey, D.   168
Cohen, E.A.   176
Cohn, K.   170
Colapietro, M.   288, 295, 296, 298–300, 302, 303, 309, 310, 314–321
Coldish, E.   255
Collins, S.   24
Collomon, J.H.   3, 5, 6, 7, 11, 15, 19–22, 25
Companion, A.L.   405
Compton, D.A.C.   217
Connor, J.A.   108, 271
Conrad, A.R.   71, 164, 172
Considine, J.L.   339
Cook, R.L.   188
Cordell, F.R.   460, 462, 469

Corfield, P.W.R.   368
Cornwell, D.C.   176
Coulson, C.A.   230, 231, 282–284, 287, 288
Cowan, D.J.   246
Cowley, A.H..   21, 64, 188, 372, 374,
Cox, A.P.   52, 56, 168, 307, 365
Cradock, S.   19, 21, 39, 44, 70, 156, 271, 332
Craig, D.P.   178
Craig, N.C.   155, 166
Crawford, R.   194
Cremer, D.   155, 183
Cremer, S.E.   23
Creswell, R.A.   135
Croatto, U.   372
Cromie, E.R.   74, 173
Cross, P.C.   463
Cross, V.R.   126, 190
Cruickshank, D.W.J.   163, 183, 191, 196, 197
Csákvári, B.   19, 106
Cubicciotti, D.D.   384, 395, 406, 440, 443
Cucinella, S.   351
Cunico, R.F.   23
Cuolt, J.P. 94, 95
Curl, R.F.   98, 111, 150, 409, 472
Curtiss, L.A.   403–405, 442–445, 447
Cusachs, L.C.   406
Cyvin, B.N.   22, 126, 130, 131, 157, 190, 271, 297, 312, 315, 316, 443
Cyvin, S.J.   16, 17, 22, 70, 149, 157, 171, 211–213, 261, 262, 271, 297, 312, 314–316, 401, 415, 417, 440, 443

D

Dai, Y.   469
Dain, C.J.   11, 12, 341
Dakkouri, M.   19, 22, 37,   151, 152, 331
D'Alessio, L.   398, 399
Dallinga, G.   101, 102, 157
Damiani, D.   170, 361
Danielson, D.D.   6, 74, 150
Danilova, T.G.   422
D'Antonio, P.   101, 102, 116, 117, 122, 155, 156, 182
Darmadi, A.   157
Dashevskii, V.G.   72, 85, 255
Dass, S.C.   95, 98, 155
Davidovits, P.   394
Davidson, P.J.   334
Davies, A.G.   365
Davis, H.   433, 434
Davis, M.   233, 234
Davis, M.I.   139, 174, 246, 255, 358

Day, M.C.   342, 344
De Lucia, F.C.   149
Decius, J.C.   463
Dédier, J.   106
DeFrees, D.   359
Dehmer, J.L.   406
Dekker, J.   358
Del Piero, G.   351
Demaison, J.   305–307
DeMaria, G.   398, 399
DeMeijere, A.   247, 254, 256, 264, 265
Demidov, A.V.   409, 432, 433
Demuth, R.   71, 165
Den Otter, G.J.   317
Dendl, G.   123
DeNeui, R.J.   213, 214
Derissen, J.L.   101, 103, 108, 109, 118–120,
    157, 251, 256, 270
Des Marteau, D.D.   40, 44, 122, 131, 132, 166,
    183, 192
Desclaux, J.-P.   355
Dewan, J.C.   24
Dewar, M.J.S.   12, 16, 370
D'Eye, R.W.M.   429
Di Renzo, F.   283, 302
Diehl, P.   299–301
Diem, M.   16
Dienes, G.J.   395–397, 442
Dillen, J.   50, 138, 140, 235–237
Diodati, F.P.   110, 122, 182, 183, 471
Dittebrandt, C.   79, 179
Dixon, D.   155
Dixon, D.A.   12, 76, 188
Dixon, W.   136
Dofferhoff, G.M.T.   113
Dolbier, W.R., Jr.   157
Dolzine, T.W.   358
Domenicano, A.   100, 114, 281–292, 295–310,
    313–321
Domnin, I.N.   256, 265
Dorofeeva, O.V.   11, 12, 15, 16, 22, 58, 85,
    235–237, 389, 416, 417
Douglas, A.E.   460
Doun, S.   193
Doun, S.K.   19, 20, 132
Dowater, I.C.   170
Downs, A.J.   11, 12, 178, 187, 337, 341, 343,
    350
Doyns, Sister V.J.   158
Dozzi, G.   350, 351
Drago, R.S.   175
Drake, J.E.   161, 162, 328, 331
Drake, M.C.   384, 399, 400, 422
Dreizler, H.   123

Dreizler, H.Z.   29, 185, 188
Drew, D.A.   348, 362–366, 370
Driessen, R.A.J.   117, 121
Du Mont, W.-W.   372, 374
Duckett, J.A.   39, 164
Dudley, F.B.   194
Duffy, D.   341
Duisenberg, A.J.M.   372
Dulova, V.G.   331
Dunitz, J.D.   231, 232, 273
Durig, J.R.   19, 21, 39, 71, 172, 173, 328, 343,
    468
Dyatkina, M.E.   422
Dyke, J.M.   429
Dyke, T.R.   431
Dyngeseth, S.   114, 120, 121
Dzhaparidze, K.G.   17

E

Eades, R.A.   76
Earnshaw, A.   406, 408, 423–425, 431
Ebsworth, E.A.V.   19, 21, 44, 47, 64, 68, 71,
    76, 108, 109, 112, 156, 171, 345, 361, 362,
    368
Eckert-Maksic, M   159
Eddy, L.   415, 416
Edwards, A.J.   435
Egawa, T.   231
Egorov, M.P.   332
Ehrenson, S.   289
Eijck, B.P.   119
Eisch, J.J.   339
Eisenstein, O.   365
Elfimova, T.L.   49
Eliel, E.L.   215, 216, 227, 233, 258
Ellengsen, B.H.   97, 98
Elliott, N.   401, 402
El'natov, Y.I.   65, 66
Emons, von H.H.   398
Engle, P.S.   472
Enjalbert, R.   343, 346
Eriksson, A.   415
Ermolayeva, L.I.   3, 4, 402, 405, 406, 445
Erokhin, E.V.   440, 442
Eujen, J.   162, 163
Eujen, R.   328, 330, 356
Evans, W.J.   376
Evdokimov, V.V.   69
Ewart, I.C.   307
Ewbank, J.D.   114, 115, 182, 233, 234
Ewing, V.   16, 106, 163
Exner, O.   126, 130, 131, 290, 296, 308, 315

Eyman, D.P.   336, 341
Eysel, H.H.   44, 131, 132, 192
Ezhov, Y.S.   3, 398–400, 403, 429, 430, 434

F

Faber, D.H.   101, 103
Faegri, K., Jr.   368, 372, 374, 375
Falconer, W.E.   422, 431, 437, 438
Fan, K.   465
Fanning, M.O.   15, 16
Fauvet, G.   299
Favero, P.G.   170
Fayad, N.K.   429
Fellegvári, I.   100, 114, 266
Fernholt, L.   23, 24, 37, 52, 65, 95, 96, 109,
    113, 137, 152, 153, 157, 158, 233, 234, 297,
    312, 315, 316, 339, 340, 343, 344, 346, 353,
    355–358, 367, 372, 374, 375
Ferretti, L.   361
Fickes, M.G.   406–408
Fikke, M.N.   137
Fild, M.   174
Finder, C.J.   86
Fink, M.J.   28, 151, 152, 193, 213, 396, 397
Fischer, B.   372, 373
Fischer, K.J.   21
Fitzpatrick, N.J.   15, 16
Fitzwater, S.   215–218, 220, 221, 274
Fjeldberg, T.   6, 20, 21, 25, 27, 47, 334–336,
    339, 340, 359
Flood, E.   157, 472
Flygare, W.H.   468
Fogarasi, G.   28, 305, 306, 347, 457, 462, 463,
    465
Fong, G.D.   65, 67
Foord, A.   4, 19, 20, 40, 163, 197, 253
Forgaard, F.R.   337, 341–343, 351, 448
Forster, A.M.   187
Forte, C.   302, 303
Forti, P.   170
Franczek, F.R.   342, 344
Franklin, J.L.   432
Fransen, H.F.   406, 408
Frantsen, E.B.   261
Fraser, A.   19, 271
Fraser, T.E.   76
Frasson, E.   368
Fredin, L.S.   27
Freeman, J.M.   70, 71, 105, 160, 161, 163, 164,
    170, 172, 328
Freiberg, L.A.   234
French, R.J.   175

Frey, R.A.   389
Friedman, A.H.   415
Fries, W.   339, 340
Friesen, D.   152, 153
Fristrom, R.M.   190
Fritsch, F.N.   230
Fritz, G.   27, 163
Fujii, H.   163, 327, 328, 408
Fukuyama, T.   50, 51, 114, 115, 156, 157, 166,
    230, 231, 244, 254, 257, 259, 262, 266, 271,
    275

G

Gabes, W.   328
Gadünz, A.   184
Galiakberov, R.M.   69, 86
Gallaher, K.L.   15, 51, 116, 139, 152
Gallinella, E.   361
Galy, J.   343, 346
Ganter, C.   273
Garaeva, R.N.   102, 108, 110, 126, 128, 130
Gard, G.L.   132, 193, 433, 434
Gase, W.   122, 184, 471
Gasner, E.L.   438
Gasser, O.   76
Gassman, P.G.   76
Gaunt, A.D.   161
Gavin, R.M.   198, 365
Gavryuchenkov, F.G.   440
Gebhardt, K.F.   58
Gesie, H.J.   22, 50, 101–103, 105, 108, 110,
    113, 114, 125, 136, 138, 140, 152–154,
    232–237, 243–246, 252, 255, 259, 312
Gelly, M.C.L.   95
George, C.   116, 117, 122, 135, 155, 156,182,
    250
George, C.F.   101, 102
George, P.   283, 295, 298, 299, 301–306, 309,
    310
Gerasimov, Y.I.   349
Gergö, E.   19, 27, 106
Gerritsen, J.   317
Gerry, M.C.L.   39, 164, 196
Gershikov, A.G.   3, 4, 356, 393, 398–400, 402,
    405, 406, 409, 413–415, 445
Gibson, J.A.   176
Gigli, G.   399–401, 412
Gilbert, M.M.   50, 166, 431–433
Gillard, I.R.   165
Gillespie, P.   176, 181
Gillespie, R.J.   4, 148, 169, 175, 179, 287, 331,
    400, 410, 431, 438

Gillies, C.V.   158
Ginn, S.G.   4
Girbasova, N.V.   47, 332
Girichev, G.V.   394, 395, 402, 404–406, 414–416, 422, 424, 426, 429, 430, 436, 437
Giricheva, N.I.   402, 404, 405, 415, 416, 422, 424, 426, 429, 430
Givan, A.   389, 413, 414
Gleiter, R.   159
Glemser, O.   123, 124, 132, 133, 186, 190, 194
Glick, M.D.   339
Glidewell, C.   49, 50, 71, 105, 106, 108, 109, 163, 166, 168, 332, 333
Goddard, J.D.   18
Goddard, J.P.   365
Godfrey, P.D.   165
Gogstad, E.   271
Gold, L.P.   394
Goldberg, D.E.   334, 335
Golding, E.G.   68
Gole, J.L.   400
Golka, R.S.   331
Goltyapin, Y.V.   15
Golubinskaya, L.M.   79, 343, 346
Golubinskii, A.V.   12, 15, 16, 22, 28, 47, 55, 57, 58, 64, 65, 79, 86, 103, 111, 135, 263, 313, 332, 343, 346, 349
Gombler, W.   102, 108, 110, 184, 187
Goncharuk, V.P.   436, 437
Goode, M.J.   187
Gordon, A.J.   214
Gordon, M.S.   74
Gordy, W.   149, 152, 442
Gorokhov, L.N.   435
Gosh, S.N.   152
Gotkis, I.S.   435, 436
Gouch, K.M.   305, 306
Goulet, P.   196
Gömöry, P.   19, 106
Gradock, S.   106, 108, 112
Graeve, R.   214
Granberg, M.   270
Grant, D.F.   343
Gray, D.L.   213
Graybeal, J.D.   191
Green, A.R.   156
Green, D.W.   416, 417
Greenhalgh, D.A.   384, 390
Greenwood, N.N.   406, 408, 423–425, 431
Gregory, D.   151, 152
Gregory, D.C.   105, 108, 123, 126, 128, 296
Grenz, M.   372, 374
Gresser, G.   60
Griffiths, J.E.   176, 177

Grikina, O.G.   61, 62, 65, 66, 72
Grishina, L.N.   78, 85
Grobe, J.   80, 168, 172, 176–178
Gropen, O.   345, 359, 366
Gruber, J.B.   422
Guarnieri, A.   110
Guido, M.   399–401, 412
Guillory, J.P.   114, 115, 445, 446
Gulova, V.G.   331
Gulyaeva, N.A.   65, 66, 72, 85
Gundersen, G.   5, 6, 19, 27, 47, 49, 50, 56, 58, 68, 108, 109, 138, 150, 164, 166, 171, 191, 193, 270, 339, 343, 347, 348, 364, 448
Gundersen, S.   157, 158, 297, 312, 315, 316
Gupta, D.G.   194, 196
Gupta, K.D.   80, 175, 180
Gurvich, L.V.   237
Gusarov, A.V.   435, 436
Gutekunst, B.   18
Gutekunst, G.   18
Gutoewski, R.   17, 18
Günthard, H.H.   389
Günther, H.   44, 46, 48, 131, 132, 157, 165, 192, 193, 356
Gwinn, W.D.   186, 260, 468, 469

H

Haaland, A.   8, 11, 20, 21, 43, 109, 167, 184, 243–245, 259, 273, 325, 327, 332–337, 339–375, 377, 443, 445, 448
Haas, A.   157, 158, 184, 185, 187
Haas, B.   122, 153, 183
Haase, J.   77, 123, 124, 132, 133, 140, 153, 186, 194, 271–273
Hagen, K.   42, 43, 75, 95, 96, 114, 116,120, 121, 126, 137, 153, 155, 160, 161, 173, 190, 246–250, 256, 263, 264, 266, 267, 271, 273, 431–433
Hahn, E.   372, 374
Haley, L.V.   428
Halgren, T. A.   12
Hall, J.H.   12
Hall, S.M.   401, 405
Haller, K.J.   28
Halonen, L.   21
Halvorsen, S.   336, 337, 342
Ham, N.S.   339
Hambley, T.W.   370, 371
Hamill, D.   429
Hamm, R.   157
Hammaker, R.M.   40, 166
Hansen, K.W.   174, 175, 179, 182, 433

Hanson, H.P.   246
Hanusa, T.P.   376
Hanzawa, Y.   24
Harding, M.M.   361, 362, 368
Hargittai, I.   2, 8, 11, 17–19, 21, 27, 28, 37, 39,
    40, 42, 59, 72, 75, 81, 85, 99–101, 103, 105,
    106, 108, 111, 114, 123–136, 128, 130, 131,
    135, 138, 139, 150, 159, 185, 186, 188, 190,
    191, 211–213, 253, 267, 270, 284, 295, 296,
    298, 299, 301, 302, 304–307, 309, 310–321,
    327, 328, 332, 349, 367, 384, 386, 388, 390,
    393, 394, 398–400, 403, 405, 408–410,
    412–416, 421, 425, 440, 442–446
Hargittai, M.   8, 19, 52, 59, 109, 126, 150, 190,
    327, 343, 346, 383, 388, 389, 393, 394,
    398–400, 403, 405, 412–417, 421, 425, 440,
    443–447
Harmony, M.D.   468
Harris, D.H.   334
Harsányi, L.   267
Harshbarger, W.   8, 136, 151, 198
Hartford, W.D.   59, 165
Hartley, F.R.   329, 330, 336, 355
Hartley, J.G.   396, 397
Hartmann, A.O.   158
Hartmann, G.   159
Hase, Y.   399–401
Hassel, O.   139, 233, 234
Hastie, J.W.   384, 406, 416, 421, 422, 440, 442
Hauge, R.H.   384, 390, 416, 421, 422, 429
Haugen, T.   339, 343, 347, 348, 443, 445, 448
Haugen, W.   263
Hausen, H.-D.   340
Hauser, R.   149, 164
Hayashi, M.   100, 469
Hayes, E.F.   400
Hayes, R.G.   165
Häfelinger, G.   158
Hedberg, E.   48, 70, 170
Hedberg, K.   2, 6, 11, 12, 16, 21, 42, 48, 50,
    56, 59, 63, 70, 75, 79, 80, 95, 96, 106, 108,
    114–116, 121, 126, 132, 137, 150, 152, 153,
    160, 163, 165, 166, 170, 173, 175, 178, 180,
    190–194, 230, 255, 256, 262, 266, 267, 358,
    410, 411, 431–434, 469
Hedberg, L.   11, 48, 56, 75, 79, 132, 170, 173,
    175, 178, 191, 192, 198, 256, 262, 431–434,
    469
Heeg, M.J.   374
Heenan, R.K.   197, 215, 216
Hehre, W.J.   445
Heinricher, A.   443
Heitsch, C.W.   346
Helgaker, T.U.   355, 356

Hellams, K.L.   328
Helminger, P.   149
Hellwege, A.M.   36, 39, 40, 42–47, 52, 58,
    63–65, 74, 75, 79, 80
Hellwege, K.-H.   3, 5, 6, 7, 11, 15, 19–22, 25,
    36, 39, 40, 42–47, 52, 58, 63–65, 74, 75, 79,
    80
Hemmings, R.T.   328, 331
Hemple, S.   400, 403
Hencher, J.L.   42, 125, 135, 158, 161–163, 190,
    191, 327, 328, 331
Hendricks, S.B.   393, 394
Hendricksen, D.K.   468
Hengge, E.   17, 25, 27
Hennings, R.T.   161, 162
Henrickson, C.H.   336, 341
Henriksen, L.   113
Henry, B.R.   305, 306
Hersh, O.L.   165
Herzberg, G.   185, 272, 338, 359, 360
Hess, H.   349–351
Hettich, B.   189
Hewitt, T.G.   71, 139, 161, 196, 332, 342, 343,
    445
Heyder, F.   46, 165
Higgenbotham, H.K.   50, 166, 214
Hildenbrand, D.L.   384, 429, 440, 443
Hildenbrandt, R.L.   12, 17, 22, 23, 58, 85, 117,
    121, 136, 137, 156, 216, 221, 222, 235, 236,
    247, 250, 260, 267, 271, 331
Hilmo, J.   40, 43
Hinderer, A.   349, 351
Hirose, C.   133, 134, 158
Hirota, E.   3, 5–7, 11, 15, 17, 19–22, 25, 169,
    213, 409
Hitchcock, P.B.   334, 335
Hodges, H.L.   71, 172
Hoekstra, A.   261, 285, 308
Hoffmann, P.   176, 181
Hoffmann, R.   180, 188
Hofmann, P.   374
Hogel, J.   62
Hohlt, H.J.   255, 272, 273
Holmes, R.R.   177
Holywell, G.C.   65, 70, 71, 164, 169, 170, 172
Homer, G.D.   23
Homes, R.R.   174, 175
Hope, H.   21, 269
Hopf, H.   270, 271
Horlbeck, W.   398
Hom, A.   98, 273
Hougen, J.   357, 359
Houldsworth, N.J.   152
Howell, J.M.   180

Hoy, A.R.   160
Höfs, H.U.   159
Høg, J.H.   46, 83, 282, 302, 303, 307, 311, 317
Huber, K.P.   359, 360
Hudson, G.A.   108, 271
Huffman, J.C.   12, 335, 336
Hughes, L.A.   376
Huglen, R.   440, 443
Huisman, P.A.G.   45, 115, 154, 157, 246
Hunter, B.K.   428
Hunter, W.E.   372, 374
Huntley, C.M.   19, 21, 39, 70, 112, 170
Hursthouse, M.B.   365
Hutchison, D.J.   71

I

Iga, I.   139
Ignatov, S.M.   49
Ignatova, N.P.   61, 62
Iijima, K.   8, 9, 11, 57, 79, 102, 150, 431, 437
Iijima, T.   29, 102, 108, 113, 114, 116, 117,
    120, 152, 153, 184, 214, 215, 261, 263, 309,
    327, 328, 355, 356, 408
Il'enko, T.M.   19, 163
Ilsley, W.H.   340
Imaishi, H.   100
Imaizumi, S.   273
Ino, T.   395-397
Ischenko, A.A. (Ishchenko, A.A.)   174, 196,
    197, 409, 411, 412, 426-428, 432-437
Ishiyama, J.   273
Isselmann, P.H.   44
Ito, T.   402, 405, 406, 411
Ivanov, A.A.   3, 4, 398-400, 402, 405, 406,
    409, 411, 414, 415, 422, 432, 437, 443, 445,
    446
Ivashkevich, L.S.   174, 409, 411, 426-428, 432
Ivshin, V.P.   55
Iwasaki, M.   63, 410

J

Jackson, R.H.   183
Jacob, E.J.   74, 75, 151, 197, 198, 225, 433,
    434, 438, 439
Jacobsen, G.G.   256
Jahn, H.A.   427, 428, 433, 434
Jalilian, M.   343
Jandal, P.   107, 108
Janiak, C.   374
Jansen, P.   37, 39, 43
Janssen, J.   260, 268, 269
Janzen, J.   149
Jeffrey, E.A.   339, 347
Jeffrey, G.A.   103, 342, 344, 350, 351
Jemmis, E.D.   370
Jenny, S.N.   401, 405
Jensen, H.   235, 236
Jin, A.   193
Johansen, R.   345
Johns, J.W.C.   166, 338
Johnsen, F.   331
Johnson, D.R.   184
Johnson, Q.   12
Jokisaari, J.   295, 298-300
Jondal, P.   251, 252
Jones, C.E.   68, 171
Jones, M.E.   11, 12
Jones, M.O.   152
Jones, R.A.   372, 374
Jonvik, T.   5, 6, 472
Jordan, K.D.   394
Joyce, T.E.   412
Jugie, G.   343, 346
Jutzi, P.   367, 372, 374
Juvek, R.   196

K

Kagramanov, N.D.   17, 367, 409, 410
Kaiser, E.W.   422, 438
Kakinoki, J.   395-397
Kakubari, H.   153
Kakumoto, K.   395-397
Kalaichev, Y.S.   411, 442
Kalasinsky, K.S.   173, 468
Kalasinsky, V.F.   173, 468
Kalcher, K.   40
Kalinin, V.N.   16
Kallury, M.R.   18
Kálmán, A.   316
Kambora, H.   22-25, 231, 233, 234
Kanáan, A.S.   390, 416
Kapfer, C.A.   23, 331
Kapovits, I.   55, 125
Kappler, H.A.   246
Kaptev, G.S.   332
Karakida, K.   74, 135, 173
Karimov, A.S.   16
Karl, R.R.   44, 123, 124, 136, 154, 186, 246
Karle, I.L.   17, 168
Karle, J.   101, 102, 116, 117, 122, 135, 156,
    168, 182, 250
Karlsson, F.   270

Karni, M. 162
Karsh, H.H. 64, 65
Kashiwabara, K. 355, 356, 414
Kasparov, V.V. 398–400, 403
Katada, K. 151, 198
Kataeva, O.N. 64, 65, 68
Kato, C. 114
Kato, H. 469
Katritzky, A.R. 289
Katsumata, S. 108, 271
Katsyuba, S.A. 65
Kattenberg, H.W. 328
Käss, D. 122
Keidel, F.A. 282, 297
Kelley, M.H. 193
Keneshea, F.J. 406
Kennard, O. 2
Kenney, J.K. 4
Kerk, G.J.M.v.d. 372
Kerr, C.R. 342, 344
Khachkuruzov, G.A. 183
Khaikin, L.S. 17, 37, 47, 48, 61, 62, 65–69, 72,
    75, 77, 80, 163, 180, 271, 328, 330, 332,
    349
Kibardin, A.M. 40
Kiessling, D. 398
Kilb, R.W. 161, 162
Kilduff, J.E. 21, 64
Killean, R.C.G. 343
Kimura, K. 45, 151, 186, 191, 192
Kimura, M. 8, 27, 29, 101, 102, 108, 113, 114,
    150, 153, 158, 163, 169, 182, 184, 254, 261,
    263, 309, 327, 328, 345, 355, 356, 408, 411,
    413, 430, 438
Kinneging, A.J. 95
Kirby, R.G. 44
Kirchhoff, W.H. 184, 188, 409,
Kirpichenko, S.V. 135
Kirsch, G. 182, 233, 234
Kiseljev, Y.M. 429, 430
Kisin, A.V. 362
Kiss, A.I. 100, 114, 266
Kitaev, Y.P. 42, 51
Kitaigorodsky, A.I. 218, 233, 234
Kitano, M. 51, 114, 115
Kizer, K.L. 328
Klaeboe, P. 126, 153, 401–406, 443
Klebe, K.J. 115, 157
Kleemann, G. 132, 192
Kleier, D.A. 12
Klein, A.W. 271
Klemperer, W. 389, 391, 399–401, 413, 416,
    422, 431, 438
Klewe, B. 51

Kloster-Jensen, E. 312, 316
Klusacek, H. 176, 181
Klüver, H. 190
Knerr, G.D. 44, 123, 124, 126, 186, 190
Knieriem, B. 273, 274
Koch, B. 158
Kodera, S. 395–397
Kohata, K. 50, 115, 154, 157, 166, 246
Kohl, D.A. 134, 136, 217–219, 262
Kohl, F.X. 367, 372, 374
Kohler, D.A. 11
Kohler, U. 349–351
Kohrmann, J. 157
Kojima, T. 99
Kolderup Fikke, M. 158
Kolesnikov, A.U. 73
Kolesnikov, S.P. 332
Kollmann, P.A. 395
Kolomeets, V.I. 64–66, 77, 82
Kolonits, M. 27, 75, 100, 114, 126, 128, 130,
    131, 163, 190, 191, 266, 296, 305, 306, 309,
    310, 312, 315, 415, 416, 421
Kolsaker, P. 136
Komalenkova, N.G. 22
Komarov, S.A. 3, 430
Konaka, S. 4, 8, 27, 101, 114, 149, 150, 158,
    169, 182, 254, 345, 355, 356, 411, 414
Koningstein, J.A. 428
Konschin, H. 283, 302 304, 309
Koprowski, J. 19, 271
Koptev, G.S. 47
Koput, J. 168
Korenevsky, V.A. 362
Koreshkov, Y.D. 331
Korovik, I. 111
Koshimizu, E. 8, 9, 79
Kostyanovskii, R.G. 49, 65, 66
Kovar, D. 25
Kozina, M.P. 22, 24
Kozunov, V.A. 350
Krabbes, v.G. 412
Kraemer, W.P. 45
Krasnov, K.S. 394 395, 406, 415, 416, 422,
    424–426, 429, 430, 436, 437
Krasnova, T.L. 263
Krats, H. 77
Kratus, M.T. 42
Krauss, H. 162
Krebs, A. 140, 255, 259, 260, 272, 273, 332
Krishna, R. 18
Kroll, W.E. 366
Kroto, H.W. 60
Krüger, C. 374
Kubo, N. 45

Kuchitsu K.   3-7, 9, 11, 15, 19-25, 50, 51, 74,
    101, 108, 114-117, 135, 149-152, 154, 156,
    157, 161, 166, 169 173, 211-216, 218, 230,
    231, 233, 234, 242-244, 246, 249, 254,
    257-260, 262, 266 269-271, 314, 390, 405,
    445, 446
Kucsman, A.   55, 111
Kuczkowski, R.L.   65, 67, 150, 166, 173 185
Kuhler, M.   185 188
Kulikov, V.A.   56, 57 414, 415
Kumar, R.C.   44, 123, 124, 186
Kunze, K.R.   429
Kuonanoja, J.   295 298
Kurkin A.N.   64, 65, 77
Kursanov, D.N.   331
Kusch, P.   397
Kuyper, J.   44
Kuznetsov, G.N. 394
Kuznetsova, T.M.   19, 163, 253
Kück, B.   110
Kvále, E.   372
Kveseth, K.   25, 37, 85, 152, 153, 156, 262,
    266
Kuvada, K.   100

L

Labarre, J.-F.   126, 253, 443, 445
Labartkava, M.O.   263
Lancaster, J.E.   158
Lapitskij, P.V.   432
Lappert, M.F.   20, 21, 27, 334, 335, 359
Laptev, V.T.   15
Larsen, N.W.   46, 282, 283, 302 304, 307, 317
Lathan, W.A.   445
Laurenson, G.S.   11, 12, 39, 48, 70, 74, 170
    173
Laurie, V.W.   155, 160, 162, 328
Lavigne, J.   60
Lawrence, J.L.   343
Laws, E.A.   230
Lazareth, O.W.   395-397, 442
Le Van, D.   80, 172, 176
Leavitt, H.   85
Lee, G.   8, 151
Lee, J.G.   460, 464
Lee, P.L.   170
Lee, S.   370
Lee, Y.S.   328
Lee, Y.T.   370
Lees, R.M.   95
Legon, A.C.   123, 165
Lehmkuhl, H.   273, 358

Lekies, R.   184
Lentz, D.   132, 167, 191, 192
Lesiecki, M.L.   389, 399-401, 403, 405, 407,
    408
Lewis, M.   422
Leyte, J.C.   113
Li, Y.S.   39, 328, 343, 468
Lide, D.R.   150, 165, 190, 215, 222, 328, 394,
    409
Lindøy, S.   5
Lipscomb, W.N.   9, 12, 230
Lister, D.G.   46, 134, 282, 307, 317, 356
Litvinov, O.A.   40, 50, 51
Liu, C.S.   370
Ljunggren, S.J.   40
Lloyd, D.R.   108, 271
Loewenschuss, A.   389, 413, 414
Lohr, A.   2
Lorberth, J.   365
Lory, E.R.   150
Lovas, F.J.   40, 166
Lowrey, A.   313, 315
Lowrey, A.H.   101, 102, 116, 117, 121, 122,
    135, 155, 156, 182, 250
Lu, K.C.   158, 159, 256
Lucas, N.J.D.   185, 186
Lucchese, R.R.   183
Luckenbach, R.   174
Lugié, M.   271
Luong-Thi, N.T.   359
Lusztyk, J.   368-371
Lutoshkin, B.I.   442
Lüthi, H.F.   368, 372
Lüttke, W.   255, 260, 264, 268, 269, 272-273

M

Maagdenberg, A.A.J.   119, 157
Macoll, A.   178
Macdonald, E.K.   68, 108, 109, 171
Macdonald, J.N.   471
Machinek, R.   255, 272, 273
Macho, C.   175
Mack, H.G.   157, 189
Mackle, H.   161
MacLean, C.   317
Magdesieva, N.N.   136
Magnuson, D.W.   197
Mahaffy, P.G.   17, 18
Mahler, W.   175-178, 181
Maier, E.   2
Mair, H.J.   256

Majer, Z. 2
Malcev, A.A. 401, 405
Malone, J.F. 339
Maltese, M. 398, 399
Maltsev, A.K. 17, 19, 367, 409, 410
Maly, H. 334
Malysheva, T.N. 426
Mamaeva, G.I. 11, 356
Mamantov, G. 440, 443
Mangerud, M. 5, 49, 150, 164
Mann, D.E. 190, 399, 400, 403
Mannafov, T.G. 55, 108, 110, 111
Marchand, A. 106
Marciante, C. 288, 295, 296, 298, 302, 303,
    315, 316, 321
Margrave, J.L. 384, 390, 416, 421, 422, 429,
    432
Maroni, V.A. 443–445, 447
Marquarding, D. 176, 181
Marquart, J.R. 398
Marram, E.P. 404
Marsden, C.J. 49, 64, 80, 99, 103, 122, 132,
    133, 166, 168, 169, 173, 174, 180, 183–186,
    189, 193, 199, 471, 473
Marsden, H.M. 157
Marstokk, K.-M. 95–98, 273, 363
Martin, T.P. 395
Martinez, J.V. 198
Marynick, D.S. 11, 12, 370
Masamune, S. 24
Massaux, M. 299
Mastryukov, V.S. 1, 3, 11, 12, 15–17, 19, 20,
    22–25, 27, 28, 58, 79, 85,135, 235–237, 245,
    326, 327, 331, 343, 346, 349,
Mathew, S.N. 468
Matin, J. 72, 85
Matrosov, E.I. 394
Matsumura, C. 50
Mawhorter, R.J. 396, 397
Maxwell, L.R. 393, 394
Mayo, R.A. 19, 47
Mazzei, A. 350, 351
Mazzeo, P. 283
McAdam, A. 71
McAloon, K. 328
McClelland, B.W. 56, 79, 175, 178, 433
McDermott, D.P. 416, 417
McDonald, T.R.R. 350
McDonald, W.S. 339, 350
McFadden, W.H. 394
McKean, D.C. 462
McKee, M.L. 12, 16
McLean, W.C. 342, 344
McNeill, E.A. 15

McPhail, A.T. 283, 287
McPherson, A.M. 339
Medley, J.H. 342, 344
Medwid, A.R. 8, 29, 45, 64, 71, 150, 165, 169,
    170
Mehmood, T. 182
Meikle, G.D. 19, 44
Mel'nikov, N.N. 61, 62, 75
Menegus, F. 368
Mews, R. 39, 123, 124, 132, 133, 159, 186,
    188, 193, 194, 196
Meyer, R. 183
Miakshin, I.N. 196, 197, 411, 412, 435, 437
Michel, F. 299, 300
Michl, J. 28
Mijlhoff, F.C. 18, 19, 22, 44, 45, 95, 101–103,
    105, 106, 108, 110, 112, 113, 115, 117, 121,
    152–154, 157, 185, 186, 233, 234, 246, 259,
    261, 312
Miki, H. 395–397
Miles, S.J. 334, 335
Millen, D.J. 123, 165
Miller, R.C. 397
Mills, I.M. 21, 160, 464
Milne, A. 395
Milvitskaya, E.M. 22, 24
Minkwitz, R. 165, 175, 177, 178, 183, 334
Mislow, K. 214
Mitchell, K.A.R. 412
Mochaklov, V.I. 261, 262
Moffitt, W.E. 230, 231
Mohamad, A.B. 39
Mohammadi, M.A. 134
Mohanty, A.K. 191
Mole, T. 339, 347
Molnár, Z. 415, 416
Mom, V. 45, 95, 154, 246
Monaghan, J.J. 71, 161, 164, 172, 222, 223
Montero, S. 217
Montgomery, L.K. 17, 18, 209, 234–236, 263
Moody, D.C. 12
Mooney, R.C.L. 429
Moore, J.H. 105, 108, 123, 126, 128, 296
Morgan, G.L. 353, 354, 356
Morino, Y. 17, 109, 114, 169, 173, 182, 262,
    266, 270, 271, 328, 402, 405, 406, 408, 411,
    426–428
Moritani, T. 169, 173
Morris, A. 429
Morrison, G.A. 215, 216, 227, 233, 258
Mosley, V.M. 393, 394
Motzfeldt, T. 118, 119, 372, 374
Moutran, R. 19, 163, 253
Mozzi, R.L. 350, 351

Møllendal, H. (Möllendal, H.) 85, 95–98, 273, 363
Muecke, T.W. 255
Muenter, A.A. 431
Muenter, J.S. 406, 407, 438
Muetterties, E.L. 175–178, 180, 181
Mukmenev, E.T. 72
Murakami, S. 24
Murata, Y. 408
Murchison, C. 198
Murdoch, J.D. 109, 332
Murphy, W.F. 217
Murray, E.K. 47
Murray-Rust, P. 283–286, 288, 289–292, 295, 297–306, 308, 313, 316
Murty, A.N. 98
Murty, K.S.R. 191
Mustoe, F.J. 125, 161–163, 327, 328, 331
Mu-tao, L. 139
Mutter, R. 2
Mürdoch, J.D. 39, 40

N

Naegele, W. 366
Nagashima, M 163, 327, 328
Nagel, B. 126, 253
Náhlovská, Z. 121, 136
Náhlovský, B. 121, 136
Nair, K.P.R. 126, 130, 131, 190
Nakamura, Y. 328, 408
Nakata, M. 50, 115, 157, 230, 231, 246, 257
Narath, A. 468
Nasarenko, A.Y. 409
Naumov, V.A. 40, 42, 47, 50, 51, 64–69, 72, 75, 77, 78, 85, 93, 102, 205, 108, 110, 125, 126, 128, 130, 253, 355, 402
Nazarenko, I.I. 165
Nedjak, S.V. 401, 405
Nefedov, O.M. 17, 19, 332, 367, 409, 410
Negrebetskii, V.V. 61, 62
Nehl, H. 273, 358
Nemukhin, A.V. 3, 4
Nery, H. 299, 300
Nesterov, V.Y. 65, 67, 253
Neumann, W.P. 334
Newton, M.G. 86
Nibler, J.W. 399, 400, 401, 403, 407, 408
Nielsen, C.J. 126, 266
Nielsen, E.W. 363
Niepel, H. 45, 165
Nikitin, V.S. 271, 328
Nikolaev, A.N. 411
Nilsson, J.E. 353, 354, 356

Nipan, M.E. 263
Niu, Z. 61, 62, 462
Nixon, J.F. 60
Noakes, T.J. 40
Nor, O. 37, 271
Nordman, C.E. 346
Nordtømme, T. 266
Norrestam, R. 288, 289, 292, 302–306, 316
Norton, B.G. 71, 164, 172
Nosberger, P. 299, 300
Novak, D.P. 11, 343, 350, 352, 364, 443, 445, 448
Novikov, G.I. 440
Novikov, V.P. 54, 64–66, 77, 82, 163, 296, 302, 303, 311, 313, 328, 343, 346
Novikova, Z.S. 64, 65, 77
Nugent, K.W. 370, 371
Nyburg, S.C. 18
Nygaard, L. 83, 282, 283, 287, 295, 297–299, 301, 307
Nyholm, R.S. 148, 169, 175, 178, 179

O

Oberhammer, H. 19, 25, 27, 28, 39, 40, 43–46, 58, 60, 64, 71, 77, 79, 80, 85, 102, 108, 110, 122–124, 126, 128, 131–133, 147, 149, 151, 152, 154, 156–169, 172, 175–196, 250, 256, 263, 271, 328, 330, 347, 356, 470, 473
Odom, J.D. 71, 172, 343
Offenbach, J.L. 27
Ofitserov, E.H. 69
Ogasawara, M. 261, 263
Ogden, J.S. 401, 405
Oka, T. 109
Okije, K. 134
Olbrich, G. 334
Oliver, J.P. 339, 340, 358, 368
Oppenheim, V.D. 17
Oppermann, H. 412
Orgel, L.E. 178
Orville-Thomas, W.J. 211–213, 314
Osina, E.L. 22, 23
Oskam, A. 328
Ostoja Starzewski, K.A. 43
Ostropikov, V.V. 415, 416
Ottar, B. 233, 234
Ouzounis, K. 349–351
Overend, J. 4, 198
Oyamada, T. 327, 328
Oyanagi, K. 101, 108, 151, 154, 246
Østensen, H. 95, 98, 273
Øye, H.A. 440, 443

P

Page, E.M.   431
Palmer, K.J.   401, 402
Panattoni, C.   368, 372
Panchenko, Y.N.   261, 262
Pandey, G.K.   29
Pang, F.   295, 298, 305, 306, 457, 462
Pankrushev, Y.A.   53–55, 302, 310, 312, 313, 315
Papatheodorou, G.N.   401–403, 405, 406, 440, 443–445, 447
Pappalardo, G.C.   105, 108, 126, 128
Parameswaran, T.   428
Parent, C.R.   196
Párkányi, L.   316
Parris, G.E.   364
Parry, G.S.   350, 351
Parry, R.W.   346
Patai, S.   21, 37, 329, 330, 336, 355
Patton, J.V.   6, 150
Paul, I.C.   21, 37
Paul, J.W., Jr.   174
Paulen, G.   114, 117, 261, 262, 266, 270
Pauling, L.   9, 44, 45, 151, 161, 169, 233, 234, 332, 333, 336, 425
Pawelke, G.   43, 45, 46, 149, 156, 164, 165, 167, 250
Pearson, R.   40, 166
Pedersen, T.   282, 283, 287, 295, 297, 298, 307
Peixoto, E.M.A.   137, 139, 267
Pela Ceccarini, G.   295, 298
Pelissier, M.   443, 445
Pence, D.T.   155
Penionzhkevich, N.P.   54–56, 81, 296, 299, 300, 302, 310, 311, 313, 315
Penn, R.E.   95
Penner, G.H.   302, 303
Pépin, C.   60
Perego, G.   351
Permanand, R.R.   471
Perov, P.A.   401, 405
Pervov, V.S.   436
Pete, B.   100, 114, 266
Peters, E.M.   132, 189, 191, 192, 254
Peterson, G.E.   177
Peterson, S.H.   198
Petrov, K.P.   73, 411, 442
Petrov, V.M.   414, 415, 422, 426, 429, 430, 436, 437
Petrova, V.N.   426, 430, 436, 437
Petrunin, A.B.   15
Pfafferott, G.   58, 85, 162
Pfohl, S.   176, 181

Pfund, R.A.   273
Pickardt, J.   372, 374
Pierce, L.   158, 161, 162, 165
Piero, G.D.   350, 351
Pilcher, G.   329, 330, 336, 355
Pinder, P.M.   71, 148
Piper, L.G.   155, 166
Pirzadeh, T.   2
Pitz, C.L.   79
Pitzer, K.S.   198, 355
Plaats, G.   119
Planie, M.C.   101, 102
Plant, C.   471
Plato, V.   59, 165
Pongor, G.   465
Poon, Y.C.   18
Popenko, N.I.   402, 405, 406, 422, 445
Popik, M.V.   54, 55, 110, 111, 293, 296, 302, 310, 311, 313, 315
Popik, N.I.   54
Pople, J.A.   103, 445, 458
Potalone, G.   100, 114, 283, 288, 295, 296, 298, 299, 301–305, 307, 309, 310, 314–321
Porter, R.F.   8, 43, 149–151, 164, 167, 256, 401, 402, 415
Poulin, D.D.   175–177
Powel, D.L.   153
Powell, F.X.   184, 409
Power, P.P.   21, 358
Pozdeev, N.M.   136
Pötter, B.   188, 189
Preiss, H.   432
Prenzel, H.   175
Priebe, H.   98, 273
Pritchard, R.G.   20, 21, 134, 158
Pronicheva, L.D.   125
Przhevalskii, I.N.   183
Puddephatt, R.J.   359
Pudovik, A.N.   64, 69
Pudovik, M.A.   72, 85
Pulay, P.   173, 295, 298, 305, 306, 457, 458, 460, 462–465, 468, 472, 474
Pulkkinen, A.   295, 298
Purt, A.W.   65, 66
Pyckhout, W.   101, 103, 251
Pyyköö, P.   355

R

Rademacher, P.   58
Radloff, P.L.   406
Radom, L.   183
Raevskii, O.A.   64–66, 77, 82

Rambidi, N.G.   56, 57, 73, 394, 398–400, 402–405, 414, 415, 424, 425, 429, 430, 442, 443
Ramirez, F.   176, 181
Ramme, K.J.   184, 187
Rankin, D.W.H.   11, 12, 19, 21, 25, 27, 39, 40, 44, 47–50, 58, 64–66, 68, 70, 71, 74–77, 80, 99, 105–109, 112, 156, 163, 164, 167–171, 173, 174, 180, 187, 263, 271, 332, 337, 341, 343, 345, 350, 361, 362, 368
Rao, R.S.   355, 356
Rastrup-Andersen, J.   282, 283, 287, 295, 297, 298, 307
Raw, T.T.   158
Reader, W.   4, 20
Reichman, S.   198
Rempfer, B.   161, 162
Renes, G.   18, 44, 45, 95, 101, 103, 106, 112, 115, 117, 121, 152–154, 157, 246, 259, 312
Reuter, D.   469
Rhine, W.E.   339
Rice, D.A.   431
Richards, W.G.   405
Richardson, T.J.   437, 438
Riddell, F.G.   49, 99
Ridgen, J.S.   182
Riethmiller, S.   443
Riffel, H.   349–351
Riley, P.E.   188
Rippon, D.M.   428
Ritter, D.M.   11
Riva di Sanseverino, L.   302, 309
Robert, J.B.   72, 85
Robertson, A.   19, 25, 164, 167
Robertson, H.E.   21, 27, 47, 64, 65, 75, 76, 106–109, 171, 187, 345, 350
Robiette, A.G.   28, 39, 49, 50, 71, 105, 106, 108, 149, 155, 160, 163, 164, 166, 168, 197, 211–213, 314, 332
Robinet, G.   80, 126, 180, 253
Roch, B.S.   334
Rodler, M.   99
Rodom, L.   103
Rodwell, W.R.   183
Roelandt, F.F.   161
Rogers, M.T.   196
Rogers, R.D.   342, 344
Rogowski, R.   188
Rohmann, J.   175, 177, 178
Rolinski, E.J.   412
Romanov, G.V.   64, 196, 197, 398–400, 411, 412, 426–428, 430–432, 434, 435, 437
Romers, C.   113
Ron, A.   413, 414

Ronova, I.A.   163, 368
Rooij, J.   112
Roon, P.H.   119
Rosenblatt, G.M.   384, 399, 400, 422
Rossiter, B.W.   9
Roussy, G.   299, 300, 305–307
Rozsondai, B.   17, 19, 21, 25, 28, 105, 106, 108, 123, 126, 128, 163, 253, 270, 295, 296, 310, 332
Rubio, J.   365
Rudchenko, V.F.   49
Rudolph, R.W.   12, 68, 171
Ruelle, P.   61
Rummens, F.H.A.   259
Rundle, R.E.   179, 335, 353, 354
Rustad, S.   18, 20, 137, 197
Ryan, R.R.   150, 198
Rypdal, K.   377
Rytter, E.   401–406, 443, 445, 447
Rzepa, H.S.   370

S

Sachse, H.   233
Sadova, N.I.   15, 24, 27, 35, 52–56, 75, 136, 245, 263, 296, 299, 300, 302, 303, 310–313, 315, 327, 349
Saebo, S. (Saebø, S.)   6, 51, 474
Saito, S.   123, 182
Samdal, S.   51, 52, 55, 74, 75, 103, 108, 113–115, 156–158, 247, 251, 252, 293, 296, 297, 311–317, 343, 345, 355, 356, 364, 365, 410, 411
Sandnes, T.W.   135
Sarvin, A.P.   434
Sasaki, Y.   45
Sastry, K.V.L.N.   95, 98
Satge, J.   334
Sattler, E.   27, 163
Saunders, J.K.   339
Schaber, H.   395
Schachtschneider, J.H.   216
Schack, C.   39
Schack, C.J.   167
Schaefer, H.F.   18, 183, 370, 391, 409
Schaefer, T.   302, 304, 305
Schaeffer, R.   12
Schardey, A.   175
Scharfenberg, P.   126, 295, 296, 315
Schatte, G.   189
Schäfer, H.   440
Schäfer, L.   114–116, 182, 233–235, 370, 395, 462, 472
Schei, H.   23, 58, 114, 248, 267, 271

Schei, S.H.   37, 101, 103, 120, 121, 155,
247–252, 316, 317
Schepper, L.   288, 289, 292, 302, 303, 304–306,
316
Schiele, M.   19
Schilling, B.E.R.   21, 334, 335, 363–365, 367,
372, 374, 375
Schindler, G.   347
Schirdewahn, H.G.   2
Schlessinger, H.I.   339
Schleyer, P.v.R.   359, 370
Schlosser, K.   184, 185
Schmidbaur, H.   75, 76, 347
Schmidling, D.G.   44, 336
Schmidt, M.W.   74
Schmiedekamp, C.W.   151, 152
Schmutzler, R.   64, 169, 175–178, 180–182
Schnepp, O.   413, 414
Schnidpeter, A.   62
Schnökel, v.H.   401, 405
Scholer, F.R.   15
Schomaker, V.   2, 11, 12, 20, 196, 231, 232,
255, 388, 402, 412, 413, 438
Schoonmaker, R.C.   415
Schreiner, F.   198
Schrem, G.   58
Schrumpf, G.   264, 271
Schubert, W.K.   235
Schultz, G.   17, 19, 27, 42, 55, 72, 85, 99, 100,
106, 111, 114, 125, 126, 130, 131, 135, 138,
139, 190, 266, 284, 295, 296, 298, 299, 301,
302, 304–310, 312, 314, 315, 317–321, 367,
409, 410, 415, 416
Schumberger, F.   37
Schuster, H.G.   25
Schwarz, W.   340
Schweizer, W.B.   273
Schwendeman, R.H.   68, 169–171
Schwoch, D.   12
Sebastian, R.   302, 304, 305
Seel, F.   187
Segal, G.A.   16
Seip, H.M.   5–7, 17, 23–25, 27, 101, 103,
107–109, 113, 136–138, 158, 163, 233, 234,
247, 251, 252, 256, 260, 270, 438
Seip, R.   5–7, 19–21, 27, 37, 39, 40, 49, 51, 52,
64, 75, 76, 95, 98, 101, 103, 108, 109, 126,
130, 131, 135, 138, 150, 153, 155–158, 163,
164, 170, 184, 190, 247–250, 262, 267, 270,
271, 273, 274, 284, 296, 305, 306, 334, 336,
339, 340, 343, 346, 353, 355, 356, 364, 365,
367, 372, 374, 375, 438
Seiter, U.   2
Selig, H.   438

Sellers, H.   3, 4
Sellers, H.L.   243–245, 252, 395
Sellers, S.F.   157
Semashko, V.N.   75, 77, 78, 85
Sen, K.D.   73
Senda, Y.   273
Seppelt, K.   39, 126, 128, 130, 132, 133,
188–192, 194, 195
Sereda, S.V.   332
Sergeyev, N.M.   362
Serke, I.   125
Seshadri, K.S.   399, 400
Shaidulin, S.A.   72, 75, 78, 253
Shakhparonov, M.I.   440
Shakir, R.   359
Shapkin, A.A.   136
Shaposhnikova, O.K.   426
Sharma, E.   460
Shatrukov, L.F.   77, 125
Shcherbak, G.A.   136
Shcherbik, L.K.   331
Shcherbinin, V.V.   163
Shearer, H.M.M.   368
Sheldrick, G.M.   28, 49, 50, 71, 105, 106, 108,
109, 122, 149, 159, 163, 166, 168, 184, 185,
332, 336, 351
Sheldrick, W.S.   28, 62, 149, 336, 351
Shen, Q.   17, 19, 20, 22, 23, 58, 75, 85, 106,
111, 112, 120, 121, 134, 135, 158, 161, 162,
173, 247–252, 328, 331, 336, 402–405, 445,
446
Sheridan, J.   134, 356, 365
Shibata, S.   8, 9, 57, 79, 102, 150, 365, 431
Shi-cheng Chang   40
Shidulin, S.A.   65, 67
Shimanouchi, T.   116, 117, 254
Shirk, A.E.   389, 405
Shirk, J.S.   389, 405
Shishkov, I.F.   49, 53, 54, 296, 302, 303, 311
Shoji, H.   409
Shreeve, J.M.   44, 80, 123, 124, 126, 132, 157,
175, 180, 182, 186, 190, 193–196
Shustorovich, E.M.   193
Shvetsov-Shilovskii, N.I.   61, 62
Sidorov, L.N.   440
Sille, K.   339
Sim, G.A.   211–213, 283, 287
Simmons, N.P.C.   60
Simon, A.   132, 189, 191, 192, 254
Sinyashin, O.G.   68
Siu, A.K.Q.   400
Sjøgren, C.E.   401–406, 443, 445, 447
Skancke, A.   157, 158, 297, 312, 315, 316, 468,
469

Skancke, P.N. 231, 232
Skinner, H.A. 329, 330, 336, 355
Slater, J.C. 9
Slater, R.C. 406–408
Smart, B.E. 157, 189
Smart, J.B. 358
Smirnov, V.V. 72
Smith, D.E. 345
Smith, D.F. 196
Smith, D.W. 438
Smith, G.S. 12
Smith, J.E. 169, 170
Smith, J.G. 74, 185, 186
Smith, M.B. 338
Smith, R.S. 353, 355, 356
Smith, Z. 17, 25, 27, 134–136, 138, 158
Snow, A.I. 353, 354
Snow, J.T. 24
Snow, M.R. 370, 371
Snyder, R.G. 216
So, S.P. 405
Sobolev, G.A. 398
Sokolkov, S.V. 54, 55, 111, 293, 296, 313
Skolov, V.B. 412, 435, 437
Solc, V.B. 440
Solomonik, V.G. 394, 395, 406
Solouki, B. 132, 192
Southern, J.F. 235
Souza, G.G.B. 328
Sørensen, G.O. 83, 282, 299, 301, 307
Søvik, O.I. 247–249
Speed, C.S. 358
Spek, A.L. 358, 372
Spelbos, A. 18, 44, 101, 103, 106, 154, 246
Spiridonov, V.P. (Szpiridonov, V.P.) 3, 4, 11,
    57, 174, 196, 197, 356, 363, 384, 391, 393,
    394, 398–400, 402, 405, 406, 409, 411–415,
    422, 426–428, 430–437, 440, 442, 443, 445,
    446
Špirko, V. 45
Spoliti, M. 398, 399, 412
Spreter, C. 2
St. Denis, J. 358
Stahl, I. 193
Stanko, L.V. 15
Stanley, A.E. 468
Starck, B. 2
Starowieyski, K.B. 368, 369
Stauffer, J.L. 389, 391, 399, 401, 404, 413, 416
Stedman, D. 410
Steer, I.A. 4, 20, 336
Steger, B. 167, 168, 172, 175
Steinhauser, S. 349, 351
Steinnes, O. 120, 121

Stelzer, O. 64–66, 169
Stern, R.C. 406–408
Stevens, J.F. 472
Stevens, L.G. 339
Stevens, R.M. 230
Stevenson, D.P. 20, 196, 402
Stevie, F.A. 437, 438
Stewart, C.A. 372, 374
Stigliani, W.M. 307
Stobart, S.R. 362
Stock, A. 2
Stoicheff, B.P. 18, 158, 355, 356
Stokkeland, O. 343, 345, 349, 350, 352
Stolevik, R. (Stølevik, R.) 65, 153, 154, 156,
    246–249, 260, 271, 273
Stone, F.G.A. 326, 377
Stosick, A.J. 11
Strand, T.G. 37, 39, 40, 43, 99, 121, 193, 246,
    271, 411, 412, 432–437
Strauss, H.L. 27
Strähle, J. 43
Streib, W.E. 335, 336
Streitwieser, A. 370
Strelkov, S.A. 22, 25, 135
Strey, G. 464
Struchkov, Y.T. 17, 49, 61, 62, 331, 332, 368
Stucky, G.D. 339, 353, 357
Stull, D.R. 401, 402
Su, L.S. 71, 117, 172
Subbotina, N.Y. 415, 416
Suga, H. 254
Sukhoverkhov, V.F. 196, 197, 411, 412, 437
Sullivan, J.F. 39
Surtees, J.R. 339
Suslova, E.N. 135
Sutter, D. 185, 188
Sutton, L.E. 161, 178, 211–213
Swepston, P.N. 395
Swick, D.A. 17
Switkes, E. 230
Sydnes, L.K. 260, 262
Syshchikov, Y.N. 65, 66, 82
Szabó, Z.G. 296, 305, 306
Székely, T. 19, 126, 128
Szobota, J.S. 188
Szöke, A. 126, 128, 130

T

Taft, R.W. 289
Takabayashi, F. 22, 25, 231, 270
Takahata, Y. 399–401
Takemura, M. 8, 27, 101, 150, 158, 182, 345

Takeo, H.   150, 409
Takeo, J.   50
Tamagawa, K.   8, 27, 101, 136, 150, 158, 182,
    271, 309, 345
Tanaka, T.   409
Tanimoto, M.   270, 271
Tarasenko, N.A.   17, 106
Tatevskii, V.M.   402
Taugbøl, K.   372, 374, 375
Taylor, M.J.   443
Taylor, P.R.   183
Tecle, B.   340, 368
Teller, E.   427, 428, 433, 434
Ter Brake, J.H.M.   45, 243, 245, 259
The, K.I.   175–177
Thiele, K.-H.   357
Thingstad, Ø.   246, 248, 271
Thom, E.   154
Thomas, E.C.   328
Thomas, J.   27, 163
Thomas, K.M.   334, 335
Thomas, P.D.P.   11, 337, 341
Thompson, H.B.   197, 198, 225
Thompson, K.R.   421
Thorne, A.J.   20, 21, 27, 334, 335, 359
Thornton, C.G.   151
Tiemann, E.   384
Todd, M.R.   48, 49, 70, 77, 99, 170, 174
Tokue, I.   156, 243, 244, 254, 259
Tolles, W.M.   186
Tombo, G.M.R.   273
Tomita, T.   401–406
Tomlinson, A.J.   175–177
Toneman, L.H.   101, 102, 157
Toney, J.   357
Topsom, R.D.   289
Torgrimsen, T.   103, 107, 108, 156, 247, 251,
    252
Torring, T.   384
Torrini, I.   309, 315, 319, 320
Touseev, N.I.   57
Trachtman, M.   283, 295, 298, 289, 301–306
Traetteberg, M.   16, 37, 40, 43, 95, 98, 106,
    114, 117, 135, 163, 239, 254–267, 269–274
Trambarulo, R.   152
Tremmel, J.   17, 367, 384, 386, 388, 389,
    398–400, 403, 408–410, 412–417, 421, 425,
    442
Trickle, I.R.   429
Trinquier, G.   334
Trombetti, A.   166
Trongmo, Ø.   247–249
Trotter, J.   302
Tsay, Y.-H.   374

Tsirel'nikov, V.I.   426
Tsolis, E.A.   176, 181
Tsuchiya, S.   29, 102, 108, 120, 184
Tuck, D.G.   135, 158
Tulyakova, T.F.   75
Turner, E.S.   99
Turner, J.B.   328
Turner, R.   355, 356
Turova, N.Y.   350
Tuzova, L.L.   47, 64, 65, 68, 69, 72
Tyblewski, M.   183, 273
Tyler, J.K.   46, 282, 307, 317, 365
Typke, V.   19, 37, 151, 152, 186, 272

U

Uehara, H.   426–428
Ugarov, V.V.   56, 57, 73, 411, 414, 415, 425,
    442
Ugi, I.   176, 181
Ukayi, T.   402, 405, 406, 411
Ulbrecht, V.   108, 197, 271
Urevig, D.S.   16
Ustynyuk, Y.A.   362, 368
Utkin, A.N.   402, 404, 405, 414, 415, 422, 426

V

Vaciago, A.   282–288, 290, 295, 308, 313, 317
Vajda, E.   19, 27, 75, 126, 128, 130, 163, 267,
    328, 388, 398–400, 403, 409, 412, 413, 416,
    442
Van Alsenoy, C.   101, 103, 113, 114
Van den Enden, L.   22, 101, 103, 105, 113,
    114, 243–245, 252, 259
Van der Draii, R.K.   157
Van der Plaats, G.   157
Van der Vorn, P.C.   175
Van des Does, H.   45
Van Eijck, B.P.   157
Van Hemelrijk, D.   243–245, 252, 259
Van Laere, E.   125
Van Nuffel, P.   101, 103, 113, 114
Van Roon, P.H.   157
Van Schaick, E.J.H.   312
Van Schaick, E.J.M.   101, 103, 152–154, 246
Vargha, A.   465
Vasile, M.J.   437, 438
Vasil'ev, A.F.   61, 62, 75
Väänänen, T.   295, 298
Veazy, S.E.   442
Velichko, F.K.   10

Veniaminov, N.N.   19, 22, 163, 253, 368
Veracini, C.   299, 301–303
Vereshchagin, A.N.   49
Vilkins, C.J.   75
Vilkov, L.V.   3, 11, 12, 15–17, 20, 22–25, 27, 28, 35, 37, 47–49, 52–56, 58, 61, 62, 64–66, 68, 69, 72, 75, 77, 79–82, 85, 86, 103, 106, 110, 111, 124, 135, 136, 163, 165, 180, 188, 245, 263, 271, 293, 296, 299, 300, 302, 303, 310–313, 315, 327, 328, 330–332, 343, 346, 349, 356
Vinogradov, V.S.   425
Vledder, H.J.   113
Volden, H.V.   21, 334, 359, 364, 372, 373, 376, 377
Volpin, M.E.   331
Von Carlowitz, M.   185, 187, 469, 470
Von Carlowitz, S.   173, 176, 177, 468, 469
Voronkov, M.G.   135
Vorren, O.   153
Vos, A.   285, 308
Vrieland, G.E.   401, 402

W

Wade, K.   12, 13, 368
Wagner, Z.   19, 106
Wahl, G.H.   214
Walker, M.L.   188
Walker, N.P.C.   365
Wall, F.T.   16
Walsh, A.D.   264, 265, 287, 400
Wang, S.-M.   370
Waring, S.   56, 168
Wartic, T.   339
Waterfeld, A.   132, 194, 196
Watson, D.G.   2
Watta, B.   334
Wehrlein, U.   162
Wehrung, T.   158
Weidlein, J.   109, 336, 339, 340, 343–345, 348–352, 366, 372, 373
Weinstock, B.   198, 438
Weiss, E.   353, 368
Weissenberger, A.   9
Welch, A.J.   345
Welch, D.O.   395–397, 442
Welcman, N.   158
Well, F.P.   113
Weller, F.   365
Wells, A.F.   17
Wells, C.   356

Wells, J.W.   422
Werder, R.D.   389
West, R.   24, 28
Wharton, L.   399–401
Wheeler, V.L.   155, 166
White, D.   399, 400, 403
Whitehead, G.   368
Whitt, C.D.   357
Wierl, R.   2
Wieser, J.D.   221, 222, 235, 236, 260, 263, 328
Wilcox, C.F.   264
Wildman, T.A.   302, 304–306
Wilkins, C.J.   115, 157, 173
Wilkinson, G.   326, 353, 377
Willadsen, T.   136
Williams, R.E.   12
Williamson, S.M.   168
Willis, R.E., Jr.   196
Willner, H.   102, 108, 110, 176, 184–187, 189, 470
Wilson, C.A.   234, 263
Wilson, E.B.   166, 188, 463
Wilson, P.W.   409
Winnewisser, G.   95
Wiser, J.D.   12
Wong, C.   370
Wong, C.-H.   370, 388, 412, 413
Wong, T.C.   62, 63
Wong-Ng, W.   18
Wolfel, V.   175, 177, 178

X

Xie, Y.   465

Y

Yagola, A.G.   82
Yamada, T.   102, 150
Yamada, Y.   8
Yamamoto, O.   338
Yamamoto, S.   230, 231, 257
Yanagisawa, M.   338
Yarkov, A.V.   65, 66, 77, 82
Yavari, P.   152
Yokozeki, A.   95, 98, 122, 148, 151, 152, 154, 157, 159, 330
Yoshioka, Y.   18
Yow, H.   179
Yow, H.Y.   68, 171

Z

Zakharkin, L.I. 16

Zakzhevskii, V.G. 395

Zanjanchi, M.A. 152

Zaripov, N.M. 47, 55, 64–66, 68, 69, 72, 85, 86, 103, 108, 110, 111, 125, 255

Zasorin, E.Z. 3, 4, 57, 384, 393–395, 402, 404–406, 409, 422, 424–426, 429, 430, 440, 442, 445

Zavgorodnii, V.S. 19, 47, 271, 328, 330, 332

Zav'yalov, A.P. 78, 85

Zebelman, D. 113, 116, 117

Zeil, W. 28, 58, 77, 123, 124, 128, 132, 133, 153, 157, 173, 186, 194, 271, 347, 468

Zelei, B. 21, 310

Zeller, E.E. 401, 402

Zenkin, A.A. 37

Zhigach, A.F. 3, 15

Zhigareva, G.G. 16

Ziatdinova, R.N. 64, 105, 126, 253

Zil'berg, I.Y. 75

Zimmer-Gasser, B. 75

Zozulin, A.J. 343

Zuckerman, J.J. 374

FORMULA INDEX

A

AlB_3H_{12} 11
$AlBr_3$ 403, 445, 447
$AlBr_3H_3N$ 443, 445, 447
$AlCl_3$ 384, 386, 389, 403–406, 440, 445, 447, 448
$AlCl_3H_3N$ 443, 445, 447
$AlCl_3N_2$ 389
$AlCl_4Cs$ 442
$AlCl_4K$ 442
$AlCl_4Rb$ 442
$AlCsF_4$ 442
AlF_3 404, 405, 445
AlF_3H_3N 443, 445, 447
AlF_4K 440, 442, 443
AlF_4Li 442, 443
AlF_4Na 440, 442
AlF_4Rb 442
AlH_2 338
AlH_3 346, 448
AlH_5O 345
AlI_3 403, 447
AlN 350, 351
Al_2Br_6 405, 445
Al_2Cl_6 393, 405, 443
Al_2F_6 403, 405
Al_2I_6 405
Al_4F_4 347
$AsBr_3$ 410, 411
$AsCl_2F_3$ 175
$AsCl_3$ 411
AsF_3 169, 411
AsF_5 174, 175, 411, 433
AsH_9Si_3 29, 163
AsI_3 411
AuF_5 437
Au_2F_{10} 437
Au_3F_{15} 437, 438

B

BBr_3 2, 5
BCl_3 2, 5

$BCsO_2$ 3, 4
BF_2HO 150
$BF_2H_6NSi_2$ 28, 149, 150
BF_2N 28
BF_3 2, 4, 5, 102, 149, 150
BF_3H_3P 150
BF_7Si_2 149, 164
BI_3 5
BKO_2 3
$BLiO_2$ 3
$BNSi_2$ 28
$BNaO_2$ 3
BO_2Rb 3, 4
BO_2Tl 3, 4
B_2BeH_8 11
B_2Br_4 6, 9
B_2ClH_5 11
$B_2Cl_2S_3$ 7
B_2Cl_4 6, 9
B_2D_6 11
B_2F_4 6, 9, 150
$B_2F_6H_4P_2$ 150
B_2GaH_9 11
B_2H_6 2, 9–11
B_2H_7N 11
B_2I_4 6, 9
B_2O_3 3, 4
B_2S_3 3
$B_3F_3H_3N_3$ 151
$B_3H_3O_3$ 7
$B_3H_6N_3$ 2, 7, 8, 49, 151
$B_3H_{12}Ti$ 11
B_4H_{10} 11–13
$B_4H_{16}Zr$ 11
B_5H_9 11–13
$B_5H_{11}Si$ 12
$B_{10}H_{14}$ 11
$BaBr_2$ 400, 401
$BaCl_2$ 398–401
BaF_2 400, 401
BaI_2 398–401
$BeBr_2$ 400, 401
$BeCl_2$ 363, 400, 401
BeF_2 400, 401
BeF_3K 440, 441

BeF_3Li 440, 441
BeF_3Na 440, 441
BeI_2 400, 401
BrF 196, 197
$BrFO_3$ 197
BrF_2PS 74, 173
BrF_3 197
BrF_5 197
Br_2Ca 386, 398–403
Br_2Cd 414. 415
Br_2Co 416, 421
Br_2Fe 416, 421
Br_2Ge 17, 409
Br_2Mg 400, 401
Br_2Mn 386, 416, 421
Br_2Ni 415, 416, 421
Br_2Pb 409
Br_2Si 17, 409
Br_2Sn 409
Br_2Sr 399–401
Br_2Zn 413, 414
Br_3Ce 424
Br_3Dy 424
Br_3Er 424
Br_3Eu 424
Br_3Ga 403, 446
Br_3GaH_3N 443, 446
Br_3Gd 422. 424
Br_3Ho 424
Br_3La 422, 424
Br_3Lu 422, 424
Br_3Nd 424
Br_3OP 73, 74
Br_3P 63, 68
Br_3PS 73, 74
Br_3Pm 424
Br_3Pr 424
Br_3Re 425
Br_3Sb 411
Br_3Sm 424
Br_3Tb 424
Br_3Tm 424
Br_3Yb 424
Br_4Ge 328, 329, 356
Br_4Hf 426
Br_4Mo 430, 431
Br_4Si 17
Br_4Th 429, 430
Br_4Ti 426, 428
Br_4U 429, 430
Br_4V 426–428
Br_4Zr 426
Br_5Nb 432
Br_5Ta 432, 433

Br_6Ga_2 403, 405
Br_9Re_3 425

C

$CBrF_3$ 151, 152
$CBrF_3S$ 184
$CBrN$ 37
$CBrN_3O_6$ 53
CBr_2 409
CBr_3NO_2 53
$CClFO$ 157
$CClFS$ 157
$CClF_3$ 151, 152
$CClF_3O$ 182
$CClF_3O_2$ 122, 183
$CClF_3O_2S$ 126, 127, 190
$CClF_3S$ 110, 184
$CClN$ 37
$CClNO$ 38–40
$CClNO_3S$ 39, 126, 127
$CClNS$ 38
$CClN_3O_6$ 53
CCl_2 409
$CCl_2F_6Si_2$ 163, 164
CCl_2NO_2P 39, 77
$CCl_2N_2O_4$ 52, 53
CCl_3F_3Ge 162
CCl_3N 40
CCl_3NOSi 19, 39
CCl_3NO_2 52, 53
CCl_4F_3P 177, 178
CCl_4O_2S 126, 127
CCl_4S 110
CD_4 209, 213, 214
CFN 37
CFP 60
CF_2 409
CF_2NOP 39, 70, 171
CF_2NP 65, 169
CF_2NPS 39, 70, 171
CF_2NPSe 38, 39, 70
CF_2N_2 42, 158
CF_2O 115, 156, 157, 166
CF_2S 156, 157
CF_2Se 156, 157
CF_3I 151, 152
CF_3N 40, 166, 193
CF_3NO 44, 168
CF_3NOSi 19, 39, 163, 164
CF_3NO_2 53, 168
CF_3N_3 39, 167
CF_4 151, 152, 160

$CF_4N_2P_2$ 40, 70, 171
CF_4O 110, 182
CF_4O_2 122, 183
CF_4S 110, 184
CF_5N 46, 165, 166
CF_5NOS 38, 39, 132, 133, 194
CF_5NOSe 38, 39, 132, 133, 194, 195
CF_5NOTe 38, 39, 132, 133, 194
CF_5NS 44, 123, 124, 185, 186
CF_6Ge 162
CF_6Si 162
CF_7Ns 195
CF_7P 80, 176, 177
CF_8S 193
$CF_{12}S_2$ 194
$CHBrClF$ 151
$CHCl_4P_2$ 64
$CHDO_2$ 119
$CHFO$ 115, 157
CHF_3 151, 152
CHF_3O_2 122, 183
CHF_3O_3S 126, 130, 190
$CHF_3O_6S_3$ 126, 191
CHF_3S 99, 184, 189
CHN 37, 460, 464
$CHNO$ 38, 39
$CHNS$ 38, 39
CHN_3O_6 53
CHP 60
CH_2ClNO_2 52, 53
CH_2Cl_3OP 75
CH_2Cl_3P 65
CH_2Cl_3PS 75
$CH_2Cl_4O_2P_2$ 77
$CH_2Cl_4P_2$ 64, 65
CH_2Cl_4Si 19
$CH_2F_4P_2S_2$ 77, 174
CH_2F_4S 131, 132, 191, 192, 195
$CH_2F_6Si_2$ 27, 163, 164
CH_2N_2 158
CH_2O 114
CH_2O_2 117–119
CH_3BrHg 356
CH_3Br_2PS 68
CH_3Br_3Ge 328, 329, 356
CH_3ClHg 356
CH_3ClO 182
CH_3ClO_2S 126, 127
CH_3ClO_3S 126, 127, 130, 131
CH_3ClS 110
CH_3Cl_2OP 67, 68, 75
CH_3Cl_2OPS 77
$CH_3Cl_2O_2P$ 77

CH_3Cl_2P 64
CH_3Cl_2PS 68
$CH_3Cl_2PS_2$ 77
CH_3Cl_3Ge 162, 328, 329, 356
CH_3Cl_3Sn 328, 329, 356
CH_3F 151, 160
CH_3FO_2S 126, 190
CH_3FO_3S 126, 130, 131, 190
CH_3F_2N 165
CH_3F_2OP 67, 68, 171, 173, 468
CH_3F_2PS 68, 108, 109, 171, 172
CH_3F_3Ge 161, 162, 328, 329, 331, 356
CH_3F_3OSi 105, 106, 163, 164
CH_3F_3Si 161, 162, 164
$CH_3F_4NP_2$ 48, 70, 170
CH_3F_4NS 44, 131, 132, 193, 195
CH_3F_4P 79, 80, 178–180
CH_3GeNO 39
CH_3HgI 356
CH_3Li 359
CH_3N 40, 166
CH_3NO 44, 51, 114, 115, 168
CH_3NOS 44
CH_3NOSi 19, 39, 163, 164
CH_3NO_2 52, 53, 168
CH_3NO_3 56
CH_3NSSi 19, 39
CH_3N_3 38, 39, 167, 168
$CH_3N_3O_4$ 55
CH_3OP 67
CH_4 209, 213–215, 220–223
CH_4F_2NP 70, 170
$CH_4N_2O_2$ 55
CH_4O 95, 103
CH_4O_2 103, 183
CH_4O_2Si 112
CH_4S 99, 189
CH_5N 45
CH_5NO 49, 99
CH_5P 64, 76
$CH_6Cl_2Si_2$ 163
CH_6Ge 162, 328–330, 362
$CH_6Ge_2N_2$ 40
$CH_6N_2Si_2$ 40
CH_6OSi 106, 163
CH_6Si 18, 162, 164
CH_6Sn 328, 329
CH_8Si_2 27, 28, 163, 164
CH_9NSi_2 47
$CH_{11}AlB$ 341
$CH_{11}AsB_{10}$ 16
$CH_{11}B_5$ 12
$CH_{11}B_{10}P$ 16
CIN 37

CI$_2$ 409
CN$_4$ 37, 39
CN$_4$O$_8$ 53
C$_2$BrCl   271
C$_2$BrI   271
C$_2$Br$_4$ 246
C$_2$Br$_4$N$_2$ 41, 42
C$_2$ClF$_3$O   121
C$_2$ClF$_6$NS   44, 123, 124, 186
C$_2$Cl$_2$F$_2$ 155
C$_2$Cl$_2$N$_2$O$_2$Si   19, 39
C$_2$Cl$_2$O$_2$ 121
C$_2$Cl$_3$F$_6$P   177, 178
C$_2$Cl$_4$ 246
C$_2$Cl$_6$O$_2$S   125, 126
C$_2$D$_6$ 209, 214
C$_2$F$_3$NO   167
C$_2$F$_3$NS   167
C$_2$F$_3$NSe   184
C$_2$F$_3$N$_2$S$_2$ 159
C$_2$F$_4$ 154, 155, 166, 246
C$_2$F$_4$O   121, 158
C$_2$F$_4$S$_2$ 135, 158, 195
C$_2$F$_4$Se$_2$ 158
C$_2$F$_5$N   167
C$_2$F$_5$P   168, 169, 172
C$_2$F$_6$ 152
C$_2$F$_6$Hg   356
C$_2$F$_6$NO   168
C$_2$F$_6$N$_2$ 43, 167
C$_2$F$_6$O   102, 182, 183
C$_2$F$_6$OS   123, 186
C$_2$F$_6$O$_2$ 122, 183, 184, 471
C$_2$F$_6$O$_2$S   125, 126, 190
C$_2$F$_6$S   102, 108, 184, 185, 187–189, 192, 472
C$_2$F$_6$S$_2$ 122, 185
C$_2$F$_6$Se   109, 184, 187
C$_2$F$_6$Se$_2$ 122, 185
C$_2$F$_6$Te   184, 187
C$_2$F$_7$N   46, 165
C$_2$F$_8$Ge   162
C$_2$F$_8$OS   131, 132, 187, 193
C$_2$F$_8$S   186, 187, 193
C$_2$F$_8$Se   187
C$_2$F$_8$Te   187
C$_2$F$_9$P   176, 177, 187
C$_2$HBr$_3$O   121
C$_2$HClF$_3$N   167
C$_2$HCl$_3$O   121
C$_2$HF$_3$ 154, 155, 246
C$_2$HF$_3$O$_2$ 119, 157
C$_2$HF$_4$N   167
C$_2$HF$_4$P   79, 179, 271
C$_2$HF$_5$ 152

C$_2$HF$_6$NO   49, 166, 168
C$_2$H$_2$BrClO   121
C$_2$H$_2$Br$_2$ 246
C$_2$H$_2$Br$_2$O   121
C$_2$H$_2$Cl$_2$ 246
C$_2$H$_2$Cl$_2$Ge   331
C$_2$H$_2$Cl$_2$O   120, 121
C$_2$H$_2$D$_2$O$_4$ 119
C$_2$H$_2$F$_2$ 154, 155, 246
C$_2$H$_2$F$_2$O$_2$ 119, 120, 157
C$_2$H$_2$F$_4$ 152, 153
C$_2$H$_2$I$_2$Ge   331
C$_2$H$_2$N$_2$O   51
C$_2$H$_2$O   270
C$_2$H$_2$O$_2$ 114
C$_2$H$_2$O$_3$ 113
C$_2$H$_2$O$_4$ 121, 122
C$_2$H$_3$Br   246
C$_2$H$_3$BrO   120
C$_2$H$_3$Cl 66, 246
C$_2$H$_3$ClO   66, 114, 115, 120
C$_2$H$_3$ClO$_2$ 111, 112, 119
C$_2$H$_3$ClO$_2$S   126–128, 253
C$_2$H$_3$ClSi   19, 271
C$_2$H$_3$Cl$_2$OP   75, 253
C$_2$H$_3$Cl$_2$P   65, 67, 253
C$_2$H$_3$Cl$_3$Si   20
C$_2$H$_3$F   154, 155, 246
C$_2$H$_3$FO   120
C$_2$H$_3$FO$_2$ 119, 157
C$_2$H$_3$FS   471
C$_2$H$_3$F$_2$P   65, 67
C$_2$H$_3$F$_3$ 152, 153
C$_2$H$_3$F$_3$Hg   356
C$_2$H$_3$F$_3$N$_2$ 43
C$_2$H$_3$IO   120
C$_2$H$_3$N   37
C$_2$H$_3$NO   38, 39, 167, 168
C$_2$H$_3$NS   38, 39, 167, 168
C$_2$H$_3$N$_2$P   62
C$_2$H$_3$P   60
C$_2$H$_4$ 154, 155, 230, 243, 249, 257, 258, 260, 269
C$_2$H$_4$AsBrO$_2$ 86
C$_2$H$_4$AsBrS$_2$ 86
C$_2$H$_4$AsClO$_2$ 86
C$_2$H$_4$ClN   58
C$_2$H$_4$ClNO   51, 52
C$_2$H$_4$ClOPS   77
C$_2$H$_4$ClO$_2$P   72, 85, 86
C$_2$H$_4$ClO$_3$P   78, 85
C$_2$H$_4$ClPS$_2$ 72, 85
C$_2$H$_4$ClPS$_3$ 78, 85
C$_2$H$_4$Cl$_2$ 210

$C_2H_4Cl_2O$ 101, 102
C_2H_4FNO 51, 52, 157
$C_2H_4F_2$ 152, 153
C_2H_4Ge 328, 330
C_2H_4INO 51, 52
$C_2H_4N_2$ 41, 42
C_2H_4O 103, 105, 113, 114, 116, 119, 133, 158
C_2H_4OS 112, 123
$C_2H_4O_2$ 95, 111, 112, 117–120
$C_2H_4O_3$ 136
$C_2H_4O_3S$ 124, 125
$C_2H_4O_4$ 118, 119
C_2H_4S 133, 134, 471
C_2H_4Si 18
$C_2H_5B_3$ 14–16
$C_2H_5BrO_2S$ 126, 129
$C_2H_5ClN_2O_2$ 55
C_2H_5ClO 95, 96, 101, 102
C_2H_5ClS 102, 108
$C_2H_5Cl_2P$ 65
C_2H_5F 152
C_2H_5FO 95, 96, 153
C_2H_5N 58, 81
C_2H_5NO 51
C_2H_5NS 52
C_2H_6 26, 95, 152, 153, 155, 209, 210, 214, 220–223
C_2H_6AlCl 350
C_2H_6AlF 347
$C_2H_6AsF_3Si$ 163
$C_2H_6BCl_2N$ 48
$C_2H_6BF_3O$ 8, 102, 150
$C_2H_6B_2S_3$ 7
C_2H_6Be 353, 354, 356, 357, 359, 360, 362, 377
$C_2H_6Be_2Ge$ 328, 329, 331
$C_2H_6ClNO_2S$ 125–128, 130
C_2H_6Cd 355, 356, 359, 360
$C_2H_6ClNO_2S$ 48
$C_2H_6ClO_2PS$ 76, 77
$C_2H_6Cl_2Ge$ 328, 329, 331, 356
$C_2H_6Cl_2NOP$ 77
$C_2H_6Cl_2NP$ 47, 69, 70
$C_2H_6Cl_2Si$ 19, 21, 22
$C_2H_6Cl_2Sn$ 328, 329, 356
$C_2H_6Cl_3NSi$ 47
$C_2H_6FO_2P$ 76, 77
$C_2H_6F_2Ge$ 161, 162, 328, 329, 331, 356
$C_2H_6F_2NP$ 47, 69, 70, 170
$C_2H_6F_2Si$ 161
$C_2H_6F_3NS$ 187, 188
$C_2H_6F_3NSi$ 47, 163, 164
$C_2H_6F_3P$ 80, 178, 179, 181, 182
$C_2H_6F_4NP$ 48, 80, 180
$C_2H_6F_6N_2P_2$ 180, 181

C_2H_6Ge 328, 335
C_2H_6GeOS 112
C_2H_6Hg 355, 356, 359, 360
C_2H_6Mg 353, 357, 360, 364
C_2H_6NO 168
$C_2H_6N_2$ 43, 167, 168, 471, 472
$C_2H_6N_2O$ 44
$C_2H_6N_2O_2$ 55
$C_2H_6N_2S$ 44
C_2H_6O 8, 27, 49, 56, 57, 94, 95, 101–103, 105, 109, 134, 135, 141, 150, 171, 341, 345, 346, 357
C_2H_6OS 123, 124, 186
C_2H_6OSSi 112
$C_2H_6O_2$ 95, 97, 122, 183
$C_2H_6O_2S$ 126, 129, 190
$C_2H_6O_2S_4Si_4$ 28
$C_2H_6O_2Si$ 112
$C_2H_6O_4S$ 126, 130, 131
C_2H_6S 29, 100, 102, 107–109, 134, 135, 138, 341
$C_2H_6S_2$ 99, 122, 135, 185, 188
C_2H_6Se 29, 109
$C_2H_6Se_2$ 122
C_2H_6Si 18
C_2H_6Te 109
C_2H_6Zn 355–357, 359, 360, 365
C_2H_7Al 337, 339
$C_2H_7BN_4$ 7
C_2H_7N 27, 45, 48
C_2H_7NO 49
C_2H_7NOS 44, 128
$C_2H_7NO_2$ 49
$C_2H_7N_2$ 58
C_2H_7P 64
C_2H_8AlN 349
C_2H_8GaN 350
C_2H_8Ge 328, 329
$C_2N_8N_2$ 49, 50
$C_2H_8N_2S$ 44, 128
C_2H_8Si 21
C_2H_8Sn 328, 329
C_2H_9AsSi 163
C_2H_9NSi 19, 47, 163, 164
C_2H_9PSi 71
$C_2H_{10}AlB$ 11, 337, 341
$C_2H_{10}BGa$ 11, 341
$C_2H_{10}B_2$ 10, 11
$C_2H_{10}B_8$ 14–16
$C_2H_{10}OSi_2$ 106, 107
$C_2H_{11}B_2N$ 48
$C_2H_{12}B_{10}$ 3, 14–16
$C_2H_{14}AlB_3$ 341
$C_2H_{14}B_3Ga$ 341

C_2N_2 37
C_2N_4 37, 43
C_3AsF_9 170
C_3ClN 37, 271
$C_3ClN_3O_3Si$ 19, 39
$C_3Cl_2F_6$ 153
$C_3Cl_2F_9P$ 177, 178
C_3Cl_4 256
C_3Cl_6O 117
C_3D_6 230
$C_3F_3N_3$ 158
C_3F_6 156, 250
C_3F_6O 117, 134, 158
C_3F_9N 45, 46, 80, 165, 166
$C_3F_9NS_3$ 48, 185
$C_3F_9NSe_3$ 185
C_3F_9OP 173
C_3F_9P 64, 80, 169, 172
$C_3F_{11}P$ 80, 176, 177
$C_3HCl_3F_2$ 155, 156, 247, 250
C_3HF_6N 40
C_3H_2O 271
C_3H_3ClO 120, 121, 266
$C_3H_3F_3$ 156
$C_3H_3F_3O$ 117
$C_3H_3F_3Si$ 156, 163
C_3H_3N 37
C_3H_3NS 471
$C_3H_3N_2OP$ 62
$C_3H_3N_3$ 158
C_3H_4 255, 257, 270
C_3H_4BrCl 247, 248
$C_3H_4Br_2$ 247,248
C_3H_4ClF 155, 156, 247, 249, 250
$C_3H_4Cl_2$ 247–250, 468, 469
$C_3H_4Cl_3N$ 40
C_3H_4O 114, 115, 266
$C_3H_4O_2$ 465
$C_3H_4O_3$ 113
C_3H_5Br 247–249
C_3H_5BrO 134
$C_3H_5BrO_2S$ 126, 128, 129, 252, 253
C_3H_5Cl 247–249
C_3H_5ClO 120, 121, 134
$C_3H_5F_3Si$ 19, 163, 253
$C_3H_5NO_2$ 40, 41, 99
C_3H_6 133, 134, 156, 157, 209, 230, 231, 243, 244, 257, 264
C_3H_6ClNO 52
$C_3H_6Cl_2Si$ 21, 22
$C_3H_6F_2$ 153
C_3H_6NP 65
$C_3H_6N_2$ 38
$C_3H_6N_2O_2$ 55, 56

$C_3H_6N_2O_4$ 52, 53
C_3H_6O 57, 98, 101, 103, 105, 106, 113, 114, 116, 117, 135, 140, 141, 251, 252
C_3H_6OS 123
$C_3H_6OS_2$ 112, 113
$C_3H_6O_2$ 118, 119
$C_3H_6O_2S$ 126, 128, 129, 252, 253
$C_3H_6O_3$ 95–97, 112, 113, 139, 141
$C_3H_6O_3S$ 125
C_3H_6S 103, 108, 135, 251, 471
$C_2H_6S_3$ 112, 113
C_3H_6Si 19, 20, 271
$C_3H_7AsS_2$ 86
C_3H_7ClNOP 72
C_3H_7ClO 469
$C_3H_7Cl_2P$ 64
C_3H_7F 153
C_3H_7FO 96
C_3H_7N 57, 58, 468
C_3H_7NO 51, 52
$C_3H_7NO_2$ 52, 53
$C_3H_7O_3P$ 72
C_3H_8 27, 65, 153, 209, 215, 220–223
$C_3H_8BrClSi$ 20
$C_3H_8N_2$ 58
C_3H_8O 101
$C_3H_8O_2$ 95, 101, 103, 104
C_3H_8S 107, 108
C_3H_8Si 17–19, 21, 22, 163, 253
C_3H_9Al 335–339, 341–344, 353, 355, 357, 377, 448
$C_3H_9AlCl_3N$ 343, 443, 445, 447, 448
C_3H_9AlN 350
C_3H_9AlS 350, 353
C_3H_9As 29, 170
C_3H_9B 5, 335, 336, 338
$C_3H_9BBr_3N$ 8, 59
$C_3H_9BBr_3P$ 8, 9, 79
$C_3H_9BCl_3N$ 8, 59
$C_3H_9BCl_3P$ 8, 9, 79
$C_3H_9BF_3N$ 8, 59, 150
$C_3H_9BI_3N$ 8, 59
$C_3H_9BI_3P$ 8, 79
C_3H_9BO 5
$C_3H_9BO_2$ 5
$C_3H_9BO_3$ 5
C_3H_9BS 5
$C_3H_9BS_2$ 5
$C_3H_9BS_3$ 5
C_3H_9BrGe 328, 329, 356
C_3H_9ClGe 328, 329, 356
C_3H_9ClSn 328, 329, 356
$C_3H_9Cl_3GaP$ 343, 346
C_3H_9FGe 161

C_3H_9FSi 161
$C_3H_9F_2P$ 80, 178, 179
C_3H_9Ga 335, 336, 338, 339, 357, 377
C_3H_9In 335, 336, 338
C_3H_9N 8, 45, 150, 165, 166, 341, 342, 344–346, 357, 445, 448
C_3H_9NO 49, 97, 98
$C_3H_9NO_2S$ 48, 126, 128, 130
$C_3H_9N_3Si$ 19, 38, 39
C_3H_9OP 75, 79, 173
$C_3H_9O_3P$ 68
$C_3H_9O_4P$ 76, 77
C_3H_9P 8, 29, 64, 66, 67, 75, 79, 80, 341, 344
C_3H_9PS 75
$C_3H_9PS_3$ 68, 69
C_3H_9PSe 75
C_3H_9Tl 335, 336, 338
$C_3H_{10}BN$ 5, 49, 150
$C_3H_{10}Ge$ 328, 329
$C_3H_{10}Sn$ 328, 329
$C_3H_{11}BN_2$ 5, 49, 150
$C_3H_{11}NSi$ 47
$C_3H_{12}AlN$ 343, 448
$C_3H_{12}BN$ 2, 8, 59
$C_3H_{12}BN_3$ 5, 8, 49
$C_3H_{12}GaN$ 343
$C_3H_{13}NSi_2$ 47
$C_3H_{15}NSi_3$ 47
C_3N_2O 37
C_3N_3P 65
C_3O_2 270
$C_4Cl_2O_3$ 267
C_4Cl_6 261
$C_4F_2O_3$ 137
$C_4F_4O_3$ 137, 158
C_4F_6 156, 256, 261
$C_4F_6O_3$ 113
$C_4F_6S_2$ 135, 158
C_4F_7NO 157
C_4F_9I 154
$C_4F_{10}P_2$ 172
$C_4F_{12}Ge$ 160, 163, 328, 330
$C_4F_{12}N_2$ 50, 166, 172
$C_4F_{12}P_2$ 70, 71, 172
$C_4F_{12}Sn$ 160, 163, 328, 330
C_4HF_9 154
C_4HF_9O 95, 98, 154
$C_4H_2Cl_2O_2$ 266
$C_4H_2F_6$ 155, 156, 250
$C_4H_2O_3$ 137, 267
C_4H_3BrO 136
C_4H_3BrS 136
C_4H_3Cl 270
C_4H_3ClO 136

$C_4H_3ClO_2S_2$ 126, 127
C_4H_3ClS 136
$C_4H_3Cl_2PS$ 65, 67
$C_4H_3CoO_4Si$ 19
$C_4H_3NO_2$ 267
C_4H_4 270
$C_4H_4Cl_2$ 256, 261
$C_4H_4Cl_2OSi$ 263
$C_4H_4Cl_4Ge_2$ 331
$C_4H_4Ge_2I_4$ 331
$C_4H_4N_2$ 37
C_4H_4O 136
$C_4H_4O_2$ 114, 115, 139, 140, 266
$C_4H_4O_3$ 137, 158
C_4H_4S 136
C_4H_4Se 136
C_4H_5ClO 121, 266
C_4H_5N 37, 38, 248, 249, 465
C_4H_6 255, 257, 262, 271
$C_4H_6BF_6N$ 149, 150
C_4H_6ClOP 75, 78, 253
$C_4H_6Cl_2$ 248, 250
$C_4H_6Cl_2Si$ 21
$C_4H_6N_3$ 45
C_4H_6O 98, 101, 103, 105, 114, 115, 136, 270
$C_4H_6O_2$ 116, 117, 134
$C_4H_6O_2S$ 126, 128, 253
$C_4H_6O_3$ 113
C_4H_6S 107–109, 270
$C_4H_6S_2$ 108, 271
C_4H_7Br 248
C_4H_7Cl 248
C_4H_7ClNOP 72
$C_4H_7NO_2$ 116
C_4H_8 22, 25, 27, 57, 157, 209, 231, 242–245, 252, 254, 259, 274
C_4H_8AsCl 86
$C_4H_8B_2$ 11
C_4H_8ClN 58
$C_4H_8Cl_2Si$ 21, 22
$C_4H_8F_2Si$ 22
$C_4H_8F_3P$ 181, 182
C_4H_8Ge 331
$C_4H_8NO_2$ 51
$C_4H_8N_2$ 42
C_4H_8O 57, 85, 95, 98, 101, 103, 105, 114, 116, 117, 135, 136, 141, 251, 252
C_4H_8OS 139
$C_4H_8O_2$ 138, 139
$C_4H_8O_3S$ 124, 125
$C_4H_8O_4$ 118, 140, 141
C_4H_8S 85, 136
$C_4H_8S_2$ 107, 108, 138, 251
C_4H_8Se 85, 136

C_4H_8Si 21, 22
C_4H_9ClSi 20
$C_4H_9Cl_2P$ 64
C_4H_9D 222
C_4H_9F 153
$C_4H_9F_2P$ 64, 169
C_4H_9N 57, 58, 85
C_4H_9OP 65–67
C_4H_9NOSi 19, 39
$C_4H_9NO_2$ 52, 53
C_4H_9NSSi 39
C_4H_9NSi 19
C_4H_9P 86
C_4H_{10} 209, 210, 215, 216, 218, 220–223, 274
$C_4H_{10}BClN_2$ 7
$C_4H_{10}Be$ 353
$C_4H_{10}ClN_2P$ 72, 85
$C_4H_{10}ClP$ 64, 65
$C_4H_{10}Cl_2Si$ 19
$C_4H_{10}Ge$ 331
$C_4H_{10}N_2$ 51, 58
$C_4H_{10}O_2$ 101
$C_4H_{10}O_3$ 101, 103, 104
$C_4H_{10}SSi$ 135
$C_4H_{10}Si$ 17, 21, 22, 85
$C_4H_{10}Zn$ 356
$C_4H_{11}N$ 45
$C_4H_{11}NO$ 95
$C_4H_{11}P$ 76
$C_4H_{12}AlN$ 349
$C_4H_{12}AlNa$ 342, 344
$C_4H_{12}Al_2Cl_2$ 350, 352
$C_4H_{12}AuP$ 359
$C_4H_{12}ClN_2P$ 47, 69, 70
$C_4H_{12}Cl_2OSi_2$ 106, 107
$C_4H_{12}Cl_2Si_2$ 25
$C_4H_{12}F_2N_2S$ 188
$C_4H_{12}Ge$ 160, 161, 163, 327–329, 356, 362, 377
$C_4H_{12}N_2$ 50
$C_4H_{12}N_2OS$ 48, 124, 188
$C_4H_{12}N_2O_2S$ 48, 126, 128, 130
$C_4H_{12}N_2S$ 48
$C_4H_{12}OSi$ 19, 105, 106
$C_4H_{12}O_2Si_2$ 28
$C_4H_{12}O_3Si$ 19
$C_4H_{12}O_4Si$ 18, 105, 106
$C_4H_{12}P_2$ 71
$C_4H_{12}Pb$ 327, 328
$C_4H_{12}Si$ 18, 19, 161
$C_4H_{12}Si_2$ 18
$C_4H_{12}Sn$ 160, 163, 327–330
$C_4H_{13}NSi$ 47
$C_4H_{13}PSi$ 76

$C_4H_{14}Al$ 339, 341
$C_4H_{14}OSi_2$ 106, 107
$C_4H_{15}NSi_2$ 47
$C_4H_{16}B_{10}$ 15
$C_4H_{20}FNO_4$ 342, 344
$C_4H_{22}B_{20}$ 16
$C_4N_4O_4Si$ 18, 39
$C_5F_3O_5ReSi$ 19
C_5F_8 157, 256
C_5F_{12} 153, 154, 160
$C_5F_{12}S_4$ 184
$C_5H_2F_6O_2$ 116, 117
C_5H_3ClOS 266
$C_5H_3ClO_2$ 266
$C_5H_3MnO_5Si$ 19
$C_5H_3O_5ReSi$ 19
C_5H_4OS 266
$C_5H_4O_2$ 114, 115, 266, 267
C_5H_5As 62, 63
C_5H_5BeBr 364
C_5H_5BeCl 363, 367
C_5H_5ClGe 367
$C_5H_5Cl_3$ 247
$C_5H_5F_3O_2$ 116, 117
C_5H_5In 365, 368, 374, 377
C_5H_5N 62, 63
C_5H_5P 62, 63
C_5H_5Tl 365, 368, 374
C_5H_6 271, 361
C_5H_6Be 363
C_5H_6OS 100
C_5H_7ClO 266
$C_5H_7N_2OP$ 61, 62
C_5H_8 157, 240, 243, 254, 255, 261–264
C_5H_8Ge 331, 361, 362, 368
$C_5H_8N_2$ 42
$C_5H_8N_6O$ 135
C_5H_8O 121, 135, 274
C_5H_8OS 138
$C_5H_8O_2$ 116, 117
C_5H_8Si 263
C_5H_9Al 339, 340
C_5H_9BBe 364
C_5H_9Cl 85
C_5H_9ClSi 19
C_5H_9N 60
C_5H_9P 60
C_5H_{10} 22, 27, 57, 84, 85, 197, 209, 232, 233, 243, 254, 259, 438
$C_5H_{10}Cl_2Si$ 21, 22
$C_5H_{10}F_3P$ 181, 182
$C_5H_{10}O$ 138, 141, 274
$C_5H_{10}O_3S$ 125
$C_5H_{10}Si$ 271

$C_5H_{10}Sn$ 271, 328
$C_5H_{11}N$ 58
C_5H_{12} 160, 209, 217–223, 225
$C_5H_{12}BClN_2$ 6, 7
$C_5H_{12}BF_3N_2$ 149, 150
$C_5H_{12}N_2$ 51
$C_5H_{12}N_2O$ 52
$C_5H_{12}N_2S$ 52
$C_5H_{12}OSi$ 19, 106, 251, 252
$C_5H_{12}O_2$ 101, 103
$C_5H_{12}O_3$ 101, 103
$C_5H_{12}O_4$ 101, 103, 104
$C_5H_{12}Si$ 18, 21–23
$C_5H_{12}Sn$ 330
$C_5H_{13}Ga$ 340
$C_5H_{13}In$ 340
$C_5H_{14}N$ 45
$C_5H_{14}P_2$ 64, 65
$C_5H_{15}AlO$ 341, 343, 345–347, 350
$C_5H_{15}AlS$ 109, 341, 343, 346, 353
$C_5H_{15}GaO$ 343, 346
$C_5H_{15}NSi$ 345
$C_5H_{15}NZn$ 357
$C_5H_{17}NSi_2$ 47
C_5N_4 37
$C_6ClF_5O_2S$ 126–128
$C_6Cl_4O_2$ 267
$C_6F_4O_2$ 267
$C_6F_5Cl_2N$ 40
$C_6H_2F_4$ 316, 317
$C_6H_3BrN_2O_4$ 54
$C_6H_3Br_3$ 313, 314
$C_6H_3ClN_2O_4$ 54
$C_6H_3Cl_3$ 313, 314
$C_6H_3F_3$ 313, 314, 317
C_6H_3ISn 271, 328
$C_6H_3N_3O_6$ 54, 313, 314
$C_6H_4BrNO_2$ 53–55, 81, 83, 311, 312
$C_6H_4BrNO_2Se$ 55, 111
$C_6H_4Br_2$ 271, 310–312
$C_6H_4ClNO_2$ 53–55, 81, 83, 311, 312
$C_6H_4ClNO_2S$ 55, 111
$C_6H_4ClO_2P$ 72
$C_6H_4ClO_3P$ 78
$C_6H_4Cl_2$ 310–312
$C_6H_4F_2$ 310–312, 316, 317
$C_6H_4FeO_4$ 358
$C_6H_4N_2O_4$ 54, 83, 310–312
$C_6H_4O_2$ 267
$C_6H_4O_3S$ 124, 125
C_6H_5Br 55, 271, 296, 307
C_6H_5BrHg 356
C_6H_5BrSe 110, 111
C_6H_5Cl 54, 284, 289, 290, 296, 299, 300, 307

$C_6H_5ClO_2S$ 126–128
C_6H_5ClS 110, 111
C_6H_5ClSe 110
$C_6H_5Cl_2OP$ 75
$C_6H_5Cl_2P$ 65, 66
C_6H_5F 282, 284, 289, 290, 295, 296, 298, 299, 305, 307, 308, 316
$C_6H_5F_2P$ 65, 66
$C_6H_5F_3Si$ 19
$C_6H_5F_4P$ 79, 179, 180
C_6H_5Li 284
$C_6H_5NO_2$ 53, 54, 81, 83, 289, 290, 296, 302, 303, 307, 308
$C_6H_5N_3O_4$ 316
C_6H_6 271, 279, 281, 283, 284, 290, 307, 309, 311, 312, 314
$C_6H_6N_2O_2$ 53, 54, 288, 311, 312, 316
C_6H_6O 283, 289, 290, 296, 302, 304, 307, 308
$C_6H_6O_2$ 100, 309–312, 318, 321
$C_6H_6S_2$ 100, 310–312
C_6H_7ClFN 298
C_6H_7ClSi 20
C_6H_7N 46, 55, 282, 289, 290, 307, 308, 317
C_6H_7P 65, 66
C_6H_8 140, 254, 256, 261–263
C_6H_8Be 362–365, 369
C_6H_8Hg 365
C_6H_8Mg 364
$C_6H_8N^+$ 289, 290
$C_6H_8N_2$ 310–312, 317, 318, 320, 321
C_6H_8OS 138
C_6H_8OSi 105
$C_6H_8O_2$ 117
C_6H_8Si 282, 297, 307
C_6H_8Zn 364, 365, 368, 372, 377
C_6H_9B 4, 5, 20
C_6H_9Cl 256
C_6H_9Ga 339
C_6H_9N 85
$C_6H_9O_3P$ 68
C_6H_{10} 60, 254, 255, 262, 271
$C_6H_{10}N_2$ 42
$C_6H_{10}O$ 135, 138
C_6H_{12} 22, 23, 27, 86, 138, 157, 182, 209, 233, 234, 243–245, 255, 259
$C_6H_{12}AlN$ 339
$C_6H_{12}ClO_2P$ 72, 85
$C_6H_{12}N_2$ 58
$C_6H_{12}N_3P$ 69
$C_6H_{12}O$ 140
$C_6H_{12}O_3$ 139
$C_6H_{12}O_4$ 118
C_6H_{14} 209, 217, 219, 220, 221, 225–227
$C_6H_{14}Be$ 353

$C_6H_{14}O$ 101
$C_6H_{14}Si$ 19
$C_6H_{14}Zn$ 356
$C_6H_{15}Al$ 338
$C_6H_{15}AlO$ 347
$C_6H_{15}O_3P$ 68
$C_6H_{16}Si$ 19
$C_6H_{18}AlN$ 341–344, 346, 350, 351, 448
$C_6H_{18}AlN_3$ 350
$C_6H_{18}AlP$ 341, 343, 344
$C_6H_{18}Al_2$ 353
$C_6H_{18}Al_2S_2$ 350, 352
$C_6H_{18}BN_3$ 48, 49
$C_6H_{18}F_2N_3P$ 80, 180
$C_6H_{18}GaN$ 343
$C_6H_{18}GaP$ 79, 343
$C_6H_{18}Ge_2O$ 106, 332, 333
$C_6H_{18}NPSi$ 43, 76
$C_6H_{18}N_2Si_2$ 28
$C_6H_{18}N_3P$ 47, 69
$C_6H_{18}OSi_2$ 106
$C_6H_{18}OSn_2$ 332, 333
$C_6H_{18}O_2Si_2$ 122
$C_6H_{18}O_3Si_3$ 28
$C_6H_{18}Si_2$ 25
$C_6H_{19}NSi_2$ 27, 47
$C_6H_{21}AlN$ 346
$C_6H_{21}N_3Si_3$ 28
$C_6H_{24}Al_3N_3$ 351
C_6N_4 269
$C_7H_4Cl_2O_2$ 315
$C_7H_4N_2O_6$ 315
C_7H_5ClO 114, 115
$C_7H_5ClO_2$ 300
$C_7H_5FO_2$ 298
$C_7H_5F_3$ 296, 307
C_7H_5N 282, 289, 290, 296, 299, 301, 305, 307, 316, 317
C_7H_6Be 364, 365
C_7H_6O 261
$C_7H_6O_2$ 114, 289, 290
C_7H_7O 263
C_7H_8 261, 263, 264, 289, 290, 296, 305–307
C_7H_8O 95, 98, 101, 103, 105, 283, 289, 290, 308
$C_7H_8O_2S$ 126, 129, 289, 290, 296, 307
C_7H_8S 103, 108
C_7H_8Se 103, 105
C_7H_{10} 262–264
$C_7H_{11}Al$ 362, 366–368, 377
$C_7H_{12}O$ 101, 135
$C_7H_{13}ClSi$ 23
$C_7H_{13}N$ 58, 59
C_7H_{14} 209, 235–237

$C_7H_{14}Si$ 23
$C_7H_{15}NO_3Si$ 17
C_7H_{16} 209, 217–221
$C_7H_{16}O_2Si$ 105, 106
$C_7H_{18}P_2$ 76
$C_7H_{19}LiSi_2$ 21, 359
$C_7H_{20}Si$ 20, 21
$C_7H_{20}Si_2$ 21, 27
$C_8F_{18}S_6$ 185
$C_8H_4N_2$ 38, 310–312, 317–320
C_8H_4Si 18
C_8H_4Sn 271, 328, 330
C_8H_6 307
$C_8H_6Cl_2$ 249
$C_8H_6O_2$ 309–312
C_8H_7N 37, 38
C_8H_8 263, 283
C_8H_8O 289, 290
$C_8H_8O_2$ 101, 103, 311
C_8H_8S 108
C_8H_9ClNOP 72
C_8H_9N 40
C_8H_9NO 289, 290
C_8H_{10} 269, 296, 307, 310–312
$C_8H_{10}ClN_2P$ 72
$C_8H_{10}Si$ 23
$C_8H_{11}Al$ 339
$C_8H_{11}AlO$ 347
$C_8H_{11}N$ 45, 46, 289, 290, 308
$C_8H_{11}P$ 65, 66, 83
C_8H_{12} 254, 256, 261, 263, 270, 272, 273
$C_8H_{12}O_3$ 137
$C_8H_{12}Si$ 18, 20
$C_8H_{12}Sn$ 328
C_8H_{14} 23, 255, 257, 258, 262
$C_8H_{14}Ge$ 362, 366, 368
$C_8H_{14}Si$ 362, 366
$C_8H_{14}Sn$ 362, 366, 368
$C_8H_{14}Zn$ 274, 358
C_8H_{16} 209, 235–237
$C_8H_{16}Ge_2$ 331
$C_8H_{16}Si$ 23
C_8H_{18} 209, 224, 226
$C_8H_{18}Be$ 353–356
$C_8H_{18}ClP$ 64
$C_8H_{18}FP$ 64, 169
$C_8H_{18}F_2Si$ 161
$C_8H_{18}Ge_2O$ 332, 333
$C_8H_{18}N_2Zn$ 358
$C_8H_{18}OSi_2$ 21, 270
$C_8H_{18}O_2$ 122
$C_8H_{18}O_2Zn$ 357
$C_8H_{18}S_2Zn$ 358
$C_8H_{18}Sn_2$ 328, 330

$C_8H_{24}Al_2N_2$ 351
$C_8H_{24}Al_4F_4$ 347, 348
$C_8H_{24}N_4Sn$ 47
$C_8H_{24}OSi$ 347
$C_8H_{24}Si_3$ 25
$C_8H_{24}Si_4$ 25
C_9H_7N 460
$C_9H_9N_2P$ 61, 62
$C_9H_{10}O$ 312
$C_9H_{10}O_2$ 315
C_9H_{12} 269, 313, 314
C_9H_{14} 272
C_9H_{16} 258
C_9H_{20} 209, 229, 230
$C_9H_{21}AlO_3$ 350
$C_9H_{27}AlO$ 347, 348
$C_9H_{27}NSn_3$ 47, 332, 333
$C_{10}F_{18}$ 157
$C_{10}H_8N_2S$ 105, 108, 140
$C_{10}H_{10}$ 256, 265
$C_{10}H_{10}Be$ 369–372
$C_{10}H_{10}Ge$ 372, 374, 375
$C_{10}H_{10}Mg$ 364, 368, 369, 372, 374
$C_{10}H_{10}Ni$ 368
$C_{10}H_{10}Pb$ 372, 374, 377
$C_{10}H_{10}Sn$ 372, 374–376
$C_{10}H_{10}Zn$ 372
$C_{10}H_{11}Al$ 330
$C_{10}H_{11}O$ 312
$C_{10}H_{12}N_3P$ 62
$C_{10}H_{12}O$ 312
$C_{10}H_{12}O_2$ 267
$C_{10}H_{14}BTi$ 11
$C_{10}H_{15}ClGe$ 367, 372, 374
$C_{10}H_{16}$ 256
$C_{10}H_{16}Mg$ 364, 365, 368
$C_{10}H_{16}S$ 140, 272
$C_{10}H_{18}$ 262
$C_{10}H_{18}Ge$ 331
$C_{10}H_{18}Si$ 23
$C_{10}H_{18}Zn$ 274, 358, 359
$C_{10}H_{20}$ 209, 235, 236
$C_{10}H_{20}N_2$ 40, 41
$C_{10}H_{20}N_4$ 259, 269
$C_{10}H_{22}Mg$ 353, 355–357, 359, 360, 365
$C_{10}H_{22}Mn$ 377
$C_{10}H_{22}O_2Zn$ 357
$C_{10}H_{29}PSi_3$ 21, 64
$C_{11}H_5Cl_5Hg$ 365
$C_{11}H_{10}N_3P$ 61, 62
$C_{11}H_{22}Si_2$ 21
$C_{12}D_{10}$ 297, 307
$C_{12}F_{10}$ 158
$C_{12}F_{12}Sn$ 163, 328

$C_{12}H_8Br_2$ 312
$C_{12}H_8Cl_2$ 312
$C_{12}H_8F_2$ 158, 312
$C_{12}H_8S_2$ 139, 140
$C_{12}H_9Cl$ 297, 307, 312
$C_{12}H_9F$ 157, 297, 307, 312
$C_{12}H_{10}$ 157, 158, 289, 290, 297, 307
$C_{12}H_{10}AlN$ 350
$C_{12}H_{10}Hg$ 356
$C_{12}H_{10}N_2$ 43
$C_{12}H_{10}O$ 105
$C_{12}H_{10}OS$ 123
$C_{12}H_{10}O_2S$ 126, 128
$C_{12}H_{10}S$ 105, 108, 140, 296, 307
$C_{12}H_{11}P$ 65, 66
$C_{12}H_{14}Ge$ 372, 374, 375
$C_{12}H_{14}Sn$ 372, 374, 375
$C_{12}H_{22}Si_2$ 21, 310–312
$C_{12}H_{27}Al$ 339
$C_{12}H_{27}OP$ 75
$C_{12}H_{27}P$ 64, 169
$C_{12}H_{28}NP$ 76
$C_{12}H_{28}Si$ 19, 20
$C_{12}H_{30}Al_2O_2$ 347, 349
$C_{12}H_{32}Al_4N_4$ 351
$C_{12}H_{36}Al_2N_6$ 350, 351
$C_{12}H_{36}BeN_2Si_4$ 21, 47, 48
$C_{12}H_{36}CdN_2Si_4$ 21, 47, 48
$C_{12}H_{36}GeN_2Si_4$ 21, 47, 48
$C_{12}H_{36}HgN_2Si_4$ 21, 47, 48
$C_{12}H_{36}Li_2N_2Si_4$ 21
$C_{12}H_{36}MgN_2Si_4$ 21, 47, 48
$C_{12}H_{36}N_2PbSi_4$ 21, 47, 48
$C_{12}H_{36}N_2Si_4Sn$ 21, 47, 48
$C_{12}H_{36}N_2Si_4Zn$ 21, 47, 48, 358
$C_{12}H_{36}Si_5$ 24, 25
$C_{13}H_{11}N$ 40
$C_{13}H_{28}$ 209, 227, 228
$C_{13}H_{36}Si_4$ 21
$C_{14}H_{12}$ 261
$C_{14}H_{28}$ 234, 235
$C_{14}H_{38}GeSi_4$ 21, 334, 335
$C_{14}H_{38}PbSi_4$ 334
$C_{14}H_{38}Si_4Sn$ 21, 334, 335
$C_{15}H_{15}F$ 158
$C_{16}H_{20}$ 269
$C_{16}H_{22}Al_2O_2$ 347
$C_{16}H_{26}SiZn$ 372, 373
$C_{16}H_{32}BeN_2$ 357
$C_{16}H_{32}MgN_2$ 357
$C_{16}H_{34}$ 209, 210, 217, 218, 220, 221, 234, 235
$C_{18}H_{15}N$ 45, 46
$C_{18}H_{32}$ 259
$C_{18}H_{36}$ 259, 260

508   FORMULA INDEX

$C_{18}H_{48}Al_6N_6$ 351
$C_{18}H_{54}AlN_3Si_6$ 351
$C_{18}H_{54}CeN_3Si_6$ 21, 47, 48
$C_{18}H_{54}N_3PrSi_6$ 21, 47, 48
$C_{18}H_{54}N_3ScSi_6$ 21, 47, 48
$C_{20}H_{15}Al$ 339
$C_{20}H_{30}Ca$ 376, 377
$C_{20}H_{30}Ge$ 372, 374, 375, 377
$C_{20}H_{30}Mg$ 376, 377
$C_{20}H_{30}Pb$ 372, 374
$C_{20}H_{30}Sm$ 376
$C_{20}H_{30}Sn$ 374, 375
$C_{20}H_{30}Yb$ 376, 377
$C_{20}H_{30}Zn$ 372–374
$C_{24}H_{20}Sn$ 328, 330
$C_{24}H_{33}Al_3O_3$ 347
$C_{26}H_{29}NO$ 240
$C_{27}H_{63}Al_3O_9$ 350
$C_{31}H_{25}N$ 285
$C_{70}H_{50}Sn$ 374, 375
$CaCl_2$ 386, 398–403
CaF_2 399–401
CaI_2 386, 398–403
$CeCl_3$ 424
CeF_4 430
CeI_3 424
$ClCs$ 397
$ClCu$ 386, 388, 412, 413
ClF 196
$ClFO_2$ 196
$ClFO_3$ 196
ClF_2N 46, 165
ClF_2NOS 44, 190
ClF_2NS 44, 123, 124, 186, 190
ClF_2P 169
ClF_2PS 74, 173
ClF_3 196
ClF_3O 196
ClF_4P 175
ClF_5 196, 197
ClF_5S 132, 133, 193
ClK 397
$ClNOS$ 44
ClN_3 39
$ClNa$ 395–397
$ClRb$ 397
Cl_2Co 415–417, 419–421
Cl_2Cr 389, 416–418
Cl_2F_3P 175
$Cl_2F_8N_2S_2$ 132, 133, 195
Cl_2Fe 386, 389, 416, 417, 419–421
Cl_2Ge 17, 367, 409
Cl_2Hg 356, 414
Cl_2Mg 398–401, 403

Cl_2Mn 415–418, 421
Cl_2Na_2 395–397
Cl_2Ni 415–417, 421
Cl_2O_2S 125–127
Cl_2Pb 409
Cl_2Si 17, 409
Cl_2Sn 409
Cl_2Sr 398–401
Cl_2V 386, 387, 389, 416–418
Cl_2Zn 413, 414
Cl_3Cu_3 388, 393, 413
Cl_3Dy 424
Cl_3Er 424
Cl_3Eu 424
Cl_3F_2P 175
Cl_3Fe 384, 386, 388, 389, 424
Cl_3Ga 403, 446
Cl_3GaH_3N 443, 446
Cl_3Gd 422, 424
Cl_3Ho 422, 424
Cl_3La 422, 424
Cl_3Lu 422, 424
Cl_3N 46, 410
Cl_3Nd 424
Cl_3OP 73–75, 78, 79
Cl_3P 63, 65, 67, 68, 73, 79, 410
Cl_3PS 73–75
Cl_3Pm 424
Cl_3Pr 422, 424
Cl_3Sb 411
Cl_3Sm 424
Cl_3Tb 422, 424
Cl_3Tm 424
Cl_3Yb 424
Cl_4Cu_4 388, 413
Cl_4FP 175
Cl_4Fe_2 393
Cl_4Ge 328, 329, 356, 408
Cl_4Hf 426
Cl_4InTl 440
Cl_4KY 440
Cl_4Mo 430, 431
Cl_4NV 43
Cl_4OW 431
Cl_4SW 431
Cl_4SeW 431
Cl_4Si 16, 18, 20, 408
Cl_4Sn 328, 329, 256, 408
Cl_4Th 429, 430
Cl_4Ti 426, 428
Cl_4U 429, 430
Cl_4V 416, 426–428, 433
Cl_4W 430, 431
Cl_4Zr 426

Cl$_4$Mo 434–436
Cl$_5$Nb 432, 433
Cl$_5$P 79, 175, 178, 433
Cl$_5$Sb 411
Cl$_5$Ta 432, 433
Cl$_5$W 434
Cl$_6$Fe$_2$ 425
Cl$_6$Ga$_2$ 405
Cl$_6$N$_3$P$_3$ 78
Cl$_6$OSi$_2$ 106
Cl$_6$Si$_2$ 17, 25
Cl$_8$Si$_3$ 25
CrF$_5$ 433–435
CsNO$_2$ 56
CsNO$_3$ 57
CsO$_3$P 73
CuI 412, 413
CuN$_2$O$_6$ 57
Cu$_3$I$_3$ 412, 413

D

D$_4$P$_2$ 71
DyI$_3$ 424

E

ErI$_3$ 424
EuI$_3$ 424

F

FGeH$_3$ 161
FH 149
FH$_2$N 165
FH$_2$NO$_2$S 128, 130, 190
FH$_3$Si 160
FLi 394, 442
FNO 165
FNO$_2$ 165
FNS 188
FTl 395, 406, 407
F$_2$ 196
F$_2$Ge 409
F$_2$GeH$_3$PS 68, 108, 109, 171, 172
F$_2$Ge$_2$H$_6$P$_2$ 70, 71
F$_2$HN 165
F$_2$HOP 74, 173, 174

F$_2$HPS 173
F$_2$HPSe 74, 173, 174
F$_2$H$_2$NP 70, 170
F$_2$H$_2$Si 160
F$_2$H$_5$N$_2$P 80, 180
F$_2$H$_6$NPSi$_2$ 48, 70, 170
F$_2$IPS 74
F$_2$Kr 198
F$_2$Li$_2$ 394, 395
F$_2$Mg 398–401, 403
F$_2$Mn 415, 416, 421
F$_2$N 165
F$_2$N$_2$ 43, 166, 193
F$_2$Ni 415, 416, 421
F$_2$O 182, 183
F$_2$OS 185, 186
F$_2$OSe 190
F$_2$O$_2$ 183–185
F$_2$O$_2$S 126, 189, 190
F$_2$O$_2$Se 190
F$_2$O$_2$S$_2$ 183, 190, 191
F$_2$O$_8$S$_3$ 190, 191
F$_2$Pb 409
F$_2$S 184, 185, 187, 470
F$_2$S$_2$ 185, 186, 188
F$_2$Si 409
F$_2$Sn 409
F$_2$Sr 399–401
F$_2$Ti 416
F$_2$Tl$_2$ 406–408
F$_2$Xe 198
F$_2$HSi 160
F$_3$H$_2$PSi 71, 163
F$_3$H$_4$N$_2$P 80, 180
F$_3$N 46, 165, 166, 169
F$_3$NO 59, 165
F$_3$NS 188
F$_3$OP 73, 74, 78, 79, 173, 192
F$_3$P 63, 65, 68, 79, 169, 172, 173, 192
F$_3$PS 73, 74, 173, 174
F$_3$Sc 422
F$_4$OP$_2$ 68, 69
F$_4$Ge 161
F$_4$GeH$_3$NP$_2$ 48, 70, 170
F$_4$HNP$_2$ 70, 170
F$_4$HNS 131, 193
F$_4$H$_3$NP$_2$Si 70, 170
F$_4$H$_3$NP$_2$Sn 48
F$_4$Hf 426
F$_4$MoO 437
F$_4$N$_2$ 50, 166, 172
F$_4$OP$_2$ 171, 174, 183
F$_4$OP$_2$S$_2$ 77, 78, 174, 183
F$_4$OS 131, 132, 191–193, 195

F_4OXe 198
F_4P_2 70, 71, 172
F_4P_2S 68, 69, 108, 109, 171, 172
F_4P_2Se 68, 69, 171
F_4S 186, 187, 191, 192, 430, 431
F_4S_2 187, 188, 470, 471
F_4Se 186
F_4Si 18, 160, 161
F_4Th 429, 430
F_4Ti 426
F_4U 429, 430
F_4Xe 198
F_4Zr 426
F_5I 197
F_5IO 197
F_5Mo 436
F_5NS 44, 131, 132, 191, 192, 195
F_5Nb 412, 435
F_5P 79, 174, 175, 177–179, 186, 433
F_5Sb 411, 412
F_5Ta 412, 435
F_5V 431, 433
F_6Ir 438
F_6Mo 438
F_6NP_3 48, 70, 170
$F_6N_3P_3$ 78, 174
F_6Np 438
F_6OS 194
F_6OSi_2 28, 106, 163, 183
F_6Os 438
F_6Pu 438
F_6Re 438
F_6S 132, 192, 193, 195
F_6Se 193–195
F_6Si_2 25, 164
F_6Te 193–195
F_6U 438
F_6W 438
F_6Xe 198
F_7I 197
F_7NS 132, 133, 194
F_7Re 438, 439
$F_8O_2Se_2$ 195
$F_8O_2Te_2$ 195
F_9NS_2 195
$F_{10}HNS_2$ 194
$F_{10}OS_2$ 132, 133, 183, 194
$F_{10}OSe_2$ 132, 183, 194, 195
$F_{10}OTe_2$ 132, 183, 194
$F_{11}NS_2$ 194
$F_{15}Mo_3$ 437, 438
$F_{15}Nb_3$ 412, 436–438
$F_{15}Sb_3$ 412, 436, 438
$F_{15}Ta_3$ 412, 436–438

G

GaI_3 405, 406
GdI_3 422, 424
GeH_3N_3 39
GeH_4 328, 329
GeI_2 409
Ge_2H_6O 29, 106, 332, 333, 350
Ge_2H_6S 29, 108, 332, 333, 350
Ge_2H_6Se 109, 332, 333
Ge_3H_9N 29, 47, 48, 332, 333, 350
Ge_3H_9P 29, 71, 332, 350

H

HN_3 38, 39
H_2B 338
H_2F_2NP 47
H_2N_2 166
H_2O_2 183
H_2Se 109
H_3N 45, 46, 445–448
H_3NO 49
H_3NOSSi 19, 44
H_3N_3Si 19, 39
H_3P 63, 172
H_4N_2 50, 56, 166, 172
H_4P_2 70, 71, 172
H_3Sn 328, 329
H_5NS 188
H_5P 79
H_5PSi 71, 163
H_6OSi_2 16, 27, 28, 105, 106, 163, 182, 183
H_6SSi_2 29, 108
H_6SeSi_2 29, 209
H_6Si_2 16, 25, 26, 28, 164
H_6Si_3 24
H_7NSi_2 27, 28, 47
H_8Si_3 25
H_8Si_4 24
H_9NSi_3 16, 47
H_9PSi_3 29, 71
$H_{10}Si_5$ 17, 25, 27
$H_{12}Li_2N_2Si_4$ 59
$H_{12}N_2Si_4$ 50
$H_{12}Si_6$ 24, 25, 27
HfI_4 426
HgI_2 356, 414, 415
HoI_3 424

I

I_2Mg 400, 401
I_2Pb 409

I_2Si 409
I_2Sn 409
I_2Sr 398–401, 403
I_2Zn 413, 414
I_3La 424
I_3Lu 422, 424
I_3Nd 422, 424
I_3P 63
I_3Pm 424
I_3Pr 422, 424
I_3Sb 411
I_3Sm 424
I_3Tb 424
I_3Tm 424
I_3Yb 424
I_4Ti 426
I_4Zr 426

N

NO_2 56
NO_2Rb 56

NO_3Rb 57
NO_3Tl 57
N_2 37, 38
N_2O_4 56
N_2O_5 56
N_4 38
$N_4O_{12}Sn$ 57
$N_4O_{12}Zr$ 57
NaO_3P 73

O

OS 185
O_2 183
$O_{10}P_4$ 78, 79

S

S_2 185